NUCLEIC ACIDS IN MEDICINAL CHEMISTRY AND CHEMICAL BIOLOGY

Wiley Series in Drug Discovery and Development

Binghe Wang, Series Editor

A complete list of the titles in this series appears at the end of this volume.

NUCLEIC ACIDS IN MEDICINAL CHEMISTRY AND CHEMICAL BIOLOGY

DRUG DEVELOPMENT AND CLINICAL APPLICATIONS

Edited by

LIHE ZHANG
Peking University
Beijing, China

XINJING TANG
Peking University
Beijing, China

ZHEN XI
Nankai University
Tianjin, China

JYOTI CHATTOPADHYAYA
University of Uppsala
Uppsala, Sweden

WILEY

This edition first published 2023

Registered Office(s)
John Wiley & Sons, Inc., 111 River Street, Hoboken, NJ 07030, USA

Editorial Office
John Wiley & Sons, Inc., 111 River Street, Hoboken, NJ 07030, USA

For details of our global editorial offices, customer services, and more information about Wiley products visit us at www.wiley.com.

Wiley also publishes its books in a variety of electronic formats and by print-on-demand. Some content that appears in standard print versions of this book may not be available in other formats.

Library of Congress Cataloging-in-Publication Data
Names: Zhang, Lihe, editor. | Tang, Xinjing, editor. | Xi, Zhen, editor. | Chattopadhyaya, Jyoti, editor.
Title: Nucleic acids in medicinal chemistry and chemical biology: drug development and clinical applications / Lihe Zhang (Peking University, Beijing, China), Xinjing Tang (Peking University, Beijing, China), Zhen Xi (Nankai University, Tianjin, China), Jyoti Chattopadhyaya (University of Uppsala, Uppsala, Sweden).
Description: Hoboken, NJ : Wiley, 2023. | Series: Wiley series in drug discovery and development | Includes bibliographical references and index.
Identifiers: LCCN 2022033410 (print) | LCCN 2022033411 (ebook) | ISBN 9781119692744 (hardcover) | ISBN 9781119692751 (adobe pdf) | ISBN 9781119692782 (epub)
Subjects: LCSH: Nucleic acids–Therapeutic use. | Pharmaceutical chemistry.
Classification: LCC RM666.N87 N835 2023 (print) | LCC RM666.N87 (ebook) | DDC 615.1/9–dc23/eng/20221005
LC record available at https://lccn.loc.gov/2022033410
LC ebook record available at https://lccn.loc.gov/2022033411

Cover Design: Wiley
Cover Image: Courtesy of Xiaoxuan Su and Yu Zhang

Set in 10.25/12pt TimesLTStd by Straive, Pondicherry, India

CONTENTS

PREFACE

As is often the case, research progress involving a given class of compounds can be limited by its availability. As a consequence of extensive investigations carried out by many laboratories, it is now possible to acquire (modified) DNAs and RNAs of diverse structure commercially, or by following detailed preparative descriptions in the literature. This certainly now includes the nucleic acids that encode many proteins, facilitating the enhanced study of both classes of molecules. The increased accessibility of such molecules is clear from the range of topics addressed in the current volume.

One would expect advances of understanding in many areas of inquiry to be enabled by ready access to nucleic acids and molecules with which they interact. The chapters in this volume certainly provide this increased level of insight into biological systems and their modes of function. Perhaps somewhat more surprising is the large increase in the number of applications reflected in the individual chapters of this volume. These now extend well beyond the development of new therapeutic strategies to a great variety of new tools to interrogate detailed elements of biological and biochemical processes. A number of these applications are introduced briefly below as part of this preface to illustrate the expanded reach of current studies in nucleic acids.

The current volume addresses several aspects of ongoing research activities pertinent to the medicinal chemistry of nucleic acids. Three chapters deal with nucleoside and nucleotide analogues as physiological mediators. Among these is Chapter 1, which deals with nucleoside analogues that function in biological systems as phosphorylated species. Such nucleoside analogues are generally not internalized by cells when phosphorylated due to the accompanying negative charge and often cannot be phosphorylated efficiently intracellularly. An alternative is the administration of neutral phosphate and phosphonate prodrugs. These can be unmasked within cells using any of several strategies. Chapter 2 discusses the structures of the cyclic dinucleotides and the mechanisms that they use to produce an immune response. This includes the STING (stimulator of interferon) pathway and the identification of hSTING protein, which has 379 amino acids and for which 3 structural domains have been defined. There are potential applications in anticancer therapy (via IFN, cytotoxic T-lymphocytes, and natural killer cells). The cyclic dinucleotides are good candidates for adjuvant chemotherapy with numerous therapeutic agents. Cyclic dinucleotides are also potentially useful for the treatment of autoimmune diseases resulting from STING gain-of-function mutations. Chapter 3 discusses cyclic ADP-ribose and NAADP, two Ca-mobilizing messengers. They have numerous biological functions, including Ca trafficking processes and the transfer of mitochondria from human astrocytes to glioma cells. The chapter also discusses their enzymatic synthesis (involving ADP-ribosyl cyclase and CD38). Type III CD38 may also be involved in the endogenous production of cyclic ADP-ribose, while type II CD38 is implicated in the endogenous production of NAADP. Recently, it was shown that protein SARM1, not at all similar to CD38, is expressed in neuronal tissue that biosynthesizes both mediators.

Two additional chapters are also concerned with medicinal chemistry. These include Chapter 14, which details the use of structurally distinct forms of G-quadruplexes as targets for selective binding by small molecules. The small molecules include derivatives of indoles and bis-indoles, which have been functionalized with side chains that can be protonated to enable specific types of G-tetrad interactions. Recently discovered ligands include copper metallo-crowns and trisubstituted fan-shaped trinuclear Pt(II) complexes. Additionally, carbohydrates have been used to functionalize G-tetrad binders, imparting increased affinity and selectivity, as well as improved solubility. Finally, Chapter 8 describes the use of DNA-encoded chemical libraries as a strategy for drug discovery. There is a surprisingly large number of strategies that can be employed to utilize DNA encoding for the discovery of small molecules potentially useful as drugs, and this chapter cites several small-molecule clinical candidates discovered from DNA-encoded libraries.

Of continuing interest to the nucleic acids community are the numerous studies that employ molecules capable of interacting selectively with nucleic acids. These studies have provided very important insights in the past and promise to continue to do so. Chapter 5 discusses work that has been done to define and utilize compounds that can bind to chromosomal DNA in a sequence-specific fashion. As noted, intervening at the level of cellular DNA is very attractive conceptually because the number of target molecules is minimal. The authors compare several natural, mostly protein-based systems (including CRISPR/Cas9, TALENs, ZNF, and TFOs) to synthetic peptide nucleic acids (PNA) and pyrrole-imidazole polyamides, noting key differences and issues. The chapter notes recent initiatives, including targeted mutagenesis, homologous recombination, and the use of PNA for dsDNA invasion, as well as potential applications (e.g. gene editing, gene modulation via epigenetics, and PNA regulation of gene structure). Chapter 12 provides a highly informative summary of the basics of gene editing with CRISPR/Cas9 for therapeutic applications. It also includes the details of a clinical trial to counter the effects of HIV infection by ablating the *CCR5* gene. Yet another example of molecules that interact selectively with nucleic acids is provided in Chapter 10, which deals with the fluorescence hybridization imaging of neurological disease mRNAs in live neuronal cells. MAOs occur in the mitochondrial outer membrane. The imaging probes included MAOA-specific PNA dodecamers, separated by an N-terminal spacer to a μ-opioid receptor targeting peptide, with a spacer and fluorophore at the C-terminus. These were delivered into SH-SY5Y neuroblastoma cells through μ-opioid receptor-mediated endocytosis.

Two strategies that have been pursued for a number of years continue to receive substantial attention. One of these, oligonucleotide therapeutics, has arguably become even more active in the past several years. The comprehensive review of this topic in Chapter 11, comparing ASOs and siRNAs, includes a discussion of adjuvant delivery by GalNAc and lipids. Antibodies have also been used as conjugates. Oligonucleotide drugs can work by active mechanisms involving degradation of endogenous mRNA target sequences (by RNase H or argonaute-2). An additional mechanism involves physical blocking, e.g. by translation arrest or modulation of RNA processing. Finally, Chapter 13 describes the properties functions and applications of aptamers, single-stranded DNA or RNA that can bind selectively to a specific target. Aptamers occur in nature as riboswitches and can be selected to mediate enzyme inhibition, growth factor neutralization, immune function, and antiviral activity.

The remaining four chapters deal with modified nucleic acids. There are numerous modified nucleosides that appear both in RNA and DNA, and these have a variety of functions. High density CpG methylation of DNA promoter regions can block transcription factor binding and diminish gene expression. Likewise, post-transcriptional modification of mRNA can affect its stability and expression efficiency. There is a strong link of epigenetic modifications and disease, as well established for cancers. Chapter 7 outlines techniques used to detect and characterize such modifications. Also of interest is Chapter 6, which describes the preparation of modified genomes and organisms. Such modified genomes contain new structural features, such as the elimination of repetitive sequences to improve stability, the minimization of genome size, and recoding (simplification) of genomes. The crystallization of nucleic acid structures and their characterization by X-ray crystallography is greatly facilitated by the introduction of Se atoms at specific positions. As described in Chapter 4, there are now methods for effecting substitutions throughout the entire DNA oligonucleotide structure. Finally, the study of DNA origami has become very popular and has led to a variety of techniques for preparing highly specialized structures. Chapter 9 focuses on the preparation of nucleic acid-based nanocarriers that can be used for drug delivery, including three-dimensional cages. These include bio-orthogonal conjugation-induced functionalization of DNA nanocages. RGD-conjugated DNA structures showed high binding affinity to HeLa cells, and a cell-penetrating peptide (CPP)-modified DNA nanostructure developed to facilitate endocytosis-mediated uptake into target cells.

Sidney M. Hecht
Arizona State University

CONTRIBUTORS

JUNBIAO CHANG College of Chemistry, Zhengzhou University, Zhengzhou, China; School of Pharmaceutical Sciences, Zhengzhou University, Zhengzhou, China

CEN CHEN SeNA Research Institute, Chengdu, Sichuan, China

HONGKUI DENG School of Basic Medical Sciences, State Key Laboratory of Natural and Biomimetic Drugs, Peking University Health Science Center and the MOE Key Laboratory of Cell Proliferation and Differentiation, College of Life Sciences, Peking-Tsinghua Center for Life Sciences, Peking University, Beijing, China

SANGWOO HAN Department of Chemical Engineering, University of Seoul, Seoul, Republic of Korea

ZHEN HUANG SeNA Research Institute, Chengdu, Sichuan, China; College of Life Sciences, Sichuan University, Chengdu, Sichuan, China; Department of Biochemistry, Szostak-CDHT Large Nucleic Acids Institute, Chengdu, Sichuan, China

YOONBIN JI Department of Chemical Engineering, University of Seoul, Seoul, Republic of Korea

DAJEONG KIM Department of Chemical Engineering, University of Seoul, Seoul, Republic of Korea

Hon Cheung Lee State Key Laboratory of Chemical Oncogenomics, Key Laboratory of Chemical Genomics, Peking University Shenzhen Graduate School, Shenzhen, China

Jong Bum Lee Department of Chemical Engineering, University of Seoul, Seoul, Republic of Korea

Xiaoyu Li Department of Chemistry and State Key Laboratory of Synthetic Chemistry, The University of Hong Kong, Hong Kong, China; Laboratory for Synthetic Chemistry and Chemical Biology, Health@InnoHK Program by Innovation and Technology Commission, Hong Kong, China

Yizhou Li Chongqing Key Laboratory of Natural Product Synthesis and Drug Research, Chemical Biology Research Center, School of Pharmaceutical Sciences, Chongqing University, Chongqing, China

Yue Liu Frontiers Science Center for Synthetic Biology and Key Laboratory of Systems Bioengineering of Ministry of Education, School of Chemical Engineering and Technology, Tianjin University, Tianjin, PR China

Yulin Liu School of Basic Medical Sciences, State Key Laboratory of Natural and Biomimetic Drugs, Peking University Health Science Center and the MOE Key Laboratory of Cell Proliferation and Differentiation, College of Life Sciences, Peking-Tsinghua Center for Life Sciences, Peking University, Beijing, China

Dejun Ma Department of Chemical Biology and State Key Laboratory of Elemento-Organic Chemistry, College of Chemistry, Nankai University, Tianjin, China

Sunghyun Moon Department of Chemical Engineering, University of Seoul, Seoul, Republic of Korea

Manlio Palumbo Department of Pharmaceutical and Pharmacological Sciences, University of Padova, Padova, Italy

Yufei Pan State Key Laboratory of Natural and Biomimetic Drugs, School of Pharmaceutical Sciences, Peking University, Beijing, China

Ganesh N. Pandian Institute for Integrated Cell-Material Sciences (iCeMS), Kyoto University, Kyoto, Japan

Shuang Peng College of Chemistry and Molecular Sciences, Key Laboratory of Biomedical Polymers of Ministry of Education, Hubei Province Key Laboratory of Allergy and Immunology, Wuhan University, Wuhan, Hubei, China

Claudia Sissi Department of Pharmaceutical and Pharmacological Sciences, University of Padova, Padova, Italy

Hiroshi Sugiyama Institute for Integrated Cell-Material Sciences (iCeMS), Kyoto University, Kyoto, Japan; Department of Chemistry, Graduate School of Science, Kyoto University, Kyoto, Japan

Shicheng Sun The Murdoch Children's Research Institute, Department of Paediatrics, University of Melbourne, Parkville, VIC, Australia

Weihong Tan The Cancer Hospital of the University of Chinese Academy of Sciences (Zhejiang Cancer Hospital), Institute of Basic Medicine and Cancer (IBMC), Chinese Academy of Sciences, Hangzhou, Zhejiang, China; Molecular Science and Biomedicine Laboratory, State Key Laboratory of Chemo/Biosensing and Chemometrics, College of Chemistry and Chemical Engineering, College of Biology, Aptamer Engineering Center of Hunan Province, Hunan University, Changsha, Hunan, China

Zhenghua Wang Department of Chemical Biology and State Key Laboratory of Elemento-Organic Chemistry, College of Chemistry, Nankai University, Tianjin, China

Xiaocheng Weng College of Chemistry and Molecular Sciences, Key Laboratory of Biomedical Polymers of Ministry of Education, Hubei Province Key Laboratory of Allergy and Immunology, Wuhan University, Wuhan, Hubei, China

Eric Wickstrom Laboratory of Molecular Design, Thomas Jefferson University, Philadelphia, PA, USA

Zhen Xi Department of Chemical Biology and State Key Laboratory of Elemento-Organic Chemistry, College of Chemistry, Nankai University, Tianjin, China

Ze Xiong Xie Frontiers Science Center for Synthetic Biology and Key Laboratory of Systems Bioengineering of Ministry of Education, School of Chemical Engineering and Technology, Tianjin University, Tianjin, PR China

Fukang Yang SeNA Research Institute, Chengdu, Sichuan, China

Shilian Yang Chongqing Key Laboratory of Natural Product Synthesis and Drug Research, Chemical Biology Research Center, School of Pharmaceutical Sciences, Chongqing University, Chongqing, China

Zhenjun Yang State Key Laboratory of Natural and Biomimetic Drugs, School of Pharmaceutical Sciences, Peking University, Beijing, China

Wenquan Yu College of Chemistry, Zhengzhou University, Zhengzhou, China

Zutao Yu Institute for Integrated Cell-Material Sciences (iCeMS), Kyoto University, Kyoto, Japan

Kexin Yuan College of Chemistry and Molecular Sciences, Key Laboratory of Biomedical Polymers of Ministry of Education, Hubei Province Key Laboratory of Allergy and Immunology, Wuhan University, Wuhan, Hubei, China

Ying Jin Yuan Frontiers Science Center for Synthetic Biology and Key Laboratory of Systems Bioengineering of Ministry of Education, School of Chemical Engineering and Technology, Tianjin University, Tianjin, PR China

Liqin Zhang State Key Laboratory of Natural and Biomimetic Drugs, School of Pharmaceutical Sciences, Peking University, Beijing, China

Guixian Zhao Chongqing Key Laboratory of Natural Product Synthesis and Drug Research, Chemical Biology Research Center, School of Pharmaceutical Sciences, Chongqing University, Chongqing, China

Yong Juan Zhao Ciechanover Institute of Precision and Regenerative Medicine, School of Life and Health Sciences, The Chinese University of Hong Kong, Shenzhen, China

Zilong Zhao Molecular Science and Biomedicine Laboratory, State Key Laboratory of Chemo/Biosensing and Chemometrics, College of Chemistry and Chemical Engineering, College of Biology, Aptamer Engineering Center of Hunan Province, Hunan University, Changsha, Hunan, China

Xiang Zhou College of Chemistry and Molecular Sciences, Key Laboratory of Biomedical Polymers of Ministry of Education, Hubei Province Key Laboratory of Allergy and Immunology, Wuhan University, Wuhan, Hubei, China

Xinyang Zhou State Key Laboratory of Natural and Biomimetic Drugs, School of Pharmaceutical Sciences, Peking University, Beijing, China; School of Criminal Investigation & Forensic Science, People's Public Security University of China, Beijing, China

Rui Ying Zhu Frontiers Science Center for Synthetic Biology and Key Laboratory of Systems Bioengineering of Ministry of Education, School of Chemical Engineering and Technology, Tianjin University, Tianjin, PR China

Guangrong Zou College of Chemistry and Molecular Sciences, Key Laboratory of Biomedical Polymers of Ministry of Education, Hubei Province Key Laboratory of Allergy and Immunology, Wuhan University, Wuhan, Hubei, China

FOREWORD

NUCLEIC ACID-BASED THERAPEUTICS: A TOUR DE FORCE

As the serial editor, it gives me special pleasure to write a foreword for this volume focused on nucleic acids in the *Wiley Series in Drug Discovery and Develop* to showcase past accomplishments and envision future opportunities. This is the second edition of a monograph published in 2011. This new edition includes both updates of some earlier chapters and new chapters that discuss various aspects of nucleic acid-based therapeutics. The foreword by Professor Sidney Hecht provides comprehensive highlights of this new edition. Thus, this foreword aims to complement by providing some context and background for this monograph. First, the *Wiley Series in Drug Discovery and Develop* published its inaugural book in 2004. Since then, about 30 volumes have been published covering a broad range of subjects that impact drug discovery and development. The monograph on nucleic acid is especially important because of the central roles that this field has played in advancing biomedical science. Since the historical work by many on DNA structures culminating with the proposal of the double helix structure for DNA by Watson and Crick in the 1950s, discoveries of milestone nature have continued in the nucleic acid field. The human genome project was a monumental undertaking that has fundamentally transformed the biomedical field. Further, the nucleic acid field has been recognized with the Nobel Prize no less than seven times in the past 20 years in areas including telomeres and chromosomal protection, RNA interference, split genes, genome editing, DNA repairs, DNA chemistry (PCR and site-directed mutagenesis), and ribosomal structures and functions. With all the advancements in the nucleic acid field as the background, the first edition of this book was a tremendous success, which led to the commissioning of the second edition nine years later. Furthermore, the nucleic acid field has seen much progress including CRISPR gene editing, antisense therapy, DNA-encoded library, and nanoparticle-based delivery, which need to be reflected in a new edition, especially in relation to their application in developing new therapeutics.

In terms of medicinal chemistry work in the field of nucleosides, nucleotides, and nucleic acids, there has been an impressive number of success stories from nucleoside-based small-molecule drugs. However, oligonucleotides and nucleic acids are somewhat different. On one hand, they present exciting opportunities in precision and specificity in addressing the various underlying molecular mechanisms of diseases. On the other hand, the development of oligonucleotide-/nucleic acid-based therapeutics also faces unique challenges in delivery. Nevertheless, it is exciting to see the tremendous progress that has been made along this line. One transformative example is the recent success of mRNA-based vaccines for the SARS-CoV-2 virus, which are poised to rescue the world from the possibly catastrophic consequence of a pandemic that at times seemed to be out of control. The success of these mRNA-based vaccines also helps to break the myth of "drug-likeness" that has hovered over the field of nucleic acid-based therapeutics for a long time. At a time when nucleic acid-based approaches are positioned to bring a large number of unique solutions to issues of human health and suffering, this volume timely focuses on the medicinal chemistry and therapeutic application aspects of oligonucleotides and nucleic acids including important drug delivery approaches. It is quite a *tour de force* covering past accomplishments and, yet more importantly, it envisions future opportunities for much more to come. It is hoped that this volume will help to stimulate increasing interests in this field in order to accelerate the development of nucleic acid-based therapeutics.

On a personal note, the lead editor, Professor Zhang, and his wife, Professor Peiying Zhang, were my professors when I was an undergraduate student at Peking University Health Sciences Center (then Beijing Medical College). Thus, it is with special personal gratitude that I write this foreword to express my appreciation of Professor Zhang's efforts in leading a group of four editors in organizing this book and very importantly of his contributions to the development of the nucleic acid chemistry field in China and worldwide.

With much progress made in the past few decades, the time has come for nucleic acids to be a major force in the development of therapeutic options with precision and specificity!

Binghe Wang
April 11, 2021, On Lake Allatoona

INTRODUCTION

In 2011, we invited some scientists to contribute their research progress in areas of nucleic acid medicinal chemistry and published a collection of these articles, named by ***Medicinal Chemistry in Nucleic Acids*** by ***Wiley.*** As nine years passed since the publishing of above book, and many new developments have been achieved in nucleic acid medicinal chemistry and chemical biology, we think that it's right time to publish a new edition in this field.

Nucleoside, nucleotide, and nucleobase analogues have been widely utilized for the treatment of viral pathogens, neoplasms, and in anticancer chemotherapy for decades. Nucleoside analogues are known to be incorporated into DNA and RNA, producing the replication and transcription inhibition, respectively. It is believed that nucleoside transporter proteins are responsible for cellular uptake of these nucleoside derivatives. Once they enter cells, some kinases phosphorylate nucleoside analogues to the corresponding mono-, di-, and triphosphates. Nucleoside triphosphate analogues are regarded as the active cytotoxic form that leads to cell apoptosis as these are substrates for DNA polymerases and can be incorporated into DNA. When they integrated into an elongate viral DNA chain, the nucleotide inhibitor works as a chain terminator to terminate viral DNA synthesis [1–4]. Moreover, certain nucleoside 5-O-triphosphate analogues also can be incorporated into RNA to inhibit RNA synthesis. Currently, thousands of nucleoside and nucleotide derivatives are synthesized and their synthetic strategy is highlighted [5].

The attractive progress in this field would be new anti-HBV and anti-HCV drugs approved for market. The first-generation anti-HBV drug is Entecavir, which was approved by FDA in 2005 [6]. Tenofovir (TFV) was originally described as anti-HIV agent in 1993 under the name of (R)-PMPA. Then many TFV derivatives were synthesized through modification of phosphate moiety such as tenofovir disoproxil fumarate (TDF). Tenofovir alafenamide (TAF) can be considered a new prodrug of TFV with enhanced delivery to lymphoid and hepatocytes. Its active diphosphate metabolite can target RNA-dependent DNA polymerase (reverse transcriptase) of both HIV and HBV. TAF is equally potent as an antiretrovirus agent at a 30-fold lower dose than TDF. TAF (Vamlidy), approved by FDA in 2016, is not only described for the treatment and prevention of HIV infections but also shows promise for the treatment of HBV infections [7] (Figure 1).

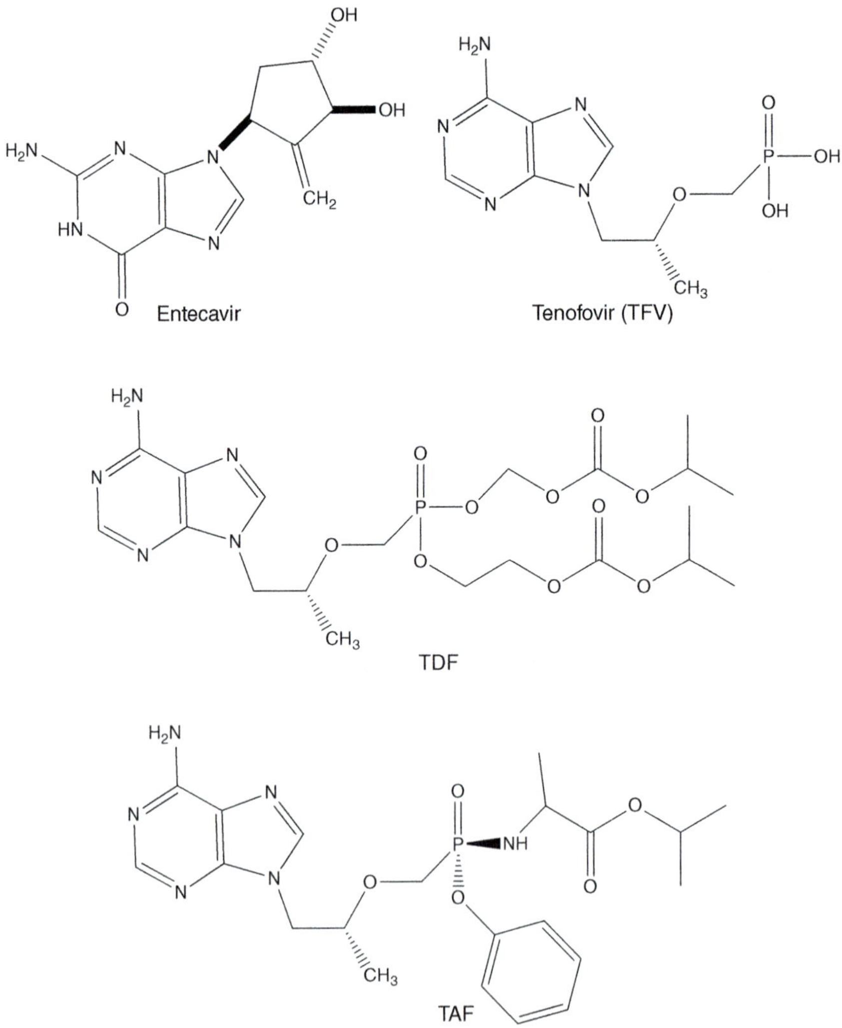

Figure 1 Structures of entecavir, tenofovir (TFV), and its derivatives TDV and TAF.

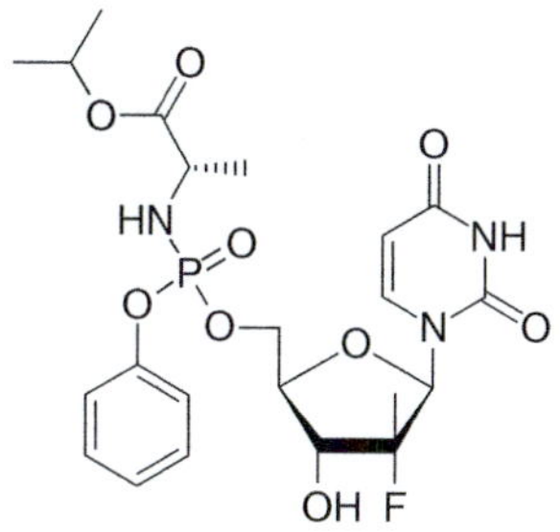

Figure 2 Structure of Sofosbuvir.

HCV is a potentially life-threatening disease and may lead to progressive liver damage and liver failure consequently. The first generation of anti-HCV drug is sofosbuvir. It undergoes the intracellular metabolism to its pharmacologically active form, uridine analogue triphosphate (GS-461203). This active metabolite is then incorporated into HCV RNA by NS5B RNA-dependent RNA polymerase and leads to the chain termination of HCV RNA. *In vitro*, the polymerase activity of the recombinant NS5B from several different HCV genotypes can be inhibited by GS-461203. This magic anti-HCV drug is a fixed-dose combination tablet containing sofosbuvir and NS5A/B protease inhibitors velpatasvir and voxilapavir for oral administration [8]. Another fixed dosage combination tablet named Harnovi in market, where NS5A/B protease inhibitor ledipasvir is used instead of velpatasvir and voxilapavir (Figure 2).

Oligonucleotides targeting RNA have attracted substantial interest from both academic institutions and pharmaceutical companies. Antisense, Aptamer, and siRNA oligonucleotides are designed based on different mechanisms and developed to be new agents for the sequence specific gene silencing of target cellular mRNAs or selective interaction with target proteins. And the potential of mRNA technology for rapid vaccine development is valuable in light of COVID-19 pandemic [9].

The first antisense oligonucleotide drug is Vitravene, which was developed by Isis Pharmaceutics and approved by FDA in 1998 for the specific treatment of CMV Retinitis. Followed by Vitravene, an aptamer drug (Pegaptanib) was approved by FDA in 2003 targeting of VEGF (vascular endothelial growth factor) for the treatment of diabetic macular edema (DME). However, the *in vivo* therapeutic potency of the aptamer drug is crucially limited by their inherent physicochemical characteristics which may affect pharmacokinetic properties. These effects result in metabolic instability, rapid renal filtration, rapid distribution from the plasma compartment into the tissues (for example the liver or spleen), nonspecific immune activation, and polyanionic effects [10]. Pegaptanib has been superseded by VEGF-specific monoclonal antibodies that show improved therapeutic effects. Another antisense oligonucleotide Kynamro that targets messenger RNA for apolipoprotein B was developed and approved by FDA (2013) to treat homozygous familial hypercholesterolemia, FoFH.

The wide adoption of RNA-based therapeutics across the biopharmaceutical industry takes a few more years. Recent research achievements have begun to open a new area for RNA-based therapeutics, especially new delivery systems are developed for oligonucleotide drugs, such as *N*-acetylgalactosamine (GalNAc) conjugation [11] and lipid nanoparticle for the delivery drugs into hepatocytes. And several candidates were further developed for the treatment of hepatic viruses, liver-centric genetic diseases, and cardiometabolic disorders. Drug formulations of RNA-based therapeutics have made significant progress including pharmacodynamics-related challenges in targeting specificity, off-target RNAi activity and immune-mediated cytotoxicity, and pharmacokinetics-related challenges in systemic circulation, cellular uptake, and endosomal escape [12, 13]. In 2016, two antisense oligonucleotides, Eteplirsen (Exondys 510) and Spinraza (Nusinersen) were also approved by FDA for the treatment of Duchenne muscular dystrophy (DMD) and spinal muscular atrophy (SMA), respectively.

RNA interference (RNAi) is a hot field in nucleic acid drug development by the use of small inhibitory double-stranded RNA(siRNA) to target degradation of the sequence-specific cellular mRNAs, and as a result to silencing gene expression. With the more recent development of RNAi in mammalian systems, investigators have opened a new therapeutic approaches in human genetics and/or infectious diseases. Patisiran is the first RNAi therapeutics, approved by FDA in 2018, targeting transthyretin (TTR) for the treatment of polyneuropathy in adult patients with hereditary transthyretin-mediated (hATTR) amyloidosis, a progressive, debilitating, chronic, and often fatal disease. Another antisense oligonucleotide Givosiran for the treatment of hATTR has also been approved in 2019. Very recently, Novartis receives positive opinion for Leqvio (Inclisiran) from Committee for Medicinal Products for Humen Use (CHMP) of European Medicines Agency (EMA), a potential first-in-class siRNA for the treatment of high cholesterol. Inclisiran has already been approved by FDA now. In 23 November 2020, FDA approved another siRNA drug, Oxlumo (Lumasiran), for the treatment of primary hyperoxaluria type 1 (PH1) to lower oxalate in urine. Oxlumo works by degrading hydroxyacid oxidase 1(HAO1) messenger RNA and reducing the synthesis of glycolate oxidase, which inhibits hepatic production of oxalate, the toxic metabolite responsible for the clinical manifestations of PH1. Up to now, 431 RNA-based therapeutics are in different stage for clinic trials [9]. A blowout growth of RNA-based therapeutics will be expected in the future drug market.

Nucleic acid detection has shown a wide range of applications, including diagnostics, biosensing and bioimaging, affinity isolation, biomarker discovery. Recent advances in the sequence-specific detection of nucleic acids now provide a number of options for rapid and cost-effective diagnostics, such as nanomaterial-based nucleic acid detection approaches [14] and point-of-care CRISPR/Cas nucleic acid detection [15]. With the more recent development of sgRNA in CRISPR/Cas systems, investigators attempt the development of new therapeutic approaches in human genetics and/or infectious diseases. On another hand, precision medicine and gene therapy became a very hot field. Recent technological and analytical advances in genomics have made it possible to rapidly identify and interpret the genetic variation underlying a single patient's disease. The current COVID-19 pandemic presents a serious public health crisis, and a better understanding of the spread scope and pathway of the virus would be aided by more widespread testing. Nucleic-acid-based tests currently offer the most sensitive and early detection of COVID-19 [16].

RNA epigenetics is another hot research area in RNA chemical biology [17]. Many research contributions are still trying to figure out how and why around 170 different chemical modifications impact RNA biology. Now, increasing evidences suggest that RNA methylation plays the key role in diseases from cancers to infectious diseases. RNA methylation to form N^6-methyladenosine (m^6A) in mRNA accounts for the most abundant mRNA internal modification and has emerged as a widespread regulatory mechanism that controls gene expression in diverse physiological processes. For DNA, RNA, and histones alike, three factors, writers, readers and erasers, are most important for these regulation process: "writers" add post-synthesis modifications that alter the molecular structure to either recruit or repel "readers." Readers interpret those marks to alter transcription in the case of DNA and histones, or to modulate translation and degradation in the case of RNA. And "erasers" remove the modifications, restoring the unaltered functions of the source materials. For histone biology, drug developers have already made inroads against all three components of the writer–reader–eraser paradigm. The same things are going to do in RNAs, small-molecule inhibitors of the METTL3–METTL14 complex, which regulates epigenetic marks on RNA, have been developed for clinic trials [18]. On the another hand, many RNAs fold into structures that can be selectively targeted with small molecules, which makes it possible to develop sequence-based design of ligands targeting RNAs [19].

Drugs are often divided into the broad classes of small organic molecules and "biologics," the latter of which are most often medicinal preparations of proteins such as natural and recombinant antibodies. In general, drug discovery is the identification of novel active chemical compounds that typically target a desired protein using a biological assay. RNA-based therapeutics, targeting directly to the desired mRNA, will open a new field for the clinical medicine as well as medicinal chemistry. In this book, 14 research groups are invited to contribute the recent progresses in nucleic acid medicinal chemistry and chemical biology. I would like to take this opportunity to express my most sincere thanks for their contributions and also for the hard work of ***Wiley*** Publishers.

Li He Zhang
Xin Jing Tang
Beijing, China
31 December 2020

REFERENCES

1. Smal, C., Vertommen, D., Bertrand, L. et al. (2006) Identification of in vivo phosphorylation sites on human 2′-deoxycytidine kinase. Role of Ser-74 in the control of enzyme activity. *J. Biol. Chem.* 281 (8): 4887–4893.
2. Arner, E. S. and Eriksson, S. (1995) Mammalian deoxyribonucleoside kinases. *Pharmacol. Ther.* 67 (2): 155–186.
3. Saada-Reisch, A. (2004) Deoxyribonucleoside kinases in mitochondrial DNA depletion. *Nucleosides Nucleotides Nucleic Acids* 23 (8–9): 1205–1215.
4. Knecht, W., Petersen, G. E., Munch-Petersen, B. J. (2002) Deoxyribonucleoside kinases belonging to the thymidine kinase 2 (TK2)-like group vary significantly in substrate specificity, kinetics and feed-back regulation. *J. Mol. Biol.* 315 (4): 529–540.
5. Shelton, J., Lu, X., Hollenbaugh, J. A. et al. (2016) Metabolism, biochemical actions, and chemical synthesis of anticancer nucleosides, nucleotides, and base analogs. *Chem. Rev.* 116: 14379–14455.
6. Opio, C. K., Lee, W. M., Kirkpatrick, P. (2005) Entecavir. *Nat. Rev. Drug Discov.* 4: 535–536.
7. De Clercq, E. (2016) Tenofovir alafenamide (TAF) as the successor of tenofovir disoproxil fumarate (TDF). *Biochem. Pharmacol.* 119: 1–7.
8. Heo Y. -A. and Deeks, E. D. (2018) Sofosbuvir/Velpatasvir/Voxilaprevir: a review in chronic hepatitis C. *Drugs* 78: 577–587.
9. Wang, F., Zuroske, T., Watts, J. K. (2020) RNA therapeutics on the rise. *Nat. Rev. Drug Discov.* 19: 441–442
10. Zhou, J. and Rossi, J. (2017) Aptamers as targeted therapeutics: current potential and challenges. *Nat. Rev. Drug Discov.* 16: 181–202
11. Jayaprakash, K. N., Jennifer, L. S., Willoughby, A.C. et al. (2014) Multivalent N-acetylgalactosamine-conjugated siRNA localizes in hepatocytes and elicits robust RNAi-mediated gene silencing. *J. Am. Chem. Soc.* 136: 16958–16961
12. Dowdy, S. F. (2017) Overcoming cellular barriers for RNA therapeutics. *Nat. Biotechenol.* 35(3): 222–229
13. Setten, R. L., Rossi, J. J., Han, S. P. (2019) The current state and future directions of RNAi- based therapeutics. *Nat. Rev. Drug Discov.* 18: 421–446
14. Smith, S. J., Nemr, C. R., Kelley, S. O. (2017) Chemistry-driven approaches for ultrasensitive nucleic acid detection. *J. Am. Chem. Soc.* 139: 1020–1028
15. van Dongen, J. E., Berendsen, J. T. W., Steenbergen, R. D. M. (2020) Point-of-care CRISPR/Cas nucleic acid detection: Recent advances, challenges and opportunities. *Biosens. Bioelectron.* 166: 112445.
16. Esbin, M. N., Whitney, O. N., Chong, S. et al. (2020) Overcoming the bottleneck to widespread testing: a rapid review of nucleic acid testing approaches for COVID-19 detection. *RNA*, 26:771–783
17. Zaccara, S., Ries, R. J., Jaffrey, S. R. (2019) Reading, writing and erasing mRNA methylation. *Nat. Rev. Mol. Cell Biol.* 20: 608–624
18. Cully, M. (2019) Chemical inhibitors make their RNA epigenetic mark. *Nature* 18: 892–894
19. Costales, M. G., Childs-Disney, J. L., Haniff, H. S. (2020) How we think about targeting RNA with small molecules, *J. Med. Chem.* 63(17): 8880–8900

CHAPTER 1

DESIGN, SYNTHESIS, AND APPLICATIONS OF NUCLEOSIDE PHOSPHATE AND PHOSPHONATE PRODRUGS

JUNBIAO CHANG[1,2] AND WENQUAN YU[1]
[1] College of Chemistry, Zhengzhou University, Zhengzhou, China
[2] School of Pharmaceutical Sciences, Zhengzhou University, Zhengzhou, China

1.1. INTRODUCTION

For decades, numerous nucleoside analogs have been developed as drugs or drug candidates [1, 2] for the treatment of various cancers and infections by different viruses, such as human immunodeficiency virus (HIV), hepatitis B virus (HBV), hepatitis C virus (HCV), herpes simplex virus (HSV), and human cytomegalovirus (HCMV). In cells, nucleoside analog molecules are converted to the corresponding phosphates by nucleoside or nucleotide kinases [3, 4] and then are incorporated into viral or tumor DNA/RNA where they serve as chain terminators. For example, the anti-HIV nucleoside drug azvudine (FNC) [4–6] is further phosphorylated *in vivo* to the active nucleoside triphosphate (FNC-TP, Figure 1.1), which inhibits the viral replication. However, the structural differences between the nucleoside analogs and the natural nucleosides may affect the phosphorylation processes, thus decreasing the pharmacological activity of the compounds [7, 8]. Due to their poor chemical stability and high polarity, the corresponding phosphates themselves cannot be used directly as drugs

Nucleic Acids in Medicinal Chemistry and Chemical Biology: Drug Development and Clinical Applications, First Edition. Edited by Lihe Zhang, Xinjing Tang, Zhen Xi, and Jyoti Chattopadhyaya.

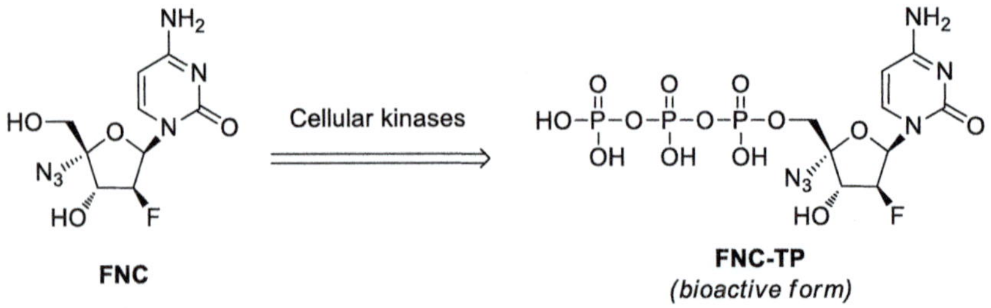

Figure 1.1 Formation *in vivo* of the active triphosphate (TP) of FNC.

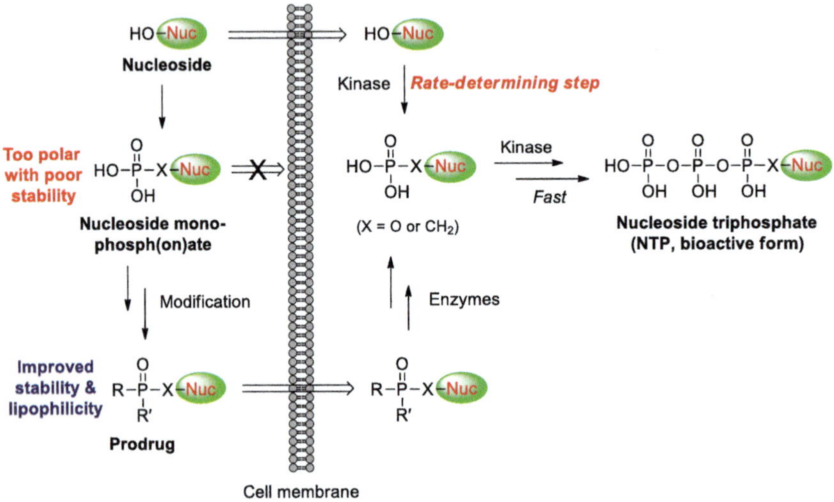

Figure 1.2 Design and mechanism of action of nucleoside phosph(on)ate prodrugs.

(Figure 1.2). To address these issues, diverse prodrugs of nucleoside (nucleotide) phosph(on)ate analogs have been designed and synthesized in drug discovery [3, 9–14].

Introduction of phosph(on)ate esters as prodrugs (Figure 1.2) has emerged as a very useful tool in the design and discovery of nucleoside and nucleotide analog drugs for the treatment of cancers and viral infectious diseases. The features of a prodrug may include (1) improved chemical stability; (2) increased lipophilicity for better bioavailability; (3) oral availability (the parent compound may often be administered only by injection); and (4) an improved therapeutic effect with reduced toxicity *via* targeted drug delivery. In recent years, considerable effort has been devoted to the design, synthesis, and biological evaluation of nucleoside phosph(on)ate prodrugs and more than twenty drug candidates have entered clinical development. Of these, several prodrugs have received FDA approval for clinical use. These include

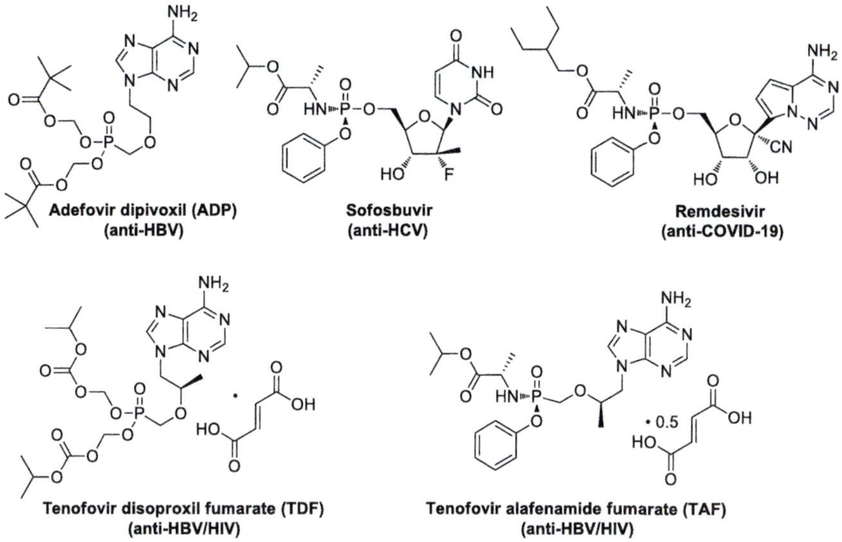

Figure 1.3 Nucleoside phosph(on)ate prodrugs in clinical use.

ADP, TDF, sofosbuvir, TAF, and remdesivir (Figure 1.3). In this chapter, four main classes of nucleoside phosph(on)ate prodrugs and prodrug candidates are discussed: carbonyloxymethyl diester prodrugs, alkoxyalkyl monoester prodrugs, cyclic 1-aryl-1,3-propanyl ester (HepDirect) prodrugs, and phosphoramidate/phosphonamidate prodrugs.

1.2. NUCLEOSIDE PHOSPH(ON)ATE PRODRUGS

1.2.1. Carbonyloxymethyl Diester Prodrugs

Two nucleotide prodrugs developed *via* this strategy, ADP (trade names: Preveon and Hepsera) and TDF (trade name: Viread), have been approved by FDA for the treatment of viral infectious diseases. Both the acyclic nucleoside phosphonate drugs have a carbonyloxymethyl diester moiety. Specifically, ADP contains a bis(pivaloyloxymethyl) (POM) substructure and TDF bears a bis(isopropyloxymethyl carbonate) group (POC). These carbonyloxymethyl (e.g., POM) and alkyloxycarbonyloxyalkyl (e.g., POC) diester units can increase the oral bioavailability and overall systemic exposure when compared to the parent nucleotide molecules. Activation of this carbonyloxymethyl ester prodrug involves the esterase-catalyzed cleavage of the first carbonate ester group followed by chemical degradation to form an unstable POM- or POC-monoester (**2**). The monoester intermediate then undergoes a second degradation cycle to afford the expected nucleoside monophosph(on)ate (**1**)

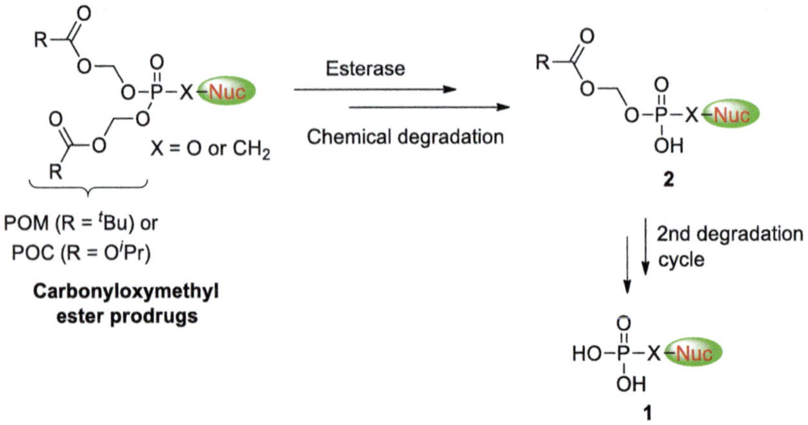

Figure 1.4 Activation of carbonyloxymethyl ester prodrugs.

(Figure 1.4). In addition to the above two marketed nucleotide prodrugs, besifovir (**LB80380**) is a bis(POM)-type prodrug candidate of an acyclic purine nucleotide analog and has entered clinical development for the treatment of HBV infection.

1.2.1.1. Adefovir Dipivoxil Adefovir (ADV), also known as 9-(2-phosphonomethoxyethyl)adenine (PMEA), is an acyclic nucleotide analog of adenosine monophosphate. ADV was first reported by De Clercq *et al.* [15] as having potent antiviral activity against HIV and other retroviruses [16, 17]. To increase the lipophilicity and intestinal permeability of the parent compound (ADV) for oral administration, various prodrugs of adefovir have been designed. However, the previously designed prodrugs with simple alkyl diester or amide groups failed, probably due to their inefficient degradation to adefovir *in vivo* [18], and the monoesters showed poor oral bioavailability because of the unmasked negative charge. Eventually, ADP (bis-POM PMEA, Figure 1.5) designed by masking the phosphonic acid

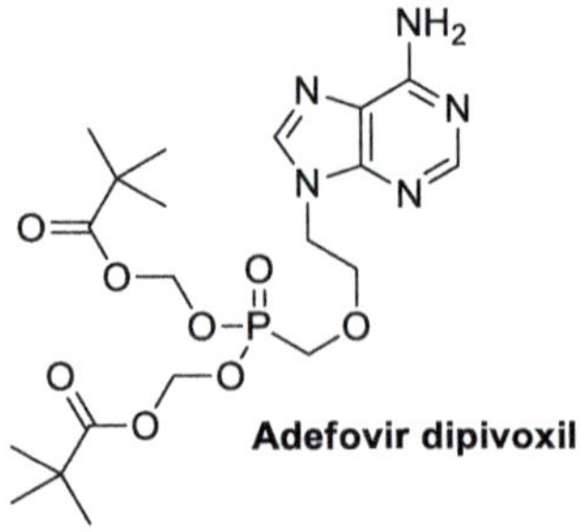

Figure 1.5 Structure of adefovir dipivoxil.

moiety with two POM groups was identified and was found to have favorable lipophilicity and intestinal permeability. It can be conveniently metabolized to the parent nucleotide adefovir *in vivo*, leading to high oral bioavailability. ADP was initially developed for the treatment of HIV infections (AIDS) [19, 20], but failed due to its toxicity. Further development demonstrated that this ADV prodrug can significantly reduce the HBV viral load at a nontoxic dose [21, 22]. ADP (Preveon and Hepsera) received FDA approval for the treatment of chronic HBV infection *via* oral administration as a reverse transcriptase (RT) inhibitor.

1.2.1.2. Tenofovir Disoproxil Fumarate Tenofovir, (*R*)-9-(2-phosphonylmethoxypropyl)adenine (PMPA), was designed by incorporation of a methyl group into the side chain of adefovir. As a nucleotide analog reverse transcriptase inhibitor (nRTI), PMPA exhibits good inhibitory activity against HIV and other retroviruses [23]. However, like other nucleotide analogs, the parent nucleotide analog suffers from poor oral bioavailability. To improve the oral delivery and also avoid the side effects [24] from pivalic acid derived from ADP, new prodrugs were designed and evaluated. Among these, the bis(isopropyloxycarbonyloxym ethyl) ester (bis-POC) prodrug (tenofovir disoproxil) displayed favorable solubility, stability, and improved oral bioavailability compared to the parent nucleotide and thus was selected for clinical development [25, 26]. TDF (Figure 1.6) has been approved by the FDA for the treatment of HIV/AIDS as well as chronic hepatitis B under the trade name Viread.

1.2.1.3. Besifovir Besifovir (**LB80380**, Figure 1.7) is a bis(POM) phosphonate prodrug of a 2-aminopurine nucleotide analog and is in clinical trials for the treatment of HBV infection [27]. **LB80380** displayed potent anti-HBV activity against wild-type and drug-resistant strains [28]. In the liver, the prodrug will be rapidly degraded to release the parent nucleotide, which is then oxidized at the C6-position of the purine moiety followed by phosphorylation to form the bioactive metabolites that serve as inhibitors of HBV DNA replication [27, 29].

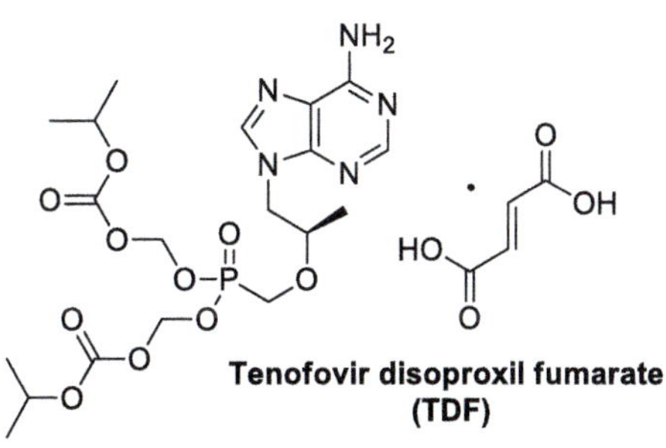

Figure 1.6 Structure of tenofovir disoproxil fumarate (TDF).

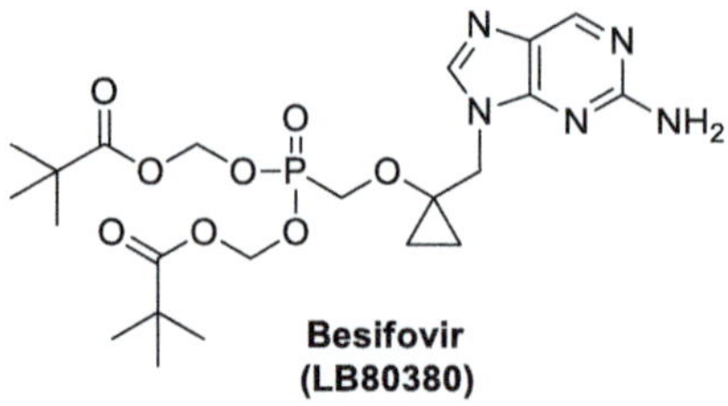

Figure 1.7 Structure of besifovir (**LB80380**).

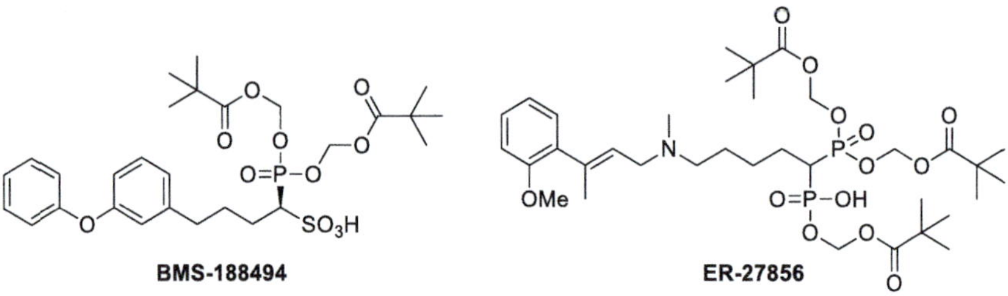

Figure 1.8 Structure of non-nucleoside prodrugs **BMS-188494** and **ER-27856**.

Application of the carbonyloxymethyl ester prodrug strategy to non-nucleoside compounds led to the discovery of **BMS-188494**, a bis(POM)-substituted squalene synthase inhibitor (Figure 1.8), which is in clinical trials for the treatment of hypercholesterolemia [30]. Another squalene synthase inhibitor bearing three POM groups (**ER-27856**) (Figure 1.8) was demonstrated to have a cholesterol-lowering effect [31] and inhibitory activity on the growth of *Trypanosoma cruzi* [32].

1.2.2. Alkoxyalkyl Monoester Prodrugs

In most of the phosph(on)ate prodrugs, both oxygens of the phosph(on)ate groups are fully masked, but in this prodrug approach, only one of them is masked and the other remains as a free OH group. The concept underlying the alkoxyalkyl monoester prodrug design is to mimic the structural features of the natural lysophosphatidylcholine (LPC). Further structural optimization of the masking groups led to hexadecyloxypropyl (HDP) and octadecyloxyethyl (ODE) derivatives, two important prodrug moieties. These prodrugs are orally available, and they may be delivered into the targeted tissues through the same uptake pathway in the small intestine as LPC. After reaching the desired tissue, the prodrug is cleaved by specific cellular enzymes such as phospholipase C to release the nucleoside monophosph(on)ate (**1**) (Figure 1.9). The drug candidates brincidofovir (**CMX-001**), **CMX-157**, and fozivudine tidoxil belong to the alkoxyalkyl monoester prodrug class.

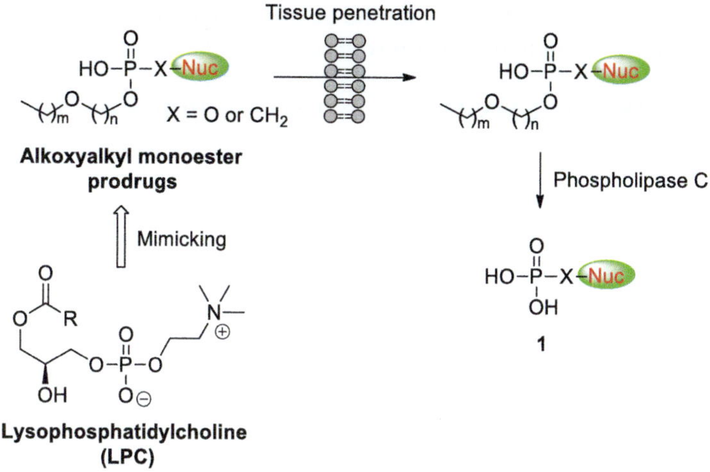

Figure 1.9 Design and activation of alkoxyalkyl monoester prodrugs.

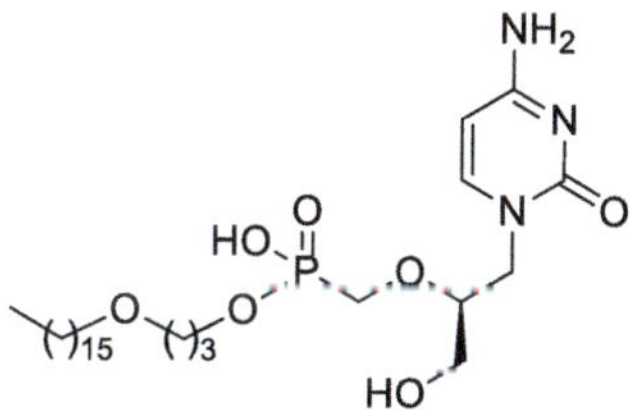

Brincidofovir (CMX-001)

Figure 1.10 Structure of brincidofovir (**CMX-001**).

1.2.2.1. Brincidofovir Cidofovir (CDV) is an FDA-approved injectable antiviral agent targeted to treat smallpox infections and complications caused by smallpox vaccines. To improve the lipophilicity of CDV and enhance the uptake of the orally administered drug in the small intestine, the prodrug brincidofovir (**CMX-001**, Figure 1.10) was designed with a lipid HDP monoester group linked to the phosphonate moiety [33]. As expected, **CMX-001** is more lipophilic than the parent compound and could be absorbed by active transport in the small intestine, resulting in a more rapid accumulation in the target cell cytoplasm. After being degraded to the parent nucleotide (CDV), it is further phosphorylated by host cell nucleoside kinases into cidofovir diphosphates, which serve as viral DNA polymerase inhibitors. **CMX-001** displays broad-spectrum antiviral activity against adenoviruses, smallpox, cytomegalovirus (CMV), papillomavirus, polyomavirus, and orthopoxviruses [34]. As an orally available drug candidate, brincidofovir has entered clinical development for the treatment of smallpox infections and complications related to the smallpox vaccine.

1.2.2.2. CMX-157 **CMX-157** (Figure 1.11) is a HDP monoester prodrug candidate of nucleotide analog tenofovir and is in clinical trials [35] as a potential treatment for HIV-1 and HBV infections. A lipophilic HDP group has been incorporated into this prodrug to mimic natural LPC for efficient drug delivery through the LPC uptake pathway and thus to improve oral availability, efficacy, and the toxicity profile of tenofovir [35, 36]. As expected, **CMX-157** displays > 300-fold better antiviral activity than the parent compound and is active against most of the HIV-1 virus subtypes [36].

1.2.2.3. Fozivudine Tidoxil Fozivudine tidoxil (Figure 1.12) is a prodrug of zidovudine (AZT) developed *via* the alkoxyalkyl monoester prodrug strategy with a similar thioether lipid moiety instead of the HDP group. Fozivudine tidoxil has progressed into phase II clinical trials for the treatment of patients with HIV infection.

1.2.3. Cyclic 1-Aryl-1,3-propanyl Ester (HepDirect) Prodrugs

A HepDirect prodrug approach [8, 10, 37, 38] was developed by Metabasis Therapeutics in the early 2000s. Prodrugs from this work contain cyclic 1-aryl-1,3-propanyl ester moieties, which possess high plasma and tissue stability. They can be metabolized by cytochrome P_{450} (CYP) enzymes that are

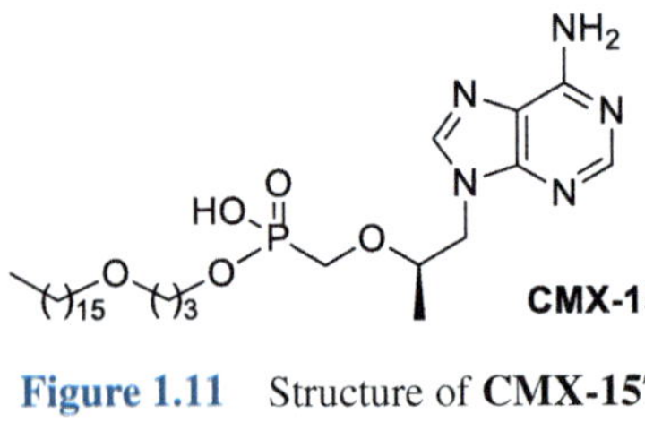

CMX-157

Figure 1.11 Structure of **CMX-157**.

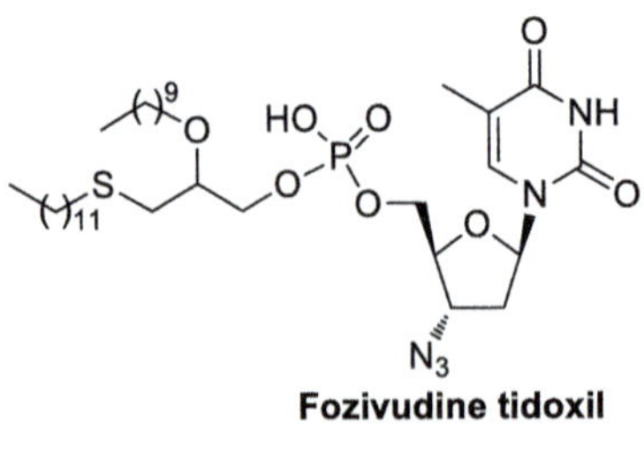

Fozivudine tidoxil

Figure 1.12 Structure of fozivudine tidoxil.

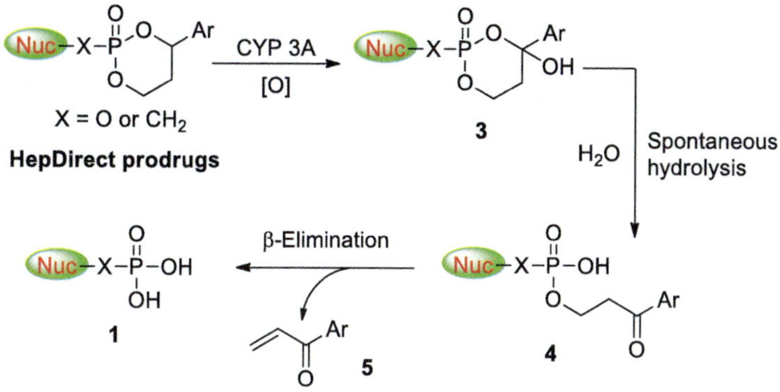

Figure 1.13 Activation of HepDirect prodrugs.

expressed predominantly in the liver to achieve liver-targeted drug delivery. The CYP-catalyzed degradation of the prodrugs results in oxidative hydroxylation at the benzylic position (**3**). Subsequently, ring-opening of the intermediate (**3**) followed by β-elimination liberates the expected nucleoside phosph(on)ate (**1**) (Figure 1.13). The resulting vinyl ketone (**5**) may further conjugate with glutathione to form less toxic metabolites. Nucleoside or nucleotide prodrug candidates developed *via* this strategy, which are in clinical trials, include pradefovir (**MB06866**) and **MB07133**.

1.2.3.1. Pradefovir Pradefovir (**MB06866**, Figure 1.14) is a HepDirect prodrug candidate of adefovir bearing a 3-chlorophenyl group. It was developed for the treatment of hepatitis B virus infection [38].

The previously marketed anti-HBV drug ADP is a carbonyloxymethyl diester prodrug of the acyclic nucleotide adefovir. ADP was originally developed as an antiviral agent against HIV, but failed due to side effects. Although it received FDA approval for the treatment of chronic hepatitis B, ADP is only used at submaximally efficacious doses on account of its renal toxicity. To improve the therapeutic index, cyclic 1-aryl-1,3-propanyl ester (HepDirect) prodrugs of the parent compound were designed and synthesized to deliver the bioactive

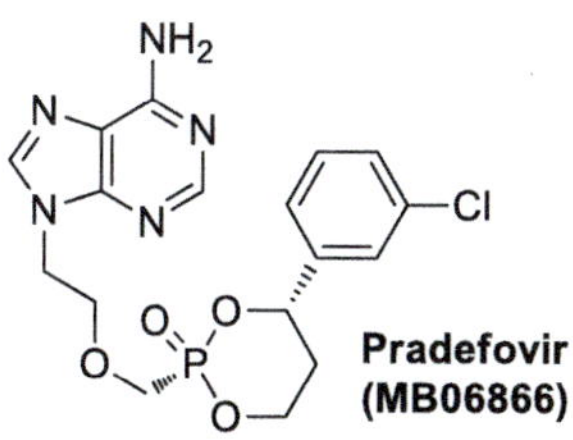

Figure 1.14 Structure of pradefovir (**MB06866**).

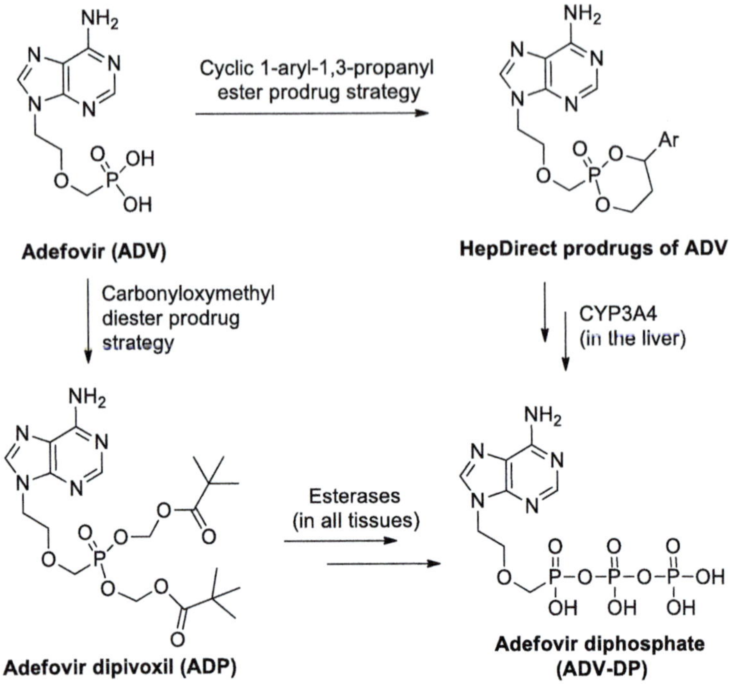

Figure 1.15 Design rationale of HepDirect prodrugs of adefovir.

adefovir diphosphate (ADP-DP) selectively to the liver (Figure 1.15) and thus reduce side effects by decreasing the exposure to the kidney [38].

Among the HepDirect prodrugs of ADV, the analog bearing a 3-chlorophenyl group displayed a high activation rate in microsomes, in primary hepatocytes, and orally administered in rats. Further studies suggested that the stereochemistry has different influences on the absorption and activation of the prodrug. Among the four stereoisomers, the racemic *cis*-isomers showed a high rate of activation but the racemic *trans*-isomers were not activated in human microsomes. The *cis*-isomers also gave a 12-fold improvement in the liver/kidney exposure ratio compared to ADP in rats *via* oral administration. The *cis*-(2*R*,4*S*) prodrug of the *cis*-diastereomers is the better-activated compound than the other *cis*-isomer. Evaluation of the *cis*-diastereomers led to the selection of the *cis*-(2*R*,4*S*) prodrug isomer in the form of the mesylate salt (**MB06866**) as the drug candidate for further development (Figure 1.16). Pradefovir (**MB06866**, previously known as remofovir) displayed favorable oral bioavailability in rats and dogs and has been advanced into clinical trials for the treatment of hepatitis B patients.

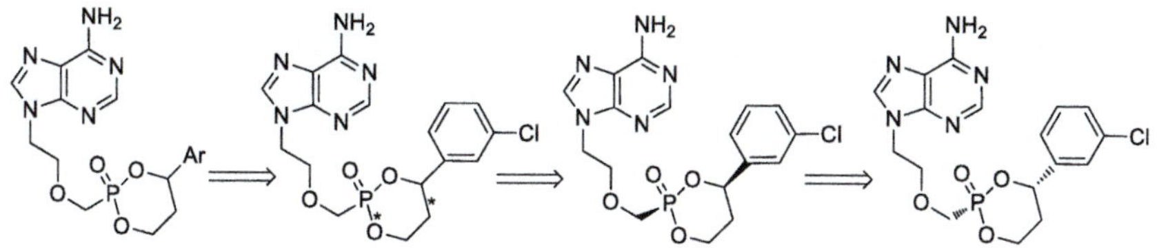

Figure 1.16 Lead optimization for the discovery of pradefovir (**MB06866**).

1.2.3.2. MB07133 Cytarabine is a pyrimidine nucleoside analog in clinical use for the treatment of acute myelocytic leukemia [39]. Like many nucleoside analogs, cytarabine has a limited phosphorylation rate forming the bioactive nucleoside triphosphate (NTP) *in vivo*, and the higher doses necessary to achieve the therapeutic NTP level may cause side effects, such as myelosuppression [40]. Application of the HepDirect prodrug strategy to this cytidine analog led to **MB07133** (Figure 1.17), a prodrug candidate to treat hepatocellular carcinoma (HCC) [40]. This prodrug with its 1-(4-pyridyl)-1,3-propanyl ester group exhibited profiles in activation, stability and solubility favorable for parenteral administration during the lead optimization. Compared to the parent nucleoside, **MB07133** can achieve much higher levels of the active cytarabine triphosphate in the liver than in the bone marrow and the plasma [40].

The HepDirect prodrug strategy was also involved in the discovery of non-nucleoside drug candidates. For example, **MB07811** is a liver-activated prodrug candidate of a phosphonic acid (PA) thyroid hormone receptor (TR) agonist and is used for the treatment of hyperlipidemia [41] (Figure 1.18). It displays potent lipid-lowering activity in a cholesterol-fed rat (CFR) model and good oral bioavailability ($F = 40\%$) in rats.

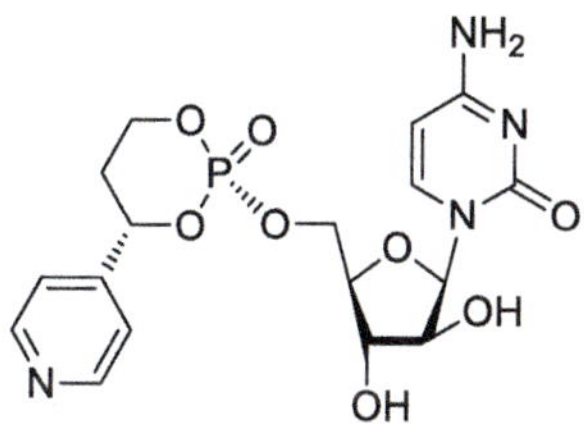

MB07133

Figure 1.17 Structure of **MB07133**.

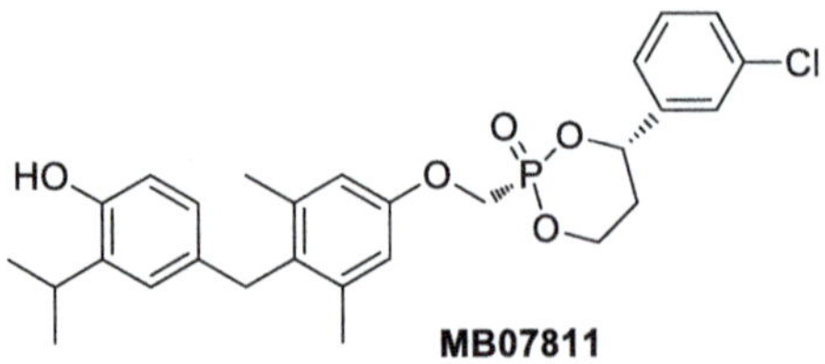

Figure 1.18 Structure of non-nucleoside prodrug **MB07811**.

1.2.4. Phosphoramidate and Phosphonamidate Prodrugs

In the early 1990s, McGuigan *et al.* developed the aryloxy phosphoramidate prodrug approach, also known as ProTide (PROdrug + nucleoTIDE) technology, to deliver nucleoside monophosphates into cells. Generally, in ProTide prodrugs, there is an amino acid alkyl ester and an aryloxy group connected to a phosphorus atom. During the development of this type of prodrug, a variety of related prodrugs were designed and synthesized [3], including di(halo)alkyl phosphates, (halo)alkyloxy phosphoramidates, phosphorodiamidates, diaryl phosphates, and alkyloxy phosphor-amidates (ProTides). Among these, the ProTide prodrugs are superior to other related drugs, as they not only can significantly increase or even reveal the biological activity of the parent nucleos(t)ides but also are more convenient to synthesize. To date, the ProTide technology has become a powerful tool for medicinal chemists in the discovery and development of new nucleos(t)ide prodrug candidates. This prodrug approach has led to three FDA-approved drugs, sofosbuvir (trade name: Sovaldi), TAF (trade name: Vemlidy), and remdesivir (trade name: Veklury), as well as a number of drug candidates for clinical development.

The mechanism of action of ProTide prodrugs [3, 42, 43] has been shown to be as illustrated in Figure 1.19. After entrance into cells, the metabolism of this prodrug starts with hydrolysis of the carboxylic ester group by an esterase

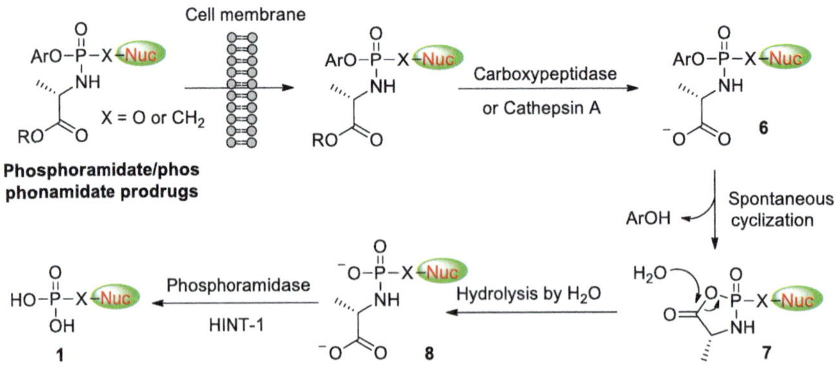

Figure 1.19 Activation of phosphoramidate/phosphonamidate prodrugs.

or cathepsin A to generate a carboxylate intermediate (**6**). Intramolecular cyclization of carboxylate **6** leads to a five-membered cyclic intermediate (**7**) by releasing a phenol molecule. Then, water-mediated ring-opening of compound **7** forms a diacid intermediate (**8**) and finally, cleavage of the amino acid moiety from compound **8** by intracellular phosphoramidase or the histidine triad nucleotide-binding protein 1 (HINT-1) produces the nucleoside monophosphate (**1**).

1.2.4.1. Sofosbuvir Sofosbuvir (**PSI-7977**, Sovaldi, Figure 1.20) is the first marketed prodrug developed *via* the ProTide approach and is used to treat patients infected with hepatitis C virus (HCV). Sofosbuvir is on the World Health Organization (WHO) model list of essential medicines [44]. After cellular uptake into hepatocytes, the prodrug undergoes the metabolism process illustrated in Figure 1.19 to form the pharmacologically active nucleoside triphosphate (NTP) [45]. The bioactive NTP will exert its action on HCV NS5B polymerase, which is responsible for the viral genome replication. By inhibiting HCV NS5B polymerase, sofosbuvir causes RNA chain termination and eventually stops the HCV replication [46].

The design rationale of the prodrug **PSI-7977** [46] is illustrated in Figure 1.21. Previous studies indicated that although its triphosphate (**9-TP**) is a potent HCV NS5B inhibitor [47–49], the parent nucleoside, 2′-fluoro-2′-*C*-methyluridine (**9**), is inactive in HCV replicon assays. Metabolism studies demonstrated that 2′-fluoro-2′-*C*-methylcytidine (**10**) can be metabolized to both the active triphosphate (**10-TP**) and the inactive 2′-fluoro-2′-*C*-methyluridine (**9**). Metabolism of the cytidine monophosphate (**10-MP**) leads to the monophosphate of uridine (**9-MP**), which could be further phosphorylated to the active uridine triphosphate (**9-TP**). However, 2′-fluoro-2′-*C*-methyluridine (**9**) cannot be directly metabolized into its monophosphate (**9-MP**) and this could be the reason why the uridine (**9**) itself is inactive. With a design based on these results, the nucleoside monophosphate (**9-MP**) prodrug may overcome the inefficient monophosphorylation step of the parent compound (**9**) to achieve high levels of the desired triphosphate (**9-TP**) in the liver.

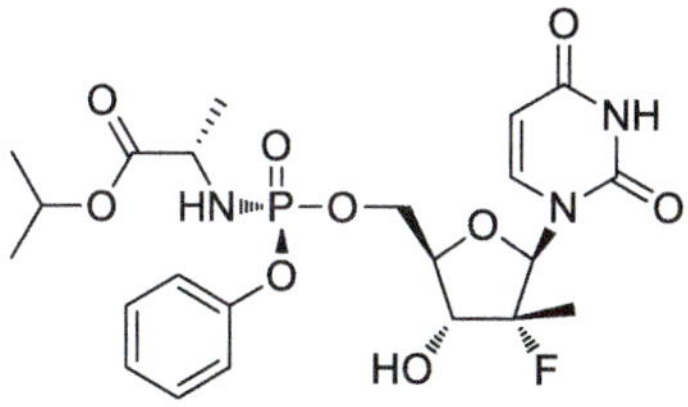

Sofosbuvir
(PSI-7977)

Figure 1.20 Structure of sofosbuvir (**PSI-7977**).

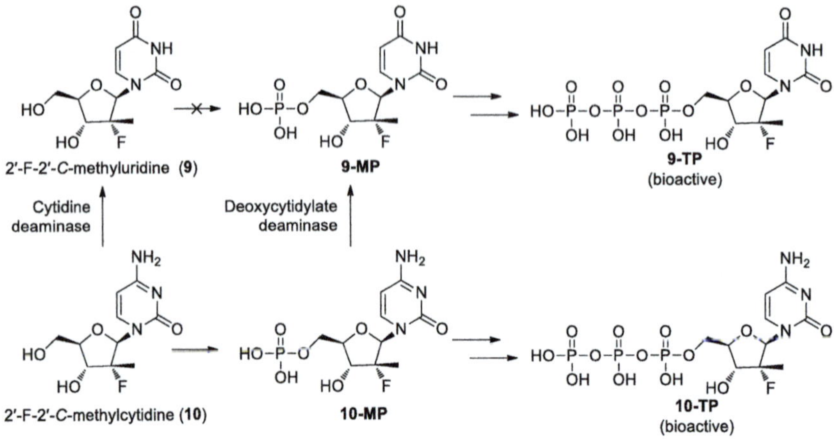

Figure 1.21 The design rationale of sofosbuvir (**PSI-7977**).

1.2.4.2. Tenofovir Alafenamide Fumarate Tenofovir alafenamide fumarate (TAF, **GS-7340**, Figure 1.22) is a new, orally available prodrug of the acyclic nucleotide analog tenofovir developed *via* the ProTide strategy. This prodrug has been approved by FDA [50] as a nucleotide reverse transcriptase inhibitor (Vemlidy) for the treatment of HIV-1 infection and chronic hepatitis B. Compared to the previously approved prodrug TDF, TAF exhibits significantly increased antiviral activity with higher stability *in vivo* and better lymphoid-tissue distribution [51]. The higher intracellular levels but lower plasma levels of TAF could be the reason for the fewer TAF-associated side effects compared to TDF [52]. After the cellular uptake, this ProTide is metabolized into the pharmacologically active tenofovir diphosphate (DP), which then results in the DNA chain termination, thus inhibiting the viral replication [53].

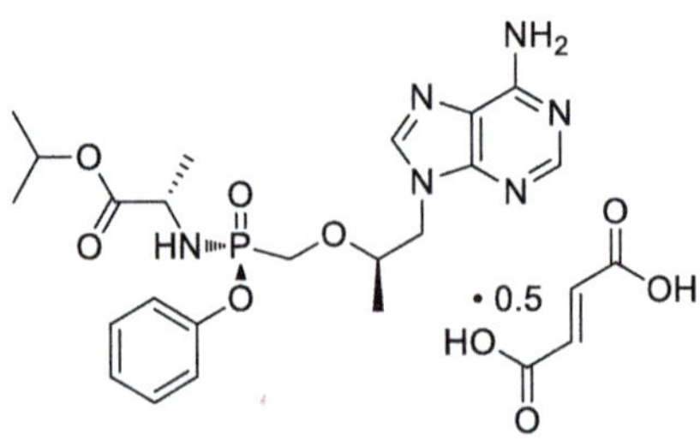

Tenofovir alafenamide fumarate (TAF)

Figure 1.22 Structure of tenofovir alafenamide fumarate (TAF).

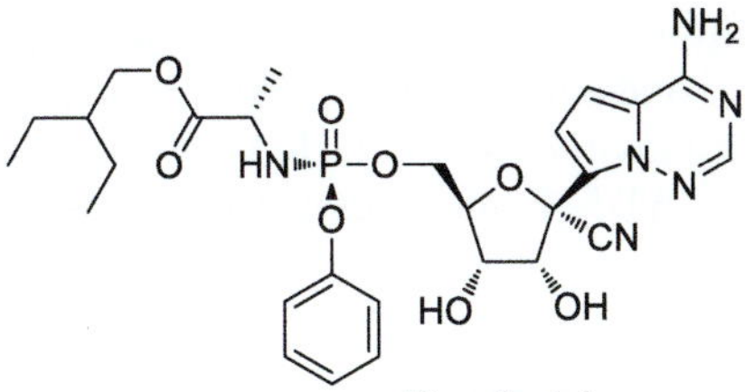

Remdesivir
(GS-5734)

Figure 1.23 Structure of remdesivir (**GS-5734**).

1.2.4.3. Remdesivir Remdesivir (**GS-5734**, Figure 1.23) is a *C*-nucleoside, 1′-cyano-4-aza-7,9-dideaza adenosine prodrug bearing a phenyl 2-ethylbutyl L-alanine ester phosphoramidate moiety. This ProTide prodrug (Veklury) was first developed as a potential antiviral agent for the treatment of Ebola *via* intravenous administration [54, 55]. **GS-5734** also displayed a broad spectrum antiviral activity against RNA viruses including arenaviruses, coronaviruses, and filoviruses. During the COVID-19 pandemic, it was originally granted an FDA Emergency Use Authorization (EUA) for use in patients with suspected or confirmed COVID-19 in May 2020 and received full approval in October 2020 to treat COVID-19. In addition, remdesivir has been approved or authorized in some 50 countries as an emergency treatment of COVID-19.

1.2.4.4. AT-527 **AT-527** (Figure 1.24) is a hemisulfate salt of a 2′-fluoro-2′-*C*-methylguanosine phosphoramidate prodrug. As a novel HCV NS5B polymerase inhibitor, **AT-527** possesses potent *in vitro* anti-HCV activity with favorable preclinical profiles [56, 57]. This ProTide prodrug displayed about 10-fold higher antiviral activity than sofosbuvir against laboratory strains as well as clinical isolates of HCV genotypes 1–5. It is fully active against S282T-resistant variants with up to 58-fold higher potency than sofosbuvir. No *in vitro* toxicity was observed at concentrations up to 100 μM. **AT-527** was predominantly metabolized into the active nucleoside triphosphate (NTP) in

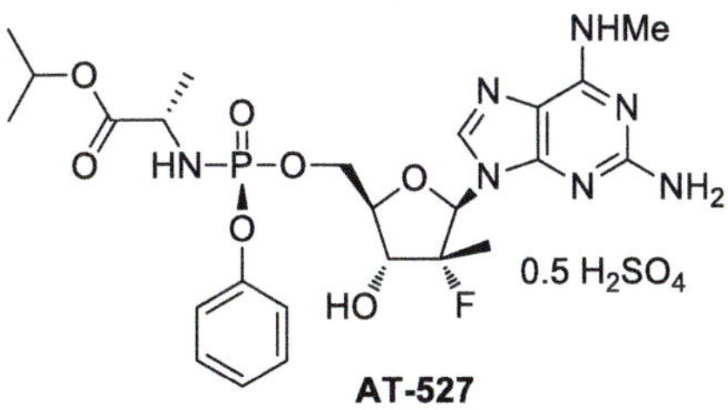

AT-527

Figure 1.24 Structure of **AT-527**.

hepatocytes from multiple species. Consistently, it led to high levels of NTP in rats and monkeys after oral administration. This prodrug has exhibited clinical tolerability and potent antiviral activity in HCV-infected patients with a favorable human PK profile. **AT-527** interferes with viral RNA polymerase, a key enzyme in the replication of enveloped, positive single-stranded RNA viruses, including flaviviruses and coronaviruses. It has also demonstrated antiviral activity against SARS-CoV-2 infection and is currently under evaluation in clinical development for the treatment of COVID-19.

1.2.4.5. GS-6620 **GS-6620** (Figure 1.25) was the first *C*-nucleoside derivative reported as an HCV polymerase inhibitor with antiviral response in hepatitis C patients [58, 59]. It is a ProTide prodrug of 1′-cyano-2′-*C*-methyl 4-aza-7,9-dideaza adenosine with the 3′-OH masked by the isobutyryl group. **GS-6620** has also been developed as a potential treatment for other infectious diseases caused by viruses such as Ebola [60, 61].

1.2.4.6. NUC-1031 **NUC-1031** (Figure 1.26) is a ProTide prodrug of gemcitabine and a new promising anticancer reagent [62]. The activation of this prodrug is less dependent on deoxycytidine kinase and nucleoside transporters than the nucleoside gemcitabine. **NUC-1031** is resistant to cytidine

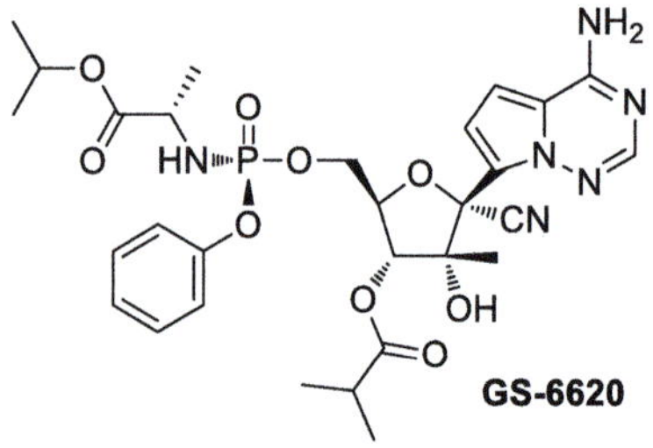

GS-6620

Figure 1.25 Structure of **GS-6620**.

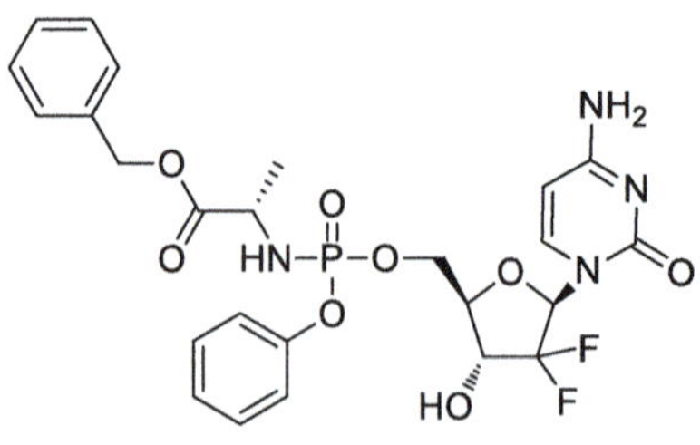

NUC-1031

Figure 1.26 Structure of **NUC-1031**.

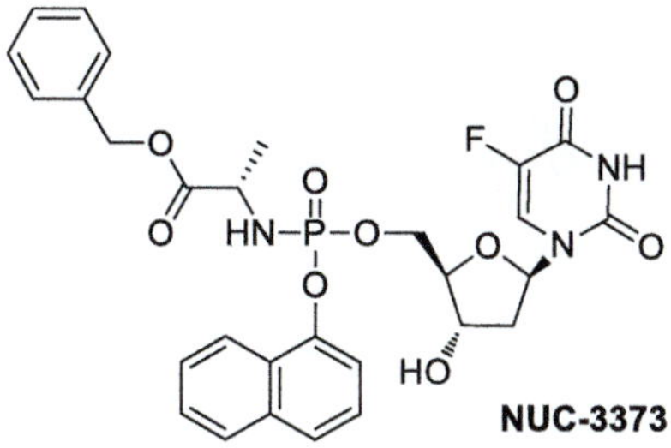

NUC-3373

Figure 1.27 Structure of **NUC-3373**.

deaminase-mediated degradation and displays a significant inhibitory effect on tumor growth in pancreatic cancer xenograft models.

1.2.4.7. NUC-3373 **NUC-3373** (Figure 1.27) is another potential anticancer agent developed *via* the ProTide technology. It is a phosphoramidate prodrug of 5-fluoro-2′-deoxyuridine (FDUR) and was discovered by McGuigan *et al.* [63] In an extensive structure–activity relationship (SAR) investigation of 39 ProTide analogs, the amino acid ester region exhibited little contribution to the activity of this type prodrug. Replacement of the phenol moiety in the ProTide with 1-naphthalenol significantly improved the potency of the compound. Eventually, the *O*-naphthyl L-alanine benzyl ester ProTide (**NUC-3373**) was selected for further development.

NUC-3373 is the first drug candidate reported to overcome the key drug resistance associated with 5-fluorouracil (5-FU) and FDUR [64]. Thymidine kinase inhibition resulted in the reductions in the cytotoxic activity of the ProTide prodrug by a factor of only 4 and of the parent nucleoside FDUR by a factor of 136. Nucleoside transport inhibition mildly decreased cytotoxicity of **NUC-3373** (*vs* a 63-fold reduction in the cytotoxicity of FUDR). In an assessment of sensitivity to dihydropyrimidine dehydrogenase (DPD) degradation, **NUC-3373** concentration was not affected while the 5-FU concentration was significantly increased in an experiment with 5-FU. These results indicate that the prodrug is more independent of thymidine kinase and nucleoside transporters and also is not a substrate for DPD metabolism. **NUC-3373** displayed greater inhibitory effect on tumor growth (47%) than 5-FU (25%) in a colorectal cancer xenograft model.

1.2.4.8. Stampidine Stavudine is an FDA-approved anti-HIV/AIDS drug. The first phosphorylation of this nucleoside analog, however, is rate-limiting during the formation of the bioactive triphosphate metabolite [65]. Application of the ProTide approach to stavudine led to the discovery of stampidine (Figure 1.28), a ProTide prodrug bearing a *para*-bromophenyl L-alanine methyl ester moiety [66]. Stampidine displayed potent antiviral activity against both the wild-type HIV-1 strains and drug-resistant viral strains [66, 67]. Incorporation

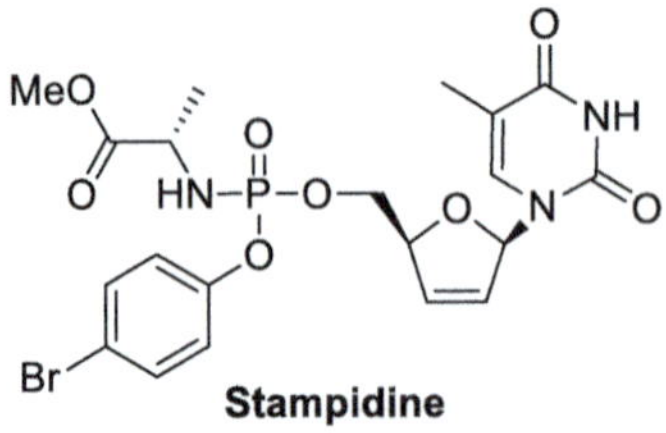

Figure 1.28 Structure of stampidine.

of an electron-withdrawing atom, bromine into the *O*-phenyl group can accelerate the metabolism of the ProTide to form the active triphosphate species. Stampidine is well tolerated in mice and rats at a dose of up to 500 mg/kg *via* intraperitoneal or oral administration [68]. This prodrug has been advanced into clinical development stages and is also considered to be a promising agent for pre-exposure prophylaxis against HIV infection [69].

1.2.4.9. Thymectacin Thymectacin, also known as **NB-1011** [70], (Figure 1.29) is a ProTide prodrug of brivudine (BVdU), an antiviral drug used to treat herpes zoster (shingles). Interestingly, this prodrug possesses potent anticancer activity and targets tumor cells with high levels of thymidylate synthase (TS) [71]. **NB-1011** displayed 10-fold higher cytotoxicity to 5-FU resistant and TS overexpressing colorectal tumor cells than to healthy cells. It is metabolized by TS into the nucleoside monophosphate (BVdU-MP), and BVdU-MP can compete with deoxyuridine monophosphate, the natural substrate, to bind to TS. The binding between BVdU-MP and TS is reversible, and thus TS can further convert BVdU-MP into cytotoxic metabolites. This prodrug has progressed into clinical trials for the treatment of colon cancer [72].

1.2.4.10. BMS-986094 **BMS-986094** (also known as **INX-08189**, Figure 1.30) is a double prodrug of anti-HCV 2′-β-*C*-methyl guanosine analog with a phosphoramidate group and a 6-*O*-methoxy purine base [73–75]. Both

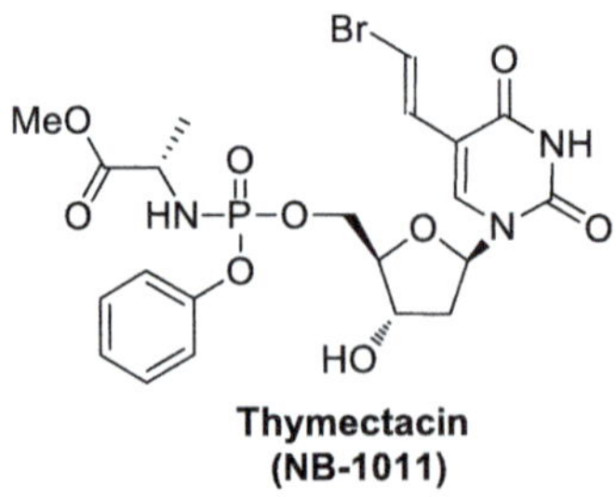

Figure 1.29 Structure of thymectacin.

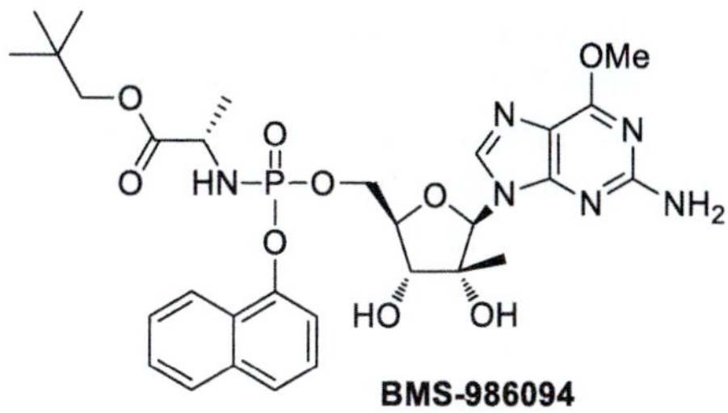

BMS-986094
(INX-08189)

Figure 1.30 Structure of **BMS-986094** (**INX-08189**).

the prodrug moieties can increase the lipophilicity of the parent guanosine monophosphate. **INX-08189** displays a 500-fold better antiviral activity (EC_{50} = 10 nM) against the HCV genotype 1b subgenomic replicon than the parent compound, which is consistent with the intracellular levels of the active 2′-*C*-methyl guanosine triphosphate in primary human hepatocytes. Separated single isomers of the prodrug exhibited similar anti-HCV activity and both isomers were degraded by enzymes, showing similar half-lives. Therefore, the prodrug **INX-08189** was selected for clinical development to treat chronic HCV infection.

1.2.4.11. PSI-353661 **PSI-353661** (Figure 1.31) is also a double prodrug containing ProTide and 6-*O*-methoxy moieties for the treatment of HCV infection [76]. The parent nucleoside 2′-deoxy-2′-fluoro-2′-*C*-methylguanosine displayed only weak anti-HCV active (EC_{90} = 69.2 μM) in a replicon assay although its triphosphate is a potent HCV NS5B polymerase inhibitor (IC_{50} = 5.94 μM). Most of these prodrugs exhibited significantly improved antiviral activity (> 1000-fold) compared to the parent nucleoside. Among them, **PSI-353661** demonstrated high potency against both the wild type and the S282T-resistant replicon (EC_{90} = 0.008~0.011 μM) with no cytotoxicity at 100 μM concentration.

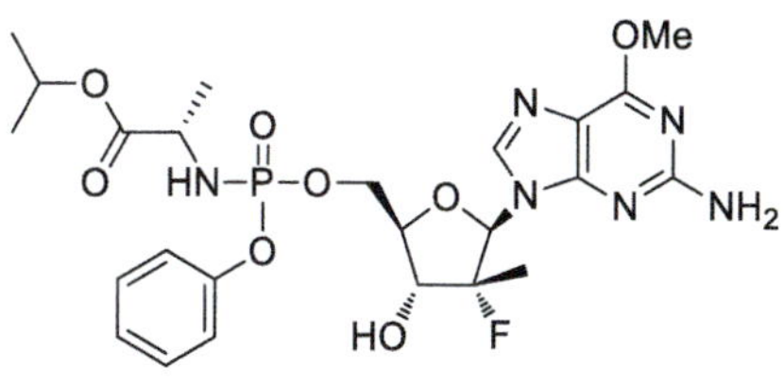

PSI-353661

Figure 1.31 Structure of **PSI-353661**.

1.2.4.12. GS-9131 The parent compound of **GS-9131**, (5-(6-amino-purin-9-yl)-4-fluoro-2,5-dihydro-furan-2-yloxymethyl)phosphonic acid (2′-Fd4AP, **GS-9148**) is a novel and potent nucleotide HIV-1 reverse transcriptase (RT) inhibitor against multiple NRTI-resistant HIV-1 variants [77, 78]. **GS-9131** (Figure 1.32) is an orally bioavailable ProTide prodrug of 2′-Fd4AP bearing an ethylalaninyl phosphoramidate moiety [78–80]. This phosphoramidate prodrug was designed as a substrate for the lysosomal carboxypeptidase (cathepsin A), which is highly expressed in peripheral blood mononuclear cells (PBMC). It also displayed favorable intestinal and hepatic stabilities, and thus can reach lymphoid cells efficiently after oral administration, achieving high levels of the bioactive diphosphate of the parent nucleotide.

1.2.4.13. Uprifosbuvir Uprifosbuvir (**MK-3682/IDX21437**, Figure 1.33) is a potent antiviral ProTide prodrug candidate of 2′-chloro-2′-methyluridine [81, 82]. Unlike most S_P-configuration- and L-amino acid-based ProTide prodrugs, **MK-3682** is a R_P-isomer prodrug bearing a D-alanine isopropyl ester moiety [81] and has entered phase III clinical trials as a pan-genotypic inhibitor of HCV NS5B for the treatment of patients with HCV [83–85].

1.2.4.14. CL-096 Our laboratory reported a phosphoramidate prodrug of β-D-2′-deoxy-2′-α-fluoro-2′-β-*C*-(fluoromethyl)uridine [86–88] (**CL-096,** Figure 1.34). This novel difluoro uridine phosphoramidate prodrug possesses potent antiviral activity against HCV genotypes 1b, 1a, 2a, and S282T

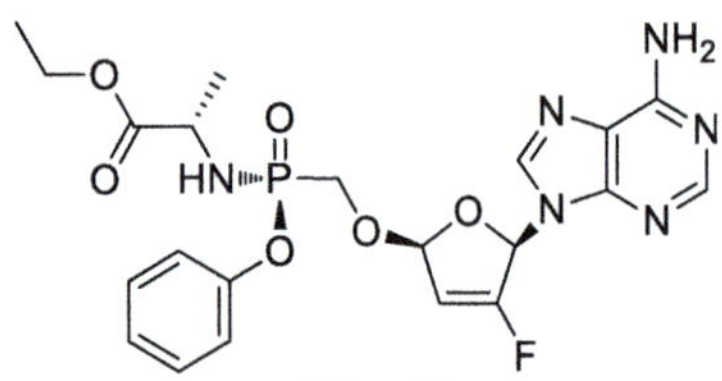

GS-9131

Figure 1.32 Structure of **GS-9131**.

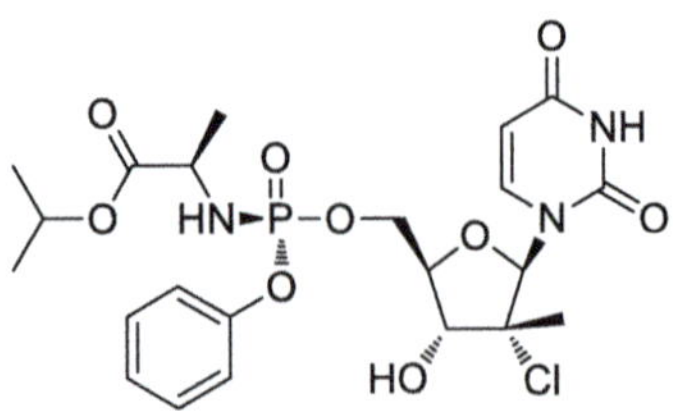

Uprifosbuvir
(MK-3682)

Figure 1.33 Structure of uprifosbuvir (**MK-3682**).

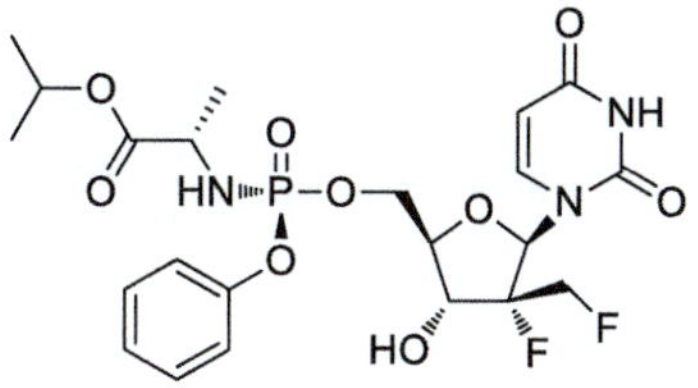

Figure 1.34 Structure of **CL-096**.

(EC_{50} = 0.18–1.13 μM) with no cytotoxicity observed at the highest tested concentration (1000 μM). **CL-096** is orally available, is well tolerated at a dose of 4 g/kg in mice, and produces a high level of the corresponding nucleoside triphosphate (NTP) in rat livers.

1.2.4.15. IDX-184 **IDX-184** (Figure 1.35) is a guanosine analog prodrug developed as a potent NS5B RNA polymerase inhibitor [89, 90] for the treatment of HCV infection. Unlike most ProTides, this phosphoramidate prodrug contains an *N*-benzyl group instead of the amino acid ester moiety and a masked 2-mercaptoethyl group instead of the *O*-aryl moiety. This drug candidate has progressed into clinical development and has displayed reasonable effectiveness in early clinical trials. In addition, **IDX-184** also showed antiviral activity against emerging viruses such as the Zika virus [91, 92], MERS-CoV [93], and SARS CoV-2 [94, 95].

1.2.4.16. CL-206 Our group also developed a series of new phosphoramidate prodrugs by replacing the aryl group in ProTide with a substituted benzyl moiety. These novel phosphoramidates have a liver-targeting feature similar to that of HepDirect prodrugs, as the cleavage of the benzyl group requires cytochrome P450 (CYP) enzymes in the liver. Application of this prodrug approach to 5-fluoro-2′-deoxyuridine (FDUR) led to the discovery of **CL-206** (Figure 1.36) as a candidate drug for the treatment of hepatocellular carcinoma (HCC) [87, 96]. Biological evaluation including stability, activation rate, and cytostatic activity on a cancer cell line (SMMC7721) demonstrated that the optimal *O*-benzyl moiety is an *ortho*-methylbenzyl group. Compared to the racemic prodrug (**13**),

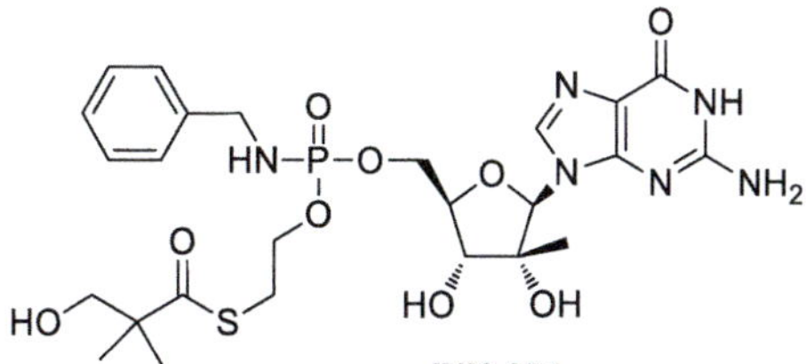

INX-184

Figure 1.35 Structure of **IDX-184**.

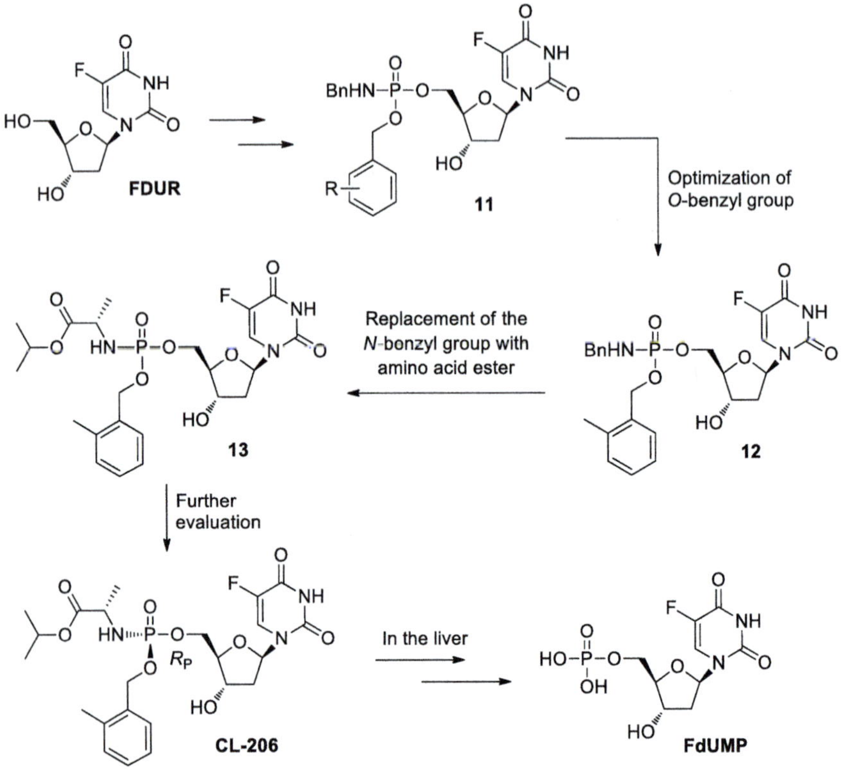

Figure 1.36 Discovery of **CL-206**, a new phosphoramidate prodrug with an *O*-benzyl group.

the R_p-enantiomer (**CL-206**) displayed equally good anticancer activity with slightly lower toxicity on a normal liver cell line (LO2). As expected, metabolism studies in rats revealed that **CL-206** can deliver the desired nucleoside monophosphate (FdUMP) to the liver with a high liver-targeting index. This prodrug is well-tolerated at an oral dose up to 3 g/kg in mice and exhibits a good inhibitory effect on tumor growth in a mouse xenograft model. These results suggest that **CL-206** is a potential new agent for anti-HCC therapy *via* oral administration.

1.3. SYNTHESIS OF NUCLEOSIDE PHOSPH(ON)ATE PRODRUGS

1.3.1. Synthesis of Carbonyloxymethyl Diester Prodrugs

Generally, there are three approaches to the development of nucleoside phosphate diester prodrugs bearing carbonyloxymethyl groups. Taking the preparation of bis(POM) nucleoside phosphates as an example, the synthetic methods include [3] (A) direct reactions of a nucleoside analog with the bis(POM)-phosphorochloridate in the presence of base, or with bis(POM)-phosphate *via*

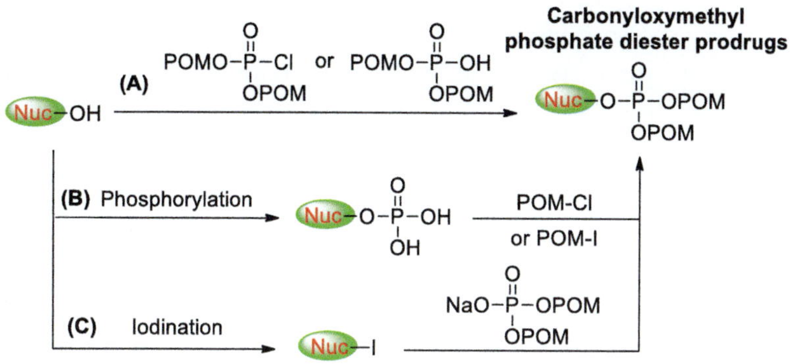

Scheme 1.1 Synthesis of carbonyloxymethyl phosphate diester prodrugs.

Mitsunobu coupling; (B) phosphorylation of a nucleoside followed by the reactions with a carbonyloxymethyl chloride or iodide; (C) iodination of the hydroxyl group in a nucleoside followed by replacement of the iodo group with a bis(POM)-phosphate salt (Scheme 1.1).

The preparation of nucleotide carbonyloxymethyl diester prodrugs is quite straightforward. Starting with nucleotide analogs protected with simple dialkyl groups [3], the expected prodrugs can be synthesized *via* a direct reaction with carbonyloxymethyl chloride reagents in the presence of sodium iodide or through phosphonate deprotection followed by the reaction with carbonyloxymethyl chlorides (Scheme 1.2).

1.3.2. Synthesis of Alkoxyalkyl Monoester Prodrugs

The synthesis of nucleoside alkoxyalkyl phosphate monoester prodrugs can be achieved from the corresponding parent compound *via* the following pathways [3]: (A) Mitsunobu coupling with an alkoxyalkyl phosphate reagent followed by deprotection, if necessary; (B) sequential phosphorylation, coupling with an alkoxyalkyl phosphate and deprotection; (C) coupling with an

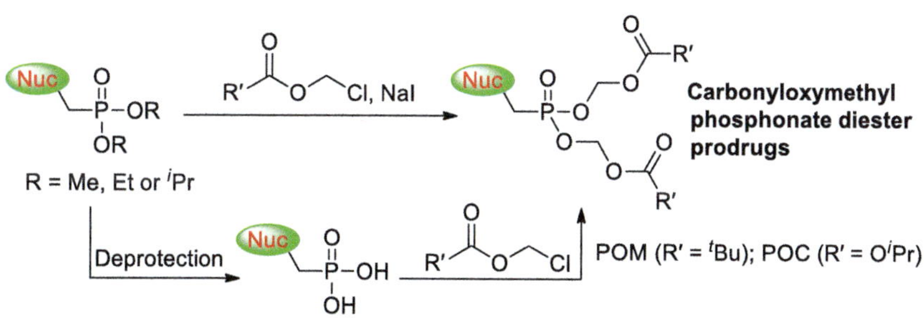

Scheme 1.2 Synthesis of carbonyloxymethyl phosphonate diester prodrugs.

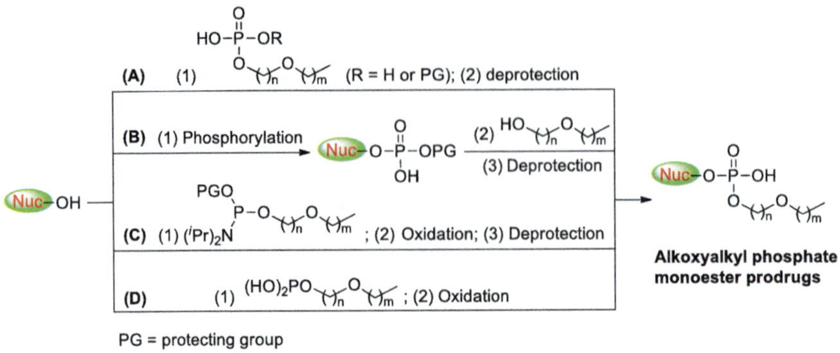

Scheme 1.3 Synthesis of alkoxyalkyl phosphate monoester prodrugs.

alkoxyalkyl phosphoramidite reagent followed by oxidation and deprotection; (D) condensation with a phosphite and then oxidation (Scheme 1.3).

Nucleotide alkoxyalkyl monoester prodrugs can be prepared [3] from the reaction of the nucleoside phosphonic acid with an alkoxyalkyl alcohol in the presence of *N,N'*-dicyclohexylcarbodiimide (DCC) or *via* Mitsunobu coupling. The parent nucleotide can also be directly alkylated by an alkoxyalkyl bromide or first activated through monochlorination followed by substitution with an alkoxyalkyl alcohol to access the prodrug. This kind of prodrug can also be synthesized by selective hydrolysis of the corresponding nucleotide alkoxyalkyl diester intermediate (Scheme 1.4).

Hostetler *et al.* reported a concise synthesis of the (*S*)-3-hydroxy-2-(phosphono-methoxy)propyl (HPMP) adenine prodrug (**16**) by direct introduction of the monoalkoxyalkyl phosphonate monoester moiety with a tosylate intermediate already containing the HDP chain (**15**) [97] (Scheme 1.5). Brincidofovir (**CMX-001**), a prodrug of cidofovir, can be prepared *via* sequential intramolecular cyclization (**17**) of the parent nucleotide, alkylation with alkoxyalkyl bromide producing **18**, and finally hydrolytic ring-opening [98] (Scheme 1.6).

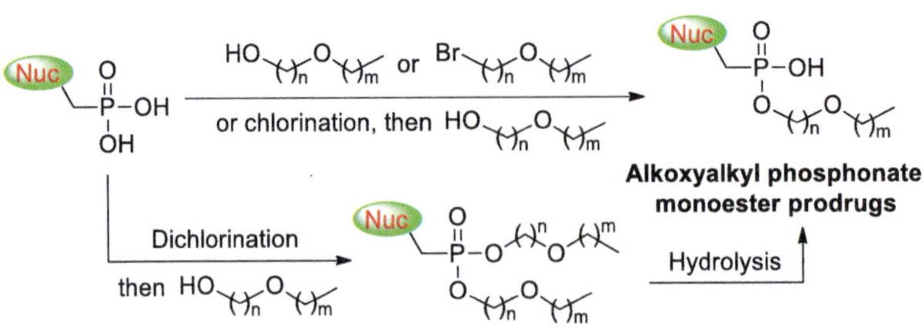

Scheme 1.4 Synthesis of alkoxyalkyl phosphonate monoester prodrugs.

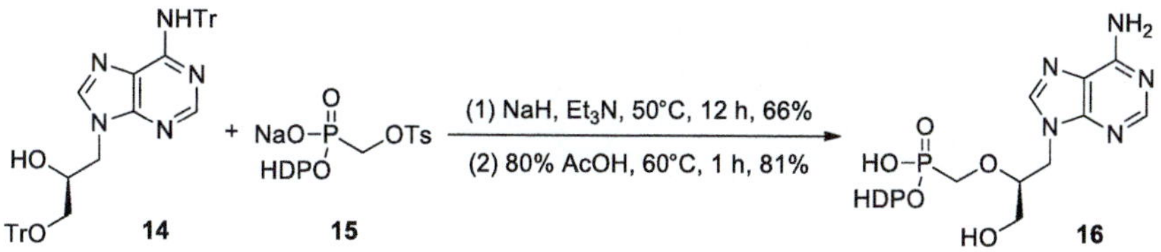

Scheme 1.5 Synthesis of HPMP-adenine prodrug (**16**).

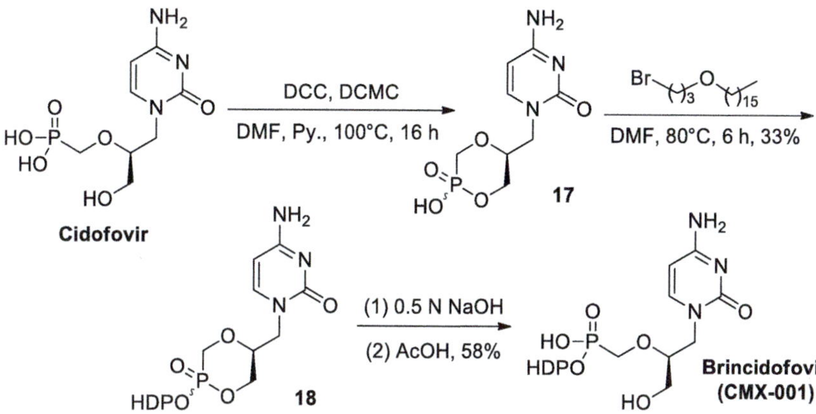

Scheme 1.6 Synthesis of prodrug brincidofovir (**CMX-001**)

1.3.3. Synthesis of HepDirect Prodrugs

HepDirect phosphate prodrugs [8] can conveniently be synthesized by coupling the parent nucleoside with pre-prepared *p*-nitrophenyl phosphate intermediates or P(III) phosphoramidite intermediates followed by oxidation. Preparation of HepDirect phosphonate prodrugs [38] can be achieved through direct coupling with 1-arylpropane-1,3-diol (Scheme 1.7).

Because of there are two chiral centers in the HepDirect prodrugs moiety, there could be four diastereoisomers. Using enantiomerically pure propane-1,3-diols as starting materials will lead to the formation of only two of the diastereoisomers, either the *cis*- or the *trans*-isomers. Under optimized reaction conditions, pradefovir and the other *cis*-isomer can even be synthesized enantioselectively from the corresponding diols **19** and **20**, respectively [38] (Scheme 1.8).

1.3.4. Synthesis of Phosphoramidate and Phosphonamidate Prodrugs

Generally, the phosphoramidate prodrugs can be accessed by any of three pathways [3]: (A) pre-formation of a phosphoramidate intermediate already containing an aryl, amino acid ester and a leaving group, followed by substitution of

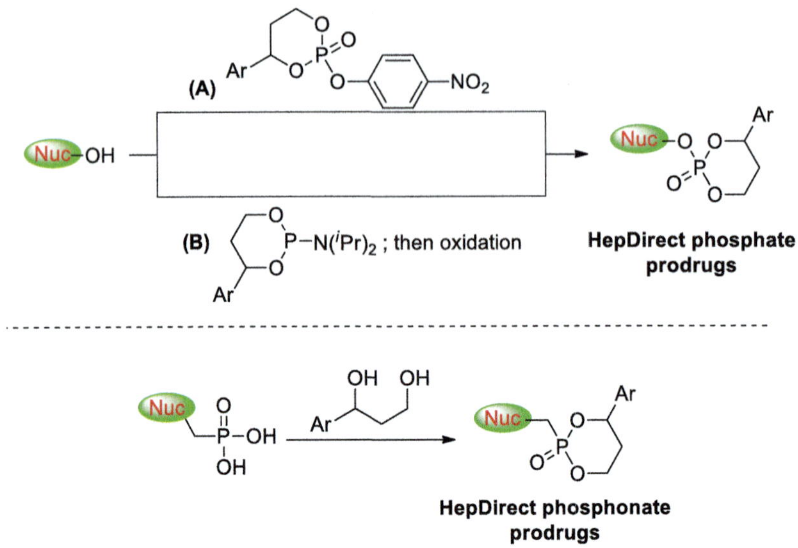

Scheme 1.7 Synthesis of HepDirect phosph(on)ate prodrugs.

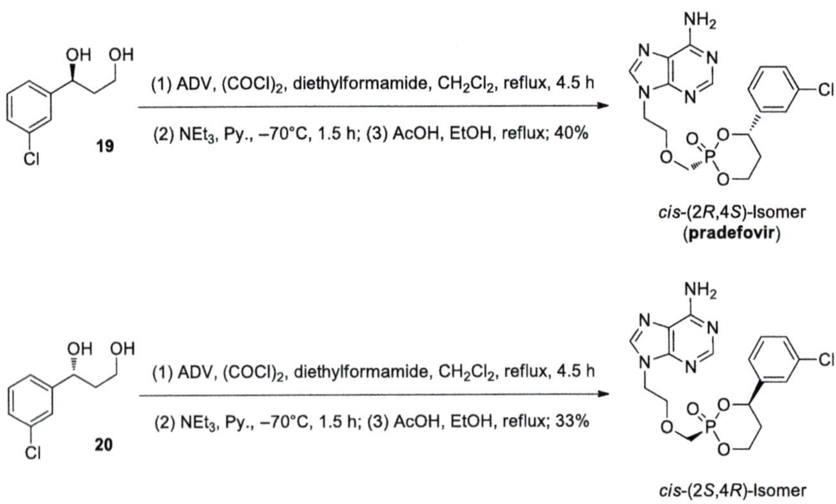

Scheme 1.8 Enantioselective synthesis of pradefovir and the other *cis*-isomer.

the parent nucleoside in the presence of *N*-methylimidazole (NMI) or *tert*-butylmagnesium chloride (^{t}BuMgCl); (B) sequential substitution of a diarylphosphite [O=PH(OAr)$_2$] by the nucleoside and oxidative amination with an amino acid ester; (C) coupling of the nucleoside with an aryl phosphorodichloridate, followed by reaction with an amino acid ester (Scheme 1.9).

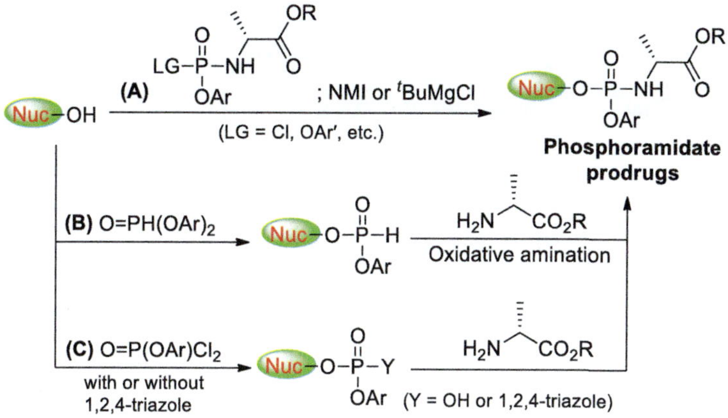

Scheme 1.9 Synthesis of phosphoramidate prodrugs.

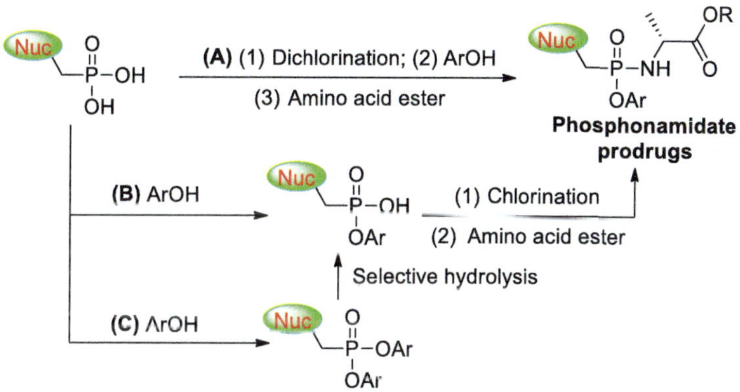

Scheme 1.10 Synthesis of phosphonamidate prodrugs.

The preparation of phosphonamidate prodrugs can be achieved [3] by (A) dichlorination of the parent nucleotide and subsequent substitution with first a phenol and then an amino acid ester; or (B and C) chlorination of a pre-formed nucleotide monoaryl phosphonate followed by coupling with an amino acid ester (Scheme 1.10).

In most cases, the S_P and R_P isomers of phosphoramidate and phosphonamidate prodrugs display different biological profiles, and this promotes the development of diastereoselective approaches to access these types of prodrugs. One of the most frequently used methods is use of pre-formed chiral aryloxy phosphoramidate intermediates (see Method A in Scheme 1.9). For example, Ross *et al.* [99] reported the synthesis of sofosbuvir (**PSI-7977**) using chiral phosphoramidate reagents with substituted phenols as leaving groups (Scheme 1.11). Such phosphoramidate reagents can be readily prepared from commercially

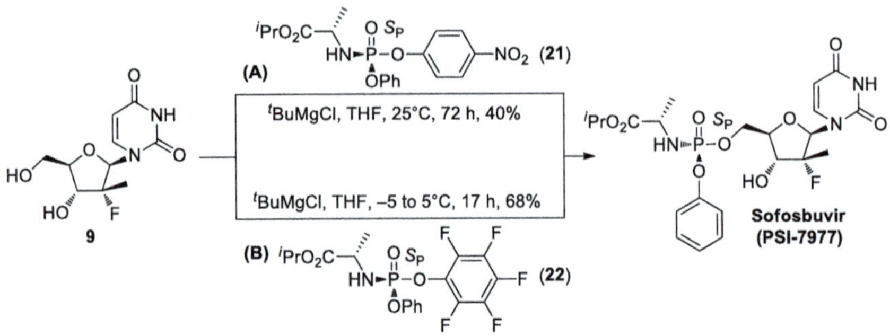

Scheme 1.11 Asymmetric synthesis of sofosbuvir (**PSI-7977**).

available substituted aryl dichlorophosphates by sequential reactions with phenol and then with an amino acid ester. The reaction of 2′-fluoro-2′-methyluridine with the S_P-isomer of *p*-nitrophenyl intermediate **21** in the presence of ^{t}BuMgCl followed by two recrystallizations from CH_2Cl_2 produced **PSI-7977** in 40% yield (Scheme 1.11A). Replacement of the *p*-nitrophenyl group in the phosphoramidate reagent with a pentafluorophenyl group (**22**) resulted in the formation of the desired prodrug in a better yield after two recrystallizations (68%, Scheme 1.11B).

As the parent nucleotide tenofovir already possesses a phosphonic acid group, the chiral ProTide prodrug TAF (**GS-7340**) cannot be prepared using the above strategy. Chapman *et al.* developed a practical method for the kilogram-scale preparation of the free base of this prodrug [100]. Starting with the parent nucleoside tenofovir, a DCC-mediated coupling reaction with phenol in *N*-methyl pyrrolidone (NMP) selectively masked the first hydroxyl group (**23**). The remaining hydroxyl group is chlorinated with sulfurous dichloride ($SOCl_2$) and then further substituted by the amino acid isopropyl ester moiety. Finally, the diastereomerically pure prodrug (**24**) was isolated by simulated moving bed chromatography (Scheme 1.12).

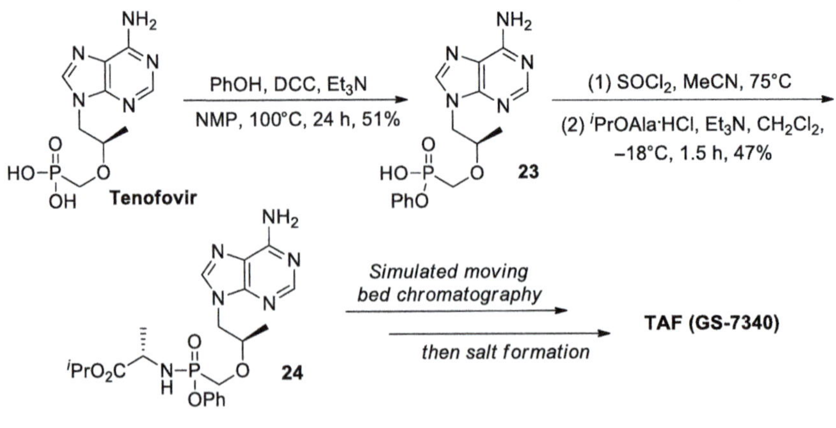

Scheme 1.12 Preparation of tenofovir alafenamide fumarate (**GS-7340**).

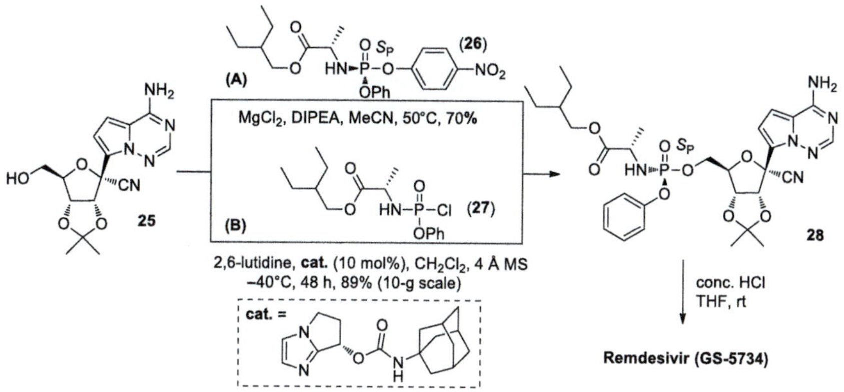

Scheme 1.13 Asymmetric synthesis of remdesivir (**GS-5734**).

Remdesivir (**GS-5734**) was previously synthesized [101] employing the single S_P-isomer of a *p*-nitrophenoxy phosphoramidate precursor (**26**) with the protected parent adenine *C*-nucleoside (**25**) in $MgCl_2$ and *N*,*N*-diisopropylethylamine (DIPEA) followed by deprotection (Scheme 1.13A). Recenty, Wang *et al.* [102] reported a catalytic asymmetric synthesis of this anti-COVID-19 drug in 96% yield from the P-racemic phosphoryl chloride (**27**) and the protected parent nucleoside (**25**) with excellent stereoselectivity (22 : 1 S_P:R_P). This dynamic kinetic asymmetric transformation (DyKAT) was achieved by utilization of a chiral imidazole catalyst. This catalyst contains a bicyclic imidazole skeleton and an adamantinyl-substituted carbamate group, which are both required for the DyKAT (Scheme 1.13B).

DiRocco *et al.* [82] reported a multifunctional catalyst for stereoselective synthesis of phosphoramidate prodrugs such as uprifosbuvir (**MK-3682**) from the P-racemic phosphoryl chloride (**30**). The rational catalyst design from mechanistic studies and computational modeling led to a stereoselective synthesis up to 99 : 1 (Scheme 1.14).

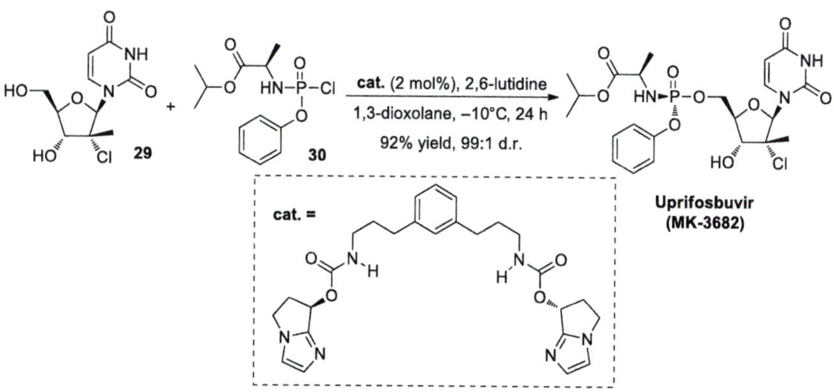

Scheme 1.14 Catalytic asymmetric synthesis of uprifosbuvir (**MK-3682**).

1.4. CONCLUSION

Prodrug strategies for phosphates or phosphonates have significantly facilitated the drug discovery and development of nucleotide and nucleoside analogs. They can increase the potency of a parent compound and even reveal the activity of an inactive nucleos(t)ide analog. The phosph(on)ate prodrugs usually display better bioavailability than the associated parent molecule mainly because they have improved lipophilicity. Some nucleos(t)ides, traditionally administered only by injection, may be developed into orally available drugs through these prodrug approaches. However, there are still some challenges, such as efficient methods for enantioselective phosph(on)ate prodrug synthesis and tissue-targeted drug delivery. Given the advantages and achievements of phosph(on)ate prodrugs, this synthetic branch may develop potential drug candidates against emerging virus infections for which no treatment is currently available, and it will further explore the applications of the phosph(on)ate prodrug strategies in non-nucleoside drug discovery.

ABBREVIATIONS

3TC	lamivudine
5-FU	5-fluorouracil
ADP	adefovir dipivoxil
ADV	adefovir
AIDS	acquired immune deficiency syndrome
AZT	zidovudine
BVdU	brivudine
CC_{50}	50% cytotoxic concentration
CDV	cidofovir
CFR	cholesterol-fed rat
COVID-19	corona virus disease 2019
CYP	cytochrome P450
DCC	*N,N′*-dicyclohexylcarbodiimide
DIPEA	*N,N*-diisopropylethylamine
DNA	deoxyribonucleic acid
DP	diphosphate
DPD	dihydropyrimidine dehydrogenase
DyKAT	dynamic kinetic asymmetric transformation
EC_{50}	50% effective concentration
EC_{90}	90% effective concentration
EUA	Emergency Use Authorization
FDA	the US Food and Drug Administration
FdUMP	5-fluoro-2′-deoxyuridine monophosphate
FDUR	5-fluoro-2′-deoxyuridine

FNC	azvudine
HBV	hepatitis B virus
HCC	hepatocellular carcinoma
HCMV	human cytomegalovirus
HCV	hepatitis C virus
HDP	hexadecyloxypropyl
HINT-1	histidine triad nucleotide-binding protein 1
HIV	human immunodeficiency virus
HPMP	(*S*)-3-hydroxy-2-(phosphonomethoxy)propyl
HSV	herpes simplex virus
IC_{50}	50% inhibitory concentration
LPC	lysophosphatidylcholine
MERS-CoV	Middle East respiratory syndrome coronavirus
MP	monophosphate
NMI	*N*-methylimidazole
NMP	*N*-methyl pyrrolidone
nRTI	nucleotide analog reverse transcriptase inhibitor
NS5B	non-structural protein 5B
NTP	nucleoside triphosphate
ODE	octadecyloxyethyl
PA	phosphonic acid
PBMC	peripheral blood mononuclear cell
PG	protecting group
PK	pharmacokinetics
PMEA	9-(2-phosphonomethoxyethyl)adenine
PMPA	(*R*)-9-(2-phosphonylmethoxypropyl)adenine
POC	isopropyloxymethyl
POM	pivaloyloxymethyl
RT	reverse transcriptase
SAR	structure–activity relationship
SARS-CoV-2	severe acute respiratory syndrome coronavirus 2
$t_{1/2}$	half-life
TAF	tenofovir alafenamide fumarate
TDF	tenofovir disoproxil fumarate
TP	triphosphate
TR	thyroid hormone receptor
TS	thymidylate synthase
WHO	World Health Organization

ACKNOWLEDGMENT

We would like to thank the National Natural Science Foundation of China (Nos. U1804283 and 81773570) for financial support.

REFERENCES

1. De Clercq, E.; Holy, A. Antiviral activity of aliphatic nucleoside analogs: structure-function relationship. *J. Med. Chem.* 1979, *22*, 510–513.
2. Jordheim, L. P.; Durantel, D.; Zoulim, F.; Dumontet, C. Advances in the development of nucleoside and nucleotide analogues for cancer and viral diseases. *Nat. Rev. Drug Discov.* 2013, *12*, 447–464.
3. Pradere, U.; Garnier-Amblard, E. C.; Coats, S. J.; Amblard, F.; Schinazi, R. F. Synthesis of nucleoside phosphate and phosphonate prodrugs. *Chem. Rev.* 2014, *114*, 9154–9218.
4. Sun, L.; Peng, Y.; Yu, W.; Zhang, Y.; Liang, L.; Song, C.; Hou, J.; Qiao, Y.; Wang, Q.; Chen, J.; Wu, M.; Zhang, D.; Li, E.; Han, Z.; Zhao, Q.; Jin, X.; Zhang, B.; Huang, Z.; Chai, J.; Wang, J.-H.; Chang, J. Mechanistic insight into antiretroviral potency of 2′-Deoxy-2′-β-fluoro-4′-azidocytidine (FNC) with a long-lasting effect on HIV-1 prevention. *J. Med. Chem.* 2020, *63*, 8554–8566.
5. Chang, J. 2′-Fluoro-4′-substituted nucleosides, the preparation and use. Eur. Patent EP2177527B, December 18, 2013; US Patent US8835615B2, September 16, 2014.
6. Chang, J.; Bao, X.; Wang, Q.; Guo, X.; Wang, W.; Qi, X. 2′-Fluoro-4′-substituted nucleosides, the preparation and use. CN Patent ZL200710137548.0, August 26, 2009.
7. Boyer, S. H.; Sun, Z. L.; Jiang, H. J.; Esterbrook, J.; Gomez-Galeno, J. E.; Craigo, W.; Reddy, K. R.; Ugarkar, B. G.; MacKenna, D. A.; Erion, M. D. Synthesis and characterization of a novel liver-targeted prodrug of cytosine-1-beta-D-arabinofuranoside monophosphate for the treatment of hepatocellular carcinoma. *J. Med. Chem.* 2006, *49*, 7711–7720.
8. Erion, M. D.; Reddy, K. R.; Boyer, S. H.; Matelich, M. C.; Gornez-Galeno, J.; Lemus, R. H.; Ugarkar, B. G.; Colby, T. J.; Schanzer, J.; van Poelje, P. D. Design, synthesis, and characterization of a series of cytochrome P_{450} 3A-activated prodrugs (HepDirect prodrugs) useful for targeting phosph(on)ate-based drugs to the liver. *J. Am. Chem. Soc.* 2004, *126*, 5154–5163.
9. Wagner, C. R.; Iyer, V. V.; McIntee, E. J. Pronucleotides: toward the in vivo delivery of antiviral and anticancer nucleotides. *Med. Res. Rev.* 2000, *20*, 417–451.
10. Erion, M. D.; van Poelje, P. D.; MacKenna, D. A.; Colby, T. J.; Montag, A. C.; Fujitaki, J. M.; Linemeyer, D. L.; Bullough, D. A. Liver-targeted drug delivery using HepDirect prodrugs. *J. Pharmacol. Exp. Ther.* 2005, *312*, 554–560.
11. Hecker, S. J.; Erion, M. D. Prodrugs of phosphates and phosphonates. *J. Med. Chem.* 2008, *51*, 2328–2345.
12. Thornton, P. J.; Kadri, H.; Miccoli, A.; Mehellou, Y. Nucleoside phosphate and phosphonate prodrug clinical candidates. *J. Med. Chem.* 2016, *59*, 10400–10410.
13. Nie, B.; Jin, C.; Zhong, W.; Ren, Q.; Zhang, Y.; Zhang, J. Application and recent progress of phosphoramidate prodrugs strategies and protide technology in drug discovery. *Chinese J. Org. Chem.* 2017, *37*, 2818–2840.
14. Mehellou, Y.; Rattan, H. S.; Balzarini, J. The protide prodrug technology: from the concept to the clinic. *J. Med. Chem.* 2018, *61*, 2211–2226.
15. Clercq, E. D.; Holý, A.; Rosenberg, I.; Sakuma, T.; Balzarini, J.; Maudgal, P. C. A novel selective broad-spectrum anti-DNA virus agent. *Nature* 1986, *323*, 464–467.

16. Pauwels, R.; Balzarini, J.; Schols, D.; Baba, M.; Desmyter, J.; Rosenberg, I.; Holy, A.; De Clercq, E. Phosphonylmethoxyethyl purine derivatives, a new class of anti-human immunodeficiency virus agents. *Antimicrob. Agents Chemother.* 1988, *32*, 1025–1030.
17. Balzarini, J.; Naesens, L.; Herdewijn, P.; Rosenberg, I.; Holy, A.; Pauwels, R.; Baba, M.; Johns, D. G.; De Clercq, E. Marked in vivo antiretrovirus activity of 9-(2-phosphonylmethoxyethyl)adenine, a selective anti-human immunodeficiency virus agent. *Proc. Natl. Acad. Sci. U. S. A.* 1989, *86*, 332–336.
18. Starrett, J. E., Jr.; Tortolani, D. R.; Russell, J.; Hitchcock, M. J. M.; Whiterock, V.; Martin, J. C.; Mansuri, M. M. Synthesis, oral bioavailability determination, and in vitro evaluation of prodrugs of the antiviral agent 9-[2-(phosphonomethoxy) ethyl]adenine (PMEA). *J. Med. Chem.* 1994, *37*, 1857–1864.
19. Naesens, L.; Balzarini, J.; Bischofberger, N.; De Clercq, E. Antiretroviral activity and pharmacokinetics in mice of oral bis(pivaloyloxymethyl)-9-(2-phosphonylmethoxyethyl)adenine, the bis(pivaloyloxymethyl) ester prodrug of 9-(2-phosphonylmethoxyethyl)adenine. *Antimicrob. Agents Chemother.* 1996, *40*, 22–28.
20. De Clercq, E. Clinical potential of the acyclic nucleoside phosphonates cidofovir, adefovir, and tenofovir in treatment of DNA virus and retrovirus infections. *Clin. Microbiol. Rev.* 2003, *16*, 569–596.
21. Hadziyannis, S. J.; Tassopoulos, N. C.; Heathcote, E. J.; Chang, T.-T.; Kitis, G.; Rizzetto, M.; Marcellin, P.; Lim, S. G.; Goodman, Z.; Wulfsohn, M. S.; Xiong, S.; Fry, J.; Brosgart, C. L. Adefovir dipivoxil for the treatment of hepatitis B e antigen–negative chronic hepatitis B. *N. Engl. J. Med.* 2003, *348*, 800–807.
22. Marcellin, P.; Chang, T.-T.; Lim, S. G.; Tong, M. J.; Sievert, W.; Shiffman, M. L.; Jeffers, L.; Goodman, Z.; Wulfsohn, M. S.; Xiong, S.; Fry, J.; Brosgart, C. L. Adefovir dipivoxil for the treatment of hepatitis B e antigen–positive chronic hepatitis B. *N. Engl. J. Med.* 2003, *348*, 808–816.
23. Balzarini, J.; Holy, A.; Jindrich, J.; Naesens, L.; Snoeck, R.; Schols, D.; De Clercq, E. Differential antiherpesvirus and antiretrovirus effects of the (S) and (R) enantiomers of acyclic nucleoside phosphonates: potent and selective in vitro and in vivo antiretrovirus activities of (R)-9-(2-phosphonomethoxypropyl)-2,6-diaminopurine. *Antimicrob. Agents Chemother.* 1993, *37*, 332–338.
24. Barditch-Crovo, P.; Toole, J.; Hendrix, C. W.; Cundy, K. C.; Ebeling, D.; Jaffe, H. S.; Lietman, P. S. Anti-human immunodeficiency virus (HIV) activity, safety, and pharmacokinetics of adefovir dipivoxil (9-[2-(bis-pivaloyloxymethyl)-phosphonylmetho xyethyl]adenine) in HIV-infected patients. *J. Infect. Dis.* 1997, *176*, 406–413.
25. Naesens, L.; Bischofberger, N.; Augustijns, P.; Annaert, P.; Van den Mooter, G.; Arimilli, M. N.; Kim, C. U.; De Clercq, E. Antiretroviral efficacy and pharmacokinetics of oral bis(isopropyloxycarbonyloxymethyl)9-(2-phosphonylmethoxypropyl) adenine in mice. *Antimicrob. Agents Chemother.* 1998, *42*, 1568–1573.
26. Shaw, J.-P.; Sueoka, C. M.; Oliyai, R.; Lee, W. A.; Arimilli, M. N.; Kim, C. U.; Cundy, K. C. Metabolism and pharmacokinetics of novel oral prodrugs of 9-[(*R*)-2-(phosphonomethoxy)propyl]adenine (PMPA) in dogs. *Pharm. Res.* 1997, *14*, 1824–1829.
27. Choi, J.-R.; Cho, D.-G.; Roh, K. Y.; Hwang, J.-T.; Ahn, S.; Jang, H. S.; Cho, W.-Y.; Kim, K. W.; Cho, Y.-G.; Kim, J.; Kim, Y.-Z. A novel class of phosphonate nucleosides.

9-[(1-phosphonomethoxycyclopropyl)methyl]guanine as a potent and selective anti-HBV agent. *J. Med. Chem.* 2004, *47*, 2864–2869.

28. Fung, J.; Lai, C.-L.; Yuen, M.-F. LB80380: a promising new drug for the treatment of chronic hepatitis B. *Expert Opin. Investig. Drugs* 2008, *17*, 1581–1588.
29. Yuen, M.-F.; Lee, S.-H.; Kang, H.-M.; Kim, C. R.; Kim, J.; Ngai, V.; Lai, C.-L. Pharmacokinetics of LB80331 and LB80317 following oral administration of LB80380, a new antiviral agent for chronic hepatitis B (CHB), in healthy adult subjects, CHB patients, and mice. *Antimicrob. Agents Chemother.* 2009, *53*, 1779–1785.
30. Dickson, J. K.; Biller, S. A.; Magnin, D. R.; Petrillo, E. W.; Hillyer, J. W.; Hsieh, D. C.; Lan, S.-J.; Rinehart, J. K.; Gregg, R. E.; Harrity, T. W.; Jolibois, K. G.; Kalinowski, S. S.; Kunselman, L. K.; Mookhtiar, K. A.; Ciosek, C. P. Orally active squalene synthase inhibitors: bis((acyloxy)alkyl) prodrugs of the α-phosphonosulfonic acid moiety. *J. Med. Chem.* 1996, *39*, 661–664.
31. Hiyoshi, H.; Yanagimachi, M.; Ito, M.; Ohtsuka, I.; Yoshida, I.; Saeki, T.; Tanaka, H. Effect of ER-27856, a novel squalene synthase inhibitor, on plasma cholesterol in rhesus monkeys: comparison with 3-hydroxy-3-methylglutaryl-CoA reductase inhibitors. *J. Lipid Res.* 2000, *41*, 1136–1144.
32. Braga, M. V.; Urbina, J. A.; de Souza, W. Effects of squalene synthase inhibitors on the growth and ultrastructure of *Trypanosoma cruzi*. *Int. J. Antimicrob. Agents* 2004, *24*, 72–78.
33. Ciesla, S. L.; Trahan, J.; Wan, W. B.; Beadle, J. R.; Aldern, K. A.; Painter, G. R.; Hostetler, K. Y. Esterification of cidofovir with alkoxyalkanols increases oral bioavailability and diminishes drug accumulation in kidney. *Antivir. Res.* 2003, *59*, 163–171.
34. Florescu, D. F.; Keck, M. A. Development of CMX001 (Brincidofovir) for the treatment of serious diseases or conditions caused by dsDNA viruses. *Expert Rev. Anti-Infect. Ther.* 2014, *12*, 1171–1178.
35. Painter, G. R.; Almond, M. R.; Trost, L. C.; Lampert, B. M.; Neyts, J.; De Clercq, E.; Korba, B. E.; Aldern, K. A.; Beadle, J. R.; Hostetler, K. Y. Evaluation of hexadecyloxypropyl-9-*R*-[2-(phosphonomethoxy)propyl]- adenine, CMX157, as a potential treatment for human immunodeficiency virus type 1 and hepatitis B virus infections. *Antimicrob. Agents Chemother.* 2007, *51*, 3505–3509.
36. Lanier, E. R.; Ptak, R. G.; Lampert, B. M.; Keilholz, L.; Hartman, T.; Buckheit, R. W.; Mankowski, M. K.; Osterling, M. C.; Almond, M. R.; Painter, G. R. Development of hexadecyloxypropyl tenofovir (CMX157) for treatment of infection caused by wild-type and nucleoside/nucleotide-resistant HIV. *Antimicrob. Agents Chemother.* 2010, *54*, 2901–2909.
37. Erion, M. D.; Bullough, D. A.; Lin, C. C.; Hong, Z. HepDirect prodrugs for targeting nucleotide-based antiviral drugs to the liver. *Curr. Opin. Investig. D.* 2006, *7*, 109–117.
38. Reddy, K. R.; Matelich, M. C.; Ugarkar, B. G.; Gomez-Galeno, J. E.; Dare, J.; Ollis, K.; Sun, Z.; Craigo, W.; Colby, T. J.; Fujitaki, J. M.; Boyer, S. H.; van Poelje, P. D.; Erion, M. D. Pradefovir: a prodrug that targets adefovir to the liver for the treatment of hepatitis B. *J. Med. Chem.* 2008, *51*, 666–676.
39. Dombret, H.; Gardin, C. An update of current treatments for adult acute myeloid leukemia. *Blood* 2016, *127*, 53–61.

40. Boyer, S. H.; Sun, Z.; Jiang, H.; Esterbrook, J.; Gómez-Galeno, J. E.; Craigo, W.; Reddy, K. R.; Ugarkar, B. G.; MacKenna, D. A.; Erion, M. D. Synthesis and characterization of a novel liver-targeted prodrug of cytosine-1-β-d-arabinofuranoside monophosphate for the treatment of hepatocellular carcinoma. *J. Med. Chem.* 2006, *49*, 7711–7720.
41. Boyer, S. H.; Jiang, H.; Jacintho, J. D.; Reddy, M. V.; Li, H.; Li, W.; Godwin, J. L.; Schulz, W. G.; Cable, E. E.; Hou, J.; Wu, R.; Fujitaki, J. M.; Hecker, S. J.; Erion, M. D. Synthesis and biological evaluation of a series of liver-selective phosphonic acid thyroid hormone receptor agonists and their prodrugs. *J. Med. Chem.* 2008, *51*, 7075–7093.
42. Thornton, P. J.; Kadri, H.; Miccoli, A.; Mehellou, Y. Nucleoside phosphate and phosphonate prodrug clinical candidates. *J. Med. Chem.* 2016, *59*, 10400–10410.
43. Mehellou, Y.; Rattan, H. S.; Balzarini, J. The protide prodrug technology: from the concept to the clinic. *J. Med. Chem.* 2018, *61*, 2211–2226.
44. World Health Organization, *World Health Organization Model List of Essential Medicines: 21st List 2019*; World Health Organization: Geneva, 2019.
45. Kattakuzhy, S.; Levy, R.; Kottilil, S. Sofosbuvir for treatment of chronic hepatitis C. *Hepatol. Int.* 2015, *9*, 161–173.
46. Sofia, M. J.; Bao, D.; Chang, W.; Du, J.; Nagarathnam, D.; Rachakonda, S.; Reddy, P. G.; Ross, B. S.; Wang, P.; Zhang, H.-R.; Bansal, S.; Espiritu, C.; Keilman, M.; Lam, A. M.; Steuer, H. M. M.; Niu, C.; Otto, M. J.; Furman, P. A. Discovery of a β-d-2′-deoxy-2′-α-fluoro-2′-β-*C*-methyluridine nucleotide prodrug (PSI-7977) for the treatment of hepatitis C virus. *J. Med. Chem.* 2010, *53*, 7202–7218.
47. Ma, H.; Jiang, W.-R.; Robledo, N.; Leveque, V.; Ali, S.; Lara-Jaime, T.; Masjedizadeh, M.; Smith, D. B.; Cammack, N.; Klumpp, K.; Symons, J. Characterization of the metabolic activation of hepatitis C virus nucleoside inhibitor β-D-2′-deoxy-2′-fluoro-2′-*C*-methylcytidine (PSI-6130) and identification of a novel active 5′-triphosphate species. *J. Biol. Chem.* 2007, *282*, 29812–29820.
48. Murakami, E.; Bao, H.; Ramesh, M.; McBrayer, T. R.; Whitaker, T.; Micolochick Steuer, H. M.; Schinazi, R. F.; Stuyver, L. J.; Obikhod, A.; Otto, M. J.; Furman, P. A. Mechanism of activation of β-D-2′-deoxy-2′-fluoro-2′-*C*-methylcytidine and inhibition of hepatitis C virus NS5B RNA polymerase. *Antimicrob. Agents Chemother.* 2007, *51*, 503–509.
49. Murakami, E.; Niu, C.; Bao, H.; Micolochick Steuer, H. M.; Whitaker, T.; Nachman, T.; Sofia, M. A.; Wang, P.; Otto, M. J.; Furman, P. A. The mechanism of action of β-D-2′-deoxy-2′-fluoro-2′-*C*-methylcytidine involves a second metabolic pathway leading to β-D-2′-deoxy-2′-fluoro-2′-*C*-methyluridine 5′-triphosphate, a potent inhibitor of the hepatitis C virus RNA-dependent RNA polymerase. *Antimicrob. Agents Chemother.* 2008, *52*, 458–464.
50. De Clercq, E. Tenofovir alafenamide (TAF) as the successor of tenofovir disoproxil fumarate (TDF). *Biochem. Pharmacol.* 2016, *119*, 1–7.
51. Eisenberg, E. J.; He, G.-X.; Lee, W. A. Metabolism of GS-7340, a novel phenyl monophosphoramidate intracellular prodrug of PMPA, in blood. *Nucleos. Nucleot. Nucl.* 2001, *20*, 1091–1098.
52. Gibson, A. K.; Shah, B. M.; Nambiar, P. H.; Schafer, J. J. Tenofovir alafenamide. *Ann. Pharmacother.* 2016, *50*, 942–952.

53. Menéndez-Arias, L.; Álvarez, M.; Pacheco, B. Nucleoside/nucleotide analog inhibitors of hepatitis B virus polymerase: mechanism of action and resistance. *Curr. Opin. Virol.* 2014, *8*, 1–9.
54. Warren, T. K.; Jordan, R.; Lo, M. K.; Ray, A. S.; Mackman, R. L.; Soloveva, V.; Siegel, D.; Perron, M.; Bannister, R.; Hui, H. C.; Larson, N.; Strickley, R.; Wells, J.; Stuthman, K. S.; Van Tongeren, S. A.; Garza, N. L.; Donnelly, G.; Shurtleff, A. C.; Retterer, C. J.; Gharaibeh, D.; Zamani, R.; Kenny, T.; Eaton, B. P.; Grimes, E.; Welch, L. S.; Gomba, L.; Wilhelmsen, C. L.; Nichols, D. K.; Nuss, J. E.; Nagle, E. R.; Kugelman, J. R.; Palacios, G.; Doerffler, E.; Neville, S.; Carra, E.; Clarke, M. O.; Zhang, L.; Lew, W.; Ross, B.; Wang, Q.; Chun, K.; Wolfe, L.; Babusis, D.; Park, Y.; Stray, K. M.; Trancheva, I.; Feng, J. Y.; Barauskas, O.; Xu, Y.; Wong, P.; Braun, M. R.; Flint, M.; McMullan, L. K.; Chen, S.-S.; Fearns, R.; Swaminathan, S.; Mayers, D. L.; Spiropoulou, C. F.; Lee, W. A.; Nichol, S. T.; Cihlar, T.; Bavari, S. Therapeutic efficacy of the small molecule GS-5734 against Ebola virus in rhesus monkeys. *Nature* 2016, *531*, 381–385.
55. Malin, J. J.; Suárez, I.; Priesner, V.; Fätkenheuer, G.; Rybniker, J. Remdesivir against COVID-19 and other viral diseases. *Clin. Microbiol. Rev.* 2020, *34*, e00162-20.
56. Berliba, E.; Bogus, M.; Vanhoutte, F.; Berghmans, P.-J.; Good, S. S.; Moussa, A.; Pietropaolo, K.; Murphy, R. L.; Zhou, X.-J.; Sommadossi, J.-P. Safety, pharmacokinetics, and antiviral activity of AT-527, a novel purine nucleotide prodrug, in hepatitis C virus-infected subjects with or without cirrhosis. *Antimicrob. Agents Chemother.* 2019, *63*, e01201-19.
57. Good, S. S.; Moussa, A.; Zhou, X.-J.; Pietropaolo, K.; Sommadossi, J.-P. Preclinical evaluation of AT-527, a novel guanosine nucleotide prodrug with potent, pangenotypic activity against hepatitis C virus. *PLoS One* 2020, *15*, e0227104.
58. Cho, A.; Zhang, L.; Xu, J.; Lee, R.; Butler, T.; Metobo, S.; Aktoudianakis, V.; Lew, W.; Ye, H.; Clarke, M.; Doerffler, E.; Byun, D.; Wang, T.; Babusis, D.; Carey, A. C.; German, P.; Sauer, D.; Zhong, W.; Rossi, S.; Fenaux, M.; McHutchison, J. G.; Perry, J.; Feng, J.; Ray, A. S.; Kim, C. U. Discovery of the first *C*-nucleoside HCV polymerase inhibitor (GS-6620) with demonstrated antiviral response in HCV infected patients. *J. Med. Chem.* 2014, *57*, 1812–1825.
59. Feng, J. Y.; Cheng, G.; Perry, J.; Barauskas, O.; Xu, Y.; Fenaux, M.; Eng, S.; Tirunagari, N.; Peng, B.; Yu, M.; Tian, Y.; Lee, Y.-J.; Stepan, G.; Lagpacan, L. L.; Jin, D.; Hung, M.; Ku, K. S.; Han, B.; Kitrinos, K.; Perron, M.; Birkus, G.; Wong, K. A.; Zhong, W.; Kim, C. U.; Carey, A.; Cho, A.; Ray, A. S. Inhibition of hepatitis C virus replication by GS-6620, a potent *C*-nucleoside monophosphate prodrug. *Antimicrob. Agents Chemother.* 2014, *58*, 1930–1942.
60. De Clercq, E. *C*-nucleosides to be revisited. *J. Med. Chem.* 2016, *59*, 2301–2311.
61. De Clercq, E. New nucleoside analogues for the treatment of hemorrhagic fever virus infections. *Chem. Asian J.* 2019, *14*, 3962–3968.
62. Slusarczyk, M.; Lopez, M. H.; Balzarini, J.; Mason, M.; Jiang, W. G.; Blagden, S.; Thompson, E.; Ghazaly, E.; McGuigan, C. Application of ProTide technology to gemcitabine: a successful approach to overcome the key cancer resistance mechanisms leads to a new agent (NUC-1031) in clinical development. *J. Med. Chem.* 2014, *57*, 1531–1542.
63. McGuigan, C.; Murziani, P.; Slusarczyk, M.; Gonczy, B.; Vande Voorde, J.; Liekens, S.; Balzarini, J. Phosphoramidate ProTides of the anticancer agent FUDR

successfully deliver the preformed bioactive monophosphate in cells and confer advantage over the parent nucleoside. *J. Med. Chem.* 2011, *54*, 7247–7258.
64. Ghazaly, E. A.; Slusarczyk, M.; McGuigan, C.; Harrison, D.; Blagden, S. P. Abstract B46: NUC-3373: a novel pyrimidine nucleotide analogue that overcomes key cancer drug resistance limiting patient survival. *Mol. Cancer Ther.* 2015, *14*, B46.
65. Balzarini, J.; Herdewijn, P.; De Clercq, E. Differential patterns of intracellular metabolism of 2′,3′-didehydro-2′,3′-dideoxythymidine and 3′-azido-2′,3′-dideoxy thymidine, two potent anti-human immunodeficiency virus compounds. *J. Biol. Chem.* 1989, *264*, 6127–6133.
66. Venkatachalam, T. K.; Tai, H.-L.; Vig, R.; Chen, C.-L.; Jan, S.-T.; Uckun, F. M. Enhancing effects of a mono-bromo substitution at the para position of the phenyl moiety on the metabolism and anti-HIV activity of D4T-phenyl methoxyalaninyl phosphate derivatives. *Bioorg. Med. Chem. Lett.* 1998, *8*, 3121–3126.
67. Uckun, F. M.; Pendergrass, S.; Venkatachalam, T. K.; Qazi, S.; Richman, D. Stampidine is a potent inhibitor of zidovudine- and nucleoside analog reverse transcriptase inhibitor-resistant primary clinical human immunodeficiency virus type 1 isolates with thymidine analog mutations. *Antimicrob. Agents Chemother.* 2002, *46*, 3613–3616.
68. Uckun, F. M.; Chen, C. L.; Lisowski, E.; Mitcheltree, G. C.; Venkatachalam, T. K.; Erbeck, D.; Chen, H.; Waurzyniak, B. Toxicity and pharmacokinetics of stampidine in mice and rats. *Arzneim. Forsch.* 2003, *53*, 357–367.
69. Uckun, F. M.; Cahn, P.; Qazi, S.; D'Cruz, O. Stampidine as a promising antiretroviral drug candidate for pre-exposure prophylaxis against sexually transmitted HIV/AIDS. *Expert Opin. Investig. Drugs* 2012, *21*, 489–500.
70. Lackey, D. B.; Groziak, M. P.; Sergeeva, M.; Beryt, M.; Boyer, C.; Stroud, R. M.; Sayre, P.; Park, J. W.; Johnston, P.; Slamon, D.; Shepard, H. M.; Pegram, M. Enzyme-catalyzed therapeutic agent (ECTA) design: activation of the antitumor ECTA compound NB1011 by thymidylate synthase. *Biochem. Pharmacol.* 2001, *61*, 179–189.
71. Boyer, C. R.; Karjian, P. L.; Wahl, G. M.; Pegram, M.; Neuteboom, S. T. Nucleoside transport inhibitors, dipyridamole and p-nitrobenzylthioinosine, selectively potentiate the antitumor activity of NB1011. *Anti-Cancer Drugs* 2002, *13*, 29–36.
72. Wilson, R. H. Novel therapeutic developments other than EGFR and VEGF inhibition in colorectal cancer. *Oncologist* 2006, *11*, 1018–1024.
73. McGuigan, C.; Madela, K.; Aljarah, M.; Gilles, A.; Brancale, A.; Zonta, N.; Chamberlain, S.; Vernachio, J.; Hutchins, J.; Hall, A.; Ames, B.; Gorovits, E.; Ganguly, B.; Kolykhalov, A.; Wang, J.; Muhammad, J.; Patti, J. M.; Henson, G. Design, synthesis and evaluation of a novel double pro-drug: INX-08189. A new clinical candidate for hepatitis C virus. *Bioorg. Med. Chem. Lett.* 2010, *20*, 4850–4854.
74. Simmons, B.; Liu, Z.; Klapars, A.; Bellomo, A.; Silverman, S. M. Mechanism-based solution to the ProTide synthesis problem: selective access to sofosbuvir, acelarin, and INX-08189. *Org. Lett.* 2017, *19*, 2218–2221.
75. Vernachio, J. H.; Bleiman, B.; Bryant, K. D.; Chamberlain, S.; Hunley, D.; Hutchins, J.; Ames, B.; Gorovits, E.; Ganguly, B.; Hall, A.; Kolykhalov, A.; Liu, Y.; Muhammad, J.; Raja, N.; Walters, C. R.; Wang, J.; Williams, K.; Patti, J. M.; Henson, G.; Madela, K.; Aljarah, M.; Gilles, A.; McGuigan, C. INX-08189, a

phosphoramidate prodrug of 6-*O*-methyl-2′-*C*-methyl guanosine, is a potent inhibitor of hepatitis C virus replication with excellent pharmacokinetic and pharmacodynamic properties. *Antimicrob. Agents Chemother.* 2011, *55*, 1843–1851.

76. Chang, W.; Bao, D.; Chun, B. K.; Naduthambi, D.; Nagarathnam, D.; Rachakonda, S.; Reddy, P. G.; Ross, B. S.; Zhang, H. R.; Bansal, S.; Espiritu, C. L.; Keilman, M.; Lam, A. M.; Niu, C.; Steuer, H. M.; Furman, P. A.; Otto, M. J.; Sofia, M. J. Discovery of PSI-353661, a novel purine nucleotide prodrug for the treatment of HCV infection. *ACS Med. Chem. Lett.* 2011, *2*, 130–135.

77. Boojamra, C. G.; Mackman, R. L.; Markevitch, D. Y.; Prasad, V.; Ray, A. S.; Douglas, J.; Grant, D.; Kim, C. U.; Cihlar, T. Synthesis and anti-HIV activity of GS-9148 (2′-Fd4AP), a novel nucleoside phosphonate HIV reverse transcriptase inhibitor. *Bioorg. Med. Chem. Lett.* 2008, *18*, 1120–1123.

78. Cihlar, T.; Ray, A. S.; Boojamra, C. G.; Zhang, L.; Hui, H.; Laflamme, G.; Vela, J. E.; Grant, D.; Chen, J.; Myrick, F.; White, K. L.; Gao, Y.; Lin, K.-Y.; Douglas, J. L.; Parkin, N. T.; Carey, A.; Pakdaman, R.; Mackman, R. L. Design and profiling of GS-9148, a novel nucleotide analog active against nucleoside-resistant variants of human immunodeficiency virus type 1, and its orally bioavailable phosphonoamidate prodrug, GS-9131. *Antimicrob. Agents Chemother.* 2008, *52*, 655–665.

79. Ray, A. S.; Vela, J. E.; Boojamra, C. G.; Zhang, L.; Hui, H.; Callebaut, C.; Stray, K.; Lin, K.-Y.; Gao, Y.; Mackman, R. L.; Cihlar, T. Intracellular metabolism of the nucleotide prodrug GS-9131, a potent anti-human immunodeficiency virus agent. *Antimicrob. Agents Chemother.* 2008, *52*, 648–654.

80. Mackman, R. L.; Ray, A. S.; Hui, H. C.; Zhang, L.; Birkus, G.; Boojamra, C. G.; Desai, M. C.; Douglas, J. L.; Gao, Y.; Grant, D.; Laflamme, G.; Lin, K. Y.; Markevitch, D. Y.; Mishra, R.; McDermott, M.; Pakdaman, R.; Petrakovsky, O. V.; Vela, J. E.; Cihlar, T. Discovery of GS-9131: design, synthesis and optimization of amidate prodrugs of the novel nucleoside phosphonate HIV reverse transcriptase (RT) inhibitor GS-9148. *Bioorg. Med. Chem.* 2010, *18*, 3606–3617.

81. Alexandre, F.-R.; Badaroux, E.; Bilello, J. P.; Bot, S.; Bouisset, T.; Brandt, G.; Cappelle, S.; Chapron, C.; Chaves, D.; Convard, T.; Counor, C.; Da Costa, D.; Dukhan, D.; Gay, M.; Gosselin, G.; Griffon, J.-F.; Gupta, K.; Hernandez-Santiago, B.; La Colla, M.; Lioure, M.-P.; Milhau, J.; Paparin, J.-L.; Peyronnet, J.; Parsy, C.; Pierra Rouvière, C.; Rahali, H.; Rahali, R.; Salanson, A.; Seifer, M.; Serra, I.; Standring, D.; Surleraux, D.; Dousson, C. B. The discovery of IDX21437: design, synthesis and antiviral evaluation of 2′-α-chloro-2′-β-*C*-methyl branched uridine pronucleotides as potent liver-targeted HCV polymerase inhibitors. *Bioorg. Med. Chem. Lett.* 2017, *27*, 4323–4330.

82. DiRocco, D. A.; Ji, Y.; Sherer, E. C.; Klapars, A.; Reibarkh, M.; Dropinski, J.; Mathew, R.; Maligres, P.; Hyde, A. M.; Limanto, J.; Brunskill, A.; Ruck, R. T.; Campeau, L.-C.; Davies, I. W. A multifunctional catalyst that stereoselectively assembles prodrugs. *Science* 2017, *356*, 426–430.

83. Borgia, G.; Maraolo, A. E.; Nappa, S.; Gentile, I.; Buonomo, A. R. NS5B polymerase inhibitors in phase II clinical trials for HCV infection. *Expert Opin. Investig. Drugs* 2018, *27*, 243–250.

84. Lawitz, E.; Gane, E.; Feld, J. J.; Buti, M.; Foster, G. R.; Rabinovitz, M.; Burnevich, E.; Katchman, H.; Tomasiewicz, K.; Lahser, F.; Jackson, B.; Shaughnessy, M.; Klopfer, S.; Yeh, W. W.; Robertson, M. N.; Hanna, G. J.; Barr, E.; Platt, H. L.; on behalf of the C-BREEZE-2 Study Investigators. Efficacy and safety of a two-drug direct-acting

antiviral agent regimen ruzasvir 180 mg and uprifosbuvir 450 mg for 12 weeks in adults with chronic hepatitis C virus genotype 1, 2, 3, 4, 5 or 6. *J. Viral Hepat.* 2019, *26*, 1127–1138.
85. Soriano, V.; Fernandez-Montero, J. V.; de Mendoza, C.; Benitez-Gutierrez, L.; Peña, J. M.; Arias, A.; Barreiro, P. Treatment of hepatitis C with new fixed dose combinations. *Expert. Opin. Pharmacother.* 2017, *18*, 1235–1242.
86. Chang, J.; Huang, Q. Preparation of 2'-fluoro-2'-(fluoromethyl)-2'-deoxynucleosi des and phosphate prodrugs. CN Patent, ZL201310665447.6, November 30, 2016.
87. Chang, J.; Huang, Q.; Liu, R.; Wang, Z. Double-liver-targeting phosphoramidate and phosphonoamidate prodrugs. US Patent, US9156874B2, October 13, 2015.
88. Li, E.; Wang, Y.; Yu, W.; Lv, Z.; Peng, Y.; Liu, B.; Li, S.; Ho, W.; Wang, Q.; Li, H.; Chang, J. Synthesis and biological evaluation of a novel β-D-2′-deoxy-2′-α-fluoro-2′-β-*C*-(fluoromethyl)uridine phosphoramidate prodrug for the treatment of hepatitis C virus infection. *Eur. J. Med. Chem.* 2018, *143*, 107–113.
89. Abdo, A. E. Novel guanosine derivatives as anti-HCV NS5b polymerase: a QSAR and molecular docking study. *Med. Chem.* 2019, *15*, 130–137.
90. Elfiky, A. A.; Elshemey, W. M. IDX-184 is a superior HCV direct-acting antiviral drug: a QSAR study. *Med. Res. Rev.* 2016, *25*, 1005–1008.
91. Elfiky, A. A. Zika viral polymerase inhibition using anti-HCV drugs both in market and under clinical trials. *J. Med. Virol.* 2016, *88*, 2044–2051.
92. Elfiky, A. A.; Ismail, A. M. Molecular docking revealed the binding of nucleotide/side inhibitors to Zika viral polymerase solved structures. *SAR QSAR Environ. Res.* 2018, *29*, 409–418.
93. Elfiky, A. A.; Mahdy, S. M.; Elshemey, W. M. Quantitative structure-activity relationship and molecular docking revealed a potency of anti-hepatitis C virus drugs against human corona viruses. *J. Med. Virol.* 2017, *89*, 1040–1047.
94. Elfiky, A. A. Anti-HCV, nucleotide inhibitors, repurposing against COVID-19. *Life Sci.* 2020, *248*, 117477.
95. Elfiky, A. A. Ribavirin, remdesivir, sofosbuvir, galidesivir, and tenofovir against SARS-CoV-2 RNA dependent RNA polymerase (RdRp): a molecular docking study. *Life Sci.* 2020, *253*, 117592.
96. Peng, Y.; Yu, W.; Li, E.; Kang, J.; Wang, Y.; Yang, Q.; Liu, B.; Zhang, J.; Li, L.; Wu, J.; Jiang, J.; Wang, Q.; Chang, J. Discovery of an orally active and liver-targeted Prodrug of 5-fluoro-2′-deoxyuridine for the treatment of hepatocellular carcinoma. *J. Med. Chem.* 2016, *59*, 3661–3670.
97. Beadle, J. R.; Wan, W. B.; Ciesla, S. L.; Keith, K. A.; Hartline, C.; Kern, E. R.; Hostetler, K. Y. Synthesis and antiviral evaluation of alkoxyalkyl derivatives of 9-(*S*)-(3-hydroxy-2-phosphonomethoxypropyl)adenine against cytomegalovirus and orthopoxviruses. *J. Med. Chem.* 2006, *49*, 2010–2015.
98. Kern, E. R.; Hartline, C.; Harden, E.; Keith, K.; Rodriguez, N.; Beadle, J. R.; Hostetler, K. Y. Enhanced inhibition of orthopoxvirus replication in vitro by alkoxyalkyl esters of cidofovir and cyclic cidofovir. *Antimicrob. Agents Chemother.* 2002, *46*, 991–995.
99. Ross, B. S.; Reddy, P. G.; Zhang, H. R.; Rachakonda, S.; Sofia, M. J. Synthesis of diastereomerically pure nucleotide phosphoramidates. *J. Organomet. Chem.* 2011, *76*, 8311–8319.
100. Chapman, H.; Kernan, M.; Prisbe, E.; Rohloff, J.; Sparacino, M.; Terhorst, T.; Yu, R. Practical synthesis, separation, and stereochemical assignment of the PMPA pro-drug GS-7340. *Nucleos. Nucleot. Nucl.* 2001, *20*, 621–628.

101. Siegel, D.; Hui, H. C.; Doerffler, E.; Clarke, M. O.; Chun, K.; Zhang, L.; Neville, S.; Carra, E.; Lew, W.; Ross, B.; Wang, Q.; Wolfe, L.; Jordan, R.; Soloveva, V.; Knox, J.; Perry, J.; Perron, M.; Stray, K. M.; Barauskas, O.; Feng, J. Y.; Xu, Y.; Lee, G.; Rheingold, A. L.; Ray, A. S.; Bannister, R.; Strickley, R.; Swaminathan, S.; Lee, W. A.; Bavari, S.; Cihlar, T.; Lo, M. K.; Warren, T. K.; Mackman, R. L. Discovery and synthesis of a phosphoramidate prodrug of a pyrrolo[2,1-f][triazin-4-amino] adenine *C*-nucleoside (GS-5734) for the treatment of Ebola and emerging viruses. *J. Med. Chem.* 2017, *60*, 1648–1661.
102. Wang, M.; Zhang, L.; Huo, X.; Zhang, Z.; Yuan, Q.; Li, P.; Chen, J.; Zou, Y.; Wu, Z.; Zhang, W. Catalytic asymmetric synthesis of the anti-COVID-19 drug remdesivir. *Angew. Chem. Int. Ed.* 2020, *59*, 20814–20819.

CHAPTER 2

CYCLIC DINUCLEOTIDES: A NEW-GENERATION DRUG FOR IMMUNE THERAPY

ZHENGHUA WANG, DEJUN MA, AND ZHEN XI
Department of Chemical Biology and State Key Laboratory of Elemento-Organic Chemistry, College of Chemistry, Nankai University, Tianjin, China

2.1. STRUCTURE OF CDNs

CDNs were constituted of two nucleosides connected by two phosphate diester bonds. As shown in Figure 2.1, CDNs could be divided into three types based on the cyclic-phosphate diester backbone. The 3′-3′ CDNs were constituted of two nucleosides with two 3′-5′ phosphate diester bonds. The 2′-3′ CDNs were constituted of two nucleosides with one 2′-5′ and one 3′-5′ phosphate diester bond. The 2′-2′ CDNs were constituted of two nucleosides with two 2′-5′ phosphate diester bonds.

Until now, all of the CDNs found in bacteria are 3′-3′ CDNs, such as **X-1** (c[G(3′,5′)G(3′,5′)], 3′,3′-c-di-GMP) [4], **X-2** (c[A(3′,5′)A(3′,5′)], 3′-3′-c-di-AMP) [5], **X-3** (c[G(3′,5′)A(3′,5′)], 3′,3′-cGAMP) [6]. 2′-3′ CDNs were enriched in mammals like **X-5** (c[G(2′,5′)A(3′,5′)]), which was also termed 2′,3′-cGAMP [7–11]. The 2′-5′ phosphodiester linked the 2′-OH of GMP and 5′-OH of AMP and the 3′-5′ phosphodiester linked the 3′-OH of AMP and 5′-OH of GMP. However, no 2′-2′ CDNs have been found in natural organisms so far.

Nucleic Acids in Medicinal Chemistry and Chemical Biology: Drug Development and Clinical Applications, First Edition. Edited by Lihe Zhang, Xinjing Tang, Zhen Xi, and Jyoti Chattopadhyaya.

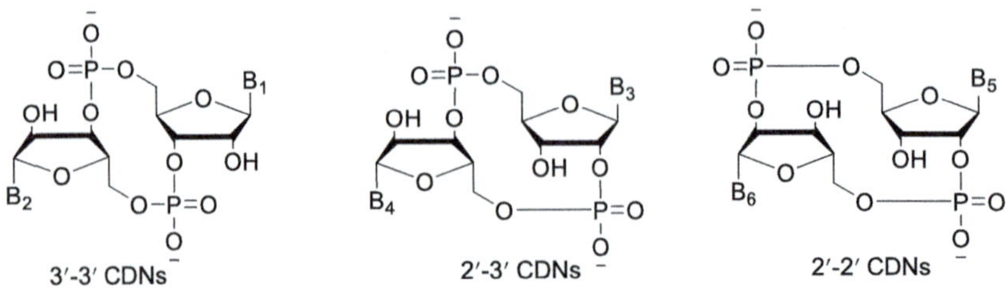

X-1 B_1=B_2=G
X-2 B_1=B_2=A
X-3 B_1=G, B_2=A
X-4 B_1=A, B_2=U

X-5 B_3=G, B_4=A

X-6 B_5=G, B_6=U
X-7 B_5=I, B_6=U

Figure 2.1 General structure of CDNs.

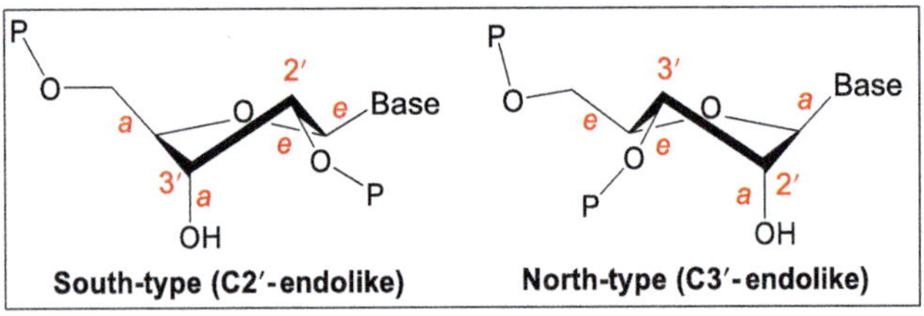

Figure 2.2 Ribose conformation of CDNs.

The conformational space of CDNs was investigated. As shown in Figure 2.2, the ribose moieties predominantly adopted the North-type conformation (C3′-endolike) in 3′-3′ CDNs through the combination of computational and NMR methods [12]. When the ribose moieties was attached by 2′-5′ phosphodiester linkage, the ribose was prone to be a South-type (C2′-ednolike) conformation [13]. For 2′-3′ cGAMP, the ribose of guanosine was prone to be a South-type, while the sugar of adenosine was prone to be a North-type conformation. The ribose moieties and cyclic di-phosphonate backbone of 3′-3′ CDNs constituted of different base type adopted the similar conformational space [14]. The *syn*-glyosidic bond could encounter less steric clash on the C2′-endolike ribose than on the C3′-endolike ribose. The glyosidic bonds of North-type ribose tended to give the *anti*-conformation, while the South-type ribose was more prone to forming the *syn*-glycosyl conformation [13]. The N2-amino group of G base would form electrostatic interaction with phosphate group leading to favorably adopting the *syn*-glycosyl conformation in 2′-H- and F-modified 3′,3′-c-di-GMP [12]. Hence, the smallest ring in the 3′-3′ CDNs generated the highest rigid ribose conformation, while the largest ring in 2′-2′ CDNs rendered the highest flexibility. The conformation space of CDNs with different backbone would affect its physiological activity and bioactivity [13–16].

2.2. cGAS-cGAMP-STING PATHWAY

Cyclic GMP-AMP synthase (cGAS) has been identified as a new dsDNA sensor in cytoplasm in mammalian cells [17]. During the infection, pattern recognition receptors (PRRs) act as key sensors of the innate immune system, which recognized distinct pathogen-associated molecular patterns (PAMPs) such as lipids, lipoproteins, nucleic acids, and damaged-associated molecular patterns (DAMP) such as the damaged mitochondria and nuclei. The identified PRR family mainly included TLRs, RIRs, NLR, and others. Foreign dsDNA from bacteria or virus and self-dsDNA leaked from the nucleus or mitochondria under some pressure could be detected by cGAS. After activation, cGAS could synthesize 2′,3′-cGAMP with the substrates of GTP and ATP (Figure 2.3) [9].Until now, there was no evidence that cGAS had DNA sequence requirements in sensing dsDNA. However, cGAS, especially the hcGAS, was prone to bind longer dsDNA to form a condensed model. When hcGAS bind to dsDNA, hcGAS and dsDNA could form the highly ordered complex with the molar ratio of 2n : 2 (Figure 2.4). The ladder-like cGAS dimer located between two dsDNA in the head-to-head and tail-to-tail mode to stabilize the cGAS-dsDNA complex structure, which endowed the cGAS the better activity [18]. When the cGAS-dsDNA complex was added with HMGB/TFAM [19–22], the longer dsDNA could form a U-sharp structure with the more potential to form liquid-like droplet.

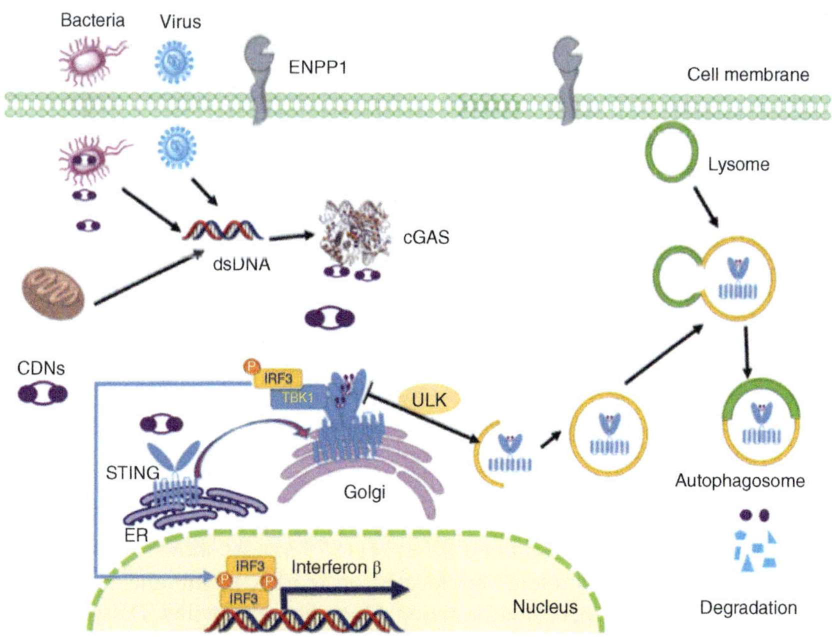

Figure 2.3 dsDNA-induced cGAS-STING pathway.

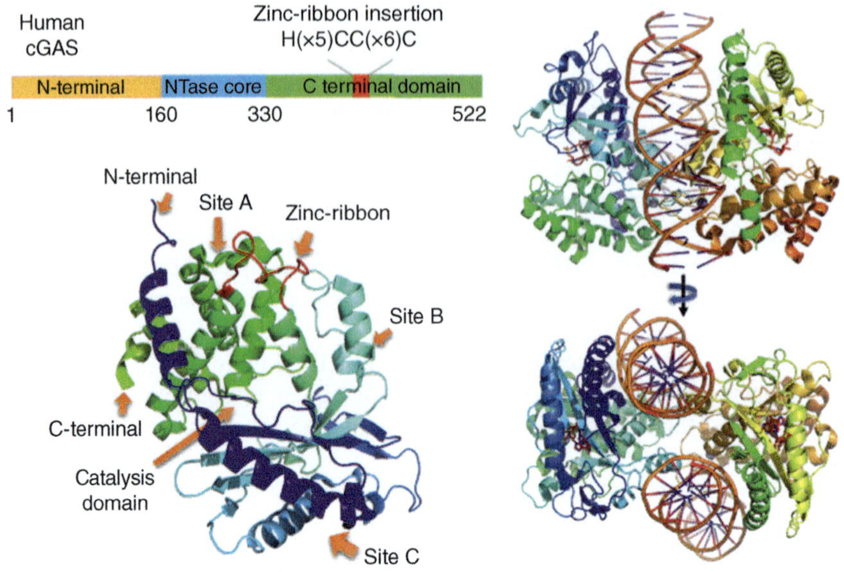

Figure 2.4 Molecular basis of dsDNA sensing by cGAS.

The liquid–liquid phase-separation effects dramatically enhance the activation of cGAS, in which Zn^{2+} plays an important role during the droplet formation [23]. Furthermore, a third DNA binding site (the site-C interface of hcGAS) was essential to DNA condensate formation A (Figure 2.4) [24]. Both the formation of ladder-like complex and droplet by phase separation would promote the condensation and raise the local concentration of cGAS/dsDNA, resulting in the enhancement of the activity of cGAS. Once the cGAS was activated, two distinct chemical reactions were involved in the generation of 2′,3′-cGAMP. Firstly, cGAS bound ATP and GTP as substrates and transferred ATP to the 2′-OH of GTP to form the linear intermediate pppG(2′,5′)pA. Subsequently, the linear pppG(2′,5′)pA was taken as a new substrate and the GTP moiety intramolecularly transferred to the 3′-OH of the adenosine nucleoside phosphate to form 2′,3′-cGAMP.

Stimulator of interferon genes (STING), also known as TMEM173, MPYS, MITA, or ERIS, was recognized as a direct receptor of CDNs (Figure 2.3) [8, 25–29]. The full-length hSTING protein was composed of 379 amino acids (379 aa) including transmembrane (TM) domains (1–139 aa) located in the endoplasmic reticulum (ER) membrane, the CDNs binding domain (CBD) (159–340 aa), and C-terminal tail (CTT) (341–379 aa). As shown in Figure 2.5, the inactivated STING protein was located on the ER membrane as a dimer, exhibiting a flexible "open" conformation. Upon bounded with CDNs, the CTT was released from a self-inhibited conformation [26]. The two C-terminal part of STING would twist by 180° [30] and shift to a "closed" conformation with

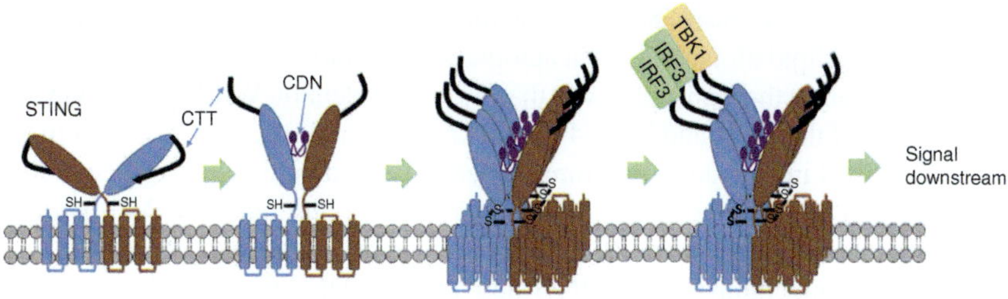

Figure 2.5 Activation mode of STING stimulated by CDN.

four ordered sheet lid covering on the binding pocket. After the palmitoylation of C81/89 residues [31, 32] and the formation of intermolecular disulfide bound of C148 residues [33], the STING dimers would polymerized and translocated to Golgi. The CTT scaffold contributed to the recruitment and activation of TBK1 and IRF3, which were then phosphorylated [34, 35]. The phosphorylated IRF-3 was finally translocated into the nucleic to induce the expression of type I interferon and other inflammatory cytokines.

Although cGAS-cGAMP-STING pathway has been fully revealed, there were also some controversy about the molecular mechanism of the CDNs-STING activation. As reported, the crystal complex of wild-type hSTING with 3′,3′-c-di-GMP in an "open" conformation was still able to activate wild-type hSTING, which indicated that the "closed" conformation was not essential to activation [36]. However, V147L/C148A hSTING double mutant was still able to be activated by cGAMP to levels comparable to wild-type STING, which indicated that the disulfide bound of C148 was also not essential [33]. Until now, no crystal complex of wild-type hSTING containing CTT was reported. How the CTT functioned when stimulated by CDNs was also unknown. Although a STING-deficient cell-line overexpressing TM-deficient STING could not undergo STING signaling when CDNs were added, the well-defined self-assembled tetrameric complex of CDNs with TM-deficient STING protein could effectively trigger the STING signaling pathway even in STING-deficient cell lines [37]. Hence, further investigation of STING activation by CDNs would be necessary in the future work.

In addition to cGAS-cGAMP-STING pathway, CDNs could also induce autophagy through STING-dependent but IRF-3-independent pathway (Figure 2.3). After treatment with CDNs, STING was firstly phosphorylated by serine/threonine UNC-51-like kinase (ULK1/ATG1) to suppress IRF3 function. When released from its repressor AMP-activated protein kinase (AMPK), ULK1 was activated. Activation of AMPK/ULK1 and STING with CDNs prevented the sustained activation of STING-dependent proinflammatory genes, which otherwise led to inflammation and autoimmune disorders [38, 39]. Upon bound by cGAMP, STING translocated to the endoplasmic reticulum–Golgi intermediate compartment (ERGIC) and the Golgi in a process that was dependent on the COP-II complex and ARF GTPases. STING-containing ERGIC then served as a membrane source for LC3 lipidation involved in autophagosome biogenesis. cGAMP induced the LC3 lipidation through a pathway that was dependent on WIPI2 and ATG5 but independent of the ULK and VPS34-beclin kinase complexes [40]. The dual functions of STING in activating the immune response and autophagy suggested that STING could be involved in the innate immune response and cGAMP-induced autophagy mediated the clearance of DNA and viruses in the cytosol [40–42].

As for the mechanism of STING activation, CDNs would induce a large conformational change of hSTING dimer (Figure 2.5). As shown in Figure 2.6, the two hSTING dimer could undergo a ~20 Å inward rotation and the two high flexible regions formed four ordered sheet lid covered by the CBD pocket to ensure the deeper binding site for 2′,3′-cGAMP [43]. The ITC binding assay have exhibited that 2′,3′-cGAMP had the higher binding affinity (nM) than bacterial CDNs [15]. In general, the base group would form the stacking interaction with Y167 and R238 residues when CDNs bound to the CBD of hSTING protein (Figure 2.6). The E260 formed the hydrogen binding with the exocyclic-NH_2

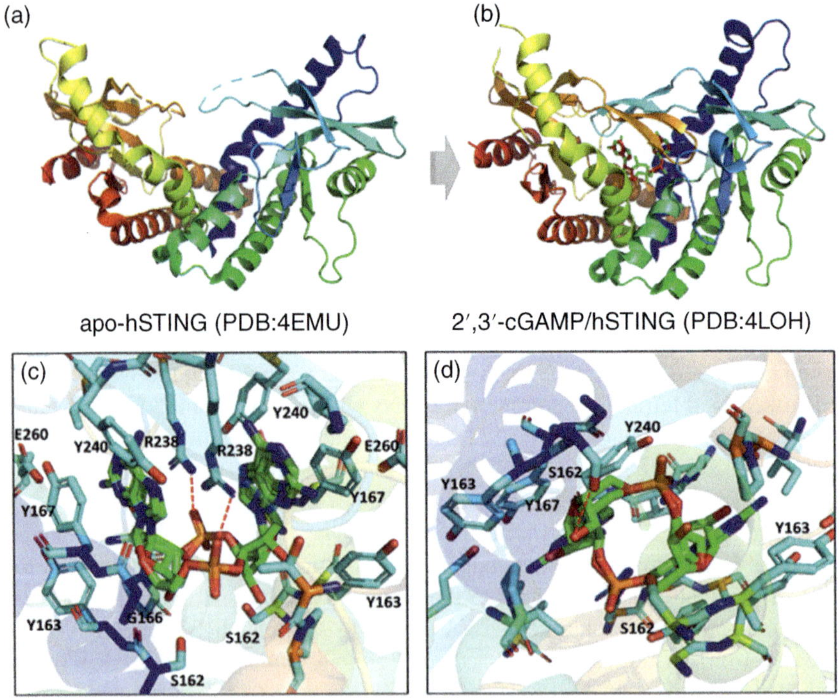

Figure 2.6 Molecular basis of hSTING activation by CDNs. (a) The structure of apo-hSTING (PDB:4EMU); (b) The complex structure of 2′,3′-cGAMP/hSTING (PDB:4LOH); (c, d) The interaction mode of CDNs with different residues in the CBD of hSTING protein.

of G base. The S162 residues could form hydrogen binding with the OH group of ribose. The R238 residues would form direct hydrogen binding with phosphonate groups. All of these interactions stabilized the structure of hSTING-CDNs complexes. According to the previous reports, CDNs with purine rings (G, A and I) would have more potential to activate hSTING pathway because the purine rings had more favor to form stacking reactions with the R238 and Y167 residues of hSTING protein. However, Xi *et al.* have demonstrated that CDNs with pyrimidine rings could also activate hSTING-dependent immune responses, especially the **X-6** (c[G(2′,5′)U(2′,5′)], 2′,2′-cGUMP) [14] and **X-7** (c[I(2′,5′)U(2′,5′)], 2′,2′-cIUMP). CDNs with 2′-5′ phosphodiester linkages had more potential to activate hSTING due to the 2′-5′ linked phosphonate group tending to form hydrogen bond with R232 residue of hSTING protein.

Although CDNs could stimulate STING, the existence of single nucleotide polymorphisms (SNPs) of hSTING genes in 1000 human Genome Project database impeded the application of CDNs as a common immune activator. Different hSTING variants showed different responses to different CDNs. As reported, hSTING contained five major variants, such as 57.9% WT, 13.7% R232H, 1.5% R293Q, 5.2% G230A-R293Q (denoted as AQ), and 20.4% R71H-G230A-R293Q (denoted as HAQ) in 98.7% of human population (Figure 2.7) [44]. Xi's group [14] also characterized these five major variants' activity stimulated by CDNs through IFN-β promoter responsive luciferase reporter assay. The results exhibited that WT could be activated by most CDNs, while the HAQ and AQ variants showed relatively weak response to part of CDNs. Both R232H and R293Q lost the response to a big part of CDNs. Intriguingly, the endogenous 2′,3′-cGAMP could activate all of these variants. The full-length cryo-EM structure of hSTING showed that R71 was buried into the N-terminal transmembrane domain and R293 was located in the stem region [30]. Both of these two sites could not directly interact with the CBD, any of whose mutation interfered with the translocation or aggregation of hSTING to initiate the IFN-β pathway. It might be more important for R293Q mutation to affect the rotation of each hSTING protomer induced by CDN binding [30]. Both G230 and R232 were located in the highly flexible regions

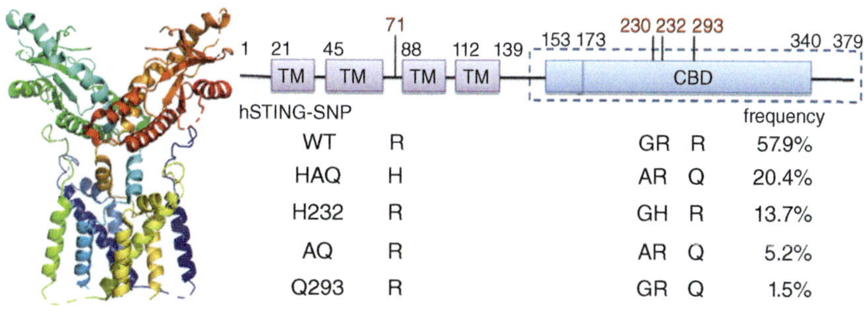

Figure 2.7 SNP distribution of hSTING protein.

of hSTING to form the ordered β-sheets covered on the CBD when bound to CDNs. hSTING including G230A mutation favored the closed conformation in the presence of some CDNs, which would increase the binding affinity of hSTING protein to CDNs [44]. R232 residue could form hydrogen binding interaction with the phosphate group of CDNs [10], which was absent in R232H mutation [43]. As a result, R232H mutation decreased the activation of hSTING and provided an explanation of the weak activation to hSTING-R232H by the part of CDNs [14].

Different hSTING genotype was widely spread in different human populations. 31% Americans were R232/R232, which was less than Europeans (50%). However, most of East Asian were R232/HAQ (34%) and only 22% East Asian were R232/R232 [44]. Jin *et al.* [45] reported that the HAQ was a kind of null allele as it had the low expression and transcriptional level in B cell lines. Ruiz-Moreno *et al.* [46] also reported that the production of type I IFNs in people containing HAQ allele was severely impaired and pro-inflammatory cytokines were easily stimulated by *L. pneumophila*, bacterial DNA, and CDNs. Human populations containing R293H allele responded weakly to bacterial CDNs but had no any response to *L. pneumophila* and DNA. Wu *et al.* found that Chinese populations containing HAQ and R232H alleles were likely associated with high risk to the development of esophageal squamous cell carcinoma (ESCC) [47]. It was still unknown whether the loss of function of hSTING mutation caused diseases.

Besides the STING acceptor, the mouse aldo-keto reductase family 1 member C13 (AKRF1MC13), termed as reductase controlling NF-κB (RECON) protein, could also bind to CDNs [48, 49]. Different from the STING pathway, the RECON protein regulated the NF-κB pathway. The RECON protein act as an inhibitor of NF-κB. Once the 3′,3′-c-di-AMP secreted by *L. monocytogenes* bound to RECON protein, the oxidoreductase activity of RECON was disrupted to activate NF-κB, which mediated the resistance of bacterial infection. In fact, RECON had selectivity to CDNs. The bacterial CDNs like 3′,3′-c-di-AMP and 3′,3′-cGAMP could bind and inhibit RECON, while 3′,3′-c-di-GMP and mammalian endogenous 2′,3′-cGAMP could not inhibit RECON. Kranzusch *et al.* found that the 3′,3′-cAUMP could also bind to RECON protein [50]. They also identified a newly cyclic trinucleotides cAAG could also bind to RECON and inhibit RECON activity. Fan *et al.* [51] firstly found that the endoplasmic reticulum (ER) membrane adaptor (ERAdP) could also directly bind with 3′,3′-c-di-AMP. When *L. monocytogenes* infected, the secreted 3′,3′-c-di-AMP could bound to the C-terminal of ERAdP, followed by the dimerization of ERAdP and the subsequent activation of kinase TAK1. It would result in the activation of NF-κB pathway to induce the production of pro-inflammation cytokines to control the bacterial infection. Whether other CDNs could also bound and activate ERAdP remains unknown. The CDNs-RECON and CDNs-ERAdP pathway might act as a role in defending bacterial pathogens when cells had low or no expression of STING protein.

2.3. DEGRADATION OF CDNs IN ENPP1 AND SERUM

CDNs were more stable in water solution than linear nucleotides. As mentioned above, the highly rigid ribose conformation of the macrocyclic phosphodiester backbone prevented CDNs from the nonspecific hydrolases. As reported, two phosphodiester bonds in CDNs could be degraded by some esterases like SVPD or nP1, while CDNs were more resistant to PDE12 [7, 52, 53]. In contrast with esterases, ecto-nucleotide pyrophosphatase/phosphodiesterase (ENPP1) was the only identified hydrolase of CDNs in cells [53]. Nureki *et al.* [54] found that 2′,3′-cGAMP was transformed into pApG intermediates (PDB: 6AEK) based on the co-crystal structure of 3′,3′-cGAMP with ENPP1 (PDB: 6AEL). The ENPP1 firstly degraded 2′-5′ phosphodiester bond rather than the 3′-5′ phosphodiester bond (Figure 2.8a). Xi's group also demonstrated that the 3′-3′ CDNs showed the higher stability than 2′-2′ CDNs and 2′-3′ CDNs toward ENPP1 [14]. There was no much difference between 2′-3′ CDNs and 2′-2′ CDNs toward ENPP1.

The stability of CDNs in serum was also studied [14]. The *in vitro* nuclease stability assay showed that the 3′-3′ CDNs have the highest serum stability, while 2′-2′ CDNs showed the lowest serum stability, which implied that the 2′-5′ phosphodiester linkage had lower stability than 3′-5′ phosphodiester linkage in CDNs. The stability of CDNs in both serum and ENPP1 was determined by the phosphodiester backbone. The base type contributed little to the stability of CDNs. It suggested that CDNs with more stable 2′-5′ phosphodiester substitution, such as phosphorothioate, would be better suited to the clinical applications [53]. The inhibition of ENPP1 would also enhance the CDN-STING-dependent immune response. Some ENPP1 inhibitors have been developed as the indirect STING agonists for the antitumor therapy [55, 56]. In cells, ENPP1 was not found in cytosol but located on the plasma membrane and ER membrane. This raised the question of how the negatively charged 2′,3′-

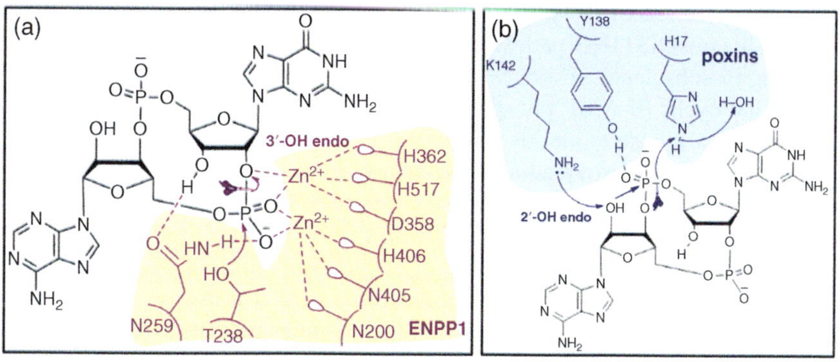

Figure 2.8 Proposed degradation mechanism of 2′,3′-cGAMP by (a) ENPP1 and (b) poxins.

cGAMP was transported into the membrane transmembrane from the cytoplasm to the extracellular catalysis site of ENPP1. Interestingly, deletion of *Sting* increased the cGAMP levels, suggesting that STING also participated in the clearance of cGAMP [57]. There was another hypothesis that the CDNs was degraded along with the STING protein by endosome (Figure 2.3). The mechanism of STING-mediated cGAMP clearance required further investigation.

In addition to ENPP1, Kranzusch's group [58] have identified that poxvirus immune nucleases (poxins) also could hydrolyze 2′,3′-cGAMP to restrict the STING-mediated signaling pathway to help the virus evade innate immunity. But different to ENPP1 needing Ca^{2+}, the degradation of 2′,3′-cGAMP by poxins was free from metal ions. As shown in Figure 2.8b, the poxins firstly degraded the 3′-5′ phosphodiester bond, which was also different from ENPP1. Poxins shared no any homology with other nucleases and phosphodiesterases, but degraded 2′,3′-cGAMP specifically. Since Poxvirus was commonly used as a vector for vaccination, gene therapy, and cancer treatments, poxins might affect the immune reaction in the poxvirus-mediated therapy [59, 60].

2.4. CDN-BASED THERAPY FOR ANTI-INFECTION TREATMENT

When bacteria or virus infected the host, the cGAS-STING pathway dominated the immune responses. As CDNs were firstly found in bacteria, the primary function of STING responded to bacterial CDNs quickly. After the infection by bacteria, CDNs synthesized by bacteria would be transferred into host cell and activate STING-dependent immune signal pathway to defend bacterial infection. For example, *Mycobacterium tuberculosis* (Mtb) infection caused serious morbidity and mortality. After Mtb infected hosts, bacterial DNA would be secreted into host cells through ESX-1 system and then initiated the cGAS-STING immune response [61–63]. When mice were infected by Mtb strain with the overexpression of disA, the excess of the secreted 3′,3′-c-di-AMP would activate the host hSTING pathway and elicit increased macrophage autophagy, resulting in substantial virulence attenuation in mice [42]. However, the deletion of the cyclic di-AMP phosphodiesterase gene (cnpB) in Mtb reduced the virulence in a mouse model of infection [64]. Van Dis *et al.* [65] also demonstrated that STING provided significant protection when CDNs was used as adjuvants against Mtb and the mucosal administration of this vaccine elicited a Th17 immune response that correlated with enhanced protection. As mentioned above, the RECON and ERadP receptor could also bind to the 3′,3′-c-di-AMP secreted by *L. monocytogenes* and initiated the NF-kB pathway against the *L. monocytogenes* infection [48, 51]. All of these studies indicated that the exogenous CDNs were promising antibacterial agents.

Besides STING activation, cGAS could be also directly activated by dsDNA from various sources, including bacteria, DNA viruses, and retroviruses.

In the past decades, a number of studies have reported that the cGAS-STING pathway play an important role during the infection of DNA viruses and retroviruses including adenovirus papillomaviruses HIV, HSV, and so on [66–74]. CDNs could also transport in a cell-to-cell manner to restrict viruses through gap junction [75], viral particles [76, 77], or via LRRC8 volume-regulated anion channels [68]. Some viruses also have acquired effective mechanisms for disrupting the cGAS-cGAMP-STING signaling pathway to block the host defense. Zika virus (ZIKV) infection triggered the activation of NLRP3 inflammasome, which could be further enhanced by viral nonstructural protein NS1 to promote the cleavage of cGAS to reduce the immune responses [78]. Similarly, human dengue viruses virus 2 (DENV2) could encode protease NS2B3, which could directly cleave hSTING protein, reduce type I interferon production, and boost viral titers in humans [79, 80]. The DENV NS2B protease cofactor targeted the cGAS for lysosomal degradation to avoid the detection of mitochondrial DNA during infection [81]. Poxvirus immune nucleases (poxins) was identified as a family of 2′,3′-cGAMP-degrading enzymes, which could cleave 2′,3′-cGAMP to restrict STING-dependent signaling [58]. The deep exploitation of the cGAS–STING pathway in maintaining the immune defenses and eliminating viral infection further enhanced the understanding of CDNs in the treatment of infectious diseases.

2.5. CDN-BASED THERAPY FOR ANTI-TUMOR TREATMENT

Immune therapy including antibody, cell-based, and cancer vaccine have been viewed as an efficient strategy for cancer treatment [82]. In particular, anti CTLA-4 monoclonal antibody, anti-PD-1/L1 monoclonal antibody, and cell-based therapy like CAR-T therapy has entered into the market for many years. As a small molecular drug, CDNs could directly promote innate immune responses in the tumor microenvironment to enhance the overexpression of type I IFN and other cytokines. Those cytokines would promote the maturation and activation of DC cells to activate tumor-specific T cells to kill cancer cells, especially cytosolic T lymphocyte and natural killer cells (Figure 2.9). CDNs have been developed as a strong immune initiator showing the big potential in clinical applications.

Earlier in 2005, Karaolis *et al.* [83] have already found that 3′,3′-c-di-GMP could inhibit both basal and growth factor and 3′,3′-c-di-GMP promoted cell cycle arrest in the lymphoblastic $CD4^+$ Jurkat cell line. This was the first report of cancer treatment with CDNs. 3′,3′-c-di-GMP was also safe to the normal kidney cells and human neuroblastoma cells even if the concentration of CDNs was 50 μM. The treatment with CDNs could keep cancers in remission [84]. As shown in Table 2.1, the combinational therapies showed more potency than monotherapy in clinical trials. CDNs could enhance the immune response

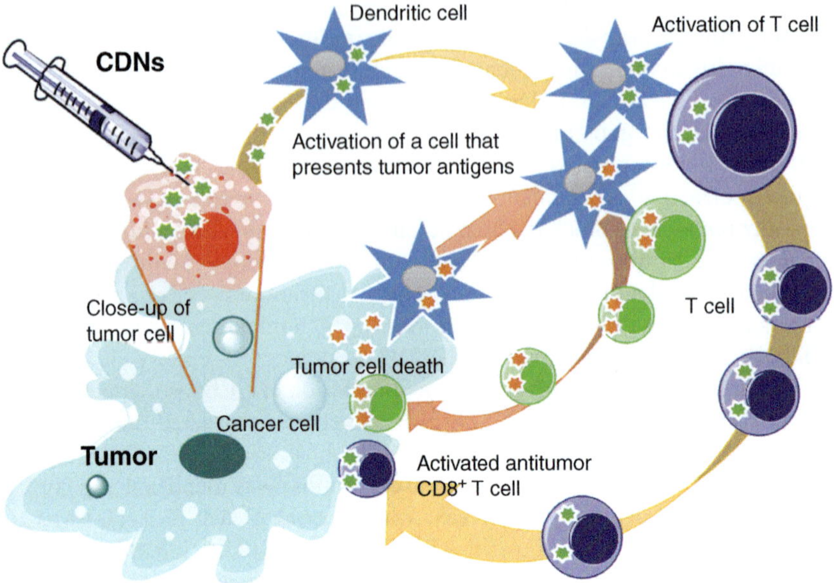

Figure 2.9 CDN-based therapy for anti-tumor treatment.

TABLE 2.1 Combinational therapies based on CDNs for antitumor treatment

No	CDNs	Combination	Tumor	Ref.
1	2′,3′-cGAMP	5-FU	Lung cancer	[85]
2	2′,3′-cGAMP	Anti-PD-1/anti-CTLA-4	Melanoma	[86]
3	2′,3′-cGAMP	Anti-PD-L1	Melanoma	[87]
4	2′,3′-cGAMP	Radiotherapy	Colon cancer	[88]
5	ADU-S100	Carboplatin /anti-PD-1	Ovarian cancer	[89]
6	ADU-S100	Anti-PD-1	Melanoma	[90]
7	ADU-S100	Anti-PD-1/anti-CTLA-4	Melanoma	[91]
8	ADU-V16	PancVAX PancVAX/anti-PD-1 PancVAX/anti-OX40	Pancreatic cancer	[92]
9	c-di-GMP	LM-Mb	Metastatic Breast Cancer	[93]
10	c-di-GMP	CAR-T	Pancreatic cancer	[94]
11	c-di-GMP	OVA/CpG	Thymoma	[95]
12	cGAMP or RR-CDA	VEGFR2-blocking antibody	Lung cancer	[96]

through the production of cytokines and proinflammatory factors, leading to the activation and differentiation of DC and Mφ cells. The latter would further enhance the infiltration and activation of T cells, especially $CD8^+$ and $CD4^+$ T cells [97, 98].

Based on these effects, CDNs would be a good choice of immune adjuvants combined with other therapies for the cancer treatment [99, 100]. Since the STING agonist could induce immune tolerogenic state by inhibitory signals like IDO [101] and PD-1/PD-L1 [89], which protected cancer cells from immune attachment and promoted tumor outgrowth [102], the combinational treatment of CDNs with anti-PD-1/PD-L1 monoclonal antibody showed better therapeutic effect than single treatment [89, 90]. The clinical trials also showed that the monotreatment of CDNs did not give much expected effect. Recently, a series of studies demonstrated that the combination of CDNs with other therapies such as cancer vaccine [100], immune checkpoint blockade (PD-1/PD-L1 [90], CTLA-4 [86]), and chimeric antigen [90, 92] have fully demonstrated the efficiency of CDNs in the anticancer treatment.

Therapeutic T cells that were programmed to express chimeric antigen receptors (CAR-T cells) for hematological malignancies has produced impressive results in clinical trials. Tyrel T. Smith *et al.* [94] demonstrated that the co-delivery of CDNs with CAR-T cells could stimulate immune responses to eliminate tumor cells that were not recognized by the adoptively transferred lymphocytes. Neoantigens, expressed on the tumor cells caused by tumor genome mutations, were an attractive immune target because their selective expression on tumors might minimize immune tolerance as well as the risk of autoimmunity [103, 104]. The combination of neoantigen-targeted vaccine with CDNs could robustly induce tumor-specific and polyfunctional effector T cells resulting in the enhancement of antitumor activity [92]. The combination of CDNs and small-molecule chemotherapy could also improve the antitumor activity. CDNs could also reduce side effects from 5-FU toxicity, which might be attributed to the activation of dendritic cells [85]. The co-administration of CDNs with cisplatin could also overcome the negative impact of cisplatin by inflaming tumor microenvironment [105].

There were several reports indicating that the combination of CDNs with HER2 or VEGFR inhibitors could also enhance the therapeutic effect. Xu *et al.* [106] found that the HER2 act as a negative regulator of hSTING-mediated immune response. They found that the HER2 could associate with STING and recruit AKT1, which could directly phosphorylate TBK1. As a result, the STING-TBK1 interaction and TBK1 K63-linked ubiquitination were prevented, which disrupted the STING-IRF3 pathway in both antiviral and antitumor processes. After the treatment of HER2 inhibitor like lapatinib, it augmented damage-induced cellular senescence and apoptosis and enhanced STING-mediated antiviral and antitumor immunity. Type I IFN signaling was negatively regulated by VEGF signaling, which could reprogram vascular involve multiple tumor microenvironment. Yang *et al.* [96] found that the combination of CDN with VEGFR inhibitor (DC101) and ICB antibody would dramatically inhibit the LLC (lewis lung cancer) proliferation. All of these findings suggested that the CDNs played an important role in the antitumor treatment, especially in the combinatorial therapy, which exhibited the high potential in clinical treatment.

cGAS-STING pathway is initially activated by cytoplasmic dsDNA. Therapies that directly or indirectly targeted to DNA such as radiotherapy [88, 107], PARP targeted inhibitors [108–111], DNA topoisomerase blockade [112, 113], and carboplatin [89] would cause the accumulation of dsDNA, resulting in the stimulation of the cGAS-STING immune pathway. Studies have showed that the combination of CDNs with these therapies also showed better anticancer efficiency than monotherapy.

Until now, several CDN-based drugs have entered into clinical researches for cancer therapy [114]. The first one, ADU-S100 (also MIW-815, Figure 2.10), developed by ADURO-biotech, now have entered into phase II clinical study. Both of the MSD and ADURO have disclosed the phase I clinical study results of MK1454 and ADU-S100 separately. Data from phase Ib combination therapy of ADU-S100 plus spartalizumab (PDR001, anti-PD-1 mono-antibody developed by Novartis) was generally well tolerated with no dose-limiting toxicity in patients having solid tumors or lymphomas. With some exceptions, partial responses have been observed in patients having PD-1-naive triple-negative breast cancer, PD-1-relapsed or refractory melanoma [115]. The preliminary data from a Phase I clinical trial evaluating MK-1454 has revealed that no remissions were observed in its monotherapy cohort and six of 34 patients responded to MK-1454 plus Keytruda [116]. Despite the disappointing monotherapy results with MK-1454, the observations of several robust antitumor responses in combination therapy still encouraged MSD to continue further studies. Generally, the single treatment of MK1454 or ADU-S100 exerted no significant improvement to the tumor suppression. Meanwhile, the combination therapy of MK1454 or ADU-S100 with immune checkpoint inhibitors also showed the limited antitumor efficacy. Due to the low response rate of ADU-S100 in patients, Novartis have removed ADU-S100 from its portfolio so far. Except ADU-S100, the chemical structure of other drugs are undisclosed. As shown in Table 2.2, STING agonists from other companies such as GSK, BMS (ex by IFM), and Springbank also have entered into clinical trial studies.

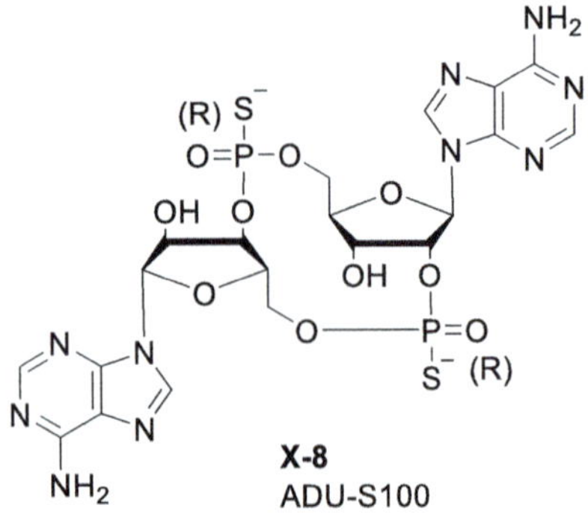

X-8
ADU-S100

Figure 2.10 Structure of ADU-S100.

TABLE 2.2 Pipeline of STING targeted drugs

Company	Drug	Combination	Target	Indication	Stage	NCT number	Time	Delivery	Partner
		+/– Ipilimumab /Yervoy		Advanced/Met astatic solid	Phase I	NCT02675439	2016.03–2020.12		
Aduro	ADU-S100 (MIW-815)	PDR001	CDN-STING agonist	Tumors or lymphomas	Phase I	NCT03172936	2017.09–2020.12	IT	Novartis
		Pembrolizumab/ Keytruda		Head and neck cancer	Phase II	NCT03937141	2019.05–2022.01		
Merck&Co	MK1454	+/– Pembrolizumab	CDN-STING agonist	Advanced/Met astatic solid tumors or lymphomas	Phase I	NCT03010176	2017.02–2021.10	IT	
GSK	GSK3745417	+/– Pembrolizumab	STING agonist	Advanced solid tumors	Phase I	NCT03843359	2019.03–2022.07	IV	
Spring bank	SB 11285	+/– Nivolumab/ Opdivo	STING agonist	Advanced solid tumors	Phase I	NCT04096638	2019.09–2020.11	IV/IT	
			ADC Drug		Preclinical				
Bristol Myers Squibb	BMS-986301	+/– Nivolumab/ Ipilimumab	STING agonist	Advanced solid tumors	Phase I	NCT03956680	2019.02–2022.11	IT	IFM
NIMBUS			STING agonist	Advanced solid tumors	Discovery				
AbbVie	MAVU-104 (Mavupharam)		ENPP1-inhibitor	Advanced solid tumors	Preclinical				
Stimunity			STING agonist Virus Like Particle	Infection and cancer	Discovery				
Synlogic	SYNB1891	Atezolizumab	STING agonist E. Coli	Advanced solid tumors	IND-enabling				ROCHE
Curadev	CRD100 CRD5500		Non-nucleotide STING agonist and ADC Trastuzumab	Solid tumors	Discovery				Takeda

2.6. CDN-BASED THERAPY FOR AUTOIMMUNE TREATMENT

Immunological tolerance was defined as unresponsiveness to an antigen that was induced by the previous exposure to that antigen. Immunological tolerance included the central tolerance and the peripheral tolerance. If the immunological tolerance was disrupted, the immune reactions was activated to attack self-antigens and then caused the autoimmunity, which was viewed as the autoimmune diseases. In fact, the cGAS-STING act as an important role through recognizing the dsDNA or infections in the immune responses. Once the cGAS-STING immune system was over-activated, it could also induce the autoimmune disease. The dysfunction of the cGAS-STING immune system included the accumulation of dsDNA to induce cGAS to produce excess cGAMP and hSTING mutation to increase the IFN production (Figure 2.11).

The accumulation of dsDNA to induce cGAS to produce excess cGAMP could be attributed to the dysfunction of nucleases in cells. Three prime repair exonuclease 1 (Trex1) was an enzyme that was encoded by the TREX1 gene in humans [117, 118]. TREX1 encoded the major 3′-5′ DNA exonuclease in human cells. Mutations of TREX1 could cause the autoimmune diseases like Aicardi-Goutieres syndrome (AGS), familial chilblain lupus (FCL), systemic lupus erythematosus (SLE), retinal vasculopathy, and cerebral leukodystrophy (RVCL). Loss of function of TREX1 resulted in the accumulation of dsDNA in cytoplasm that cannot be degraded. In Trex1$^{-/-}$ mouse model, the cGAMP level

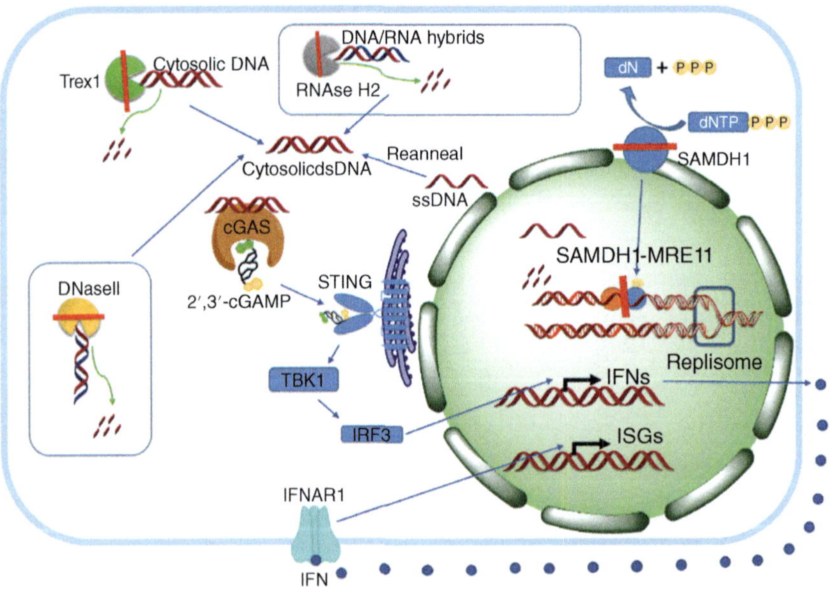

Figure 2.11 cGAS-STING pathway and autoimmune diseases.

was dramatically increased and resulted in high level of interferon-stimulated genes (ISGs) and inflammatory cytokines [57, 119, 120].

DNaseII was also a major cellular nuclease localized in the lysosome, which was responsible for the clearance of DNA from dead cells. In DNaseII-deficient mice, the indigestion of DNA induced the overexpression of IFN caused lethal autoimmunity [57]. These results were consistent with the findings that the fetal liver macrophages in DNase $II^{-/-}$ mice expressed IFN-β and TNF-α and the cytosolic DNA activated the innate immunity through a TLR-independent system. Recently, Wang *et al.* [121] found that the accumulation of dsDNA in cytoplasm contributed to fibroblast-like synoviocytes (FLS)-mediated rheumatoid synovial inflammation in rheumatoid arthritis (RA) patients through cGAS-STING axis pathway.

Ribonuclease H2 (RNase H2) could also degrade cellular DNA:RNA heteroduplexes, which cleaved and initiated the removal of single ribonucleotides embedded in DNA. RNase $H2b^{-/-}$ led to the increased proinflammatory response and the activation of ISGs. It suggested that the dysfunction of RNase H2 caused the autoimmunity diseases, which was dependent on the cGAS-STING pathway [122].

SAMHD1 act as dNTP hydrolase to maintain balanced dNTP pools [123]. After CycA-CDK phosphorylated SAMHD1 on T597 site, the phosphorylated SAMHD1 stimulated the exonuclease activity of MRE11. When the mutation or the deficiency of SAMHD1 occurred, RECQ1 DNA helicase displaced the nascent DNA and formed the flap that could be cleaved by the MRE11. The ssDNA fragment then escaped into the cytoplasm and activated cGAS-STING pathway [123, 124]. It raised the question how ssDNA activated cGAS.

As mentioned above, gain-of-function mutation of hSTING would also induce some autoimmune diseases. The activating mutations in the TMEM173 gene led to a newly classified rare auto-inflammatory disease called STING-associated vasculopathy with onset in infancy (SAVI) [125]. These gain-of-mutations of hSTING included G166E, V155M, N154S, V155M, V147L, V147M, and C206Y, R284G, R281Q, all of which showed the higher IFN-β production activity [126–129]. In the absence of CDNs stimulation, these gain-of-function mutation models even showed more IFN-β production than the health one. Until now, more than 30 cases have been reported to have SAVI all over the world [130]. JAK1/2 inhibitors like ruxolitinib [131] and baricitinib [132, 133] were the possible treatments for SAVI.

Due to the over activation of cGAS-STING pathway, which correlated well with the autoimmune diseases, the development of the inhibitor of cGAS or STING would be a choice of treatment for the autoimmune diseases. As shown in Figure 2.12, there were several inhibitors targeted to cGAS or STING. Quine compounds, such as **X-9** (**quinacrine** (QC)) [134], could block the interaction of cGAS and dsDNA. As these compounds could intercalate dsDNA, these non-specific interactions might have side effects in cells. Benzimidazole compound such as **X-10** (**RU.521**) [135] and **X-11** (**G150**) [136] could inhibit the synthesis

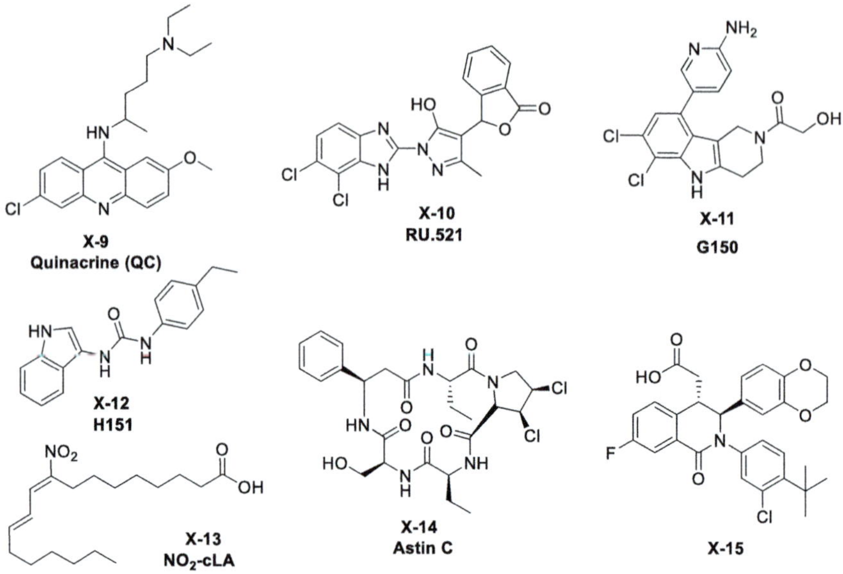

Figure 2.12 Inhibitors of cGAS-hSTING pathway.

of cGAMP by cGAS. The crystal structure showed that these compounds could bind to the catalysis pocket instantly at the interaction surface with dsDNA. **X-10** showed the poor activity to hcGAS, while the **X-11** showed high activity to inhibit hcGAS with EC_{50} value of 0.9 μM. Recent study showed that a very common used drug aspirin could directly inhibit the activity of cGAS through the acetylation of cGAS at Lys384, Lys394, and Lys414 residues. However, the *in vitro* dosage of aspirin was 4 mM and the daily dose of aspirin above 3 mM in serum would also cause salicylate poisoning. These new findings also provided a new method for the treatment of autoimmune diseases.

Compound **X-12 (H151)** [31] was identified as a covalent inhibitor of hSTING. It could form the covalent bound with C91 of hSTING through the interruption of the palmitoylation of hSTING and then block the assembly of multimeric complexes at Golgi, which was essential to STING activation. The mouse experiments showed that the treatment of **X-12** could markedly reduce systemic cytokine responses in both CMA-pretreated and *Trex*$^{-/-}$ mouse model. Nitro-fatty acids (NO_2-FAs) compound **X-13 (NO_2-cLA)** [137] could also react with C91 and C81 through the additional reaction to inhibit the palmitoylation of hSTING to interrupt the immune reactions. Natural cyclopeptide compound **X-14 (Astin C)** [138] isolated from the Chinese traditional plant *Aster tataricus* could completely block the binding of CDNs to STING proteins and suppress the recruitment of IRF3 onto the STING signalosome. Compound **X-15** also act as the competitive inhibitor of STING, which showed good oral exposure and slow binding kinetics to decrease

STING-mediated cytokine production [139]. These lead compounds would have more potential in hSTING-mediated autoimmune disease treatment.

2.7. CDNs DELIVERY

CDNs were constructed by two nucleotides linked by two phosphodiester bonds. Because of the negatively charged phosphate group, CDNs penetrated the cell membrane in an inefficient way, which limited the widespread distribution of CDNs in the administrated body. In order to increase the distribution of CDNs in bodies, several delivery systems have been developed (Figure 2.13).

Adenovirus has been used to express the CDN synthase for CDN synthesis in cells and activate the immune system. The incorporation of the *Vibrio cholerae* DGC gene (3′,3′-c-di-GMP synthase) into a replicating-deficient adenovirus vaccine produced a number of 3′,3′-c-di-GMP in mammalian cells [140]. Furthermore, the combination of OVA- and DGC-contained adenovirus vaccine significantly improved the responses of OVA-specific T cells and OVA-specific humoral B cells. In addition, the coadministration of the DGC-adenovirus vaccine with another adenovirus vector containing the

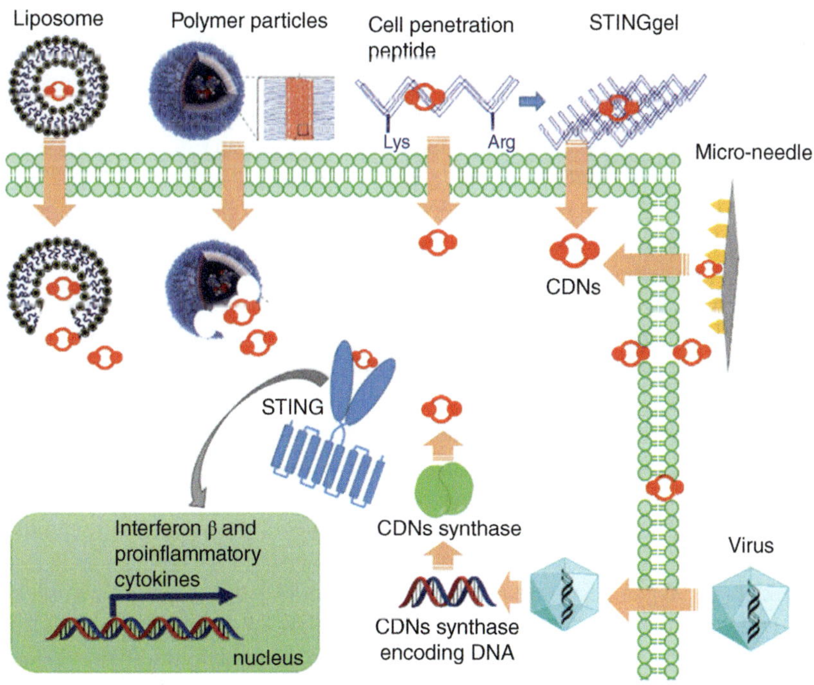

Figure 2.13 Drug delivery systems of CDNs.

HIV-1-derived Gag Ag or *Clostridium difficile*–derived toxin B resulted in significant inhibitory effects on the induction of Gag and toxin B–specific adaptive immune responses [141]. Due to the potential risk of usage, adenovirus-based delivery of CDNs was still limited in the practical utilization.

To enhance the cellular endosome escape of conventional liposomes like lipofectamine 2000, Harashima *et al.* [142] developed YSK05 lipids, which could functionalize at pH = 6.4 with high fusogenic activity and high ability of endosome escape. The 3′,3′-c-di-GMP/YSK05 liposome exhibited good cell uptaken ability and high level of 3′,3′-c-di-GMP releasing. The *in vivo* treatment of B16-F10 melanomas mice model with 3′,3′-c-di-GMP/YSK05 liposome induced the MHC-I nonrestricted anticancer immunity mediated by NK cells [143]. These results revealed that YSK05-lipids could be used for the delivery of CDNs in immunotherapy [141].

As small molecules (<1 kDa) easily penetrated blood vessels instead of lymph or draining lymph nodes, PEGylated lipid nanoparticles system was developed to enhance the efficacy and safety of CDNs uptaken in draining lymph nodes (dLNs) APCs [144]. Meanwhile, the PEG-NP-cdGMP could promoted the substantial immune responses to enhance the efficacy of immune therapy. Junkins *et al.* [145] have reported the utilization of acid-sensitive (pH<5) acetylated dextran (Ace-DEX) polymeric microparticles (MPs) for the delivery of CDNs passively into APCs. With the help of the Ace-DEX MPs, the production of type I IFN and antibody titers were dramatically enhanced by 1000-fold and 10000-fold, respectively. In addition, the HA adjuvanted with the MPs could provide long-term immune responses to the infection of lethal influenza. Wilson *et al.* exhibited the utilization of biodegradable poly(beta-amino ester) (PBAE) nanoparticles for cytosolic delivery to elicit the immune response [146]. Shae *et al.* [147] recently also reported STING-activating nanoparticles (NPs) through the fragment combination for the cytosolic delivery of 2′,3′-cGAMP. This STING-NP was constructed with cationic 2-(diethylamino) ethyl methacrylate (DEAEMA) groups which relied on pH–sensitive, hydrophobic butyl methacrylate (BMA) moieties to increase endosomal escape ability. Pyridyl disulfide ethyl methacrylate (PDSMA) groups could make the crosslinking of polymer chains through the thiol reaction and PEG helped to neutralize the particle surface. The STING-NP increased the penetration of CDNs in the tumor microenvironment and sentinel LN to enhance the immunotherapeutic efficacy of 2′,3′-cGAMP through both intravenous and intratumoral administration.

Cationic peptide as cell penetration peptide, such as lysine and arginine residues, could penetrate the cell membrane through endocytosis independent way. Yildiz *et al.* [95] demonstrated that a simple nona-Arginine peptide complexation improved the cellular uptake of CDNs resulting in the enhancement of the immune response. Leach *et al.* [148] synthesized a kind peptide hydrogel constructed with K2(SL)6K2 multi-domain peptide (MDP). This STINGel could localize and provide the controlled release of CDNs delivery.

Recently, microneedles could be also used for the delivery of CDNs into the skin to activate the Th1 pathway, which posed a great potential in the vaccine development targeted to skin-associated infection diseases [149]. An *et al.* demonstrated that the inorganic cationic silica nanoparticles (CSiNPs) could induce cell necrosis and lead to oxidative stress with their morphological changes such as membrane swelling and rupture [150]. The complexes of CSiNPs with CDNs could enhance their cellular uptake leading to the local retention with the TME to activate tumor-infiltrating APCs. Hammond *et al.* even reported the well-defined self-assembled complex of cGAMP with transmembrane (TM)-domain-deficient STING could effectively enhance the immune responses independent of the endogenous STING [37].

All of these materials could help CDNs go through the cell membrane to induce immune response for immune therapies. However, the utilization of these additional materials might increase the potential unexpected effects to host cells or tissues. These double-negatively charged CDNs could itself function as an immune stimuli. Recently, a transporter protein SLC19A1 has been found to transport CDNs into cytoplasm [151, 152]. The efficiency and selectivity to CDNs was much uncertain. A high concentration of CDNs was still needed to activate the immune response. In clinical trials, CDNs are delivered into cells via intratumoral (IT) injection, which made patients suffer from pain. Considering the high dose of CDNs in stimulating the tissue or whole body inflammation and potential danger of self-immune diseases, more safety and efficiency drug delivery systems of CDNs were still necessary to be developed for the further use.

To have a safe use of CDNs, prodrug strategies could be used to improve their biological activities, potential therapeutic efficacy, and physicochemical properties. As shown in Figure 2.14, Xi's group designed and synthesized the S-acylthioalkyl ester (SATE)- and dithioethanol (DTE)-based prodrugs of CDNs (Figure 2.14). The SATE- and DTE-CDNs prodrugs could penetrate the cell membrane freely without the help of any transfection reagents. Once entering the cells, CDN prodrugs were separately released by the cytoplasmic thioesterase or reductase. The SATE- and DTE-CDNs prodrugs helped the negatively charged CDNs pass the membrane and released CDNs in cytoplasm to induce the STING-dependent immune responses with low cytotoxicity. The results showed that these prodrugs had better antitumor activity than the clinical drug ADU-S100 in mouse model. The prodrug dramatically enhanced the bioavailability of CDNs as an immune stimulator in immunotherapies. Hence, cell-specific ligands could be also introduced to achieve the tissue or cell-targeted delivery of CDNs to avoid possible side effects.

In order to avoid the poor druggability of CDNs such as poor oral bioavailability, low cell penetration ability, and instability in circulation, non-nucleotide molecules were also screened to activate the STING pathway as the substitutes of CDNs. As shown in Figure 2.15, there were many reported

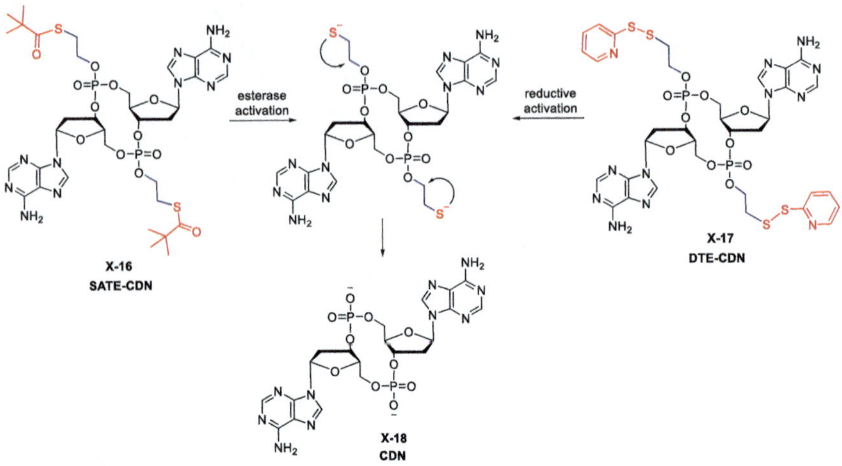

Figure 2.14 SATE- and DTE-based prodrug of CDN.

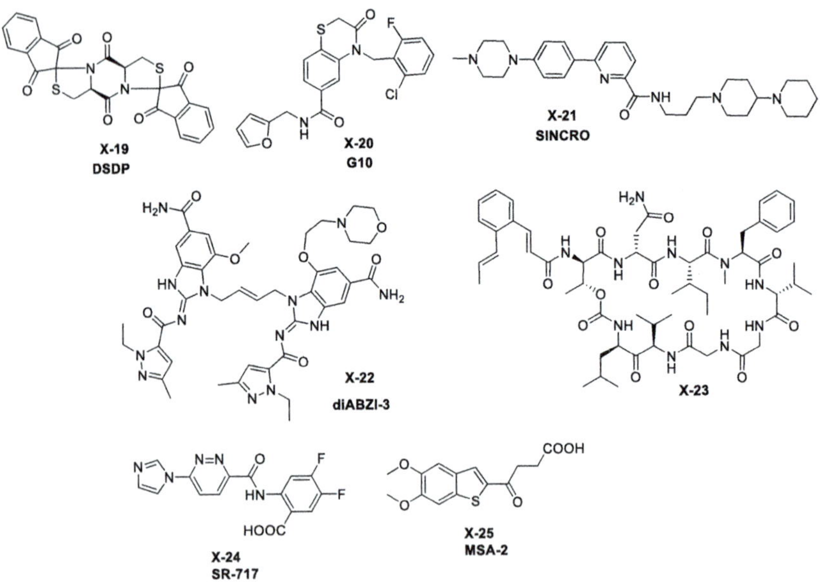

Figure 2.15 Non-nucleotide-based STING agonist.

non-CDN-based STING agonists. DMXAA was failed in phase III clinical trial because it only responded to mSTING but not to hSTING. Han *et al.* identified C7-substituted DMXAA analogues could bind to hSTING [153]. Compounds dispiro diketopiperzine (**DSDP**, **X-19**) [154] and 4-(2-chloro-6-fluorobenzyl)-N-(furan-2-ylmethyl)-3-oxo-3,4-dihydro-2H-benzo[b][1,4] thiazine-6-carboxamide (**G10**, **X-20**) were discovered through HTS, which could activate cytokine production through hSTING-IRF3 pathway and exhibited the antiviral activity [155]. Kimura *et al.* reported the compound N-{3-[(1,4′-bipiperidin)-1′-yl]propyl}-6-[4-(4-methylpiperazin-1-yl)phenyl] picolinamide **X-21** (**SINCRO**, STING-mediated interferon-inducing and cytotoxic reagent) not only could activate the production of type I interferon through STING-dependent pathway but also exhibited side-effects in untransformed cells [156]. Recently, Ramanjulu *et al.* from GSK identified small-molecule amidobenzimidazole (ABZI)-based compounds could activate hSTING through the high-throughput screening of small molecules that competed with the binding of radio-labeled 2′,3′-cGAMP [157]. These compounds could directly bind to the CBD domain. The further structural optimization showed that compound **diABZI-3** (**X-22**) would have great potential in clinical use. Shi *et al.* recently found that cinnamoyl-containing nonribosomal peptide (CCNP) compounds **X-23** could also increase IFN-β production through STING pathway through increasing the phosphorylation and nucleic localization of IRF3 instead of binding to STING directly [158]. Recently, **SR717** (**X-24**) [159] and **MSA-2** (**X-25**) [160] were also reported as STING agonists by binding to the CBD domain as dimer. **X-25** preferentially targeted tumor tissue upon extracellular acidification, which made it orally available. It indicated that **X-25** exhibited high potential in clinical application.

2.8. CDNs SYNTHESIS

2.8.1. Biosynthesis

Similar to all known natural compounds, CDNs could be directly obtained from natural sources, while the efficiency was quite low. It mainly relied on cellular enzymes to achieve the synthesis of CDNs. DncV (cGAMP synthase) could synthesize analogs of several 3′-3′ CDNs including different base combinations, 2′-modified (2′-deoxy, F, OMe, NH_2), and phosphorothioated analogs with the help of different metal ions [161]. cGAS could also be used for the biosynthesis of 2′-3′ CDNs from different NTP analogues [162, 163]. The engineered diguanylate cyclase (DGC) with the mutation in the putative regulatory I site was also utilized for the large-scale synthesis of c-di-GMP [164]. What is more, DGC immobilized in the methyltrimethoxysilane (MTMOS) and tetramethoxysilane (TMOS) gels to avoid the hindrance of the dimerization and conformational changes during the catalysis exhibited the extended lifetime and catalytic activity in comparison to the free enzyme in solution [165]. Lv *et al.* [166] firstly used engineered *Escherichia coli* with overexpression of the synthase of CDNs including DGC, DncV and cGAS and other related protein to synthesize 3′,3′-c-di-GMP, 3′,3′-cGAMP and 2′,3′-cGAMP, respectively. This method exhibited more economical and practical. But the substrate selectivity of the catalysis pocket in the synthase would limit the use for the special structure of CDNs. As known, the synthesis of 2′-2′ CDNs could not be achieved through the biosynthesis method.

2.8.2. Chemical Synthesis

With the development of DNA and RNA synthesis, the chemical synthesis of CDNs was achievable. CDNs could be synthesized through either the solid-phase or solution-phase synthesis approach.

2.8.2.1. Solid-Phase Synthesis In the 1990s, the utilization of solid-phase synthesis of the linear oligonucleotides has become the most popular approach [167, 168]. The synthesis of cyclic oligonucleotides through the solid-phase synthesis was also reported [169, 170]. Sintim *et al.* [171] reported two solid-phase-based synthetic methods of CDNs.

As shown in Scheme 2.1, commercially available phosphate CPG was started as the original compound and coupled to P-methoxy phosphoramidites, followed by the oxidation with an iodine/pyridine/H_2O mixture to give quantitatively coupled mononucleotides. In the next step, a cyanoethyl phosphoramidite was used because a selective deprotection of the methoxy-protected phosphate group was required for the formation of the cyclic dinucleotide. 1 M Na_2S_2 in DMF at room temperature was then added to selectively cleave the P-methoxy moiety without affecting the other functionalities on dinucleotides. Cyclization of the linear dinucleotide into c-di-dGMP was then accomplished using 1-mesitylenesulfonyl-3-nitro-1,2,4-triazole (MSNT) or triisopropylsulfonyl chloride and N-methylimidazole. After the selective removal of the cyclized dinucleotide with trimethylamine, the CPG solid support was treated with ammonia to return the noncyclized linear dinucleotide in over 50% yield.

In principle, an iterative cycle of cyclization followed by selective deprotection of the cyclized product would ultimately lead to the complete conversion of the linear dinucleotide into the cyclized product after several cycles. As shown

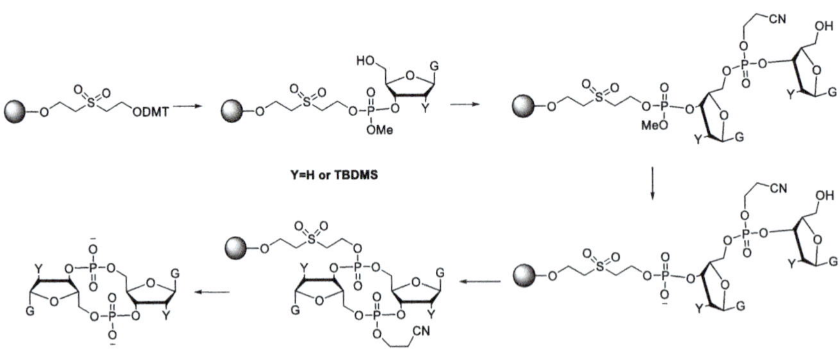

Scheme 2.1 Solid phase synthesis of CDNs - 1.

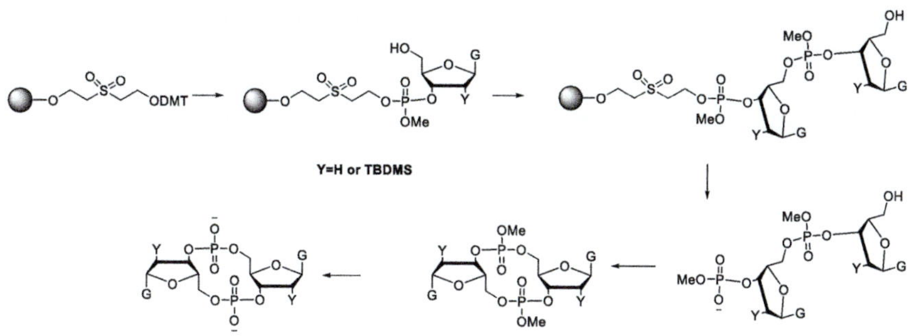

Scheme 2.2 Solid phase synthesis of CDNs - 2.

in Scheme 2.2, the same P-methoxyphosphoramidite was used for both coupling steps. The solid-supported dinucleotide was dried under vacuum and treated with triethylamine. A cyclization reagent such as MSNT was then used to give cyclized dinucleotides in greater than 95% conversion. It appeared that the cyclization of dinucleotides in solution phase was superior to solid support. Importantly, this new approach (Scheme 2.2) allowed a simple cleavage of the dinucleotide into the solution phase followed by a tandem cyclization in the same pot without any isolation. The solid-supported dinucleotide was treated with triethylamine before MSNT was added. Using this protocol, the cyclized product could be obtained in less than 24 hours. The cyclized dinucleotides were used for the next step without purification. Global deprotection of the cyclized products with 28% ammonia and the TBDMS protecting groups were removed with HF-TEA after ammonia deprotection. As shown in Scheme 2.3, 3-chloro-4-hydroxyphenyllacetic-modified CPG could also be used for the synthesis of CDNs [172, 173]. The cyclization step was carried out with MSNT. After deprotection of the phosphates with PhSH/Et_3N (R_1=Me) or Et_3N/pyridine (R_1=CNE), the cleavage from the resin with tetramethylguanidinium (TMG)/*syn*-2-pyridinecarbaldehyde oxime (PBO), and the subsequent deprotection of the nucleobases with the concentrated ammonia to give the crude CDNs. The final

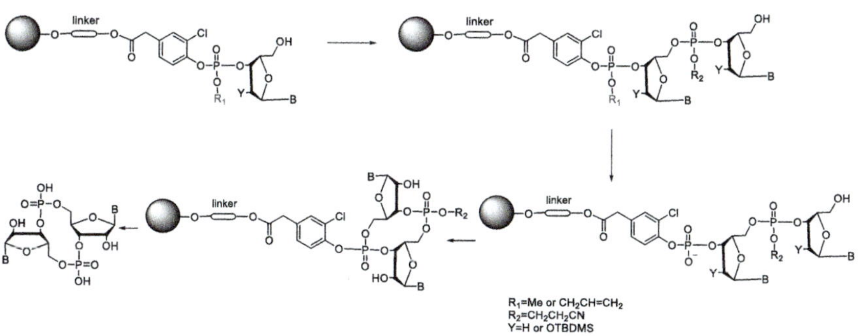

Scheme 2.3 Solid phase synthesis of CDNs – 3.

selective cleavage from the solid support provided the cyclic oligoribonucleotides in high purity after deprotection.

2.8.2.2. Solution-Phase Synthesis

2.8.2.2.1. Phosphotriester Approaches Phosphate triester approach was established in the 1970s. The core advantage of phosphate triester approach was the selectivity of protection group. As shown in Scheme 2.4 path A, both the linear and cyclization steps were achieved through phosphotriester step. At the same time, the two phosphate protection groups could be removed under different conditions and should not influence other protection groups. Cyanoethyl groups could be removed under alkaline condition. Usually, the second protection groups could select 2-chlorophenyl, nitrophenyl, and allyl groups. The phenyl group could be removed under nucleophilic reagents and base like TMG and PBO mixture [174]. The allyl groups could be removed under the mixture of $Pd_2(P(Ph)_3)_3$, $P(Ph)_3$, *n*-butylamine and formic acid [175], or NaI-acetone solution [176, 177]. The methyl group could be removed by adding Na_2S_2 [171]. When the linear dinucleotide intermediates were obtained, one protection group was removed to free the phosphonate group. Both the linear coupling and cyclization reaction were achieved with the help of 1H-tetrazole, and N-methyl imidazole cooperated with phenyl sulfonyl chloride like TPSCl and MSC. MSNT and TPSNT could be also used as the coupling reagents. After the full-protected CDNs were obtained, protection group like phosphonate, 2′- or 3′-OH, and NH_2 of base group were removed to get the targeted CDNs. Through the phosphotriester approach, all intermediates could be purified through silica-gel chromatography easily.

2.8.2.2.2. Phosphoramidate and H-Phosphonate Approach Throughout all known literatures, the phosphoramidate and H-phosphonate approach were the mostly used method for the synthesis of CDNs [178, 179]. As shown in Scheme 2.4 path B, after the linear coupling reaction, the linear dinucleotide intermediate was oxidized with tert-butyl hydroperoxide (TBHP). Next, the cyclization reaction was achieved with the help of 2-chloro-5,5-dimethyl-1,3,2-dioxaphosphorinane-2-oxide (DMOCP) and then oxidized by I_2. Through this method, the thiophosphate analogues of CDNs could be also obtained. When the oxidization reagent TBHP was changed into 3-((dimethylaminomethylidene)-amino)-3H-1,2,4-dithiazole-5-thione (DDTT), phenylacetyl disulfide (PADS), bis(3-triethoxysilylpropyl) tetrasulfide (TEST), or S8 could form P=S with two isomers (Rp and Sp), while the H-phosphonate oxidization with the sulfurization reagent like 3-*H*-1,2-benzodithiol-3-one only get the Rp diastereomer [178]. Through this method, the reaction efficient was dramatically increased and took less time. However, the *H*-phosphonate intermediates always existed as salts, which made it have high hygroscopicity and hard to purify. The coupling reaction was water sensitive, which made the reactions need a stricter condition.

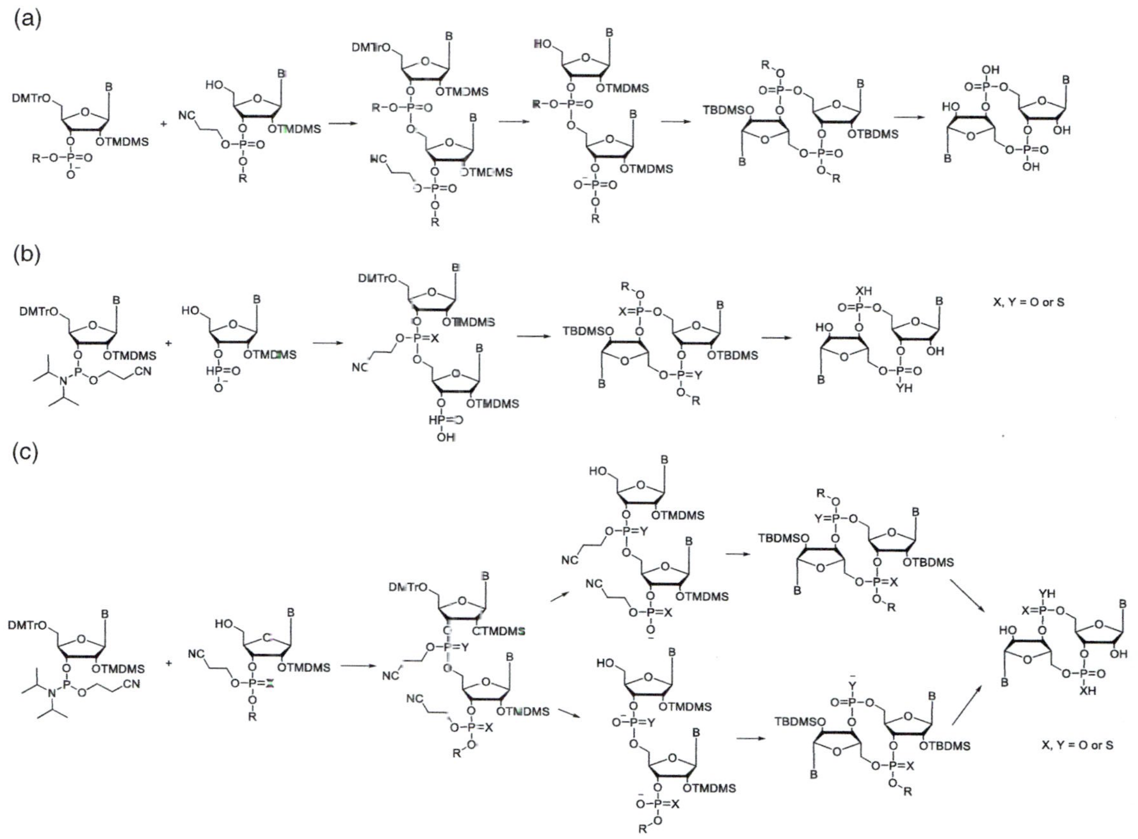

Scheme 2.4 Solution-phase chemistry for the synthesis of CDNs. (a) Phosphotriester approaches; (b) phosphoramidate and H-phosphonate approach; (c) the combination approach of phosphotriester with phosphoramidate.

2.8.2.2.3. Phosphotriester and Phosphoramidate The coupling efficiency of phosphoramidate with OH was high and the phosphotriester compounds were easily purified. As shown in Scheme 2.4 path C, the combination of phosphotriester with phosphoramidate could be also used for the synthesis of CDNs [176, 177]. Meanwhile, we could also obtain P=O or P=S substrates by adding different oxidization reagents during the linearly coupling reaction leading to the achievement of thiophosphate analogues of CDNs [52].

Thiophosphate analogues of CDNs, such as the ADU-S100 drug, has exhibited excellent therapeutic ability with the improvement of stability, lipophilicity, and activity. The traditional P=S substitution of CDNs was always achieved through P(III) approach followed by the oxidation of sulfurization reagent. As a result, different diastereoisomers could be obtained, which made the purification of pure enantiomers very difficult. As shown in Scheme 2.5, Baran *et al.* [180] developed a P(V) reagent for the stereocontrolled synthesis of CDNs, which could be used through both solution- and solid-phase chemistry.

2.9. CONCLUSION AND PERSPECTIVE

Cancer immunotherapy has been currently a powerful strategy for cancer treatment from antibodies to small-molecule immune regulators [181]. Compared

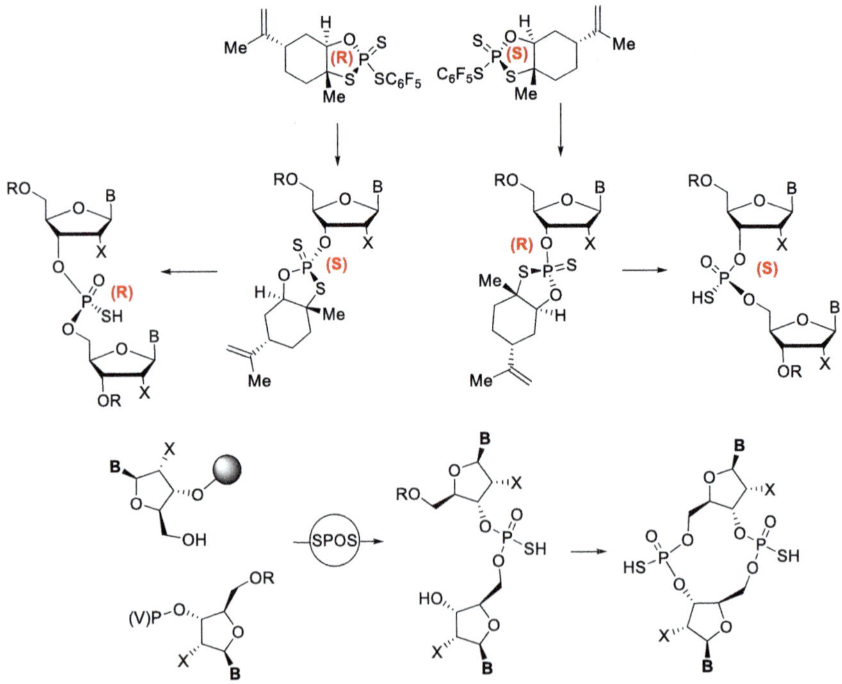

Scheme 2.5 Stereo-controlled synthesis of thiophosphate substituted CDNs.

with antibodies, small-molecule immune regulators have indicated their better pharmacokinetics. Small molecules for immunotherapies in clinical trials included checkpoint inhibitors, Toll-like receptor agonists, IDO inhibitors, chemokine receptor antagonists, STING antagonists, and so on [182–187]. CDNs were presented as good candidates for immune therapies. Nowadays, there are an increasing literatures and pipelines of CDN-based antitumor therapies. Due to the CDNs relying on stimulating the STING-dependent immune responses, CDNs could be used for a broad spectrum of cancers. However, there are still many challenges to be addressed for the further development of CDN-based STING agonists for immune therapy.

1. The molecular basis of the CDN-induced STING activation

To date, little agreement was established on how the N-terminal TM domain and CTT of STING protein functioned during the signal transduction. Further work was needed to understand full-length complex structure of STING and CDNs. When STING was activated by CDNs, CDNs induced IFN-β, NF-κB, and other cytokines, and even autophagy process, implying the existence of signaling cross-talk among different cytokines and autophagy pathways. How these pathways influenced the immune responses would help us know better about the CDN-based treatment.

2. Efficiency of CDNs in tumor microenvironment

Human population with different SNPs of hSTING gene showed different responses to CDNs' stimulation. Both the loss-of-function and the gain-of-function mutation of hSTING would cause some diseases. The hSTING expression level in tumor microenvironment would also respond differently to the stimulation of CDNs. Epigenetic modification of cGAS and STING would decrease the immune responses stimulated by CDNs [188, 189]. How to enhance the efficiency of CDNs in tumor microenvironment with STING deficiency remained the further goal.

3. Targeted delivery of CDNs

CDNs drugs were not targeted to any cells or tissues. System administration of CDNs would cause some unexpected side effects. In order to enhance the therapeutic efficiency and decrease the possible toxicity of CDNs, targeted delivery of CDNs would be a good choice. The endothelial cells as the principal cellular target of STING activation stimulated by CDNs in tumor microenvironment [86] indicated that tumor endothelial cells could be targeted for immunotherapy of melanoma by CDNs.

4. Combinational therapy

The phase I clinical trial studies showed that the administration of CDNs alone did not give any improvement to patients. The combination of CDNs with ICB induced slight responses. Many papers demonstrated that different types of combinational therapies with CDNs were efficient to the stimulation of CDNs. However, the mechanism of the combinational therapy remained unknown. During the formation of tumor, caner-driven immunosuppression led to the tumor cells having low or no responses to immune system. Future studies should attempt to identify which was the most efficient combination therapy approaches for different type of cancers.

5. Side effects of CDNs
Loss control of immune responses initiated by CDNs would cause autoimmune diseases. At the same time, excess administration dosage of CDNs would also over-elicit the immune system such as flu-like syndrome or cytokine storm side effects, which might cause the damage to tissues or bodies. Whether these side-effects could be controlled remained unknown. More researches would be necessary to determine the efficacy and safety of CDN-based treatment.

In summary, CDNs represented a new generation of small molecular drug in immune therapy for antitumor and anti-infectious treatments and in the future more efficient and safer CDN-based drugs would be waiting to be developed for cancer immunotherapy.

REFERENCES

1. Kalia, D.; Merey, G.; Nakayama, S.; Zheng, Y.; Zhou, J.; Luo, Y.; Guo, M.; Roembke, B. T.; Sintim, H. O. (2013) Nucleotide, c-di-GMP, c-di-AMP, cGMP, cAMP, (p)ppGpp signaling in bacteria and implications in pathogenesis. *Chem Soc Rev*, *42*, 305–341.
2. Zitvogel, L.; Galluzzi, L.; Kepp, O.; Smyth, M. J.; Kroemer, G. (2015) Type I interferons in anticancer immunity. *Nat Rev Immunol*, *15*, 405–414.
3. McNab, F.; Mayer-Barber, K.; Sher, A.; Wack, A.; O'Garra, A. (2015) Type I interferons in infectious disease. *Nat Rev Immunol*, *15*, 87–103.
4. Ross, P.; Weinhouse, H.; Aloni, Y.; Michaeli, D.; Weinberger-Ohana, P.; Mayer, R.; Braun, S.; de Vroom, E.; van der Marel, G. A.; van Boom, J. H. et al. (1987) Regulation of cellulose synthesis in Acetobacter xylinum by cyclic diguanylic acid. *Nature*, *325*, 279–281.
5. Witte, G.; Hartung, S.; Buttner, K.; Hopfner, K. P. (2008) Structural biochemistry of a bacterial checkpoint protein reveals diadenylate cyclase activity regulated by DNA recombination intermediates. *Mol Cell*, *30*, 167–178.
6. Davies, B. W.; Bogard, R. W.; Young, T. S.; Mekalanos, J. J. (2012) Coordinated regulation of accessory genetic elements produces cyclic di-nucleotides for V. cholerae virulence. *Cell*, *149*, 358–370.
7. Ablasser, A.; Goldeck, M.; Cavlar, T.; Deimling, T.; Witte, G.; Rohl, I.; Hopfner, K. P.; Ludwig, J.; Hornung, V. (2013) cGAS produces a 2′-5′-linked cyclic dinucleotide second messenger that activates STING. *Nature*, *498*, 380–384.
8. Diner, E. J.; Burdette, D. L.; Wilson, S. C.; Monroe, K. M.; Kellenberger, C. A.; Hyodo, M.; Hayakawa, Y.; Hammond, M. C.; Vance, R. E. (2013) The innate immune DNA sensor cGAS produces a noncanonical cyclic dinucleotide that activates human STING. *Cell Rep*, *3*, 1355–1361.
9. Gao, P.; Ascano, M.; Wu, Y.; Barchet, W.; Gaffney, B. L.; Zillinger, T.; Serganov, A. A.; Liu, Y.; Jones, R. A.; Hartmann, G. et al. (2013) Cyclic [G(2′,5′)pA(3′,5′)p] is the metazoan second messenger produced by DNA-activated cyclic GMP-AMP synthase. *Cell*, *153*, 1094–1107.

10. Zhang, X.; Shi, H.; Wu, J.; Zhang, X.; Sun, L.; Chen, C.; Chen, Z. J. (2013) Cyclic GMP-AMP containing mixed phosphodiester linkages is an endogenous high-affinity ligand for STING. *Mol Cell*, *51*, 226–235.
11. Wu, J.; Sun, L.; Chen, X.; Du, F.; Shi, H.; Chen, C.; Chen, Z. J. (2013) Cyclic GMP-AMP is an endogenous second messenger in innate immune signaling by cytosolic DNA. *Science*, *339*, 826–830.
12. Wang, B.; Wang, Z.; Javornik, U.; Xi, Z.; Plavec, J. (2017) Computational and NMR spectroscopy insights into the conformation of cyclic di-nucleotides. *Sci Rep*, *7*, 16550.
13. Che, X.; Zhang, J.; Zhu, Y.; Yang, L.; Quan, H.; Gao, Y. Q. (2016) Structural flexibility and conformation features of cyclic dinucleotides in aqueous solutions. *J Phys Chem B*, *120*, 2670–2680.
14. Wang, Z.-H.; Zhao, C.-C.; Zhang, Q.-Z.; Wang, C.-L.; Zhang, H.; Ma, D.-J.; Wang, D.-W.; Wen, X.; Li, L.-Y.; Xi, Z. (2020) Design, synthesis and systematic evaluation of all possible cyclic dinucleotides (CDNs) that activate human stimulator of interferon genes (STING) variants. *Science China Chemistry*, *63*, 534–545.
15. Shi, H.; Wu, J.; Chen, Z. J.; Chen, C. (2015) Molecular basis for the specific recognition of the metazoan cyclic GMP-AMP by the innate immune adaptor protein STING. *Proc Natl Acad Sci U S A*, *112*, 8947–8952.
16. Che, X.; Zhang, J.; Quan, H.; Yang, L.; Gao, Y. Q. (2018) CDNs-STING interaction mechanism investigations and instructions on design of CDN-derivatives. *J Phys Chem B*, *122*, 1862–1868.
17. Sun, L.; Wu, J.; Du, F.; Chen, X.; Chen, Z. J. (2013) Cyclic GMP-AMP synthase is a cytosolic DNA sensor that activates the type I interferon pathway. *Science*, *339*, 786–791.
18. Zhou, W.; Whiteley, A. T.; de Oliveira Mann, C. C.; Morehouse, B. R.; Nowak, R. P.; Fischer, E. S.; Gray, N. S.; Mekalanos, J. J.; Kranzusch, P. J. (2018) Structure of the human cGAS-DNA complex reveals enhanced control of immune surveillance. *Cell*, *174*, 300–311 e311.
19. Yanai, H.; Ban, T.; Wang, Z.; Choi, M. K.; Kawamura, T.; Negishi, H.; Nakasato, M.; Lu, Y.; Hangai, S.; Koshiba, R. et al. (2009) HMGB proteins function as universal sentinels for nucleic-acid-mediated innate immune responses. *Nature*, *462*, 99–103.
20. Rubio-Cosials, A.; Sidow, J. F.; Jimenez-Menendez, N.; Fernandez-Millan, P.; Montoya, J.; Jacobs, H. T.; Coll, M.; Bernado, P.; Sola, M. (2011) Human mitochondrial transcription factor A induces a U-turn structure in the light strand promoter. *Nat Struc Mol Biol*, *18*, 1281–1289.
21. Ngo, H. B.; Kaiser, J. T.; Chan, D. C. (2011) The mitochondrial transcription and packaging factor Tfam imposes a U-turn on mitochondrial DNA. *Nat Struc Mol Biol*, *18*, 1290–1296.
22. Ngo, H. B.; Lovely, G. A.; Phillips, R.; Chan, D. C. (2014) Distinct structural features of TFAM drive mitochondrial DNA packaging versus transcriptional activation. *Nat Commun*, *5*, 3077.
23. Du, M.; Chen, Z. J. (2018) DNA-induced liquid phase condensation of cGAS activates innate immune signaling. *Science*, *361*, 704–709.
24. Xie, W.; Lama, L.; Adura, C.; Tomita, D.; Glickman, J. F.; Tuschl, T.; Patel, D. J. (2019) Human cGAS catalytic domain has an additional DNA-binding interface that enhances enzymatic activity and liquid-phase condensation. *Proc Natl Acad Sci U S A*, *116*, 11946–11955.

25. Burdette, D. L.; Monroe, K. M.; Sotelo-Troha, K.; Iwig, J. S.; Eckert, B.; Hyodo, M.; Hayakawa, Y.; Vance, R. E. (2011) STING is a direct innate immune sensor of cyclic di-GMP. *Nature*, *478*, 515–518.
26. Yin, Q.; Tian, Y.; Kabaleeswaran, V.; Jiang, X.; Tu, D.; Eck, M. J.; Chen, Z. J.; Wu, H. (2012) Cyclic di-GMP sensing via the innate immune signaling protein STING. *Mol Cell*, *46*, 735–745.
27. Sun, W.; Li, Y.; Chen, L.; Chen, H.; You, F.; Zhou, X.; Zhou, Y.; Zhai, Z.; Chen, D.; Jiang, Z. (2009) ERIS, an endoplasmic reticulum IFN stimulator, activates innate immune signaling through dimerization. *Proc Natl Acad Sci U S A*, *106*, 8653–8658.
28. Zhong, B.; Yang, Y.; Li, S.; Wang, Y. Y.; Li, Y.; Diao, F.; Lei, C.; He, X.; Zhang, L.; Tien, P. et al. (2008) The adaptor protein MITA links virus-sensing receptors to IRF3 transcription factor activation. *Immunity*, *29*, 538–550.
29. Jin, L.; Waterman, P. M.; Jonscher, K. R.; Short, C. M.; Reisdorph, N. A.; Cambier, J. C. (2008) MPYS, a novel membrane tetraspanner, is associated with major histocompatibility complex class II and mediates transduction of apoptotic signals. *Mol Cell Biol*, *28*, 5014–5026.
30. Shang, G.; Zhang, C.; Chen, Z. J.; Bai, X. C.; Zhang, X. (2019) Cryo-EM structures of STING reveal its mechanism of activation by cyclic GMP-AMP. *Nature*, *567*, 389–393.
31. Haag, S. M.; Gulen, M. F.; Reymond, L.; Gibelin, A.; Abrami, L.; Decout, A.; Heymann, M.; van der Goot, F. G.; Turcatti, G.; Behrendt, R. et al. (2018) Targeting STING with covalent small-molecule inhibitors. *Nature*, *559*, 269–273.
32. Mukai, K.; Konno, H.; Akiba, T.; Uemura, T.; Waguri, S.; Kobayashi, T.; Barber, G. N.; Arai, H.; Taguchi, T. (2016) Activation of STING requires palmitoylation at the Golgi. *Nat Commun*, *7*, 11932.
33. Ergun, S. L.; Fernandez, D.; Weiss, T. M.; Li, L. (2019) STING polymer structure reveals mechanisms for activation, hyperactivation, and inhibition. *Cell*, *178*, 290–301.e210.
34. Zhang, C.; Shang, G.; Gui, X.; Zhang, X.; Bai, X. C.; Chen, Z. J. (2019) Structural basis of STING binding with and phosphorylation by TBK1. *Nature*, *567*, 394–398.
35. Zhao, B.; Du, F.; Xu, P.; Shu, C.; Sankaran, B.; Bell, S. L.; Liu, M.; Lei, Y.; Gao, X.; Fu, X. et al. (2019) A conserved PLPLRT/SD motif of STING mediates the recruitment and activation of TBK1. *Nature*, *569*, 718–722.
36. Shang, G.; Zhu, D.; Li, N.; Zhang, J.; Zhu, C.; Lu, D.; Liu, C.; Yu, Q.; Zhao, Y.; Xu, S. et al. (2012) Crystal structures of STING protein reveal basis for recognition of cyclic di-GMP. *Nat Struc Mol Biol*, *19*, 725–727.
37. He, Y.; Hong, C.; Yan, E. Z.; Fletcher, S. J.; Zhu, G.; Yang, M.; Li, Y.; Sun, X.; Irvine, D. J.; Li, J. et al. (2020) Self-assembled cGAMP-STINGDeltaTM signaling complex as a bioinspired platform for cGAMP delivery. *Sci Adv*, *6*, eaba7589.
38. Konno, H.; Konno, K.; Barber, G. N. (2013) Cyclic dinucleotides trigger ULK1 (ATG1) phosphorylation of STING to prevent sustained innate immune signaling. *Cell*, *155*, 688–698.
39. Konno, H.; Chinn, I. K.; Hong, D.; Orange, J. S.; Lupski, J. R.; Mendoza, A.; Pedroza, L. A.; Barber, G. N. (2018) Pro-inflammation associated with a gain-of-function mutation (R284S) in the innate immune sensor STING. *Cell Rep*, *23*, 1112–1123.

40. Gui, X.; Yang, H.; Li, T.; Tan, X.; Shi, P.; Li, M.; Du, F.; Chen, Z. J. (2019) Autophagy induction via STING trafficking is a primordial function of the cGAS pathway. *Nature*, *567*, 262–266.
41. Liu, D.; Wu, H.; Wang, C.; Li, Y.; Tian, H.; Siraj, S.; Sehgal, S. A.; Wang, X.; Wang, J.; Shang, Y. et al. (2019) STING directly activates autophagy to tune the innate immune response. *Cell Death Differ*, *26*, 1735–1749.
42. Dey, B.; Dey, R. J.; Cheung, L. S.; Pokkali, S.; Guo, H.; Lee, J. H.; Bishai, W. R. (2015) A bacterial cyclic dinucleotide activates the cytosolic surveillance pathway and mediates innate resistance to tuberculosis. *Nat Med*, *21*, 401–406.
43. Gao, P.; Ascano, M.; Zillinger, T.; Wang, W.; Dai, P.; Serganov, A. A.; Gaffney, B. L.; Shuman, S.; Jones, R. A.; Deng, L. et al. (2013) Structure-function analysis of STING activation by c[G(2′,5′)pA(3′,5′)p] and targeting by antiviral DMXAA. *Cell*, *154*, 748–762.
44. Yi, G.; Brendel, V. P.; Shu, C.; Li, P.; Palanathan, S.; Cheng Kao, C. (2013) Single nucleotide polymorphisms of human STING can affect innate immune response to cyclic dinucleotides. *PLoS One*, *8*, e77846.
45. Patel, S.; Blaauboer, S. M.; Tucker, H. R.; Mansouri, S.; Ruiz-Moreno, J. S.; Hamann, L.; Schumann, R. R.; Opitz, B.; Jin, L. (2017) The common R71H-G230A-R293Q human TMEM173 is a null Allele. *J Immunol*, *198*, 776–787.
46. Ruiz-Moreno, J. S.; Hamann, L.; Shah, J. A.; Verbon, A.; Mockenhaupt, F. P.; Puzianowska-Kuznicka, M.; Naujoks, J.; Sander, L. E.; Witzenrath, M.; Cambier, J. C. et al. (2018) The common HAQ STING variant impairs cGAS-dependent antibacterial responses and is associated with susceptibility to Legionnaires' disease in humans. *PLoS Pathog*, *14*, e1006829.
47. Wu, C.; Wang, Z.; Song, X.; Feng, X. S.; Abnet, C. C.; He, J.; Hu, N.; Zuo, X. B.; Tan, W.; Zhan, Q. et al. (2014) Joint analysis of three genome-wide association studies of esophageal squamous cell carcinoma in Chinese populations. *Nat Genet*, *46*, 1001–1006.
48. McFarland, A. P.; Luo, S.; Ahmed-Qadri, F.; Zuck, M.; Thayer, E. F.; Goo, Y. A.; Hybiske, K.; Tong, L.; Woodward, J. J. (2017) Sensing of bacterial cyclic dinucleotides by the oxidoreductase RECON promotes NF-kappaB activation and shapes a proinflammatory antibacterial state. *Immunity*, *46*, 433–445.
49. Maelfait, J.; Rehwinkel, J. (2017) RECONsidering sensing of cyclic dinucleotides. *Immunity*, *46*, 337–339.
50. Whiteley, A. T.; Eaglesham, J. B.; de Oliveira Mann, C. C.; Morehouse, B. R.; Lowey, B.; Nieminen, E. A.; Danilchanka, O.; King, D. S.; Lee, A. S. Y.; Mekalanos, J. J. et al. (2019) Bacterial cGAS-like enzymes synthesize diverse nucleotide signals. *Nature*, *567*, 194–199.
51. Xia, P.; Wang, S.; Xiong, Z.; Zhu, X.; Ye, B.; Du, Y.; Meng, S.; Qu, Y.; Liu, J.; Gao, G. et al. (2018) The ER membrane adaptor ERAdP senses the bacterial second messenger c-di-AMP and initiates anti-bacterial immunity. *Nat Immunol*, *19*, 141–150.
52. Lioux, T.; Mauny, M. A.; Lamoureux, A.; Bascoul, N.; Hays, M.; Vernejoul, F.; Baudru, A. S.; Boularan, C.; Lopes-Vicente, J.; Qushair, G. et al. (2016) Design, synthesis, and biological evaluation of novel cyclic adenosine-inosine monophosphate (cAIMP) analogs that activate stimulator of interferon genes (STING). *J Med Chem*, *59*, 10253–10267.

53. Li, L.; Yin, Q.; Kuss, P.; Maliga, Z.; Millan, J. L.; Wu, H.; Mitchison, T. J. (2014) Hydrolysis of 2′3′-cGAMP by ENPP1 and design of nonhydrolyzable analogs. *Nat Chem Biol*, *10*, 1043–1048.
54. Kato, K.; Nishimasu, H.; Oikawa, D.; Hirano, S.; Hirano, H.; Kasuya, G.; Ishitani, R.; Tokunaga, F.; Nureki, O. (2018) Structural insights into cGAMP degradation by Ecto-nucleotide pyrophosphatase phosphodiesterase 1. *Nat Commun*, *9*, 4424.
55. Kawaguchi, M.; Han, X.; Hisada, T.; Nishikawa, S.; Kano, K.; Ieda, N.; Aoki, J.; Toyama, T.; Nakagawa, H. (2019) Development of an ENPP1 fluorescence probe for inhibitor screening, cellular imaging, and prognostic assessment of malignant breast cancer. *J Med Chem*, *62*, 9254–9269.
56. Onyedibe, K. I.; Wang, M.; Sintim, H. O. (2019) ENPP1, an old enzyme with new functions, and small molecule inhibitors-A STING in the tale of ENPP1. *Molecules*, *24*, 4192.
57. Gao, D.; Li, T.; Li, X. D.; Chen, X.; Li, Q. Z.; Wight-Carter, M.; Chen, Z. J. (2015) Activation of cyclic GMP-AMP synthase by self-DNA causes autoimmune diseases. *Proc Natl Acad Sci U S A*, *112*, E5699–E5705.
58. Eaglesham, J. B.; Pan, Y.; Kupper, T. S.; Kranzusch, P. J. (2019) Viral and metazoan poxins are cGAMP-specific nucleases that restrict cGAS-STING signalling. *Nature*, *566*, 259–263.
59. Decout, A.; Ablasser, A. (2019) Virology: poxins soothe the STING. *Curr Biol*, *29*, R332–R334.
60. Dai, P.; Wang, W.; Yang, N.; Serna-Tamayo, C.; Ricca, J. M.; Zamarin, D.; Shuman, S.; Merghoub, T.; Wolchok, J. D.; Deng, L. (2017) Intratumoral delivery of inactivated modified vaccinia virus Ankara (iMVA) induces systemic antitumor immunity via STING and Batf3-dependent dendritic cells. *Sci Immunol*, *2*, eaal1713.
61. Collins, A. C.; Cai, H.; Li, T.; Franco, L. H.; Li, X. D.; Nair, V. R.; Scharn, C. R.; Stamm, C. E.; Levine, B.; Chen, Z. J. et al. (2015) Cyclic GMP-AMP synthase is an innate immune DNA sensor for mycobacterium tuberculosis. *Cell Host Microbe*, *17*, 820–828.
62. Manzanillo, P. S.; Shiloh, M. U.; Portnoy, D. A.; Cox, J. S. (2012) Mycobacterium tuberculosis activates the DNA-dependent cytosolic surveillance pathway within macrophages. *Cell Host Microbe*, *11*, 469–480.
63. Wassermann, R.; Gulen, M. F.; Sala, C.; Perin, S. G.; Lou, Y.; Rybniker, J.; Schmid-Burgk, J. L.; Schmidt, T.; Hornung, V.; Cole, S. T. et al. (2015) Mycobacterium tuberculosis differentially activates cGAS- and inflammasome-dependent intracellular immune responses through ESX-1. *Cell Host Microbe*, *17*, 799–810.
64. Yang, J.; Bai, Y.; Zhang, Y.; Gabrielle, V. D.; Jin, L.; Bai, G. (2014) Deletion of the cyclic di-AMP phosphodiesterase gene (cnpB) in Mycobacterium tuberculosis leads to reduced virulence in a mouse model of infection. *Mol Microbiol*, *93*, 65–79.
65. Van Dis, E.; Sogi, K. M.; Rae, C. S.; Sivick, K. E.; Surh, N. H.; Leong, M. L.; Kanne, D. B.; Metchette, K.; Leong, J. J.; Bruml, J. R. et al. (2018) STING-activating adjuvants elicit a Th17 immune response and protect against Mycobacterium tuberculosis infection. *Cell Rep*, *23*, 1435–1447.
66. Ritchie, C.; Li, L. (2020) cGAMP as an adjuvant in antiviral vaccines and cancer immunotherapy. *Biochemistry*, *59*, 1713–1715.
67. Lee, J. M.; Ghonime, M. G.; Cassady, K. A. (2019) STING restricts oHSV replication and spread in resistant MPNSTs but is dispensable for Basal IFN-stimulated gene upregulation. *Mol Ther Oncolytics*, *15*, 91–100.

68. Zhou, C.; Chen, X.; Planells-Cases, R.; Chu, J.; Wang, L.; Cao, L.; Li, Z.; López-Cayuqueo, K. I.; Xie, Y.; Ye, S. et al. (2020) Transfer of cGAMP into bystander cells via LRRC8 volume-regulated anion channels augments STING-mediated interferon responses and anti-viral immunity. *Immunity*, *52*, 767–781.e766.
69. Sanchez, M. V.; Ebensen, T.; Schulze, K.; Cargnelutti, D.; Blazejewska, P.; Scodeller, E. A.; Guzman, C. A. (2014) Intranasal delivery of influenza rNP adjuvanted with c-di-AMP induces strong humoral and cellular immune responses and provides protection against virus challenge. *PLoS One*, *9*, e104824.
70. Krapp, C.; Jonsson, K.; Jakobsen, M. R. (2018) STING dependent sensing – does HIV actually care? *Cytokine Growth Factor Rev*, *40*, 68–76.
71. Lee, J.; Ghonime, M. G.; Wang, R.; Cassady, K. A. (2019) The antiviral apparatus: STING and oncolytic virus restriction. *Mol Ther Oncolytics*, *13*, 7–13.
72. Lin, L. C. W.; Huang, C. Y.; Yao, B. Y.; Lin, J. C.; Agrawal, A.; Algaissi, A.; Peng, B. H.; Liu, Y. H.; Huang, P. H.; Juang, R. H. et al. (2019) Viromimetic STING agonist-loaded hollow polymeric nanoparticles for safe and effective vaccination against middle east respiratory syndrome coronavirus. *Adv Fun Mater*, *29*, 1807616.
73. Yoh, S. M.; Schneider, M.; Seifried, J.; Soonthornvacharin, S.; Akleh, R. E.; Olivieri, K. C.; De Jesus, P. D.; Ruan, C.; de Castro, E.; Ruiz, P. A. et al. (2015) PQBP1 is a proximal sensor of the cGAS-dependent innate response to HIV-1. *Cell*, *161*, 1293–1305.
74. Skouboe, M. K.; Knudsen, A.; Reinert, L. S.; Boularan, C.; Lioux, T.; Perouzel, E.; Thomsen, M. K.; Paludan, S. R. (2018) STING agonists enable antiviral cross-talk between human cells and confer protection against genital herpes in mice. *PLoS Pathog*, *14*, e1006976.
75. Ablasser, A.; Schmid-Burgk, J. L.; Hemmerling, I.; Horvath, G. L.; Schmidt, T.; Latz, E.; Hornung, V. (2013) Cell intrinsic immunity spreads to bystander cells via the intercellular transfer of cGAMP. *Nature*, *503*, 530–534.
76. Gentili, M.; Kowal, J.; Tkach, M.; Satoh, T.; Lahaye, X.; Conrad, C.; Boyron, M.; Lombard, B.; Durand, S.; Kroemer, G. et al. (2015) Transmission of innate immune signaling by packaging of cGAMP in viral particles. *Science*, *349*, 1232–1236.
77. Bridgeman, A.; Maelfait, J.; Davenne, T.; Partridge, T.; Peng, Y.; Mayer, A.; Dong, T.; Kaever, V.; Borrow, P.; Rehwinkel, J. (2015) Viruses transfer the antiviral second messenger cGAMP between cells. *Science*, *349*, 1228–1232.
78. Zheng, Y.; Liu, Q.; Wu, Y.; Ma, L.; Zhang, Z.; Liu, T.; Jin, S.; She, Y.; Li, Y. P.; Cui, J. (2018) Zika virus elicits inflammation to evade antiviral response by cleaving cGAS via NS1-caspase-1 axis. *EMBO J*, *37*, e99347.
79. Wu, Z.; Zhang, W.; Wu, Y.; Wang, T.; Wu, S.; Wang, M.; Jia, R.; Zhu, D.; Liu, M.; Zhao, X. et al. (2019) Binding of the Duck Tembusu virus protease to STING is mediated by NS2B and is crucial for STING cleavage and for impaired induction of IFN-beta. *J Immunol*, *203*, 3374–3385.
80. Stabell, A. C.; Meyerson, N. R.; Gullberg, R. C.; Gilchrist, A. R.; Webb, K. J.; Old, W. M.; Perera, R.; Sawyer, S. L. (2018) Dengue viruses cleave STING in humans but not in nonhuman primates, their presumed natural reservoir. *eLife*, *7*, e31919.
81. Aguirre, S.; Luthra, P.; Sanchez-Aparicio, M. T.; Maestre, A. M.; Patel, J.; Lamothe, F.; Fredericks, A. C.; Tripathi, S.; Zhu, T.; Pintado-Silva, J. et al. (2017) Dengue virus NS2B protein targets cGAS for degradation and prevents mitochondrial DNA sensing during infection. *Nat Microbiol*, *2*, 17037.

82. Hoos, A. (2016) Development of immuno-oncology drugs – from CTLA4 to PD1 to the next generations. *Nat Rev Drug Discov*, *15*, 235–247.
83. Karaolis, D. K.; Cheng, K.; Lipsky, M.; Elnabawi, A.; Catalano, J.; Hyodo, M.; Hayakawa, Y.; Raufman, J. P. (2005) 3′,5′-cyclic diguanylic acid (c-di-GMP) inhibits basal and growth factor-stimulated human colon cancer cell proliferation. *Biochem Biophys Res Commun*, *329*, 40–45.
84. Corrales, L.; Glickman, L. H.; McWhirter, S. M.; Kanne, D. B.; Sivick, K. E.; Katibah, G. E.; Woo, S. R.; Lemmens, E.; Banda, T.; Leong, J. J. et al. (2015) Direct activation of STING in the tumor microenvironment leads to potent and systemic tumor regression and immunity. *Cell Rep*, *11*, 1018–1030.
85. Li, T.; Cheng, H.; Yuan, H.; Xu, Q.; Shu, C.; Zhang, Y.; Xu, P.; Tan, J.; Rui, Y.; Li, P. et al. (2016) Antitumor activity of cGAMP via stimulation of cGAS-cGAMP-STING-IRF3 mediated innate immune response. *Sci Rep*, *6*, 19049.
86. Demaria, O.; De Gassart, A.; Coso, S.; Gestermann, N.; Di Domizio, J.; Flatz, L.; Gaide, O.; Michielin, O.; Hwu, P.; Petrova, T. V. et al. (2015) STING activation of tumor endothelial cells initiates spontaneous and therapeutic antitumor immunity. *Proc Natl Acad Sci U S A*, *112*, 15408–15413.
87. Wang, H.; Hu, S.; Chen, X.; Shi, H.; Chen, C.; Sun, L.; Chen, Z. J. (2017) cGAS is essential for the antitumor effect of immune checkpoint blockade. *Proc Natl Acad Sci U S A*, *114*, 1637–1642.
88. Deng, L.; Liang, H.; Xu, M.; Yang, X.; Burnette, B.; Arina, A.; Li, X. D.; Mauceri, H.; Beckett, M.; Darga, T. et al. (2014) STING-dependent cytosolic DNA sensing promotes radiation-induced type I interferon-dependent antitumor immunity in immunogenic tumors. *Immunity*, *41*, 843–852.
89. Ghaffari, A.; Peterson, N.; Khalaj, K.; Vitkin, N.; Robinson, A.; Francis, J. A.; Koti, M. (2018) STING agonist therapy in combination with PD-1 immune checkpoint blockade enhances response to carboplatin chemotherapy in high-grade serous ovarian cancer. *Br J Cancer*, *119*, 440–449.
90. Fu, J.; Kanne, D. B.; Leong, M.; Glickman, L. H.; McWhirter, S. M.; Lemmens, E.; Mechette, K.; Leong, J. J.; Lauer, P.; Liu, W. et al. (2015) STING agonist formulated cancer vaccines can cure established tumors resistant to PD-1 blockade. *Sci Transl Med*, *7*, 283ra252.
91. Sivick, K. E.; Desbien, A. L.; Glickman, L. H.; Reiner, G. L.; Corrales, L.; Surh, N. H.; Hudson, T. E.; Vu, U. T.; Francica, B. J.; Banda, T. et al. (2018) Magnitude of therapeutic STING activation determines CD8(+) T cell-mediated anti-tumor immunity. *Cell Rep*, *25*, 3074–3085.e3075.
92. Kinkead, H. L.; Hopkins, A.; Lutz, E.; Wu, A. A.; Yarchoan, M.; Cruz, K.; Woolman, S.; Vithayathil, T.; Glickman, L. H.; Ndubaku, C. O. et al. (2018) Combining STING-based neoantigen-targeted vaccine with checkpoint modulators enhances antitumor immunity in murine pancreatic cancer. *JCI Insight*, *3*, e122857.
93. Chandra, D.; Quispe-Tintaya, W.; Jahangir, A.; Asafu-Adjei, D.; Ramos, I.; Sintim, H. O.; Zhou, J.; Hayakawa, Y.; Karaolis, D. K.; Gravekamp, C. (2014) STING ligand c-di-GMP improves cancer vaccination against metastatic breast cancer. *Cancer Immunol Res*, *2*, 901–910.
94. Smith, T. T.; Moffett, H. F.; Stephan, S. B.; Opel, C. F.; Dumigan, A. G.; Jiang, X.; Pillarisetty, V. G.; Pillai, S. P. S.; Wittrup, K. D.; Stephan, M. T. (2017) Biopolymers codelivering engineered T cells and STING agonists can eliminate heterogeneous tumors. *J Clin Invest*, *127*, 2176–2191.

95. Yildiz, S.; Alpdundar, E.; Gungor, B.; Kahraman, T.; Bayyurt, B.; Gursel, I.; Gursel, M. (2015) Enhanced immunostimulatory activity of cyclic dinucleotides on mouse cells when complexed with a cell-penetrating peptide or combined with CpG. *Eur J Immunol*, *45*, 1170–1179.
96. Yang, H.; Lee, W. S.; Kong, S. J.; Kim, C. G.; Kim, J. H.; Chang, S. K.; Kim, S.; Kim, G.; Chon, H. J.; Kim, C. (2019) STING activation reprograms tumor vasculatures and synergizes with VEGFR2 blockade. *J Clin Invest*, *130*, 4350–4364.
97. Downey, C. M.; Aghaei, M.; Schwendener, R. A.; Jirik, F. R. (2014) DMXAA causes tumor site-specific vascular disruption in murine non-small cell lung cancer, and like the endogenous non-canonical cyclic dinucleotide STING agonist, 2′3′-cGAMP, induces M2 macrophage repolarization. *PLoS One*, *9*, e99988.
98. Dubensky Jr, T. W.; Kanne, D. B.; Leong, M. L. (2013) Rationale progress and development of vaccines utilizing STING-activating cyclic dinucleotide adjuvants. *Ther Adv Vaccines Immunother*, *1*, 131–143.
99. Gutjahr, A.; Papagno, L.; Nicoli, F.; Kanuma, T.; Kuse, N.; Cabral-Piccin, M. P.; Rochereau, N.; Gostick, E.; Lioux, T.; Perouzel, E. et al. (2019) The STING ligand cGAMP potentiates the efficacy of vaccine-induced CD8+ T cells. *JCI Insight*, *4*, e125107.
100. Vassilieva, E. V.; Taylor, D. W.; Compans, R. W. (2019) Combination of STING pathway agonist with saponin is an effective adjuvant in immunosenescent mice. *Front Immunol*, *10*, 3006.
101. Lemos, H.; Mohamed, E.; Huang, L.; Ou, R.; Pacholczyk, G.; Arbab, A. S.; Munn, D.; Mellor, A. L. (2016) STING promotes the growth of tumors characterized by low antigenicity via IDO activation. *Cancer Research*, *76*, 2076–2081.
102. Spranger, S.; Spaapen, R. M.; Zha, Y.; Williams, J.; Meng, Y.; Ha, T. T.; Gajewski, T. F. (2013) Up-regulation of PD-L1, IDO, and T(regs) in the melanoma tumor microenvironment is driven by CD8(+) T cells. *Sci Transl Med*, *5*, 200ra116.
103. Yarchoan, M.; Johnson III, B. A.; Lutz, E. R.; Laheru, D. A.; Jaffee, E. M. (2017) Targeting neoantigens to augment antitumour immunity. *Nat Rev Cancer*, *17*, 209–222.
104. Schumacher, T. N.; Schreiber, R. D. (2015) Neoantigens in cancer immunotherapy. *Science*, *348*, 69–74.
105. Harabuchi, S.; Kosaka, A.; Yajima, Y.; Nagata, M.; Hayashi, R.; Kumai, T.; Ohara, K.; Nagato, T.; Oikawa, K.; Ohara, M. et al. (2020) Intratumoral STING activations overcome negative impact of cisplatin on antitumor immunity by inflaming tumor microenvironment in squamous cell carcinoma. *Biochem Biophys Res Commun*, *522*, 408–414.
106. Wu, S.; Zhang, Q.; Zhang, F.; Meng, F.; Liu, S.; Zhou, R.; Wu, Q.; Li, X.; Shen, L.; Huang, J. et al. (2019) HER2 recruits AKT1 to disrupt STING signalling and suppress antiviral defence and antitumour immunity. *Nat Cell Biol*, *21*, 1027–1040.
107. Marill, J.; Mohamed Anesary, N.; Paris, S. (2019) DNA damage enhancement by radiotherapy-activated hafnium oxide nanoparticles improves cGAS-STING pathway activation in human colorectal cancer cells. *Radiother Oncol*, *141*, 262–266.
108. Shen, J.; Zhao, W.; Ju, Z.; Wang, L.; Peng, Y.; Labrie, M.; Yap, T. A.; Mills, G. B.; Peng, G. (2019) PARPi triggers the STING-dependent immune response and enhances the therapeutic efficacy of immune checkpoint blockade independent of BRCAness. *Cancer Research*, *79*, 311–319.

109. Pantelidou, C.; Sonzogni, O.; De Oliveria Taveira, M.; Mehta, A. K.; Kothari, A.; Wang, D.; Visal, T.; Li, M. K.; Pinto, J.; Castrillon, J. A. et al. (2019) PARP inhibitor efficacy depends on CD8(+) T-cell recruitment via intratumoral STING pathway activation in BRCA-deficient models of triple-negative breast cancer. *Cancer Discov*, *9*, 722–737.
110. Sen, T.; Rodriguez, B. L.; Chen, L.; Corte, C. M. D.; Morikawa, N.; Fujimoto, J.; Cristea, S.; Nguyen, T.; Diao, L.; Li, L. et al. (2019) Targeting DNA damage response promotes antitumor immunity through STING-mediated T-cell activation in small cell lung cancer. *Cancer Discov*, *9*, 646–661.
111. Reislander, T.; Lombardi, E. P.; Groelly, F. J.; Miar, A.; Porru, M.; Di Vito, S.; Wright, B.; Lockstone, H.; Biroccio, A.; Harris, A. et al. (2019) BRCA2 abrogation triggers innate immune responses potentiated by treatment with PARP inhibitors. *Nat Commun*, *10*, 3143.
112. Wang, Z.; Chen, J.; Hu, J.; Zhang, H.; Xu, F.; He, W.; Wang, X.; Li, M.; Lu, W.; Zeng, G. et al. (2019) cGAS/STING axis mediates a topoisomerase II inhibitor-induced tumor immunogenicity. *J Clin Invest*, *130*, 4850–4862.
113. Zhao, B.; Liu, P.; Fukumoto, T.; Nacarelli, T.; Fatkhutdinov, N.; Wu, S.; Lin, J.; Aird, K. M.; Tang, H. Y.; Liu, Q. et al. (2020) Topoisomerase 1 cleavage complex enables pattern recognition and inflammation during senescence. *Nat Commun*, *11*, 908.
114. Le Naour, J.; Zitvogel, L.; Galluzzi, L.; Vacchelli, E.; Kroemer, G. (2020) Trial watch: STING agonists in cancer therapy. *Oncoimmunology*, *9*.
115. Meric-Bernstam, F.; Sandhu, S.; Hamid, O.; Spreafico, A.; Kasper, S.; Dummer, R.; Shimizu, T.; Steeghs, N.; Lewis, N.; Talluto, C. et al. (2019) Phase Ib study of MIW815 (ADU-S100) in combination with spartalizumab (PDR001) in patients with advanced/metastatic solid tumors or lymphomas (NCT03172936). 2019 ASCO Annual Meeting.
116. Harrington, K. J.; Brody, J.; Ingham, M.; Strauss, J.; Cemerski, S.; Wang, M.; Tse, A.; Khilnani, A.; Marabelle, A.; Golan, T. (2018) LBA15 – preliminary results of the first-in-human (FIH) study of MK-1454, an agonist of stimulator of interferon genes (STING), as monotherapy or in combination with pembrolizumab (pembro) in patients with advanced solid tumors or lymphomas. *Ann Oncol*, *29*, viii712.
117. Rice, G. I.; Rodero, M. P.; Crow, Y. J. (2015) Human disease phenotypes associated with mutations in TREX1. *J Clin Immunol*, *35*, 235–243.
118. Crow, Y. J.; Chase, D. S.; Lowenstein Schmidt, J.; Szynkiewicz, M.; Forte, G. M.; Gornall, H. L.; Oojageer, A.; Anderson, B.; Pizzino, A.; Helman, G. et al. (2015) Characterization of human disease phenotypes associated with mutations in TREX1, RNASEH2A, RNASEH2B, RNASEH2C, SAMHD1, ADAR, and IFIH1. *Am J Med Genet A*, *167A*, 296–312.
119. Gray, E. E.; Treuting, P. M.; Woodward, J. J.; Stetson, D. B. (2015) Cutting Edge: cGAS is required for lethal autoimmune disease in the Trex1-deficient mouse model of Aicardi-Goutieres Syndrome. *J Immunol*, *195*, 1939–1943.
120. Ablasser, A.; Hemmerling, I.; Schmid-Burgk, J. L.; Behrendt, R.; Roers, A.; Hornung, V. (2014) TREX1 deficiency triggers cell-autonomous immunity in a cGAS-dependent manner. *Journal of Immunology*, *192*, 5993–5997.
121. Wang, J.; Li, R.; Lin, H.; Qiu, Q.; Lao, M.; Zeng, S.; Wang, C.; Xu, S.; Zou, Y.; Shi, M. et al. (2019) Accumulation of cytosolic dsDNA contributes to fibroblast-like synoviocytes-mediated rheumatoid arthritis synovial inflammation. *Int Immunopharmacol*, *76*, 105791.

122. Mackenzie, K. J.; Carroll, P.; Lettice, L.; Tarnauskaite, Z.; Reddy, K.; Dix, F.; Revuelta, A.; Abbondati, E.; Rigby, R. E.; Rabe, B. et al. (2016) Ribonuclease H2 mutations induce a cGAS/STING-dependent innate immune response. *EMBO J*, *35*, 831–844.
123. Coquel, F.; Silva, M. J.; Techer, H.; Zadorozhny, K.; Sharma, S.; Nieminuszczy, J.; Mettling, C.; Dardillac, E.; Barthe, A.; Schmitz, A. L. et al. (2018) SAMHD1 acts at stalled replication forks to prevent interferon induction. *Nature*, *557*, 57–61.
124. Coquel, F.; Neumayer, C.; Lin, Y. L.; Pasero, P. (2018) SAMHD1 and the innate immune response to cytosolic DNA during DNA replication. *Curr Opin Immunol*, *56*, 24–30.
125. Liu, Y.; Jesus, A. A.; Marrero, B.; Yang, D.; Ramsey, S. E.; Sanchez, G. A. M.; Tenbrock, K.; Wittkowski, H.; Jones, O. Y.; Kuehn, H. S. et al. (2014) Activated STING in a vascular and pulmonary syndrome. *N Engl J Med*, *371*, 507–518.
126. Keskitalo, S.; Haapaniemi, E.; Einarsdottir, E.; Rajamaki, K.; Heikkila, H.; Ilander, M.; Poyhonen, M.; Morgunova, E.; Hokynar, K.; Lagstrom, S. et al. (2019) Novel TMEM173 mutation and the role of disease modifying alleles. *Front Immunol*, *10*, 2770.
127. Patel, S.; Jin, L. (2019) TMEM173 variants and potential importance to human biology and disease. *Genes Immun*, *20*, 82–89.
128. Tarantino, G.; Esposito, S.; Andreozzi, L.; Bracci, B.; D'Errico, F.; Rigante, D. (2016) Lung involvement in children with hereditary autoinflammatory disorders. *Int J Mol Sci*, *17*, 2111.
129. Melki, I.; Rose, Y.; Uggenti, C.; Van Eyck, L.; Fremond, M. L.; Kitabayashi, N.; Rice, G. I.; Jenkinson, E. M.; Boulai, A.; Jeremiah, N. et al. (2017) Disease-associated mutations identify a novel region in human STING necessary for the control of type I interferon signaling. *J Allergy Clin Immunol*, *140*, 543–552. e545.
130. Tang, X.; Xu, H.; Zhou, C.; Peng, Y.; Liu, H.; Liu, J.; Li, H.; Yang, H.; Zhao, S. (2020) STING-associated vasculopathy with onset in infancy in three children with new clinical aspect and unsatisfactory therapeutic responses to tofacitinib. *J Clin Immunol*, *40*, 114–122.
131. Volpi, S.; Insalaco, A.; Caorsi, R.; Santori, E.; Messia, V.; Sacco, O.; Terheggen-Lagro, S.; Cardinale, F.; Scarselli, A.; Pastorino, C. et al. (2019) Efficacy and adverse events during Janus Kinase inhibitor treatment of SAVI syndrome. *J Clin Immunol*, *39*, 476–485.
132. Balci, S.; Ekinci, R. M. K.; de Jesus, A. A.; Goldbach-Mansky, R.; Yilmaz, M. (2020) Baricitinib experience on STING-associated vasculopathy with onset in infancy: a representative case from Turkey. *Clin Immunol*, *212*, 108273.
133. Sanchez, G. A. M.; Reinhardt, A.; Ramsey, S.; Wittkowski, H.; Hashkes, P. J.; Berkun, Y.; Schalm, S.; Murias, S.; Dare, J. A.; Brown, D. et al. (2018) JAK1/2 inhibition with baricitinib in the treatment of autoinflammatory interferonopathies. *J Clin Invest*, *128*, 3041–3052.
134. An, J.; Woodward, J. J.; Sasaki, T.; Minie, M.; Elkon, K. B. (2015) Cutting edge: antimalarial drugs inhibit IFN-beta production through blockade of cyclic GMP-AMP synthase-DNA interaction. *J Immunol*, *194*, 4089–4093.
135. Vincent, J.; Adura, C.; Gao, P.; Luz, A.; Lama, L.; Asano, Y.; Okamoto, R.; Imaeda, T.; Aida, J.; Rothamel, K. et al. (2017) Small molecule inhibition of

cGAS reduces interferon expression in primary macrophages from autoimmune mice. *Nat Commun*, *8*, 750.

136. Lama, L.; Adura, C.; Xie, W.; Tomita, D.; Kamei, T.; Kuryavyi, V.; Gogakos, T.; Steinberg, J. I.; Miller, M.; Ramos-Espiritu, L. et al. (2019) Development of human cGAS-specific small-molecule inhibitors for repression of dsDNA-triggered interferon expression. *Nat Commun*, *10*, 2261.
137. Hansen, A. L.; Buchan, G. J.; Ruhl, M.; Mukai, K.; Salvatore, S. R.; Ogawa, E.; Andersen, S. D.; Iversen, M. B.; Thielke, A. L.; Gunderstofte, C. et al. (2018) Nitro-fatty acids are formed in response to virus infection and are potent inhibitors of STING palmitoylation and signaling. *Proc Natl Acad Sci U S A*, *115*, E7768–E7775.
138. Li, S.; Hong, Z.; Wang, Z.; Li, F.; Mei, J.; Huang, L.; Lou, X.; Zhao, S.; Song, L.; Chen, W. et al. (2018) The cyclopeptide Astin C specifically inhibits the innate immune CDN sensor STING. *Cell Rep*, *25*, 3405–3421.e3407.
139. Siu, T.; Altman, M. D.; Baltus, G. A.; Childers, M.; Ellis, J. M.; Gunaydin, H.; Hatch, H.; Ho, T.; Jewell, J.; Lacey, B. M. et al. (2019) Discovery of a novel cGAMP competitive ligand of the inactive form of STING. *ACS Med Chem Lett*, *10*, 92–97.
140. Koestler, B. J.; Seregin, S. S.; Rastall, D. P.; Aldhamen, Y. A.; Godbehere, S.; Amalfitano, A.; Waters, C. M. (2014) Stimulation of innate immunity by in vivo cyclic di-GMP synthesis using adenovirus. *Clin Vaccine Immunol*, *21*, 1550–1559.
141. Alyaqoub, F. S.; Aldhamen, Y. A.; Koestler, B. J.; Bruger, E. L.; Seregin, S. S.; Pereira-Hicks, C.; Godbehere, S.; Waters, C. M.; Amalfitano, A. (2016) in vivo synthesis of cyclic-di-GMP using a recombinant adenovirus preferentially improves adaptive immune responses against extracellular antigens. *J Immunol*, *196*, 1741–1752.
142. Miyabe, H.; Hyodo, M.; Nakamura, T.; Sato, Y.; Hayakawa, Y.; Harashima, H. (2014) A new adjuvant delivery system 'cyclic di-GMP/YSK05 liposome' for cancer immunotherapy. *J Control Release*, *184*, 20–27.
143. Nakamura, T.; Miyabe, H.; Hyodo, M.; Sato, Y.; Hayakawa, Y.; Harashima, H. (2015) Liposomes loaded with a STING pathway ligand, cyclic di-GMP, enhance cancer immunotherapy against metastatic melanoma. *J Control Release*, *216*, 149–157.
144. Koshy, S. T.; Cheung, A. S.; Gu, L.; Graveline, A. R.; Mooney, D. J. (2017) Liposomal delivery enhances immune activation by STING agonists for cancer immunotherapy. *Adv Biosys*, *1*, 1600013.
145. Junkins, R. D.; Gallovic, M. D.; Johnson, B. M.; Collier, M. A.; Watkins-Schulz, R.; Cheng, N.; David, C. N.; McGee, C. E.; Sempowski, G. D.; Shterev, I. et al. (2018) A robust microparticle platform for a STING-targeted adjuvant that enhances both humoral and cellular immunity during vaccination. *J Control Release*, *270*, 1–13.
146. Wilson, D. R.; Sen, R.; Sunshine, J. C.; Pardoll, D. M.; Green, J. J.; Kim, Y. J. (2017) Biodegradable STING agonist nanoparticles for enhanced cancer immunotherapy. *Nanomedicine*, *14*, 237–246.
147. Shae, D.; Becker, K. W.; Christov, P.; Yun, D. S.; Lytton-Jean, A. K. R.; Sevimli, S.; Ascano, M.; Kelley, M.; Johnson, D. B.; Balko, J. M. et al. (2019) Endosomolytic polymersomes increase the activity of cyclic dinucleotide STING agonists to enhance cancer immunotherapy. *Nat Nanotechnol*, *14*, 269–278.

148. Leach, D. G.; Dharmaraj, N.; Piotrowski, S. L.; Lopez-Silva, T. L.; Lei, Y. L.; Sikora, A. G.; Young, S.; Hartgerink, J. D. (2018) STINGel: controlled release of a cyclic dinucleotide for enhanced cancer immunotherapy. *Biomaterials*, *163*, 67–75.
149. Shakya, A. K.; Lee, C. H.; Uddin, M. J.; Gill, H. S. (2018) Assessment of Th1/Th2 bias of STING agonists coated on microneedles for possible use in skin allergen immunotherapy. *Mol Pharm*, *15*, 5437–5443.
150. An, M.; Yu, C.; Xi, J.; Reyes, J.; Mao, G.; Wei, W. Z.; Liu, H. (2018) Induction of necrotic cell death and activation of STING in the tumor microenvironment via cationic silica nanoparticles leading to enhanced antitumor immunity. *Nanoscale*, *10*, 9311–9319.
151. Luteijn, R. D.; Zaver, S. A.; Gowen, B. G.; Wyman, S. K.; Garelis, N. E.; Onia, L.; McWhirter, S. M.; Katibah, G. E.; Corn, J. E.; Woodward, J. J. et al. (2019) SLC19A1 transports immunoreactive cyclic dinucleotides. *Nature*, *573*, 434–438.
152. Ritchie, C.; Cordova, A. F.; Hess, G. T.; Bassik, M. C.; Li, L. (2019) SLC19A1 is an importer of the immunotransmitter cGAMP. *Mol Cell*, *75*, 372–381.e375.
153. Hwang, J.; Kang, T.; Lee, J.; Choi, B. S.; Han, S. (2019) Design, synthesis, and biological evaluation of C7-functionalized DMXAA derivatives as potential human-STING agonists. *Org Biomol Chem*, *17*, 1869–1874.
154. Liu, B.; Tang, L.; Zhang, X.; Ma, J.; Sehgal, M.; Cheng, J.; Zhang, X.; Zhou, Y.; Du, Y.; Kulp, J. et al. (2017) A cell-based high throughput screening assay for the discovery of cGAS-STING pathway agonists. *Antiviral Res*, *147*, 37–46.
155. Sali, T. M.; Pryke, K. M.; Abraham, J.; Liu, A.; Archer, I.; Broeckel, R.; Staverosky, J. A.; Smith, J. L.; Al-Shammari, A.; Amsler, L. et al. (2015) Characterization of a novel human-specific STING agonist that elicits antiviral activity against emerging alphaviruses. *PLoS Pathog*, *11*, e1005324.
156. Kimura, Y.; Negishi, H.; Matsuda, A.; Endo, N.; Hangai, S.; Inoue, A.; Nishio, J.; Taniguchi, T.; Yanai, H. (2018) Novel chemical compound SINCRO with dual function in STING-type I interferon and tumor cell death pathways. *Cancer Sci*, *109*, 2687–2696.
157. Ramanjulu, J. M.; Pesiridis, G. S.; Yang, J.; Concha, N.; Singhaus, R.; Zhang, S. Y.; Tran, J. L.; Moore, P.; Lehmann, S.; Eberl, H. C. et al. (2018) Design of amidobenzimidazole STING receptor agonists with systemic activity. *Nature*, *564*, 439–443.
158. Shi, J.; Liu, C. L.; Zhang, B.; Guo, W. J.; Zhu, J.; Chang, C. Y.; Zhao, E. J.; Jiao, R. H.; Tan, R. X.; Ge, H. M. (2019) Genome mining and biosynthesis of kitacinnamycins as a STING activator. *Chemical Science*, *10*, 4839–4846.
159. Chin, E. N.; Yu, C.; Vartabedian, V. F.; Jia, Y.; Kumar, M.; Gamo, A. M.; Vernier, W.; Ali, S. H.; Kissai, M.; Lazar, D. C. et al. (2020) Antitumor activity of a systemic STING-activating non-nucleotide cGAMP mimetic. *Science*, *369*, 993–999.
160. Pan, B. S.; Perera, S. A.; Piesvaux, J. A.; Presland, J. P.; Schroeder, G. K.; Cumming, J. N.; Trotter, B. W.; Altman, M. D.; Buevich, A. V.; Cash, B. et al. (2020) An orally available non-nucleotide STING agonist with antitumor activity. *Science*, *369*, eaba6098.
161. Launer-Felty, K. D.; Strobel, S. A. (2018) Enzymatic synthesis of cyclic dinucleotide analogs by a promiscuous cyclic-AMP-GMP synthetase and analysis of cyclic dinucleotide responsive riboswitches. *Nucleic Acids Res*, *46*, 2765–2776.

162. Novotna, B.; Vanekova, L.; Zavrel, M.; Budesinsky, M.; Dejmek, M.; Smola, M.; Gutten, O.; Tehrani, Z. A.; Pimkova Polidarova, M.; Brazdova, A. et al. (2019) Enzymatic preparation of 2′-5′,3′-5′-cyclic dinucleotides, their binding properties to stimulator of interferon genes adaptor protein, and structure/activity correlations. *J Med Chem*, *62*, 10676–10690.
163. Rosenthal, K.; Becker, M.; Rolf, J.; Siedentop, R.; Hillen, M.; Nett, M.; Lutz, S. (2020) Catalytic promiscuity of cGAS: a facile enzymatic synthesis of 2′-3′-linked cyclic dinucleotides. *Chembiochem*. https://doi.org/10.1002/cbic.202000433.
164. Rao, F.; Pasunooti, S.; Ng, Y.; Zhuo, W.; Lim, L.; Liu, A. W.; Liang, Z. X. (2009) Enzymatic synthesis of c-di-GMP using a thermophilic diguanylate cyclase. *Anal Biochem*, *389*, 138–142.
165. Pasunooti, S.; Surya, W.; Tan, S. N.; Liang, Z.-X. (2010) Sol–gel immobilization of a thermophilic diguanylate cyclase for enzymatic production of cyclic-di-GMP. *J Mol Catal B: Enzym*, *67*, 98–103.
166. Lv, Y.; Sun, Q.; Wang, X.; Lu, Y.; Li, Y.; Yuan, H.; Zhu, J.; Zhu, D. (2019) Highly efficient preparation of cyclic dinucleotides via engineering of dinucleotide cyclases in *Escherichia coli*. *Front Microbiol*, *10*, 2111.
167. Beaucage, S. L. (2008) Solid-phase synthesis of siRNA oligonucleotides. *Curr Opin Drug Discov Devel*, *11*, 203–216.
168. Caruthers, M. H. (2005) New methods for synthesizing DNA, RNA, and phosphonates. *CHIMIA*, *59*, 39–42.
169. Gaglione, M.; Di Fabio, G.; Messere, A. (2012) Current methods in synthesis of cyclic oligonucleotides and analogues. *Curr Org Chem*, *16*, 1371–1389.
170. Frieden, M.; Grandas, A.; Pedroso, E. (1999) Making cyclic RNAs easily available. *Chem Commun (Camb)*, 1593–1594.
171. Kiburu, I.; Shurer, A.; Yan, L.; Sintim, H. O. (2008) A simple solid-phase synthesis of the ubiquitous bacterial signaling molecule, c-di-GMP and analogues. *Mol Biosyst*, *4*, 518–520.
172. Micura, R. (1999) Cyclic oligoribonucleotides (RNA) by solid-phase synthesis. *Chem Eur J*, *5*, 2077–2082.
173. Alazzouzi, E.; Escaja, N.; Grandas, A.; Pedroso, E. (1997) A straightforward solid-phase synthesis of cyclic oligodeoxyribonucleotides. *Angew Chem Int Ed Eng*, *36*, 1506–1508.
174. Hamoir, G.; Sonveaux, E. (1993) 3′-5′ cyclic oligothymidylic acids conformation and complexation of intercalating agents. *Bull Soc Chlm Belg*, *102*, 335–342.
175. Hayakawa, Y.; Nagata, R.; Hirata, A.; Hyodo, M.; Kawai, R. (2003) A facile synthesis of cyclic bis(3′→5′)diguanylic acid. *Tetrahedron*, *59*, 6465–6471.
176. Ching, S. M.; Tan, W. J.; Chua, K. L.; Lam, Y. (2010) Synthesis of cyclic dinucleotidic acids as potential inhibitors targeting diguanylate cyclase. *Bioorg Med Chem*, *18*, 6657–6665.
177. Shiraishi, K.; Saito-Tarashima, N.; Igata, Y.; Murakami, K.; Okamoto, Y.; Miyake, Y.; Furukawa, K.; Minakawa, N. (2017) Synthesis and evaluation of c-di-4′-thioAMP as an artificial ligand for c-di-AMP riboswitch. *Bioorg Med Chem*, *25*, 3883–3889.
178. Gaffney, B. L.; Veliath, E.; Zhao, J.; Jones, R. A. (2010) One-flask syntheses of c-di-GMP and the [Rp,Rp] and [Rp,Sp] thiophosphate analogues. *Org Lett*, *12*, 3269–3271.

179. Wang, C.; Sinn, M.; Stifel, J.; Heiler, A. C.; Sommershof, A.; Hartig, J. S. (2017) Synthesis of all possible canonical (3′-5′-Linked) cyclic dinucleotides and evaluation of riboswitch interactions and immune-stimulatory effects. *J Am Chem Soc*, *139*, 16154–16160.
180. Knouse, K. W.; deGruyter, J. N.; Schmidt, M. A.; Zheng, B.; Vantourout, J. C.; Kingston, C.; Mercer, S. E.; McDonald, I. M.; Olson, R. E.; Zhu, Y. et al. (2018) Unlocking P(V): reagents for chiral phosphorothioate synthesis. *Science*, *361*, 1234–1238.
181. Mellman, I.; Coukos, G.; Dranoff, G. (2011) Cancer immunotherapy comes of age. *Nature*, *480*, 480–489.
182. Dhanak, D.; Edwards, J. P.; Nguyen, A.; Tummino, P. J. (2017) Small-molecule targets in immuno-oncology. *Cell Chem Biol*, *24*, 1148–1160.
183. Toogood, P. L. (2017) Small molecule immuno-oncology therapeutic agents. *Bioorg Med Chem Lett*, *28*, 319–329.
184. Acurcio, R. C.; Scomparin, A.; Conniot, J.; Salvador, J. A. R.; Satchi-Fainaro, R.; Florindo, H. F.; Guedes, R. C. (2018) Structure-function analysis of immune checkpoint receptors to guide emerging anticancer immunotherapy. *J Med Chem*, *61*, 10957–10975.
185. Cheng, B.; Yuan, W. E.; Su, J.; Liu, Y.; Chen, J. (2018) Recent advances in small molecule based cancer immunotherapy. *Eur J Med Chem*, *157*, 582–598.
186. Zhu, H. F.; Li, Y. (2018) Small-molecule targets in tumor immunotherapy. *Nat Prod Bioprospect*, *8*, 297–301.
187. Adams, J. L.; Smothers, J.; Srinivasan, R.; Hoos, A. (2015) Big opportunities for small molecules in immuno-oncology. *Nat Rev Drug Discov*, *14*, 603–622.
188. Ma, F.; Lei, Y. Y.; Ding, M. G.; Luo, L. H.; Xie, Y. C.; Liu, X. L. (2020) LncRNA NEAT1 interacted with DNMT1 to regulate malignant phenotype of cancer cell and cytotoxic T cell infiltration via epigenetic inhibition of p53, cGAS, and STING in lung cancer. *Front Genet*, *11*, 250.
189. Konno, H.; Yamauchi, S.; Berglund, A.; Putney, R. M.; Mulé, J. J.; Barber, G. N. (2018) Suppression of STING signaling through epigenetic silencing and missense mutation impedes DNA damage mediated cytokine production. *Oncogene*, *37*, 2037–2051.

CHAPTER 3

ENZYMATIC AND ENDOGENOUS SYNTHESIS OF NAD(P)-DERIVED CALCIUM-MOBILIZING MESSENGERS

HON CHEUNG LEE[1] AND YONG JUAN ZHAO[2]

[1] State Key Laboratory of Chemical Oncogenomics, Key Laboratory of Chemical Genomics, Peking University Shenzhen Graduate School, Shenzhen, China

[2] School of Life and Health Sciences, School of Medicine, The Chinese University of Hong Kong, Shenzhen, China

3.1. INTRODUCTION

Ever since Ca^{2+} was found to influence the contraction of the isolated frog heart decades ago [1], it is now well established that Ca^{2+} regulates virtually all aspects of life. Modern research on Ca^{2+} is greatly aided by the development of fluorescent Ca^{2+} indicators [2], leading to the finding that inositol trisphosphate (IP_3) can release Ca^{2+} from the endoplasmic reticulum (ER) and establishing it as a messenger for calcium mobilization [3]. The mobilization of Ca^{2+} from internal stores as a major cellular signaling mechanism has since moved to the front stage.

Soon after IP_3, two more Ca^{2+}-mobilizing messengers, cyclic ADP-ribose (cADPR) and nicotinic acid adenine dinucleotide phosphate (NAADP) were discovered [4–7]. Unlike IP_3, the two new messengers are nucleotides derived from NAD and NADP, respectively. Much is now known about the signaling functions of these two new messengers. Like IP_3, cADPR also mobilizes the ER Ca^{2+}-stores, but targets a different Ca^{2+} release channel, the ryanodine receptor [8–10].

Nucleic Acids in Medicinal Chemistry and Chemical Biology: Drug Development and Clinical Applications, First Edition. Edited by Lihe Zhang, Xinjing Tang, Zhen Xi, and Jyoti Chattopadhyaya.

NAADP, on the other hand, releases Ca^{2+} from the endo-lysosomal stores via the two-pore channel [11, 12]. The finding, indeed, has germinated substantial interests on the functionality and regulation of the acidic Ca^{2+} stores in cells [13, 14].

The enzymatic synthesis of cADPR and NAADP had been a totally uncharted territory. That both are nucleotides unrelated to IP_3 indicates the enzymatic pathway have no relationship with that of IP_3. In fact, there was no known reaction that could circularize a long linear NAD molecule head to tail to a complete cyclic product, cADPR. The structure of NAADP is much more mundane with just a simple replacement of the amide group of NADP with a carboxyl group. The distinct structures of cADPR and NAADP makes the discovery that both are produced by the same enzymes that much more surprising. The remarkable advances in the understanding of the enzymatic synthesis of cADPR and NAADP, unraveling the mystery and elucidating the unknown, are the focuses of this chapter.

Furthermore, the enzymatic synthesis of cADPR and NAADP also produces ADP-ribose (ADPR), a well-known metabolite of NAD. Unexpectedly, it has also been found to have signaling functions as well, regulating the Ca^{2+}-influx through gating the transient receptor potential cation channel subfamily M member 2 (TRPM2). Whether ADPR functions as a messenger like cADPR or NAADP under physiological conditions will be discussed at the last part of this chapter.

3.2. CYCLIC ADP-RIBOSE AND NAADP

The story behind the discovery of cADPR and NAADP has been described [15]. The main impetus for the discovery was attributed to the development of a cell-free bioassay system for messengers for mobilizing the Ca^{2+}-stores using the homogenates from sea urchin egg [4, 7, 16]. It is extremely stable and simple to prepare and can be stored frozen for years without loss of sensitivity. For more than 30 years, the assay has been used routinely and has recently been amended for mass screening for antagonists for Ca^{2+}-mobilization [17].

With the assay, NAD was found to induce Ca^{2+} release from the homogenates with an initial lag, leading to the discovery of its enzymatic conversion to cADPR [4, 7]. NADP likewise induces Ca^{2+} release from the homogenates but without the delay. It was subsequently determined to be due to a contaminant, NAADP, that can be produced by an alkaline treatment of NADP also [5]. These results provided the first means to produce enough cADPR and NAADP for structural determination. The crystal structures of cADPR (PDB: 2O3Q)[6] and NAADP (PDB: 3HCT)[18] are shown in Figure 3.1.

The complete circular structure of cADPR is novel and no similar structure had ever been described before. The nicotinamide in NAD attached to the C1 of the ribose is released and linked back to the N1 of the adenine ring forming a complete circle. The structural requirement for its Ca^{2+}-releasing activity is very high. Most modifications render it inactive. A notable exception is the analog, 4″ α-azidoethyl-cyclic ADP-carbocyclic-ribose, which is active and has an azidoethyl

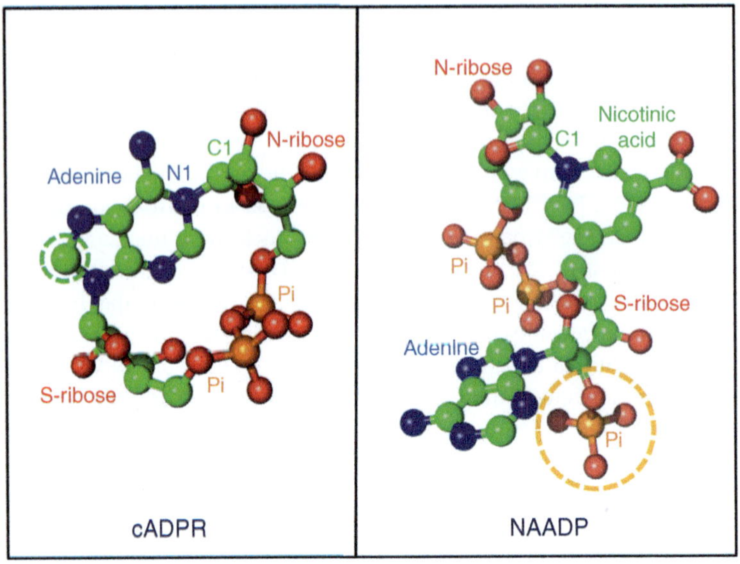

Figure 3.1 The crystal structures of cADPR and NAADP. The dotted green circle indicates the 8-position of cADPR. Attaching a bromine to that position converts cADPR to its antagonist. The two ribose groups are separately designated as Southern (S-) or Northern (N-). The dotted orange circle in NAADP is critical for its Ca^{2+}-mobilizing activity. Color coding: Carbon-green, oxygen-red, nitrogen-blue, phosphorus-orange.

group attached to the C4 of the terminal ribose [19]. This analog can potentially be useful for labeling the cADPR-receptor or use as a tool to isolate the receptor. The 8-position on the adenine ring (green circle, Figure 3.1) is of interest. Attaching an amino group or bromine turns cADPR into competitive antagonists for Ca^{2+} release [20]. The 8-Br-cADPR is cell permeant and is the most widely used diagnostic antagonist for the cADPR-induced Ca^{2+} release.

Sea urchin egg homogenates not only provide a stable system investigating Ca^{2+}-stores mobilization, leading to the discovery of cADPR, but also a specific bioassay for the first quantitative determination of endogenous cADPR in cells and tissues [21, 22]. The bioassay, however, does require the availability of urchin eggs, which can be inconvenient. The development of a fluorometric assay has since greatly facilitates the investigation of stimulus-induced changes of cADPR in cells [23].

In contrast to cADPR, the structure of NAADP is nondescript and is essentially the same as NADP, except with the nicotinamide replaced with nicotinic acid. But this simple structural change endows NAADP with potent Ca^{2+}-mobilizing activity in the nanomolar range [5]. The Pi group attached to the ribose (orange circle, Figure 3.1) is also critical, as nicotinic acid adenine dinucleotide has no Ca^{2+}-release activity [5, 24]. Virtual screening, based on 3D-shape and electrostatic similarity, identified Ned-19 as an analog of NAADP.

It turns out to be a specific and cell permeant antagonist for the NAADP-induced Ca^{2+} release and is widely used [25]. Similar to cADPR, a fluorometric enzyme-coupled assay for NAADP with sensitivity in nanomolar range has also been developed for measuring the endogenous NAADP levels [26, 27].

A wide range of cells are responsive to cADPR and/or NAADP, spanning three biological kingdoms. The physiological functions the two messengers regulate are equally wide spread [28–33]. Table 3.1 lists the biological functions shown to be regulated by cADPR and NAADP in the recent two years. Cells and systems include not only mammalian, such as neurons, but also zebrafish, brown macroalgae, and plants as well, documenting the two messengers are universal and species independent. Understanding the endogenous mechanisms for producing these two universal messengers is clearly of fundamental importance and is the emphasis of this chapter.

TABLE 3.1 Literature summary of the biological functions regulated by cADPR and NAADP.

Biological functions regulated by cADPR and NAADP	Reference
Transfer of mitochondria from human astrocytes to glioma cells	[34]
Epithelial secretory cells channeling of food allergens to lamina propria mucosal mast cells	[35]
Calcium signaling in parotid acini and control of secretion of primary saliva	[36]
Ischemia reperfusion injury in heart	[37]
Glutamate-induced calcium oscillations and nitric oxide release	[38]
Rat experimental autoimmune encephalomyelitis	[39]
Synaptically activated mGluR5 signaling	[40]
The mTORC1 signaling pathway in Paneth cells	[41]
Anticonvulsant mechanisms of the drug, phenytoin	[42]
Glucagon-like peptide-1-mediated signaling and insulin secretion from pancreatic beta-cells	[43]
Ebola virus entry	[44]
Ultraviolet B induces IL-1beta secretion	[45]
Beta-adrenergic-stimulated calcium uptake in sarcoplasmic reticulum	[46]
Reactive oxygen species generation in lymphokine-activated killer cells	[47]
The mTOR-mediated calcium signaling in pulmonary arterial myocytes	[48]
Fungal metabolite, HYTLO1, induced calcium signaling in plant	[49]
Calcium signaling during zebrafish myogenesis	[50]
The establishment of synchronized calcium activity in the primary motor neurons during zebrafish development	[51]
Calcium signaling mediating Middle East respiratory syndrome-coronavirus pseudovirus translocation through the endo-lysosomal system	[52]
Copper-induced calcium signaling in brown macroalgae	[53]
Hippocampal mGluR1-dependent long-term potentiation	[54]
Renal ischemia-reperfusion injury	[55]
Calcium microdomains upon T-cell activation	[56]

3.3. ENZYMATIC SYNTHESIS

3.3.1. Structures of ADP-Ribosyl Cyclase and CD38

The first enzyme that had ever been shown to be able to cyclize NAD into cADPR was a small soluble protein isolated from the *Aplysia* ovotestis and was named ADP-ribosyl cyclase (cyclase) [57]. Enzymatically, it is entirely different from the classical NADases, such as that from *Neurospora*, which also use NAD as a substrate but produce ADPR instead of cADPR [58, 59]. As cADPR was a newly discovered molecule, known assays for NADase could not easily distinguish ADPR from cADPR. A simple and diagnostic fluorometric assay using NGD as a substrate had been devised that can easily differentiate NADase from the cyclase and related enzymes. The latter cyclize NGD to produce fluorescent cyclic GDP-ribose, while NADases hydrolyze it to nonfluorescent GDP-ribose [58, 60].

The cyclase is a homodimer both in solution and when crystalized [61, 62]. The active site is in the middle of the molecule (Figure 3.2). NAD, the substrate,

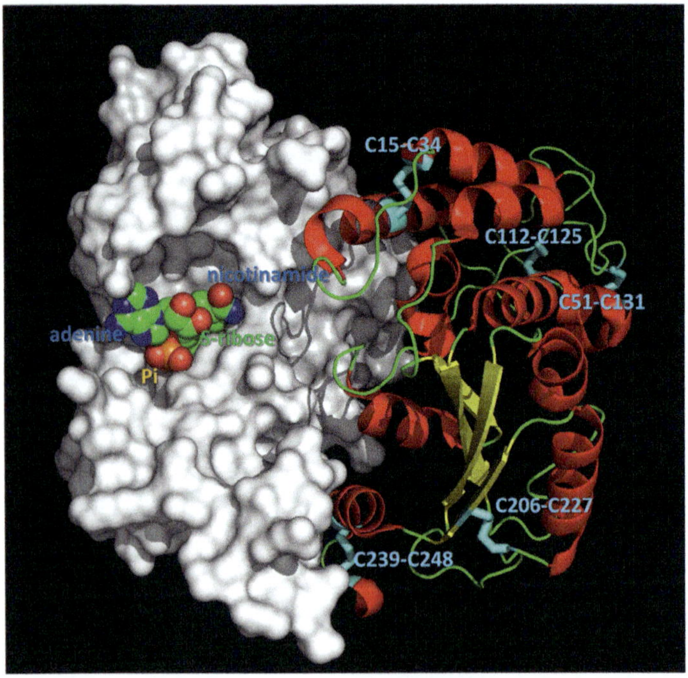

Figure 3.2 The crystal structure of the ADP-ribosyl cyclase. The cyclase is a homodimer. The secondary structure is shown for one monomer with helices shown in red, coils in green, and sheet in yellow, disulfides in cyan. The surface view is rendered for the other monomer with an NAD molecule bound at the active site in the middle of the protein. Color coding for NAD: carbon-green, oxygen-red, nitrogen-blue, phosphorus-orange.

enters into the active pocket with the nicotinamide end first, while the adenine end remains mostly out of the pocket. The recognition of the adenine group is not strict, enabling the cyclase to use many analogs of NAD, such as NGD, as substrates. The hydroxyl groups of the adenylyl ribose (S-ribose, Figure 3.1) are also pointing outward from the pocket, consistent with the cyclase can readily use NADP as substrate as well. There are five disulfides and four of them are involved in linking various helices (Figure 3.2), bestowing stability to the molecule. The catalytic residue is identified by mutagenesis and crystallography as Glu179 and its mutation to glycine renders the enzyme inactive [63]. The residue is at the bottom of the active site pocket and is close to the C1 of the terminal ribose of NAD where nicotinamide is attached.

The discovery of the cyclase provided a direct route to identify cADPR synthesizing enzymes in mammalian cells. Sequence comparison showed that CD38 has about 30% sequence identity with the cyclase [64]. CD38 is a lymphocyte surface membrane antigen and its expression changes according to the differentiation stage of the cells, but it had no known function. Different from the cyclase, CD38 is a single-pass type II transmembrane protein with a short amino(N)-tail facing the cytosol and a large carboxyl(C)-domain facing outside [65]. The C-domain, recombinantly produced and purified, indeed possesses the cADPR synthesizing activity [66] and can also cyclize NGD to produce fluorescent cGDPR [58], indicating it is enzymatically related to the cyclase. The results clearly establish that CD38 is an enzyme and not just a surface protein without function. The identification of CD38 as a cADPR synthesis enzyme had then ushered in a new frontier for investigating the signaling function of cADPR.

The close up view of the crystal structure of the catalytic C-domain is shown in Figure 3.3 (right panel) [67, 68]. The C-domain is virtually superimposable with that of the cyclase, despite only having about 30% sequence identity. Like the cyclase, the active site is also at the middle of the molecule, shown with a substrate bound (stick structures). The catalytic residue is Glu226 (magenta space-filled). CD38 has six disulfides and five of them are conserved with those of the cyclase (cyan, Figure 3.3, right panel).

3.3.2. Catalysis

How an enzyme can cyclize a long linear molecule to a cyclic product in a head to tail fashion had never been described. It is thus of considerable interest that the process can actually be captured using crystallography [68]. When CD38 was incubated with an NAD analog, arabinosyl-2′-fluoro-deoxy-nicotinamide adenine dinucleotide, the C1 of the N-ribose (Figure 3.1) formed a covalent bond with the catalytic residue, Glu226 (magenta space-filled, Figure 3.3, right panel). This anchored the substrate to the active site but allowed free movement of its adenine end. With an appropriate condition, crystals of the inhibited CD38

(a) (b)

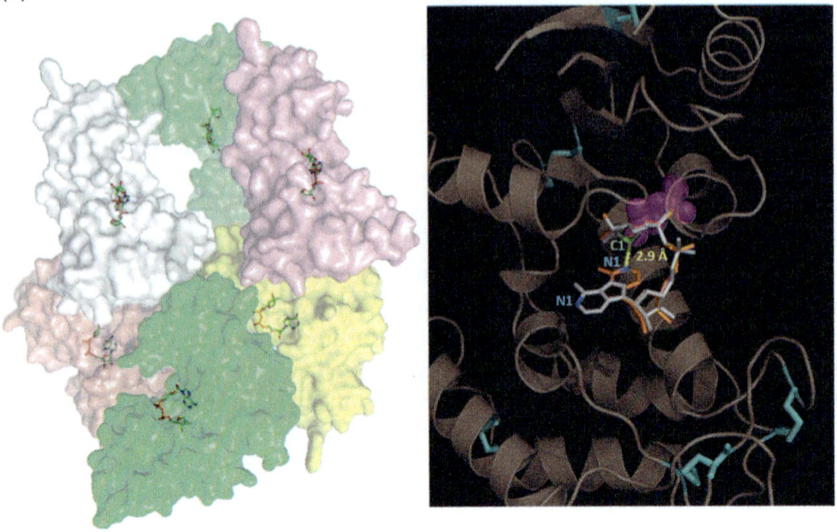

Figure 3.3 Crystallographic visualization of the cyclization of a substrate by CD38. (a) CD38 was crystalized with six molecules in one asymmetric unit. The substrate used was arabinosyl-2′-fluoro-deoxy-nicotinamide adenine dinucleotide, which one end formed a covalent linkage with the catalytic Glu226 and anchored it to the active site. The other end was free to cyclize and showed different conformation in each of the CD38 molecules, representing sub-states of the cyclization process. (b) A close-up view of the active site of CD38. Two of the substates are shown with the gray one being the most extended state. The orange one is the most folded state with C1 being 2.9Å from N1, close enough to form the cyclizing covalent bond. Disulfides are colored cyan and the Glu226 magenta.

could be obtained that have as many as six molecules in a crystallographic asymmetric unit. As shown in Figure 3.3 (left panel), the linked substrate in each of the six molecules has a different conformation, ranging from a fully extended form (gray stick, Figure 3.3, right panel) with the N1 of the adenine far apart from the C1 of the northern ribose to the completely folded form (orange stick, Figure 3.3, right panel) with the N1 and C1 as close as 2.9Å, close enough to form bonding. All six conformations of the linked substrate can be mapped by single bond rotation and a movie can be made showing the entire folding process of the cyclization of the substrate [68]. The structural results indicate a folding-back process for the adenine ring of the substrate and provide the first multiple snapshots of the process. Similar results were obtained with the cyclase. In this case, as many as twelve substrate conformations were seen in a crystallographic asymmetric unit, providing an even more detailed view of the cyclization process [69].

The discovery of the cyclase represents one of the important milestones that had greatly advanced the investigation of the signaling function of cADPR. This is because of its high cyclization efficiency, making it simple to prepare large amounts of cADPR for a variety of investigations [57]. Furthermore, the full reversibility of its cyclization reaction makes it the most critical component and essential feature of the fluorometric cycling assay mentioned above for the measurement of cADPR as well [23].

The catalytic mechanisms of the cyclase and CD38 are novel, as both are multi-functional and can use different substrates in addition to NAD to produce entirely different products. NAADP is a linear molecule like NADP, entirely different from the cyclic cADPR. It also targets a different type Ca^{2+} stores and release mechanism. Unexpectedly, NAADP is synthesized by both the cyclase and CD38 through a base-exchange reaction exchanging the nicotinamide base of NADP with nicotinic acid [70]. It should be noted that the cyclase and CD38 are entirely different from classical NADases, which do not make cADPR nor NAADP at all [71]. The catalytic similarity between the cyclase and CD38 indicates they can be grouped as members of a new family, the cyclase family.

The two members, however, do have some catalytic differences. Most notable is that CD38 possesses every high hydrolytic activity for NAD. It can also use cADPR itself as a substrate and hydrolyze it to ADPR [66]. In fact, it had been the only known enzyme that can hydrolyze cADPR, a molecule with rather unique structure. This remarkable specificity of CD38 is particularly useful as a diagnostic tool in many assays for cADPR. Only until recently, ADP-ribose/CDP-alcohol diphosphatases has been shown to be able to cleave the diphosphate bond of cADPR. The activity is inefficient but can be enhanced by mutating specific residues, providing another use tool for cADPR investigations [72].

Another important difference between the cyclase and CD38 is the amounts of cADPR produced from NAD. The cyclase produces predominantly cADPR, while CD38 produces mostly ADPR instead [66]. It is extraordinary that these dramatic catalytic differences are, in fact, determined by a single residue in the active-site pockets, Glu146 in CD38 and Phe174 in the cyclase. Making the mutations E146A and F174T inter-converted the catalytic properties of the two enzymes [73–75].

A unified mechanism that can account for all the novel characteristics and the crystal structures of the cyclase and CD38 has been described [76]. The nicotinamide end of the substrate, NAD or NADP, first enters the active-site pocket at the middle of the enzymes. Interactions with the catalytic residue, Glu226 in CD38 and Glu179 in the cyclase [63, 77], lead to the hydrolysis of the glycosidic bond and the release of the nicotinamide group, forming an intermediate. Both covalent and non-covalent intermediates have been observed using crystallography [78]. Subsequent attack by nicotinic acid leads base exchange and the formation of NAADP. Likewise, the attack by water leads to hydrolysis and the formation of ADPR. The active sites of both enzymes can also accommodate cADPR and NAADP as substrate [18, 79] and similar interactions can lead to hydrolysis and the formation of ADPR or ADP-ribose phosphate, respectively [73].

3.4. ENDOGENOUS SYNTHESIS

The elucidation of the enzymatic mechanisms of cADPR and NAADP synthesis is impressive, especially in view of the fact that they were unknown molecules that had never been described before. In contrast, how these nucleotide messengers are produced in cells has not been well understood. Ever since the catalytic activity of CD38 was identified [66], it has been the most studied enzyme in mammalian cells for the endogenous synthesis of the nucleotide messengers. Indeed, ablation of the CD38 gene results in large reduction of cADPR contents in many tissues [80] and produces multiple physiological defects in mice, documenting its importance [31, 81–83]. The difficulty lies, to a large extent, on the membrane topology of CD38 [84].

Plasma membrane forms a permeability barrier to separate the cell from the external environment and allows the maintenance of the reductive cytosol conducive for life. Bilayers also separate the cytosol from the lumen of organelles, which conditions can be regulated to be very different, but compatible with specific organellar functions. The acidic lumen of the endo-lysosomes required for degradation is an example.

The bilayer also serves to define the topology of all transmembrane proteins. Its hydrophobic interior effectively prevents the hydrophilic portions of the proteins from flipping from one side of the bilayer to the other. Each single-pass transmembrane protein thus has two opposite orientations, either facing the cytosol or the exterior of the cell that is also topologically contiguous with the organellar lumen.

3.4.1. Type III CD38: Biogenesis of cADPR

Ever since CD38 was identified as lymphocyte antigen by monoclonal antibody typing [85], it has been generally believed to be a single-pass transmembrane protein with its C-domain facing outside, a type II membrane orientation [65]. As described above, the C-domain of CD38 is also the catalytic domain (Figure 3.3). It has long been a puzzle how cytosolic NAD can have access to the catalytic domain and how cADPR, the product, can interact with the ryanodine receptor in the ER inside the cell. This topological enigma has now begun to be resolved by the recent discovery of type III CD38 with its catalytic domain facing the cytosol [84, 86].

The possibility that type III CD38 exists naturally was first proposed based on sequence analysis [32, 87]. The orientation of single-pass transmembrane proteins is thought to be determined by the positive charges at the two sides immediately adjacent to the transmembrane segment, with the more positively charged side being more likely to be facing the cytosol [88, 89]. For CD38, there are three positive residues on the amino side and two in the carboxyl side, consistent with it expressing mostly as a type II protein on cell surface. The small difference of only one positive charge suggests, however, a portion may express in an opposite orientation as a type III protein with the catalytic C-domain facing the cytosol.

Recent results indeed have documented the natural existence of both types of CD38 in cells [86, 90]. Two independent methods were used to detect endogenous type III CD38. The first one was to use specific antibodies to detect the amino-tail of the surface type III CD38 exposed to the outside [47, 86]. The second was a novel approach called duel epitopes protein identification (DepID), which employs nanobodies, single-domain antibodies [91], expressed inside cells to target two different epitopes on the C-domain of the type III CD38 that expose to the cytosol [90, 92]. Both assays show that type III CD38 is endogenously expressed in the plasma membrane and in the ER in a number of mouse and human cells [47, 86, 90].

Most importantly, type III CD38 is enzymatically active and is efficient in elevating cellular cADPR. This was first shown by expressing just the catalytic C-domain as a soluble protein in the cytosol, which indeed resulted in increase in cellular cADPR [93]. Similar increases were also seen when a constructed type III CD38, with tail charges described above mutated, was expressed [86, 94]. The endogenous type III CD38 is regulated by a cytosolic Ca^{2+}-binding regulator protein, CIB1, which binds to its C-domain [90]. Using CRISPR/Cas9 to knockout CIB1 caused decrease in cADPR levels, consistent with the important role of type III CD38 in maintaining cellular cADPR levels. That the C-domain of type III CD38 can interact specifically with cytosolic CIB1 also confirms the type III membrane topology of the enzyme.

In striking contrast, large changes, either increase or decrease, in type II CD38, has no effect on the cellular cADPR levels [95], providing further support that type III CD38 is responsible for biogenesis of cellular cADPR. The role of type II CD38 in the biogenesis of the nucleotide messengers will be discussed further below.

It is quite remarkable that type III CD38 is enzymatically active with its catalytic domain being in the cytosol, which indicates that the domain is folded correctly and with its disulfides fully formed even in the reductive environment. This was, in fact, confirmed by using disulfide-specific Western blot and mutation analyses [93]. Protein folding and disulfide formation are generally aided by cytosolic chaperones. Consistently, a complex of chaperones have been shown to associate with the catalytic domain of type III CD38 [96].

A disulfide in the mouse type III CD38 is actually responsible for regulating its activity [47]. In lymphokine-activated killer cells, interleukin-8 activates cADPR production and Ca^{2+} changes [47, 97]. The process was shown to be mediated by stimulation of NADPH oxidase 4, resulting in oxidation of Cys164 and Cys177 in the type III CD38 to form a disulfide, resulting in activating its cADPR production activity.

3.4.2. Type II CD38: Biogenesis of NAADP

That cellular cADPR production is mediated mainly by the type III CD38 represents a paradigm shift and raises the question of the functional role of the type II CD38 prominently expressed on cell surface of many cells. Being exposed to the oxidative outside environment, its disulfides are fully formed and it is enzymatically active. Evidence, however, indicates that the type II CD38 contribute minimally to cellular cADPR production [95]. Large changes in type II CD38 levels can be effected by induced endocytosis using a nanobody construct (CD71N-Nb551, Figure 3.4), but essentially no change in cellular cADPR was seen. Retention of the type II CD38 inside the ER lumen using another nanobody construct (erNb, Figure 3.4), likewise did not change the cellular cADPR.

The results suggest that the type II CD38 is tightly compartmentalize within the endo/exo-cytotic pathways (Figure 3.4). The endo-lysosomal lumen is highly

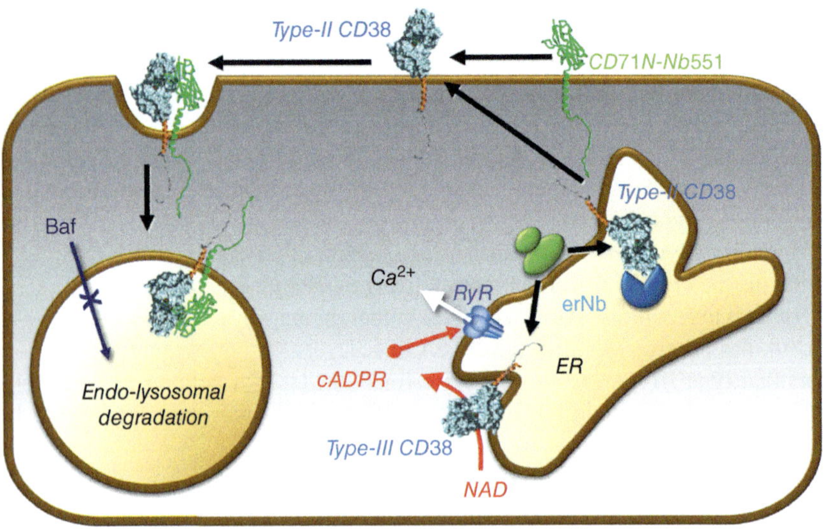

Figure 3.4 Compartmentation of the two topological types of CD38. Type III CD38 with its catalytic domain facing the cytosol and can use NAD to produce cADPR that targets the ryanodine receptor (RyR), mobilizing the endoplasmic Ca^{2+} stores (ER). The type II CD38 is tightly compartmentalized on the surface and within the endo-/exo-cytotic pathway and contribute minimally to the production of cellular cADPR. A construct based on nanobody (Nb) against CD38 effectively induces endocytosis of type II CD38 (CD71N-Nb551), while another (erNb) retains it in the ER and prevents its surface transport. Bafilomycin (Baf) alters the acidic pH of the endo-lysosomes and blocks them degradation of the internalized CD38.

acidic, well suited for the base-exchange reaction to produce NAADP by CD38, which has an optimum at pH 4 [70, 73]. Ever since the first documentation of this peculiar acidic requirement of the NAADP synthesis [70], the endo-lysosomes have been suggested to be a special cellular compartment for the biogenesis of NAADP [98]. It is a novel notion that the two forms of CD38 are topologically segregated into separate compartments so as to produce two structurally and functionally distinct Ca^{2+} messengers.

Consistent with this notion is the finding that induced endocytosis of surface type II CD38 could elevate cellular NAADP levels, especially so in the presence of nicotinic acid (NA), the co-substrate of the base-exchange reaction. Also, a lysosomal CD38 constructed and directed to express in the lysosomes efficiently elevated the cellular NAADP levels [99]. Further support comes from results obtained in mouse lymphokine-activated killer cells. As described above, interleukin-8 activated type III CD38 and cADPR production. Concomitantly, surface type II CD38 was internalized and the formation of NAADP in lysosome-related acidic organelles was observed [100].

The great enhancement of NAADP production by NA, however, indicates that the access of substrate to the extra-cytosolic compartment is still limited. This tight compartmentation suggests the possibility of a novel mechanism for signaling. Type II CD38 is constitutively active and the regulation is not on CD38 itself but on the access of substrates through activation of specific membrane transporters. For example, connexin 43 was shown to be permeable to NADP and the permeability regulated by phosphorylation, allowing controlled entry of NADP into the endo-lysosomes [100, 101],

3.5. SARM1 – A NEW ENZYME FOR BIOGENESIS OF cADPR AND NAADP

For more than two decades, CD38 had been the only known enzyme in mammalian cells capable of producing cADPR. Intriguingly, the cADPR content in brain remains substantial in CD38-knockout mice [80], suggesting the existence of a yet unidentified ADP-ribosyl cyclase in the brain. This section describes the identification of sterile alpha and toll/interleukin-1 receptor motif–containing 1 (SARM1) as such a new enzyme [102, 103].

3.5.1. Discovery of the Enzymatic Activity of SARM1

SARM1 is a protein prominently expressed in neuronal tissues. Genetic screening in *Drosophila* demonstrated SARM1 is required for axon degeneration (AxD) [104], also named Wallerian degeneration. It is an early feature of most neurodegenerative disorders and can also be induced by traumatic injury, chemotherapy, chronic diabetes, etc. (see review [105]). That NAD metabolism plays a prominent role in AxD was first suggested by analyses done in a special mice line,

Wallerian degeneration slow (Wld^s) [106]. Wld^s mice express a fusion protein containing nicotinamide mononucleotide adenylyltransferase 1 (NMNAT1) [107], a key enzyme involved in the synthesis of NAD. Subsequent works document SARM1 is an NADase [102, 103, 108] and its activation can lead to NAD depletion and AxD [108].

SARM1 contains three domains (Figure 3.5a). The armadillo repeats (ARMs) domain is composed of tandem imperfect 45 amino acid repeats, forming a superhelix that is usually mediating interactions with proteins [109], such as β-catenin [110]. The sterile alpha motif (SAM) is a small helical protein module containing around 70 amino acids. It is considered to be facilitating the formation of homo- and hetero-oligomers [111]. Indeed, Figure 3.5b shows that the purified SAM domains crystalize as a ring-shaped homo-octamer [112]. The Toll/IL-1 Receptor (TIR) domain contains around

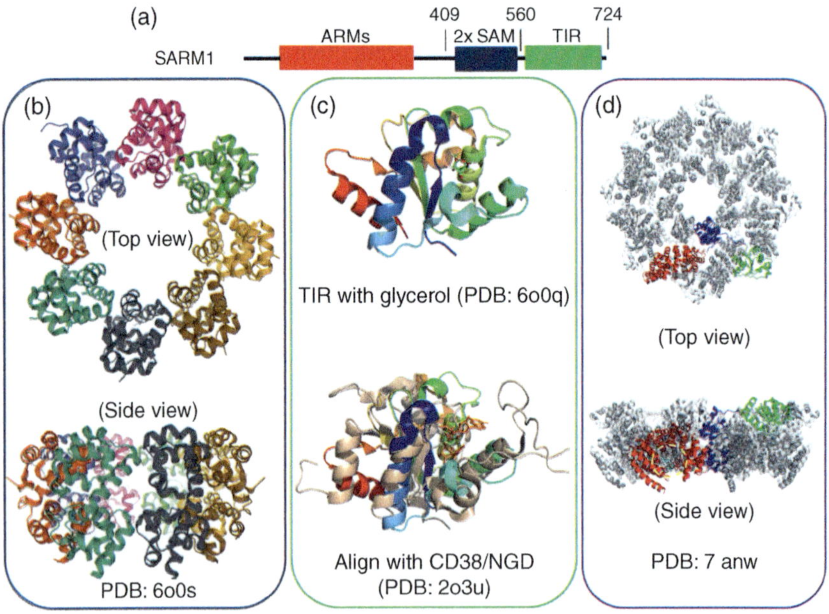

Figure 3.5 The structure of SARM1. (a) Illustration of the domains of human SARM1. (b) The cartoon presentation of the SAM domain (PDB: 6o0s). Upper panel: top view; lower panel: side view. (c) The cartoon presentation of the TIR domain (PDB: 6o0q). Upper panel: TIR complexed with glycerol; lower panel: Superposition (RMSD of 4.9 Å over 112 residues) of hSARM1TIR (rainbow) bound to glycerol (green) and the C-terminal region of the human CD38 (P118-C296, wheat, PDB: 203u) bound to NGD (orange). (d) The structure of full-length human SARM1 in a self-inhibited state complexed with NAD (PDB: 7anw). Upper panel shows in top view, while lower panel in side view. One protomer is colored in red for ARM, blue for SAM, and green for TIR. NAD molecule is in the yellow circle. The other protomers are colored in gray.

200 amino acids and its crystal structure [113] is shown in Figure 3.5c. The name derived from its presence in the Toll-like receptor and interleukin-1 receptor superfamily, which play important roles in innate immunity [114]. The TIR domain is catalytic, possessing not only the NADase and cADPR synthesis activities [102, 103] but also the production of NAADP, much like CD38 [103]. The TIR-like motifs are widely present in many proteins and have also been shown to be catalytic in several bacterial [115] and plant proteins [113, 116].

3.5.2. The Structure of the TIR Domain

SARM1 is a multi-domain and much larger protein than CD38. Even just consider the similar size TIR domain, it also has no sequence similarity with CD38. It is thus surprising that both catalyze essentially the same set of enzymatic reactions. When their secondary structures were analyzed, however, both were found to possess the features of the Rossmannoid fold [117] that was first noted in lactate dehydrogenase. The motif is composed of a β-α-β super-secondary structure and is always found in proteins that bind nucleotides, such as enzyme cofactors FAD, NAD, and NADP. This structural similarity suggests that the TIR of SARM1 and CD38 can be grouped into the (deoxy)ribohydrolase superfamily, which consists of enzymes that hydrolyze the bond between the base and the ribose in the nucleotide/nucleoside substrates [118]. Indeed, when the Rossmannoid regions of TIR and CD38 are compared, they superimpose reasonably well as shown in Figure 3.5c. This structural similarity can account for both being NAD(P)-utilizing enzymes.

More structural similarities can be seen in the ligand binding sites. In CD38, the C-2 and C-3 hydroxyl groups of the nicotinamide ribose in NAD interact with a catalytic glutamate [119]. This interaction feature is observed in other NADase as well [113]. Similar diol–glutamate interaction is seen in the crystal structures of the SARM1-TIR domains complexed with a ligand, glycerol (Figure 3.5c, PDB: 6o0r), or ribose (PDB: 6o0q). Another substrate interaction similarly seen in the crystal structure of TIR domain from plant RUN1 (PDB: 6o0w) [113] and CD38 (PDB: 2o3u) [79] is a face-to-face aromatic stacking of the adenine ring of the bound NAD(P) with a tryptophan in the active-site pocket. These similarities in substrate interactions underscore the catalytic similarity of the two enzymes, despite the lack of sequence homology.

3.5.3. Activation and Regulation of the Enzymatic Activity of SARM1

It is very unusual that two entirely different proteins can not only catalyze the same set of multi-reactions, but also exhibit the same pH dependence. Like CD38, at neutral pH, SARM1 hydrolyzes NAD to ADPR and cyclizes NAD to cADPR (Figure 3.6). At acidic pH, it catalyzes the exchange of NA with the nicotinamide group of NADP produce NAADP, just like CD38.

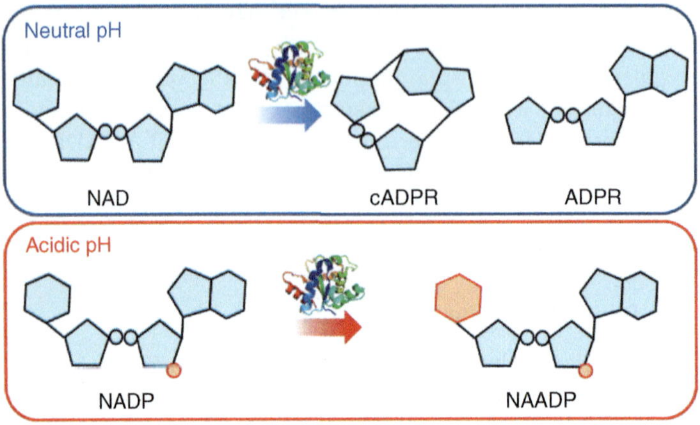

Figure 3.6 The reactions catalyzed by SARM1. All the substrates and products are shown in their simplified structures. Upper panel: the reactions of hydrolysis and cyclization at neutral pH with NAD as the substrate. The leaving nicotinamide is omitted. Lower panel: the base-exchange reaction at acidic pH. The exchanging compounds, nicotinamide and nicotinic acid, are omitted.

This peculiar acidic dependence of the base-exchange reaction has been elucidated in CD38 and is attributed to the electrostatic repulsion between the negatively charged substrate, NA, by the residues Glu146 and Asp155 at the active site pocket, which is neutralized at acidic pH [73]. Similar acidic residues may well be present at the active site of SARM1, determining its acidic dependence.

A remarkable feature of SARM1 that is distinct from CD38 is that it is an autoregulated enzyme. This was suggested first by the axon degeneration assay [120] and directly demonstrated by biochemical analyses [103]. Full-length SARM1 had very low enzymatic activity but became fully activated when the ARM domain was deleted, indicating it acts as an intra-molecular regulator of the catalytic activity of the TIR domain.

The endogenous activation mechanism of SARM1 had been investigated mainly through analyses of neuronal injury, during which the amount of NMN quickly accumulates because of the depletion of the converting enzyme, NMNAT2 [121], as summarized in Figure 3.7. It was suspected that the elevated NMN somehow activated the SARM1 and triggered axon degeneration [122]. But the direct evidence was absent until the discovery of CZ-48, a cell permeable analogue of NMN.

The story of CZ-48 started from CD38. In 2012, a series of mimetics of NMN was designed and synthesized. They formed covalent bonds with Glu226 of CD38 and inhibited its enzymatic activities [123]. Unexpectedly, when testing the cellular effects, it was observed that one of these inhibitors, sulfo-ara-F-NMN (CZ-48), could effectively elevate cellular cADPR contents in HEK293T cells

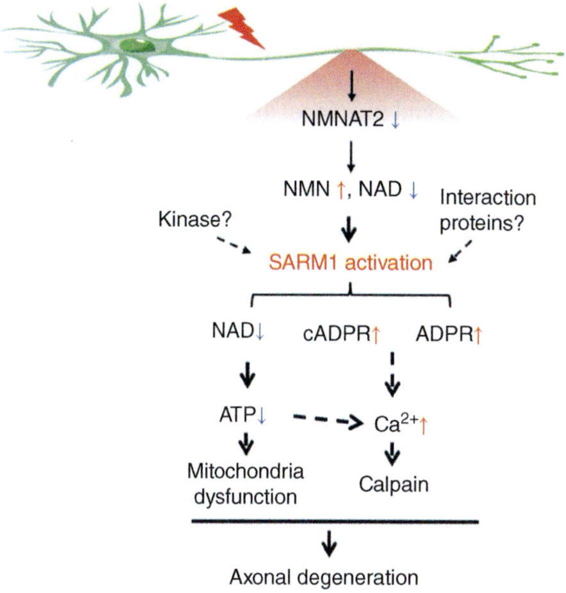

Figure 3.7 Working model of axonal degeneration. After injury or chemotherapy agent treatment, the transportation of vesicles or proteins along axons is impaired. The level of NMNAT2, a major enzyme converting NMN into NAD in axon, decreases quickly due to its short life-time, which results in the accumulation of NMN and decline of NAD, and subsequent SARM1 activation. The activated SARM1 further consumes NAD, which leads to ATP depletion and mitochondrial dysfunction, and produces cADPR and ADPR, which causes elevation of cellular Ca^{2+} levels and activation of calpain, etc. All the events finally account for the degeneration of axons. SARM1 is the central molecule because its knockout significantly delays the process. The mechanism of how kinases and interaction proteins regulate its function is to be elucidated.

not expressing CD38 [103]. The result was astonishing and strongly indicated that there is an unknown ADP-ribosyl cyclase existing in the mammalian cells. The mystery came to light when the NAD-utilizing activity of TIR domain in SARM1 was demonstrated [102]. Knocking out SARM1 eliminated the CZ-48-induced cADPR elevation, indicating the activating target of CZ-48 is SARM1 [103]. This was then directly established using isolated recombinant SARM1, showing that CZ-48 was as effective as NMN in activating its enzymatic activities. CZ-48 is thus not only an authentic mimic of NMN but also possesses the novel advantage of being cell permeant and unmetabolizable, which NMN is not, and it can activate SARM1 in live cells [103]. Both NMN and CZ-48 activate all the enzymatic activities in similar extent, suggesting that the regulators act to unmask the single catalytic site in the TIR domain auto-inhibited by the ARM domain. The discovery of CZ-48 not only proves directly that NMN is an endogenous regulator of SARM1 but also provides a valuable tool for manipulating the enzymatic activity and investigating the functions of SARM1 in live cells. Evidence suggests that its unmasking of the auto-inhibition may involve the oligomerization of the catalytic TIR domain as demonstrated in both the enzymatic [103, 113] and functional axon degeneration studies [108].

So far, NMN is the best-understood regulator of SARM1. Its cellular levels, however, are tightly maintained low by abundantly expressed NMNATs, which readily produce NAD from excess NMN. How and whether stimuli can raise NMN sufficiently to the activate SARM1 remained to be determined. Other regulation possibilities, such as kinases [124] and activating proteins [125], have been proposed, but how they modulate the enzymatic activities of SARM1 have not been clarified. Recently, the inhibitory structures of SARM1 have been solved with the CryoEM technique ([126–128], and our recently published structure [129] data). The structures revealed that NAD, one substrate of SARM1, also acts as the allosteric inhibitor, binding at the ARM domain and maintaining the inhibitory conformation (Figure 3.5d), which indicates that decrease of the cellular NAD concentration might be another trigger for SARM1 activation during AxD.

SARM1 is a ubiquitous protein widely expressed in many cell types, not just neuronal cells [103]. It is intriguing to consider the functional role of SARM1 in non-neuronal cells. Most studies on SARM1 focus on AxD because the enzyme is highly expressed in neurons. Its activation can lead to large decreases in cellular NAD levels and effect cell death and AxD [102, 103, 108]. The levels of SARM1 in non-neuronal cells are low, not sufficient to significantly alter cellular NAD when activated. It can, however, readily produce effective amounts of cADPR and NAADP for mobilizing Ca^{2+} stores [103], suggesting SARM1 may well be a Ca^{2+} signaling enzyme like CD38.

3.6. ADPR: A CA^{2+}-MOBILIZING SECOND MESSENGER?

As discussed in the previous sections, CD38 and SARM1 are signaling enzymes producing two general messengers that are documented to regulate a wide range of physiological functions. Both enzymes also produce ADPR from NAD, which is in fact the dominant product, much more than cADPR [66, 103]. The possibility that ADPR is also a messenger is worth considering.

3.6.1. The Metabolism of ADPR

ADPR is a common metabolite produced not just by CD38 and SARM1 (Figure 3.5a). Another major source of ADPR is the ADP-ribosylated proteins (Figure 3.8b). The mono-ADP-ribose (MARs) and poly-ADP-ribose (PARs) of these modified proteins are removed and converted to free ADPR by (ADP-ribosyl) hydrolases (ARHs), including poly(ADP-ribose) glycohydrolases (PARGs) and macro domains [130].

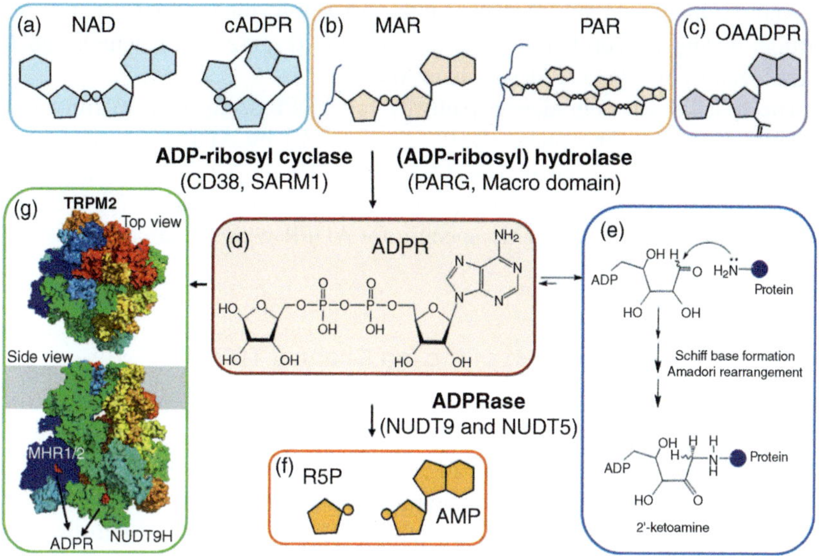

Figure 3.8 The metabolism and functions of ADPR in mammalian cells. (a)–(c) The simplified structures for the cellular source of free ADPR. NAD and cADPR are hydrolyzed by ARCs such as CD38 and SARM1; proteins modified by mono-ADPR (MAR) or poly(ADP-ribose) (PAR), free O-Acetyl-ADP-Ribose (OAADPR) are hydrolyzed by (ADP-ribosyl) hydrolases (ARHs) such as PARG and macro domain proteins. (d) The chemical structure of ADPR; (e) The reaction mechanism of non-enzymatic ADP-ribosylation of proteins. (f) The simplified structures for the ADPR degradation products, AMP and ribose 5'-phosphate, catalyzed by ADP-ribose pyrophosphatases (ADPRase) such as NUDT9 and NUDT5. (g) The top and side view of the structure of tetrameric TRPM2 (PDB:6PUS). The structures are in surface view and two molecules of ADPR are presented as spheres located in NUDT9H and MHR1/2 domains.

ADP-ribosylated proteins are produced during many of the critical and highly regulated processes, such as mitosis, gene transcription, proliferation, differentiation, and cell death. The ADP-ribosylation of proteins is catalyzed by ADP-ribosyl transferases (ARTs) with NAD as the substrate (reviewed in [131]). Human cells express 21 ARTs, including 4 ARTCs, which express at the cell surface or secreted into the extracellular space, and 17 ARTDs, among which ARTD1-6 are polymerases (PARPs) and ARTD7-17 are mono-ART [132].

ADPR itself is also a reactive compound and can chemically ADP-ribosylate proteins. Its terminal ribose can serve as a nucleophile to react with the $-NH_2$ and -SH in the lysine and cysteine residues, forming stable ketoamine through Schiff base formation and Amadori rearrangement (Figure 3.8e) [133, 134]. Proteins such as actin [135], histone [136, 137], eukaryotic elongation factor 2 [138], and defensin HNP-1 [139], are susceptible to such modification.

Another main source of ADPR is from OAADPR, the product generated by sirtuins during deacetylation of target proteins (Figure 3.8c), which can be hydrolyzed by ARHs [140] to produce ADPR.

The best-known degradation pathway for ADPR is through hydrolysis to AMP and ribose-5-phosphate (R5P) (Figure 3.8f) by ADP-ribose pyrophosphatases (ADPRases) such as NUDT5 and NUDT9, the Nudix hydrolase family members. NUDT9 mainly locates in mitochondria [141], while NUDT5 is a nuclear protein. Both are highly specific for ADPR with K_m values in the high micromolar range.

3.6.2. ADPR as an Agonist of the Ion Channel TRPM2

The most investigated signaling function of ADPR is the activation of a nonselective Ca^{2+}-permeable cation channel, TRPM2, resulting in cellular Ca^{2+} changes. TRPM2 [142, 143] plays an essential role in diverse physiological processes, such as cell migration, chemokine production, and insulin secretion, as well as in pathological conditions, such as cell death and neuronal disorders (reviewed in [144]).

Remarkable progress has been made in resolving the 3D structures of TRPM2, which is shown in Figure 3.8g [145–148]. The full-length TRPM2 has three parts, the N-terminal TRPM homology region (MHR), followed by six transmembrane segments with the putative pore-forming region, and the C-terminal region containing a coiled-coil region (CCR) and a NUDT9-homology region (NUDT9H), which shows high homology with NUDT9 and contains a Nudix box motif characteristic of a family of pyrophosphatases that accept nucleoside diphosphate substrates like ADPR [149]. In the human TRPM2, the structure (Figure 3.8g) shows two ligand-binding sites, one in MHR and the other in NUDT9H domains. The former one acts as a orthostatic ADPR-binding site while that in the NUDT9H assists channel activation [145]. Together with Ca^{2+} binding in the transmembrane domain, ADPR induces significant rearrangement of the structure and activates TRPM2 [145].

The half maximal activating concentrations (EC_{50}s) of ADPR for TRPM2 reported in the literature are highly variable, which may reflect the contribution by various conditions, such as Ca^{2+} [142], temperature [150], oxidative stress [151], and the presence of various NAD metabolites, including cADPR [152, 153] and NAADP [154]. The most cited EC_{50} values are in the order of 10–40 μM range (reviewed in [144]).

For ADPR to serve as a messenger, it is expected that its cellular levels can undergo rapid and transient increases. With so many sources and enzymes involving in ADPR metabolism, it is unclear how that can be accomplished. Indeed, rapid and significant changes in cellular ADPR levels following stimulation has not been definitively documented. This may be related to the difficulty in avoiding artifactual changes during preparation of cellular ADPR, as either acid or alkaline extraction can convert cellular NADH or NAD, respectively, to ADPR.

Furthermore, as a messenger, it is expected its stimulus-induced rise should reach its EC_{50}, which would require the hydrolysis of a major portion of the cellular NAD content, a condition that should reflect dysfunction. In this scenario, ADPR, through its activation Ca^{2+}-influx via the TRPM2, could be functioning as a signal or mediator for cell death or AxD.

REFERENCES

1. Ringer, S. (1882). Concerning the influence exerted by each of the constituents of the blood on the contraction of the ventricle. *J Physiol (London) 3*, 380–393.
2. Tsien, R. Y. (1981). A non-disruptive technique for loading calcium buffers and indicators into cells. *Nature 290*, 527–528.
3. Streb, H., Irvine, R. F., Berridge, M. J., and Schulz, I. (1983). Release of Ca^{2+} from a nonmitochondrial intracellular store in pancreatic acinar cells by inositol-1,4,5-trisphosphate. *Nature 306*, 67–69.
4. Clapper, D. L., Walseth, T. F., Dargie, P. J., and Lee, H. C. (1987). Pyridine nucleotide metabolites stimulate calcium release from sea urchin egg microsomes desensitized to inositol trisphosphate. *J Biol Chem 262*, 9561–9568.
5. Lee, H. C., and Aarhus, R. (1995). A derivative of NADP mobilizes calcium stores insensitive to inositol trisphosphate and cyclic ADP-ribose. *J Biol Chem 270*, 2152–2157.
6. Lee, H. C., Aarhus, R., and Levitt, D. (1994b). The crystal structure of cyclic ADP-ribose. *Nat Struct Biol 1*, 143–144.
7. Lee, H. C., Walseth, T. F., Bratt, G. T., Hayes, R. N., and Clapper, D. L. (1989). Structural determination of a cyclic metabolite of NAD^+ with intracellular Ca^{2+}-mobilizing activity. *J Biol Chem 264*, 1608–1615.
8. Galione, A., Lee, H. C., and Busa, W. B. (1991). Ca^{2+}-induced Ca^{2+} release in sea urchin egg homogenates: modulation by cyclic ADP-ribose. *Science 253*, 1143–1146.
9. Lee, H. C., Aarhus, R., Graeff, R., Gurnack, M. E., and Walseth, T. F. (1994a). Cyclic ADP ribose activation of the ryanodine receptor is mediated by calmodulin. *Nature 370*, 307–309.
10. Ogunbayo, O. A., Zhu, Y., Rossi, D., Sorrentino, V., Ma, J., Zhu, M. X., and Evans, A. M. (2011). Cyclic adenosine diphosphate ribose activates ryanodine receptors, whereas NAADP activates two-pore domain channels. *J Biol Chem 286*, 9136–9140.
11. Brailoiu, E., Churamani, D., Cai, X., Schrlau, M. G., Brailoiu, G. C., Gao, X., Hooper, R., Boulware, M. J., Dun, N. J., Marchant, J. S., and Patel, S. (2009). Essential requirement for two-pore channel 1 in NAADP-mediated calcium signaling. *J Cell Biol 186*, 201–209.
12. Calcraft, P. J., Ruas, M., Pan, Z., Cheng, X., Arredouani, A., Hao, X., Tang, J., Rietdorf, K., Teboul, L., Chuang, K.-T., *et al.* (2009). NAADP mobilizes calcium from acidic organelles through two-pore channels. *Nature 459*, 596–601.

13. Morgan, A. J., Davis, L. C., Ruas, M., and Galione, A. (2015). TPC: the NAADP discovery channel? *Biochem Soc Trans 43*, 384–389.
14. Patel, S., and Brailoiu, E. (2018). Thinking through acidic Ca(2+) stores. *Sci Signal 11*:eaau3342.
15. Lee, H. C. (2002). Cyclic ADP-ribose and NAADP. A story of two calcium messenger. In *Cyclic ADP-ribose and NAADP. Structures, Metabolism and Functions.*, H.C. Lee, ed. Dordrecht: Kluwer Academic Publishers, pp. 1–21.
16. Clapper, D. L., and Lee, H. C. (1985). Inositol trisphosphate induces Ca^{2+}- release from non-mitochondrial stores in sea urchin egg homogenates. *J Biol Chem 260*, 13947–13954.
17. Gunaratne, G. S., Johns, M. E., Hintz, H. M., Walseth, T. F., and Marchant, J. S. (2018a). A screening campaign in sea urchin egg homogenate as a platform for discovering modulators of NAADP-dependent Ca(2+) signaling in human cells. *Cell Calcium 75*, 42–52.
18. Zhang, H., Graeff, R., Lee, H. C., and Hao, Q. (2013). Crystal structures of human CD38 in complex with NAADP and ADPRP. *MESSENGER 2*, 44–53.
19. Sato, T., Watanabe, M., Tsuzuki, T., Takano, S., Murayama, T., Sakurai, T., Kameda, T., Arisawa, M., Fukuda, H., and Shuto, S. (2016). Design, synthesis, and identification of 4″α-azidoethyl-cyclic ADP-carbocyclic-ribose as a highly potent analogue of cyclic ADP-ribose, a Ca^{2+}-mobilizing second messenger. *J Med Chem 59*, 7282–7286.
20. Walseth, T. F., and Lee, H. C. (1993). Synthesis and characterization of antagonists of cyclic-ADP-ribose-induced Ca^{2+} release. *Biochim Biophys Acta 1178*, 235–242.
21. Walseth, T. F., Aarhus, R., Zeleznikar, R. J., Jr., and Lee, H. C. (1991). Determination of endogenous levels of cyclic ADP-ribose in rat tissues. *Biochim Biophys Acta 1094*, 113–120.
22. Walseth, T. F., Wong, L., Graeff, R. M., and Lee, H. C. (1997). A bioassay for determining endogenous levels of cyclic ADP-ribose. *Methods Enzymol 280*, 287–294.
23. Graeff, R., and Lee, H. C. (2002a). A novel cycling assay for cellular cADP-ribose with nanomolar sensitivity. *Biochem J 361*, 379–384.
24. Lee, H. C., and Aarhus, R. (1997). Structural determinants of nicotinic acid adenine dinucleotide phosphate important for its calcium-mobilizing activity. *J Biol Chem 272*, 20378–20383.
25. Naylor, E., Arredouani, A., Vasudevan, S. R., Lewis, A. M., Parkesh, R., Mizote, A., Rosen, D., Thomas, J. M., Izumi, M., Ganesan, A., *et al.* (2009). Identification of a chemical probe for NAADP by virtual screening. *Nat Chem Biol 5*, 220–226.
26. Graeff, R., and Cheung Lee, H. (2013). *An Improved Enzymatic Cycling Assay for NAADP*, Vol 2), 96–105.
27. Graeff, R., and Lee, H. C. (2002b). A novel cycling assay for nicotinic acid-adenine dinucleotide phosphate with nanomolar sensitivity. *Biochem J 367*, 163–168.
28. De Flora, A., Zocchi, E., Guida, L., Franco, L., and Bruzzone, S. (2004). Autocrine and paracrine calcium signaling by the CD38/NAD^+/cyclic ADP-ribose system. *Ann N Y Acad Sci 1028*, 176–191.
29. Galione, A., Morgan, A. J., Arredouani, A., Davis, L. C., Rietdorf, K., Ruas, M., and Parrington, J. (2010). NAADP as an intracellular messenger regulating lysosomal calcium-release channels. *Biochem Soc Trans 38*, 1424–1431.

30. Guse, A. H., and Lee, H. C. (2008). NAADP: a universal Ca^{2+} trigger. *Sci Signal 1*, re10.
31. Lee, H. C. (2001). Physiological functions of cyclic ADP-ribose and NAADP as calcium messengers. *Annu Rev Pharmacol Toxicol 41*, 317–345.
32. Lee, H. C. (2012a). Cyclic ADP-ribose and nicotinic acid adenine dinucleotide phosphate (NAADP) as messengers for calcium mobilization. *J Biol Chem 287*, 31633–31640.
33. Lee, H. C. (2012b). The cyclic ADP-ribose/NAADP/CD38-signaling pathway: past and present. *Messenger 1*, 16–33.
34. Sun, C., Liu, X., Wang, B., Wang, Z., Liu, Y., Di, C., Si, J., Li, H., Wu, Q., Xu, D., *et al.* (2019). Endocytosis-mediated mitochondrial transplantation: transferring normal human astrocytic mitochondria into glioma cells rescues aerobic respiration and enhances radiosensitivity. *Theranostics 9*, 3595–3607.
35. Noah, T. K., Knoop, K. A., McDonald, K. G., Gustafsson, J. K., Waggoner, L., Vanoni, S., Batie, M., Arora, K., Naren, A., Wang, Y. H., *et al.* (2019). IL-13-induced Intestinal secretory epithelial cell antigen passages are required for IgE-mediated food-induced anaphylaxis. *J Allergy Clin Immunol 144*, 1058–1073.e3.
36. Imbery, J. F., Iqbal, A. K., Desai, T., and Giovannucci, D. R. (2019). Role of NAADP for calcium signaling in the salivary gland. *Cell Calcium 80*, 29–37.
37. Boslett, J., Reddy, N., Alzarie, Y. A., and Zweier, J. L. (2019). Inhibition of CD38 with the thiazoloquin(az)olin(on)e 78c protects the heart against post-ischemic injury. *J Pharmacol Exp Ther*, *369*, 55–64.
38. Zuccolo, E., Kheder, D. A., Lim, D., Perna, A., Nezza, F. D., Botta, L., Scarpellino, G., Negri, S., Martinotti, S., Soda, T., *et al.* (2018). Glutamate triggers intracellular Ca(2+) oscillations and nitric oxide release by inducing NAADP- and InsP3 -dependent Ca(2+) release in mouse brain endothelial cells. *J Cell Physiol 234*, 3538–3554.
39. Zhang, B., Watt, J. M., Cordiglieri, C., Dammermann, W., Mahon, M. F., Flugel, A., Guse, A. H., and Potter, B. V. L. (2018). Small molecule antagonists of NAADP-Induced Ca(2+) release in T-lymphocytes suggest potential therapeutic agents for autoimmune disease. *Sci Rep 8*, 16775.
40. Yu, W., Sohn, J. W., Kwon, J., Lee, S. H., Kim, S., and Ho, W. K. (2018). Enhancement of dendritic persistent Na(+) currents by mGluR5 leads to an advancement of spike timing with an increase in temporal precision. *Molecular brain 11*, 67.
41. Ye, J., Yuan, K., Dai, W., Sun, K., Li, G., Tan, M., Song, W., and Yuan, Y. (2018). The mTORC1 signaling modulated by intracellular C3 activation in Paneth cells promotes intestinal epithelial regeneration during acute injury. *Int Immunopharmacol 67*, 54–61.
42. Sadeghi, L., Yekta, R., and Dehghan, G. (2018). New mechanisms of phenytoin in calcium homeostasis: competitive inhibition of CD38 in hippocampal cells. *DARU J journal of Faculty of Pharmacy, Tehran University of Medical Sciences 26*, 191–198.
43. Rahman, F. U., Park, D. R., Joe, Y., Jang, K. Y., Chung, H. T., and Kim, U. H. (2018). Critical roles of carbon monoxide and nitric oxide in Ca^{2+} signaling for insulin secretion in pancreatic islets. *Antioxid Redox Signal 30*, 560–576.
44. Penny, C. J., Vassileva, K., Jha, A., Yuan, Y., Chee, X., Yates, E., Mazzon, M., Kilpatrick, B. S., Muallem, S., Marsh, M., *et al.* (2018). Mining of Ebola virus entry inhibitors identifies approved drugs as two-pore channel pore blockers. *Biochim Biophys Acta, Mol Cell Res 1866*, 1151–1161.

45. Park, K. H., Park, D. R., Kim, Y. W., Nam, T. S., Jang, K. Y., Chung, H. T., and Kim, U. H. (2018c). The essential role of Ca(2+) signals in ultraviolet B-induced interleukin-1beta secretion in keratinocytes. *J Invest Dermatol 139*, 1362–1372.
46. Park, D. R., Nam, T. S., Kim, Y. W., Lee, S. H., and Kim, U. H. (2018b). CD38-cADPR-SERCA signaling axis determines skeletal muscle contractile force in response to beta-adrenergic stimulation. *Cell Physiol Biochem Int J Exp Cell Physiol Biochem Pharmacol 46*, 2017–2030.
47. Park, D. R., Nam, T. S., Kim, Y. W., Bae, Y. S., and Kim, U. H. (2018a). Oxidative activation of type III CD38 by NADPH oxidase-derived hydrogen peroxide in Ca(2+) signaling. *FASEB J 33*, 3404–3419.
48. Ogunbayo, O. A., Duan, J., Xiong, J., Wang, Q., Feng, X., Ma, J., Zhu, M. X., and Evans, A. M. (2018). mTORC1 controls lysosomal Ca(2+) release through the two-pore channel TPC2. *Sci Signal 11*:eaao5775.
49. Moscatiello, R., Sello, S., Ruocco, M., Barbulova, A., Cortese, E., Nigris, S., Baldan, B., Chiurazzi, M., Mariani, P., Lorito, M., and Navazio, L. (2018). The hydrophobin HYTLO1 secreted by the biocontrol fungus *Trichoderma longibrachiatum* triggers a NAADP-mediated calcium signalling pathway in *Lotus japonicus*. *Int J Mol Sci 19*, 2596.
50. Kelu, J. J., Webb, S. E., Galione, A., and Miller, A. L. (2018a). Characterization of ADP-ribosyl cyclase 1-like (ARC1-like) activity and NAADP signaling during slow muscle cell development in zebrafish embryos. *Dev Biol 445*, 211–225.
51. Kelu, J. J., Webb, S. E., Galione, A., and Miller, A. L. (2018b). TPC2-mediated Ca(2+) signaling is required for the establishment of synchronized activity in developing zebrafish primary motor neurons. *Dev Biol 438*, 57–68.
52. Gunaratne, G. S., Yang, Y., Li, F., Walseth, T. F., and Marchant, J. S. (2018b). NAADP-dependent Ca(2+) signaling regulates Middle East respiratory syndrome-coronavirus pseudovirus translocation through the endolysosomal system. *Cell Calcium 75*, 30–41.
53. Gonzalez, A., Saez, C. A., and Moenne, A. (2018). Copper-induced activation of TRPs and VDCCs triggers a calcium signature response regulating gene expression in Ectocarpus siliculosus. *PeerJ 6*, e4556.
54. Foster, W. J., Taylor, H. B. C., Padamsey, Z., Jeans, A. F., Galione, A., and Emptage, N. J. (2018). Hippocampal mGluR1-dependent long-term potentiation requires NAADP-mediated acidic store Ca(2+) signaling. *Sci Signal 11*, eaat9093.
55. Eraslan, E., Tanyeli, A., Polat, E., and Polat, E. (2018). 8-Br-cADPR, a TRPM2 ion channel antagonist, inhibits renal ischemia-reperfusion injury. *J Cell Physiol 234*, 4572–4581.
56. Diercks, B. P., Werner, R., Weidemuller, P., Czarniak, F., Hernandez, L., Lehmann, C., Rosche, A., Kruger, A., Kaufmann, U., Vaeth, M., *et al.* (2018). ORAI1, STIM1/2, and RYR1 shape subsecond Ca(2+) microdomains upon T cell activation. *Sci Signal 11*, eaat0358.
57. Lee, H. C., and Aarhus, R. (1991). ADP-ribosyl cyclase: an enzyme that cyclizes NAD^+ into a calcium-mobilizing metabolite. *Mol Biol Cell 2*, 203–209.
58. Graeff, R. M., Walseth, T. F., Fryxell, K., Branton, W. D., and Lee, H. C. (1994). Enzymatic synthesis and characterizations of cyclic GDP-ribose. A procedure for distinguishing enzymes with ADP-ribosyl cyclase activity. *J Biol Chem 269*, 30260–30267.

59. Lee, H. C., Graeff, R., and Walseth, T. F. (1995). Cyclic ADP-ribose and its metabolic enzymes. *Biochimie 77*, 345–355.
60. Graeff, R. M., Walseth, T. F., Hill, H. K., and Lee, H. C. (1996). Fluorescent analogs of cyclic ADP-ribose: synthesis, spectral characterization, and use. *Biochemistry 35*, 379–386.
61. Munshi, C., Baumann, C., Levitt, D., Bloomfield, V. A., and Lee, H. C. (1998). The homo-dimeric form of ADP-ribosyl cyclase in solution. *Biochim Biophys Acta 1388*, 428–436.
62. Prasad, G. S., McRee, D. E., Stura, E. A., Levitt, D. G., Lee, H. C., and Stout, C. D. (1996). Crystal structure of *Aplysia* ADP ribosyl cyclase, a homologue of the bifunctional ectozyme CD38. *Nat Struct Biol 3*, 957–964.
63. Munshi, C., Thiel, D. J., Mathews, I. I., Aarhus, R., Walseth, T. F., and Lee, H. C. (1999). Characterization of the active site of ADP-ribosyl cyclase. *J Biol Chem 274*, 30770–30777.
64. States, D. J., Walseth, T. F., and Lee, H. C. (1992). Similarities in amino acid sequences of *Aplysia* ADP-ribosyl cyclase and human lymphocyte antigen CD38. *Trends Biochem Sci 17*, 495.
65. Jackson, D. G., and Bell, J. I. (1990). Isolation of a cDNA encoding the human CD38 (T10) molecule, a cell surface glycoprotein with an unusual discontinuous patern of expression during lymphocyte differentiation. *J Immunol 144*, 2811–2815.
66. Howard, M., Grimaldi, J. C., Bazan, J. F., Lund, F. E., Santos-Argumedo, L., Parkhouse, R. M., Walseth, T. F., and Lee, H. C. (1993). Formation and hydrolysis of cyclic ADP-ribose catalyzed by lymphocyte antigen CD38. *Science 262*, 1056–1059.
67. Liu, Q., Kriksunov, I. A., Graeff, R., Munshi, C., Lee, H. C., and Hao, Q. (2005). Crystal structure of human CD38 extracellular domain. *Structure 13*, 1331–1339.
68. Zhang, H., Graeff, R., Chen, Z., Zhang, L., Lee, H., and Hao, Q. (2011). Dynamic conformations of the CD38-mediated NAD cyclization captured in a single crystal. *J Mol Biol 405*, 1070–1078.
69. Kotaka, M., Graeff, R., Zhang, L., Lee, H. C., and Hao, Q. (2012). Structural studies of intermediates along the cyclization pathway of *Aplysia* ADP-ribosyl cyclase. *J Mol Biol 415*, 514–526.
70. Aarhus, R., Graeff, R. M., Dickey, D. M., Walseth, T. F., and Lee, H. C. (1995). ADP-ribosyl cyclase and CD38 catalyze the synthesis of a calcium-mobilizing metabolite from NADP. *J Biol Chem 270*, 30327–30333.
71. Lee, H. C., Graeff, R. M., and Walseth, T. F. (1997). ADP-ribosyl cyclase and CD38. Multi-functional enzymes in Ca^{2+} signaling. *Adv Exp Med Biol 419*, 411–419.
72. Ribeiro, J. M., Canales, J., Cabezas, A., Rodrigues, J. R., Pinto, R. M., Lopez-Villamizar, I., Costas, M. J., and Cameselle, J. C. (2018). Specific cyclic ADP-ribose phosphohydrolase obtained by mutagenic engineering of Mn(2+)-dependent ADP-ribose/CDP-alcohol diphosphatase. *Sci Rep 8*, 1036.
73. Graeff, R., Liu, Q., Kriksunov, I. A., Hao, Q., and Lee, H. C. (2006). Acidic residues at the active sites of CD38 and ADP-ribosyl cyclase determine NAADP synthesis and hydrolysis activities. *J Biol Chem 281*, 28951–28957.
74. Graeff, R., Liu, Q., Kriksunov, I. A., Kotaka, M., Oppenheimer, N., Hao, Q., and Lee, H. C. (2009). Mechanism of cyclizing NAD to cyclic ADP-ribose by ADP-ribosyl cyclase and CD38. *J Biol Chem 284*, 27629–27636.

75. Graeff, R., Munshi, C., Aarhus, R., Johns, M., and Lee, H. C. (2001). A single residue at the active site of CD38 determines its NAD cyclizing and hydrolyzing activities. *J Biol Chem 276*, 12169–12173.
76. Lee, H. C. (1999). A unified mechanism for enzymatic synthesis of two calcium messengers, cyclic ADP-ribose and NAADP. *Biol Chem 380*, 785–793.
77. Munshi, C., Aarhus, R., Graeff, R., Walseth, T. F., Levitt, D., and Lee, H. C. (2000). Identification of the enzymatic active site of CD38 by site-directed mutagenesis. *J Biol Chem 275*, 21566–21571.
78. Liu, Q., Kriksunov, I. A., Jiang, H., Graeff, R., Lin, H., Lee, H. C., and Hao, Q. (2008). Covalent and noncovalent intermediates of an NAD utilizing enzyme, human CD38. *Chem Biol 15*, 1068–1078.
79. Liu, Q., Kriksunov, I. A., Graeff, R., Lee, H. C., and Hao, Q. (2007a). Structural basis for formation and hydrolysis of calcium messenger cyclic ADP-ribose by human CD38. *J Biol Chem 282*, 5853–5861.
80. Partida-Sanchez, S., Cockayne, D., Monard, S., Jacobson, E. L., Oppenheimer, N., Garvy, B., Kusser, K., Goodricj, S., Howard, M., Harmsen, A., *et al.* (2001a). Cyclic ADP-ribose production by CD38 regulates intracellular calcium release, extracellular calcium influx and chemotaxis in neutrophils and is required for bacterial clearance in vivo. *Nat Med 7*, 1209–1216.
81. Galione, A., and Chuang, K. T. (2020). Pyridine nucleotide metabolites and calcium release from intracellular stores. *Adv Exp Med Biol 1131*, 371–394.
82. Lee, H. C. (2003). Calcium signaling: NAADP ascends as a new messenger. *Curr Biol 13*, R186–R188.
83. Lee, H. C. (2004). Multiplicity of Ca^{2+} messengers and Ca^{2+} stores: a perspective from cyclic ADP-ribose and NAADP. *Curr Mol Med 4*, 227–237.
84. Lee, H. C., and Zhao, Y. J. (2019). Resolving the topological enigma in Ca^{2+}-signaling by cyclic ADP-ribose and NAADP. *J Biol Chem 294*, 19831–19843.
85. Reinherz, E. L., Kung, P. C., Goldstein, G., Levey, R. H., and Schlossman, S. F. (1980). Discrete stages of human intrathymic differentiation: analysis of normal thymocytes and leukemic lymphoblasts of T-cell lineage. *Proc Natl Acad Sci U S A 77*, 1588–1592.
86. Zhao, Y. J., Lam, C. M. C., and Lee, H. C. (2012). The membrane-bound enzyme CD38 exists in two opposing orientations. *Sci Signal 5*, ra67.
87. Lee, H. C. (2011). Cyclic ADP-ribose and NAADP: fraternal twin messengers for calcium signaling. *Sci China Life Sci 54*, 699–711.
88. Higy, M., Junne, T., and Spiess, M. (2004). Topogenesis of membrane proteins at the endoplasmic reticulum. *Biochemistry 43*, 12716–12722.
89. Von Heijne, G. (1989). Control of topology and mode of assembly of a polytopic membrane protein by positively charged residues. *Nature 341*, 456–458.
90. Liu, J., Zhao, Y. J., Li, W. H., Hou, Y. N., Li, T., Zhao, Z. Y., Fang, C., Li, S. L., and Lee, H. C. (2017). Cytosolic interaction of type III human CD38 with CIB1 modulates cellular cyclic ADP-ribose levels. *Proc Natl Acad Sci U S A 114*, 8283–8288.
91. Kirchhofer, A., Helma, J., Schmidthals, K., Frauer, C., Cui, S., Karcher, A., Pellis, M., Muyldermans, S., Casas-Delucchi, C. S., Cardoso, M. C., *et al.* (2010). Modulation of protein properties in living cells using nanobodies. *Nat Struct Mol Biol 17*, 133–138.
92. Li, T., Qi, S., Unger, M., Hou, Y. N., Deng, Q. W., Liu, J., Lam, C. M. C., Wang, X. W., Xin, D., Zhang, P., *et al.* (2016). Immuno-targeting the multifunctional CD38 using nanobody. *Sci Rep 6*, 27055.

93. Zhao, Y. J., Zhang, H. M., Lam, C. M. C., Hao, Q., and Lee, H. C. (2011). Cytosolic CD38 protein forms intact disulfides and is active in elevating intracellular cyclic ADP-ribose. *J Biol Chem 286*, 22170–22177.
94. Zhao, Y. J., Zhu, W. J., Wang, X. W., Zhang, L.-H., and Lee, H. C. (2014). Determinants of the membrane orientation of a calcium signaling enzyme CD38. *Biochim Biophys Acta, Mol Cell Res 1853*, 2095–2103.
95. Deng, Q. W., Zhang, J., Li, T., He, W. M., Fang, L., Lee, H. C., and Zhao, Y. J. (2019). The transferrin receptor CD71 regulates type II CD38, revealing tight topological compartmentalization of intracellular cyclic ADP-ribose production. *J Biol Chem 294*, 15293–15303.
96. Wu, Y., Zhang, J., Fang, L., Lee, H. C., and Zhao, Y. J. (2019). A cytosolic chaperone complex controls folding and degradation of type III CD38. *J Biol Chem 294*, 4247–4258.
97. Rah, S. Y., Mushtaq, M., Nam, T. S., Kim, S. H., and Kim, U. H. (2010). Generation of cyclic ADP-Ribose and nicotinic acid adenine dinucleotide phosphate by CD38 for Ca^{2+} signaling in interleukin-8-treated lymphokine-activated killer cells. *J Biol Chem 285*, 21877–21887.
98. Lee, H. C. (1997). Mechanisms of calcium signaling by cyclic ADP-ribose and NAADP. *Physiol Rev 77*, 1133–1164.
99. Fang, C., Li, T., Li, Y., Xu, G. J., Deng, Q. W., Chen, Y. J., Hou, Y. N., Lee, H. C., and Zhao, Y. J. (2018). CD38 produces nicotinic acid adenosine dinucleotide phosphate in the lysosome. *J Biol Chem 293*, 8151–8160.
100. Nam, T. S., Park, D. R., Rah, S. Y., Woo, T. G., Chung, H. T., Brenner, C., and Kim, U. H. (2020). Interleukin-8 drives CD38 to form NAADP from NADP(+) and NAAD in the endolysosomes to mobilize Ca(2+) and effect cell migration. *FASEB J 34*, 12565–12576.
101. Bruzzone, S., Franco, L., Guida, L., Zocchi, E., Contini, P., Bisso, A., Usai, C., and De Flora, A. (2001). A self-restricted CD38-connexin 43 cross-talk affects NAD(+) and cyclic ADP-ribose metabolism and regulates intracellular calcium in 3T3 fibroblasts. *J Biol Chem 276*, 48300–48308.
102. Essuman, K., Summers, D. W., Sasaki, Y., Mao, X., DiAntonio, A., and Milbrandt, J. (2017). The SARM1 toll/interleukin-1 receptor domain possesses intrinsic NAD(+) cleavage activity that promotes pathological axonal degeneration. *Neuron 93*, 1334–1343 e1335.
103. Zhao, Z. Y., Xie, X. J., Li, W. H., Liu, J., Chen, Z., Zhang, B., Li, T., Li, S. L., Lu, J. G., Zhang, L., *et al.* (2019). A cell-permeant mimetic of NMN activates SARM1 to produce cyclic ADP-ribose and induce non-apoptotic cell death. *iScience 15*, 452–466.
104. Osterloh, J. M., Yang, J., Rooney, T. M., Fox, A. N., Adalbert, R., Powell, E. H., Sheehan, A. E., Avery, M. A., Hackett, R., Logan, M. A., *et al.* (2012). dSarm/Sarm1 is required for activation of an injury-induced axon death pathway. *Science 337*, 481–484.
105. Coleman, M. P., and Hoke, A. (2020). Programmed axon degeneration: from mouse to mechanism to medicine. *Nat Rev Neurosci 21*, 183–196.
106. Lunn, E. R., Perry, V. H., Brown, M. C., Rosen, H., and Gordon, S. (1989). Absence of wallerian degeneration does not hinder regeneration in peripheral nerve. *Eur J Neurosci 1*, 27–33.
107. Mack, T. G., Reiner, M., Beirowski, B., Mi, W., Emanuelli, M., Wagner, D., Thomson, D., Gillingwater, T., Court, F., Conforti, L., *et al.* (2001). Wallerian

degeneration of injured axons and synapses is delayed by a Ube4b/Nmnat chimeric gene. *Nat Neurosci 4*, 1199–1206.

108. Gerdts, J., Brace, E. J., Sasaki, Y., DiAntonio, A., and Milbrandt, J. (2015). SARM1 activation triggers axon degeneration locally via NAD(+) destruction. *Science 348*, 453–457.
109. Hatzfeld, M. (1999). The armadillo family of structural proteins. *Int Rev Cytol 186*, 179–224.
110. Xing, Y., Takemaru, K., Liu, J., Berndt, J. D., Zheng, J. J., Moon, R. T., and Xu, W. (2008). Crystal structure of a full-length beta-catenin. *Structure 16*, 478–487.
111. Vincenzi, M., Mercurio, F. A., and Leone, M. (2020). Sam domains in multiple diseases. *Curr Med Chem 27*, 450–476.
112. Sporny, M., Guez-Haddad, J., Lebendiker, M., Ulisse, V., Volf, A., Mim, C., Isupov, M. N., and Opatowsky, Y. (2019). Structural evidence for an octameric ring arrangement of SARM1. *J Mol Biol 431*, 3591–3605.
113. Horsefield, S., Burdett, H., Zhang, X., Manik, M. K., Shi, Y., Chen, J., Qi, T., Gilley, J., Lai, J. S., Rank, M. X., *et al.* (2019). NAD(+) cleavage activity by animal and plant TIR domains in cell death pathways. *Science 365*, 793–799.
114. Luo, L., Lucas, R. M., Liu, L., and Stow, J. L. (2019). Signalling, sorting and scaffolding adaptors for Toll-like receptors. *J Cell Sci 133*, jcs239194.
115. Essuman, K., Summers, D. W., Sasaki, Y., Mao, X., Yim, A. K. Y., DiAntonio, A., and Milbrandt, J. (2018). TIR domain proteins are an ancient family of NAD(+)-consuming enzymes. *Curr Biol 28*, 421–430.e424.
116. Wan, L., Essuman, K., Anderson, R. G., Sasaki, Y., Monteiro, F., Chung, E. H., Osborne Nishimura, E., DiAntonio, A., Milbrandt, J., Dangl, J. L., and Nishimura, M. T. (2019). TIR domains of plant immune receptors are NAD(+)-cleaving enzymes that promote cell death. *Science 365*, 799–803.
117. Adams, M. J., Ford, G. C., Koekoek, R., Lentz, P. J., McPherson, A., Jr., Rossmann, M. G., Smiley, I. E., Schevitz, R. W., and Wonacott, A. J. (1970). Structure of lactate dehydrogenase at 2-8 A resolution. *Nature 227*, 1098–1103.
118. Burroughs, A. M., Zhang, D., Schaffer, D. E., Iyer, L. M., and Aravind, L. (2015). Comparative genomic analyses reveal a vast, novel network of nucleotide-centric systems in biological conflicts, immunity and signaling. *Nucleic Acids Res 43*, 10633–10654.
119. Liu, Q., Kriksunov, I. A., Graeff, R., Munshi, C., Lee, H. C., and Hao, Q. (2006). Structural basis for the mechanistic understanding of human CD38 controlled multiple catalysis. *J Biol Chem 281*, 32861–32869.
120. Summers, D. W., Gibson, D. A., DiAntonio, A., and Milbrandt, J. (2016). SARM1-specific motifs in the TIR domain enable NAD^+ loss and regulate injury-induced SARM1 activation. *Proc Natl Acad Sci U S A 113*, E6271–E6280.
121. Di Stefano, M., Nascimento-Ferreira, I., Orsomando, G., Mori, V., Gilley, J., Brown, R., Janeckova, L., Vargas, M. E., Worrell, L. A., Loreto, A., *et al.* (2015). A rise in NAD precursor nicotinamide mononucleotide (NMN) after injury promotes axon degeneration. *Cell Death Differ 22*, 731–742.
122. Loreto, A., Di Stefano, M., Gering, M., and Conforti, L. (2015). Wallerian degeneration is executed by an NMN-SARM1-dependent late Ca(2+) influx but only modestly influenced by Mitochondria. *Cell Rep 13*, 2539–2552.
123. Kwong, A. K., Chen, Z., Zhang, H., Leung, F. P., Lam, C. M., Ting, K. Y., Zhang, L., Hao, Q., Zhang, L. H., and Lee, H. C. (2012). Catalysis-based inhibitors of the calcium signaling function of CD38. *Biochemistry 51*, 555–564.

124. Murata, H., Khine, C. C., Nishikawa, A., Yamamoto, K. I., Kinoshita, R., and Sakaguchi, M. (2018). c-Jun N-terminal kinase (JNK)-mediated phosphorylation of SARM1 regulates NAD(+) cleavage activity to inhibit mitochondrial respiration. *J Biol Chem 293*, 18933–18943.
125. Sethurathinam, S., Singh, L. P., Panneerselvam, P., Byrne, B., and Ding, J. L. (2013). UXT plays dual opposing roles on SARM-induced apoptosis. *FEBS Lett 587*, 3296–3302.
126. Bratkowski, M., Xie, T., Thayer, D. A., Lad, S., Mathur, P., Yang, Y. S., Danko, G., Burdett, T. C., Danao, J., Cantor, A., *et al.* (2020). Structural and mechanistic regulation of the pro-degenerative NAD hydrolase SARM1. *Cell Rep 32*, 107999.
127. Jiang, Y., Liu, T., Lee, C. H., Chang, Q., Yang, J., and Zhang, Z. (2020). The NAD(+)-mediated self-inhibition mechanism of pro-neurodegenerative SARM1. *Nature 588*, 658–663.
128. Sporny, M., Guez-Haddad, J., Khazma, T., Yaron, A., Dessau, M., Shkolnisky, Y., Mim, C., Isupov, M. N., Zalk, R., Hons, M., and Opatowsky, Y. (2020). Structural basis for SARM1 inhibition and activation under energetic stress. *eLife 9*, e62021.
129. Li, W. H., Huang, K., Cai, Y., Wang, Q. W., Zhu, W. J., Hou, Y. N., Wang, S., Cao, S., Zhao, Z. Y., Xie, X. J., *et al.* (2021). Permeant fluorescent probes visualize the activation of SARM1 and uncover an anti-neurodegenerative drug candidate. *eLife* 10, e67381.
130. Rack, J. G. M., Palazzo, L., and Ahel, I. (2020). (ADP-ribosyl)hydrolases: structure, function, and biology. *Genes Dev 34*, 263–284.
131. Cohen, M. S., and Chang, P. (2018). Insights into the biogenesis, function, and regulation of ADP-ribosylation. *Nat Chem Biol 14*, 236–243.
132. Di Girolamo, M., and Fabrizio, G. (2019). Overview of the mammalian ADP-ribosyl-transferases clostridia toxin-like (ARTCs) family. *Biochem Pharmacol 167*, 86–96.
133. Cervantes-Laurean, D., Minter, D. E., Jacobson, E. L., and Jacobson, M. K. (1993). Protein glycation by ADP-ribose: studies of model conjugates. *Biochemistry 32*, 1528–1534.
134. Jacobson, E. L., Cervantes-Laurean, D., and Jacobson, M. K. (1994). Glycation of proteins by ADP-ribose. *Mol Cell Biochem 138*, 207–212.
135. Tsuyama, S., Inoue, Y., and Tanaka, M. (1997). ADP-ribosylated actin as part of the actin monomer pool in rat brain. *Int J Biochem Cell Biol 29*, 293–301.
136. Pashikanti, S., Boissonneault, G. A., and Cervantes-Laurean, D. (2011). ex vivo detection of histone H1 modified with advanced glycation end products. *Free Radic Biol Med 50*, 1410–1416.
137. Wondrak, G. T., Cervantes-Laurean, D., Jacobson, E. L., and Jacobson, M. K. (2000). Histone carbonylation in vivo and in vitro. *Biochem J 351 Pt 3*, 769–777.
138. Bektas, M., Nurten, R., Ergen, K., and Bermek, E. (2006). Endogenous ADP-ribosylation for eukaryotic elongation factor 2: evidence of two different sites and reactions. *Cell Biochem Funct 24*, 369–380.
139. Stevens, L. A., Levine, R. L., Gochuico, B. R., and Moss, J. (2009). ADP-ribosylation of human defensin HNP-1 results in the replacement of the modified arginine with the noncoded amino acid ornithine. *Proc Natl Acad Sci U S A 106*, 19796–19800.

140. Kasamatsu, A., Nakao, M., Smith, B. C., Comstock, L. R., Ono, T., Kato, J., Denu, J. M., and Moss, J. (2011). Hydrolysis of *O*-acetyl-ADP-ribose isomers by ADP-ribosylhydrolase 3. *J Biol Chem 286*, 21110–21117.
141. Perraud, A. L., Shen, B., Dunn, C. A., Rippe, K., Smith, M. K., Bessman, M. J., Stoddard, B. L., and Scharenberg, A. M. (2003). NUDT9, a member of the Nudix hydrolase family, is an evolutionarily conserved mitochondrial ADP-ribose pyrophosphatase. *J Biol Chem 278*, 1794–1801.
142. Perraud, A. L., Fleig, A., Dunn, C. A., Bagley, L. A., Launay, P., Schmitz, C., Stokes, A. J., Zhu, Q., Bessman, M. J., Penner, R., *et al.* (2001). ADP-ribose gating of the calcium-permeable LTRPC2 channel revealed by Nudix motif homology. *Nature 411*, 595–599.
143. Sano, Y., Inamura, K., Miyake, A., Mochizuki, S., Yokoi, H., Matsushime, H., and Furuichi, K. (2001). Immunocyte Ca^{2+} influx system mediated by LTRPC2. *Science 293*, 1327–1330.
144. Faouzi, M., and Penner, R. (2014). Trpm2. *Handb Exp Pharmacol 222*, 403–426.
145. Huang, Y., Roth, B., Lu, W., and Du, J. (2019). Ligand recognition and gating mechanism through three ligand-binding sites of human TRPM2 channel. *eLife 8*, e50175.
146. Huang, Y., Winkler, P. A., Sun, W., Lu, W., and Du, J. (2018). Architecture of the TRPM2 channel and its activation mechanism by ADP-ribose and calcium. *Nature 562*, 145–149.
147. Wang, L., Fu, T. M., Zhou, Y., Xia, S., Greka, A., and Wu, H. (2018). Structures and gating mechanism of human TRPM2. *Science 362*.
148. Zhang, Z., Toth, B., Szollosi, A., Chen, J., and Csanady, L. (2018b). Structure of a TRPM2 channel in complex with Ca(2+) explains unique gating regulation. *elife 7*, e36409.
149. Shen, B. W., Perraud, A. L., Scharenberg, A., and Stoddard, B. L. (2003). The crystal structure and mutational analysis of human NUDT9. *J Mol Biol 332*, 385–398.
150. Tan, C. H., and McNaughton, P. A. (2016). The TRPM2 ion channel is required for sensitivity to warmth. *Nature 536*, 460–463.
151. Hara, Y., Wakamori, M., Ishii, M., Maeno, E., Nishida, M., Yoshida, T., Yamada, H., Shimizu, S., Mori, E., Kudoh, J., *et al.* (2002). LTRPC2 Ca^{2+}-permeable channel activated by changes in redox status confers susceptibility to cell death. *Mol Cell 9*, 163–173.
152. Kolisek, M., Beck, A., Fleig, A., and Penner, R. (2005). Cyclic ADP-ribose and hydrogen peroxide synergize with ADP-ribose in the activation of TRPM2 channels. *Mol Cell 18*, 61–69.
153. Yu, P., Liu, Z., Yu, X., Ye, P., Liu, H., Xue, X., Yang, L., Li, Z., Wu, Y., Fang, C., *et al.* (2019). Direct gating of the TRPM2 channel by cADPR via specific interactions with the ADPR binding pocket. *Cell Rep 27*, 3684–3695.e3684.
154. Beck, A., Kolisek, M., Bagley, L. A., Fleig, A., and Penner, R. (2006). Nicotinic acid adenine dinucleotide phosphate and cyclic ADP-ribose regulate TRPM2 channels in T lymphocytes. *FASEB J 20*, 962–964.

CHAPTER 4

SELENIUM ATOM-MODIFIED NUCLEIC ACIDS FOR STRUCTURAL, FUNCTIONAL, AND DIAGNOSTIC STUDIES

CEN CHEN[1], FUKANG YANG[1], AND ZHEN HUANG[1,2,3]

[1] Department of Chemical Biology, SeNA Research Institute, Chengdu, Sichuan, China

[2] College of Life Sciences, Sichuan University, Chengdu, Sichuan, China

[3] Department of Biochemistry, Szostak-CDHT Large Nucleic Acids Institute, Chengdu, Sichuan, China

4.1. BACKGROUND

DNAs and RNAs are essential biomolecules in living organisms and chiefly responsible for genetic information storage, replication, transcription, and translation. Since they play these essential roles, nucleic acids have been considered as potential diagnostic and therapeutic targets, especially as anticancer drugs [1–8]. Moreover, they have been widely applied in biotechnology and nanotechnology [9–16]. Well-defined 3D structures of nucleic acids and their protein complexes make them fine drug targets. DNAs and RNAs can fold into many different shapes and structures in addition to classical duplexes [17–19], such as triplexes [20], quadruplexes [21, 22], Hollidays Junctions [23], and i-motifs [24], let alone the structures formed by nucleic acid–protein complexes. The 3D-structure determination of nucleic acids and nucleic acid-protein

Nucleic Acids in Medicinal Chemistry and Chemical Biology: Drug Development and Clinical Applications, First Edition. Edited by Lihe Zhang, Xinjing Tang, Zhen Xi, and Jyoti Chattopadhyaya.

complexes can significantly enhance nucleic acid-target-based anticancer drug discovery as well as selective gene silencing research.

Fortunately, in the past two decades, the development of X-ray crystallographic technologies has largely facilitated the structural and functional studies of biological macromolecules, such as nucleic acids. More and more novel structures of nucleic acids and nucleic acid–protein complexes have been determined and they have emerged as potential drug targets. At the same time, there are many promising advancements in nucleic acid-based therapeutic discovery, including antisense oligonucleotides, siRNAs, microRNAs, and aptamers. These innovative investigations make structural and functional studies of nucleic acids and their protein complexes an increasingly active research field, thereby demanding new methodologies, especially novel strategies for nucleic acid X-ray crystallography [25].

However, compared with protein crystallography, there are unique challenges in nucleic acid X-ray crystallography. One of the long-existing challenge is the poor crystallizability of nucleic acids. While a protein molecule has a variety of structural and chemical groups on its surface to enable crystal contacts, the surface of a nucleic acid molecule is full of negatively charged repetitive phosphate groups. These repelling charges and structural dynamics make the crystal packing of DNA and RNA molecules much more difficult [25]. Screening process for the crystallization buffer is usually labor-intensive, and the formation of high-quality crystals is usually rare and time-consuming. These challenges have significantly slowed down the structure determination of nucleic acids as well as their protein complexes.

To face these challenges, we have developed the selenium atom-specific modification (SAM) of nucleic acids. On the basis of our research, we have discovered that it is possible to tailor the local structures and molecular dynamics of nucleic acids and accelerate the crystallization processes of nucleic acid molecules using a single selenium atom [26–28]. Selenium atom, which is in the same elemental family of oxygen, has been introduced to various locations of nucleic acids, especially the nucleobases, backbones, and sugars, to study its unique structural, functional, and diagnostic impacts on nucleic acids [25, 29–32].

4.2. SeNA ADVANTAGES FOR CRYSTALLIZATION AND PHASING IN NUCLEIC ACID CRYSTALLOGRAPHY

As mentioned previously, crystal growth is a major challenge existing in nucleic acid X-ray crystallography due to the repulsion of negatively charged backbone and multiple conformations of nucleic acids. Besides crystallization, phase problem is another major challenge needed to overcome. During the diffraction data analysis, the application of a Fourier transformation is required to construct the electron density map of macromolecules. In the

Fourier transformation, both structure factor amplitude F and phase of the reflection α are needed for calculating the electron density map. The former can be measured during the X-ray diffraction experiments. However, the phase information is missing, though it is critical to deduce the correct electron density map and crystal structure.

There are several different strategies to address this phasing problem, such as isomorphism replacement (SIR; MIR), molecular replacement (MR), and anomalous dispersion techniques (SAD and MAD). With an available similar model, MR should be the most convenient method, which does not require a heavy-atom derivative. However, for novel structures, especially nucleic acids containing highly repetitive and stacked units [33], molecular replacement is very difficult, and heavy-atom method is necessary in order to solve the phase problem. Isomorphous replacement has been successfully used in protein crystallography by soaking and co-crystallization with heavy-metal cations. Unfortunately, it often causes random hydrolysis of the phosphate backbone, and it requires to prepare many isomorphous native crystals and heavy-atom derivatives in order to finally determine 3D structures. Compared with isomorphous replacement, MAD and SAD techniques only require one single crystal in order to obtain all the diffraction information needed for the structure determination, and they provide a good solution to the phase problem of nucleic acid structure determination.

Many heavy atoms, such as Se, Br, Hg, and Zn, can be used as the anomalous scattering center for MAD [34]. Selenium is the most popular anomalous scattering atom used for proteins, because of its ideal K absorption edge (0.9795 Å) for X-ray diffraction experiment and the minimal structure perturbation. Selenium derivatives of nucleic acids have been developed and achieved great success in facilitating crystal growth and solving phase problem, which are the two major problems in nucleic acid crystallography. During data collection, traditional 5-halogen pyrimidine derivatives can be significantly affected by radiation damage. In contrast, radiolysis has little effect on selenium-modified nucleic acids [35]. Selenium can provide similar phasing power as bromine [30] and has more modification sites (Figure 4.1) to choose than bromine. Previous studies on selenium-modified nucleic acids (SeNA) X-ray crystallography have shown that the selenium modification can facilitate the crystal growth generally [28, 36–42], without significant structure perturbations [30, 31]. SAM can also facilitate the structure determination of protein–nucleic acid complexes, helping study the functions and mechanisms of enzymes, such as RNase-H [43, 44].

A study revealed that the 2′-methylseleno modification could facilitate the crystallization of A-form DNA by destabilizing the B-form helix [45]. Due to the steric hindrance by the neighboring residues, the large methylseleno group can't be tolerated by B-helix geometry, but it fits well in the minor groove of the A-form helix, serving as the driving force of B-to-A conversion, thereby facilitating the crystallization (Figure 4.2).

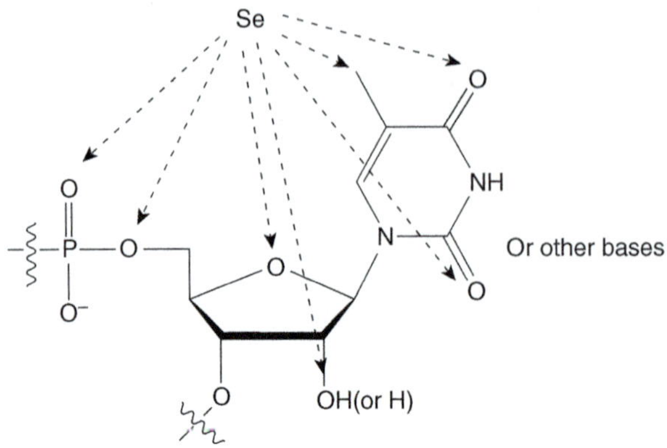

Figure 4.1 Chemical structure of selenium-modified nucleic acids (SeNA).

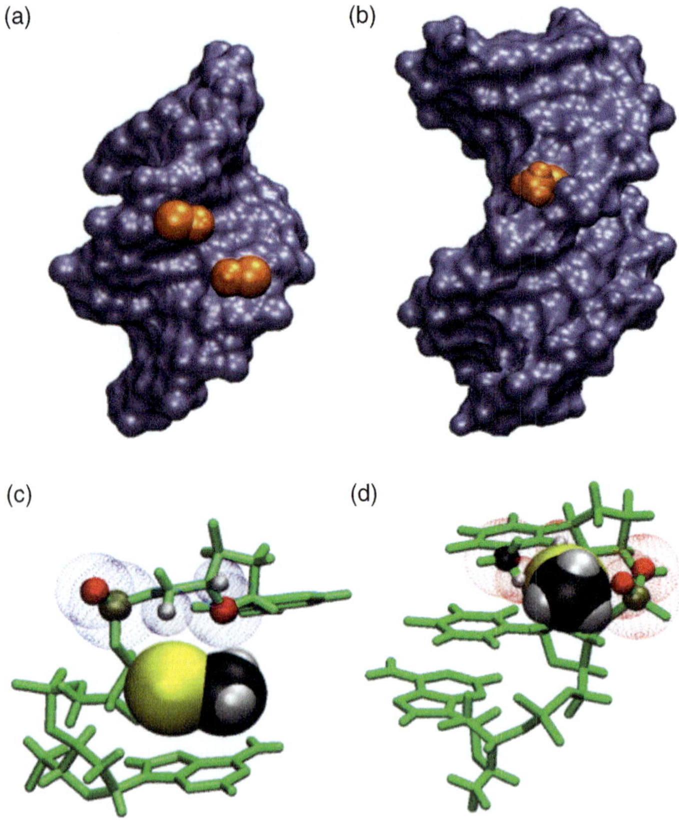

Figure 4.2 The 2′-$SeCH_3$ modifications of the ribose are implanted in A- and B-form DNA helical structures. (a) The $SeCH_3$ group in A-type helix. (b) The $SeCH_3$ group in B-type helix. (c) In A-type helix, no clashes between the $SeCH_3$ group and neighboring residue. (d) In B-type helix, the predicted potential clashes between the $SeCH_3$ group and neighboring residue. Source: Thompson et al. [45] / with permission of Taylor & Francis.

4.3. RECENT SeNA MODIFICATIONS

The studies of SeNA and SAM were started in 1998. Our research group synthesized the first selenium-derivatized phosphoramidite and triphosphate building blocks, pioneered site-specific synthesis of Se-modified nucleic acids, initiated and established the research areas of selenium nucleic acid chemical and structural biology as well as SeNA-assisted diagnostic research. Since the beginning, we have synthesized over 100 different Se-building blocks and incorporated a variety of Se-modifications into DNAs and RNAs for structural and functional studies. Our early research progresses were well reviewed [25] and will not be discussed here again. Since recent years, more novel Se-modified nucleosides, nucleotides and nucleic acids have been synthesized and investigated structurally, functionally, and diagnostically, we will focus this review on the recent Se-modification synthesis, structural and functional studies, detection investigations, such as diagnostic applications.

4.3.1. Se-Modifications on the Sugar and Phosphate

4.3.1.1. 5′-Se-Thymidine The synthesis of 5′-Se-derivatized thymidine and its phosphoramidite was reported by Zhang et al. in 2011, and the building block was then successfully incorporated into DNA oligonucleotides through solid-phase synthesis [12]. The protected Se-functionality was introduced within two steps with 71% yield in total. With the 5′-Se-moiety, the fluorescent group can be easily incorporated into DNA during the deprotection step under a mild condition. It was also observed that DNA polymerase recognized the labeled DNA well, which is useful in bioreaction monitoring and disease diagnosis.

4.3.1.2. Se-Triphosphates via One-Pot Facile Synthesis By using a protection-free one-pot triphosphates synthesis strategy [35], Lin et al. successfully synthesized nucleoside 5′-(α-P-seleno)-triphosphates (NTPαSe) and incorporate them into RNAs by T7 RNA polymerase [46]. During the triphosphate synthesis, the phosphitylating reagent was generated *in situ* without any purification and the reaction was highly regioselective at the 5′-hydroxyl group of nucleosides containing no protecting groups on the sugar or nucleobases. The selenium functionality was introduced via oxidation with 3H-1,2-benzothaselenol-3-one (BTSe) at room temperature, followed by hydrolysis (Scheme 4.1). The NTPαSe analogs can be purified by boronate affinity method using the strong interaction between the cis-diols at the 2′ and 3′ positions of NTPs and boronate to purify NTPαSe. Compared with the conventional HPLC method, the boronate affinity gives higher capacity for the large-scale synthesis with less time and steps.

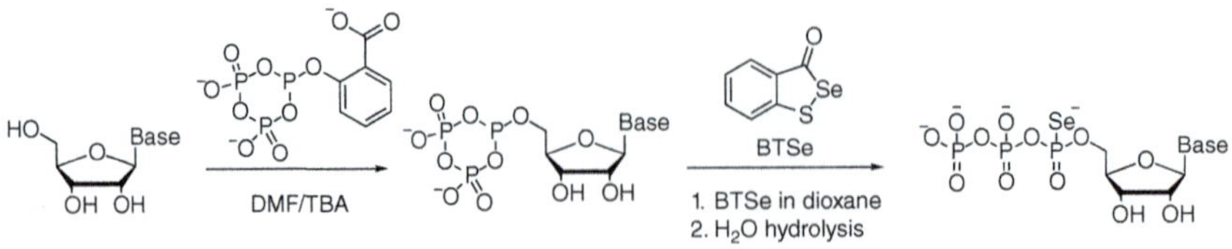

Scheme 4.1 Synthesis of NTPαSe analogs.

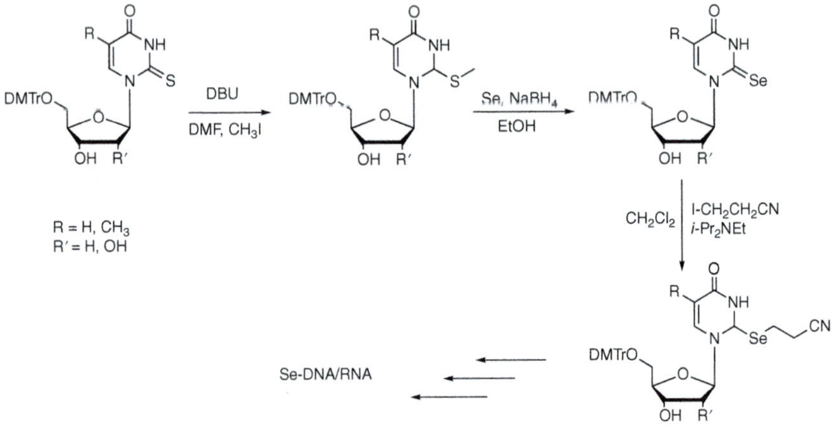

Scheme 4.2 Synthesis of 2-Se-U/T modified DNA and RNA.

4.3.2. Se-Modifications on the Nucleobases

4.3.2.1. Bases (2-Se-U and 2-Se-T) To study the high fidelity of base pairing in both DNA and RNA, 2-Se-modified thymidine and uridine were synthesized via the Huang Reaction and incorporated into oligonucleotides (Scheme 4.2) [27, 28]. Both the X-ray crystal structures of Se-DNA (5′-G-$^{2'\text{MeSe}}$dU-G-$^{2\text{Se}}$T-ACAC-3′)$_2$ and Se-RNA r (5′-GUAUA-$^{2\text{Se}}$U-AC-3′)$_2$ were solved by molecular replacement at high resolution (1.58 and 2.2 Å, respectively; Figure 4.3). The results indicated that 2-Se-modified DNA and RNA have very similar structures with the corresponding native. The 2-Se-substitution enhanced the base pair fidelity by further discouraging the T/C and T/G or U/G wobble pair in DNA and RNA, respectively, without significantly altering the native base pairing and overall duplex structure. The steric hindrance and electronic effect introduced by the bulky Se atom can discriminate against the wobble base-pair formation, hence increasing the base-pair recognizing specificity. Theoretical studies, including quantum mechanics calculation, molecular dynamics simulation, and free-energy perturbation simulation, have further confirmed the experimental conclusions [47]. The free-energy calculation indicates that both the disfavored hydrogen bonding interaction and the unfavorable solvent entropic contribution lead to the high specificity of the 2-Se-T/A pair.

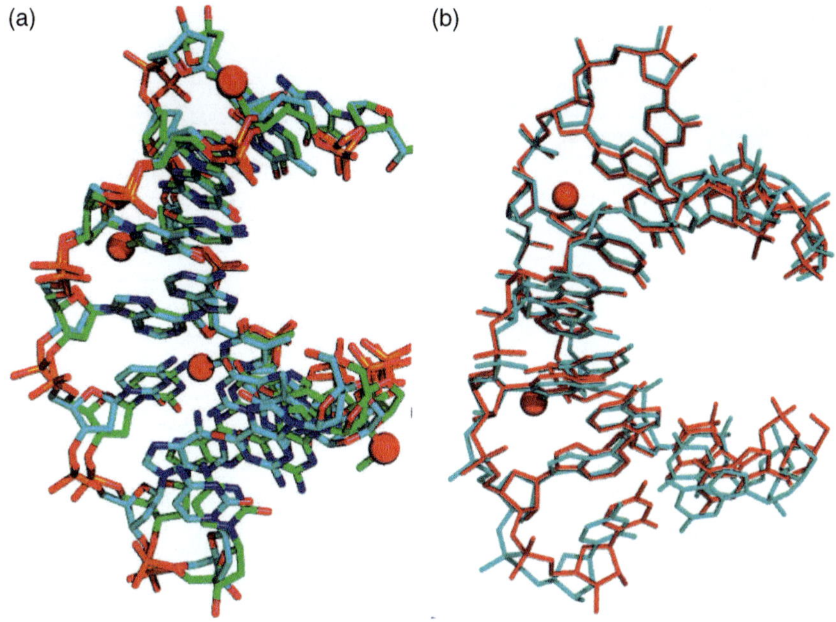

Figure 4.3 X-ray crystal structures of the Se-DNA and Se-RNA. (a) Se-DNA: (5′-G-$^{2'MeSe}$dU-G-^{2Se}T-ACAC-3′)$_2$ at high resolution of 1.58 Å. (b) Se-RNA: r(5′-GUAUA-^{2Se}U-AC-3′)$_2$ at high resolution of 2.2 Å.

4.3.2.2. 4-Se-U Sheng et al. reported the synthesis of 4-Se-uridine and Se-RNA to probe the U•U and U/A base pair [48]. The Se-modified RNA duplexes r(5′-U-^{4Se}U-CGCG-3′)$_2$ and r(5′-GUG-^{4Se}U-AAC-3′)$_2$ formed yellow crystals, and the structures were determined via SAD and molecular replacement (Figure 4.4). The Se-derivatized and native structures are virtually identical based on the crystal structure and thermostability studies, implying the RNA flexibility to accommodate the large Se atom. The comparison between the Se-modified and native structures also suggested a stronger stacking interaction between the Se-nucleobase to 3′-cytosine, strengthening the RNA duplexes. The decrease of H-bonding in U/A pair by 4-Se modification of uridine was compensated by the base-stacking increase strengthening the duplex thermostability. By introducing the selenium atomic probe to the RNA duplex junctions, the study reveals that due to the uracil stacking, each Se atom has contributed additional 0.4–0.8 kcal/mol to the RNA duplex stabilization. Interestingly, color RNA was generated by the single Se-modification shifting the uridine UV spectrum over 100 nm.

4.3.2.3. 6-Se-G 6-Se-guanosine, which can also be used as a Se-atomic probe to study the RNA secondary structures, Se-phosphoramidite, and Se-RNAs were synthesized by Salon et al. in 2013 [43]. The ^{Se}G-RNAs show yellow color and are useful for visualizing protein-RNA cocrystallization. It was observed that this Se-modification could probe the bulge and wobble structures *vs* a duplex structure.

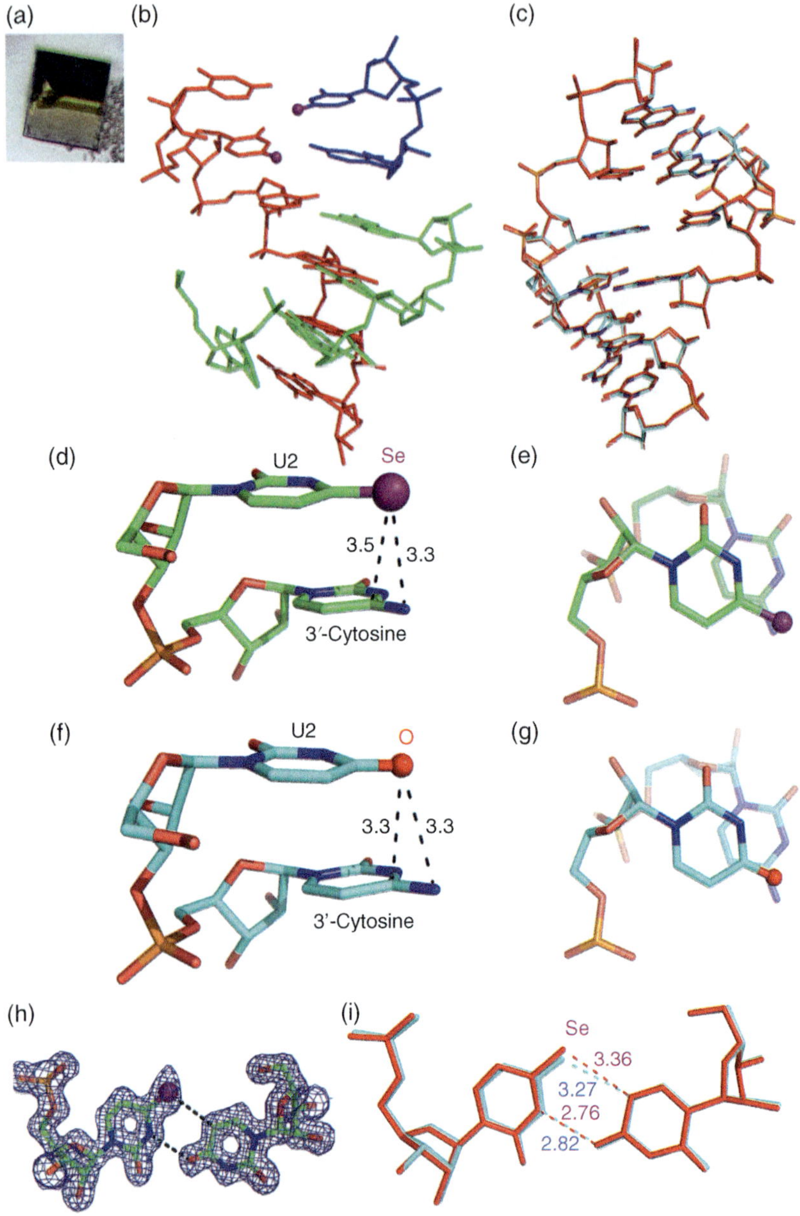

Figure 4.4 Crystal and structure of the 4-Se-U RNA hexamer, (5′-U-^{Se}U-CGCG-3′)$_2$. (a) Yellow Se-RNA crystal (0.1 × 0.1 × 0.1 mm). (b) Se-RNA duplex structure containing the Se-RNA hexamer (in red), the base-paired CGCG (in green) and the U•U-paired U-^{Se}U (in blue). (c) Superimposition of the Se-RNA and the corresponding native structure. (d) and (e) Se-modified U2 stacks on its 3′-cytosine. (f) and (g) Native U2 stacks on its 3′-cytosine. (h) Electron density map and model of the ^{Se}U•U pair. (i) Superimposition of ^{Se}U•U pair (in red) with native U•U pair (in cyan).

4.3.2.4. 5-MeSe-T, 5-MeSe-dC, and 4,5-diSe-T In addition, some other Se-modified nucleosides and nucleotides were reported. Zhang et al. synthesized the 2′-deoxy-5-methylselenyl-cytidine and incorporated it into DNA for biophysical and structural study [49]. No significant perturbation was observed, and the modified DNA duplex has similar stability to the corresponding native. Further, they reported a 4, 5-diSe-modified thymidine and DNA synthesis in 2013 [50]. An improved 5-MeSe-T synthesis was developed by our group using lithium/halogen exchange reaction, increasing the yield from 34 to 60%.

Besides chemical synthesis of Se-modified nucleosides, which is labor-intensive and usually low overall yield, Zhou et al. explored the biocatalytic synthesis by employing thermostable nucleoside phosphorylases [51]. These enzymes can recognize the Se-modified nucleobases and show excellent adaptability to some modified purines with high substrate conversion up to 97%.

4.4. SeNA-FACILITATED STRUCTURAL AND FUNCTIONAL STUDIES VIA X-RAY AND NEUTRON DIFFRACTION CRYSTALLOGRAPHY

4.4.1. SeNA-Enhanced X-ray Diffraction and Function Studies

6-Se-G-modified-RNA/DNA/RNase H complex crystallization was conducted by Salon et al. [43] The study showed that the Se-modification was remove by hydrolysis naturally during the crystallization. However, high-diffraction quality crystals were surprisingly generated, and the structure of RNA/DNA/RNase H complex was determined by the molecular replacement with high resolution (1.60 Å, PDB: 3ULD), in contrast to the corresponding native RNA complex with DNA and RNase H (2.70 Å). Further, the selenium-derivatized nucleic acids were applied to functional studies, such as RNase cleavage. The duplex of 6-Se-G-modified-DNA/RNA was complexed with RNase H, and its structure was reported with 1.80 Å resolution (PDB: 3TWH). A mechanistic study was conducted based on this high-resolution structure, which was compared with the corresponding native RNA/DNA/RNase H complex structure (PDB: 2G8U) [44]. In the native structure, the RNase H binds to the junction of two RNA/DNA duplexes. However, with the modified Se-DNA/RNA duplex, the RNase H recognized the internal position, though it had the same sequence as in the native complex structure. Moreover, the determined structure revealed that the selenium modification introduced a subtle duplex unwinding by 0.5–0.7 Å, and the scissile phosphate was shifted closer to the transition state of the catalytic hydrolysis reaction, thereby significantly accelerating the RNA cleavage by 6.2-fold faster. In the meanwhile, the scissile phosphate positioned the water nucleophile in the structure by forming a hydrogen bond (3.10 Å) between the phosphate and water molecule. These studies offered us better understandings on the guide-dependent RNA cleavage.

4.4.2. SeNA-Enhanced Neutron Diffraction and Function Studies

The SeNA for X-ray crystallography has been pioneered and developed for nucleic acids and their protein complexes, which solves the phasing problem rationally [25, 29, 52]. After addressing this problem in nucleic acid X-ray crystallography, we are taking on the remaining challenge: crystallization of nucleic acids. It is notorious that nucleic acid crystallization is much more challenging than protein crystallization, due to nucleic acid negative charges, multiple conformations, and chemical heterogeneity caused by synthesis and purification [53–56]. Moreover, it is much more difficult to grow large crystals (approximately 0.5–0.6 mm in size), required for neutron diffraction crystallography.

Generally, due to the low X-ray scattering power of hydrogen atoms, complete information about macromolecular H-bonds is hard to obtain in X-ray crystallographic structures. However, as a complementary of X-ray, neutron nucleic acid crystallography can determine the orientation of water molecule with a relatively low-resolution diffraction data (2.5–2.6 Å), because atomic nuclei instead of electron clouds scatter off the neutrons [57, 58]. Thus, even the lightest atom, hydrogen, can be observed directly. In addition, the neutrons do not cause direct radiation damage compared to X-ray diffraction, which can produce damage-free biomacromolecular structures and the data can be collected equally well at room and cryo temperatures. Joint X-ray/neutron structures of a self-complementary A-DNA oligonucleotide at both room and cryo temperatures were reported, with the 2′-SeMe modification on Cyt5 [59]. This selenium modification significantly facilitated the crystal growth and generated large DNA crystals (about 1000 folds larger in volume than the corresponding native). Interestingly, a subtle difference in the structures between these two temperatures was observed. We discovered that a proton on a backbone phosphate oxygen was replaced with a hydrated Mg^{2+}, upon cooling the crystal from the room temperature to 100 K, indicating the favorable metal binding at low temperature (Figure 4.5). Consequently, the enzymatic mechanisms and biological functions established on the 3D structures may lead to significant errors, because most of crystal and EM structures are obtained under the cryo temperature. Neutron crystallography has the advantage over X-ray crystallography in many aspects. However, the requirement of large crystal volume (generally than 1 mm^3 for biomacromolecules) has largely limited the application and development of neutron crystallography. Surprisingly, the selenium modification has dramatically reduced the volume requirement of neutron crystal diffraction to approximately 0.1 mm^3 in crystal volume (a reduction in 10 folds).

4.5. SeNA POTENTIALS IN THERAPEUTIC AND DIAGNOSTIC DEVELOPMENT

In the past, compared to proteins, nucleic acids attracted much less attention as drug targets in the structure-based drug design (SBDD), partially due to the lack of structural information on nucleic acids with drug candidates. By facilitating crystallization and improving crystal quality, SAM strategy can greatly

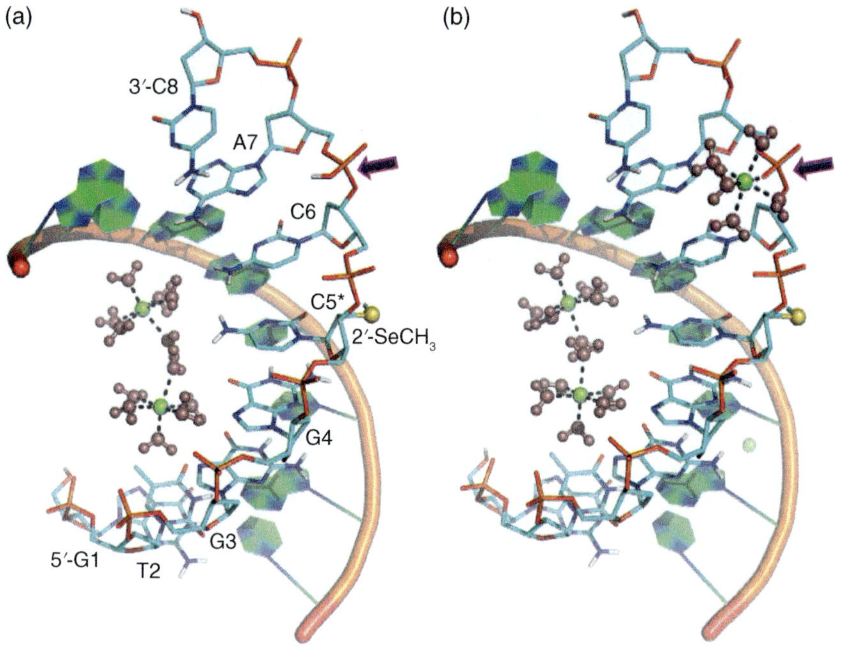

Figure 4.5 The neutron-diffraction-determined structures of the Se-DNA: d[GTGG($C^{2'MeSe}$)CAC]$_2$. (a) At room temperature; the arrow points to the Ade7 phosphate backbone with protonation. (b) At 100 K; the arrow points to the Ade7 phosphate backbone with Mg^{2+} coordination.

strengthen nucleic acid SBDD. The detailed 3D structures can help researchers understand the target structures and even the interactions between the target and small-molecule ligand. Moreover, oligonucleotides have been investigated as potential therapeutics for over 30 years, including antisense oligonucleotides (ASOs), aptamers, ribozymes, and siRNAs with approximately ten drugs approved by FDA [60]. Many obstacles, including the nuclease instability, off-target effect, poor delivery, limited bioavailability, and low affinity, have slowed down oligonucleotide-sequence-based drug discovery. Many chemical modifications, such as phosphorothioate (PS), 2′-*O*-methyl (2′-OMe), locked nucleic acids (LNAs), phosphorodiamidate morpholino oligomer (PMO), and so on [61], have been developed to overcome those challenges, by altering the structures and charges of the oligonucleotides with therapeutic potential.

Our previous studies on Se-modified nucleic acids have shown that the Se-replacement of the non-bridging oxygen on backbone phosphates can largely prevent the oligonucleotides from nuclease degradation [62, 63]. This Se-modification on siRNA may improve the RNase stability, suggesting its potential therapeutic application in the RNAi applications. Moreover, certain Se-modifications, such as 2′-MeSe, can reduce multiple conformations of nucleic acids by destabilizing other unfavorable structures [45] and can potentially increase the affinity and specificity of nucleic acid duplexes (including DNA/DNA, DNA/RNA, and RNA/RNA) by reducing entropic penalty. Further, the anticancer activities of selenium-modified nucleosides have also been reported [64]. Therefore, SAM can potentially increase the siRNA half-life and allow accurate sequence targeting, thereby better delivering to the tissues of interest.

Recently, Hu et al. reported a protection-free, one-pot strategy to synthesize the 2′-deoxynucleoside 5′-(alpha-P-seleno)-triphosphates (dNTPαSe; Scheme 4.3). With the selenium incorporation, the triphosphates (dNTPαSe) offered two diastereomers. The study showed that only one of two diastereomers could be efficiently recognized by DNA polymerases. Moreover, the Se-derivatized dNTPs could inhibit non-specific DNA polymerization caused by mis-priming, which provides a more specific strategy for nucleic acid polymerization and detection.

Our following investigations have revealed that dNTPαSe enhanced the polymerization specificity over 240-fold better than the corresponding natives and increased sensitivity up to the single-digit detection (Figure 4.6) [65]. In performing PCR or RT–PCR for detecting human papilloma virus and

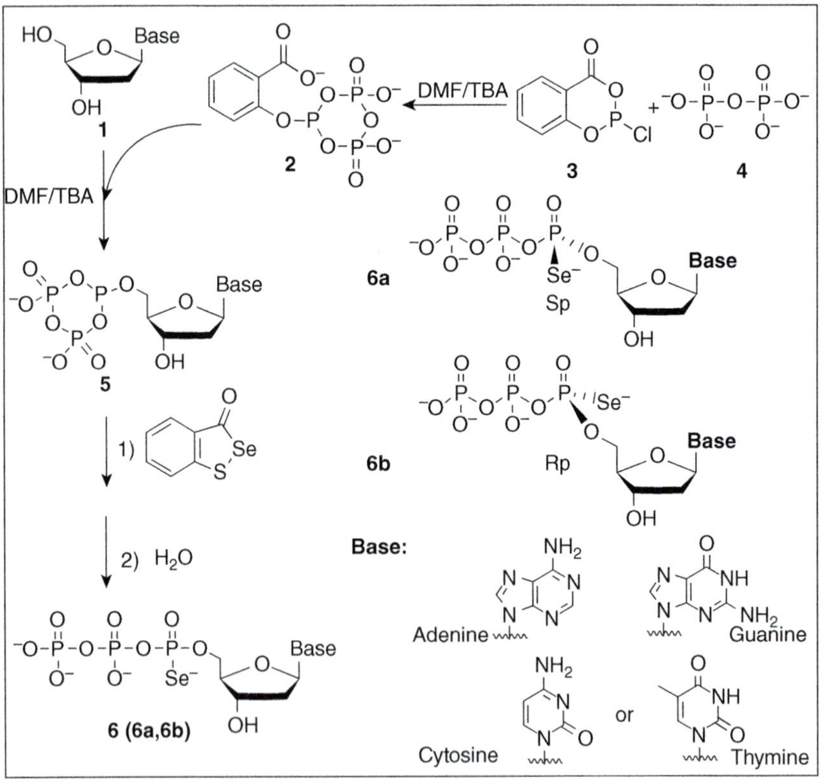

Scheme 4.3 Simple and mild synthesis of dNTPαSe analogs via the one-pot strategy.

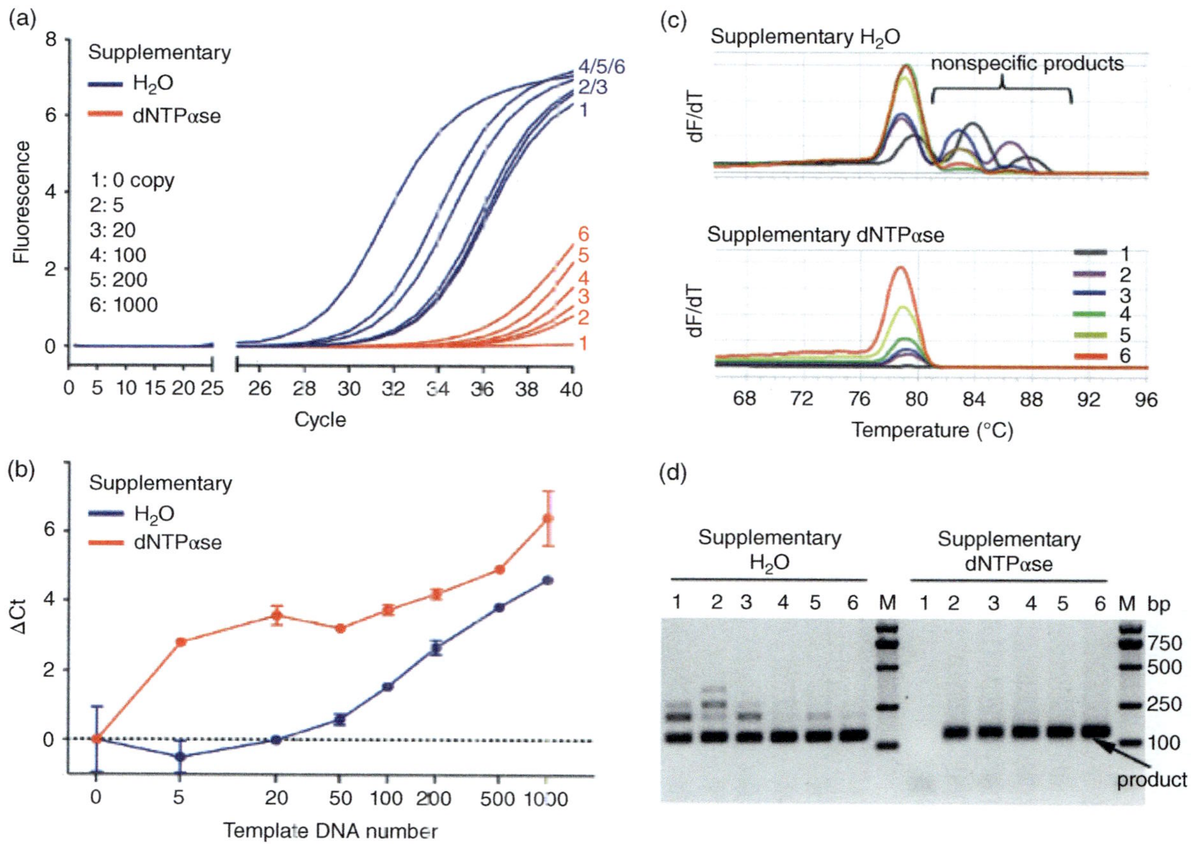

Figure 4.6 (a) Amplification curves; single-digit copies of DNA molecules were detected. (b) Relationship between ΔCt and viral DNA copies in qPCR (SYBR Green I); ΔCt is PCR Ct (Cycle of threshold) difference between those of background and sample; ΔCt <1 was defined as negative detection result. (c) High-resolution melting analyses of the corresponding reaction products in **A**. (d) Agarose gel analysis of reaction products in **A**.

COVID-19 viral RNA, dNTPαSe largely increased specificity in the presence of human genomic background DNA. Clearly, SAM can enhance specificity of DNA synthesis and polymerization significantly, and in turn it allows detection of clinical COVID-19 viral RNA in the low copy numbers (single-digit copies), while the conventional RT-qPCR doesn't. By avoiding the by-products formed with the commercial kits, the SeNA strategy is much more sensitive than the original kit, showing its great potential in disease diagnosis.

Our further study [66] accomplished by Luo et al. has demonstrated the ability of Se-RNAs as templates for reverse transcription and shown their potential as both positive and negative controls for RT–PCR kits. Overall, SeNA possess many unique properties and show great potentials in discovering therapeutics and diagnostics. However, the studies on SeNA are still limited, and the synthesis is relatively costly, which may slow down broader applications. Through further investigations and synthesis of SeNA, we expect that more progresses and applications on SAM will be accomplished and reported in the near future.

4.6. CONCLUSIONS AND PERSPECTIVES

In conclusion, SeNAs have been successfully synthesized systematically via chemical and biochemical strategies. Our research results have indicated that SAM can significantly facilitate DNA and RNA crystallization, consistent with their thermal stability study. Clearly, SAM can provide useful tools to probe the crystallization and phasing, facilitate structure determination of nucleic acids and their protein complexes, and reveal the mechanistic and functional insights into the molecular interaction and catalysis, on the basis of structure and function studies [43, 44, 67–70]. Further, because of the biostability, SeNAs have great potentials in developing diagnostics and therapeutics, such as siRNA, microRNA, aptamer, mRNA vaccine, and antisense therapeutics.

4.6.1. General Significance

In general, SeNA can facilitate crystallization, phasing and high-resolution structure determination of nucleic acids and their protein complexes, and the selenium-atom-specific modifications (SAM) can tailor dynamics of nucleic acid molecules, especially local dynamics, without causing significant structure and function perturbations. Selenium, in the same elemental family of oxygen, has been introduced to various locations of nucleic acids by replacing oxygen atom with selenium for crystallography. In addition, SAM provides the mechanistic insights into nucleic acid crystallization and inter- and intra-molecular interactions, thereby facilitating discovery of the small-molecule drugs that target nucleic acids. Further, in order to minimize nucleobase mismatches jeopardizing polymerization specificity and causing mis-diagnosis in nucleic acid detection, we have established the first synthesis of the Se-triphosphates against nucleobase mismatches. Surprisingly, SAM can significantly enhance specificity and sensitivity of the polymerization and detection and allow detection of nucleic acids (DNA and RNA) in the low copy numbers (single-digit copies), while the conventional strategies don't. Our novel SAM strategy via SeNA can allow nucleic acid polymerization and detection with much higher specificity and sensitivity.

REFERENCES

1. Opalinska, J. B.; Gewirtz, A. M., Nucleic-acid therapeutics: basic principles and recent applications. *Nat. Rev. Drug Discov.* 2002, *1* (7), 503–514.
2. Palchaudhuri, R.; Hergenrother, P. J., DNA as a target for anticancer compounds: methods to determine the mode of binding and the mechanism of action. *Curr. Opin. Biotechnol.* 2007, *18* (6), 497–503.
3. Ding, S.; Qiao, X.; Kucera, G. L.; Bierbach, U., Using a build-and-click approach for producing structural and functional diversity in DNA-targeted hybrid anticancer agents. *J. Med. Chem.* 2012, *55* (22), 10198–10203.
4. Ueberschaar, N.; Dahse, H. M.; Bretschneider, T.; Hertweck, C., Rational design of an apoptosis-inducing photoreactive DNA intercalator. *Angew. Chem. Int. Ed. Eng.* 2013, *52* (24), 6185–6189.
5. Wang, S.; Kumar, A.; Aston, K.; Nguyen, B.; Bashkin, J. K.; Boykin, D. W.; Wilson, W. D., Different thermodynamic signatures for DNA minor groove binding with changes in salt concentration and temperature. *Chem. Commun. (Camb.)* 2013, *49*, 8543–8545.
6. Chenoweth, D. M.; Meier, J. L.; Dervan, P. B., Pyrrole-imidazole polyamides distinguish between double-helical DNA and RNA. *Angew. Chem. Int. Ed. Eng.* 2013, *52*, 415–418.
7. Ranjan, N.; Davis, E.; Xue, L.; Arya, D. P., Dual recognition of the human telomeric G-quadruplex by a neomycin-anthraquinone conjugate. *Chem. Commun. (Camb.)* 2013, *49*, 5796–5798.
8. Minarini, A.; Milelli, A.; Tumiatti, V.; Rosini, M.; Lenzi, M.; Ferruzzi, L.; Turrini, E.; Hrelia, P.; Sestili, P.; Calcabrini, C.; Fimognari, C., Exploiting RNA as a new biomolecular target for synthetic polyamines. *Gene* 2013, *524*, 232–240.
9. Sassolas, A.; Leca-Bouvier, B. D.; Blum, L. J., DNA biosensors and microarrays. *Chem. Rev.* 2008, *108* (1), 109–139.
10. Xue, X.; Wang, F.; Liu, X., One-step, room temperature, colorimetric detection of mercury (Hg^{2+}) using DNA/nanoparticle conjugates. *J. Am. Chem. Soc.* 2008, *130* (11), 3244–3245.
11. Lan, T.; Furuya, K.; Lu, Y., A highly selective lead sensor based on a classic lead DNAzyme. *Chem. Commun. (Camb.)* 2010, *46* (22), 3896–3898.
12. Zhang, W.; Huang, Z., Synthesis of the 5′-se-thymidine phosphoramidite and convenient labeling of DNA oligonucleotide. *Org. Lett.* 2011, *13* (8), 2000–2003.
13. Yang, Z.; Chen, F.; Alvarado, J. B.; Benner, S. A., Amplification, mutation, and sequencing of a six-letter synthetic genetic system. *J. Am. Chem. Soc.* 2011, *133* (38), 15105–15112.

14. Cristovao, M.; Sisamakis, E.; Hingorani, M. M.; Marx, A. D.; Jung, C. P.; Rothwell, P. J.; Seidel, C. A.; Friedhoff, P., Single-molecule multiparameter fluorescence spectroscopy reveals directional MutS binding to mismatched bases in DNA. *Nucleic Acids Res.* 2012, *40* (12), 5448–5464.
15. Sha, R.; Birktoft, J. J.; Nguyen, N.; Chandrasekaran, A. R.; Zheng, J.; Zhao, X.; Mao, C.; Seeman, N. C., Self-assembled DNA crystals: the impact on resolution of 5′-phosphates and the DNA source. *Nano Lett.* 2013, *13* (2), 793–797.
16. Han, D.; Jiang, S.; Samanta, A.; Liu, Y.; Yan, H., Unidirectional scaffold-strand arrangement in DNA origami. *Angew. Chem. Int. Ed. Eng.* 2013, *52*, 9031–9034.
17. Zhang, J.; Ferre-D'Amare, A. R., Co-crystal structure of a T-box riboswitch stem I domain in complex with its cognate tRNA. *Nature* 2013, *500* (7462), 363–366.
18. Ke, A.; Zhou, K.; Ding, F.; Cate, J. H.; Doudna, J. A., A conformational switch controls hepatitis delta virus ribozyme catalysis. *Nature* 2004, *429* (6988), 201–5.
19. Doherty, E. A.; Doudna, J. A., Ribozyme structures and mechanisms. *Annu. Rev. Biophys. Biomol. Struct.* 2001, *30*, 457–475.
20. Frank-Kamenetskii, M. D.; Mirkin, S. M., Triplex DNA structures. *Annu. Rev. Biochem.* 1995, *64* (1), 65–95.
21. Davis, J. T., G-quartets 40 years later from 5′-GMP to molecular biology and supramolecular chemistry. *Angew. Chem. Int. Ed. Eng.* 2004, *43* (6), 668–698.
22. Lim, K. W.; Phan, A. T., Structural basis of DNA quadruplex-duplex junction formation. *Angew. Chem. Int. Ed. Eng.* 2013, *52*, 8566–8569.
23. Ho, P. S.; Eichman, B. F., The crystal structures of DNA Holliday junctions. *Curr. Opin. Struct. Biol.* 2001, *11* (3), 302–308.
24. Phan, A. T.; Mergny, J. L., Human telomeric DNA: G-quadruplex, i-motif and Watson–Crick double helix. *Nucleic Acids Res.* 2002, *30* (21), 4618–4625.
25. Lin, L.; Sheng, J.; Huang, Z., Nucleic acid X-ray crystallography via direct selenium derivatization. *Chem. Soc. Rev.* 2011, *40* (9), 4591–4602.
26. Pallan, P. S.; Greene, E. M.; Jicman, P. A.; Pandey, R. K.; Manoharan, M.; Rozners, E.; Egli, M., Unexpected origins of the enhanced pairing affinity of 2′-fluoro-modified RNA. *Nucleic Acids Res.* 2011, *39* (8), 3482–3495.
27. Sun, H.; Sheng, J.; Hassan, A. E.; Jiang, S.; Gan, J.; Huang, Z., Novel RNA base pair with higher specificity using single selenium atom. *Nucleic Acids Res.* 2012, *40* (11), 5171–5179.
28. Hassan, A. E.; Sheng, J.; Zhang, W.; Huang, Z., High fidelity of base pairing by 2-selenothymidine in DNA. *J. Am. Chem. Soc.* 2010, *132* (7), 2120–2121.
29. Carrasco, N.; Ginsburg, D.; Du, Q.; Huang, Z., Synthesis of selenium-derivatized nucleosides and oligonucleotides for X-ray crystallography. *Nucleos. Nucleot. Nucl. Acids* 2001, *20* (9), 1723–1734.
30. Teplova, M.; Wilds, C. J.; Wawrzak, Z.; Tereshko, V.; Du, Q.; Carrasco, N.; Huang, Z.; Egli, M., Covalent incorporation of selenium into oligonucleotides for X-ray crystal structure determination via MAD: proof of principle. *Biochimie* 2002, *84* (9), 849–858.
31. Wilds, C. J.; Pattanayek, R.; Pan, C.; Wawrzak, Z.; Egli, M., Selenium-assisted nucleic acid crystallography: use of phosphoroselenoates for MAD phasing of a DNA structure. *J. Am. Chem. Soc.* 2002, *124* (50), 14910–14916.
32. Hobartner, C.; Micura, R., Chemical synthesis of selenium-modified oligoribonucleotides and their enzymatic ligation leading to an U6 SnRNA stem-loop segment. *J. Am. Chem. Soc.* 2004, *126* (4), 1141–1149.

33. Campbell, N. H.; Parkinson, G. N., Crystallographic studies of quadruplex nucleic acids. *Methods* 2007, *43* (4), 252–263.
34. Hendrickson, W. A., Determination of macromolecular structures from anomalous diffraction of synchrotron radiation. *Science* 1991, *254* (5028), 51–58.
35. Olieric, V.; Rieder, U.; Lang, K.; Serganov, A.; Schulze-Briese, C.; Micura, R.; Dumas, P.; Ennifar, E., A fast selenium derivatization strategy for crystallization and phasing of RNA structures. *RNA* 2009, *15* (4), 707–715.
36. Hassan, A. E.; Sheng, J.; Jiang, J.; Zhang, W.; Huang, Z., Synthesis and crystallographic analysis of 5-Se-thymidine DNAs. *Org. Lett.* 2009, *11* (12), 2503–2506.
37. Jiang, J.; Sheng, J.; Carrasco, N.; Huang, Z., Selenium derivatization of nucleic acids for crystallography. *Nucleic Acids Res.* 2006, *35* (2), 477–485.
38. Moroder, H.; Kreutz, C.; Lang, K.; Serganov, A.; Micura, R., Synthesis, oxidation behavior, crystallization and structure of 2′-methylseleno guanosine containing RNAs. *J. Am. Chem. Soc.* 2006, *128* (30), 9909–9918.
39. Salon, J.; Sheng, J.; Gan, J.; Huang, Z., Synthesis and crystal structure of 2′-Se-modified guanosine containing DNA. *J. Org. Chem.* 2010, *75* (3), 637–641.
40. Salon, J.; Sheng, J.; Jiang, J.; Chen, G.; Caton-Williams, J.; Huang, Z., Oxygen replacement with selenium at the thymidine 4-position for the Se base pairing and crystal structure studies. *J. Am. Chem. Soc.* 2007, *129* (16), 4862–4863.
41. Sheng, J.; Jiang, J.; Salon, J.; Huang, Z., Synthesis of a 2′-Se-thymidine phosphoramidite and its incorporation into oligonucleotides for crystal structure study. *Org. Lett.* 2007, *9* (5), 749–752.
42. Sheng, J.; Salon, J.; Gan, J.; Huang, Z., Synthesis and crystal structure study of 2′-Se-adenosine-derivatized DNA. *Sci. China Chem.* 2010, *53* (1), 78–85.
43. Salon, J.; Gan, J.; Abdur, R.; Liu, H.; Huang, Z., Synthesis of 6-Se-guanosine RNAs for structural study. *Org. Lett.* 2013, *15* (15), 3934–3937.
44. Abdur, R.; Gerlits, O. O.; Gan, J.; Jiang, J.; Salon, J.; Kovalevsky, A. Y.; Chumanevich, A. A.; Weber, I. T.; Huang, Z., Novel complex MAD phasing and RNase H structural insights using selenium oligonucleotides. *Acta Crystallogr. D Biol. Crystallogr.* 2014, *70* (2), 354–361.
45. Thompson, R. A.; Spring, A. M.; Sheng, J.; Huang, Z.; Germann, M. W., The importance of fitting in: conformational preference of selenium 2′ modifications in nucleosides and helical structures. *J. Biomol. Struct. Dyn.* 2015, *33* (2), 289–297.
46. Lin, L.; Caton-Williams, J.; Kaur, M.; Patino, A. M.; Sheng, J.; Punetha, J.; Huang, Z., Facile synthesis of nucleoside 5′-(α-P-seleno)-triphosphates and phosphoroselenoate RNA transcription. *RNA* 2011, *17* (10), 1932–1938.
47. Christofferson, A.; Zhao, L.; Sun, H.; Huang, Z.; Huang, N., Theoretical studies of the base pair fidelity of selenium-modified DNA. *J. Phys. Chem. B* 2011, *115* (33), 10041–10048.
48. Sheng, J.; Gan, J.; Soares, A. S.; Salon, J.; Huang, Z., Structural insights of non-canonical U• U pair and Hoogsteen interaction probed with Se atom. *Nucleic Acids Res.* 2013, *41* (22), 10476–10487.
49. Zhang,W.;Sheng,J.;Hassan,A.E.;Huang,Z.,Synthesisof2′-deoxy-5-(methylselenyl) cytidine and Se-DNAs for structural and functional studies. *Chem. Asian J.* 2012, *7* (3), 476–479.
50. Zhang, W.; Hassan, E. A.; Huang, Z., Synthesis of novel di-Se-containing thymidine and Se-DNAs for structure and function studies. *Sci. China Chem.* 2013, *56* (3), 273–278.

51. Zhou, X.; Yan, W.; Zhang, C.; Yang, Z.; Neubauer, P.; Mikhailopulo, I. A.; Huang, Z., Biocatalytic synthesis of seleno-, thio-and chloro-nucleobase modified nucleosides by thermostable nucleoside phosphorylases. *Catal. Commun.* 2019, *121*, 32–37.
52. Zhang, W.; Szostak, J. W.; Huang, Z., Nucleic acid crystallization and X-ray crystallography facilitated by single selenium atom. *Front. Chem. Sci. Eng.* 2016, *10* (2), 196–202.
53. Choi, J.; Majima, T., Conformational changes of non-B DNA. *Chem. Soc. Rev.* 2011, *40* (12), 5893–5909.
54. Dock-Bregeon, A.; Moras, D.; Giegé, R., Nucleic acids and their complexes. In Arnaud Ducruix, Richard Giegé Eds., *Crystallization of Nucleic Acids and Proteins. A Practical Approach*. New York: Oxford University Press Inc. 1999, 209–243.
55. Ke, A.; Doudna, J. A., Crystallization of RNA and RNA–protein complexes. *Methods* 2004, *34* (3), 408–414.
56. Mooers, B. H., Crystallographic studies of DNA and RNA. *Methods* 2009, *47* (3), 168–176.
57. Blakeley, M. P.; Langan, P.; Niimura, N.; Podjarny, A., Neutron crystallography: opportunities, challenges, and limitations. *Curr. Opin. Struct. Biol.* 2008, *18* (5), 593–600.
58. Chatake, T., Neutron nucleic acid crystallography. In Eric Ennifar Ed. *Nucleic Acid Crystallography*, New York: Humana Press: 2016; pp 283–300.
59. Vandavasi, V. G.; Blakeley, M. P.; Keen, D. A.; Hu, L. R.; Huang, Z.; Kovalevsky, A., Temperature-induced replacement of phosphate proton with metal ion captured in neutron structures of A-DNA. *Structure* 2018, *26* (12), 1645–1650.e3.
60. Stein, C. A.; Castanotto, D., FDA-approved oligonucleotide therapies in 2017. *Mol. Ther.* 2017, *25* (5), 1069–1075.
61. Sharma, V. K.; Sharma, R. K.; Singh, S. K., Antisense oligonucleotides: modifications and clinical trials. *Med. Chem. Commun.* 2014, *5* (10), 1454–1471.
62. Carrasco, N.; Huang, Z., Enzymatic synthesis of phosphoroselenoate DNA using thymidine 5′-(α-P-seleno) triphosphate and DNA polymerase for X-ray crystallography via MAD. *J. Am. Chem. Soc.* 2004, *126* (2), 448–449.
63. Brandt, G.; Carrasco, N.; Huang, Z., Efficient substrate cleavage catalyzed by hammerhead ribozymes derivatized with selenium for X-ray crystallography. *Biochemistry* 2006, *45* (29), 8972–8977.
64. Lin, L.; Sheng, J.; Momin, R. K.; Du, Q.; Huang, Z., Facile synthesis and antitumor cell activity of Se-containing nucleosides. *Nucleos. Nucleot. Nucl. Acids* 2009, *28* (1), 56–66.
65. Hu, B.; Wang, Y.; Li, N.; Zhang, S.; Luo, G.; Huang, Z., Highly convenient and highly specific-and-sensitive PCR using Se-atom modified dNTPs. *Chem. Commun.* 2021, *57* (1), 57–60.
66. Luo, G.; Zhang, J.; Zhang, S.; Hu, B.; Hu, L.; Huang, Z., High-quality RT-PCR with chemically modified RNA controls. *Talanta* 2021, *224*, 121850.
67. Helm, M.; Petermeier, M.; Ge, B.; Fiammengo, R.; Jäschke, A., Allosterically activated Diels–Alder catalysis by a ribozyme. *J. Am. Chem. Soc.* 2005, *127* (30), 10492–10493.
68. Serganov, A.; Keiper, S.; Malinina, L.; Tereshko, V.; Skripkin, E.; Höbartner, C.; Polonskaia, A.; Phan, A. T.; Wombacher, R.; Micura, R., Structural basis for Diels–Alder ribozyme-catalyzed carbon-carbon bond formation. *Nat. Struct. Mol. Biol.* 2005, *12* (3), 218.

69. Egli, M.; Pallan, P. S.; Pattanayek, R.; Wilds, C. J.; Lubini, P.; Minasov, G.; Dobler, M.; Leumann, C. J.; Eschenmoser, A., Crystal structure of homo-DNA and nature's choice of pentose over hexose in the genetic system. *J. Am. Chem. Soc.* 2006, *128* (33), 10847–10856.
70. Salon, J.; Jiang, J.; Sheng, J.; Gerlits, O. O.; Huang, Z., Derivatization of DNAs with selenium at 6-position of guanine for function and crystal structure studies. *Nucleic Acids Res.* 2008, *36* (22), 7009–7018.

CHAPTER 5

SEQUENCE-SPECIFIC CHROMOSOMAL DNA BINDERS: AN UNFULFILLED PROMISE?

ZUTAO YU[1], GANESH N. PANDIAN[1], and HIROSHI SUGIYAMA[1,2]
[1] Institute for Integrated Cell-Material Sciences (iCeMS), Kyoto University, Kyoto, Japan
[2] Department of Chemistry, Graduate School of Science, Kyoto University, Kyoto, Japan

5.1. INTRODUCTION

Why targeting chromosomal DNA? Cells containing one single copy gene can be used indefinitely as template to generate enormous RNA copy transcripts, and those RNAs will be translated to further large copy numbers of protein. From the perspective of biological intervention, such as cellular regulation and disease therapy, controlling the gene transcription at the DNA level would be an ideal way since there is only one copy of target gene DNA sequence in a cell. On the other hand, to inhibit 1000 copies of protein, it requires at least 1000 binder and it requires all those 1000 copies are targeted by the binder, which is far more challenging. Therefore, an artificial genetic switch (DNA binding system) would be of high interest to specifically bind the target DNA sequence of a gene and then modulate its output at the DNA level. Another key advantage of targeting chromatin DNA over its translated products lies in its simplicity since the interactions between nucleic acids can be predicted by a few simple base-pairing nucleobases that are highly specific and predictable.

Nucleic Acids in Medicinal Chemistry and Chemical Biology: Drug Development and Clinical Applications, First Edition. Edited by Lihe Zhang, Xinjing Tang, Zhen Xi, and Jyoti Chattopadhyaya.

RNA-guided DNA binding system CRISPR-Cas has revolutionized research toolkit to manipulate genomic DNA on both permanent gene editing and transient epigenetic modulation [1]. While protein-based DNA binding system such as CRISPR, TALEs, and ZNF can be delivered to cells transiently or long-term through integration into the genome, treatment with small molecules allows for more finely tuned control of dosage and timing of transcriptional regulation without the introduction of genetic material. Small-molecule DNA binders with sequence specificity and programmability are so attractive and have been pursued for decades, owing to its small molecule size, tenability of activity and specificity, user-friendly, and large potential for pharmaceutical application. Moreover, synthetic DNA binders are excellent probes in the field of deciphering, reprogramming, and manipulating cell state precisely [2, 3].

There are mainly three types of synthetic chromatin DNA binders, i.e., triplex-forming oligonucleotides (TFOs), peptide nucleic acid (PNA) [3], and pyrrole–imidazole polyamides (PIPs) [4]. TFOs and PNA are nucleic acid analogues that interact with DNA following Watson–Crick and Hoogsteen bond interaction. They have been extensively explored as gene editing tool for decades and studies proved the efficacy in animal models. As natural product-inspired small polymer, PIPs interact with dsDNA via H-bond in minor groove and have been explored in transcription regulation and anticancer study. Although progress is still ongoing, none of these DNA binders reaches clinical trial until now. Several challenges impede the broad biological study and clinic application.

To address these challenges, we first overview the protein-based and synthetic DNA binding system and make side-by-side comparison. Then detailed discussion of TFO, PNA, and PIPs about structure, DNA binding mode, and biological applications are presented, with a focus on current limitation and potential strategies for future improvement.

5.2. COMPARISON OF PROTEIN-BASED AND SYNTHETIC DNA TARGETING SYSTEMS

Scientists from the areas of chemistry, biology, and related have just welcomed the award of 2020 Nobel prize in Chemistry to CRISPR-Cas9 technology. Although the characterization of Cas9 happens just less than 10 years, it has revolutionized scientists' toolkit in targeting chromatin DNA. Like TALEs and ZNF, protein-based DNA targeting system has high sequence specificity and targeting efficiency. Moreover, as RNA-guided targeting system, Cas9 is easy to program its guide RNA when targeting a new locus. Thus, scientists choose to use user-friendly Cas9 in pre-clinic study while giving up the optimization of conventional synthetic DNA binders. The potential problem is the discontinuity on the study of synthetic DNA binders.

Synthetic DNA binders, including TFOs, PNA, and PIPs, are conventionally believed to be advantages when using in clinics. As shown in Table 5.1, these

TABLE 5.1 Comparison of protein-based and synthetic DNA targeting systems

	CRISPR-Cas9	TALEs	ZNF	TFOs	PNA	PIPs
Size	163 kDa (20 bp)	12 kDa (12 bp)	12 kDa (12 bp)	4 kDa (12 bp)	4–7 kDa (12 bp)	1–2 kDa (7-bp)
Open dsDNA	√	×	×	×	√	×
Interaction mode	RNA–DNA	Protein–DNA	Protein-DNA	LNA–dsDNA triplex; major groove	Strand invasion	minor groove
Sequence specificity	High	High	High	High	High	Low-to-moderate
Programmability	Easy	Difficult	Difficult	Limited to TFO sequence	Easy	Moderate
RNAi	×	×	×	√	√	×
Binding affinity	High	High	Moderate	Moderate	High	Moderate (K_D around 10–0.5 nM)
Delivery	Transfection, genetic material	Transfection, genetic material	Transfection, genetic material	Transfection, nanoparticle	Modification, transfection, nanoparticle	Cell-uptake
Biological efficiency	High	High	High	Low	Low-to-moderate	Moderate

agents have most of the properties that protein-based systems have, albeit of difference in potency. Most significantly, they are small in size and no risk of using genetic materials. Of course, the potency could vary largely in cases like homology searching efficiency and off-target effect. A clear understanding of disadvantages is the key to improve synthetic DNA binders.

5.3. TFOs STRUCTURE VIA HOOGSTEEN BOND INTERACTION

Although anti-gene strategy whereby an oligonucleotide directly binds chromosomal DNA in the major groove has been extensively studied experimentally during the last three decades, none of them reached clinical application yet. There are mainly two classes tested until now, TFOs and strand invading oligonucleotides as represented by PNA [3, 5]. Antisense application via targeting RNA have been well summarized in other reviews.

The understanding of the working mechanism is essential for the study of their function and biological application. To function effectively as an anti-gene tool, three sequential, interconnected steps are essential, i.e., delivery and nuclease-stability, chromosomal DNA targeting, and induction of cellular responses locally such as blocking TFs binding, interfering polymerase, and inducing cellular repair. In this section, TFOs will be explained with a focus on the working mechanism, current progress, and limitation and further research opportunities.

5.3.1. TFO Structure and DNA Binding Mode

Karst Hoogsteen in 1963 observed a non-Watson–Crick hydrogen bond interaction, which laid the foundation of TFO study, and such new base interactions were named as Hoogsteen bonds subsequently [6]. Hoogsteen pairing theory has been developed and well characterized (see review [7, 8]). TFOs are inserted alongside the major groove of dsDNA with high affinity and sequence specificity through either Hoogsteen or reverse Hoogsteen interaction. TFOs have been suggested as homing device for gene manipulation for *in vivo* study.

Structural requirements influence the design and application of TFOs. Since Hoogsteen pairing in TFO is much weaker than Watson–Crick interaction in dsDNA, targeting dsDNA site has been limited to homopurine sequences to form stable TFO structure. Continuous polypurine runs in dsDNA are required, while the interruptions by pyrimidines or single base-pair mismatches can significantly destabilize the binding of TFO. Thus, TFO pairing could not be applied to random human genome sequence with a mixture of base composition.

There are two stable TFO helical motifs based on orientation and base composition, parallel/pyrimidine, and anti-parallel/purine structure (Figure 5.1). In the parallel/pyrimidine motif, TFO strand contains pyrimidine nucleotide (C and T) and interacts with the polypurine strand of dsDNA via H-bond

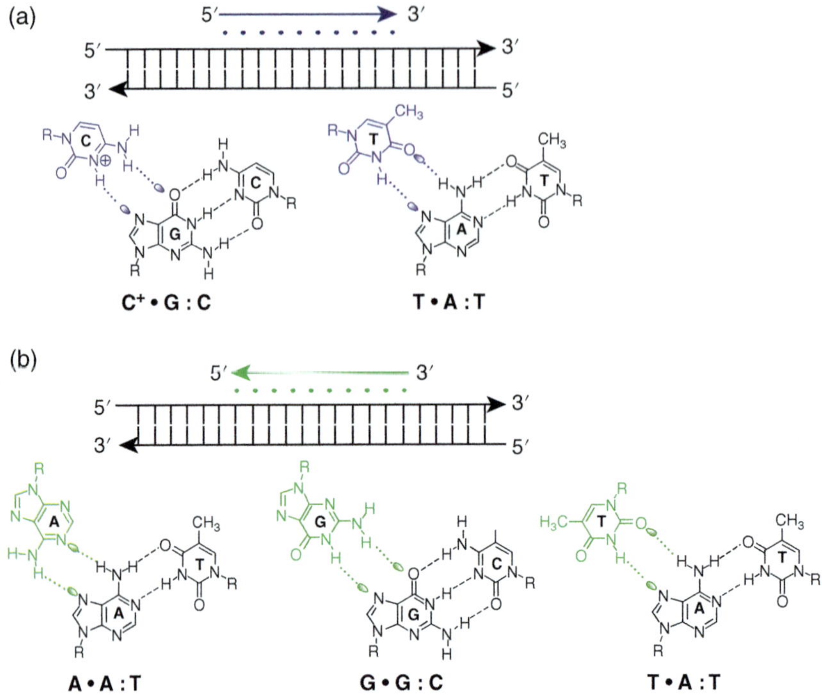

Figure 5.1 Diagram of TFO structure and Hoogsteen pairing. (a) Parallel/pyrimidine structure. The third strand (blue) is parallel to the pyrimidine-rich strand (upper strand) in terms of 5′–3′ orientation. Protonated C and T interact with upper strand via Hoogsteen bond (blue dot). (b) Antiparallel/purine structure. The third strand (green) has opposite orientation with respect to the upper strand. Hoogsteen bond interaction (green dot) are formed with A, G, and T in the third strand to stabilize triplex formation.

Hoogsteen interaction, C^+ G:C and T• A:T. C is protonated in order to form Hoogsteen H-bond interaction with G:C pair. The parallel TFO keeps same orientation with the polypurine strand of dsDNA. On the other hand, TFO containing purine nucleotide (G and A) has opposite orientation with the polypurine strand in dsDNA via reverse Hoogsteen interactions including G• G:C and A• A:T. G-rich anti-parallel/purine motif has higher stability compared with other TFOs. In addition, G and T-constituting TFOs form stable triplex mainly in antiparallel modes via reverse Hoogsteen interaction but can also build parallel modes through Hoogsteen interaction.

5.3.2. Limitations and Strategies

Chemical modifications in improving nuclease resistance have been well documented for the development of therapeutic oligonucleotides, for example RNAi and mRNA vaccine (Table 5.2) [3]. Binding affinity is important in TFOs since

TABLE 5.2 The limitations and strategies of improvement of TFOs

Category	Base modification	Sugar	Linkage and backbone
Nuclease-resistance	–	2′-O-Me; 2′-O-MOE; LNA	Phosphorothioate (PS); phosphoramidates (PN); PNA
Binding affinity	C replacement: 8-OA, 7,8-2H-OA, 6-OC, 8-NG; T replacement: 2dU, dU bis-substituted	LNA	PNA
Polyanionic repulsion of phosphate-ribose backbone	–	–	Morpholino or cationic phosphoramidate linkages such as *N,N*-diethethyleneamine (DEED) or *N,N*-dimethyl-aminopropyl amine; peptide backbone: PNA
Protonation threshold	5mC and pseudo-isoC	–	–
Inhibit G-quartet self-hybridization	2′-deoxy-6SG,7-deaza-2′-deoxyxanthosine	–	–

TFO has thermodynamic weak interaction compared with duplex DNA and has low stability of triplex helical structure, which LNA and PNA are two most studied oligonucleotide.

The destabilization poised by polyanionic repulsion cannot be ignored when a third polyanionic strand is added into the major groove of two polyanionic dsDNA. Neutral and cationic linkage/backbone are suggested to eliminate polyanionic repulsion, such as morpholino PMOs and PNA. In the parallel/pyrimidine TFO motif, protonated C is sensitive to pH and physiological pH is not applicable unless C modification incorporated. Also, G-rich TFOs in antiparallel/purine motif could be easy to self-hybridize and form stable G-quartet structure, especially in physiological high potassium environment. Thus, modified G with ablation of 7-N H-bond acceptor can effectively block self-aggregation.

There are several key issues that have not been thoroughly tackled, impeding the further application for *in vivo* study.

i. Stemming from the nature of TFO binding affinity and stability, TFO-based antigene strategy has been limited to sequence composition that only designed to target polypurine dsDNA. Therefore, TFO cannot target and modify random genomic sequence of interest with mixed insertion of pyrimidine base. Moreover, the limitation of sequence composition to polypurine largely increases the off-target rate without increasing the TFO strand length.
ii. Off-target effect has not been less evaluated in genome-wide scale. Long TFO strand will target more specific loci but can tolerate high number of

mismatches, while shorter version will encounter multiple loci for matched binding. For example, a target sequence of 16 bp long polypyrimidine TFO sequence present in more than 57 copies within the human genome [9].

iii. Rather than anti-gene effect via binding chromosomal DNA, another off-target effect originates from antisense effect through binding RNA. TFO has potent binding affinity with RNA and experimental studies are required to rule out the complexation of TFO-mediated RNAi effect.

iv. As a homing device, chromosomal accessibility significantly affects the targeting rate. Open non-chromatinized region of active transcribed gene is easier for targeting than hetero-chromatinized region in silenced gene. Also, cell cycle dependence on TFO accessibility has also been contradictory as reported in different studies. Like CRISPR-Cas9 system, increasing TFO concentration shows site-specific improvement of targeting. But the enhancement is not significant, and it would trigger high cellular stress like off-target effect.

v. The efficiency of TFO-mediated DNA mutagenesis is far lower (less than 3%) than protein-based system like CRISPR-Cas9. Low DNA mutagenesis efficiency is partially owing to the failure for inducing DNA double-strand break, and the conjugation with DNA crosslinking agent psoralen proved to improve the efficacy slightly but not dramatically. However, off-target mutation by psoralen will be a concerning issues since the mutation will be a permanent risk. Homing efficiency at target site could be attributed to the low efficiency, but no experimental evidence to confirm the binding efficiency at the target site.

5.3.3. Biological Applications

5.3.3.1. Targeted Mutagenesis To inducing DNA mutagenesis, TFO–psoralen conjugate has been studied with UV irradiation. Psoralen crosslinks with T and induces T:A to A:T transversion. Using plasmid as the acceptor of TFO binding, it showed 6% targeted mutagenesis of which 55% undergoes T:A to A:T transversion [10]. To modifying endogenous chromosome site, TFO-psoralen induced mutations at the endogenous *hprt* gene in CHO cells. It demonstrates 80% of the mutations occurred at TFO binding sites [11].

5.3.3.2. Homologous Recombination HR is a high-fidelity route for site-specific gene editing. Using simple TFOs only results in low frequency and random integration. The combination of donor DNA template and DNA strand breakage mutagen are shown to significantly enhance HR-mediated gene editing through XPA factor-dependent NER pathway. For example, AG-30 induces intramolecular HR at *supF* gene with an HR frequency of 0.37% in COS7 cells. The covalency with psoralen enhances the frequency to 0.58% [12]. In a separated study, the combination of AG30 and 51 mer donor DNA successfully induced intermolecular HR, albeit of low frequency of 0.05% in CHO derived cells [13].

5.4. PNA AS dsDNA INVASION AGENT

Through computational modeling study, PNA was firstly designed via replacing deoxyribose-phosphate backbone with polyamide bond [14]. To maintain homomorphosis to DNA in terms of scaffold bond and distance between backbone and nucleobase, PNA has a peptide-based backbone via N-(2-aminoethyl) glycine (known as aegPNA) (Figure 5.2a). Owing to charge-neutral backbones, blocking degradation by nucleases, higher binding affinity and even better specificity than natural counterparts, and without triggering of RNase H, PNA has been a heated research topic, and substantial efforts on targeting genomic DNA have invested in mechanism understanding, structural modification, and applications [15].

5.4.1. Structure and How to Invade DNA Duplex

Despite its big deviation from phosphate/ribose-based backbone in nucleic acids, PNA has very high binding affinity to be capable of forming heteroduplex and triplex structures with DNA and RNA through Watson–Crick, Hoogsteen, and reverse-Hoogsteen bond pairing. There are mainly five working modeling developed until now on PNA–DNA duplex interaction via base pairing (Figure 5.2b). PNA can form TFO structure via Hoogsteen bond interaction as explained in TFO section. Most significantly, PNA can efficiently invade dsDNA and form DNA/PNA heteroduplex and DNA/PNA/PNA triplex, which is perhaps the most widely utilized strategy by targeting genomic DNA. Five binding modes of PNA-mediated DNA invasion have been well studied, i.e., bisPNA, tail-clamp PNA (tcPNA), pseudocomplementary PNA (pcPNA), γPNA-single strand invasion, and bifacial PNA [15].

Through covalently connecting a triplex strand and a Watson–Crick complementary strand, bisPNA via triplex invasion significantly enhance the efficiency of strand invasion and binding affinity. To form stable Hoogsteen bonding complexes, long stretch of polypurine or polypyrimidine sequence is required, such as in TFO and bisPNA. Also, C needs to be protonated as H-bond donor at N3 site. To this end, tcPNA is mainly stabilized by a longer Watson–Crick PNA strand but with less length-dependence of polypurine/polypyrimidine triplex stretch [16]. Recently, a clamp-forming bis-LNA has been designed and shows efficiency for DNA invasion [17]. Another interesting LNA-based DNA invasion method is Zorro-LNA with a Z-shape construct to stabilize inter-strand stabilization of complexes, which exhibits DNA invasion ability in an *in vitro* study [18]. However, there is a lack of side-by-side comparative study with bisPNA and tcPNA.

To fully relax the sequence requirement using double-stranded invasion as shown in pcPNA, it needs to block PNA self-complementary hybridization. Introduction of modified base pairs such as $D\sim U^{s}$ and $K^{*}\sim C^{Bz}$ proved to be effective (Figure 5.2c). As the analogue of A:T pair, D pairs with T and A pairs

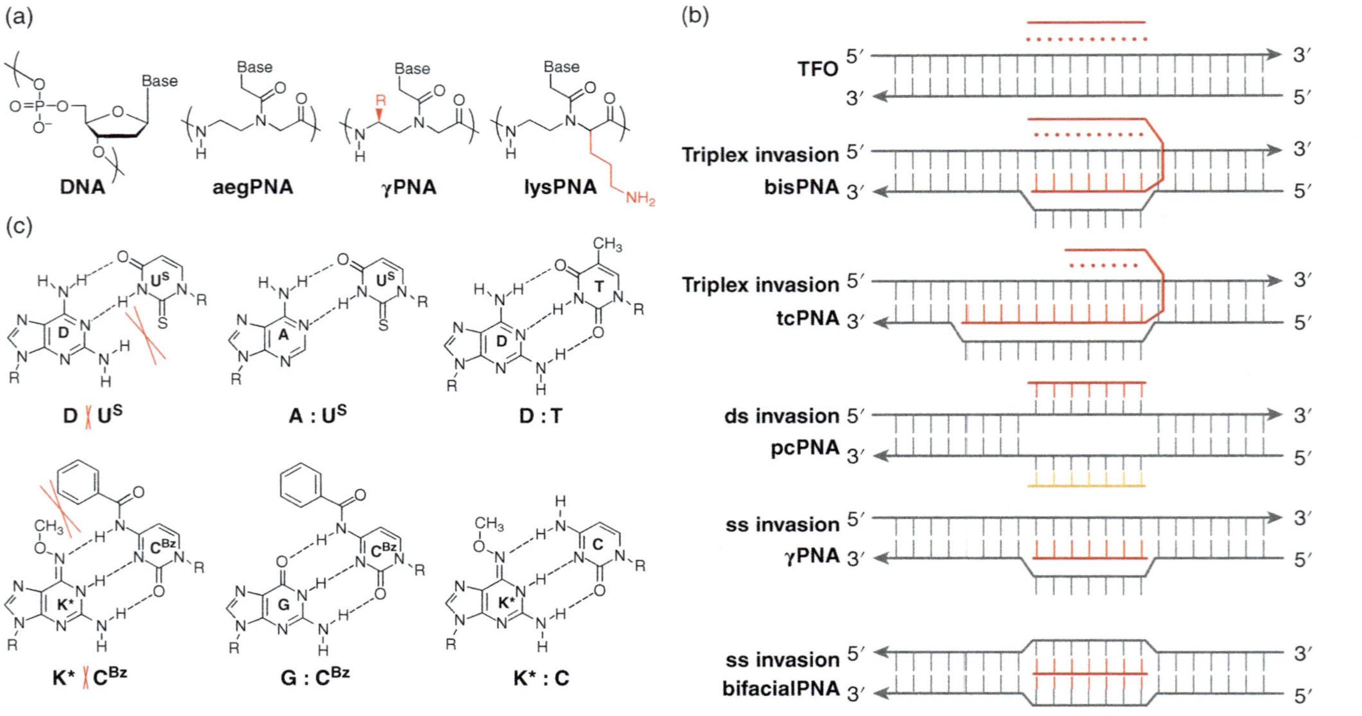

Figure 5.2 Structure and DNA binding mode of PNA. (a) Chemical structure of PNA and its modifications. (b) Schematic illustration of DNA duplex binding. Except TFO structure via triplex formation, five modes (triplex invasion bisPNA and tcPNA, double-stranded invasion pcPNA, single-stranded invasion γPNA, and single-stranded invasion bifacial PNA) require DNA duplex invasion. Watson–Crick bond showed in dash bond and Hoogsteen bond showed in dot. Triplex invasion mode using bisPNA and tcPNA requires either polypurine or polypyrimidine stretch. (c) Structure and base pairing of artificial nucleobase pairs used in pcPNA mode.

with U^s, but D cannot form stable base pair with U^s. Accordingly, pcPNA with two noncomplementary strands can cooperatively engage DNA invasion without concerns of PNA self-complementarity and sequence requirement. Alternatively, the PNA with a bulky pyrrolidine-modified backbone that cannot form stable self-pairing hybrids offers a solution for the DNA duplex invasion without the requirement of base modifications.

Single-stranded γPNA pre-organizes in helical structure and exhibits more potent binding affinity and strand invasion capacity than aegPNA. Using single-stranded γPNA proves the possibility of single-stranded invasion, since γPNA pre-organizes into a helical polymer instead of random irregular one before binding to the complementary strand. Moreover, Ly *et al.* reported efficient single-stranded invasion *in vitro* using bifacial PNA in which modified bases termed Janus bases insert into dsDNA and base pairing with bases on both sides [19].

5.4.2. Limitations and Strategies

i. **Cellular delivery:** PNA shows good efficiency to invade DNA duplex; however, poor cellular uptake and quick clearance *in vivo* hugely impede its application as therapeutics. It is extremely critical to use a cellular uptake carrier, which efficiently absorb PNA, deliver PNA inside cells, and release PNA to the nucleus. Conventional class of covalent modification is through conjugating with cellular penetrating peptide, NLS, and positive charge amino acid such as polylysine and arginine. However, the covalent linkage influences homology searching, strand invasion, or sequence specificity, and it also impedes large-scale usage for clinic use. Non-covalent carrier provides improved efficacy *in vitro* cell line study, such as electroporation, nucleofection, and microinjection, but these methods are not applicable for *in vivo* clinic study. Currently, PNA backbone modifications such as α-lysine residue termed lysPNA and mini-polyethylene-glycol at γ-site PNA improve cellular uptake and gene editing efficiency. Alternative strategy is using nanoparticle-based delivery technology, such as oncolytic adenovirus (OAds) and poly(lactic-co-glycolic acid) (PLGA) particles. As a biodegradable and biocompatible polymer, PLGA is an FDA-approved technology for *in vivo* gene editing, instead of using electroporation and nucleo-transfection [16]. These strategies also apply to TFOs delivery for *in vivo* study.
ii. **Antisense effect:** PNA is a powerful ASOs candidate for gene regulation. Actually, compared with dsDNA binding, it is more direct and efficient to bind RNA with complementary sequence. Thus, it is vital to rule out antisense effect at maximum when targeting chromosomal DNA.
iii. **Chromatin accessibility, target sequence searching, and strand invasion:** The final efficacy of PNA in living cells is a combination of a series of events, including delivery, target sequence searching, chromatin accessibility, and

DNA invasion. PNA delivery can be easily evaluated using fluorescent probe and methods to improve it has been discussed above. Chromatin accessibility can be evaluated using ATAC-seq. PNA is believed to have better accessibility to hetero-chromatinized region than protein-based targeting system, and systematical understanding of the efficiency of PNA targeting different status of chromatin will help gene targeting and optimization to be more precise. Furthermore, to initiate biological regulation, PNA needs to search the target site, usually one specific short motif from whole human genome, for example. In this scenario, the efficiency of sequencing searching is extremely critical, in the premise of sequence specificity. For instance, homology searching via Hoogsteen triplex formation is faster than strand invasion via Watson–Crick base pairing. Mechanistic study reveals that bisPNA initially forms kinetically favored PNA-dsDNA triplex complexes and then gradually start DNA invasion followed by forming thermodynamically stable invasion complexes. Thus, how to balance the triplex forming sequence and DNA invasion sequence is essential to improve the target searching and strand invasion. However, there is a lack of method to evaluate the timing and efficiency of targeting sequence searching and strand invasion *in vivo*.

iv. **Off-target mutation and genotoxicity:** One of the main obstacles for the wide application of synthetic DNA binder is off-target issue. Usually, PNA only binds a length of about 10-mer DNA sequence, high binding affinity can compromise 1- to several mismatch inside pairing site, and the triplex forming stretch requires polypurine/polypyrimidine sequence, which all cause the high rate of off-target binding. Depending on the PNA application *in vivo*, off-target mutation using PNA-mediated gene editing method posies high risk for clinic application since the editing outcome is permanent. On the one hand, it is pivotal to lower off-targeting binding through structural optimization and sequence selection. At the same time, understanding of mutation rate of off-target binding, the site of mutation, and the influence of the mutation in cellular function are also crucial.

5.4.3. Biological Applications

5.4.3.1. Gene Editing PNA-mediated gene editing is believed to be much safer than active nuclease like CRISPR, TALANE, and ZNF. dsDNA invasion via PNA first induces XPA-dependent NER and followed by HDR, both of which are high fidelity repair pathways. bisPNA/tcPNA and artificial restriction of DNA cutter with pcPNA-Ce^{IV}-EDTA complexes (ARCUT) are two mostly evaluated designs. ARCUT is shown to generate double-strand DNA break at specific loci and shows high gene-editing efficiency through HR pathway using donor DNA template [20]. Promoting HR machinery via cell-cycle synchronization at G2/M phase and inhibiting NHEJ by RNA interference enhance correct gene editing. Moreover, it clearly shows that the editing efficiency is dependent on DNA double-strand break [21]. In the future, inducing DSB at the target site can be an effective measure to enhance HR-mediated gene editing in the premise of generating low nonspecific DSB.

Recently, tcPNA-mediated gene editing demonstrates promising efficacy in blood disorder, β-thalassemia. In the thalassemia mice, γ-mini-polyethylene-glycol-tcPNA ($\gamma^{\text{-MP}}$tcPNA) coupled with PLGA nanoparticle delivery system ameliorated the disease phenotype, with sustained elevation of blood hemoglobin levels into the normal range, reversal of splenomegaly, and up to 7% b-globin gene correction, with extremely low off-target effects [16].

5.4.3.2. Gene Modulation via Epigenetic Modulator As DNA binder, PNA has been studied to modulate gene expression via covalently tethering epigenetic modulators. A 15-mer single-stranded PNA with activation domain TAT peptide and NLS domain exerts γ-globin gene activation in mouse models and human primary peripheral blood cells [22]. In an *in vitro* study, PNA demonstrated to function as histone mark presenting platform, which actively recruit epigenetic modulating complexes PRC to the specific loci [23].

5.4.3.3. PNA Regulates G4 Structure DNA G-quadruplex is a noncanonical Watson–Crick duplex structure and has been demonstrated to associate with a variety of biological processes, including transcription, replication initiation, DNA damage, and telomere function [24]. Substantial studies reveal the role of G-quadruplex dysregulation in disease development like cancer [25]. However, a clear understanding of G-quadruplex's role in living cell is not available owing to a lack of site-specific probe, which can target specific gene loci and regulate G-quadruplex formation and resolving. Applying PNA to resolve G4 structure and installing G4 structure attract lots of research focus [26]. A G-rich PNA sequence (H-KGGGTGGGGK-NH2) invades dsDNA with homology G-rich sequence and forms PNA–DNA duplex at C-rich complementary strand. Interestingly, it also forms intermolecular G-quadruplex efficiently. Further study applied this PNA to sequence-specific DN cutter via tethering Ce^{IV}-EDTA [27]. G-rich PNA can form intra- and intermolecular G-quadruplex structure [28]. Pyrazolo[3,4-d]pyrimidine guanine (PPG) is a guanine analogue that lacks a nitrogen atom at the N7 position. PPG-containing PNA oligomer has been tested to initiate dsDNA invasion and G-quadruplex installation without the concerns of forming PNA G-quadruplexes [29].

5.5. PYRROLE-IMIDAZOLE POLYAMIDES (PIPs)

PIPs represent the best-characterized class of small molecules that can be programmed to bind to any predetermined B-form DNA and regulate gene expression patterns in a transgene-free and cost-effective manner [30, 31]. PIPs as DBD have been extensively applied to regulate specific gene expression because they can bind to their cognate DNA sequences and alter their interaction with natural transcription factors (TFs), which in turn can modulate the transcriptional networks and the epigenetic landscapes that govern cell fate. More significantly, PIPs serve as a valuable DNA binding platform to achieve higher order biological functions in a programmable manner, such as, PIP-epidrug, PIP-alkylator, cooperative PIP binding system, and PIP-dye conjugates. PIPs show potent efficacy on regulating gene expression patterns in a transgene-free and cost-effective manner.

5.5.1. Structure and DNA Binding

The precursors of cell-permeable, programmable, DNA minor groove-binding PIPs are the natural antibiotics netropsin and distamycin A, a class of Py dimer/trimer molecules linked with an amide bond [32, 33]. The clarification of the crystal structure of cooperative distamycin–DNA 2:1 complexes and the incorporation of Im building block establishes the basis for the subsequent development of PIPs [34, 35]. Dervan's group first developed hairpin PIPs by incorporating aliphatic γ-aminobutyric acid (γ-turn), which nicely combined and fixed two linear binding chains [36]. Hairpin PIPs are the most used chemical architecture, which can be attributed to their ease of synthesis, unambiguous pairing rules, high binding affinity, and excellent sequence selectivity. Results from X-ray crystal studies manifested that the Im/Py pair recognizes G/C pairs and the Py/Py pair recognizes A/T or T/A; this defines the essence of their sequence specificity and binding-site programmability (Figure 5.3a) [37]. The precise hydrogen bond interactions included in G:Im are 2-NH_2···3′-N and 3-N··· amide-NH_2; for C:Py 2-CO···amide-NH_2; for A:Py 3-N···amide-NH_2; and for T:Py 2-CO···amide-NH_2 (Figure 5.3b).

The development of solid-phase peptide synthesis resulted in a surge in the design and characterization of various intricate molecular architectures (Figure 5.3c) [38, 39]. Cyclic PIPs display higher binding affinities (40-fold enhancement for 5′-TGTTA-3′) than their counterpart hairpin PIPs [40]. Despite their small reduction in binding affinity, H-pin and U-pin PIPs extend the targeting sequence to the terminal G/C-rich DNA sequence [41, 42]. Tandem hairpin structures, including head-to-head and head-to-tail styles, are applicable to the design of molecules targeting more extended sequences, such as a PIP targeting the 24-bp telomere repeat [43]. Polypurine DNA has a narrow minor groove, and a linear PIP with a 1:1 binding mode was successfully applied to bind to a GA-rich sequence [44]. The replacement at γ-turn showed significant effect on DNA binding affinity [45], orientation, sequence selectivity, solubility [46], cell uptake [47], and toxicity [48], including γ-aminobutyric acid (γ-turn), (*R*)- and (*S*)-β-amino-γ-turn, (*R*)-α-amino-γ-turn, α-diaminobutyric acid, and the respective acetylated form at amino group.

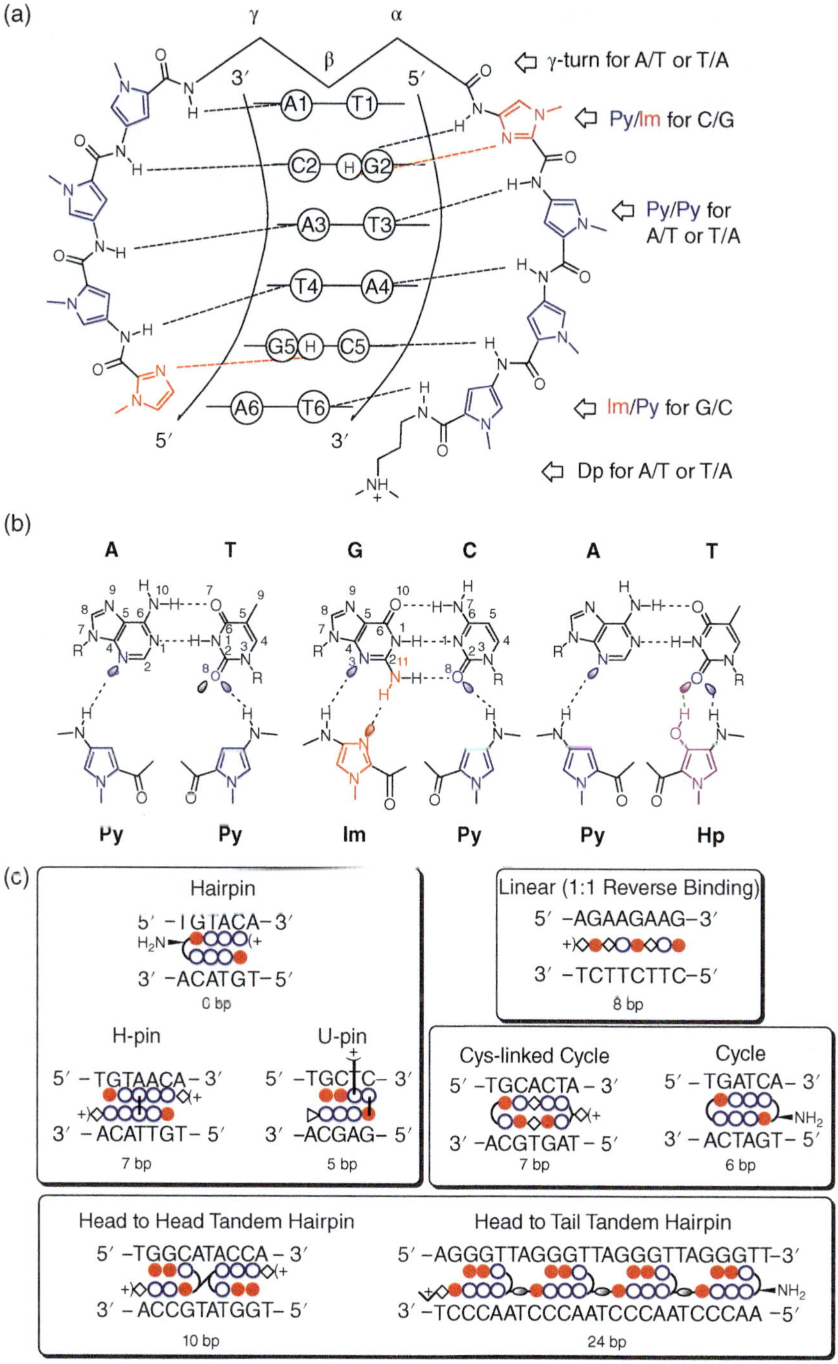

Figure 5.3 PIPs structure and DNA binding modes. (a) Schematic hydrogen interaction map of PIP-DNA. (b) Detailed hydrogen interaction and pairing rules of PIP-DNA. (c) Different types of PIP design and its binding to DNA.

5.5.2. Limitation and Strategy

To achieve ready-to-use and easy-to-use of artificial genetic switch with high sequence-selectivity and epigenetic modifying capability, substantial works are indispensable. PIPs can enter live mammalian cell lines, localize inside the nucleus, read their cognate DNA sequences, and alter basal transcription with no constraint from transfection agents [49]. Hence, they remain at the forefront of promising small-molecule-based designer drugs. While PIP-based systems show potential as sophisticated drugs, there is a need to perform substantial studies to overcome their shortcomings. Several critical impediments are remained to be tackled, especially high off-target effect, inflexible binding mode, and requirement of epigenetic modifying activity.

i. Canonical Py and Im ring cannot differentiate A/T to T/A pair, causing high off-target rate. Although 3-OH-Py can form valuable hydrogen bond with base A, it needs extra chemical protect and is chemically not so stable.
ii. The length of biologically applicable PIPs is 6–8 bp, which is suitable for blocking TFs-DNA interaction. However, for gene specific targeting, this length is far shorter, which usually needs to span at least 10–15 bp.
iii. We can design >10 bp PIPs monomer, like targeting telomere. But it has two concerns. Such PIPs cannot be uptaken by cells in the absence of transfection reagent. Meanwhile, divergent from tandem PIPs that have high rate of correct binding mode (difficult to synthesize), long hairpin PIPs would be difficult to form hairpin correctly in the minor groove, since extended bending is necessary.
iv. Another strategy is to design short PIPs monomer, but they can bind in nearby site in a cooperative manner, such as PIP-HoGu. This concept should also need to be careful since the cooperative domain might be too tedious to make it further efficient.
v. PIPs have been applied as DNA targeting platform and are equipped with epigenetic active module. This is the long-echoing direction but few breakthroughs yet. The short DNA binding sequence leads to high background noise activity, and few binding sites at the wanted gene loci cause weak anticipated biological output.
vi. Moreover, to recruit epigenetic regulator, there is only few ready-to-use small molecule until now. Also, in a specific purpose, efficient screening system to fish important potential epigenetic modulator is required. It would be necessary to learn from dCas9- [50], ZFs- [51], and TALE[52]-based activation/repression system, especially the importance of improving DNA binding specificity, and the introduction of cooperative binding system should be of significance. Lastly, further work should also focus on the selection and optimization of the suitable epi-drug that conjugated to PIPs, and an efficient assay platform that can recruit desired epigenetic enzyme but avoid the complexation of epi-drug to evaluate the efficacy of PIPs-based system must accelerate the research.

5.5.3. Biological Applications

5.5.3.1. DNA Binding Blocking Transcription Factor–DNA Interaction The primary therapeutic strategy using PIPs has been to block the interaction of transcription factor with DNA and trigger the modulation of RNAP II/III activity, which is followed by downstream gene perturbation. Crystallographic structural studies revealed allosteric perturbation of the DNA helix as the molecular basis by which minor groove-targeting PIPs disrupted the major groove-binding TFs [53]. Dervan, Gottesfeld, and colleagues demonstrated the selectivity and efficiency of PIPs as regulators of gene expression. In living cells, 8-ring hairpin PIPs were shown to interfere with 5S RNA expression by occupying specific regions of TFIII binding sites and suppressing RNAP III activity [31]. Moreover, PIPs targeting the flanking regions of the binding sites of Ets-1, LEF-1, and TBP in the HIV-1 promoter/enhancer suppressed RNAP II, thereby inhibiting virus replication in human lymphocytes by >99% [54]. Sugiyama and colleagues focused on the construction of longer designer PIPs. Such hairpin PIPs targeting longer sequences of therapeutically important genes have successfully altered the endogenous expression in different cell types of target genes including aurora kinases, TGF-β1, human lectin-like oxidized low-density lipoprotein, HER2, and EBNA1 [55–59]. Also, a groundbreaking application of PIPs targeting SOX2 was demonstrated in human-induced pluripotent stem (hiPS) cells [60].

5.5.3.2. PIP-Epigenetic Conjugates By utilizing the PIPs platform for DNA binding, artificial transcription factors have been reported with the conjugation of FD. By incorporating a variable 8-ring PIP moiety, compound library of both HDAC inhibitor conjugates (SAHA-PIP), we showed gene-specific histone acetylation and downstream gene activation in living cells [61]. The PIP-I sequence (DNA-binding sequence, 5′–WWCCWW–3′, W = A/T) exhibited the highest potential to upregulate SOX2, OCT4, and NANOG, which are responsible for iPS cell formation and maintenance (Table 5.3). Intriguingly, when SAHA was replaced with HAT activator CTB, CTB-I showed a gene activation pattern that was similar to that of SAHA-I, although it targeted a different epigenetic regulator [62]. In addition to DBD-HDAC inhibitors and DBD-HAT activators, which result in increased histone acetylation, has become one promising research field, DBD-HAT inhibitors demonstrate that these conjugates could retain their dual-target activity and exhibit potent anticancer activity. As the "reader" of histone acetylation, Brd protein binds acetylated histone to play a pivotal role in gene activation, and recruiting it to a specific gene locus has demonstrated therapeutic potential. A sequence-specific synthetic transcription elongation factor 1 (PIP–JQ1) was developed to stimulate the transcription elongation machinery, which enabled transcription across the repressive GAA repeats that silence frataxin expression in

TABLE 5.3 Biological functions of sequence-specific Epi-drugs

Comp.	Structure	Target genes	Regulated gene networks	Ref.
SAHA-I	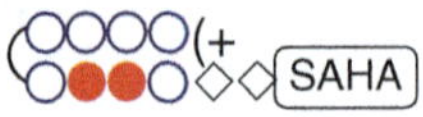	Pluripotent genes	Activate *SOX2*, *POU5F1*, and miR-302 family	[62]
SAHA-L	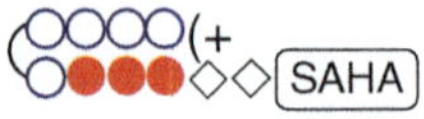	Neural genes	Induce *NESTIN*, *PAX6*, and *NGN2*	[63]
CTB-I	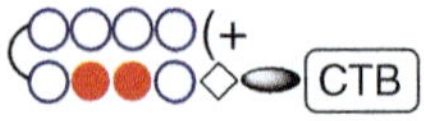	Pluripotent genes	Activate *SOX2*, *POU5F1*, and *NANOG*	[62]
C646-I	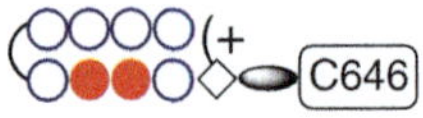	p53 targeting genes	Induce *SPATA18*, *EDA2R*, *MDM2*, *BTG2*, and *FDXR*	[64]
PIP-JQ1		*FXN1*	Specifically activate *FXN1* in FRDA cells	[65]
PIP-Bi		Sequence-selective genes	Sequence-specific relevant activation, such as *NTS* gene	[66]
PIP-NCD38		Epigenome changes	Region-selective alteration of epigenetic status	[67]

Friedreich's ataxia (FRDA). This novel design defined a potential therapeutic regimen for FRDA patients, a terminal neurodegenerative disease with no effective therapy [65].

5.5.3.3. Cooperative DNA Binding Construct PIPs show potent efficacy on regulating gene expression patterns in a transgene-free and cost-effective manner; however, several critical impediments remain, especially high off-target effect and inflexible binding mode. To this end, PIPs conjugated to Host-Guest assemblies, termed **PIP–HoGu,** has been reported, which mimic the cooperation between natural TF pairs (Table 5.4) [69]. The results consistently demonstrated that the **PIP–HoGu** system formed stable, noncovalent, cooperative complexes, thereby meeting the critical criteria to mimic a natural TF pair. Significantly, cell-based evaluation validated the capacity of **PIP–HoGu** to exhibit potent cooperative inhibitory effects on gene expression under physiological conditions by disrupting TF pair–DNA function [71]. Recently, cooperation domain in the **PIP–HoGu** system is upgraded by replacing

TABLE 5.4 Summary of representative non-covalent DNA binding systems

Non-covalent system	Property	Ref.
Host guest for PIP-Peptide 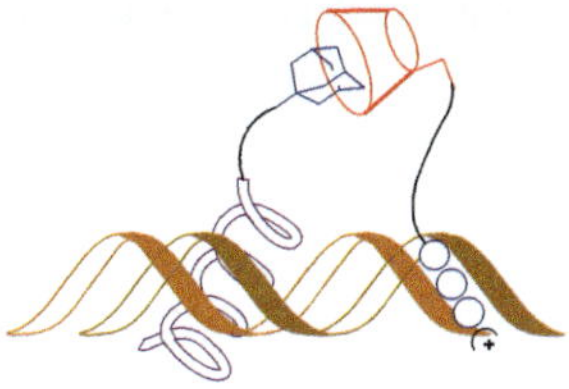	Non-covalent synergic interaction; Major and minor DNA groove binding; Need to be tested in cells	[68]
PIP-HoGu system 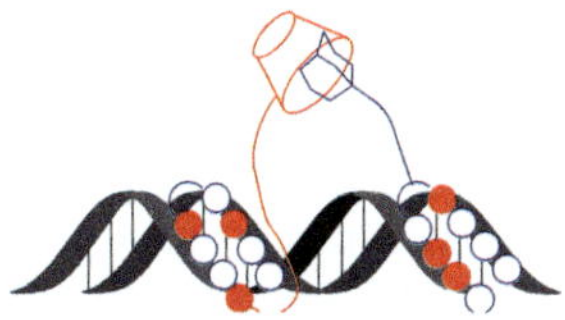	Small molecule synergic DNA binding; Proved efficacy in cells; Need to improve host-guest interaction	[69]
PIP-NaCo system 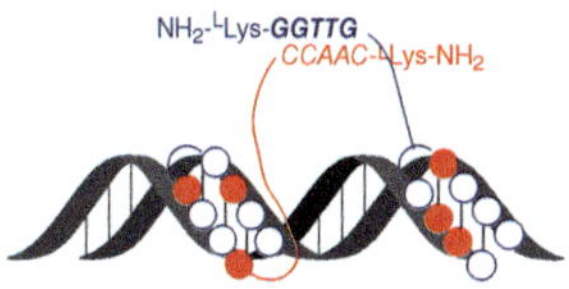	Cooperation is tunable; Orthogonality that cannot bind right-handed DNA/RNA; Synthetically labor-consuming	[70]
CB7-assisted PIP-HoGu system 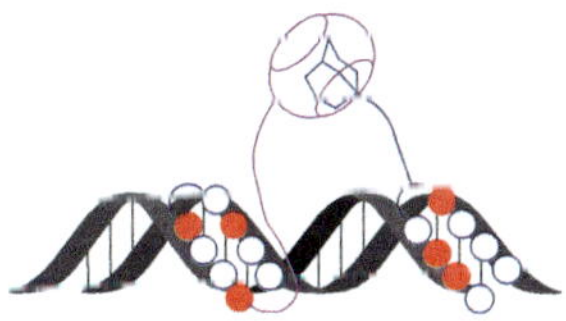	Ultratight host–guest interaction; Higher cooperation	
ePIP-HoGu system 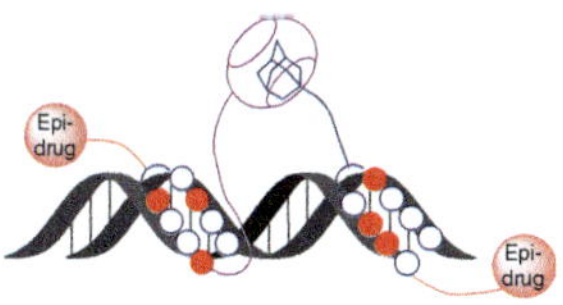	Epigenetically active synergistically DNA binding; Efficiently modify local nucleosomes with matched DNA binding sites	

cyclodextrin (Cyd) with CB7 [72, 73]. Significantly, CB7 exhibits an ultratight binding affinity to adamantane (Ada) (K_D routinely in the 10^{-9}–10^{-12} M^{-1} range), whereas Cyd–Ada has a relatively weak host–guest interaction (K_D of ca. 10^{-5} M^{-1})[73]. In another study, LH γPNA was chosen as CID because of its bio-orthogonality to the endogenous nucleic acids and its high binding affinity and sequence specificity for the partner strand [74]. **Pip–NaCo** shows cooperativity of its concerted DNA recognition highly comparable to that of the natural system, with a minimum energetics of cooperation of –3.26 kcal/mol [70]. The vertical binding mode, in which γPNA duplex is vertical to dsDNA, demonstrated further enhancement of cooperativity. The orthogonal **PIP–NaCo** platform also has the potential to provide triple to multiple hetero-binding systems to target further complicated natural TF cluster networks.

5.5.3.4. Advance DNA Targeting Construct Simple DNA binding by cooperative interaction systems induce biological effects through the disruption of TF pair binding [69]. However, in order to achieve higher level of biological efficacy and wider biological applications, such as gene activation, the next challenge is the installation of epigenetic modulator (an epi-drug) to the optimized cooperative DNA binding system [66]. A delicate design harboring all three domains was reported [75]. Dimerization domain of residues 251–281 of the yeast protein GCN4 is known to form a coiled-coil dimer motif. The activation domain used is amphipathic helix (AH) (PEFPGIELQELQELQALLQQ). Such design represented a key step toward the future design.

Based on the CB7-assisted **PIP–HoGu**, an epi-drug has been covalently tethered to **PIP–HoGu** and constructs an advanced synthetic transcription factor mimic termed **ePIP–HoGu** (**Table 5.4**) [76]. This construct is expected to be capable of cooperatively recruiting the epigenetic modifiers to the predetermined DNA locus and nearby nucleosome. Indeed, the incorporation of a cooperative dimer system into PIPs–epi-drug conjugates increase the DNA recognition length, reinforces reasonable sequence selectivity, and allows versatile binding modes. As a proof-of-concept study, the **ePIP–HoGu** system is shown to be adept at synergistically augmenting proximate histone acetylation with valuable efficiency and selectivity.

5.6. PERSPECTIVES

Synthetic chromosomal DNA binder with sequence specificity and programmability, including TFOs, PNA and PIPs, should never be out of date and could ultimately be the antidote for diseases such as genetic diseases. Numerous properties give them better advantages, such as small in size, cell permeability, sequence specificity, programmability, user friendly, and easy to modify.

In recent years, new technology especially CRISPR-Cas out-competes synthetic chromosomal DNA binder because of their high specificity and modularity. CRISPR-Cas technology has attracted huge research resources and replaces lots of other technologies in lots of applications, including gene editing in clinics, although the academic community recognizes its unavoidable off-target effect when using nuclease active format for gene editing. On the one hand, given the limited research resources in total, the long-term, intensive research investment on the area that is not so productive in short-term could be not sustainable before the breakthrough can be accumulated. The risk is the imbalance of research resources distribution could hugely slow down the optimization and development of these synthetic DNA binders. On the other hand, CRISPR-Cas system sets out a perfect example for synthetic chromosomal DNA binder to follow.

ACKNOWLEDGMENT

This work was supported by AMED [JP20am0101101 (Platform Project for Supporting Drug Discovery and Life Science Research (BINDS) to H.S.)], 19H03349 (Grants-in-Aid for Scientific Research (B)) to G P.N and 16H06356 (Grants-in-Aid for Scientific Research (S)) to H.S.], and NIH awards R01CA236350 to H.S. This chapter is to commemorate the 10th anniversary of falling in love of Dr. Yu with Dr. Ye.

REFERENCES

1. M. Jinek, K. Chylinski, I. Fonfara, M. Hauer, J.A. Doudna, E. Charpentier, A programmable dual-RNA–guided DNA endonuclease in adaptive bacterial immunity, *Science*, 337 (2012) 816–821.
2. Z. Yu, G.N. Pandian, T. Hidaka, H. Sugiyama, Therapeutic gene regulation using pyrrole-imidazole polyamides, *Adv. Drug Deliv. Rev.*, 147 (2019) 66–85.
3. C.I.E. Smith, R. Zain, Therapeutic oligonucleotides: state of the art, *Annu. Rev. Pharmacol. Toxicol.*, 59 (2019) 605–630.
4. P.B. Dervan, Molecular recognition of DNA by small molecules, *Bioorg. Med. Chem.*, 9 (2001) 2215–2235.
5. S. Buchini, C.J. Leumann, Recent improvements in antigene technology, *Curr. Opin. Chem. Biol.*, 7 (2003) 717–726.
6. K. Hoogsteen, The crystal and molecular structure of a hydrogen-bonded complex between 1-methylthymine and 9-methyladenine, *Acta Crystallogr.*, 16 (1963) 907–916.
7. M.P. Knauert, P.M. Glazer, Triplex forming oligonucleotides: sequence-specific tools for gene targeting, *Hum. Mol. Genet.*, 10 (2001) 2243–2251.
8. N.T. Thuong, C. Hélène, Sequence-specific recognition and modification of double-helical DNA by oligonucleotides, *Angew. Chem. Int. Ed.*, 32 (1993) 666–690.
9. M. Duca, P. Vekhoff, K. Oussedik, L. Halby, P.B. Arimondo, The triple helix: 50 years later, the outcome, *Nucleic Acids Res.*, 36 (2008) 5123–5138.
10. P.A. Havre, P.M. Glazer, Targeted mutagenesis of simian virus 40 DNA mediated by a triple helix-forming oligonucleotide, *J. Virol.*, 67 (1993) 7324–7331.
11. A. Majumdar, A. Khorlin, N. Dyatkina, F.L.M. Lin, J. Powell, J. Liu, Z. Fei, Y. Khripine, K.A. Watanabe, J. George, P.M. Glazer, M.M. Seidman, Targeted gene knockout mediated by triple helix forming oligonucleotides, *Nat. Gen.*, 20 (1998) 212–214.
12. Z. Luo, M.A. Macris, A.F. Faruqi, P.M. Glazer, High-frequency intrachromosomal gene conversion induced by triplex-forming oligonucleotides microinjected into mouse cells, *Proc. Natl. Acad. Sci. U.S.A.*, 97 (2000) 9003–9008.

13. M.P. Knauert, J.M. Kalish, D.C. Hegan, P.M. Glazer, Triplex-stimulated intermolecular recombination at a single-copy genomic target, *Mol. Ther.*, 14 (2006) 392–400.
14. P. Nielsen, M. Egholm, R. Berg, O. Buchardt, Sequence-selective recognition of DNA by strand displacement with a thymine-substituted polyamide, *Science*, 254 (1991) 1497–1500.
15. P. Muangkaew, T. Vilaivan, Modulation of DNA and RNA by PNA, *Bioorganic Med. Chem. Lett.*, 30 (2020) 127064.
16. R. Bahal, N. Ali McNeer, E. Quijano, Y. Liu, P. Sulkowski, A. Turchick, Y.-C. Lu, D.C. Bhunia, A. Manna, D.L. Greiner, M.A. Brehm, C.J. Cheng, F. López-Giráldez, A. Ricciardi, J. Beloor, D.S. Krause, P. Kumar, P.G. Gallagher, D.T. Braddock, W. Mark Saltzman, D.H. Ly, P.M. Glazer, in vivo correction of anaemia in β thalassemic mice by γPNA-mediated gene editing with nanoparticle delivery, *Nat. Commun.*, 7 (2016) 13304.
17. P.M.D. Moreno, S. Geny, Y.V. Pabon, H. Bergquist, E.M. Zaghloul, C.S.J. Rocha, I.I. Oprea, B. Bestas, S.E. Andaloussi, P.T. Jørgensen, E.B. Pedersen, K.E. Lundin, R. Zain, J. Wengel, C.I.E. Smith, Development of bis-locked nucleic acid (bisLNA) oligonucleotides for efficient invasion of supercoiled duplex DNA, *Nucleic Acids Res.*, 41 (2013) 3257–3273.
18. E.M. Zaghloul, A.S. Madsen, P.M.D. Moreno, I.I. Oprea, S. El-Andaloussi, B. Bestas, P. Gupta, E.B. Pedersen, K.E. Lundin, J. Wengel, C.I.E. Smith, Optimizing anti-gene oligonucleotide 'Zorro-LNA' for improved strand invasion into duplex DNA, *Nucleic Acids Res.*, 39 (2010) 1142–1154.
19. S.A. Thadke, V.M. Hridya, J.D.R. Perera, R.R. Gil, A. Mukherjee, D.H. Ly, Shape selective bifacial recognition of double helical DNA, *Commun. Chem.*, 1 (2018) 79.
20. M. Komiyama, Chemical modifications of artificial restriction DNA cutter (ARCUT) to promote its in vivo and in vitro applications, *Artificial DNA, PNA & XNA*, 5 (2014) e1112457.
21. H. Katada, T. Harumoto, N. Shigi, M. Komiyama, Chemical and biological approaches to improve the efficiency of homologous recombination in human cells mediated by artificial restriction DNA cutter, *Nucleic Acids Res.*, 40 (2012) e81.
22. J. Chen, K.R. Peterson, C. Iancu-Rubin, J.J. Bieker, Design of embedded chimeric peptide nucleic acids that efficiently enter and accurately reactivate gene expression in vivo, *Proc. Natl. Acad. Sci. U.S.A.*, 107 (2010) 16846–16851.
23. Z.Z. Brown, M.M. Müller, H.E. Kong, P.W. Lewis, T.W. Muir, Targeted histone peptides: insights into the spatial regulation of the methyltransferase PRC2 by using a surrogate of heterotypic chromatin, *Angew. Chem. Int. Ed. Engl.*, 54 (2015) 6457–6461.
24. J. Spiegel, S. Adhikari, S. Balasubramanian, The structure and function of DNA G-quadruplexes, *Trend. Chem.*, 2 (2020) 123–136.
25. S. Asamitsu, S. Obata, Z. Yu, T. Bando, H. Sugiyama, Recent progress of targeted G-quadruplex-preferred ligands toward cancer therapy, *Molecules*, 24 (2019) 429.
26. I.G. Panyutin, M.I. Onyshchenko, E.A. Englund, D.H. Appella, R.D. Neumann, Targeting DNA G-quadruplex structures with peptide nucleic acids, *Curr. Pharm. Des.*, 18 (2012) 1984–1991.
27. T. Ishizuka, J. Yang, M. Komiyama, Y. Xu, G-rich sequence-specific recognition and scission of human genome by PNA/DNA hybrid G-quadruplex formation, *Angew. Chem. Int. Ed. Engl.*, 51 (2012) 7198–7202.

28. Y. Krishnan-Ghosh, E. Stephens, S. Balasubramanian, A PNA4 quadruplex, *J. Am. Chem. Soc.*, 126 (2004) 5944–5945.
29. E.A. Englund, P. Gupta, C.M. Micklitsch, M.I. Onyshchenko, E. Remeeva, R.D. Neumann, I.G. Panyutin, D.H. Appella, PPG peptide nucleic acids that promote DNA guanine quadruplexes, *ChemBioChem*, 15 (2014) 1887–1890.
30. J.W. Trauger, E.E. Baird, P.B. Dervan, Recognition of DNA by designed ligands at subnanomolar concentrations, *Nature*, 382 (1996) 559–561.
31. J.M. Gottesfeld, L. Neely, J.W. Trauger, E.E. Baird, P.B. Dervan, Regulation of gene expression by small molecules, *Nature*, 387 (1997) 202–205.
32. A.C. Finlay, F.A. Hochstein, B.A. Sobin, F.X. Murphy, Netropsin, a new antibiotic produced by a streptomyces, *J. Am. Chem. Soc.*, 73 (1951) 341–343.
33. F. Arcamone, S. Penco, P. Orezzi, V. Nicolella, A. Pirelli, Structure and synthesis of distamycin A, *Nature*, 203 (1964) 1064.
34. J.G. Pelton, D.E. Wemmer, Structural characterization of a 2:1 distamycin A.d(CGCAAATTGGC) complex by two-dimensional NMR, *Proc. Natl. Acad. Sci. U.S.A.*, 86 (1989) 5723–5727.
35. W.S. Wade, M. Mrksich, P.B. Dervan, Binding affinities of synthetic peptides, pyridine-2-carboxamidonetropsin and 1-methylimidazole-2-carboxamidonetropsin, that form 2:1 complexes in the minor groove of double-helical DNA, *Biochemistry*, 32 (1993) 11385–11389.
36. M. Mrksich, M.E. Parks, P.B. Dervan, Hairpin peptide motif. A new class of oligopeptides for sequence-specific recognition in the minor groove of double-helical DNA, *J. Am. Chem. Soc.*, 116 (1994) 7983–7988.
37. C.L. Kielkopf, E.E. Baird, P.B. Dervan, D.C. Rees, Structural basis for G.C recognition in the DNA minor groove, *Nat. Struct. Biol.*, 5 (1998) 104–109.
38. E.E. Baird, P.B. Dervan, Solid phase synthesis of polyamides containing imidazole and pyrrole amino acids, *J. Am. Chem. Soc.*, 118 (1996) 6141–6146.
39. N.R. Wurtz, J.M. Turner, E.E. Baird, P.B. Dervan, Fmoc solid phase synthesis of polyamides containing pyrrole and imidazole amino acids, *Org. Lett.*, 3 (2001) 1201–1203.
40. J. Cho, M.E. Parks, P.B. Dervan, Cyclic polyamides for recognition in the minor groove of DNA, *Proc. Natl. Acad. Sci. U.S.A.*, 92 (1995) 10389–10392.
41. M. Mrksich, P.B. Dervan, Design of a covalent peptide heterodimer for sequence-specific recognition in the minor groove of double-helical DNA, *J. Am. Chem. Soc.*, 116 (1994) 3663–3664.
42. A. Heckel, P.B. Dervan, U-pin polyamide motif for recognition of the DNA minor groove, *Chem. Eur. J.*, 9 (2003) 3353–3366.
43. Y. Kawamoto, A. Sasaki, A. Chandran, K. Hashiya, S. Ide, T. Bando, K. Maeshima, H. Sugiyama, Targeting 24 bp within telomere repeat sequences with tandem tetramer pyrrole–imidazole polyamide probes, *J. Am. Chem. Soc.*, 138 (2016) 14100–14107.
44. A.R. Urbach, P.B. Dervan, Toward rules for 1:1 polyamide:DNA recognition, *Proc. Natl. Acad. Sci. U.S.A.*, 98 (2001) 4343–4348.
45. C. Dose, M.E. Farkas, D.M. Chenoweth, P.B. Dervan, Next generation hairpin polyamides with (R)-3,4-diaminobutyric acid turn unit, *J. Am. Chem. Soc.*, 130 (2008) 6859–6866.
46. A.E. Hargrove, J.A. Raskatov, J.L. Meier, D.C. Montgomery, P.B. Dervan, Characterization and solubilization of pyrrole-imidazole polyamide aggregates, *J. Med. Chem.*, 55 (2012) 5425–5432.

47. J.L. Meier, D.C. Montgomery, P.B. Dervan, Enhancing the cellular uptake of Py-Im polyamides through next-generation aryl turns, *Nucleic Acids Res.*, 40 (2012) 2345–2356.
48. F. Yang, N.G. Nickols, B.C. Li, J.O. Szablowski, S.R. Hamilton, J.L. Meier, C.M. Wang, P.B. Dervan, Animal toxicity of hairpin pyrrole-imidazole polyamides varies with the turn unit, *J. Med. Chem.*, 56 (2013) 7449–7457.
49. G.N. Pandian, H. Sugiyama, Nature-inspired design of smart biomaterials using the chemical biology of nucleic acids, *Bull. Chem. Soc. Jpn.*, 89 (2016) 843–868.
50. N. Matharu, S. Rattanasopha, S. Tamura, L. Maliskova, Y. Wang, A. Bernard, A. Hardin, W.L. Eckalbar, C. Vaisse, N. Ahituv, CRISPR-mediated activation of a promoter or enhancer rescues obesity caused by haploinsufficiency, *Science*, 363 (2019) eaau0629.
51. M. Garriga-Canut, C. Agustín-Pavón, F. Herrmann, A. Sánchez, M. Dierssen, C. Fillat, M. Isalan, Synthetic zinc finger repressors reduce mutant huntingtin expression in the brain of R6/2 mice, *Proc. Natl. Acad. Sci. U.S.A.*, 109 (2012) E3136–E3145.
52. M.L. Maeder, S.J. Linder, D. Reyon, J.F. Angstman, Y. Fu, J.D. Sander, J.K. Joung, Robust, synergistic regulation of human gene expression using TALE activators, *Nat. Meth.*, 10 (2013) 243.
53. D.M. Chenoweth, P.B. Dervan, Structural basis for cyclic Py-Im polyamide allosteric inhibition of nuclear receptor binding, *J. Am. Chem. Soc.*, 132 (2010) 14521–14529.
54. L.A. Dickinson, J.W. Trauger, E.E. Baird, P.B. Dervan, B.J. Graves, J.M. Gottesfeld, Inhibition of Ets-1 DNA binding and ternary complex formation between Ets-1, NF-kappaB, and DNA by a designed DNA-binding ligand, *J. Biol. Chem.*, 274 (1999) 12765–12773.
55. T. Takahashi, Y. Asami, E. Kitamura, T. Suzuki, X. Wang, J. Igarashi, A. Morohashi, Y. Shinojima, H. Kanou, K. Saito, T. Takasu, H. Nagase, Y. Harada, K. Kuroda, T. Watanabe, S. Kumamoto, T. Aoyama, Y. Matsumoto, T. Bando, H. Sugiyama, C. Yoshida-Noro, N. Fukuda, N. Hayashi, Development of pyrrole-imidazole polyamide for specific regulation of human aurora kinase-A and -B gene expression, *Chem. Biol.*, 15 (2008) 829–841.
56. T. Ueno, N. Fukuda, A. Tsunemi, E.H. Yao, H. Matsuda, K. Tahira, T. Matsumoto, K. Matsumoto, Y. Matsumoto, H. Nagase, H. Sugiyama, T. Sawamura, A novel gene silencer, pyrrole-imidazole polyamide targeting human lectin-like oxidized low-density lipoprotein receptor-1 gene improves endothelial cell function, *J. Hypertens.*, 27 (2009) 508–516.
57. T. Suzuki, Y. Asami, T. Takahashi, X. Wang, T. Watanabe, T. Bando, H. Sugiyama, N. Fukuda, H. Nagase, Development of a molecule-recognized promoter DNA sequence for inhibition of HER2 expression, *J. Antibiot.*, 62 (2009) 339–341.
58. H. Matsuda, N. Fukuda, T. Ueno, M. Katakawa, X. Wang, T. Watanabe, S. Matsui, T. Aoyama, K. Saito, T. Bando, Y. Matsumoto, H. Nagase, K. Matsumoto, H. Sugiyama, Transcriptional inhibition of progressive renal disease by gene silencing pyrrole-imidazole polyamide targeting of the transforming growth factor-beta1 promoter, *Kidney Int.*, 79 (2011) 46–56.
59. A. Yasuda, K. Noguchi, M. Minoshima, G. Kashiwazaki, T. Kanda, K. Katayama, J. Mitsuhashi, T. Bando, H. Sugiyama, Y. Sugimoto, DNA ligand designed to antagonize EBNA1 represses Epstein–Barr virus-induced immortalization, *Cancer Sci.*, 102 (2011) 2221–2230.

60. P.W. Burridge, E. Matsa, P. Shukla, Z.C. Lin, J.M. Churko, A.D. Ebert, F. Lan, S. Diecke, B. Huber, N.M. Mordwinkin, J.R. Plews, O.J. Abilez, B. Cui, J.D. Gold, J.C. Wu, Chemically defined generation of human cardiomyocytes, *Nat. Meth.*, 11 (2014) 855–860.
61. G.N. Pandian, Y. Nakano, S. Sato, H. Morinaga, T. Bando, H. Nagase, H. Sugiyama, A synthetic small molecule for rapid induction of multiple pluripotency genes in mouse embryonic fibroblasts, *Sci. Rep.*, 2 (2012) e544.
62. L. Han, G.N. Pandian, A. Chandran, S. Sato, J. Taniguchi, G. Kashiwazaki, Y. Sawatani, K. Hashiya, T. Bando, Y. Xu, X. Qian, H. Sugiyama, A synthetic DNA-binding domain guides distinct chromatin-modifying small molecules to activate an identical gene network, *Angew. Chem. Int. Ed. Engl.*, 54 (2015) 8700–8703.
63. Y. Wei, G.N. Pandian, T. Zou, J. Taniguchi, S. Sato, G. Kashiwazaki, T. Vaijayanthi, T. Hidaka, T. Bando, H. Sugiyama, A multi-target small molecule for targeted transcriptional activation of therapeutically significant nervous system genes, *ChemistryOpen*, 5 (2016) 517–521.
64. Z. Yu, J. Taniguchi, Y. Wei, G.N. Pandian, K. Hashiya, T. Bando, H. Sugiyama, Antiproliferative and apoptotic activities of sequence-specific histone acetyltransferase inhibitors, *Eur. J. Med. Chem.*, 138 (2017) 320–327.
65. G.S. Erwin, M.P. Grieshop, A. Ali, J. Qi, M. Lawlor, D. Kumar, I. Ahmad, A. McNally, N. Teider, K. Worringer, R. Sivasankaran, D.N. Syed, A. Eguchi, M. Ashraf, J. Jeffery, M. Xu, P.M.C. Park, H. Mukhtar, A.K. Srivastava, M. Faruq, J.E. Bradner, A.Z. Ansari, Synthetic transcription elongation factors license transcription across repressive chromatin, *Science*, 358 (2017) 1617–1622.
66. J. Taniguchi, Y. Feng, G.N. Pandian, F. Hashiya, T. Hidaka, K. Hashiya, S. Park, T. Bando, S. Ito, H. Sugiyama, Biomimetic artificial epigenetic code for targeted acetylation of histones, *J. Am. Chem. Soc.*, 140 (2018) 7108–7115.
67. K. Alagarswamy, K.I. Shinohara, S. Takayanagi, M. Fukuyo, A. Okabe, B. Rahmutulla, N. Yoda, R. Qin, N. Shiga, M. Sugiura, H. Sato, K. Kita, T. Suzuki, T. Nemoto, A. Kaneda, Region-specific alteration of histone modification by LSD1 inhibitor conjugated with pyrrole-imidazole polyamide, *Oncotarget*, 9 (2018) 29316–29335.
68. J.B. Blanco, V.I. Dodero, M.E. Vázquez, M. Mosquera, L. Castedo, J.L. Mascareñas, Sequence-specific DNA binding by noncovalent peptide–tripyrrole conjugates, *Angew. Chem. Int. Ed. Engl.*, 45 (2006) 8210–8214.
69. Z. Yu, C. Guo, Y. Wei, K. Hashiya, T. Bando, H. Sugiyama, Pip-HoGu: an artificial assembly with cooperative DNA recognition capable of mimicking transcription factor pairs, *J. Am. Chem. Soc.*, 140 (2018) 2426–2429.
70. Z. Yu, W.C. Hsieh, S. Asamitsu, K. Hashiya, T. Bando, D.H. Ly, H. Sugiyama, Orthogonal gammaPNA dimerization domains empower DNA binders with cooperativity and versatility mimicking that of transcription factor pairs, *Chem. Eur. J.*, 24 (2018) 14183–14188.
71. N.G. Nickols, J.O. Szablowski, A.E. Hargrove, B.C. Li, J.A. Raskatov, P.B. Dervan, Activity of a Py-Im polyamide targeted to the estrogen response element, *Mol. Cancer Ther.*, 12 (2013) 675–684.
72. S. Liu, C. Ruspic, P. Mukhopadhyay, S. Chakrabarti, P.Y. Zavalij, L. Isaacs, The Cucurbit[n]uril family: prime components for self-sorting systems, *J. Am. Chem. Soc.*, 127 (2005) 15959–15967.

73. W.S. Jeon, K. Moon, S.H. Park, H. Chun, Y.H. Ko, J.Y. Lee, E.S. Lee, S. Samal, N. Selvapalam, M.V. Rekharsky, V. Sindelar, D. Sobransingh, Y. Inoue, A.E. Kaifer, K. Kim, Complexation of ferrocene derivatives by the Cucurbit[7]uril host: a comparative study of the Cucurbituril and cyclodextrin host families, *J. Am. Chem. Soc.*, 127 (2005) 12984–12989.
74. I. Sacui, W.-C. Hsieh, A. Manna, B. Sahu, D.H. Ly, Gamma peptide nucleic acids: as orthogonal nucleic acid recognition codes for organizing molecular self-assembly, *J. Am. Chem. Soc.*, 137 (2015) 8603–8610.
75. A.K. Mapp, A.Z. Ansari, M. Ptashne, P.B. Dervan, Activation of gene expression by small molecule transcription factors, *Proc. Natl. Acad. Sci. U.S.A.*, 97 (2000) 3930–3935.
76. Z. Yu, M. Ai, S.K. Samanta, F. Hashiya, J. Taniguchi, S. Asamitsu, S. Ikeda, K. Hashiya, T. Bando, G.N. Pandian, L. Isaacs, H. Sugiyama, A synthetic transcription factor pair mimic for precise recruitment of an epigenetic modifier to the targeted DNA locus, *ChemComm*, 56 (2020) 2296–2299.

CHAPTER 6

DNA SYNTHESIS AND SYNTHETIC GENOME

ZE-XIONG XIE, RUI YING ZHU, YUE LIU, AND YING JIN YUAN
Frontiers Science Center for Synthetic Biology and Key Laboratory of Systems Bioengineering of Ministry of Education, School of Chemical Engineering and Technology, Tianjin University, Tianjin, PR China

6.1. INTRODUCTION

In 1828, the success of urea synthesis from ammonium chloride was a revolution breaking the boundary of organic and inorganic. Since then, scientists have been working toward synthesizing various bioactive organic molecules, including polypeptides, proteins, and nucleic acids. The discovery of DNA double helix structure led to the central dogma, which states the importance of genetic information stored in genomic DNA [1, 2]. Synthetic genomics is a nascent field in which scientists genetically modify pre-existing life forms, or artificially synthesize new genomes [3].

Genome synthesis is consisted of multiple steps, including design, assembly, transplantation, bug mapping, and debugging. The synthetic genome is designed from natural sequences with customized features, chemically assembled from small oligonucleotides *in vivo* or *in vitro*, and then transplanted into recipient cells. In so doing, the biological system is organized by the synthetic genome in all aspects including the maintenance of cell fitness, genomic stability, and the disruption of unnecessary metabolic processes [4, 5]. Thus, the customized synthetic systems can be optimized to generate a variety of products including drugs, vaccines, and chemical stocks [6] (Figure 6.1).

Nucleic Acids in Medicinal Chemistry and Chemical Biology: Drug Development and Clinical Applications, First Edition. Edited by Lihe Zhang, Xinjing Tang, Zhen Xi, and Jyoti Chattopadhyaya.

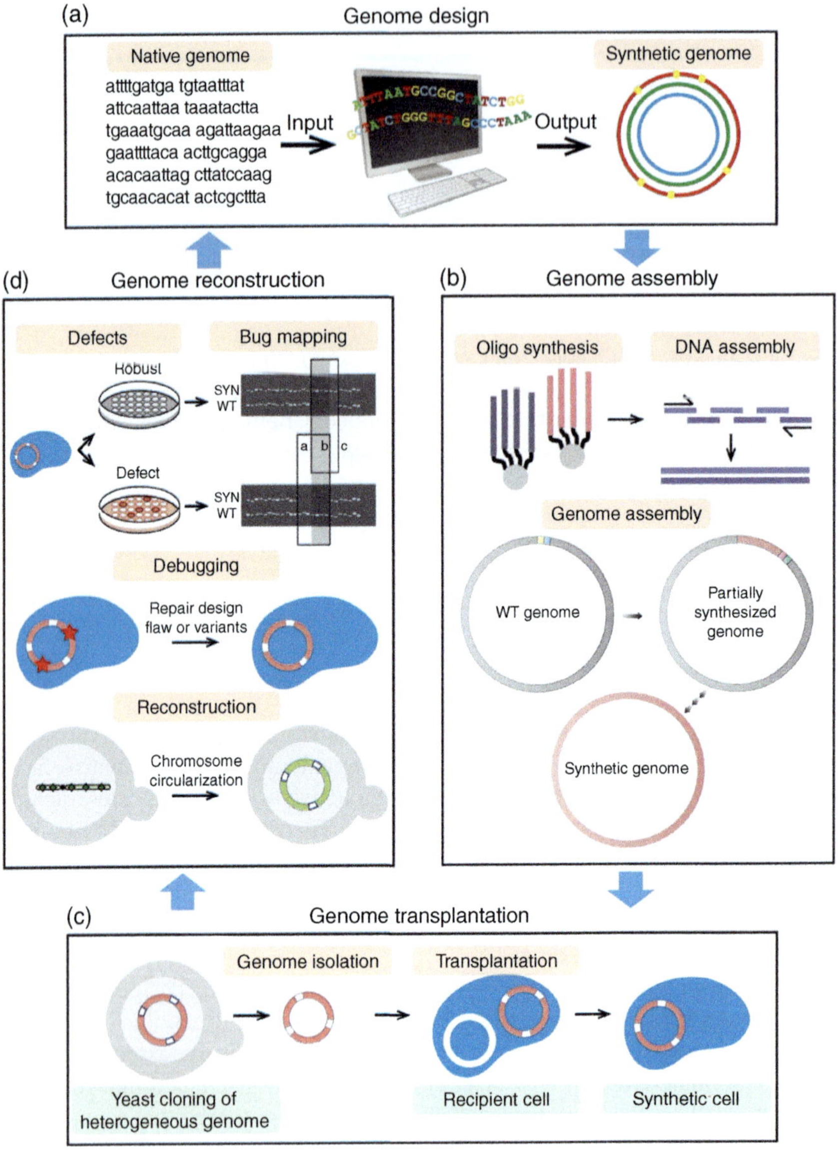

Figure 6.1 **Genome synthesis.** (a) The synthetic genome is designed from native genome sequence with the aid of artificial intelligence. (b) The whole synthetic genome is assembled from DNA segments and chemically produced oligonucleotides. (c) The assembled genome is further transplanted to recipient cells to maintain cell fitness of synthetic cell. (d) Genome sequence errors leading to growth defects and unexpected cell functions are mapped and corrected to generated variable synthetic cells.

6.2. DESIGN A SYNTHETIC GENOME

In the first stage of synthetic genomics, scientists aimed to test whether a chemically synthesized genome can maintain cell fitness. Thus, the sequence of synthetic genome was minimally altered from the native one. In the last decade, more customer features have been introduced for various purposes, for example to minimize the genome size, to recode codons, and to endow the genome new functions (Figure 6.2).

6.2.1. The First Synthetic Genome with Minimal Design

The first synthetic genome, a 1.08-Mb *Mycoplasma mycoides* genome, namely JCVI-syn1.0, was synthesized in 2010 [7]. To ensure robustness of cellular growth, the research group designed the genomic sequence with a principle not to alter the native sequence significantly. Only a few "watermark" sequences and gene deletions were included in their design [7]. JCVI-syn1.0 upon assembly in yeast was transplanted into a *Mycoplasma capricolum* recipient cell followed by knockout of the *M. capricolum* native genome, creating a new *M. mycoides* strain. The synthetic strain exhibited the cell morphology and other phenotypes of *M. mycoides* and was fully functional in self-replication.

6.2.2. Synthetic Genome with Customer Features

Upon early success, scientists became brave to introduce more customer features to synthetic genomes. For instance, the synthetic yeast genome project (Sc2.0) redesigned all 16 chromosomes of *Saccharomyces cerevisiae* with repetitive sequences deleted to increase genome stability, and with customized genetic elements such as loxPsym sites inserted to improve genome flexibility [8].

Repetitive genomic sequences are related to genomic instability, such as telomeres and transposons [9]. Complicated abnormal recombination events frequently occur at repetitive genomic sequences, leading to deletions, duplications, and chromosomal translocations [10, 11]. To increase the genomic stability, repetitive sequences including transposons and subtelomeres were designed to be removed. Although introns are rare in *S. cerevisiae*, they were also excluded from the design. All tRNA genes were translocated to a neochromosome, and the native telomeres were replaced by universal telomere caps (UTC) [4, 5, 12]. In addition, PCRTags were designed in open reading frames (ORF) by recoding the degenerate codons with synonymous substitutions [4, 8, 13–16]. Similarly, watermarks were also introduced in the synthetic *M. genitalium* and *Escherichia coli* genomes as identities of synthetic genomes [7, 13, 16].

With loxPsym recombination sites designed in 3'UTR of nonessential genes throughout the genome [8], the synthetic yeast genome could be rearranged upon the expression of a Cre recombinase [4, 15]. The genome scale rearrangement

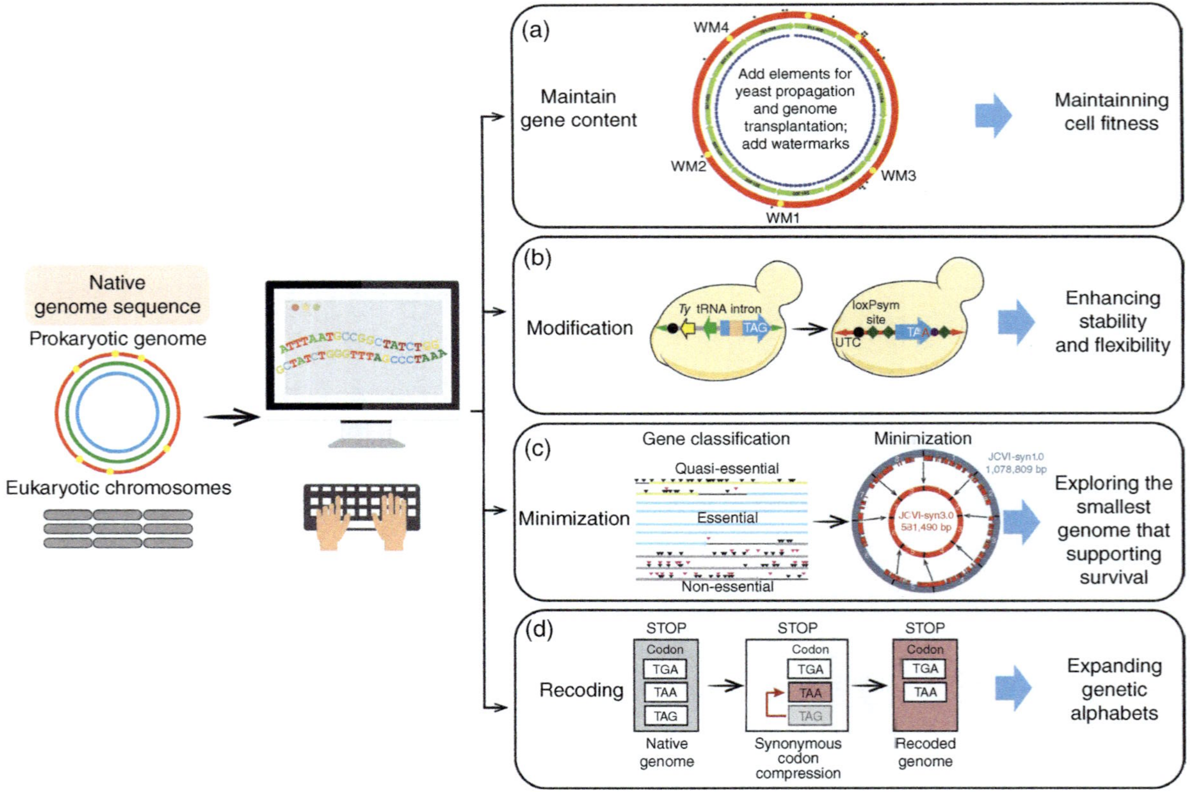

Figure 6.2 **Design a synthetic genome.** The native sequence of prokaryotic genome or eukaryotic chromosomes is redesigned for different purpose. (a) Gene content is maintained to generated the first functional synthetic genome JCVI-syn1.0, in which watermarks (WM) are incorporated to identify the synthetic genome from native one. (b) Customer features are introduced to the synthetic yeast genome to enhance the genomic stability and flexibility. (c) JCVI-syn1.0 genome is redesigned to generate the minimized genome, JCVI-syn3.0, to explore the smallest genome content supporting survival. (d) The synthetic *E. coli* genome is recoded by replacing codons with synonymous substitutions, creating an organism with a reduced-codon genome.

can generate chromosome number variations (CNVs) and structural variations (SVs) to promote genome evolution, which is called synthetic chromosome recombination and modification by loxP-mediated evolution (SCRaMbLE). The SCRaMbLE system could be used to rapidly generate yeast libraries, containing diverse genomes and exhibiting different phenotypes [17]. Investigation of these libraries could help understand the mechanism of genomic evolution while screening of these libraries could result in specific strains for high yield of chemical production [15, 18–26].

6.2.3. Minimization of Synthetic Genome

Organisms differ greatly from each other in genome sizes, which range from megabase (Mb) to gigabase (Gb). This prompted scientists to inquire about the minimal genome size that supports basic cellular life and reproduction [27, 28]. Appropriate genomic simplification saves energy and substrates that can be tuned for other utilizations [29]. By deleting nonessential sequences, genome can be "top-down" simplified. More knowledge such as synthetic lethality has accumulated from previous attempts for the purpose of constructing a minimized genome [13, 30–38]. The current record of minimal genome size was achieved by tailoring JCVI-1.0, resulting in JCVI-3.0, which has been deleted for 428 nonessential genes with remaining 473 genes and 531 kb in length [13].

6.2.4. Recoding the Genome

The genetic code is highly redundant, as there are 64 genetic codons and only 20 amino acids. Synthetic recoding of genomes by the removal of targeted sense codons could potentially save codons for the incorporation of unnatural amino acids into proteins of interest, which may have a broad range of applications including use in mechanistic study of enzymes, modifications of therapeutic proteins, preventing synthetic genome from escape, and resisting viral infection [4, 16, 39–46].

The "bottom-up" genome synthesis facilitated recoding in genome-wide scale and replacing the stop codon TAG with TAA has been successfully applied to both prokaryotic and eukaryotic genomes [4, 8, 14, 16, 47–52]. However, an attempt to remove 7 codons with 62,214 occurrences from the *E. coli* genome was only of partial success [16]. The whole genome was divided into 87 segments with each 50-kb in length. The targeted 7 codons were substituted by synonymous codons segment by segment; 55 of the segments exhibited functions, while the remaining 32 were lethal with the mechanism unexplored. In 2019, the first totally synthesized *E. coli* genome was completed, in which 3 codons (the TAG stop codon and two serine codons (TCG, TCA)) were recoded [47]. The 3.97-Mb synthetic *E.coli* genome with 61-codons exhibited near wild-type fitness.

6.3. SYNTHESIZE AND ASSEMBLE A GENOME FROM SCRATCH

Chemically synthesis of a yeast alanine tRNA was reported in 1970 [53]. Since then, tremendous progresses have been made in nucleic acid synthesis. Nowadays the design and assembly of ultra-long DNAs are practical, exemplified by the synthetic genomes of *M. mycoides* [7, 13], *E. coli* [16, 47], and *S. cerevisiae* [8, 14, 48–52].

Following the computer-aided design, a synthetic genome could be constructed starting from oligonucleotides, which are chemically synthesized from nucleotide monomers. Oligonucleotides are assembled into double-stranded DNA fragments and eventually the entire genome by various DNA assembly methods (Figure 6.3). The synthetic genome could be assessed by DNA sequencing, genome organization, RNA transcription, protein translation, and phenotypes such as cell morphology and growth rate.

6.3.1. Oligonucleotide Design and Synthesis

The computer programs such as GeneDesign considering the distribution of restriction sites and extreme GC contents are generally used to split the designed genomic sequence into numerous oligonucleotides [54, 55]. Each such oligonucleotide being dozens of bases in length is then chemically synthesized and assembled hierarchically following "oligonucleotide—short DNA segment—medium DNA segment—long DNA segment" [4, 7, 13, 14, 16, 48, 51, 56–60].

The chemical synthesis of oligonucleotides uses a column-based strategy and requires iterative cycles. Each four-step cycle, including the deprotection, coupling, oxidation, and capping, adds one base to the growing end of a oligonucleotide chain with its other end attached to a solid support matrix (see review [61, 62]). This process is highly amenable to automation and forms the basis for oligonucleotide synthesizers, which can synthesize 96–1536 distinct oligonucleotides simultaneously up to 100 nt with error rates at or below 0.5% [63].

Alternatively, microarray oligonucleotide synthesis has recently emerged and been developed (see review [64, 65]). In particular, light-based or ink-jet printing-based microarray oligonucleotide synthesis is competitive in balancing length, low cost and high accuracy [66, 67]. Microarray-based oligonucleotides are typically synthesized at femtomolar scales and up to 200–300 nt in length, which saves a large quantity of reagents and results in overall lower cost [68–70].

6.3.2. Genome Assembly

One to two dozens of overlapping chemically synthesized oligonucleotides are first assembled by polymerase cycling reaction (PCA) into short DNA segments [51, 56, 57, 71]. Adjacent short DNA fragments contain overlapping sequences. Thus, short DNA segments can be further assembled to form long DNA segments by *in vitro* assembly (DNA polymerase/endonuclease/exonuclease based) or *in vivo*

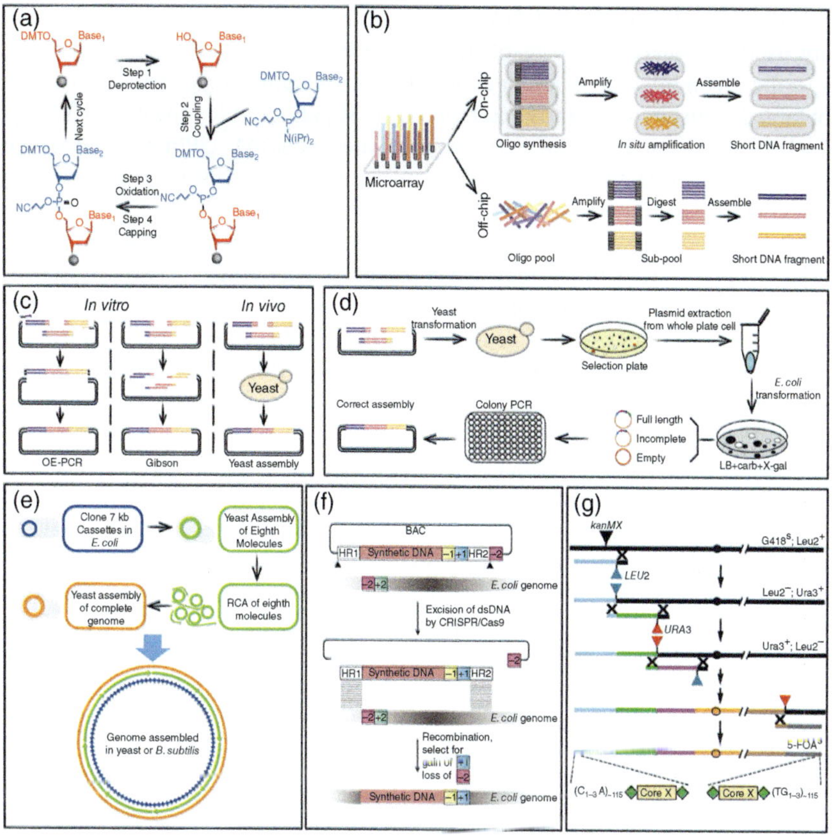

Figure 6.3 **Assembly of an entire synthetic genome from scratch.** (a) Oligonucleotide is chemically synthesized by using column-based strategy. (b) Microarray-based oligonucleotide synthesis and assembly of short DNA segments. (c) Short DNA segments are assembled *in vivo* and *in vitro*. (d) The pipeline of RADOM assembly. (e) The heterogeneous synthetic genome is assembled in naturally recombinogenic organisms like *S. cerevisiae* and *B. subtilis*. (f) The native genome is *in situ* replaced by long synthetic DNA segments iteratively. (g) The synthetic yeast chromosome is assembled *in situ* segment by segment with endogenous HR.

assembly (homologous recombination (HR)/site-specific recombination based) (see review [72]). The commonly used *in vitro* assembly includes overlap-extension PCR, Gibson isothermal assembly, and Uracil Specific Excision Reaction (USER) assembly, while Rapid Assembly of DNA Overlapping Multifragments (RADOM) is a typical *in vivo* method [7, 14, 47, 51, 58–60]. The RADOM employs blue–white screening facilitating the isolation of colonies with correct assembly [60] and becomes popular in yeast genome synthesis. RADOM has been part of the syllabus of the "Build-A-Genome" undergraduate course at Tianjin University and Johns Hopkins University [51].

Subsequently, synthetic DNA segments ranging from 2 to 100 kb in length are used to replace the corresponding endogenous sequences *in vivo* by site-specific recombination or endogenous HR until a complete synthetic genome is obtained. Clustered regularly interspaced short palindromic repeats (CRISPR)/CRISPR-associated protein 9 (Cas9)-mediated genomic editing to introduce double-stranded DNA breaks is used to increase the rate of site-specific recombination [7, 14, 47].

Synthetic genomes can be assembled *in vitro* such as the mouse mitochondrial genome. Alternatively, they can be assembled heterologously in a host cell. *S. cerevisiae* and *Bacillus subtilis* are naturally recombinogenic organisms and are hence used as chassis for genome assembly. For example, the *M. mycoides* JCVI-1.0 was finally assembled in yeast [7], while the rice chloroplastic genome was assembled in *B. subtilis* [73, 74].

Synthetic genomes can be assembled *in situ* as well. Native genomic sequences are replaced segment by segment. Briefly, the synthetic DNA segment was first cloned into a bacterial artificial chromosome (BAC) and transformed into *E. coli* where it was excised from BAC using a CRISPR/Cas9 system. The resulting linear DNA fragment was incorporated into the genome by λ-red-mediated recombination [16, 47, 75]. In order to accelerate the whole genome assembly, semisynthetic genomes were separately assembled in distinct strains. Upon bacterial conjugation, a complete synthetic *E. coli* genome was obtained [47, 76]. Similarly, yeast synthetic chromosome was assembled *in situ* segment by segment but with endogenous HR instead of site-specific recombination. [8, 14, 48–52]. The lengths of segments for replacement ranged from 30 to 60 kb. Alternatively, one to two dozens of synthetic DNAs with medium lengths (~2–4 kb) were simultaneously transformed into yeast cells. To accelerate completion of whole synthetic yeast genome, the synthesis of 16 individual chromosomes was allocated within a world-wide team [12, 77]. So far, six chromosomes have been synthesized, namely synthetic yeast chromosome II (synII) [49], synthetic yeast chromosome III (synIII) [14], synthetic yeast chromosome V (synV) [51], synthetic yeast chromosome VI (synVI) [48], synthetic yeast chromosome X (synX) [50], and synthetic yeast chromosome XII (synXII) [52].

6.3.3. Test of the Synthetic Genome

Genome synthesis is a tedious work and the paramount objective is to generate a functional genome that maintains cell fitness. However, lack of knowledge about genomic information results in design flaws, which together with frequently occurring errors during synthesis could cause lethality [4, 5, 13, 51]. The synthesis of JCVI-syn1.0 lasted 15 years and cost 40 million US dollars [78]. It was noticed that a single-base variant had once delayed the process for months [7], highlighting the importance of prompt functional evaluation of synthetic genomes.

The presence of a synthetic genome could be detected by PCR watermarks, pulsed-field gel electrophoresis (PFGE), or southern blot [7, 8, 13, 14, 16, 51, 79], whereas errors generated during synthesis including single-nucleotide variation (SNV), short insertion/deletion (InDel), structural variation (SV), and copy number variation (CNV) are mainly detected by whole genome sequencing (WGS) [7, 13, 14, 16, 48–52].

Viability is the gold standard and the most obvious phenotype to test the function of a designer genome in the cell, and it is essential for any subsequent engineering, evolution, and applications of synthetic cells [7, 13, 16]. Other phenotypes including growth rate, colony size, cell morphology, and nucleolar morphology are often tested [7, 13, 14, 16, 30, 48–52, 80]. In addition, DNA replication was analyzed using fluorescence-activated cell sorting (FACS) and replication profiling to gain insights into chromosomal modifications [14, 49, 52]. Synthetic genomes may differ from native ones in 3D structures, which in turn may affect cellular functions [80]. Therefore approaches such as Hi-C contact experiments were performed, which generate Hi-C maps in illustration of the three-dimensional conformation of the synthetic genome [80]. The deletion and insertion of specific large fragments led to conformational changes, but no phenotypic differences were detected [14, 52, 80]. Transcriptomic, proteomic, and metabolomic profiles have also been used to reveal subtly altered functions of some genomic elements [7, 13, 16, 48–52]. The aforementioned analyses taken together provided thorough understanding of the functions of synthetic genomes and could be used to link altered functions to specific design flaws and errors, a process named bug mapping described in more details in a later section.

6.4. GENOME TRANSPLANTATION

Synthetic genomes assembled *in vitro* or heterologously in a host cell need to be transplanted. Synthetic genomes assembled *in situ* in some cases also need transplantation. For example, the yeast genome synthesis project involves synthesis of multiple chromosomes done separately in different laboratories. Genome transplantation is a process in which an intact genome is moved from one species to another (Figure 6.4). Success of genome transplantation has been reported for prokaryotic and eukaryotic single-cell microorganisms [81, 82]. Transplantation of large, Gb-sized animal and plant genomes require new tools and methods [83] with high transfer efficiency, low cytotoxicity, minimal impact on normal physiology, easy to use, and repeatable. Some new methods being tested for the transplantation of super large genome will be discussed in later paragraph in this section. Generally, these methods are not mature with efficiency to be improved [84].

6.4.1. Transplantation of Synthetic Genome to Cells

The first step of genome transplantation oftentimes involves the extraction of the synthetic genome from the donor cells. To minimize DNA breakage caused by

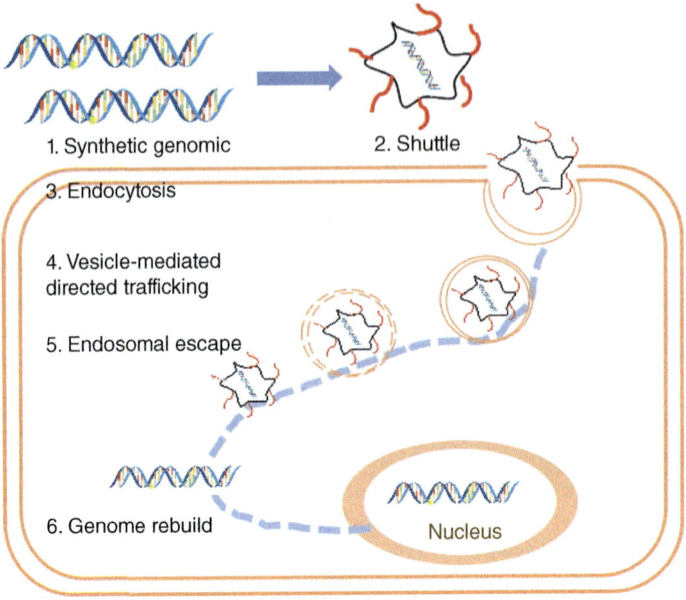

Figure 6.4 Genome transplantation.

shear forces during genome extraction, donor cells are suspended in agarose solutions and immobilized upon agarose solidification. Immobilized cells are treated with an enzyme mixture followed by dialysis to remove cellular components while intact genome remains trapped in the agarose. The completely deproteinated genomic DNA is then isolated and used for transplantation into the recipient cells by a modified polyethylene glycol (PEG)-mediated transformation strategy. Recently, the *M. mycoides* genome was successfully transplanted into a related species, *M. capricolum* [7, 81] and the spheroplasts of yeast strains VL6-48N and W303a [82]. The *M. mycoides* synthetic genome could also be isolated from yeast and transplanted back to *M. mycoides* recipient cell [85] (Figure 6.5).

In Sc2.0 project, the 16 chromosomes are first constructed in discrete strains. The "endoduplication intercross" strategy is applied to transplant and consolidate synthetic chromosomes into one strain. The GAL1 promoter is inserted adjacent to centromeres of specific native chromosomes to disrupt the chromosome segregation. For example, GAL1 promoter was inserted adjacent to the centromere of wild-type chromosome V (wtV) in the strain carrying synX, and the centromere of wild-type chromosome X (wtX) in the strain carrying synV. Then, the two strains of opposite mating types were mated to generate a diploid strain. The diploid strain was treated with galactose to simultaneously eliminate wtV and wtX. After sporulation and tetrad dissection, a haploid strain carrying both synV and synX was isolated [86]. A haploid strain carrying synIII/synVI/synIXR has been also generated similarly [4, 86] (Figure 6.6).

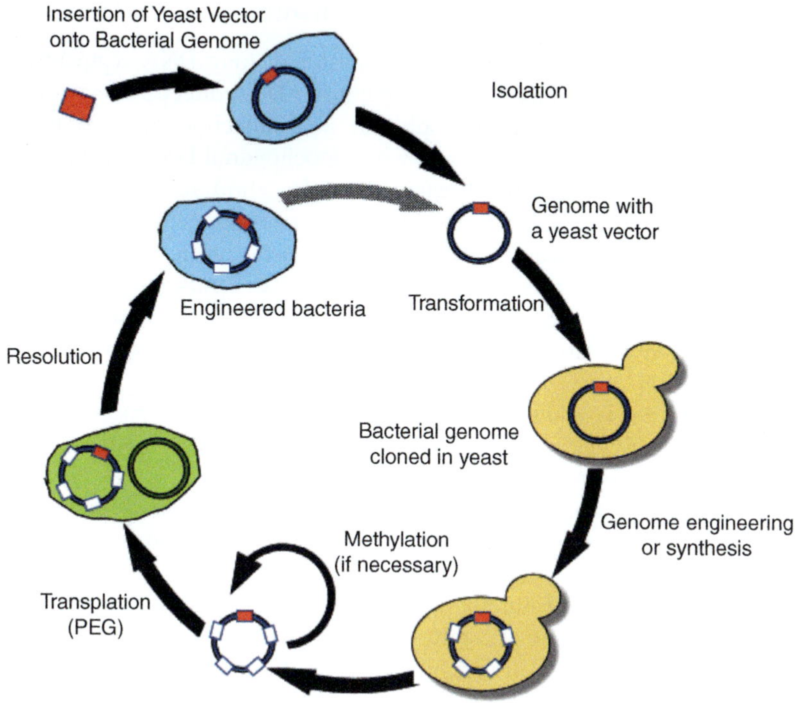

Figure 6.5 Genome transplantation between yeast and *M. mycoides*.

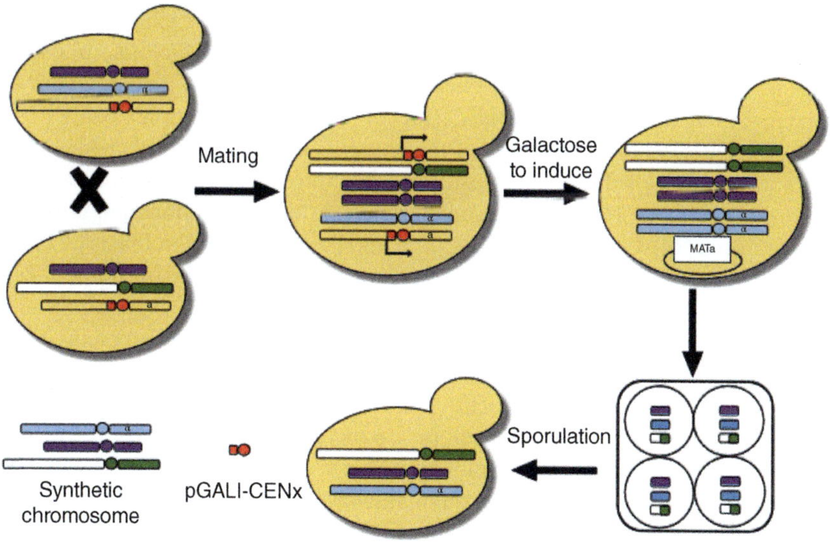

Figure 6.6 Consolidation synthetic yeast chromosome into one strain.

6.4.2. Transformation of Long DNA Segments

There are diverse methods used for the transfer of long DNA segments into mammalian cells. Microinjection is one to transfer DNA directly into cytoplasm or nucleus [87] (Figure 6.7a). Cytoplasmic microinjection has been used to transfer mitochondrial DNA into cells with mitochondrial DNA mutations that cause non-Mendelian genetic diseases through cybrid formation [88, 89]. Pronuclear microinjection has been used to generate transgenic mice by transferring a 60-Mb murine satellite DNA-based artificial chromosome (SATAC) [90]. Both microinjection methods have their limits: cytoplasm microinjection is less effective because cytoplasmic nucleases can degrade DNA; pronuclear injection is a laborious procedure because only one cell can be injected at a time [87, 91].

Electroporation is another method [92–94] (Figure 6.7a). Electroporation can be used to efficiently transfer DNA to a variety of cells. A 101-kb human gene *HPRT1* was assembled in yeast and delivered to mouse cells by electroporation [95]. The electroporation efficiency decreases notably as the size of synthetic DNA increases, and hence it has not been reported in intact genome transplantation.

Bacterial conjugation naturally occurs and is also employed to transfer long DNA segments in synthetic genomics (Figure 6.7c). The connection between a donor cell and a recipient cell is regulated by fertility factors. Formation of sex pili mediates the transfer of single-stranded DNAs from the donor cell to the recipient cell [96]. The *E. coli* genome was synthesized by 8 segments and the conjugation-based method was applied to assemble them into an intact genome [47].

Chemical transfection methods are widely used to transfer exogenous DNA into mammalian cells (Figure 6.7a). The chemicals used to facilitate transfection are diverse. Generally, these chemicals are positively charged and work as carriers of negatively charged DNA. The net charge of the complex being positive contributes to the attachment of the complex to the negatively charged cell membrane and the subsequent transfer of DNA into the cells through endocytosis or phagocytosis [97]. A hygromycin-resistant 60-Mb mammalian artificial chromosome (MAC) has been transfected into hamster lung fibroblasts (V79-4) and murine connective tissue cells (LM (TK-)) by using cationic liposomes [98].

PEGs are highly hydrated polymers used to mediate cell aggregation and fusion, generating hybrid cells [99–102]. PEG-mediated cell fusion method has been used to generate hybrids of *S. cerevisiae* protoplasts and mammalian cells, realizing the transfer of synthetic genomes with lengths from 100 kb to 1 Mb [84]. As the PEG-mediated cell fusion method unnecessitates DNA crossing cytoplasmic membranes, its transfer efficiency is not affected by the size of synthetic DNA. It also avoids hydrodynamic shear forces damaging DNA molecules, which could happen when genomic DNA is isolated from cells prior to transfer.

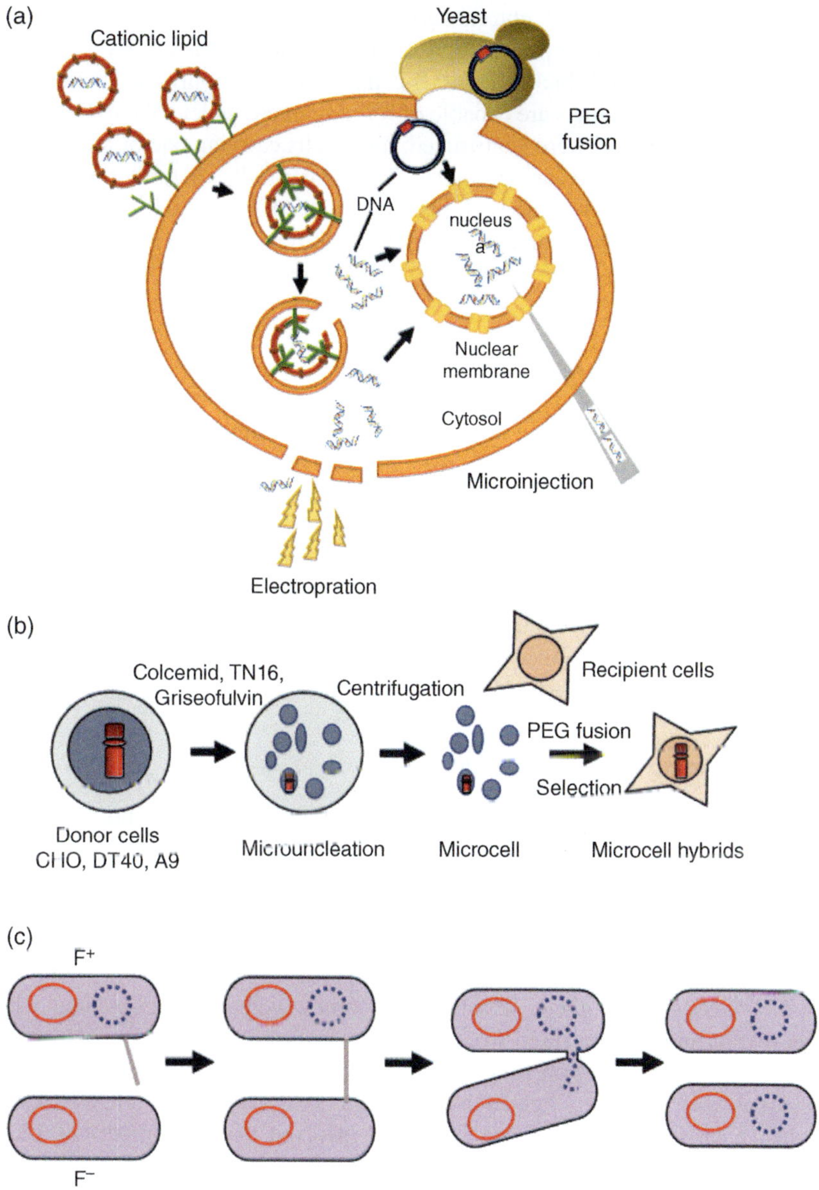

Figure 6.7 **The strategies used for long DNA segments transfer.** (a) Cationic lipid, PEG fusion, electroporation, and microinjection strategy. (b) MMCT strategy. (c) Bacterial conjugation strategy.

The microcell-mediated chromosome transfer (MMCT) is a method for chromosome-level transfer (Figure 6.7b). Chinese hamster ovary (CHO) cell line, mouse A9 cell line, and chicken DT40 B-cell line are commonly used donor cells in MMCT. These cell lines are capable of forming micronuclei. When these cells containing micronuclei were centrifuged, microcells carrying one or a few chromosomes are formed and can be isolated. These microcells are further fused with recipient cells by PEG-mediated cell fusion mentioned above [103, 104]. With the MMCT method, a chromosome in A9 cell line was transferred to a mouse embryonic cell and the trans-chromosomic humanized mice were bred [105, 106].

6.5. BUG MAPPING AND DEBUGGING

Limited biological knowledge results in flaws of genomic design that affect cell fitness [107]. DNA sequence errors are inevitable during genome synthesis. The bugs on synthetic genomes, arise from both design and construction, could lead to compromised growth and even lethality making the intended evaluation of designer features difficult. Genomic bug mapping and debugging aims to locate and correct design flaws or sequence errors on synthetic genomes. Cells carrying bug-free genomes should undergo normal DNA replication and maintain fitness, and knowledge gained from the process of genome debugging is valuable for future genomic design [108] (Figure 6.8a).

6.5.1. Mapping Genomic Bugs

Errors on synthetic genomes could occur in different stages of genome assembly. Various methods including PFGE, WGS, PCR and qPCR are employed to verify the synthetic genome. The watermarks [7, 13, 109], introduced restriction sites [8, 14, 48–52], and recoded sequences [8, 14, 16, 47–52] could be used as references for precise bug mapping in WGS, PCR, and qPCR. These methods reveal different types of errors including SNVs, CNVs, and SVs [7, 8, 14, 48–52]. CNVs on synthetic yeast chromosomes were often associated with cohesive termini that resulted from restriction enzyme digestion of DNA constructs and introduced loxPsym sites [50, 51]. For example, a medium DNA construct was tandemly ligated by a short cohesive-end-like "GCGGCGC" sequence on synV [51]. On synX, a CNV involved three fragments of 12.8, 27.7, 5.6 kb in length, respectively, which were ligated by loxPsym sites [50].

The defective growth phenotypes of a synthetic strain could be due to aforementioned variations or the synthetic DNA sequences as designed but with flaws. Various methods have been applied to locate bugs from numerous synthetic sequences and variations and will be discussed in the following paragraphs in details.

The early version of synthesized *M. mycoides* genome was not able to sustain the viability of the recipient cell [7]. In the tedious bug mapping process, different

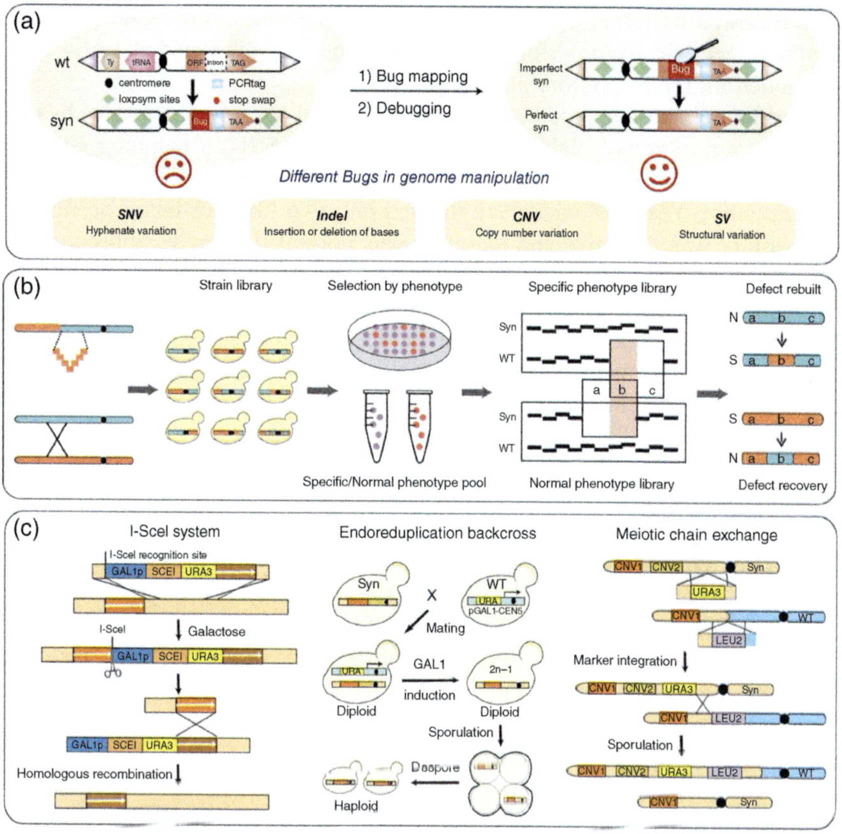

Figure 6.8 Genome bug mapping and debugging. (a) Mapping and debugging design flaws or sequence variations on synthetic genomes can recover cell fitness. (b) PoPM bug mapping strategy. (c) I-SceI-mediated HR, "endoduplication backcross" and meiotic crossover debugging methods.

combinations of 11 pre-assembled DNA fragments, each 100-kb in length, were used to construct semisynthetic genomes. All these semisynthetic genomes were transplanted into recipient cells and screened for lethality. One 100-kb fragment was associated with the lethal phenotype. Similarly, the lethal error was further narrowed down to a 20-kb synthetic segment and eventually was linked to a frameshift single bp deletion in *dnaA*, an essential gene for chromosomal replication.

Backcross of the synthetic strain with wild-type strain of the opposite mating type followed by tetrad dissection results in various meiotic recombinated spores, which could be utilized to map bugs [48, 51]. The yeast strain carrying an early version of synVI exhibited a respiratory deficient [48]. To identify the bug, this strain was backcrossed with the wild-type strain of the opposite mating type followed by tetrad dissection. Colony PCRs priming the PCRTags in the synthetic genome were performed. It was found that the recoded *PRE4* gene accounted for the respiratory growth defect [48].

Both aforementioned methods need to be carried out for colonies one by one and thus are extremely labor-intensive. A pooled PCRTag mapping (PoPM) strategy was recently developed, which enabled rapid bug mapping on synX [50] (Figure 6.8b). Yeast colonies were divided into two libraries based on fitness. One library was defective and the other was normal in growth. Subsequently, both libraries were tested with predesigned PCRTags on synthetic and wild-type chromosomes. The synthetic elements only detected in the normal library but not in the defective library were candidate-defective loci. Using the PoPM strategy, loci-caused growth defects were quickly pinpointed, where synonymously recoding of essential *FIP1* gene, deletion of *tR(CCU)J* gene, and loxPsym site insertion to the 3'UTR of *YJR120W* were identified [50].

6.5.2. Debugging

In the yeast synthetic genome project, SVs and CNVs were considered problematic and so were firstly fixed. WGS revealed an extra copy of two synV sequences (291, 217–299, 980 bp and 305, 355–325, 235 bp) that are misintegrated on other chromosome(s), which affects the fitness of the synV strain [51]. The "endoduplication backcross" strategy was applied to remove the extra copy (Figure 6.8c). The synV strain was backcrossed with the wild-type strain, engineered with the insertion of a GAL1 promoter adjacent to the centromere of wtV. Then, wtV was ejected from the diploid strain grown with galactose. After sporulation and tetrad dissection, haploid colonies were screened for the presence of synV and the extra copy of the two synV sequences removed. Extra copies of synthetic sequences present on the same synthetic chromosome were also encountered. In this case, meiotic crossover or I-SceI-mediated HR was used for debugging [50, 52]. An example of meiotic crossover debugging was reported for synX (Figure 6.8c). An early version of synX strain carrying such a CNV was crossed with a strain of opposite mating type harboring a semi-synX chromosome without the CNV [50]. Following sporulation and tetrad dissection, a haploid strain carrying a complete synX without the CNV was isolated as a result of meiotic recombination. The other debugging method, I-SceI-mediated HR, can be used only for removing the extra copy in proximity (Figure 6.8c). An example for this is the usage of the I-SceI endonuclease in repairing the CNV on synII during cell mitosis [49]. A cassette containing an I-SceI recognition site was incorporated between the original synthetic region and the duplicated region. The expression of I-SceI endonuclease generated a double-strand break (DSB) at the I-SceI recognition site, promoting HR between the original synthetic region and the duplicated region resulting in a corrected synII.

In addition to CNVs and SVs, numerous SNVs and InDels occur during genome assembly. Even a very small number of nucleotide changes could, in principle, affect cell fitness. Thus, correcting all SNVs and InDels to generate a synthetic chromosome exactly as designed is crucial to evaluate design principles [51]. Toward this end, a combination of three methods, the HR-based gene replacement, CRISPR/Cas9–mediated genomic editing, and dual-labeling correction, was used and all SNVs and InDels involving 3,331 bp on synV were successfully repaired [51].

The traditional HR-based gene replacement method was firstly applied to correct variations on synV [51]. A variation was first replaced by a selective marker, and then the selective marker was replaced by donor DNA with a correct sequence. This method requires two steps of homologous recombination and selection and is laborious. Then a CRISPR/Cas9-mediated one-step method was developed. In this method, a DSB at the targeted region was generated by CRISPR/Cas9 [51]. Subsequently, SNVs and InDels were eliminated by HR of donor DNAs at the DSB [51, 110]. However, the CRISPR/Cas9-mediated one-step method relies on the presence of NGG in target sequences and therefore is not applicable to the entire synthetic yeast genome. Other problems of this method include variable efficiencies and off-target cleavage. To overcome the aforementioned problems, a dual-labeling correction strategy was developed to completely repair remaining errors on synV [51]. A doubly selective cassette containing both an antibiotic gene and an auxotroph gene was first incorporated adjacent to a specific target, minimizing false positive rates by double selection. Meanwhile, a protospacer and a PAM site were also designed in the selective cassette. The protospacer was then cut by Cas9 to increase the HR efficiency, and a pool of donor DNAs covering multiple SNVs or InDels was co-transformed for multiplex correction [51]. The isolated clone with most correction events was subjected to the next round of correction. After 24 rounds, all SNVs and InDels, including two new mutations, were corrected resulting in synV that was perfectly matching the designed sequence. This synV strain was as robust as the wild-type strain [5, 51].

Aforementioned methods also allow the repair of design flaws. Typically, synthetic sequences at the defective region are reverted to corresponding native sequences [48, 52]. For example, qPCR analysis revealed the expression of *ATP2* in synX strain was significantly lower than the wild-type level. By analyzing the designed sequence, the authors found that the inserted loxPsym site downstream of the neighboring gene *YJR120W* was hypothesized to alter the *ATP2* promoter function. The *ATP2* expression was restored by deleting this loxPsym site and the fitness of the strain was recovered [50]. Another example is the PCRTag in *FIP1* of synX, which was designed by synonymously recoding native sequences. The PCRTag happened to resemble the binding site of the transcriptional repressor Rap1p, which presumably caused the downregulation of *FIP1* expression in synX strain. The growth rate was recovered by reverting only two base pairs presumably enough to disrupt putatively Rap1p binding [50].

6.5.3. Ring Chromosome

Eukaryotic nuclear chromosomes are linear with rare exceptions. Pathological circular "ring" chromosomes are linked to human diseases, including epilepsy [111, 112], intellectual delay [112], various dysmorphic features [113], leukemia [114], and microcephaly [115, 116]. By contrast, a chromosome circularization-mediated therapy for genomic disorders involving loss of the ring chromosome carrying disease-linked mutations and the replication of the homologous chromosome has been proposed [117]. More than one-third of the yeast genes are orthologous to their human counterparts [118] and yeast can maintain ring structure chromosomes [8, 119, 120]. Therefore, yeast could be developed as a model for ring chromosome analysis.

Customer synthetic yeast chromosomes were recently cyclized by using a HR or a CRISPR/Cas9 strategy without altering the gene contents [51, 86]. To generate the ring derivative of synV (ring_synV) and wtV (ring_wtV), specific regions adjacent to telomeres were selected for modifications to generate homologous arms. Then, a DNA cassette was constructed containing a selective marker in the middle and two homologous arms on both ends pairing with the homologous arms near the left and right telomeres, respectively. The DNA cassette was transformed into yeast strains to generate ring chromosome Vs and to delete telomeres on both ends of chromosome Vs by HR [51]. CRISPR/Cas9 was used to significantly increase the efficiency of chromosome circularization. The specific regions were cut by Cas9 prior to HR [86].

Ring chromosomes expanded the scope of synthetic genomics and provided a model for the studies of genomic rearrangement, ring chromosome evolution, and human ring chromosome disorders [48, 51]. LoxPsym sites were added to the 3'UTR of nonessential genes on ring_synV, allowing for Cre-mediated site-specific recombination [25]. Cells carrying ring_synV were induced for SCRaMbLE resulting in aneuploidy of chromosome I, III, VI, XII, XIII, and ring_synV. The neochromosome of rearranged ring_synV contained a lot more unbalanced variations, including duplications and insertions than the SCRaMbLEd linear form. Novel SVs detected on the rearranged ring_synV promoted the biosynthesis of PDV, suggesting that one can use the circularization of synthetic chromosome and SCRaMbLE to induce genome structural changes and to alter specific phenotypes including increased yields of fermentation products.

ACKNOWLEDGMENTS

This work was funded by the National Key Research and Development Program of China (2020YFC1316500), the National Natural Science Foundation of China (21907074) and the Natural Science Foundation of Tianjin (20JCQNJC02090).

REFERENCES

1. Watson JD, Crick FH. Molecular structure of nucleic acids; a structure for deoxyribose nucleic acid. Nature. 1953;171(4356):737–8.
2. Crick F. Central dogma of molecular biology. Nature. 1970;227(5258):561–3.
3. Rugnetta M. Synthetic biology. Encyclopedia Britannica; 2016 [cited 2021]. Available from: https://www.britannica.com/science/synthetic-biology.
4. Richardson SM, Mitchell LA, Stracquadanio G, Yang K, Dymond JS, DiCarlo JE, et al. Design of a synthetic yeast genome. Science. 2017;355(6329):1040–4.
5. Xie ZX, Liu D, Li BZ, Zhao M, Zeng BX, Wu Y, et al. Design and chemical synthesis of eukaryotic chromosomes. Chem Soc Rev. 2017;46(23):7191–207.
6. Li S, Li Y, Smolke CD. Strategies for microbial synthesis of high-value phytochemicals. Nat Chem. 2018;10(4):395–404.
7. Gibson DG, Glass JI, Lartigue C, Noskov VN, Chuang RY, Algire MA, et al. Creation of a bacterial cell controlled by a chemically synthesized genome. Science. 2010;329(5987):52–6.
8. Dymond JS, Richardson SM, Coombes CE, Babatz T, Muller H, Annaluru N, et al. Synthetic chromosome arms function in yeast and generate phenotypic diversity by design. Nature. 2011;477(7365):471–6.
9. Bailey JA, Yavor AM, Massa HF, Trask BJ, Eichler EE. Segmental duplications: organization and impact within the current human genome project assembly. Genome Res. 2001;11(6):1005–17.
10. Mefford HC, Linardopoulou E, Coil D, van den Engh G, Trask BJ. Comparative sequencing of a multicopy subtelomeric region containing olfactory receptor genes reveals multiple interactions between non-homologous chromosomes. Hum Mol Genet. 2001;10(21):2363–72.
11. van Overveld PG, Lemmers RJ, Deidda G, Sandkuijl L, Padberg GW, Frants RR, et al. Interchromosomal repeat array interactions between chromosomes 4 and 10: a model for subtelomeric plasticity. Hum Mol Genet. 2000;9(19):2879–84.
12. Pennisi E. Building the ultimate yeast genome. Science. 2014;343(6178):1426–9.
13. Hutchison CA, III, Chuang RY, Noskov VN, Assad-Garcia N, Deerinck TJ, Ellisman MH, et al. Design and synthesis of a minimal bacterial genome. Science. 2016;351(6280):aad6253.
14. Annaluru N, Muller H, Mitchell LA, Ramalingam S, Stracquadanio G, Richardson SM, et al. Total synthesis of a functional designer eukaryotic chromosome. Science. 2014;344(6179):55–8.
15. Dymond J, Boeke J. The Saccharomyces cerevisiae SCRaMbLE system and genome minimization. Bioeng Bugs. 2012;3(3):168–71.
16. Ostrov N, Landon M, Guell M, Kuznetsov G, Teramoto J, Cervantes N, et al. Design, synthesis, and testing toward a 57-codon genome. Science. 2016;353(6301):819–22.
17. Chen S, Xie ZX, Yuan YJ. Discovering and genotyping genomic structural variations by yeast genome synthesis and inducible evolution. FEMS Yeast Res. 2020;20(2):foaa012.
18. Blount BA, Gowers GF, Ho JCH, Ledesma-Amaro R, Jovicevic D, McKiernan RM, et al. Rapid host strain improvement by in vivo rearrangement of a synthetic yeast chromosome. Nat Commun. 2018;9(1):1932.

19. Jia B, Wu Y, Li BZ, Mitchell LA, Liu H, Pan S, et al. Precise control of SCRaMbLE in synthetic haploid and diploid yeast. Nat Commun. 2018;9(1):1933.
20. Liu W, Luo Z, Wang Y, Pham NT, Tuck L, Perez-Pi I, et al. Rapid pathway prototyping and engineering using in vitro and in vivo synthetic genome SCRaMbLE-in methods. Nat Commun. 2018;9(1):1936.
21. Luo Z, Wang L, Wang Y, Zhang W, Guo Y, Shen Y, et al. Identifying and characterizing SCRaMbLEd synthetic yeast using ReSCuES. Nat Commun. 2018;9(1):1930.
22. Shen MJ, Wu Y, Yang K, Li Y, Xu H, Zhang H, et al. Heterozygous diploid and interspecies SCRaMbLEing. Nat Commun. 2018;9(1):1934.
23. Steensels J, Gorkovskiy A, Verstrepen KJ. SCRaMbLEing to understand and exploit structural variation in genomes. Nat Commun. 2018;9(1):1937.
24. Wang J, Jia B, Xie ZX, Li YX, Yuan YJ. Improving prodeoxyviolacein production via multiplex SCRaMbLE iterative cycles. Front Chem Sci Eng. 2018;12(4):806–14.
25. Wang J, Xie ZX, Ma Y, Chen XR, Huang YQ, He B, et al. Ring synthetic chromosome V SCRaMbLE. Nat Commun. 2018;9(1):3783.
26. Wu Y, Zhu RY, Mitchell LA, Ma L, Liu R, Zhao M, et al. in vitro DNA SCRaMbLE. Nat Commun. 2018;9(1):1935.
27. Hutchison CA, Peterson SN, Gill SR, Cline RT, White O, Fraser CM, et al. Global transposon mutagenesis and a minimal Mycoplasma genome. Science. 1999;286(5447):2165–9.
28. Mushegian AR, Koonin EV. A minimal gene set for cellular life derived by comparison of complete bacterial genomes. Proc Natl Acad Sci U S A. 1996;93(19):10268–73.
29. Glass JI, Assad-Garcia N, Alperovich N, Yooseph S, Lewis MR, Maruf M, et al. Essential genes of a minimal bacterium. Proc Natl Acad Sci U S A. 2006;103(2):425–30.
30. Giaever G, Chu AM, Ni L, Connelly C, Riles L, Veronneau S, et al. Functional profiling of the *Saccharomyces cerevisiae* genome. Nature. 2002;418(6896):387–91.
31. Posfai G, Plunkett G, III, Feher T, Frisch D, Keil GM, Umenhoffer K, et al. Emergent properties of reduced-genome *Escherichia coli*. Science. 2006;312(5776):1044–6.
32. Umenhoffer K, Feher T, Baliko G, Ayaydin F, Posfai J, Blattner FR, et al. Reduced evolvability of Escherichia coli MDS42, an IS-less cellular chassis for molecular and synthetic biology applications. Microb Cell Fact. 2010;9:38.
33. Csorgo B, Feher T, Timar E, Blattner FR, Posfai G. Low-mutation-rate, reduced-genome *Escherichia coli*: an improved host for faithful maintenance of engineered genetic constructs. Microb Cell Fact. 2012;11:11.
34. Sugiyama M, Nakazawa T, Murakami K, Sumiya T, Nakamura A, Kaneko Y, et al. PCR-mediated one-step deletion of targeted chromosomal regions in haploid *Saccharomyces cerevisiae*. Appl Microb Biotechnol. 2008;80(3):545–53.
35. Murakami K, Tao E, Ito Y, Sugiyama M, Kaneko Y, Harashima S, et al. Large scale deletions in the *Saccharomyces cerevisiae* genome create strains with altered regulation of carbon metabolism. Appl Microb Biotechnol. 2007;75(3):589–97.
36. Hirashima K, Iwaki T, Takegawa K, Giga-Hama Y, Tohda H. A simple and effective chromosome modification method for large-scale deletion of genome sequences and identification of essential genes in fission yeast. Nucleic Acids Res. 2006;34(2):e11.

37. Giga-Hama Y, Tohda H, Takegawa K, Kumagai H. Schizosaccharomyces pombe minimum genome factory. Biotechnol Appl Biochem. 2007;46(Pt 3):147–55.
38. Venetz JE, Del Medico L, Wolfle A, Schachle P, Bucher Y, Appert D, et al. Chemical synthesis rewriting of a bacterial genome to achieve design flexibility and biological functionality. Proc Natl Acad Sci U S A. 2019;116(16):8070–9.
39. Lajoie MJ, Kosuri S, Mosberg JA, Gregg CJ, Zhang D, Church GM. Probing the limits of genetic recoding in essential genes. Science. 2013;342(6156):361–3.
40. Park HS, Hohn MJ, Umehara T, Guo LT, Osborne EM, Benner J, et al. Expanding the genetic code of *Escherichia coli* with phosphoserine. Science. 2011;333(6046):1151–4.
41. Xie J, Schultz PG. A chemical toolkit for proteins – an expanded genetic code. Nat Rev Mol Cell Biol. 2006;7(10):775–82.
42. Isaacs FJ, Carr PA, Wang HH, Lajoie MJ, Sterling B, Kraal L, et al. Precise manipulation of chromosomes in vivo enables genome-wide codon replacement. Science. 2011;333(6040):348–53.
43. Salim HM, Ring KL, Cavalcanti AR. Patterns of codon usage in two ciliates that reassign the genetic code: *Tetrahymena thermophila* and *Paramecium tetraurelia*. Protist. 2008;159(2):283–98.
44. Lajoie MJ, Rovner AJ, Goodman DB, Aerni HR, Haimovich AD, Kuznetsov G, et al. Genomically recoded organisms expand biological functions. Science. 2013;342(6156):357–60.
45. Mandell DJ, Lajoie MJ, Mee MT, Takeuchi R, Kuznetsov G, Norville JE, et al. Biocontainment of genetically modified organisms by synthetic protein design. Nature. 2015;518(7537):55–60.
46. Rovner AJ, Haimovich AD, Katz SR, Li Z, Grome MW, Gassaway BM, et al. Recoded organisms engineered to depend on synthetic amino acids. Nature. 2015;518(7537):89–93.
47. Fredens J, Wang K, de la Torre D, Funke LFH, Robertson WE, Christova Y, et al. Total synthesis of Escherichia coli with a recoded genome. Nature. 2019;569(7757):514–8.
48. Mitchell LA, Wang A, Stracquadanio G, Kuang Z, Wang X, Yang K, et al. Synthesis, debugging, and effects of synthetic chromosome consolidation: synVI and beyond. Science. 2017;355(6329):eaaf4831.
49. Shen Y, Wang Y, Chen T, Gao F, Gong J, Abramczyk D, et al. Deep functional analysis of synII, a 770-kilobase synthetic yeast chromosome. Science. 2017;355(6329):eaaf4791.
50. Wu Y, Li BZ, Zhao M, Mitchell LA, Xie ZX, Lin QH, et al. Bug mapping and fitness testing of chemically synthesized chromosome X. Science. 2017;355(6329):eaaf4706.
51. Xie ZX, Li BZ, Mitchell LA, Wu Y, Qi X, Jin Z, et al. "Perfect" designer chromosome V and behavior of a ring derivative. Science. 2017;355(6329):eaaf4704.
52. Zhang W, Zhao G, Luo Z, Lin Y, Wang L, Guo Y, et al. Engineering the ribosomal DNA in a megabase synthetic chromosome. Science. 2017;355(6329):eaaf3981.
53. Agarwal KL, Buchi H, Caruthers MH, Gupta N, Khorana HG, Kleppe K, et al. Total synthesis of the gene for an alanine transfer ribonucleic acid from yeast. Nature. 1970;227(5253):27–34.
54. Richardson SM, Wheelan SJ, Yarrington RM, Boeke JD. GeneDesign: rapid, automated design of multikilobase synthetic genes. Genome Res. 2006;16(4):550–6.

55. Richardson SM, Nunley PW, Yarrington RM, Boeke JD, Bader JS. GeneDesign 3.0 is an updated synthetic biology toolkit. Nucleic Acids Res. 2010;38(8): 2603–6.
56. Barnett RW, Erfle H. Rapid generation of DNA fragments by PCR amplification of crude, synthetic oligonucleotides. Nucleic Acids Res. 1990;18(10):3094.
57. Ciccarelli RB, Gunyuzlu P, Huang J, Scott C, Oakes FT. Construction of synthetic genes using PCR after automated DNA synthesis of their entire top and bottom strands. Nucleic Acids Res. 1991;19(21):6007–13.
58. Dymond JS, Scheifele LZ, Richardson S, Lee P, Chandrasegaran S, Bader JS, et al. Teaching synthetic biology, bioinformatics and engineering to undergraduates: the interdisciplinary Build-a-Genome course. Genetics. 2009;181(1):13–21.
59. Gibson DG, Young L, Chuang RY, Venter JC, Hutchison CA, III, Smith HO. Enzymatic assembly of DNA molecules up to several hundred kilobases. Nat Methods. 2009;6(5):343–5.
60. Lin Q, Jia B, Mitchell LA, Luo J, Yang K, Zeller KI, et al. RADOM, an efficient in vivo method for assembling designed DNA fragments up to 10 kb long in *Saccharomyces cerevisiae*. ACS Synth Biol. 2015;4(3):213–20.
61. Beaucage SL, Caruthers MH. Deoxynucleoside phosphoramidites—a new class of key intermediates for deoxypolynucleotide synthesis. Tetrahedron Lett. 1981;22(20):1859–62.
62. Kosuri S, Church GM. Large-scale de novo DNA synthesis: technologies and applications. Nat Methods. 2014;11(5):499–507.
63. Livesay EA, Liu YH, Luebke KJ, Irick J, Belosludtsev Y, Rayner S, et al. A scalable high-throughput chemical synthesizer. Genome Res. 2002;12(12):1950–60.
64. Tian J, Gong H, Sheng N, Zhou X, Gulari E, Gao X, et al. Accurate multiplex gene synthesis from programmable DNA microchips. Nature. 2004;432(7020):1050–4.
65. Kosuri S, Eroshenko N, Leproust EM, Super M, Way J, Li JB, et al. Scalable gene synthesis by selective amplification of DNA pools from high-fidelity microchips. Nat Biotechnol. 2010;28(12):1295–9.
66. Gao X, LeProust E, Zhang H, Srivannavit O, Gulari E, Yu P, et al. A flexible light-directed DNA chip synthesis gated by deprotection using solution photogenerated acids. Nucleic Acids Res. 2001;29(22):4744–50.
67. Warner JR, Reeder PJ, Karimpour-Fard A, Woodruff LB, Gill RT. Rapid profiling of a microbial genome using mixtures of barcoded oligonucleotides. Nat Biotechnol. 2010;28(8):856–62.
68. Hughes RA, Ellington AD. Synthetic DNA synthesis and assembly: putting the synthetic in synthetic biology. Cold Spring Harb Perspect Biol. 2017;9(1):a023812.
69. LeProust EM, Peck BJ, Spirin K, McCuen HB, Moore B, Namsaraev E, et al. Synthesis of high-quality libraries of long (150mer) oligonucleotides by a novel depurination controlled process. Nucleic Acids Res. 2010;38(8):2522–40.
70. Ma S, Tang N, Tian J. DNA synthesis, assembly and applications in synthetic biology. Curr Opin Chem Biol. 2012;16(3–4):260–7.
71. Stemmer WP, Crameri A, Ha KD, Brennan TM, Heyneker HL. Single-step assembly of a gene and entire plasmid from large numbers of oligodeoxyribonucleotides. Gene. 1995;164(1):49–53.
72. Casini A, Storch M, Baldwin GS, Ellis T. Bricks and blueprints: methods and standards for DNA assembly. Nat Rev Mol Cell Biol. 2015;16(9):568–76.

73. Itaya M, Fujita K, Kuroki A, Tsuge K. Bottom-up genome assembly using the *Bacillus subtilis* genome vector. Nat Methods. 2008;5(1):41–3.
74. Gibson DG, Smith HO, Hutchison CA, III, Venter JC, Merryman C. Chemical synthesis of the mouse mitochondrial genome. Nat Methods. 2010;7(11):901–3.
75. Wang K, Fredens J, Brunner SF, Kim SH, Chia T, Chin JW. Defining synonymous codon compression schemes by genome recoding. Nature. 2016;539(7627): 59–64.
76. Ma NJ, Moonan DW, Isaacs FJ. Precise manipulation of bacterial chromosomes by conjugative assembly genome engineering. Nat Protoc. 2014;9(10):2285–300.
77. Sliva A, Yang H, Boeke JD, Mathews DJ. Freedom and responsibility in synthetic genomics: the synthetic yeast project. Genetics. 2015;200(4):1021–8.
78. Baker M. The next step for the synthetic genome. Nature. 2011;473(7347): 403–408.
79. Cello J, Paul AV, Wimmer E. Chemical synthesis of poliovirus cDNA: generation of infectious virus in the absence of natural template. Science. 2002;297(5583):1016–8.
80. Mercy G, Mozziconacci J, Scolari VF, Yang K, Zhao G, Thierry A, et al. 3D organization of synthetic and scrambled chromosomes. Science. 2017;355(6329):eaaf4597.
81. Lartigue C, Glass JI, Alperovich N, Pieper R, Parmar PP, Hutchison CA, III, et al. Genome transplantation in bacteria: changing one species to another. Science. 2007;317(5838):632–8.
82. Karas BJ, Jablanovic J, Sun L, Ma L, Goldgof GM, Stam J, et al. Direct transfer of whole genomes from bacteria to yeast. Nat Methods. 2013;10(5):410–2.
83. Boeke JD, Church G, Hessel A, Kelley NJ, Arkin A, Cai Y, et al. The genome project-write. Science. 2016;353(6295):126–7.
84. Brown DM, Chan YA, Desai PJ, Grzesik P, Oldfield LM, Vashee S, et al. Efficient size-independent chromosome delivery from yeast to cultured cell lines. Nucleic Acids Res. 2017;45(7):e50.
85. Lartigue C, Vashee S, Algire MA, Chuang RY, Benders GA, Ma L, et al. Creating bacterial strains from genomes that have been cloned and engineered in yeast. Science. 2009;325(5948):1693–6.
86. Xie ZX, Mitchell LA, Liu HM, Li BZ, Liu D, Agmon N, et al. Rapid and efficient CRISPR/Cas9-based mating-type switching of *Saccharomyces cerevisiae*. G3: Genes, Genomes, Genet. 2018;8(1):173–83.
87. Jinturkar KA, Rathi MN, Misra A. In Misra A (Ed.), Gene delivery using physical methods. Challenges in Delivery of Therapeutic Genomics and Proteomics: Elsevier; 2011. p. 83–126.
88. Barber MA. A technic for the inoculation of bacteria and other substances into living cells. J Infect Dis. 1911;8(3):348–60.
89. Kagawa Y, Inoki Y, Endo H. Gene therapy by mitochondrial transfer. Adv Drug Deliv Rev. 2001;49(1-2):107–19.
90. Borowski AH, Leung JD, Van Der Kaa J, Hengst S, Platenburg GJ, Pieper FR, et al. Generation of transgenic mice and germline transmission of a mammalian artificial chromosome introduced into embryos by pronuclear microinjection. Chromosome Res. 2000;8(3):183–91.
91. Capecchi MR. High efficiency transformation by direct microinjection of DNA into cultured mammalian cells. Cell. 1980;22(2):479–88.
92. Neumann E, Schaefer-Ridder M, Wang Y, Hofschneider P. Gene transfer into mouse lyoma cells by electroporation in high electric fields. EMBO J. 1982;1(7):841–5.

93. Tsong TY. Electroporation of cell membranes. In Neumann E, Sowers AE, Jordan CA (Eds.), Electroporation and Electrofusion in Cell Biology: Springer Science + Business Media, LLC; 1989. pp. 149–63.
94. Tsong TY. On electroporation of cell membranes and some related phenomena. J Electroanal Chem Interfacial Electrochem. 1990;299(3):271–95.
95. Mitchell LA, McCulloch LH, Pinglay S, Berger H, Bosco N, Brosh R, et al. De novo assembly, delivery and expression of a 101 kb human gene in mouse cells. Genetics. 2021;218(1):iyab038.
96. Clewell DB. Bacterial Conjugation: Springer Science & Business Media; 2013.
97. Kim TK, Eberwine JH. Mammalian cell transfection: the present and the future. Anal Bioanal Chem. 2010;397(8):3173–8.
98. De Jong G, Telenius A, Vanderbyl S, Meitz A, Drayer J. Efficient in-vitro transfer of a 60-Mb mammalian artificial chromosome into murine and hamster cells using cationic lipids and dendrimers. Chromosome Res. 2001;9(6):475–85.
99. Bilang R, Klöti A, Schrott M, Potrykus I. PEG-mediated direct gene transfer and electroporation. In Gelvin SB, Schilperoort RA (Eds.), Plant Molecular Biology Manual: Springer; 1994. p. 1–16.
100. Maas C, Werr W. Mechanism and optimized conditions for PEG mediated DNA transfection into plant protoplasts. Plant Cell Rep. 1989;8(3):148–51.
101. Ahkong Q, Fisher D, Tampion W, Lucy J. Mechanisms of cell fusion. Nature. 1975;253(5488):194–5.
102. Pells S. Nuclear Reprogramming: Methods and Protocols: Springer Science & Business Media; 2006.
103. Uno N, Abe S, Oshimura M, Kazuki Y. Combinations of chromosome transfer and genome editing for the development of cell/animal models of human disease and humanized animal models. J Hum Genet. 2018;63(2):145–56.
104. Meaburn KJ, Parris CN, Bridger JM. The manipulation of chromosomes by mankind: the uses of microcell-mediated chromosome transfer. Chromosoma. 2005;114(4):263–74.
105. Kazuki Y, Kobayashi K, Aueviriyavit S, Oshima T, Kuroiwa Y, Tsukazaki Y, et al. Trans-chromosomic mice containing a human CYP3A cluster for prediction of xenobiotic metabolism in humans. Hum Mol Genet. 2013;22(3):578–92.
106. Kugoh H, Mitsuya K, Meguro M, Shigenami K, Schulz TC, Oshimura M. Mouse A9 cells containing single human chromosomes for analysis of genomic imprinting. DNA Res. 1999;6(3):165–72.
107. Collins JJ. How best to build a cell. Nature. 2014;509(7499):155–7.
108. Coradini AL, Hull CB, Ehrenreich IM. Building genomes to understand biology. Nat Commun. 2020;11(1):1–11.
109. Gibson DG, Benders GA, Andrews-Pfannkoch C, Denisova EA, Baden-Tillson H, Zaveri J, et al. Complete chemical synthesis, assembly, and cloning of a Mycoplasma genitalium genome. Science. 2008;319(5867):1215–20.
110. Cong L, Ran FA, Cox D, Lin S, Barretto R, Habib N, et al. Multiplex genome engineering using CRISPR/Cas systems. Science. 2013;339(6121):819–23.
111. Vaudano AE, Ruggieri A, Vignoli A, Canevini MP, Meletti S. Emerging neuroimaging contribution to the diagnosis and management of the ring chromosome 20 syndrome. Epilepsy Behav. 2015;45:155–63.
112. Specchio N, Trivisano M, Serino D, Cappelletti S, Carotenuto A, Claps D, et al. Epilepsy in ring 14 chromosome syndrome. Epilepsy Behav. 2012;25(4):585–92.

113. la Cour Sibbesen E, Jespersgaard C, Alosi D, Bisgaard AM, Tumer Z. Ring chromosome 9 in a girl with developmental delay and dysmorphic features: case report and review of the literature. Am J Med Genet A. 2013;161A(6):1447–52.
114. Sivendran S, Gruenstein S, Malone AK, Najfeld V. Ring chromosome 18 abnormality in acute myelogenous leukemia: the clinical dilemma. J Hematol Oncol. 2010;3:25.
115. Ciuladaite Z, Burnyte B, Vanseviciute D, Dagyte E, Kucinskas V, Utkus A. Clinical, cytogenetic and molecular study of a case of ring chromosome 10. Mol Cytogenet. 2015;8(1):29.
116. Severino M, Accogli A, Gimelli G, Rossi A, Kotzeva S, Di Rocco M, et al. Clinico-radiological and molecular characterization of a child with ring chromosome 2 presenting growth failure, microcephaly, kidney and brain malformations. Mol Cytogenet. 2015;8:17.
117. Kim T, Bershteyn M, Wynshaw-Boris A. Chromosome therapy. Correction of large chromosomal aberrations by inducing ring chromosomes in induced pluripotent stem cells (iPSCs). Nucleus. 2014;5(5):391–5.
118. Kachroo AH, Laurent JM, Yellman CM, Meyer AG, Wilke CO, Marcotte EM. Systematic humanization of yeast genes reveals conserved functions and genetic modularity. Science. 2015;348(6237):921–5.
119. Strathern JN, Newlon CS, Herskowitz I, Hicks JB. Isolation of a circular derivative of yeast chromosome III: implications for the mechanism of mating type interconversion. Cell. 1979;18(2):309–19.
120. Clarke L, Carbon J. Isolation of a yeast centromere and construction of functional small circular chromosomes. Nature. 1980;287(5782):504–9.

CHAPTER 7

METHODS FOR MAPPING OF NUCLEIC ACIDS EPIGENETIC MODIFICATIONS AND ITS CLINIC APPLICATIONS

XIAOCHENG WENG, SHUANG PENG, GUANGRONG ZOU, KEXIN YUAN, and XIANG ZHOU
College of Chemistry and Molecular Sciences, Key Laboratory of Biomedical Polymers of Ministry of Education, Hubei Province Key Laboratory of Allergy and Immunology, Wuhan University, Wuhan, Hubei, China

7.1. INTRODUCTION

It is well known that all cells in one organism share same genetic information but have different RNA expression subsets. At a certain time, only part of genes is expressed and executes its functions. How the cell chooses to activate some specific genes is the big question that have plagued the scientists for a long time. Recent breakthroughs in the epigenetics researches help people to understand the regulation of gene expression to a certain extent. All the three classes of fundamental macromolecules, DNA, RNA and proteins, will be chemically modified after synthesis. These post-synthesis chemical modifications, also named as epigenetic modifications, are involved in the control and adaptability during virtually all biological processes. The post-translation modifications (PTM) are mainly occurred in histones. The acetylation, methylation, phosphorylation, ubiquitination, SUMOylation, and other modifications in histone tails are tightly linked with transcriptional activation or inhibition, forming a

Nucleic Acids in Medicinal Chemistry and Chemical Biology: Drug Development and Clinical Applications, First Edition. Edited by Lihe Zhang, Xinjing Tang, Zhen Xi, and Jyoti Chattopadhyaya.

collection of histone code. In the recent 10 years, the epigenetics in nucleic acid is one of the most rapidly growing field in biology. So far there are more than 17 types of chemical modifications discovered in DNA, and even more than 160 modifications identified in RNA [1].

From the molecule perspective, it is undoubted that the extra-chemical groups and covalent bonds of the epigenetic modifications will disturb the communication of nucleic acids and proteins, or change the spatial structure of biomacromolecules, nucleosomes, and even chromosomes. Finally, it will dominate the gene expression status that even produce the effects at phenotypic level. For instance, the high-density methylation in CpG regions of promoters will block the binding of transcription factors to induce transcriptional silencing. At early stages of carcinogenesis, DNA methylation in the promoter region of tumor suppressor genes are often occurred [2]. The abundance of messenger RNA (mRNA) is not only determined by transcription events, the post-transcription modifications in mRNA also affect its stability or translation efficiency.

We all knew that the chemical modifications are not added to nucleic acid randomly. Only a little partial bases will be modified at some point and the modifications are dynamic with the change of biological events. To investigate the detail biological function of chemical modifications in nucleic acid, it is necessary to know which bases have been modified, the distribution pattern of the modifications, and its dynamic rule. The abundance of chemical modifications in nucleic acids is now easily assessed by ultrasensitive detection methods such as high-performance liquid chromatography–mass spectrometry (LC–MS). However, it is difficult to accurately profile the genome-wide distribution of epigenetic modifications, although plenty of techniques have been developed. Indeed, the studies of lots of potential modifications are still in near-dormant state.

The abnormal status of epigenetic modifications is related to many kinds of diseases [3]. 5-methylcytosine (5mC) and its oxidation product 5-hydroxylmethylcytosine (5hmC) are widely studied and reported as cancer biomarkers. The rapid advance of these two cytosine analogies are not just because of its high abundance in genome, but also due to the developed reliable and sensitive detection technologies. With the ongoing development and optimization of mapping methods, the mechanism of gene expression regulation and the role in disease of epigenetic modifications will be better understood. In this chapter, we summarize the recent progress in methods for detection and mapping of nucleic acids epigenetic modifications and some clinic applications.

7.2. METHODS FOR MAPPING OF EPIGENETIC MODIFICATIONS IN DNA

7.2.1. 5-Methylcytosine (5mC)

In addition to the four canonical bases adenine (A), guanine (G), cytosine (C), and thymine (T), 5-methylcytosine (5mC) as the fifth base in DNA has been studied most in-depth so far. 5mC contains a methyl group at the C5 position of cytosine and DNA methyltransferases (DNMTs) are acknowledged to be responsible for the generation and maintenance in genomes. As for its high content in living organisms (about 1% of all DNA bases [4]) and vital biological function, such as inhibition of transcription [5], causing the formation of heterochromatin [6], many methods have been developed to detect the 5mC in the genome.

7.2.1.1. Bisulfite Sequencing (BS-Seq) of 5mC Sodium bisulfite as one of the traditional chemical treatment reagents was widely used to detect 5mC at the single base resolution. Clark *et al.* firstly found ordinary cytosine (C) could be converted to uracil (U), while 5-methylcytosine (5mC) remained unchanged because of its steric and electron-giving effect of methyl at the C5 after treating with sodium bisulfite [7]. Following DNA sequencing, all the Cs were deaminated and read as Ts, while any other remaining Cs was speculated to 5mC. By comparing the DNA sequence differences before and after treatment, the mutation sites could be determined [8]. Although a degree of DNA degradation cannot be ignored with this approach, the reaction of sodium bisulfite with DNA itself is highly efficient and specific. Thus, this method is still the most effective way to distinguish the 5-methylcytosine from unmethylated cytosine (Figure 7.1).

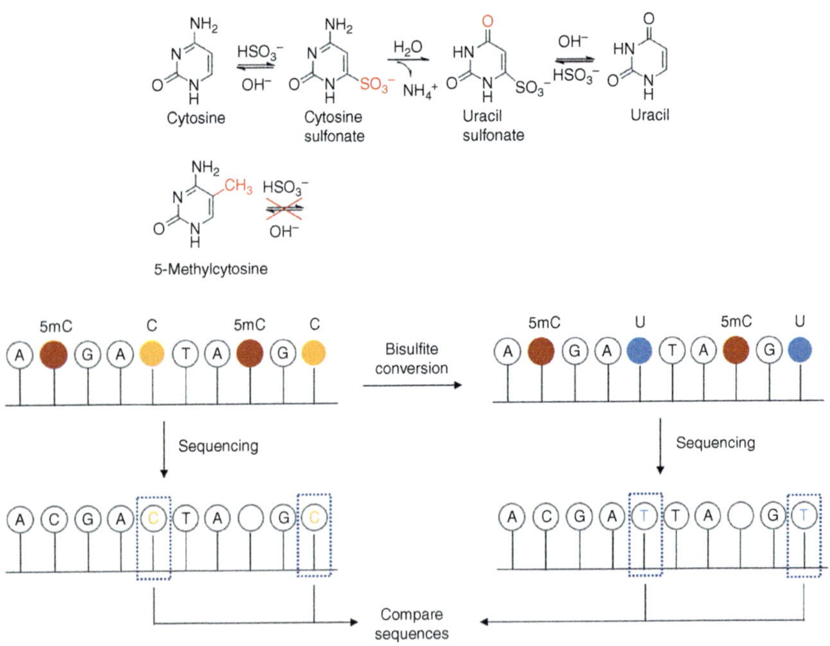

Figure 7.1 Mechanism of sodium bisulfite deamination of cytosine to uracil and schematic of BS-Seq procedure.

7.2.1.2. PyroSequencing – The Gold Standard for 5mC It is also very critical to quantify the ratio of 5-methylcytosine to cytosine in the specific 5mC site. In order to obtain more accurate DNA methylation status of specific gene fragments, scientists combined sodium bisulfite treatment with pyrosequencing to obtain the accurate quantitation information of 5mC sites. In general, after bisulfite treatment, all samples were amplified by DNA polymerase with only one kind of dNTP in each round of reaction. As for the extension reaction would produce pyrophosphate (PPi), which could bind to 5′-phosphoacyl sulfate (APS) to produce ATP under the action of ATP sulfate enzyme. Luciferase could prompt the ATP to combine with luciferin, thus producing oxidized luciferin. Therefore, specific peaks could be obtained for each round of reaction and then the sequence information of the DNA to be measured could also be acquired. The sites information of 5mC could be gained by comparing the samples with and without sodium bisulfite treatment after pyrosequencing [9]. The high sensitivity and reliability of pyrosequencing makes it the gold standard for 5mC quantitative detection.

7.2.1.3. Methylation-Specific PCR (MSP) Currently, another method based on sodium bisulfite, which named methylation-specific PCR (MSP) has been widely used in clinical practice [10]. Different from BGS technology, MSP technology requires designing two pairs of primers, methylation-specific primers and non-methylation-specific primers based on the conditions of methylation and unmethylation in advance. Because of the C to U conversion after bisulfite treatment, T_m value between the two pair of primers combined with the template is changed to affect the annealing and amplification processes. If only methylation-specific primers produced PCR products with correct length at the corresponding annealing temperature, the site could be considered as in the methylation state. Otherwise, it was unmethylated. Although this approach is economical and practical that widely used today, the primer design is quite complicate. And the false positive or false negative results are difficult to avoid if the annealing temperature is not selected correctly.

7.2.1.4. 5mC-Specific Whole-Genome Amplification Method (5mC-WGA) Although bisulfite methods play the major role for insight into the functionality of DNA methylation, the harsh condition in bisulfite treatment degrades lots of DNA, which hampers its performance in low-input samples. Recently, He *et al.* used DNA polymerase phi29 and DNMT1 to realize the 5mC-retained isothermal amplification by at least 100-fold, which avoid the loss of the 5mC information during PCR step (Figure 7.2). The amplified products were then subjected to bisulfite treatment, PCR amplification with predesigned primers, and sequencing to obtain the 5mC sites in the whole genome [11]. This methylation-retained amplification approach could enable facile the detection of 5mC in clinical samples such as DNA from biopsy or cell-free DNA (cfDNA) in plasma in the future.

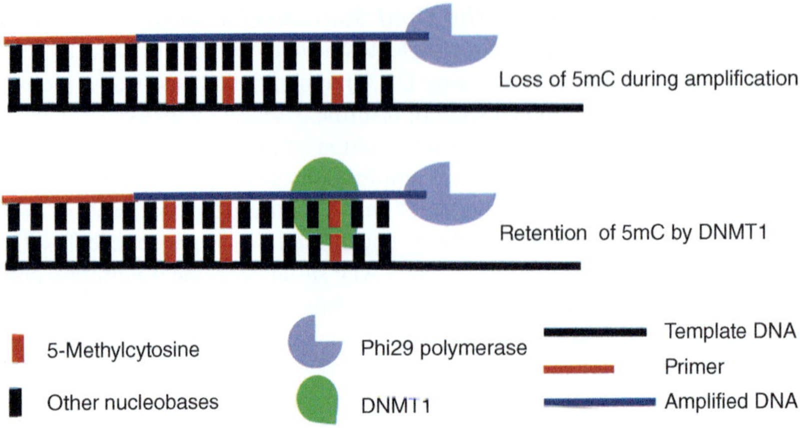

Figure 7.2 The principle of 5mC-retained whole-genome amplification (5mC-WGA).

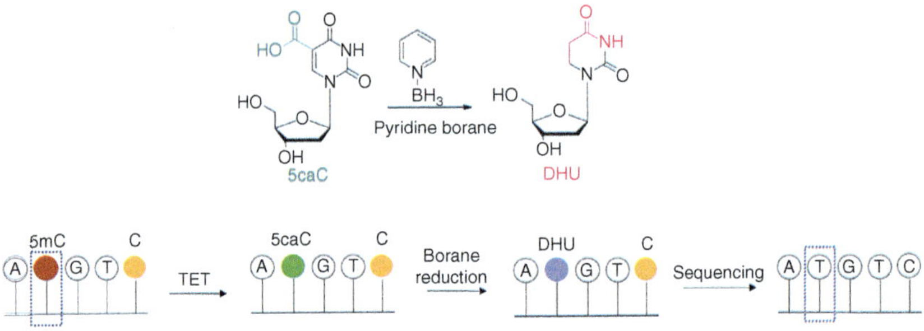

Figure 7.3 Scheme of TET-assisted pyridine borane sequencing of 5mC (TAPS).

7.2.1.5. TET-Assisted Pyridine Borane Sequencing (TAPS) Because of the severe DNA degradation and sequence complexity in bioinformatic analysis due to all cytosine conversion to thymine in bisulfite sequencing, the bisulfite-free 5-methylcytosine detection methods are urgently needed, especially for clinical application. Based on the oxidation of ten-eleven translocation enzyme (TET) and pyridine borane reduction, Song *et al.* presented a bisulfite-free and base-level-resolution sequencing method for detection of 5mC, named TET-assisted pyridine borane sequencing (TAPS) [12]. As shown in Figure 7.3, they firstly used TET enzymes to oxidize 5mC to 5caC. Then pyridine borane was applied to reduce 5caC to dihydrouracil (DHU). During the process of PCR, The DHU signal would be converted to the uracil signal, enabling a C-to-T transition of 5mC. The TAPS technology has no effect on unmodified cytosine and is not destructive, which can retain DNA fragments longer than 10 KB base length.

7.2.1.6. Detection of 5mC by Selective Oxidation Besides bisulfite treatment, other chemicals also exhibit the ability to selectively oxidize 5mC. Okamoto *et al.* found that 5mC not C could be selective oxidized by osmium complex in the presence of dipyridyl. The DNA sequence would break at the 5mC site after the oxidized products were treated with hot piperidine. Then the 5mC site information could be determined by sequencing or gel electrophoresis [13]. Except for this method, Carell *et al.* reported another similar approach [14]. The mixtures of V^V species or of $NaIO_4$ in the presence of LiBr allowed the desired conversion of the C5–C6 double bond of 5mC to be achieved with unprecedented selectivity in mixed DNA sequences. Then the 5mC site could also be confirmed by sequencing or gel electrophoresis after hot piperidine treatment. Furthermore, Zhou *et al.* have designed and synthesized two N-halogeno-N-sodiobenzenesulfonamide reagents for probing 5mC in DNA sequences [15]. In this method, N-sodio-N-bromo-m-nitrobenzenesulfonamide could realize the oxidation of 5mC and common C, while N-sodio-N-chloro-p-nitrobenzenesulfonamide could only oxidize 5mC. The numbers and sites of 5mC could also be calculated by subtraction method after hot piperidine treatment and gel electrophoresis analysis.

7.2.1.7. SMRT and Nanopore Sequencing The third-generation sequencing exhibits its attractive future and critical implication in genome biology for its longer reads capability, and also shows the potential in DNA methylation study. Turner *et al.* described a single-molecule, real-time (SMRT) sequencing method of 5mC [16]. They applied DNA polymerases to catalyze the incorporation of fluorescently labeled nucleotides into complementary nucleic acid strands. The arrival times and durations of the resulting fluorescence pulses yield information of polymerase kinetics and allow direct detection of modified nucleotides in the DNA template. This method is amenable to long read lengths and will likely enable mapping of methylation patterns in even highly repetitive genomic regions.

The DNA degradation after bisulfite treatment lead to the poor performance in the third-generation sequencing. However, with the long DNA strand after pyridine borane reduction, TAPS method [12] is compatible with both Oxford Nanopore [17] and PacBio Single-Molecule Real-Time (SMRT) sequencing, providing a new strategy and many research possibilities for whole-genome long-read methylation profiles.

7.2.2. 5-Hydroxymethylcytosine (5hmC)

5-hydroxymethylcytosine (5hmC), considered as the sixth base in genome, is present in different amounts among different cells and tissues [18]. In addition to being an intermediate of the process of DNA demethylation by TET enzymes, 5hmC is also a stable DNA modification involved in essential biological processes [19]. Besides, studies have shown that 5hmC could also be a hall marker of cancer for the diagnosis of diseases [20]. As the hydroxymethyl group of 5-hydroxymethyl cytosine can be oxidized easily, most methods for 5hmC detection usually involve oxidation.

7.2.2.1. Oxidation-Assisted Bisulfite Sequencing After the identification of 5hmC in genome, people realized that conventional BS-Seq methods cannot distinguish 5mC and 5hmC. To solve this problem, a new method based on BS-seq with the additional chemical oxidation was developed to detect 5hmC. For instance, Balasubramanian *et al.* adopted chemical molecule $KRuO_4$ to specifically oxidize 5hmC to 5fC, which could convert to U after the sodium bisulfite treatment. In contrast, the 5mC cannot be oxidized to 5fC, which still retain the C form after bisulfite treatment and amplification. By comparing the sequencing results with or without $KRuO_4$ oxidation after bisulfite sequencing, the 5hmC sites could be separately identified (Figure 7.4) [21].

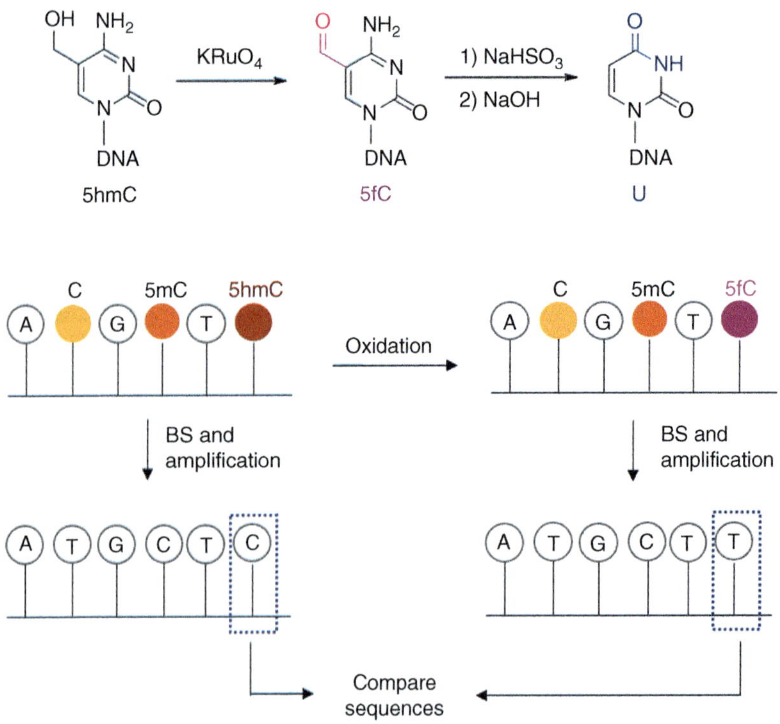

Figure 7.4 Scheme demonstrating of oxidation-assisted bisulfite sequencing.

7.2.2.2. β-Glucosyltransferase-Based Methods for 5hmC Profiling
By using β-glucosyltransferase, He *et al.* have developed a series method for 5hmC profiling genome-wide. Firstly, as early as in 2011, they used the β-glucosyltransferase to realize the selective labeling of the 5hmC by introducing a glucose containing an azide group onto the 5hmC. After click chemistry to introduce a biotin group, the 5hmC contained DNA strand that could be enriched by streptavidin beads. Therefore, by this method, genome-wide distribution of 5hmC could be determined with high efficiency and reproducibility [22]. In next year, they applied β-glucosyltransferase, TET proteins, and bisulfite treatment to develop a new method named "TAB-Seq" for 5hmC analysis at base resolution [23]. Briefly, a glucose group were introduced onto the 5hmC by using β-glucosyltransferase, thus the product β-glucosyl-5-hydroxymethylcytosine (5gmC) could protect 5hmC from further TET oxidation. Instead, 5mC would be oxidized to 5caC by TET proteins gradually. So, during the bisulfite sequencing, only 5hmC was read as C, while C, 5mC, and 5caC were converted into T. Hence, TAB-seq could realize the single-base resolution detection of 5-hydroxymethylcytosine in genome-wide. Although these two approaches have provided vital distribution and functional information of 5hmC, both of them need a large number of genomic DNA as starting material. They were not suited with the studies from rare cell populations, such as malignant stem cells, homogeneous neuronal cells, cell-free DNA and so on. To solve this problem, in 2016, they further exploit an improved approach named "Nano-hmC-Seal" [24], which used the transposase to achieve the incorporation of sequencing compatible adaptors in a single tagmentation step. This method could realize the highly sensitive and robust 5hmC sequencing from only ~1000 cells (Figure 7.5).

Recently, He *et al.* have exploited a new approach "Jump-seq" to capture and amplify 5hmC in genomic DNA based on β-glucosyltransferase [25]. The principle was mainly based on the connection of a hairpin DNA oligonucleotide carrying an alkyne group to the azide-modified 5hmC via "click" chemistry. Then the primer extension started from the hairpin motif to the modified 5hmC site and then continued to "land" on genomic DNA. These 5hmC sites could be identified from genomic DNA sequences immediately spanning the 5-prime junction. This technology with completely different strategy could also realize the detection of 5hmC at single-base resolution.

7.2.2.3. Base-Resolution Analysis of 5hmC by Bisulfite-Free Chemical Conversion The base-resolution analysis of 5hmC is extremely important for the biological function investigation. However, all above methods need bisulfite sequencing whose harsh condition will damage the integrity of genomic DNA. Soon afterward, scientists developed some bisulfite-free methods by new chemical conversion reactions for base-resolution 5hmC detection. Okamoto *et al.* reported a new method for sequencing of 5hmC by peroxotungstate-mediated oxidation [26]. After oxidation, 5hmC could be converted into

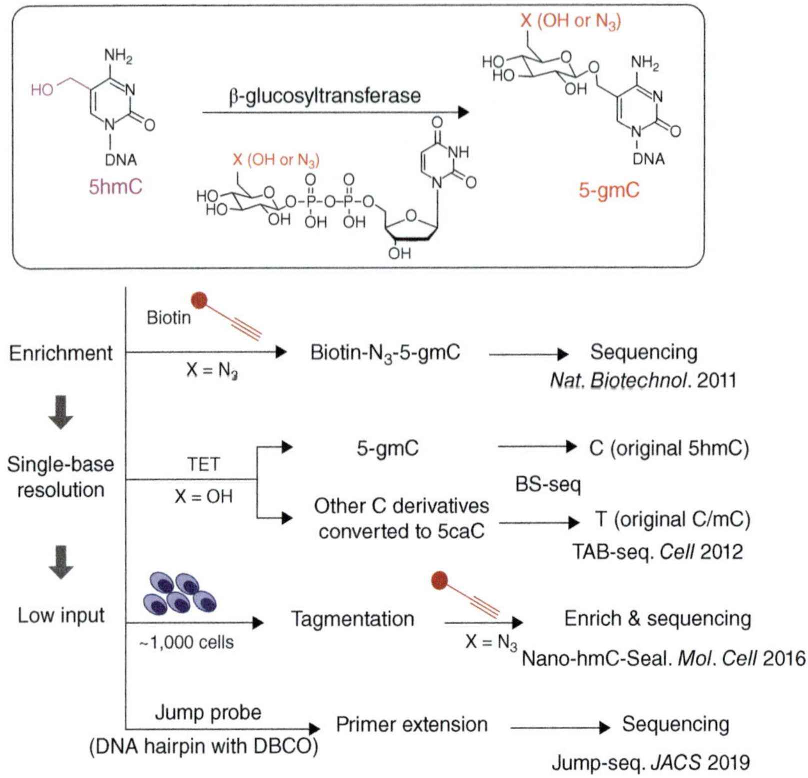

Figure 7.5 Scheme of β-glucosyltransferase based methods for 5hmC Profiling.

trihydroxylated thymine (thT), which tolerated the incorporation of dATP as a substrate in the process of DNA polymerase elongation. By comparing the results of sanger sequencing before and after the oxidation, 5hmC sites on single-stranded DNAs could be discriminated from unmethylated cytosines.

Furthermore, almost simultaneously, two methods combined the oxidation and specific 5fC probe labeling were reported for 5hmC mapping genome-wide. In these methods, as shown in Figure 7.5, 5hmC was firstly oxidized to 5-formylcytosine (5fC) by $KRuO_4$. Then two different 5fC probes were introduced to induce C-to-T transition during the PCR process. Yi *et al.* used the compound "AI" to realize the chemical labeling and further enrichment of 5fC by biotin-mediated pull-down method. This method named "hmC-CATCH" only required nanoscale input genome DNA samples in a cost-effective manner [27]. Instead, Zhou *et al.* used another 5fC specific probe "azi-BP" to induce the chemical conversion reaction for bisulfite-free, base-resolution detection of 5hmC [28]. To exclude the interference of the inherent 5fC, they blocked the 5fC by hydroxylamine in advance so that it could not be further labeled with

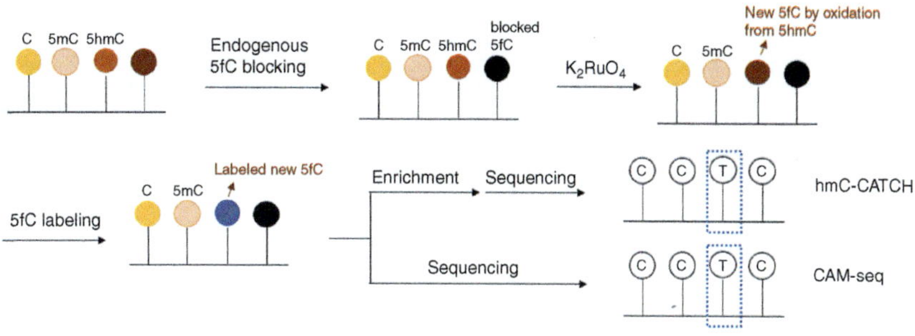

Figure 7.6 Scheme of hmC-CATCH or CAM-seq.

"azi-BP." This method termed "CAM-seq" could provide an alternative for the whole-genome mapping of 5hmC (Figure 7.6).

7.2.3. 5-Formylcytosine (5fC)

Although 5fC also is one of the oxidation products of 5mC in DNA demethylation process, it was found in many cell types and all major organs, indicating its potential biological significance in gene regulation. Since 5fC is present at quite low level, it is approximately 10-fold to 100-fold lower than that of 5hmC. Thus, the enrichment by chemical-, enzyme- or antibody-based methods for 5fC are required to reduce the sequencing depth. Several highly sensitive and selective methods for genome-wide detection of 5fC have been developed, especially some strategies can achieve the 5fC sites identification at single-base resolution.

7.2.3.1. 5fC Mapping by Biotin-Labeled Hydroxylamine Derivatives In 2012, Balasubramanian *et al.* employed a chemical compound "ARP" to realize the genome-wide mapping of 5fC, which is the first effort to profile 5fC in genome-wide [29]. The probe "ARP" contained a hydroxylamine group for reacting with the aldehyde group of 5fC and biotin tag for enrichment by streptavidin beads. Even though the reaction condition between 5fC and "ARP" was mild, this method is unable to identify accurate location information of 5fC.

7.2.3.2. fC-Seal To identify 5fC at the base resolution, He *et al.* developed a new way by coupling chemical reduction with biotin tagging [30]. They used a mild reducing agent $NaBH_4$ to reduce 5fC to 5hmC. The newly generated 5hmC from 5fC could be enriched for sequencing by 5hmC-selective method, which named "hMe-Seal." Besides the endogenous 5hmC was blocked by regular glucose based on βGT-catalyzed glucosylation. The author also resolved 5fC at base resolution by hydroxylamine-based protection from bisulfite-mediated deamination.

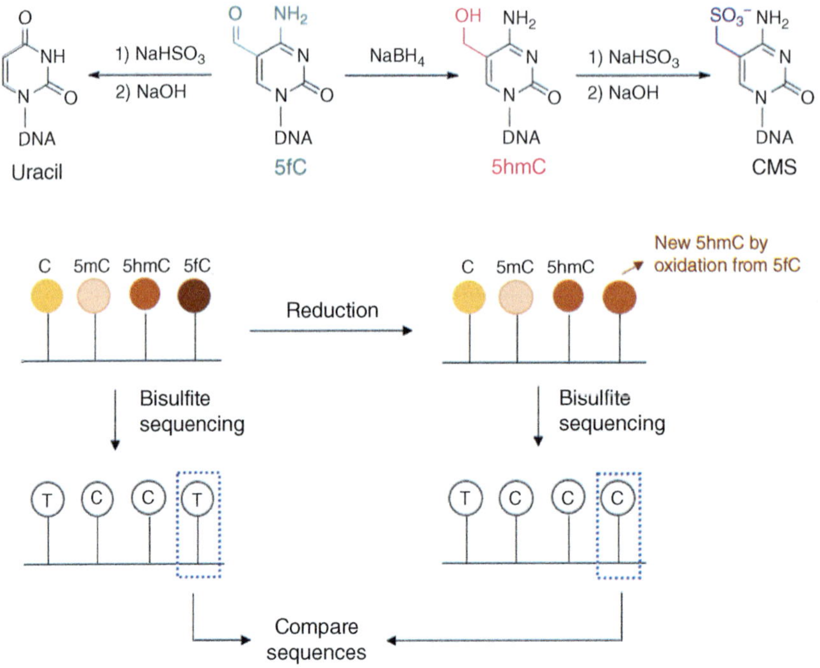

Figure 7.7 Outline of reduced bisulfite sequencing to map 5fC in DNA (redBS-Seq).

7.2.3.3. redBS-Seq In 2014, Balasubramanian *et al.* reported another new method, reduced bisulfite sequencing (redBS-Seq) to quantitatively detect 5fC at single-base resolution (Figure 7.7) [31]. Firstly, they treated the sample with sodium borohydride to reduce 5fC to 5hmC. By comparing the bisulfite sequencing result of the sample with or without reduction, the 5fC sites could be picked out. This experiment revealed the comparable levels of 5fC to 5hmC and 5mC. By combing the oxidative and bisulfite sequencing method, it could help to profile the whole genome mapping of 5fC and understand its functions.

7.2.3.4. fC-CET Although there are already some approaches can detect 5fC at the single-base resolution, most methods are complicated which need compare two rounds of sequencing or cause seriously DNA degradation by bisulfite treatment. Therefore, it is still significant to develop new straightforward technology for genome-wide detection of 5fC. Due to the high chemical activity of aldehyde group in 5fC, Yi *et al.* presented the "fC-CET," a bisulfite-free method based on chemical labeling and mutation analysis for 5fC (Figure 7.8) [32]. They applied an azido derivative of 1,3-indandione ("AI") and realized the complete conversion of 5fC to "5fC-AI" under very mild reaction

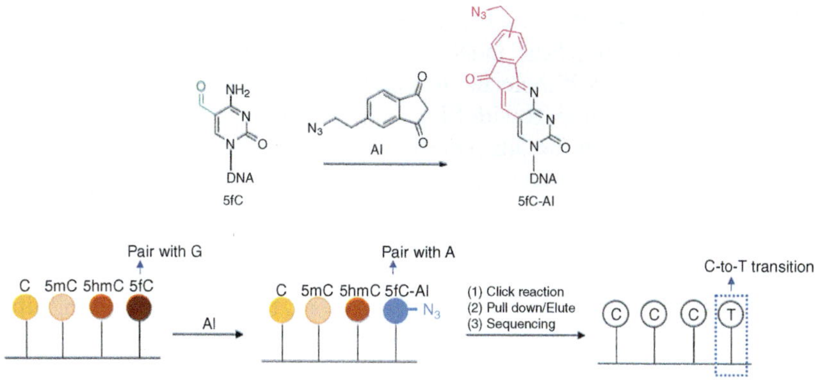

Figure 7.8 Chemical molecule "AI"-mediated cyclization labeling of 5fC and scheme of "fC-CET." Genomic DNA was sequentially labeled with "AI," conjugated to biotin for pulldown enrichment and ligated to adaptors for high-throughput sequencing. C-to-T transitions were specifically searched for to define 5fC sites in the whole genome.

conditions. The "5fC-AI" adduct could be paired with A, thus the reaction between "AI" and 5fC could cause the C to T transition of 5fC during PCR. Besides this method could achieve selective enrichment of 5fC because the AI-labeled 5fC could couple a cleavable biotin handle via click chemistry. However, the compound "AI" has poor solubility in water, which may limit its application such as for the using in few cells.

7.2.3.5. CLEVER-Seq To improve the limitation of solubility of probe in fC-CET, Yi *et al.* discovered a water-soluble small-molecule malononitrile, which can react with 5fC selectively. Based on malononitrile, they developed a new chemical-labeling-enabled C-to-T conversion sequencing method named CLEVER-seq [33]. Similar with the reaction in fC-CET, the malononitrile labeled 5fC was read as a dT during the PCR process, that means C-to-T conversion in 5fC sites. In general, by simple comparison the sequencing results before and after the malononitrile reaction with 5fC containing samples, the 5fC sites could be identified selectively. This method could be applied for detecting the single-cell 5fC landscape at single-base resolution.

7.2.4. 5-Carboxylcytosine (5caC)

TET family proteins can oxidize 5mC stepwise to 5hmC, 5fC, and 5caC. As the final oxidized derivative, the study of 5caC is lagging far behind with other modified cytosine bases. The main reason is the low content of 5caC in genome (5caC is roughly 5 ppm per cytosine) [34]. Another reason is the relative lower chemical reactivity of carboxy group in 5caC than 5fC. Nonetheless, there are still some chemical biological methods for detection of 5caC by coupling reaction.

7.2.4.1. CAB-Seq He *et al.* exploit a chemical modification-based bisulfite sequencing for detecting 5caC in DNA [34]. As shown in Figure 7.8, they firstly applied a chemical small molecule named EDC to label 5caC. The carboxylic acid group of 5caC can react with EDC to form an o-acylisourea ester intermediate, which could then be attacked by a primary amine. The author further designed a biotin-modified amine, which could realize the enrichment of 5caC by streptavidin-coated beads. What's more, even though 5-carboxylic cytosine was attached to an amide protective group, it still read as C during the bisulfite sequencing. While for the 5caC without labeling, it would converted to uracil after bisulfite treatment and read as T. Therefore, by comparing the bisulfite sequencing results of 5caC before and after EDC-catalyzed coupling reaction, the specific sites information of 5caC could be obtained (Figure 7.9).

7.2.4.2. DIPCAB-Seq Due to the quite low abundance of 5caC, bisulfite sequencing method cannot realize the direct, whole-genome detection of 5caC. Thus, He *et al.* developed an improved method based on "CAB-Seq" named "DIPCAB-seq" to generate the genome-wide, single-base resolution maps for 5caC [35]. Different from "CAB-Seq," before sequencing, they chose DNA immunoprecipitation for achieving the pre-enrichment of 5caC. Then, the 5caC

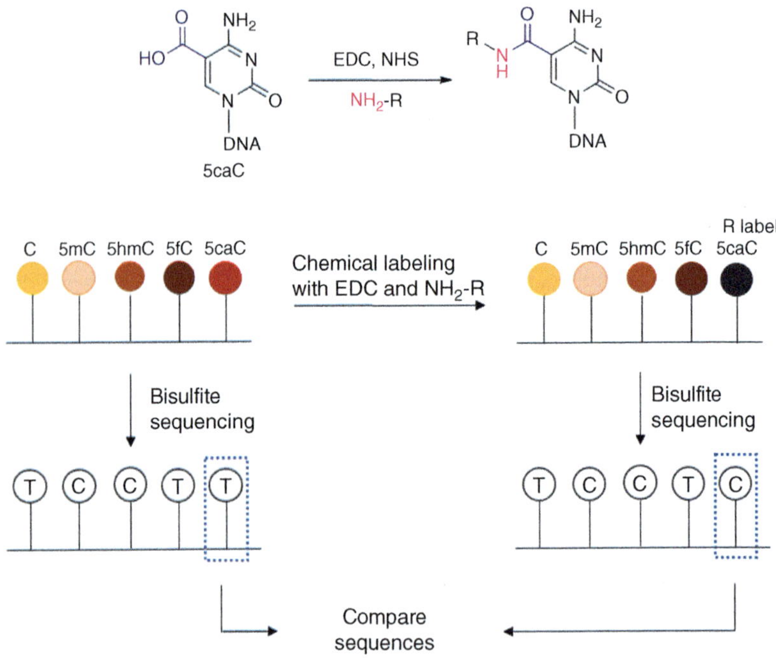

Figure 7.9 Detection method of EDC-catalyzed chemical labeling of 5caC with bisulfite sequencing.

sites were confirmed by sequencing the samples after enrichment by "CAB-Seq." This method could generate both profiling and single-base resolution maps of 5caC with more than 65% mapping efficiency.

7.2.5. 5-Formyluracil (5fU)

It was reported that the content level of 5-formyluracil and 5fC is similar in genome [36], but it is difficult to distinguish the 5fC and 5fU because these two modified bases exhibit almost same chemical properties due to aldehyde functional group. Many researchers devoted themselves to looking for effective means to detect 5fU excluding the interference of 5fC.

7.2.5.1. fU-Seq Up to now, although there were some methods for detecting 5fU based on fluorescence or mass spectrum [37], the genome-wide profiling of 5fU remains a challenge. Zhou *et al.* have drawn the first genome-wide map of 5fU in living tissues and assessed its potential roles [38]. They selected an azido-modified reagent named "azi-BIAN," which could get reaction with 5fU. The reaction products "azi-biaU" could be captured by copper-free click chemistry with commercial dibenzocyclooctyne (DBCO)-modified biotin. Thus, the genome-wide distribution of 5fU could be determined selectively (Figure 7.10). However, this approach cannot identify the specific sites of 5fU. Therefore, it is necessary to develop newly means for detecting the 5fU sites at the single-base resolution.

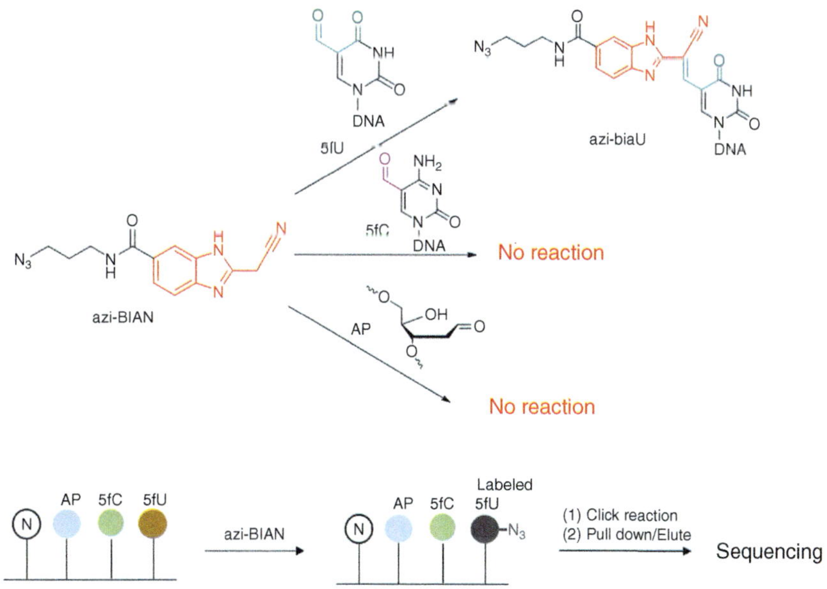

Figure 7.10 Scheme of "fU-Seq" based on chemical labeling.

7.2.6. Deoxyuridine (dU)

Deoxyuridine, as a special nucleic acid base in DNA, can be generated from cytosine deamination by AID/APOBEC family proteins [39]. In living cells, dU can be specifically recognized by four DNA glycosylase, UNG [40], SMUG1 [41], TDG [42], and MBD4 [43]. In view of the importance of dU in biology, such as inhibiting the thymidine synthase biosynthesis [44], generating the AP sites [45], it is necessary to develop the detection methods for qualitative and quantitative detection of dU sites.

7.2.6.1. "dU-Seq" Yi *et al.* applied the "dU-seq" method to determine dU in the human genome (Figure 7.11) [46]. Firstly, genomic DNA was treated by UDG enzyme to remove dU from DNA specifically. A vacancy was formed after the genomic DNA was further treated with AP endonuclease. Through the action of DNA polymerase and ligase, biotin-modified nucleotides were introduced on the vacancy. The biotin-labeled DNA strands were then enriched by streptavidin, and the dU sites information could be obtained from library construction and high-throughput sequencing. By this assay, they found that dU sites were highly enriched in centromeres, indicating the potential role of dU in-chromosome segregation.

7.2.7. N6-Methyldeoxyadenosine (6mA)

N6-methyldeoxyadenosine (6mA), which has long been considered as a vital DNA modification preserved in prokaryotes for host defense function. Until 2015, the presence of 6mA in eukaryotes were confirmed from three independent researches [47–49]. Although further studies reported the potential roles of 6mA in a series of life processes or diseases [50, 51], the main function or even the existence of 6mA in eukaryotes is still in the controversy. Unlike cytosine derivatives produced by TET oxidation, the source of 6mA in genome is still unknown. What's more, the accurate detection of 6mA, especially the single-base resolution method, is critical to answer this depute and understand its biological functions.

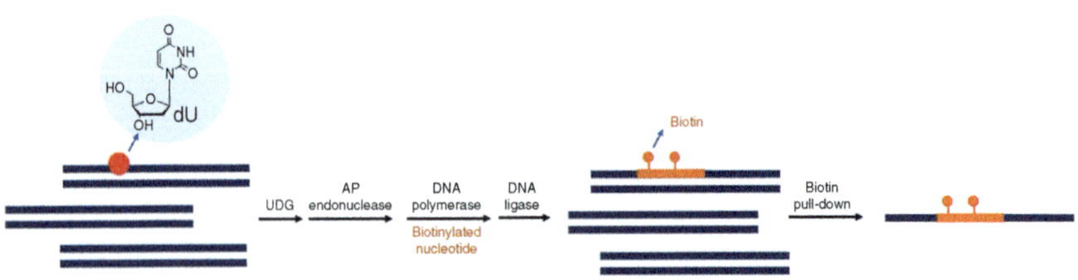

Figure 7.11 Scheme of dU-seq based on selective labeling and pull-down technology.

7.2.7.1. 6mA-CLIP-Exo In 2015, He *et al.* firstly developed an antibody-based profiling method to acquire the distribution information in a genome-wide scale. To achieve a much higher resolution, they adopt a 6mA-CLIP-exo strategy based on photo-crosslinking followed by exonuclease digestion to detect 6mA [52]. In addition, they developed 6mA-RE-seq using two methylation-sensitive restriction enzymes CviII and DpnII to genome-wide identify 6mA sites in single-base resolution. Both the two endonucleases could retain these methylation sites C(6mA)TG and G(6mA)TC and cleavage CATG and GATC sites selectively at the same time. Although the method was optimized further and could realize the 6mA detection in low abundance and limited sample [53], many 6mA sites information in the sequences outside the restriction enzyme recognition sites would be lost.

7.2.7.2. 6mA ChIP-Exo Considering the small size of the human mitochondrial genome, which is 16,569 base pairs length, He *et al.* applied a simplified 6mA ChIP-exo assay based on previous report [54] to map 6mA in mitochondrial DNA (mtDNA) at a relatively high resolution [55]. They used anti-6mA antibody to capture 6mA containing DNA. The antibody-DNA complex was then cross-linked by the irradiation of UV. Then the samples were incubated with Dynabeads Protein A or Protein G, combined with adaptor. After kinase reaction, fill-in reaction and lambda exonuclease digestion, DNA was released from beads and prepared for splint ligation. The final ligation products could be amplified by PCR and analyzed by sequencing to find out the specific 6mA sites. By this method, the author found that METTL4 could mediate the 6mA methylation of mammalian mtDNA. Besides they also proved 6mA affected the transcription, replication, and activity of mitochondrial.

7.3. METHODS FOR MAPPING OF EPIGENETIC MODIFICATIONS IN RNA

In the past 10 years, RNA epigenetic study has become increasingly popular (Figure 7.12). In view of the desire of researchers to investigate the biological functions of exact RNA modification, a transcriptome-wide analyzation method with high resolution is going to be more and more crucial. At the first place, the most classical detection methods for RNA modifications are thin layer

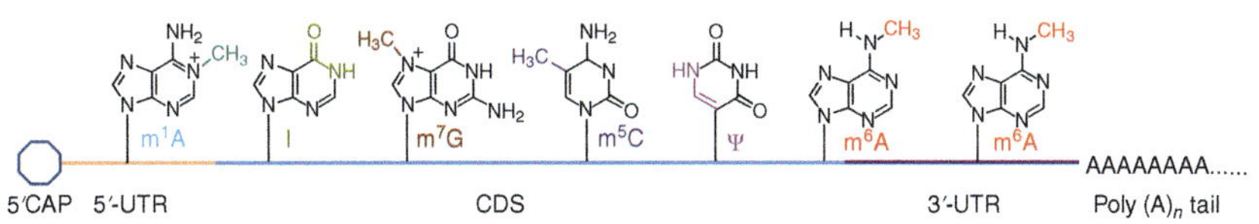

Figure 7.12 RNA epigenetic modifications in mRNA.

chromatography (TLC), HPLC-MS, and antibodies recognition. After exploration for years, an increasing number of simplified, adapted protocols were reported, assisting for the development of epigenetic studies.

7.3.1. N6-Methyldeoxyadenosine (m^6A) in RNA

As the most abundant internal modification on eukaryote mRNAs, N6-methyladenosine (m^6A) plays a significant role in RNA metabolism and function. Most m^6A marks are identified within the consensus motif DRACH (D = A, G or U; R = A or G; H = A, U or C) [56]. Although DRACH motif is ubiquitous in mRNA, merely a small fraction of them are transformed to m^6A *in vivo*. Notably, m^6A methylation is inclined to be enriched around the transcription start sites (TSS), stop codon, and the last exons [52, 57, 58]. m^6A is a dynamic RNA modification, which means that it is reversible in cellular conditions [59]. As is demonstrated so far, m^6A RNA modification is found to be mediated by a multiprotein methyltransferase complex (termed "writer"), which comprises of METTL3, METTL14, and additional protein subunits (including WTAP, KIAA1429, VIRMA, HAKAI, Zc3h13, and RBM15/15B) [60–62]. Several component of this complex (e.g., METTL3) has been identified to bind S-adenosyl-L-methionine (SAM) [63]. Meanwhile, other components are capable to interact with the METTL3-METTL14 core complex. Fe^{II}- and α-ketoglutarate-dependent fat mass and obesity-associated (FTO) protein is the first m^6A mRNA demethylase [64–66]. FTO and ALKBH5 belong to a broader family of AlkB dioxygenases, hence it's not surprising that ALKBH5 was also found to be a m^6A demethylase [67]. As far, m^6A base modification has been identified to regulate mRNA biological activities by two major approaches: blocking the protein-RNA interactions, or being recognized by m^6A binding proteins (called "readers") [68]. HNRNPC, HNRNPA2B1, YTHDF1/2, and elF3 have been reported as m^6A readers, involved in mRNA degradation, pre-mRNA processing, RNA splicing, RNA maturation, as well as many other RNA metabolism events [69–74]. Moreover, various cellular stresses, including UV light, tend to induce a transcriptome-wide redistribution of m^6A, resulting in a dynamic m^6A profile *in vivo* [75, 76]. In order to better elucidate the regulation process by m^6A, convenient, economical, and effective sequencing techniques are welcomed.

7.3.1.1. m^6A-Seq or MeRIP-Seq With the intention of uncovering the transcriptome-wide localization of m^6A, scientists have developed methods to get a precise look into the m^6A sites. Based on m^6A-specific immunoprecipitation and high-throughput sequencing, m^6A-seq compile human and mouse RNA methylomes and demonstrate the conservation of m^6A modification on distinguished species. In m^6A-seq, RNA was fragmented into 100 nt oligonucleotides and then immunoprecipitated by an anti-m^6A affinity-purified antibody. Libraries were constructed from immunoprecipitated fragments and subjected to massively parallel sequencing [69]. At the same year, MeRIP-seq employed an analogous strategy, deciphering a m^{6}A profile, which revealed the enrichment of m^{6}A sites near stop codons and in 3′ UTRs [57]. At present, these two methods based on the immunoprecipitation by anti-m^{6}A antibody are still the main research methods for m^{6}A mapping for the characteristics of simplicity and convenience.

7.3.1.2. PA-m^{6}A-Seq and miCLIP Immunoprecipitation-based sequencing techniques cannot accomplish single-base detection. To increase the resolution of m^{6}A-seq, scientists developed methods according to the strategy of photoactivatable ribonucleoside-enhanced crosslinking and immunoprecipitation (PAR-CLIP). PAR-CLIP is a photo-crosslinking-based method, which incorporate a photoactivatable ribonucleoside (4-thiouridine or 6-thioguanosine) into messenger RNA. Notably, 4SU or 6SG covalently crosslinks with nearby aromatic amino acid residues in RNA-binding proteins upon 365 nm UV irradiation, enabling the precise detection of a specific modified base. The adapted method, named photo-crosslinking-assisted m^{6}A-sequencing (PA-m^{6}A-seq), was adapted to map 6mA. PA-m^{6}A-seq efficiently rallies the accuracy of the methylation site assignments and offers a higher resolution transcriptome-wide mammalian m^{6}A map (~ 23 nt) [77]. However, m^{6}A sites out of DRACH motif are prone to be neglected in PA-m^{6}A-seq. m^{6}A-CLIP and m^{6}A individual-nucleotide-resolution crosslinking and immunoprecipitation (miCLIP) utilize UV-induced antibody-RNA crosslinking immunoprecipitation (UV-CLIP) as well. In these methods, 254-nm UV light is subjected to induce crosslinked RNA. However, in contrast to PA-m^{6}A-seq, both m^{6}A-CLIP and miCLIP take advantage of proteinase K to retrieve the crosslinked RNA. The protein–RNA crosslinking sites lead to patterned mutational or truncation profiles during RT, thereby revealing the precise position of m^{6}A (Figure 7.13) [58, 78].

7.3.1.3. m^{6}A-LAIC-Seq The pre-enrichment step will cruelly impact the hidden information of m^{6}A quantification in RNAs. Therefore, the m^{6}A stoichiometry remains largely unknown. In 2016, a method called m^{6}A-level and isoform-characterization sequencing (m^{6}A-LAIC-seq) was developed to uncover the m^{6}A stoichiometry [79]. In m^{6}A-LAIC-seq, treatment with excessive antibody is handled to ensure pulling down all m^{6}A containing RNA. Then m^{6}A levels per gene is quantified by the ratio of RNA abundances using the following formula (eluate) / (eluate + supernatant). As pre-enrichment of m^{6}A leads to the incapability of collected data representing the m^{6}A stoichiometry. m^{6}A-LAIC-Seq was designed to quantify the m^{6}A stoichiometry on a transcriptome-wide scale by comparing RNA abundance in m^{6}A-positive (eluate of immunoprecipitation) and m^{6}A-negative (supernatant) fractions.

7.3.1.4. DART-Seq Although the immunoprecipitation method based on anti-m^{6}A antibody has played a major role for m^{6}A identification, the antibody-free m^{6}A profiling method is still needed for the bias of antibody recognition and limitation of discrimination of related modifications (such as m^{6}Am). YTH family proteins were identified as m6A binding protein due to the capability for

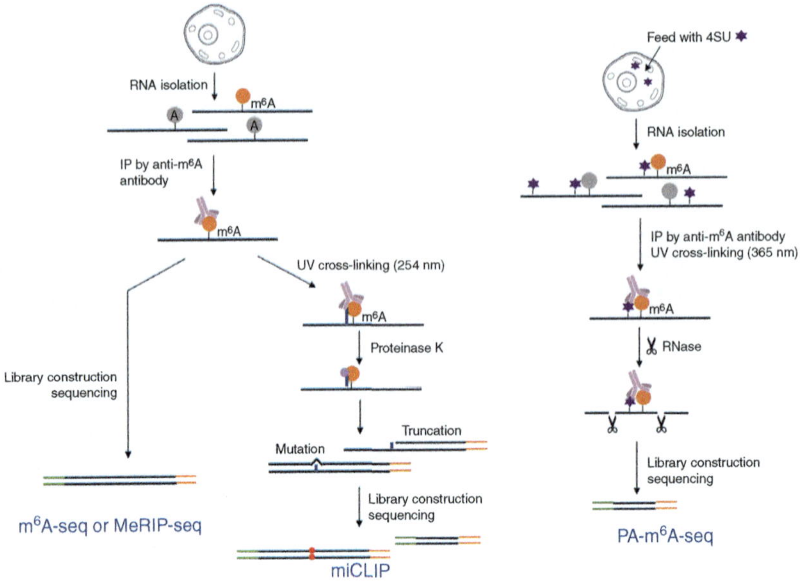

Figure 7.13 Scheme of m^{6}A-seq, PA-m^{6}A-seq and miCLIP based on anti-m^{6}A antibody.

m6A recognition of YTH domain in these family proteins. Combining cytidine deaminase APOBEC1 and m^{6}A-binding YTH domain, an antibody-free method, named DART-Seq (deamination adjacent to RNA modification targets), was developed in 2019 [80]. APOBEC1-YTH initiates C to U deamination at sites adjacent to m^{6}A residues. The transformation is later detected by standard RNA-Seq. DART-Seq requires little sample (as little as 10 ng of total RNA). Meanwhile, it won't recognize the structurally similar cap modification, m^{6}Am. Obviously, the affinity and specificity of the YTH domain to m^{6}A modification may affect the detection accuracy of the targets.

7.3.1.5. m^{6}A-REF Seq Besides YTH domain protein, the RNA-endoribonuclease MazF was applied to fulfill m^{6}A detection. The method m^{6}A-REF-seq or MAZTER-seq was developed via the m^{6}A selectivity of MazF, which can simply detect m^{6}A at single-base resolution in transcriptome-wide. In these methods, MazF can cleave the unmethylated ACA motif but not (m^{6}A) CA. After treating, the methylated (m^{6}A)CA motifs remain intact and is able to be detected by next-generation sequencing [81, 82]. Since there is no need to do immunoprecipitation, m^{6}A-REF-seq or MAZTER-seq is quantitative to validate the methylation status and abundance of detected m^{6}A sites. However, the ACA motif of MazF recognition target is not the main conservative motif and only the ~16–25% of total m^{6}A sites located within its recognition range. The directed evolution of MazF to eliminate its sequence selectivity may be very useful to expand its detected targets.

7.3.1.6. m⁶A Identification Methods by Chemical Probe The methyl group in m^6A is chemically inert, so it is difficult to discover the chemical reagent to label m^6A directly. Despite this, there are several m^6A identification methods combining the advantages of chemistry and biology. For instance, 4SedTTP, with the substitution of O4 of dTTP with selenium, result in m^6A-induced reverse transcription (RT) stop because of the steric hindrance of selenium atom with larger radius [83]. More recently, a metabolic labeling method m^6A-label-seq was designed to map mRNA m^6A. A methionine analog Se-allyl-l-selenohomocysteine is capable to substitute m^6A with a^6A. Using m^6A-label-seq, researchers has mapped mRNA m^6A sites in human HeLa, HEK293T, and mouse H2.35 cell lines [84]. Another chemical labeling method, FTO-assisted m^6A selective chemical labeling method (termed m^6A-SEAL, Figure 7.14), has been developed. Firstly, m^6A is oxidized to hm^6A by FTO, then hm^6A is

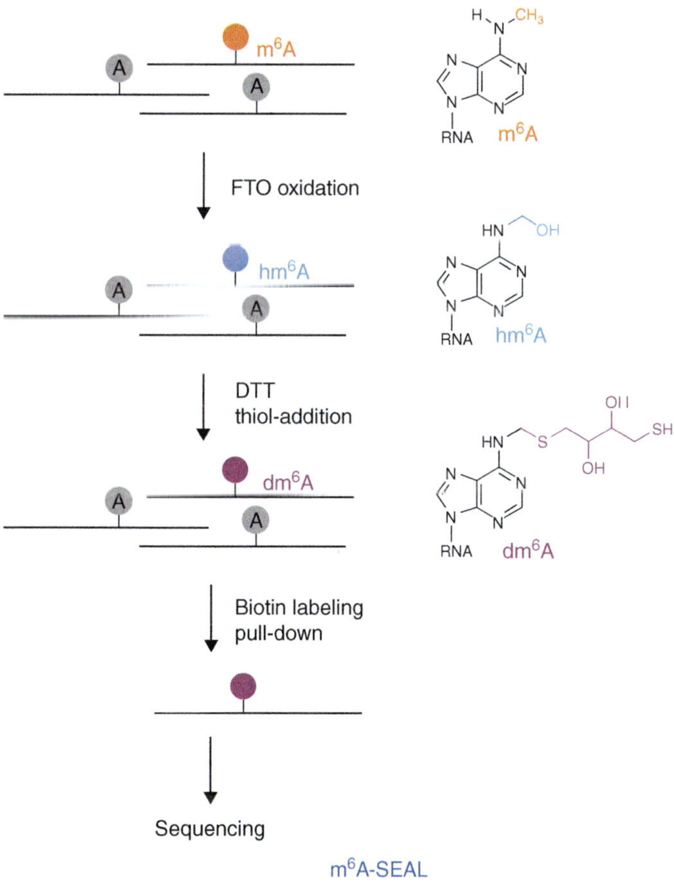

Figure 7.14 Scheme of m^6A-SEAL assisted by FTO oxidation.

transferred to more stable N6–dithiolsitolmethyladenosine (dm^6A) with a DTT-mediated thiol-addition reaction [85]. For all chemical labeling methods, labeling efficiency or the operation conditions are needed to be improved. Hence, new methods are still desired to facilitate the study of m^6A.

7.3.2. N6, 2′-*O*-Dimethyladenosine (m^6Am) in RNA

m^6Am is a cap-related modification, while it's one of the most prevalent modified bases found at the 5′ end of mRNA, adjacent to the 7-methylguanosine cap [86]. m^6Am is a dynamic modification that influences cellular mRNA fate. Methyltransferase PCIF1 can methylate 2′-*O*-methylated adenosine (Am) into m^6Am [87, 88]. Meanwhile, obesity-associated protein (FTO) reverts m^6Am into adenosine, like m^6A. Moreover, it is observed that the existence of m^6Am leads to increasing stability of mRNA [89]. As anti-m^6A is competent to recognize m^6Am as well, m^6A-seq and miCLIP technology provide m^6Am sites near the transcription start site (TSS) as well [78, 89, 90]. These truncation sites in 5′ UTR tend to occur in BCA (B = C/U/G) motifs rather than in the canonical RRACH motifs for m^6A. Recently, a transcriptome-wide, exonuclease-assisted sequencing strategy named m^6Am-Exo-Seq is developed. m^6Am-Exo-Seq employs a 5′ exonuclease to deplete the internal m^6A residues, then capped RNA 5′ end fragments are de-capped to expose the m^6Am loci. m^6Am-containing fragments are immunoprecipitated afterward and analyzed by high-throughput sequencing [87].

7.3.3. N1-Methyladenosine (m^1A) in RNA

N1-methyladenosine (m^1A) is a prevalent modification in stable noncoding RNAs including tRNA and rRNA. It was also found to be present in mammalian mRNA in 2016, and m^1A within the coding sequence can interfere with translation. m^1A is reversible in response to stimulation, indicating a dynamic m^1A methylome [91, 92].

7.3.3.1. m^1A-Seq A transcriptome-wide, antibody-based approach to localize m^1A sites termed m^1A-seq was developed. Since m^1A can rearrange to m^6A under alkaline conditions (Dimroth rearrangement), m^1A-seq was coupled with an orthogonal chemical method based on Dimroth rearrangement to obtain high-resolution m^1A maps in transcriptome-wide [91].

7.3.3.2. m^1A-ID-Seq Commonly relying on m^1A immunoprecipitation, m^1A-specific RNA immunoprecipitation with demethylase-assisted RNA sequencing (m^1A-ID-seq) utilizes the ability of m^1A to terminate reverse transcription [92]. In m^1A-ID-seq, a m^1A demethylase, E. coli AlkB, was selected, assisting m^1A reverts to adenosine. AlkB-assisted removal of m^1A from the immunoprecipitated RNA would result in the accumulation of full-length cDNAs. Consequently, demethylase-sensitive and high-confidence m^1A sites are detected by comparing the sequencing profiles of the (+) and (−) demethylase samples.

7.3.3.3. m^1A-MAP Because m^1A modification blocks the Watson–Crick base pairing, it may cause both truncation and mis-incorporation during cDNA synthesis. Additionally, mutations are more expedient over stops in NGS, as the latter is subjected to background noise resulting from unintended stops at secondary or tertiary RNA structures, fragmentation biases, and RNA degradation. Hence, TGIRT was selected from several commercially available RTases, performing distinguishing read-through efficiency and comparatively high mutation frequency at m^1A sites, and the approach is termed mis-incorporation-assisted profiling of m^1A, or m^1A-MAP [93].

7.3.3.4. Adapted m^1A-Seq Another single-nucleotide resolution protocol adapted m^1A-seq, combining with the usage of two different RTase, TGIRT and SuperScript III (SS). Respectively, TGIRT-treated cDNA possesses higher possibility to produce misincorporation on m^1A sites, while SuperScript III incorporated one tends to demonstrate truncation patterns. In addition, NGS-based single-nucleotide resolution methods characteristically rely on reading signatures, comprising of mutations and truncations, specifically introduced at the modification site by a reverse transcriptase (RT). As reading signatures initiated by a reverse transcriptase, comprising of mutations and stops, lay the groundwork for NGS-based approaches, m^1A sites are indicated through comparison of the analytical data drawn by either mis-incorporated or truncated cDNA [94]. Recently, m^1A-IP-seq and m^1A-quant-seq developed RTase by directed evolution that can read-through m^1A and generate mutation signatures, suggesting plentiful m^1A sites in human mRNA transcriptome with possible low stoichiometry [95].

7.3.4. 5-Methylcytosine (m^5C) in RNA

Alike its counterpart in DNA, m^5C modification is enriched in CG-rich regions and in regions immediately downstream of translation initiation sites. m^5C has conserved, tissue-specific, and dynamic features across mammalian transcriptomes. Moreover, m^5C formation in mRNAs is mainly catalyzed by the RNA methyltransferase NSUN2 [96]. The cysteine 321 of NSUN2, a transferase of m^5C, combines cytosine pyrimidine ring, inducing the substitution of C to m^5C. m^5C is specifically recognized by the mRNA export adaptor ALYREF, YBX1 as shown by *in vitro* and *in vivo* studies [97]. In addition, ten-eleven translocation (TET) family of enzymes has been demonstrated to oxidize m^5C to hm^5C in RNA, leading to considerable levels of hm^5C in RNA of mammalian cells and tissues [98]. m^5C are found to play essential roles in maternal mRNA stability [99].

7.3.4.1. Bisulfite Sequencing As the wide application of bisulfite sequencing in detection of DNA 5mC base modification. Scientists were inspired to adopt bisulfite sequencing in m^5C sequencing. Bisulfite treatment will deaminate unmethylated cytosine to uracil in single-strand RNA, leaving m^5C unconverted. Bisulfite sequencing provides base-resolution information of m^5C, mapping widespread m^5C sites in both coding and noncoding RNAs. However, bisulfite treatment provides sequential thermal acidic and alkaline conditions that lead to severe degradation of RNA. Furthermore, RNA bisulfite sequencing may produce false positives due to the incomplete conversion of unmethylated cytosine [100].

7.3.4.2. Aza-IP Other methods for sequencing m^5C in RNA are based on immunoprecipitation, using anti-m^5C antibodies to extract m^5C-containing RNA. However, these methods are confronting with challenges of low resolution and loss of quantitative information. A cytidine analog 5-aza-C is competent to block the m^5C transferase by breaking away from its target RNA. Taking advantage of this feature, Aza-IP performs immunoprecipitation after m^5C methyltransferase specifically recognize the target RNA, then the constructed library is subsequently sequenced [101]. An advanced approach exploits the catalytic mechanisms of the m^5C methyltransferase. As the mutation of cysteine at position 271(C271A) stabilizes the protein–RNA complex, miCLIP employs direct immunoprecipitation of the Myc-tagged C271A NSUN2 without UV crosslinking, and the protein–RNA complex can be detected by radiolabeling [102]. In conclusion, Aza-IP and miCLIP are bisulfite-independent and can pre-enrich m^5C targets. However, these protocols express redundant methyltransferase, the mis-targeting of which may lead to false positives [1].

7.3.4.3. WO-Seq and TAWO-Seq Many other m^5C-related problems and solutions are likewise inspired by their DNA counterparts. Peroxotungstate oxidation was firstly developed for selective oxidation of 5hmC in DNA to trihydroxylatedthymine(thT), which is a thymine derivative and will induce C-to-T transition in DNA after PCR. Because of the low conversion rate of the reaction in double-stranded DNA, peroxotungstate oxidation is not suitable for DNA samples. However, it could be ideal to detect hm^5C in single-stranded RNA, termed as peroxotungstate oxidation sequencing (WO-Seq). WO-Seq has been further combined with NgTET1 oxidation to successfully detect m^5C sites in RNA, which named TET-assisted WO-Seq (TAWO-Seq, Figure 7.15) [103].

7.3.5. N7-Methylguanosine (m^7G) in RNA

N7-methylguanosine (m^7G) is an evolutionarily conserved type of modification that is known to be present in the 5′ cap structure of eukaryotic mRNA [104]. m^7G is the first co-transcriptional 5′-terminal modification of nascent mRNA. This modification is catalyzed by Trm8/Trm82 complex in yeast and

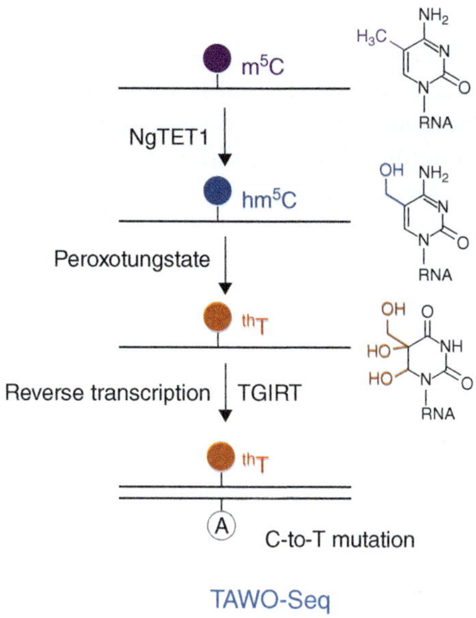

Figure 7.15 Scheme of TAWO-seq based on peroxotungstate oxidation.

METTL1/WDR4 complex in human [105]. Previous studies also demonstrated that the 5′-m^{7}G-cap structure is essential for pre-mRNA splicing, mRNA stability, nuclear export, and efficient translation [106].

Strategies to detect m^{7}G modification mostly couple enzymatic digestion with later analyzation of degradation product nucleosides, including mass spectrometry, capillary electrophoresis, and thin-layer chromatography. For instance, a method developed in 2018 applies the distinguished activities toward internal and 5′-terminal m^{7}G by S1 nuclease and phosphodiesterase I. Afterward, the differential enzymatic digestion in conjugation with liquid chromatography-tandem mass spectrometry analysis, revealing the m^{7}G stoichiometry and indicating the existence of internal m^{7}G modification in mRNA [107]. Other studies have employed m^{7}G antibody-based or chemical-labeling strategy, revealing much more widespread m^{7}G RNA modifications in mammals [108, 109].

7.3.6. Pseudouridine in RNA

Pseudouridine (ψ) is an isomer of uridine. Ψ is the most abundant modified nucleoside in noncoding RNA, enhances the function of transfer RNA and ribosomal RNA by stabilizing RNA structure. Pseudouridylation is catalyzed by pseudouridine synthases. ψ can be catalyzed by two types of pseudouridine synthases (PUSs): either the RNA-dependent PUSs that require box H/ACA RNA to act as a guide for substrate recognition or the RNA-independent PUSs that require no cofactor to catalyze ψ formation [110]. While pseudouridylation is known to have functions in ribosome biogenesis, protein synthesis, mRNA stability, translation accuracy, and pre-mRNA splicing [111, 112], the exploration for specific roles of most individual pseudouridines are limited by sequencing methods. Pseudouridine does not differ from uridine in its base-pairing specificity, but the isomerism between these two nucleotides can affect the stability of base pairs, base-stacking interactions, and RNA tertiary structure.

7.3.6.1. ψ-Seq, Pseudo-Seq, PSI-Seq, and CeU-Seq Several chemical labeling methods take advantage of the ability of N-cyclohexyl-N′-b-(4-methylmorpholinium) ethylcarbodiimide (CMC) to selectively modify ψ containing RNA. Under physiological conditions, CMC selectively acylates ψ at the N1 and N3 positions. Although CMC acylates guanosine (G) at the N1 position and uracil (U) at the N3 position as well, U-CMC or G-CMC adducts are reversible under alkaline hydrolysis (pH = 10.4), while ψ-CMC adducts is endurable to this condition. Notably, ψ-CMC adducts efficiently block reverse transcription (RT), making pseudouridylated residues detectable as a distinct stop. CMC/RT-based methods comprise of Pseudo-seq, PSI-seq, ψ-seq, and CeU-seq [113–116]. All methods have all independently confirmed the ubiquity of pseudouridylation in a diverse population of RNA species, including mRNA. When it comes to the consensus and difference between the above methods, the library preparation protocols possess subtle diversity (Figure 7.16) [117]. Pseudo-seq and PSI-seq first fragment polyA selected RNAs to a uniform size range and for truncated cDNAs following CMC treatment and reverse transcription. On the other hand, CeU-seq pre-enriches the ψ-containing RNA via biotin pulldown featured by ~15- to 20-fold, while ψ-seq employs no pre-enrichment steps during sample preparation, relying on a merely bioinformatic approach to reduce background [117].

In order to obtain all-around pseudouridine information, scientists tried from a brand-new approach. Unlike the existing methods that rely on RT stops caused by Ψ-CMC, they utilized new conditions of RT that allow read-through of Ψ-CMC, simultaneously generating mutation and deletion in cDNA [110]. It is reported that Superscript II is competent to read through ψ-CMC adduct in the presence of Mn^{2+}. A shift of melting temperature could be found in the melting curves, merely resulting from the Mn^{2+} induced misincorporation in cDNA. Importantly, it is reported that the modification level quantitatively influences the degree of melting curve alteration.

7.3.7. N4-Acetylcytidine (ac^4C) in RNA

N4-acetylcytidine (ac^4C) is an mRNA modification that is catalyzed by the acetyltransferase NAT10 [118]. As an oxidized form of cytidine, ac^4C remains comparatively unexplored. Initially described in the bacterial $tRNA^{met}$ anticodon, ac^4C was subsequently detected in eukaryotic serine and leucine tRNAs and 18S rRNA [119]. In contrast to other RNA modifications, ac^4C is rather

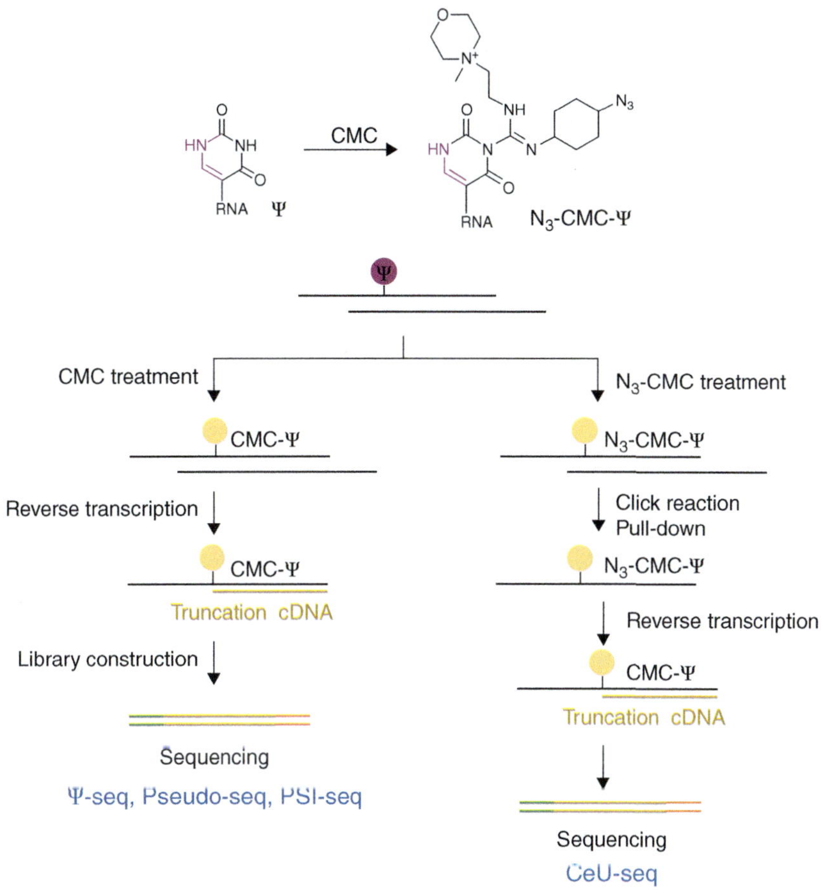

Figure 7.16 Scheme of mapping methods for pseudouridine in transcriptome-wide based on CMC chemical labeling.

nonredundant. Previous studies have detected a considerable level of ac^4C locis in mRNA. Aiming to investigate mRNA ac^4C localization and function, acRIP-seq roots in anti-ac^4C antibody strategy together with next-generation sequencing, for the first time uncovering the existence of ac^4C in mRNA [120]. It is inferable that the presence of ac^4C in mRNA enhances stability and translation efficiency.

7.3.8. Inosine in RNA

The deamination process under the action of adenosine deaminases acting on RNA (ADARs) [121], in which adenosine (A) is deaminized to inosine(I), is the most prevalent post-transcriptional modification event in higher eukaryotes.

Mainly occurring on the RNA double-stranded region, A-to-I editing frequently occurs in Alu elements in untranslated regions and introns [122]. A-to-I editing plays numerous roles in modulating gene expression, including recoding codons, altering alternative splicing, and regulating miRNA biogenesis and function. Occurring at the protein coding region, the editing process lead to the generation of different protein sequences; if the editing site appears in the non-coding region, it will significantly cause effect to the target specificity of the corresponding siRNA (small interfering RNA) and miRNA (microRNA).

7.3.8.1. Direct Detection of Mismatch Sites The conversion from adenosine to inosine directly changes the hydrogen bonding mode of the nucleobases involved, which results in the fact that inosine produced after editing is ready to interact with cytidine (C). As a result, the editing site will be recorded as guanosine (G) instead of the original adenosine. Hence, a strategy to identify A-to-I editing sites is to detect the A-to-G mismatch sites by analyzing cDNA sequences and the corresponding genomic sequences [123]. Although the strategy allows for transcriptome-wide single-nucleotide analysis, the major challenge is the false-positive results initiated by single-nucleotide polymorphisms (SNPs), somatic mutations, pseudogenes, and sequencing errors. In order to reduce background noises, it is usually necessary to utilize obtained knowledge about editing modes: such as clusters of putative editing sites or the existence of double-stranded RNA (dsRNA) structures to limit the search scope.

7.3.8.2. ICE-Seq To map the inosine profile from another prospective, a chemical-labeling method has been established. Inosine chemical erasing (ICE) takes advantage of Michael addition between acrylonitrile and inosine, giving rise to the reverse transcription truncation in cDNA, afterward adapted to be accompanied by high-throughput sequencing in 2014 [124, 125]. Notably, inosine-containing RNA is cyanoethylated to form N1-cyanoethylinosine (ce^1I) by acrylonitrile. Therefore, ICE-seq is coupled with orthogonal technologies as well (Figure 7.17). However, current inosine sequencing techniques still requires large amounts of RNA samples, expensive instrumentation, and high sequencing depth.

7.3.8.3. EndoVIPER-Seq Very recently, an inosine-cleaving enzyme, Escherichia coli Endonuclease V (eEndoV), has been identified to effectively bind A-to-I editing transcriptome without cleavage RNA when Ca^{2+} is present. Making full use of the high affinity and selectivity for inosine of eEndoV, EndoVIPER-seq (Endonuclease V inosine precipitation enrichment sequencing) is capable of overcoming many of the limitations stated above and severely improving the detection and characterization of A-to-I RNA editing in complex samples [126]. In addition, EndoVIPER is also an outstanding example where a base modification reader is repurposed to binding and enriching edited RNA.

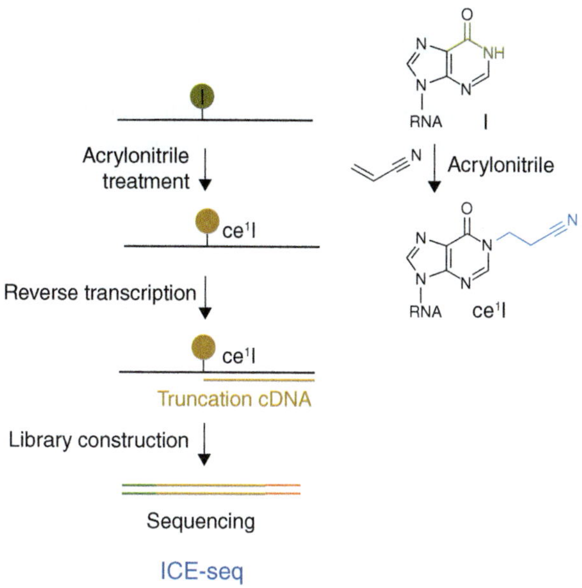

Figure 7.17 Scheme of ICE-seq based on chemical labeling.

7.3.9. Nm in RNA

Methylation of the ribose 2′-hydroxyl (Nm) exists at the second and third nucleotide in many mRNAs and as an abundant modification in tRNA and rRNA. Initially, RNA 2-*O*-methylation was shown to play a role in RNA stabilization under extreme temperatures and pH, since the substitution of the 2-OH by 2-*O*-Me considerably reduces the reactivity of the oxygen atom and thus prevents a nucleophilic attack and subsequent cleavage of the adjacent phosphodiester bond [127]. The 2′-hydroxyl group frequently participates in contacts forming higher order RNA structures.

Compared with unmodified nucleosides, the presence of 2′-OMe stabilizes 3′ end ribose of an RNA nucleoside. Nm sites have higher resistance to alkali-mediated hydrolysis, and this feature can inspire a new way to identify Nm [127]. Recently, Nm-seq was developed by iterative oxidation–elimination–dephosphorylation cycles, which can excise 2′-hydroxylated nucleotides in the 3′–5′ direction and stop at Nm sites, providing a sensitive method for mapping Nms at single-base resolution [128].

In general, identification of RNA modification abundance, sequence context and cellular dynamics is crucially based on the sequencing of these entities. Conclusively, a sensitive, quantitative, and single-base sequencing method is required for all the time [129].

7.4. THE CLINIC APPLICATIONS OF DETECTION OF NUCLEIC ACIDS EPIGENETIC MODIFICATIONS

The development of cancer is driven by the accumulation of dozens of changes that affect the structure and function of the genome. In this process, genetic changes are as important as epigenetic changes. The former disrupts the normal pattern of gene expression and sometimes leads to abnormal expression of constitutively active proteins, while the latter disrupts mechanisms such as transcriptional control, leading to the silencing or abnormal activation of cancer-related genes. Both types of changes are heritable at the cellular level, thus contributing to the clonal expansion of cancer cells [130].

The diagnosis, early detection, and prognosis of cancer have gained critical time for cancer treatment and improved the cure rate of cancer. Therefore, finding stable and specific biomarkers for early diagnosis of disease is the key to clinical diagnosis. These markers can improve the early prevention, diagnosis, and prognosis of diseases. In addition, through markers, it is also possible to predict more effective treatment methods based on the characteristics of patients, which is a well-known personalized medicine [131]. The development of next-generation sequencing technology and bioinformatics analysis technology provides the possibility for comprehensive analysis and discovery of disease-specific biomarkers from numerous nucleic acids. For example, analysis of circulating free nucleic acids in the blood provides a basis for disease diagnosis, detection of dynamic changes in circulating free nucleic acids to monitor changes in diseases. The epigenetic analysis of cell-free DNA (cfDNA) has the potential for early detection and disease monitoring, because epigenetic modifications are tissue-specific and reflect the dynamic process of cancer progression [132]. In the past 20 years, the research of genetic variation has great prospects in personalized precision medicine. And researchers found that the state of the disease cannot be explained by a single genetic variation, and the role of specific genes may be overestimated [58]. It gradually became a consensus that many diseases such as cancer are not only a genetic disease, but also the epigenetic one. Thus, the application of epigenetic modification detection for cancer diagnostics or prevention is coming to clinics.

7.4.1. DNA Methylation Detection for Disease Diagnosis and Prognosis

Thus far, detection of CpG methylation at specific loci is the most successful epigenetic biomarker, because DNA methylation is the epigenetic modification with the highest content, it is the first modified base to be studied. In this section, we will comprehensively review the application of DNA methylation detection in clinical diagnosis of human diseases, containing oncology, neurology, immunology, and pregnancy-related diseases (Table 7.1).

TABLE 7.1 Diagnostic and prognostic of DNA methylation from liquid biopsy/tissues biopsy in human diseases

Disease	Type of sample	Approach	Gene	References
		Oncology		
Colorectal cancer(CRC)	Plasma	Real-time PCR	SEPT9 methylation	[133, 134]
	Stool	ColoSure™ test, fecal immunochemical test (FIT)	VIM methylation; BMP3 and NDRG4 methylation	[135, 136]
	Primary tumor biopsy samples	Pyrosequencing	DERL3 methylation; TP53TG1 methylation	[137, 138]
	Paired adenoma and control mucosa samples	Methylome	CpG methylation at a panel of sequences	[139]
Lung cancer	Pleural effusion samples	real-time PCR	SHOX2 and PTGER4 methylation	[140]
	Plasma	real-time PCR	SHOX2 methylation	[141]
Prostate cancer	Primary tissue biopsy samples	ConfirmMDx	GSTP1, APC, and RASSF1 methylation	[142]
Bladder cancer	Urine	SNaPshot™ methylation analysis	TWIST1, ONECUT2, and OTX1 methylation	[143]
		Neurology		
Parkinson disease (PD)	PBMCs from sporadic PD patients and controls	Bisulfite sequencing; real-time PCR	SNCA methylation	[144]
Alzheimer disease (AD) and frontotemporal dementia (FTD)	PBMCs from controls, AD, and FTD donors	Pyrosequencing of bisulfite converted DNA.	PIN1 methylation	[145]
		Immunology		
Systemic lupus erythematosus (SLE)	PBMCs from active SLE and healthy donors	Bisulfite-treated DNA was analyzed by the HumanMethylation27 BeadChip	IL10 and IL1R2 methylation	[146]

(Continued)

TABLE 7.1 (Continued)

Disease	Type of sample	Approach	Gene	References
		Immunology		
Multiple sclerosis (MS)	Whole blood from patients with MS and healthy donors	Bisulfite-PCR and pyrosequencing	Global CpG methylation	[147]
	cfDNA from plasma or serum from MS and healthy donors	bisulfite-PCR; MiSeq Reagent Kit v2	Methylation at 3′ UTR of MBP and unannotated locus cg10809560	[148]
		Pregnancy-related diseases		
Fetal DNA marker	plasma	Bisulfite sequencing; Real-time PCR; Methylation-sensitive high-resolution melting (MS-HRM) analysis	SERPINB5 hypermethylated ccf-fetal DNA; RASSF1A, APC and PRKCDBP, MEST and SNRPN hypermethylated ccf-fetal DNA	[149–152]
Fetal Rh status	plasma	RT-PCR; A Multiplex PCR	RASSF1A Hypermethylated ccf-fetal DNA	[153, 154]
Trisomy 21	Blood cell	Microfluidics digital PCR analysis	HLCS	[155]
Trisomy 18	plasma	Methylation-specific PCR(MSP)	SERPINB5	[156]
Preeclampsia	plasma	Bisulfite sequencing; Real-time PCR	SERPINB5; RASSF1A hypermethylated ccf-fetal DNA	[149, 157]

7.4.1.1. DNA Methylation Markers for Cancers Diagnosis For various human diseases, especially cancer, there is an increasing need for better biomarkers for early detection of diseases. Cancer is a multistep process in which a plethora of molecular alterations synergistically contribute to the initiation and progression of the tumors. Epigenetic changes could be drivers of early disruption of cellular homeostasis in precancerous and cancerous cells through the regulation of tumors-associated genes, as well as by modulating genomic instability [158]. Cancer-associated DNA methylation have been widely explored for their applicability as cancer biomarkers (Table 7.1).

7.4.1.2. DNA Methylation Markers for Non-Cancers Diagnosis Except cancer, epigenetic regulation is critical for the normal development and functioning of the human brain. Indeed, DNA methylation have been frequently detected in neurological diseases, such as Parkinson disease (PD) [144] and Alzheimer disease (AD) [145]. Beyond human brain, epigenetic regulation is fundamental for the differentiation of immune cells. Genome-wide epigenomic maps have been generated for inflammatory rheumatic diseases, including systemic lupus erythematosus (SLE) [146] or for autoimmune diseases such as multiple sclerosis (MS) [147]. Additionally, DNA methylation also can be as biomarker for pregnancy-related diseases (Table 7.1).

7.4.2. 5-Hydroxymethylcytosines Clinical Detection

5-hydroxymethylcytosines (5hmC) is an important DNA epigenetic modification, which is widely regarded as a sign of gene expression in an activated state. The distribution of 5hmC is specific and can be observed in normal tissues and tumor tissues. It is currently believed that this modification is closely related to gene regulation and tumor pathogenesis. Its sensitive changes in whole-genome level have been repeatedly verified by various tumors, confirming that it can be used as a marker for early tumor diagnosis. 5hmc is formed by the oxidation of 5-methylcytosine (5mC) through Ten-eleven translocation. At present, studies have confirmed that the expression level of 5hmc in the normal genome is relatively stable and has a unique appearance modification effect, not just an intermediate product of 5mC metabolism [159, 160].

7.4.2.1. Liquid Biopsy for Cancers: Circulating Cell-Free DNA (cfDNA) Cell-free DNA (cfDNA) is free in body fluids outside the cell, which can be specifically produced during the necrosis and apoptosis of tumor cells, and has important research significance in the early diagnosis of tumors. Although there have been many studies in the diagnosis of various tumors around the detection of cfDNA mutations and methylation in recent years, looking for highly specific hypermethylation sites and researching more sensitive and specific early tumor minimally invasive detection methods is still an urgent need for a breakthrough. 5-Hydroxymethylcytosine (5hmC), as an apparent modification of DNA that has been recognized in recent years, has also been considered to be closely related to the occurrence and development of tumors in various studies (Table 7.2). However, cfDNA 5hmC sequencing technology has been developed and applied early tumor diagnosis and testing are still in the verification stage.

7.4.2.2. Multiple Cancer Quake and co-authors reveal the cancer-specific 5hmC distribution on cell-free DNA from blood samples, laying a good foundation for this technology in the early diagnosis of tumors, recurrence monitoring, and possible individualized treatment in the future [162]. In the same year, He and co-authors also published similar results. The expanded number of clinical samples in the study and related mouse model experiments further demonstrated that the detection of free DNA 5hmC signal is useful for early diagnosis and early screening of tumors and has important clinical significance [161].

7.4.2.3. Hepatocellular Carcinoma (HCC) Liver cancer, as one of the common malignant tumors, is easy to be ignored by patients because of the lack of obvious early symptoms. When it is discovered, the patient is already in the middle and advanced stages of liver cancer. Moreover, liver cancer can be completely cured in the early stage, but the cure rate of the disease in the middle and late stages is less than 5%.

One Chinese group carried out a large-scale, multicenter clinical study. The researchers used the 5hmC detection technology 5hmC-Seal to capture the 5hmC sequence in blood cfDNA, measured and analyzed the 5hmC level, and found significant differences between the liver cancer group and the normal group. The modification level differed by 5hmC gene locus. Through bioinformatics analysis and using the differential genes between liver cancer samples and healthy samples to establish a 5hmC liver cancer diagnostic model, it is concluded that 5hmC can be used as a marker for liver cancer staging and prognosis. They selected 917 candidate marker genes in the training group. The 5hmC modification of these genes was different between the liver cancer group and the healthy control group, as well as the liver cancer group and chronic hepatitis B patients or patients with a history of liver cirrhosis; then they were further screened by statistical analysis methods. Combination of two markers. According to the weighted diagnostic score of 32 markers, it can distinguish early liver cancer and non-liver cancer groups, as well as early liver cancer and healthy control groups [163].

One study evaluated the potential prognostic value of 5hmC-Seal technique for HCC, suggesting profiled 5hmC loci in cfDNA were enriched with liver tissue-derived regulatory markers (e.g., H3K4me1) and accurately distinguished HCC patients from controls. The result shows that this new approach in clinic could do early detection of those patients and surveillance of the disease conveniently [171].

TABLE 7.2 Diagnostic and prognostic of 5hmc from liquid biopsy/tissues biopsy in human diseases

Disease	Type of sample	Approach	Applications	References
Multiple cancers (Colorectal and gastric cancers)	Blood samples (cfDNA)	5hmC-Seal sequencing	Diagnostic and prognostic	[161]
Multiple cancers (lung cancer, HCC, pancreatic cancer, glioblastoma, gastric cancer)	Blood samples (cfDNA)	Modified 5hmC-Seal sequencing	Prediction cancer types and stages	[162]
Hepatocellular carcinoma	Blood samples (cfDNA)	5hmC-Seal sequencing	Early detection	[163]
Colorectal cancer	Blood samples (cfDNA)	5hmC-Seal sequencing	Early detection, diagnosis of precancerous adenoma	[164]
Glioblastoma, colorectal cancer, multiple myeloma, chronic lymphocytic leukemia, and pancreatic cancer	Tissues	Fluorescence-based assay for high-throughput epigenetic quantification	Diagnostics and monitoring of disease progression	[165]
Colon and Various types of blood	Tissues and blood	Fluorescence based single-molecule method for global quantification of 5hmC	Clinical diagnosing	[166]
Breast cancer	Tissue	Electrochemical immunosensor	Ultrasensitive 5-hmC detection	[167]
Esophageal cancer	Blood (cfDNA)	Nano-hmC-Seal method	Diagnosis	[168]
Non-small-cell lung cancer	Blood (cfDNA)	Sensitive 5hmC-Seal method	Diagnosis	[169]
Multiple myeloma	Plasma (cfDNA)	Nano-hmC-Seal	Diagnosis and prognosis	[170]
Hepatocellular carcinoma	Blood (cfDNA)	5hmC-Seal	Early detection of and prognostic	[171]

7.4.2.4. Multiple Myeloma (MM) The common feature of multiple myeloma (MM) is that its epigenome is severely destroyed, and gene-specific DNA hypermethylation is associated with clinical aggressive subtypes and prognosis. A research group applied nano-hmC-Seal sequencing method to profile 5hmC in plasma cfDNA samples, showing that cfDNA, 5hmC loci were enriched within gene bodies, while depleted in CpG islands. Furthermore, the result suggested that 5hmC signatures in cfDNA showed the potential to distinguish between MM patients and its two precursor conditions, and also to separate MM patients at the time of diagnosis according to relapse status [170].

7.4.2.5. Non-Small-Cell Lung Cancer (NSCLC) NSCLC accounts for 80% of all types of lung cancer. However, due to the lack of effective early diagnosis tools, the diagnosis has reached the advanced stage, which makes the treatment difficult. A research team analyzed the distribution and content change characteristics of 5hmc in cfDNA and found that 5hmc in cfDNA can be used as a biomarker for the diagnosis of NSCLC and healthy individuals and promote the development of the diagnosis and prognosis of NSCLC. In this study, they used the whole-genome 5hmc sequencing method to map the cfDNA of patients with NSCLC and healthy controls, and found that the gene bodies and promoter regions in the blood samples of tumor patients detected a significant increase of 5hmC, and found 6 potential biomarkers from 66 patients and 67 healthy controls [169].

7.4.2.6. Esophageal Cancer (EC) EC is a malignant tumor with very high morbidity and mortality, ranking sixth in the world in mortality. One study used the nano-hmc-Seal method to analyze the 5hmc distribution map of cfDNA of EC patients, revealing the whole-genome change level of 5hmc in cfDNA in the plasma of EC patients, and identified 5hmc biomarkers, which will provide a theoretical basis for EC liquid biopsy.

7.5. CONCLUSION AND FUTURE PROSPECT

As the cost of sequencing decreases and the sensitivity of detection technology increases, it should be admitted that great progresses have been made in the mapping of nucleic acid epigenetic modifications. However, there are more or less limitation in each existed technology. In order to further study the biological functions of these modified bases and their application in disease diagnosis, prognosis assessment and treatment, high-throughput sequencing methods with low input, single base resolution and high reproducibility are urgently needed. Due to lack of reliable detection method, some modifications with low abundance were fell to controversy and hard to be followed by other scientists. There is still a long way to expand the types and impact of epigenetic modifications in nucleic acid.

Of course, the good news is that the modifications of 5mC and 5hmC have exhibited its value in disease diagnosis or other applications. Despite slow uptake in the liquid biopsy tests for noninvasive detection of cancer, the investors still keep the faith and put lots of attention to this field. Recent study showed that the mutations accumulate in the normal somatic cells, which limited the prospects of current main methods for liquid biopsy by detection of genetic variations in disease DNA. By contrast, the large-scale epigenetic alterations provide the potential ability to detect and classify the disease in early stage.

So far, the nucleic acid epigenetics in clinic application are only related to DNA modification. Actually, the alteration of m^6A modification in RNA is involved in almost all vital bioprocesses, including cancer development and other disease. Thus, the expansion of RNA methylation detection for disease diagnosis and prognosis is promising for clinical trials. In addition, the present studies are mainly focusing on the diagnosis and treatment of cancer disease. Other major diseases such as viral infections also related to the epigenetic modification change either in the nucleic acid of virus or in the host cells. The clinic researches for epigenetic alternation for viral infection will provide a new prospect for viral intervention and treatment.

REFERENCES

1. Zhao LY, Song J, Liu Y, Song CX, *et al.* (2020) Mapping the epigenetic modifications of DNA and RNA *Protein Cell* 11(11):792–808.
2. Dawson MA & Kouzarides T (2012) Cancer epigenetics: from mechanism to therapy. *Cell* 150(1):12–27.
3. Cavalli G & Heard E (2019) Advances in epigenetics link genetics to the environment and disease. *Nature* 571(7766):489–499.
4. Bird AP (1987) CPG islands as gene markers in the vertebrate nucleus. *Trends Genet* 3(12):342–347.
5. Weber M, Hellmann I, Stadler MB, Ramos L, *et al.* (2007) Distribution, silencing potential and evolutionary impact of promoter DNA methylation in the human genome. *Nat Genet* 39(4):457–466.
6. Myant K, Termanis A, Sundaram AY, Boe T, *et al.* (2011) LSH and G9a/GLP complex are required for developmentally programmed DNA methylation. *Genome Res* 21(1):83–94.
7. Martin R (1994) On the relationship between preferred termination codon contexts and nonsense suppression in human-cells. *Nucleic Acids Res* 22(1):15–19.
8. Shawn J. Cokus SF, Zhang X, Chen, Z, *et al.* (2008) Shotgun bisulfite sequencing of the *Arabidopsis* genome reveals DNA methylation patterning. *Nature* 452(7184): 215–219.
9. Tost J & Gut IG (2007) DNA methylation analysis by pyrosequencing. *Nat Protoc* 2(9):2265–2275.

10. Herman JG, Graff JR, Myohanen S, Nelkin BD, *et al.* (1996) Methylation-specific PCR: a novel PCR assay for methylation status of CpG islands. *Proc Natl Acad Sci U S A* 93(18):9821–9826.
11. Liu C, Cui X, Zhao BS, Narkhede P, *et al.* (2020) DNA 5-methylcytosine-specific amplification and sequencing. *J Am Chem Soc* 142(10):4539–4543.
12. Liu Y, Siejka-Zielinska P, Velikova G, Bi Y, *et al.* (2019) Bisulfite-free direct detection of 5-methylcytosine and 5-hydroxymethylcytosine at base resolution. *Nat Biotechnol* 37(4):424–429.
13. Okamoto A, Tainaka K, & Kamei T (2006) Sequence-selective osmium oxidation of DNA: efficient distinction between 5-methylcytosine and cytosine. *Org Biomol Chem* 4(9):1638–1640.
14. Bareyt S & Carell T (2008) Selective detection of 5-methylcytosine sites in DNA. *Angew Chem Int Ed Engl* 47(1):181–184.
15. Wang T, Hong T, Tang T, Zhai Q, *et al.* (2013) Application of *N*-halogeno-*N*-sodiobenzenesulfonamide reagents to the selective detection of 5-methylcytosine in DNA sequences. *J Am Chem Soc* 135(4):1240–1243.
16. Flusberg BA, Webster DR, Lee JH, Travers KJ, *et al.* (2010) Direct detection of DNA methylation during single-molecule, real-time sequencing. *Nat Methods* 7(6):461–465.
17. Liu Q, Fang L, Yu G, Wang D, *et al.* (2019) Detection of DNA base modifications by deep recurrent neural network on Oxford Nanopore sequencing data. *Nat Commun* 10(1):2449.
18. Kriaucionis S & Heintz N (2009) The nuclear DNA base 5-hydroxymethylcytosine is present in Purkinje neurons and the brain. *Science* 324(5929):929–930.
19. Bachman M, Uribe-Lewis S, Yang X, Williams M, *et al.* (2014) 5-Hydroxymethylcytosine is a predominantly stable DNA modification. *Nat Chem* 6(12):1049–1055.
20. Vasanthakumar A & Godley LA (2015) 5-hydroxymethylcytosine in cancer: significance in diagnosis and therapy. *Cancer Gene Ther* 208(5):167–177.
21. Booth MJ, Branco MR, Ficz G, Oxley D, *et al.* (2012) Quantitative sequencing of 5-methylcytosine and 5-hydroxymethylcytosine at single-base resolution. *Science* 336(6083):934–937.
22. Song CX, Szulwach KE, Fu Y, Dai Q, *et al.* (2011) Selective chemical labeling reveals the genome-wide distribution of 5-hydroxymethylcytosine. *Nat Biotechnol* 29(1):68–72.
23. Yu M, Hon GC, Szulwach KE, Song CX, *et al.* (2012) Base-resolution analysis of 5-hydroxymethylcytosine in the mammalian genome. *Cell* 149(6):1368–1380.
24. Han D, Lu X, Shih AH, Nie J, *et al.* (2016) A highly sensitive and robust method for genome-wide 5hmC profiling of rare cell populations. *Mol Cell* 63(4):711–719.
25. Hu L, Liu Y, Han S, Yang L, *et al.* (2019) Jump-seq: Genome-wide capture and amplification of 5-hydroxymethylcytosine sites. *J Am Chem Soc* 141(22): 8694–8697.
26. Hayashi G, Koyama K, Shiota H, Kamio A, *et al.* (2016) Base-resolution analysis of 5-hydroxymethylcytosine by one-pot bisulfite-free chemical conversion with peroxotungstate. *J Am Chem Soc* 138(43):14178–14181.
27. Zeng H, He B, Xia B, Bai D, *et al.* (2018) Bisulfite-free, nanoscale analysis of 5-hydroxymethylcytosine at single base resolution. *J Am Chem Soc* 140(41): 13190–13194.

28. Wang Y, Liu C, Zhang X, Yang W, *et al.* (2018) Gene specific-loci quantitative and single-base resolution analysis of 5-formylcytosine by compound-mediated polymerase chain reaction. *Chem Sci* 9(15):3723–3728.
29. Raiber EA, Beraldi D, Ficz G, Burgess HE, *et al.* (2012) Genome-wide distribution of 5-formylcytosine in embryonic stem cells is associated with transcription and depends on thymine DNA glycosylase. *Genome Biol* 13(8):R69.
30. Song CX, Szulwach KE, Dai Q, Fu Y, *et al.* (2013) Genome-wide profiling of 5-formylcytosine reveals its roles in epigenetic priming. *Cell* 153(3):678–691.
31. Booth MJ, Marsico G, Bachman M, Beraldi D, *et al.* (2014) Quantitative sequencing of 5-formylcytosine in DNA at single-base resolution. *Nat Chem* 6(5):435–440.
32. Xia B, Han D, Lu X, Sun Z, *et al.* (2015) Bisulfite-free, base-resolution analysis of 5-formylcytosine at the genome scale. *Nat Methods* 12(11):1047–1050.
33. Zhu C, Gao Y, Guo H, Xia B, *et al.* (2017) Single-cell 5-formylcytosine landscapes of mammalian early embryos and ESCs at single-base resolution. *Cell Stem Cell* 20(5):720–731.e725.
34. Lu X, Song CX, Szulwach K, Wang Z, *et al.* (2013) Chemical modification-assisted bisulfite sequencing (CAB-Seq) for 5-carboxylcytosine detection in DNA. *J Am Chem Soc* 135(25):9315–9317.
35. Lu X, Han D, Zhao BS, Song CX, *et al.* (2015) Base-resolution maps of 5-formylcytosine and 5-carboxylcytosine reveal genome-wide DNA demethylation dynamics. *Cell Res* 25(3):386–389.
36. Liu C, Zou G, Peng S, Wang Y, *et al.* (2018) 5-Formyluracil as a multifunctional building block in biosensor designs. *Angew Chem Int Ed Engl* 57(31):9689–9693.
37. Jiang HP, Liu T, Guo N, Yu L, *et al.* (2017) Determination of formylated DNA and RNA by chemical labeling combined with mass spectrometry analysis. *Anal Chim Acta* 981:1–10.
38. Tian X, Sun B, Chen C, Gao C, *et al.* (2018) Circulating tumor DNA 5-hydroxymethylcytosine as a novel diagnostic biomarker for esophageal cancer. *Cell Res* 28(5):597–600.
39. Petersen-Mahrt SK, Harris RS, & Neuberger MS (2002) AID mutates *E. coli* suggesting a DNA deamination mechanism for antibody diversification. *Nature* 418:99–103.
40. Slupphaug G, Eftedal I, Kavli B, Bharati S, *et al.* (1995) Properties of a recombinant human uracil-DNA glycosylase from the *UNG* gene and evidence that *UNG* encodes the major uracil-DNA glycosylase. *Biochemistry* 34(1):128–138.
41. Nilsen H, Haushalter KA, Robins P, Barnes DE, *et al.* (2001) Excision of deaminated cytosine from the vertebrate genome: role of the SMUG1 uracil-DNA glycosylase. *EMBO J* 20(15):4278–4286.
42. Hardeland U, Bentele M, Lettieri T, Steinacher R, *et al.* (2001) Thymine DNA glycosylase. Prog Nucleic Acid Res Mol Biol, 68:235–253.
43. Bellacosa A (2001) Role of wihI (wfhR) gene in DNA repair and human cancer. *J Cell Physiol* 187:137–144.
44. Mauritz R, van Groeningen CJ, Smid K, Jansen G, *et al.* (2007) Thymidylate synthase and dihydropyrimidine dehydrogenase mRNA expression after administration of 5-fluorouracil to patients with colorectal cancer. *Int J Cancer* 120(12):2609–2612.
45. Tinkelenberg BA, Hansbury MJ, & Ladner RD (2002) dUTPase and uracil-DNA glycosylase are central modulators of antifolate toxicity in *Saccharomyces cerevisiae*. *Cancer Res* 62:4909–4915.

46. Shu X, Liu M, Lu Z, Zhu C, *et al.* (2018) Genome-wide mapping reveals that deoxyuridine is enriched in the human centromeric DNA. *Nat Chem Biol* 14(7):680–687.
47. Heyn H & Esteller M (2015) An adenine code for DNA: a second life for N6-methyladenine. *Cell* 161(4):710–713.
48. Luo GZ, Blanco MA, Greer EL, He C, *et al.* (2015) DNA N(6)-methyladenine: a new epigenetic mark in eukaryotes? *Nat Rev Mol Cell Biol* 16(12):705–710.
49. Summerer D (2015) N(6) -methyladenine: a potential epigenetic mark in eukaryotic genomes. *Angew Chem Int Ed Engl* 54(37):10714–10716.
50. Collier J, McAdams HH, & Shapiro L (2007) A DNA methylation ratchet governs progression through a bacterial cell cycle. *Proc Natl Acad Sci U S A* 104(43): 17111–17116.
51. Messer W, & Noyer-Weidner M (1988) Timing and targeting: the biological functions of dam methylation in *E. coli*. *Cell* 54:735–737.
52. Fu Y, Luo GZ, Chen K, Deng X, *et al.* (2015) N6-methyldeoxyadenosine marks active transcription start sites in *Chlamydomonas*. *Cell* 161(4):879–892.
53. Luo GZ, Wang F, Weng X, Chen K, *et al.* (2016) Characterization of eukaryotic DNA N(6)-methyladenine by a highly sensitive restriction enzyme-assisted sequencing. *Nat Commun* 7:11301.
54. Rossi MJ, Lai WKM, & Pugh BF (2018) Simplified ChIP-exo assays. *Nat Commun* 9(1):2842.
55. Hao Z, Wu T, Cui X, Zhu P, *et al.* (2020) N(6)-deoxyadenosine methylation in mammalian mitochondrial DNA. *Mol Cell* 78(3):382–395.e388.
56. Joan JE, Miceli SM, Roberts RJ, & Manley JL (1990) Sequence specificity of the human mRNA N6-adenosine methylase *in vitro*. *Nucleic Acids Res* 18(19):5735.
57. Meyer KD, Saletore Y, Zumbo P, Elemento O, *et al.* (2012) Comprehensive analysis of mRNA methylation reveals enrichment in 3′ UTRs and near stop codons. *Cell* 149(7):1635–1646.
58. Roadmap Epigenomics C, Kundaje A, Meuleman W, Ernst J, *et al.* (2015) Integrative analysis of 111 reference human epigenomes. *Nature* 518(7539):317–330.
59. Lichinchi G, Gao S, Saletore Y, Gonzalez GM, *et al.* (2016) Dynamics of the human and viral m(6)A RNA methylomes during HIV-1 infection of T cells. *Nat Microbiol* 1:16011.
60. Ping XL, Sun BF, Wang L, Xiao W, *et al.* (2014) Mammalian WTAP is a regulatory subunit of the RNA N6-methyladenosine methyltransferase. *Cell Res* 24(2):177–189.
61. Schwartz S, Mumbach MR, Jovanovic M, Wang T, *et al.* (2014) Perturbation of m^6A writers reveals two distinct classes of mRNA methylation at internal and 5′ sites. *Cell Rep* 8(1):284–296.
62. Yue Y, Liu J, Cui X, Cao J, *et al.* (2018) VIRMA mediates preferential m(6)A mRNA methylation in 3′UTR and near stop codon and associates with alternative polyadenylation. *Cell Discov* 4:10.
63. Pendleton KE, Chen B, Liu K, Hunter OV, *et al.* (2017) The U6 snRNA m(6)A methyltransferase METTL16 regulates SAM synthetase intron retention. *Cell* 169(5):824–835.e814.
64. Jia G, Fu Y, Zhao X, Dai Q, *et al.* (2011) N6-methyladenosine in nuclear RNA is a major substrate of the obesity-associated FTO. *Nat Chem Biol* 7(12):885–887.

65. Fu Y, Jia G, Pang X, Wang RN, *et al.* (2013) FTO-mediated formation of N6-hydroxymethyladenosine and N6-formyladenosine in mammalian RNA. *Nat Commun* 4:1798.
66. Zhao X, Yang Y, Sun BF, Shi Y, *et al.* (2014) FTO-dependent demethylation of N6-methyladenosine regulates mRNA splicing and is required for adipogenesis. *Cell Res* 24(12):1403–1419.
67. Zheng G, Dahl JA, Niu Y, Fedorcsak P, *et al.* (2013) ALKBH5 is a mammalian RNA demethylase that impacts RNA metabolism and mouse fertility. *Mol Cell* 49(1):18–29.
68. Cao GLH-B, Yin Z, Flavell RA (2016) Recent advances in dynamic m^{6}A RNA modification. *Open Biol* 6:160003.
69. Dominissini D, Moshitch-Moshkovitz S, Schwartz S, Salmon-Divon M, *et al.* (2012) Topology of the human and mouse m^{6}A RNA methylomes revealed by m^{6}A-seq. *Nature* 485(7397):201–206.
70. Wang Y, Li Y, Toth JI, Petroski MD, *et al.* (2014) N6-methyladenosine modification destabilizes developmental regulators in embryonic stem cells. *Nat Cell Biol* 16(2):191–198.
71. Louloupi A, Ntini E, Conrad T, & Orom UAV (2018) Transient *N*-6-methyladenosine transcriptome sequencing reveals a regulatory role of m^{6}A in splicing efficiency. *Cell Rep* 23(12):3429–3437.
72. Alarcon CR, Goodarzi H, Lee H, Liu X, *et al.* (2015) HNRNPA2B1 is a mediator of m(6)A-dependent nuclear RNA processing events. *Cell* 162(6):1299–1308.
73. Liu N, Dai Q, Zheng G, He C, *et al.* (2015) N(6)-methyladenosine-dependent RNA structural switches regulate RNA-protein interactions. *Nature* 518(7540):560–564.
74. Xiao W, Adhikari S, Dahal U, Chen YS, *et al.* (2016) Nuclear m(6)A reader YTHDC1 regulates mRNA splicing. *Mol Cell* 61(4):507–519.
75. Zhou J, Wan J, Gao X, Zhang X, *et al.* (2015) Dynamic m(6)A mRNA methylation directs translational control of heat shock response. *Nature* 526(7574):591–594.
76. Xiang Y, Laurent B, Hsu CH, Nachtergaele S, *et al.* (2017) RNA m(6)A methylation regulates the ultraviolet-induced DNA damage response. *Nature* 543(7646):573–576.
77. Chen K, Lu Z, Wang X, Fu Y, *et al.* (2015) High-resolution N(6) -methyladenosine (m(6)A) map using photo-crosslinking-assisted m(6)A sequencing. *Angew Chem Int Ed Engl* 54(5):1587–1590.
78. Linder B, Grozhik AV, Olarerin-George AO, Meydan C, *et al.* (2015) Single-nucleotide-resolution mapping of m^{6}A and m^{6}Am throughout the transcriptome. *Nat Methods* 12(8):767–772.
79. Molinie B, Wang J, Lim KS, Hillebrand R, *et al.* (2016) m(6)A-LAIC-seq reveals the census and complexity of the m(6)A epitranscriptome. *Nat Methods* 13(8): 692–698.
80. Meyer KD (2019) DART-seq: an antibody-free method for global m(6)A detection. *Nat Methods* 16(12):1275–1280.
81. Zhang Z, Chen L-Q, Zhao Y-L, Yang C-G *et al.* (2019) Single-base mapping of m^{6}A by an antibody-independent method. *Sci Adv* 5:eaax0250.
82. Garcia-Campos MA, Edelheit S, Toth U, Safra M, *et al.* (2019) Deciphering the "m(6) A Code" via antibody-independent quantitative profiling. *Cell* 178(3):731–747.e716.
83. Hong T, Yuan Y, Chen Z, Xi K, *et al.* (2018) Precise antibody-independent m^{6}A identification via 4SedTTP-involved and FTO-assisted strategy at single-nucleotide resolution. *J Am Chem Soc* 140(18):5886–5889.

84. Shu X, Cao J, Cheng M, Xiang S, *et al.* (2020) A metabolic labeling method detects m(6)A transcriptome-wide at single base resolution. *Nat Chem Biol* 16(8):887–895.
85. Wang Y, Xiao Y, Dong S, Yu Q, *et al.* (2020) Antibody-free enzyme-assisted chemical approach for detection of N(6)-methyladenosine. *Nat Chem Biol* 16(8):896–903.
86. Sun H, Zhang M, Li K, Bai D, *et al.* (2019) Cap-specific, terminal N(6)-methylation by a mammalian m(6)Am methyltransferase. *Cell Res* 29(1):80–82.
87. Sendinc E, Valle-Garcia D, Dhall A, Chen H, *et al.* (2019) PCIF1 Catalyzes m^6Am mRNA methylation to regulate gene expression. *Mol Cell* 75(3):620–630.e9.
88. Akichika S, Hirano S, Shichino Y, Suzuki T, *et al.* (2019) Cap-specific terminal *N*(6)-methylation of RNA by an RNA polymerase II-associated methyltransferase. *Science* 363(6423):eaav0080.
89. Mauer J, Luo X, Blanjoie A, Jiao X, *et al.* (2017) Reversible methylation of m(6) Am in the 5′ cap controls mRNA stability. *Nature* 541(7637):371–375.
90. Boulias K, Toczydlowska-Socha D, Hawley BR, Liberman N, *et al.* (2019) Identification of the m(6)Am methyltransferase PCIF1 reveals the location and functions of m(6)Am in the transcriptome. *Mol Cell* 75(3):631–643.e8.
91. Dominissini D, Nachtergaele S, Moshitch-Moshkovitz S, Peer E, *et al.* (2016) The dynamic *N*(1)-methyladenosine methylome in eukaryotic messenger RNA. *Nature* 530(7591):441–446.
92. Li X, Xiong X, Wang K, Wang L, *et al.* (2016) Transcriptome-wide mapping reveals reversible and dynamic *N*(1)-methyladenosine methylome. *Nat Chem Biol* 12(5):311–316.
93. Li X, Xiong X, Zhang M, Wang K, *et al.* (2017) Base-resolution mapping reveals distinct m(1)A methylome in nuclear- and mitochondrial-encoded transcripts. *Mol Cell* 68(5):993–1005.e1009.
94. Safra M, Sas-Chen A, Nir R, Winkler R, *et al.* (2017) The m^1A landscape on cytosolic and mitochondrial mRNA at single-base resolution. *Nature* 551(7679):251–255.
95. Zhou H, Rauch S, Dai Q, Cui X, *et al.* (2019) Evolution of a reverse transcriptase to map *N*(1)-methyladenosine in human messenger RNA. *Nat Methods* 16(12): 1281–1288.
96. Squires JE, Patel HR, Nousch M, Sibbritt T, *et al.* (2012) Widespread occurrence of 5-methylcytosine in human coding and non-coding RNA. *Nucleic Acids Res* 40(11):5023–5033.
97. Yang X, Yang Y, Sun BF, Chen YS, *et al.* (2017) 5-methylcytosine promotes mRNA export – NSUN2 as the methyltransferase and ALYREF as an m(5)C reader. *Cell Res* 27(5):606–625.
98. Fu L, Guerrero CR, Zhong N, Amato NJ, *et al.* (2014) Tet-mediated formation of 5-hydroxymethylcytosine in RNA. *J Am Chem Soc* 136(33):11582–11585.
99. Yang Y, Wang L, Han X, Yang WL, *et al.* (2019) RNA 5-methylcytosine facilitates the maternal-to-zygotic transition by preventing maternal mRNA decay. *Mol Cell* 75(6):1188–1202.e1111.
100. Schaefer M, Pollex T, Hanna K, & Lyko F (2009) RNA cytosine methylation analysis by bisulfite sequencing. *Nucleic Acids Res* 37(2):e12.
101. Khoddami V & Cairns BR (2013) Identification of direct targets and modified bases of RNA cytosine methyltransferases. *Nat Biotechnol* 31(5):458.
102. Hussain S, Sajini AA, Blanco S, Dietmann S, *et al.* (2013) NSun2-mediated cytosine-5 methylation of vault noncoding RNA determines its processing into regulatory small RNAs. *Cell Rep* 4(2):255–261.

103. Yuan F, Bi Y, Siejka-Zielinska P, Zhou YL, *et al.* (2019) Bisulfite-free and base-resolution analysis of 5-methylcytidine and 5-hydroxymethylcytidine in RNA with peroxotungstate. *Chem Commun (Camb)* 55(16):2328–2331.
104. Furuichi Y, LaFiandra A, & Shatkin AJ (1977) 5′ -Terminal structure and mRNA stability. *Nature* 266(17):235–239.
105. Alexandrov A, Martzen MR, & Phizicky EM (2002) Two proteins that form a complex are required for 7-methylguanosine modification of yeast tRNA. *RNA* 8(10):1253–1266.
106. Cowling VH (2009) Regulation of mRNA cap methylation. *Biochem J* 425(2): 295–302.
107. Chu JM, Ye TT, Ma CJ, Lan MD, *et al.* (2018) Existence of internal N^7-methylguanosine modification in mRNA determined by differential enzyme treatment coupled with mass spectrometry analysis. *ACS Chem Biol* 13(12): 3243–3250.
108. Lin S, Liu Q, Lelyveld VS, Choe J, *et al.* (2018) Mettl1/Wdr4-mediated m(7)G tRNA methylome is required for normal mRNA translation and embryonic stem cell self-renewal and differentiation. *Mol Cell* 71(2):244–255.e245.
109. Malbec L, Zhang T, Chen YS, Zhang Y, *et al.* (2019) Dynamic methylome of internal mRNA *N*(7)-methylguanosine and its regulatory role in translation. *Cell Res* 29(11):927–941.
110. Lei Z & Yi C (2017) A radiolabeling-free, qPCR-based method for locus-specific pseudouridine detection. *Angew Chem Int Ed Engl* 56(47):14878–14882.
111. Kariko K, Muramatsu H, Welsh FA, Ludwig J, *et al.* (2008) Incorporation of pseudouridine into mRNA yields superior nonimmunogenic vector with increased translational capacity and biological stability. *Mol Ther* 16(11):1833–1840.
112. Baudin-Baillieu A, Fabret C, Liang XH, Piekna-Przybylska D, *et al.* (2009) Nucleotide modifications in three functionally important regions of the *Saccharomyces cerevisiae* ribosome affect translation accuracy. *Nucleic Acids Res* 37(22):7665–7677.
113. Carlile TM, Rojas-Duran MF, Zinshteyn B, Shin H, *et al.* (2014) Pseudouridine profiling reveals regulated mRNA pseudouridylation in yeast and human cells. *Nature* 515(7525):143–146.
114. Lovejoy AF, Riordan DP, & Brown PO (2014) Transcriptome-wide mapping of pseudouridines: pseudouridine synthases modify specific mRNAs in S. cerevisiae. *PLoS One* 9(10):e110799.
115. Schwartz S, Bernstein DA, Mumbach MR, Jovanovic M, *et al.* (2014) Transcriptome-wide mapping reveals widespread dynamic-regulated pseudouridylation of ncRNA and mRNA. *Cell* 159(1):148–162.
116. Li X, Zhu P, Ma S, Song J, *et al.* (2015) Chemical pulldown reveals dynamic pseudouridylation of the mammalian transcriptome. *Nat Chem Biol* 11(8):592–597.
117. Zaringhalam M & Papavasiliou FN (2016) Pseudouridylation meets next-generation sequencing. *Methods* 107:63–72.
118. Sharma S, Langhendries JL, Watzinger P, Kotter P, *et al.* (2015) Yeast Kre33 and human NAT10 are conserved 18S rRNA cytosine acetyltransferases that modify tRNAs assisted by the adaptor Tan1/THUMPD1. *Nucleic Acids Res* 43(4): 2242–2258.
119. Boccaletto P, Machnicka MA, Purta E, Piatkowski P, *et al.* (2018) MODOMICS: a database of RNA modification pathways. 2017 update. *Nucleic Acids Res* 46(D1): D303–D307.

120. Arango D, Sturgill D, Alhusaini N, Dillman AA, *et al.* (2018) Acetylation of cytidine in mRNA promotes translation efficiency. *Cell* 175(7):1872–1886.e1824.
121. Nishikura K (2010) Functions and regulation of RNA editing by ADAR deaminases. *Annu Rev Biochem* 79:321–349.
122. Peng Z, Cheng Y, Tan BC, Kang L, *et al.* (2012) Comprehensive analysis of RNA-Seq data reveals extensive RNA editing in a human transcriptome. *Nat Biotechnol* 30(3):253–260.
123. Bahn JH, Lee JH, Li G, Greer C, *et al.* (2012) Accurate identification of A-to-I RNA editing in human by transcriptome sequencing. *Genome Res* 22(1):142–150.
124. Sakurai M, Yano T, Kawabata H, Ueda H, *et al.* (2010) Inosine cyanoethylation identifies A-to-I RNA editing sites in the human transcriptome. *Nat Chem Biol* 6(10):733–740.
125. Sakurai M, Ueda H, Yano T, Okada S, *et al.* (2014) A biochemical landscape of A-to-I RNA editing in the human brain transcriptome. *Genome Res* 24(3): 522–534.
126. Knutson SD, Arthur RA, Johnston HR, & Heemstra JM (2020) Selective enrichment of A-to-I edited transcripts from cellular RNA using endonuclease V. *J Am Chem Soc* 142(11):5241–5251.
127. Marchand V, Blanloeil-Oillo F, Helm M, & Motorin Y (2016) Illumina-based RiboMethSeq approach for mapping of 2′-*O*-Me residues in RNA. *Nucleic Acids Res* 44(16):e135.
128. Dai Q, Moshitch-Moshkovitz S, Han D, Kol N, *et al.* (2017) Nm-seq maps 2′-*O*-methylation sites in human mRNA with base precision. *Nat Methods* 14(7): 695–698.
129. Roundtree IA, Evans ME, Pan T, & He C (2017) Dynamic RNA modifications in gene expression regulation. *Cell* 169(7):1187–1200.
130. Herceg Z & Hainaut P (2007) Genetic and epigenetic alterations as biomarkers for cancer detection, diagnosis and prognosis. *Mol Oncol* 1(1):26–41.
131. Heyn H & Esteller M (2012) DNA methylation profiling in the clinic: applications and challenges. *Nature rReviews Genetics* 13(10):679–692.
132. Zeng C, Stroup EK, Zhang Z, Chiu BC, *et al.* (2019) Towards precision medicine: advances in 5-hydroxymethylcytosine cancer biomarker discovery in liquid biopsy. *Cancer Commun* 39(1):12.
133. Lamb YN & Dhillon S (2017) Epi proColon((R)) 2.0 CE: a blood-based screening test for colorectal cancer. *Mol Diagn Ther* 21(2):225–232.
134. Warren JD, Xiong W, Bunker AM, Vaughn CP, *et al.* (2011) Septin 9 methylated DNA is a sensitive and specific blood test for colorectal cancer. *BMC Med* 9:133.
135. Imperiale TF, Ransohoff DF, Itzkowitz SH, Levin TR, *et al.* (2014) Multitarget stool DNA testing for colorectal-cancer screening. *N Engl J Med* 370(14): 1287–1297.
136. Ned RM, Melillo S, & Marrone M (2011) Fecal DNA testing for colorectal cancer screening: the ColoSure test. *PLoS Curr* 3:RRN1220.
137. Lopez-Serra P, Marcilla M, Villanueva A, Ramos-Fernandez A, *et al.* (2014) A DERL3-associated defect in the degradation of SLC2A1 mediates the Warburg effect. *Nat Commun* 5:3608.
138. Diaz-Lagares A, Crujeiras AB, Lopez-Serra P, Soler M, *et al.* (2016) Epigenetic inactivation of the p53-induced long noncoding RNA TP53 target 1 in human cancer. *Proc Natl Acad Sci U S A* 113(47):E7535–E7544.

139. Bormann F, Rodriguez-Paredes M, Lasitschka F, Edelmann D, *et al.* (2018) Cell-of-origin DNA methylation signatures are maintained during colorectal carcinogenesis. *Cell Rep* 23(11):3407–3418.
140. Peter Ilse SB, Pomjanski N, Fink C, & Schramm M (2013) SHOX2 DNA methylation is a tumour marker in pleural effusions. *Cancer Genom Proteom* 10:217–224.
141. Kneip C, Schmidt B, Seegebarth A, Weickmann S, *et al.* (2011) SHOX2 DNA methylation is a biomarker for the diagnosis of lung cancer in plasma. *J Thorac Oncol* 6(10):1632–1638.
142. Wojno KJ, Costa FJ, Cornell RJ, Small JD, *et al.* (2014) Reduced rate of repeated prostate biopsies observed in confirmMDx clinical utility field study. *Am Health Drug Benefits* 7:129–134.
143. van Kessel KE, Van Neste L, Lurkin I, Zwarthoff EC, *et al.* (2016) Evaluation of an epigenetic profile for the detection of bladder cancer in patients with hematuria. *J Urol* 195(3):601–607.
144. Ai SX, Xu Q, Hu YC, Song CY, *et al.* (2014) Hypomethylation of SNCA in blood of patients with sporadic Parkinson's disease. *J Neurol Sci* 337(1–2):123–128.
145. Ferri E, Arosio B, D'Addario C, Galimberti D, *et al.* (2016) Gene promoter methylation and expression of Pin1 differ between patients with frontotemporal dementia and Alzheimer's disease. *J Neurol Sci* 362:283–286.
146. Lin SY, Hsieh SC, Lin YC, Lee CN, *et al.* (2012) A whole genome methylation analysis of systemic lupus erythematosus: hypomethylation of the IL10 and IL1R2 promoters is associated with disease activity. *Genes Immun* 13(3):214–220.
147. Neven KY, Piola M, Angelici L, Cortini F, *et al.* (2016) Repetitive element hypermethylation in multiple sclerosis patients. *BMC Genet* 17(1):84.
148. Lehmann-Werman R, Neiman D, Zemmour H, Moss J, *et al.* (2016) Identification of tissue-specific cell death using methylation patterns of circulating DNA. *Proc Natl Acad Sci U S A* 113(13):E1826–E1834.
149. Chim SSC, Tong YK, Chiu RWK, Lau TK, *et al.* (2005) Detection of the placental epigenetic signature of the maspin gene in maternal plasma. *Proc Natl Acad Sci U S A* 102(41):14753–14758.
150. Chan KC, Ding C, Gerovassili A, Yeung SW, *et al.* (2006) Hypermethylated RASSF1A in maternal plasma: a universal fetal DNA marker that improves the reliability of noninvasive prenatal diagnosis. *Clin Chem* 52(12):2211–2218.
151. Rahat B, Mahajan A, Bagga R, Hamid A, *et al.* (2017) Epigenetic modifications at DMRs of placental genes are subjected to variations in normal gestation, pathological conditions and folate supplementation. *Sci Rep* 7:40774.
152. Rahat B, Thakur S, Hamid A, Bagga R, *et al.* (2016) Association of aberrant methylation at promoter regions of tumor suppressor genes with placental pathologies. *Epigenomics* 8(6):767–787.
153. Hyland CA, Gardener GJ, Davies H, Ahvenainen M, *et al.* (2009) Evaluation of non-invasive prenatal RHD genotyping of the fetus. *Med J Aust* 191(1):21.
154. Tounta G, Vrettou C, Kolialexi A, Papantoniou N, *et al.* (2011) A multiplex PCR for non-invasive fetal RHD genotyping using cell-free fetal DNA. *in vivo (Athens, Greece)* 25(3):411–417.
155. Tong YK, Jin S, Chiu RW, Ding C, *et al.* (2010) Noninvasive prenatal detection of trisomy 21 by an epigenetic-genetic chromosome-dosage approach. *Clin Chem* 56(1):90–98.

156. Tong YK, Ding C, Chiu RW, Gerovassili A, *et al.* (2006) Noninvasive prenatal detection of fetal trisomy 18 by epigenetic allelic ratio analysis in maternal plasma: theoretical and empirical considerations. *Clin Chem* 52(12):2194–2202.
157. Tsui DW, Chan KC, Chim SS, Chan LW, *et al.* (2007) Quantitative aberrations of hypermethylated RASSF1A gene sequences in maternal plasma in pre-eclampsia. *Prenat Diagn* 27(13):1212–1218.
158. Berdasco M & Esteller M (2019) Clinical epigenetics: seizing opportunities for translation. *Nature Rev Genet* 20(2):109–127.
159. Ito S, Shen L, Dai Q, Wu SC, *et al.* (2011) Tet proteins can convert 5-methylcytosine to 5-formylcytosine and 5-carboxylcytosine. *Science* 333(6047):1300–1303.
160. Tahiliani M, Koh KP, Shen Y, Pastor WA, *et al.* (2009) Conversion of 5-methylcytosine to 5-hydroxymethylcytosine in mammalian DNA by MLL partner TET1. *Science* 324(5929):930–935.
161. Li W, Zhang X, Lu X, You L, *et al.* (2017) 5-Hydroxymethylcytosine signatures in circulating cell-free DNA as diagnostic biomarkers for human cancers. *Cell Res* 27(10):1243–1257.
162. Song CX, Yin S, Ma L, Wheeler A, *et al.* (2017) 5-Hydroxymethylcytosine signatures in cell-free DNA provide information about tumor types and stages. *Cell Res* 27(10):1231–1242.
163. Cai J, Chen L, Zhang Z, Zhang X, *et al.* (2019) Genome-wide mapping of 5-hydroxymethylcytosines in circulating cell-free DNA as a non-invasive approach for early detection of hepatocellular carcinoma. *Gut* 68(12):2195–2205.
164. Gao P, Lin S, Cai M, Zhu Y, *et al.* (2019) 5-Hydroxymethylcytosine profiling from genomic and cell-free DNA for colorectal cancers patients. *J Cell Mol Med* 23(5):3530–3537.
165. Margalit S, Avraham S, Shahal T, Michaeli Y, *et al.* (2020) 5-Hydroxymethylcytosine as a clinical biomarker: fluorescence-based assay for high-throughput epigenetic quantification in human tissues. *Int J Cancer* 146(1):115–122.
166. Gilat N, Tabachnik T, Shwartz A, Shahal T, *et al.* (2017) Single-molecule quantification of 5-hydroxymethylcytosine for diagnosis of blood and colon cancers. *Clin Epigenetics* 9:70.
167. Yang Z, Jiang W, Liu F, Zhou Y, *et al.* (2015) A novel electrochemical immunosensor for the quantitative detection of 5-hydroxymethylcytosine in genomic DNA of breast cancer tissue. *Chem Commun* 51(78):14671–14673.
168. Simon-Yarza T, Mielcarek A, Couvreur P, & Serre C (2018) Nanoparticles of metal-organic frameworks: on the road to in vivo efficacy in biomedicine. *Adv Mater* 30(37):1707365.
169. Zhang J, Han X, Gao C, Xing Y, *et al.* (2018) 5-Hydroxymethylome in circulating cell-free DNA as a potential biomarker for non-small-cell lung cancer. *Genom Proteom Bioinform* 16(3):187–199.
170. Chiu BCH, You Q, Zhang Z, Stepniak E, *et al.* (2017) 5-Hydroxymethylcytosine of circulating cell-free DNA in plasma: a novel non-invasive marker for progression and prognosis in multiple myeloma. *Blood* 130:4336.
171. Cai J, Chen L, Zhang Z, Zhang X, *et al.* (2018) 5-Hydroxymethylcytosines from circulating cell-free DNA as diagnostic and prognostic. *bioRxiv*.

CHAPTER 8

DNA-ENCODED CHEMICAL LIBRARIES USED FOR DRUG DISCOVERY

GUIXIAN ZHAO[1], SHILIAN YANG[1], XIAOYU LI[2,3], and YIZHOU LI[1]
[1] Chongqing Key Laboratory of Natural Product Synthesis and Drug Research, Chemical Biology Research Center, School of Pharmaceutical Sciences, Chongqing University, Chongqing, China
[2] Department of Chemistry and State Key Laboratory of Synthetic Chemistry, The University of Hong Kong, Hong Kong, China
[3] Laboratory for Synthetic Chemistry and Chemical Biology, Health@InnoHK Program by Innovation and Technology Commission, Hong Kong, China

8.1. INTRODUCTION

The isolation of potent and specific hit compounds against a biological target or targets of interest is the starting point for any drug development programs. Historically, hit compounds were typically identified through chemical optimization of a known bioactive molecules and mainly focused on "druggable" target family, such as kinase, enzymes, ion channels, and G-protein-coupled receptors (GPCRs). With the recent advances in omics technologies (e.g., DNA and RNA sequencing), more and more targets have been identified for drug discovery campaign, resulted in urgent needs for developing new approaches and novel technologies, which are able to reliably deliver hit compounds. These new strategies are able to rapidly and inexpensively access large chemical space in an integrated fashion, to interrogate the currently "undruggable" targets, providing starting points for further medical chemistry optimization. Generally, the technological approaches for isolation of hit compounds can be

Nucleic Acids in Medicinal Chemistry and Chemical Biology: Drug Development and Clinical Applications, First Edition. Edited by Lihe Zhang, Xinjing Tang, Zhen Xi, and Jyoti Chattopadhyaya.

classified into rational design paths and on screening-based paths. The rationally design approach requires structural information of a given target and/or the binding properties; however, it is extremely a challenging task to design a ligand with the desired binding properties because of the dynamic natures of the target protein itself and the complexity of the biological milieu around the target [1]. Currently the popular and practical approach in drug discovery is brute-force screening large-scale collection of compounds (called "chemical libraries") against the target to provide starting promising point for further development [2].

In traditional high-throughput screening (HTS) procedure, compounds from chemical libraries (typically around several millions in size) can be screened against the target individually with the support of expensive compound synthesis and sophisticated automated systems [3]. Screening at such a scale requires huge investments in compound collection, screening robotics, consumable logistics, and assay technologies, which is not often available to researchers in academia or from small companies. In sharp contrast, the isolation of therapeutic biomolecules (e.g., peptides and antibodies) with desired biological functions by various display technology platforms of encoded libraries, such as phage display [4–6], mRNA display [7], yeast surface display [8], ribosome display [9], is typically inexpensive and rapid. Notably, based on the evolution principle, these display technology platforms apply the natural transcription and translation machinery not only to generate huge of biomolecules but also to establish the physical linkage between yielding ligands' phenotype (chemical diversity) and their encoding genotype (genetic information, e.g., DNA and RNA), thereby maintaining the ability to isolate specific binders from encoded libraries by being *selected* in a whole mixed collection, instead of being *screened* one-by one against the biological targets.

Reliance on the biological machinery, the downside of display technology platforms is their yielding ligands mostly limited to peptides and biomacromolecules; however, it is worth mentioning that considerable progress has been made to expend broader chemical diversity by incorporating unnatural amino acids and other elements [10, 11] or by introducing post-translational chemical modification [12]. There is great scientific and industrial interest and demand in developing technology platforms that could harness the advantages from both biological display libraries and synthetic chemical libraries. In the last around three decades, DNA-encoded chemical library (DEL) has revolutionized traditional high-throughput screening, by coupling the power of genetic encoding with chemical synthesis, allowing the generation and selection of synthetic molecules with large chemical diversity. The productivity of DEL in discovering preclinical and clinical drug candidates has been confirmed (Figure 8.1). One famous example of a clinical candidate that originated from DEL is the first-in-class receptor-interacting protein-1 (RIP1) kinase inhibitor GSK2982772, as a Phase II clinical candidate to treat multiple inflammatory diseases such as psoriasis, rheumatoid arthritis, and ulcerative colitis [13]; notably, the chemical

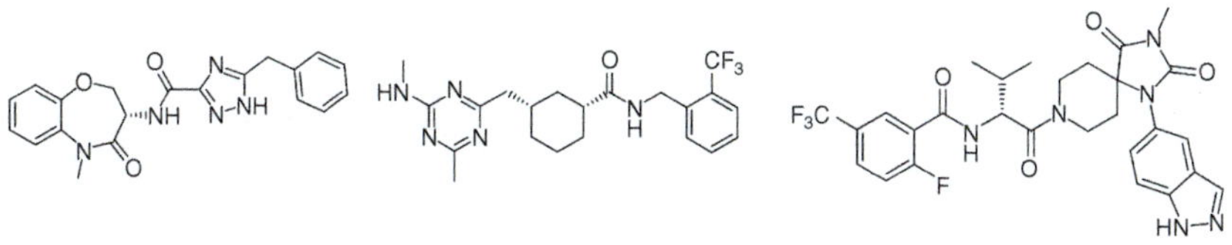

Figure 8.1 Clinical drug candidates discovered from DNA-encoded chemical library (DEL).

structure of clinical candidate, which derived with a novel benzo[*b*][1,4] oxazepin-4ones structure, has only a few atoms difference from the initial hits from DEL selection [14]. Another prominent example is GSK2256294, a soluble epoxide hydrolase (sEH) inhibitor also developed by GSK, which is currently being tested in Phase II clinical trials for treating various diseases including diabetes mellitus, endocrine system diseases, and glucose metabolism disorders [15]. Recently, X-Chem announced a novel Autotaxin inhibitor X-165, developed from hit compounds isolated from the DEL selection, which is currently being approved into Phase I clinical trials for fibrotic disease treatment [16]. These examples have demonstrated the potential of DEL in finding hit compounds with drug-like properties that can be readily developed into high-quality drug candidates.

As shown in Figure 8.2a, in a DEL, each compound individually is covalently liked with a distinctive DNA fragment, serving as the identifier for the chemical structure of the compound. As each compound members are tagged with individual DNA barcodes, the entire library with different compounds can be prepared and stored as a pool mixture, later selected simultaneously. The conventional DEL selection protocol is usually interrogated on the affinity capture procedures with an immobilized target similar to phage display. After low-affinity binders are washed away, the selected bound compounds are eluted from the target. Next, the chemical identity and relative enrichment of the selected compounds can be determined after PCR amplification, followed by high-throughput DNA sequencing (HTDS) to read the encoding DNA codes (Figure 8.2a). Thanks to the high encoding capacity of DNA tags, the sensitivity of PCR, and the ultrahigh throughput DNA sequencing, DEL technology enables the construction and selection of compound collections of very large size (typically from hundreds of millions to billions) and delivery hit compounds in an efficient and economical manner. By taking the advantage of integrating synthetic structures, DELs gain more chemical diversity than biological display libraries while still presenting the phenotype–genotype connection; DEL is a highly efficient technology that has provided large chemical diversity for drug discovery. Today, DEL has already been broadly employed by big pharmaceutical companies and employed in numerous drug discovery projects [17].

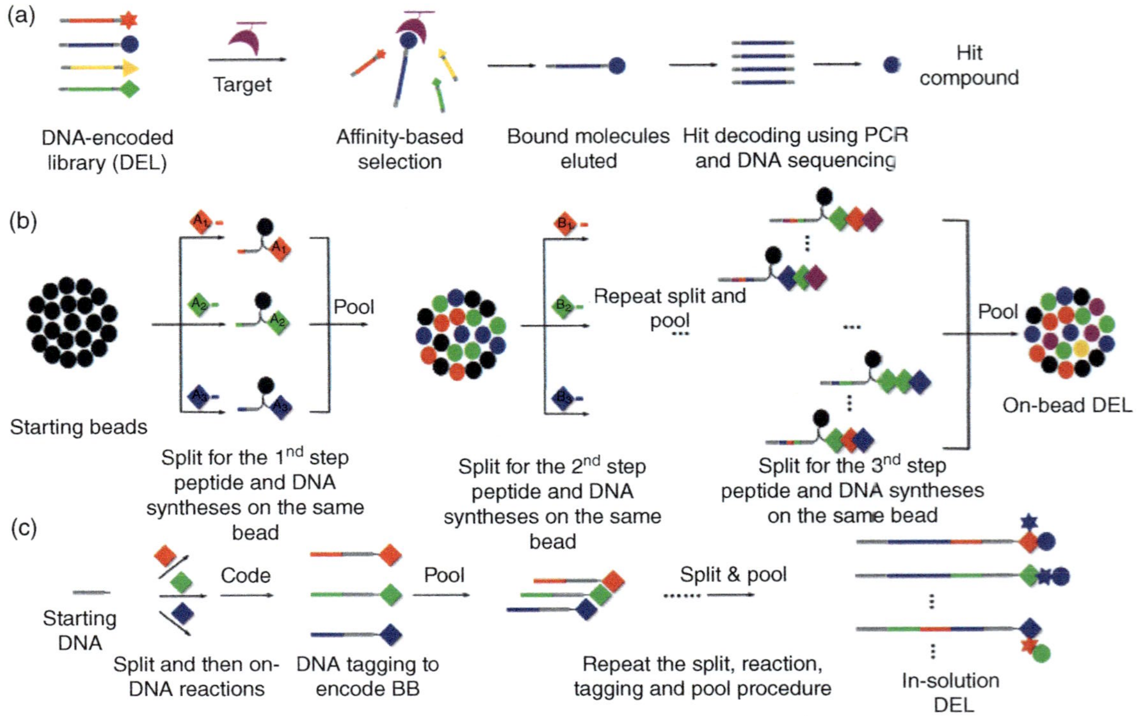

Figure 8.2 (a) The general scheme for DNA-encoded chemical library (DEL). (b) On-bead DEL by recoding each step of the chemical reactions with DNA syntheses. (c) In-solution DEL by recoding with a tagging step. (d) In-solution DEL by DNA-directing syntheses. (e) The universal template strategy. (f) The DNA routing strategy for DEL encoding.

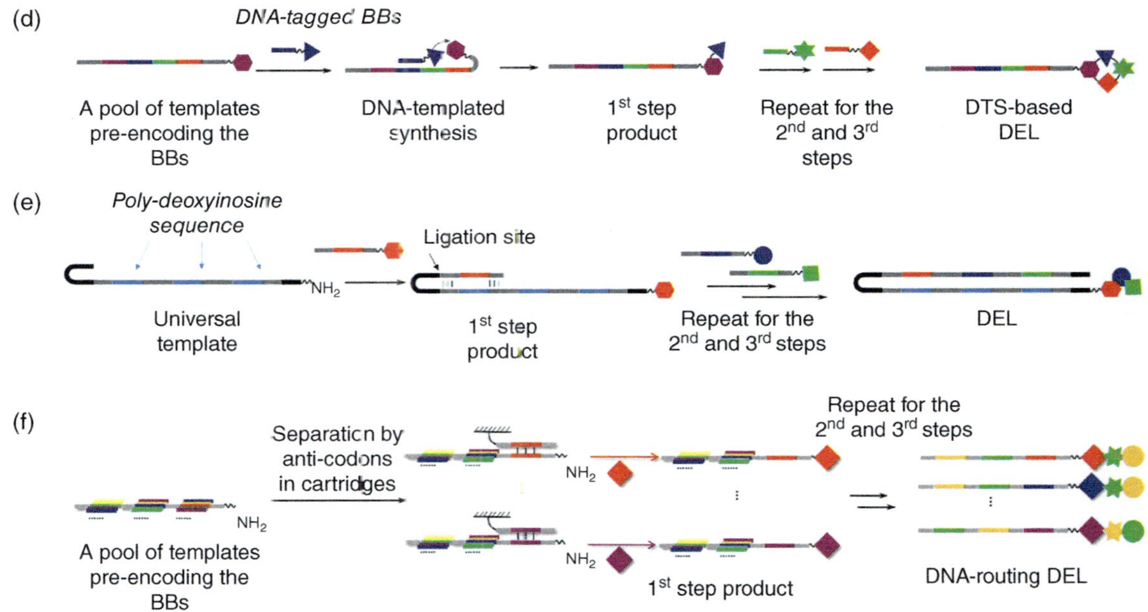

Figure 8.2 (*Continued*)

In 1992, Brenner and Lerner first postulated the original concept of encoding chemical synthetic structures with DNA barcodes [18]. In this seminal work, they theorized to encode peptide synthesis on solid support by using DNA molecules as tags. Shortly afterward, Needels and co-workers [19], as well as Janda and co-workers [20], translated Brenner and Lerner concept into reality with the synthesis of DNA-encoded peptide libraries and subjected their libraries to a selection and hit decoding process. In 1995, Kinoshita and Nishigaki reported an enzymatic oligonucleotide tagging method for the possibility of recoding chemical structures produced from a combinatorial fashion [21]. These initial reports have built up the fundamental technological features for further DEL evolution. In 2004, Neri [22], Liu [23], Harbury [24], and their respective co-workers independently reported different encoding methods for DEL synthesis and selection without beads. Since then, following the three landmark papers, great progresses have been achieved on the developments for DEL construction, selection, decoding, and hit validation. In the last few years, along with progressing advancements in DEL methodology development, DEL has been broadly exploited in drug discovery for hit generation and beyond.

Many seminal reviews, commentaries, highlight articles, and books from different prospective have appeared and then have covered every component of DEL [3, 17, 25–30]. Here in this chapter, we will first discuss the synthetic factors of the DEL, including encoding methods, DEL-compatible reactions, and chemical diversity space; then we will discuss the progressive development of selection methods. We will also briefly discuss hit decoding methods and summarize the recent hits delivered from DELs in drug discovery programs; and finally, we will present our opinions on the present challenges and directions for DEL in the future of drug discovery.

8.2. ENCODING METHODS OF DELs

Researchers have developed various types of encoding methods to build DELs. Each of the reported technique offers unique benefits and outcomes, which should be carefully considered when choosing an encoding method as the first step to build a DEL (Table 8.1). In principle, it is straightforward to create a library with small member size (typically from tens to hundreds), by tethering a DNA barcode with each compound individually. Yang, Lerner and co-workers reported an end-products labeling strategy and made a DEL pool by coupling of different nature products (some from traditional Chinese medicines), FDA-approved drugs, and compounds in clinical testing with individual DNA tags [51]. However, this one-by-one coupling process apparently becomes a laborious and impractical task for construction of a library with large member size (typically from hundreds of millions to billions). Despite the difference in the encoding process, most methods are based on the classic combinatorial chemistry strategy with "split-and-pool" manner to obtain large-scale chemical

TABLE 8.1 Summary and comparison of DEL encoding methods

Encoding methods		Features and advantages	Limitations	Refs.
DNA-recorded library	By chemical synthesis of DNA tags on solid phase	A pioneering method in initial DEL reports	Requires chemistries compatible with both peptide and DNA solid-phase syntheses	[17–19]
	By enzymatic or chemical ligation	The most widely used method; robust and extensively validated in practice; straightforward encoding procedure.	Requires the pre-preparation of dsDNA tags and buffer exchange between the enzymatic ligation and chemical synthesis step.	[31]
	By enzymatic polymerization	Also widely used and validated; only requires a single-stranded DNA to initiate the encoding	Requires buffer exchange between the enzymatic ligation and chemical synthesis step; requires high polymerase fidelity.	[32–35]
	By chemical ligation	No buffer exchange needed and no enzymatic reactions are involved	Chemical encoding forms tags with unnatural linkages not compatible with PCR and sequencing, extra processing is required	[36]
DNA-directed library	DNA-templated synthesis (DTS)	Simple library synthesis; proximity effect promotes BB reactions	Requires the pre-preparation of a template pool and careful codon design to avoid mismatch; requires bi-functional BBs	[23, 37, 38]
DNA-directed library	Universal template	Similar to DTS with only one template needed; simple codon design	Requires bi-functional BBs and special cleavable linker to release BBs	[39]
	DNA routing	Mono-functional BBs; potential for library evolution; compatible with non-aqueous condition	Requires sophisticated microfluidics instruments and codon design	[24, 40, 41]
Dual pharmacophore library	ESAC	High library purity; simple library synthesis; fragment-based discovery	Relatively small library size, requires fragment linking and optimization after selection	[22, 34, 42, 43]
	Dynamic libraries	Similar to ESAC but with features of DCL; compatible with soluble targets	Relatively small library size, requires fragment linking and optimization after selection	[44–47]
Direct DN ABB conjugate assembly		No need for template pool; no mismatch issue; simple codon design	Requires bi-functional BBs and special library architecture design	[48, 49]
PNA-encoded library		Higher chemical stability; compatible with organic reaction conditions	Not directly compatible with PCR and sequencing techniques	[26, 49, 50]

diversity through just a few synthetic steps. For demonstrative purpose, we could consider a stepwise assembly of a DEL of $1000^{100} = 10^5$ compounds, generated by the reaction of two sets of small chemical structures termed as building blocks (BBs): 100 sets of building blocks (BBs) and 1000 different BBs in each set, a 100-step "*split-and-pool*" synthesis rapidly deliveries a library of 1000^{100} compounds.

8.2.1. DNA-Recorded Library

With DNA-recording methods, each synthesis step is followed by an additional DNA encoding step to write the synthesis history for individual chemical diversity (i.e., specific BB). The early works by Brenner, Janda, and Gallop belong to this category [18–20]. The distinctive oligonucleotide sequences were synthesized directly on beads in parallel to the polypeptide synthesis (Figure 8.2b); therefore, these methods are only suitable for chemistry that is compatible with both solid-phase oligonucleotide and peptide synthesis. Later on, researchers have developed solution–phase-recorded DELs by tagging the DNA fragments during the *split-and-pool* procedure (Figure 8.2c).

In solution phase, encoding process could be accomplished by enzymatic DNA tagging. An extensively used approach, originally established by GSK (formerly Praecis Pharmaceuticals), is the direct ligation of DNA duplex with a "headpiece" DNA, which is a sticky-ended double-stranded DNA (dsDNA) with a DNA hairpin structure and an amino group serving as the starting diversity site for DEL-compatible reaction [31]. Briefly, the HP dsDNA is split into different wells and conjugated with each BBs in spatial separation manner, before respective dsDNA tags are added to each well for recoding. The solutions of all wells are pooled and re-split for next round of chemical reaction/tagging to install and encode another set of BBs. In fact, due to its design robustness and extensive applications, this "headpiece" approach has almost become the *de facto* DEL encoding protocols in industry, as well as for the researchers who just entered this field and wish to prepare their DELs [25]. Recently, Paegel, Kodadek, and co-workers have reported a new solid-phase DNA-recoded library, where dsDNA barcodes were enzymatically ligated to the pre-installed dsDNA fragments on beads to record the compound synthesis [24, 52]. Moreover, the Neri group has developed a method in which individual single-stranded DNA (ssDNA) tags are hybridized with the constant region on the sub-pool library intermediates, which conjugated with BBs and then polymerase-mediated primer extension to form the full-duplex DNA and therefore encode the chemical synthesis [32–35]. In addition, chemical ligation methods, such as Click ligation and photo-crosslinking, have also been developed to avoid the buffer exchange step between the chemical reaction and the enzymatic ligation step; however, chemical ligation does not generate native DNA structures, which requires further DNA translation steps to be compatible with PCR amplification [36].

8.2.2. DNA-Directed Library

In an alternative approach, called DNA-templated synthesis (DTS), pre-assembles a pool of DNA templates containing all the encoding DNA sequences for different combinations. Specifically, for a library with 1000^{100} compounds (100 sets of BBs and 1000 different BBs in each set), a pool of 1000^{100} DNAs with 1000 × 100 sequence combinations must be arranged before library construction, which can be achieved by using the "*split-and-pool*" approach with automated DNA synthesis and the commercial service is also provided. Next, the DNA templates directs the hybridization only with the DNA-conjugated BBs carrying complementary base sequences and the intermolecular chemical reactions spontaneously delivery the synthetic products. Usually this process is repeated to incorporate three to four sets of BBs to generate the final library (Figure 8.2d). The sequence specificity of DNA hybridization controls that only BBs tagged with the complementary DNA react with the BB/intermediate on the corresponding DNA template based on duplex formation. Therefore, parallel reactions can be conducted in the same well without spatial separation and libraries containing many different compounds can be constructed as a whole collection. The Liu group has explored and systematically optimized the DTS technology [23], which is practically for macrocyclic fumaramide DELs [37]. Recently Liu and co-workers have developed second-generation DTS with extensive optimizations and significantly improved synthetic efficiency and reaction scale, which has facilitated its industrial applications [38].

DNA-directed library takes advantage of feasible synthesis and the relatively high effective molarity between the reacting BBs [53]. In comparison with DNA-recorded libraries, however, DNA-directed library has two major limitations: (1) it requires extremely sequence design and comprehensive optimization to provide DNA hybridization fidelity and avoid mismatches, which is not straightforward for large libraries; and (2) it requires BBs carrying at least two reactivity groups, one for conjugation of DNA react and the other for DTS reaction. Also, the preparation of the rDNA react with BBs via cleavable linker conjugation requires considerable synthetic and purification efforts.

The first limitation has been addressed by Li and co-workers with developing an ingenious strategy to improve DNA template DEL, which using a single "universal template" ingrained with deoxyinosines, a nucleobase who can non-selectively pair with any of the four canonical nucleobases (Figure 8.2e) [39]. Therefore, the universal template is able to hybridize with any DNA sequences carrying the encoded BBs and then direct the synthesis. After BB delivery to reaction site, the initial DNA strand carrying BBs is linked with the universal template by enzymatic ligation to record the library synthesis history. This approach enjoys the benefit of the high-effective molarity of DTS but no sequence-specific DNA hybridization fidelity required; in this strategy, the universal template serves as a "nano-reactor" for nondiscriminatory access of BBs assembly in a combinatorial fashion. As a solution to the second limitation,

Harbury and co-workers developed a DNA-routing strategy to physically partition DNA template pool into sub-population on individual codon bases (Figure 8.2f) [24, 40, 41]; after the same group improved this strategy by accommodating with a mesofluidic devices, results with significant improvements in efficiency and scale, making the technology more suitable and mature for industrial drug discovery applications [54, 55].

8.2.3. Dual-Pharmacophore Library

Although many DELs are synthesized and encoded with dsDNAs, only one single compound (no matter how complex) is displayed on the DNA tags, and the compound is created through multistep reactions between the BBs. Such library can be termed "single-pharmacophore library." By contrast, "dual-pharmacophore library" featuring with the display of pairs of compounds at the extremity of two complementary DNA strands can be produced by resembling of two sets of sub-libraries (Figure 8.3a).

One prominent example is Encoded Self-Assembling Chemical (ESAC) libraries, which has been pioneered and systematically developed by the Neri group [22, 34, 42, 43]. As a matter of fact, the ESAC libraries are not synthesized, but assembled by combinatorial DNA hybridization. First, each pharmacophore (BBs) is conjugated with one DNA fragment, which carries a constant region that can hybridize with another set of pharmacophores. At the tail end of the DNA, a coding region is located. Second, no chemical reactions involve between the pairing BBs, and while DNA strand hybridization bringing the pharmacophores in proximity, the library is simply formed through the combinatorial self-assembly. Therefore, an exciting feature of ESAC library (generally for dual-pharmacophore library) presents the possibility that both pharmacophores, conjugated at the extremities of hybridized strands, synergize a "chelate effect," by enabling the simultaneous binding to adjacent nonoverlapping binding sites. In principle, ESAC is in full analogy to fragment-based drug discovery (FBDD), where small fragments that bind the target cooperatively are characterized before being evolved into high-affinity binders through fragment linking and growing [2]. ESAC services as a unique and complementary method to explore the utilities of FBDD.

In the first-generation of ESAC, the encoding information of the two pharmacophores are separated on the two different DNA strands; therefore, the selection of ESAC libraries was only able to isolate individual pharmacophore fragments, limiting ESAC mostly to be used for affinity maturation of known hit compounds. To address this limitation, the Neri group later developed an innovative approach that incorporates the encoding information of the two pharmacophores into a single DNA strand using a *d-spacer* to avoid base pair mismatches and transferring codes by polymerase extension strategy (Figure 8.3b), therefore synergistic binding combinations from both pharmacophores, instead of individual fragments, can be identified [34]. The second-generation of ESAC

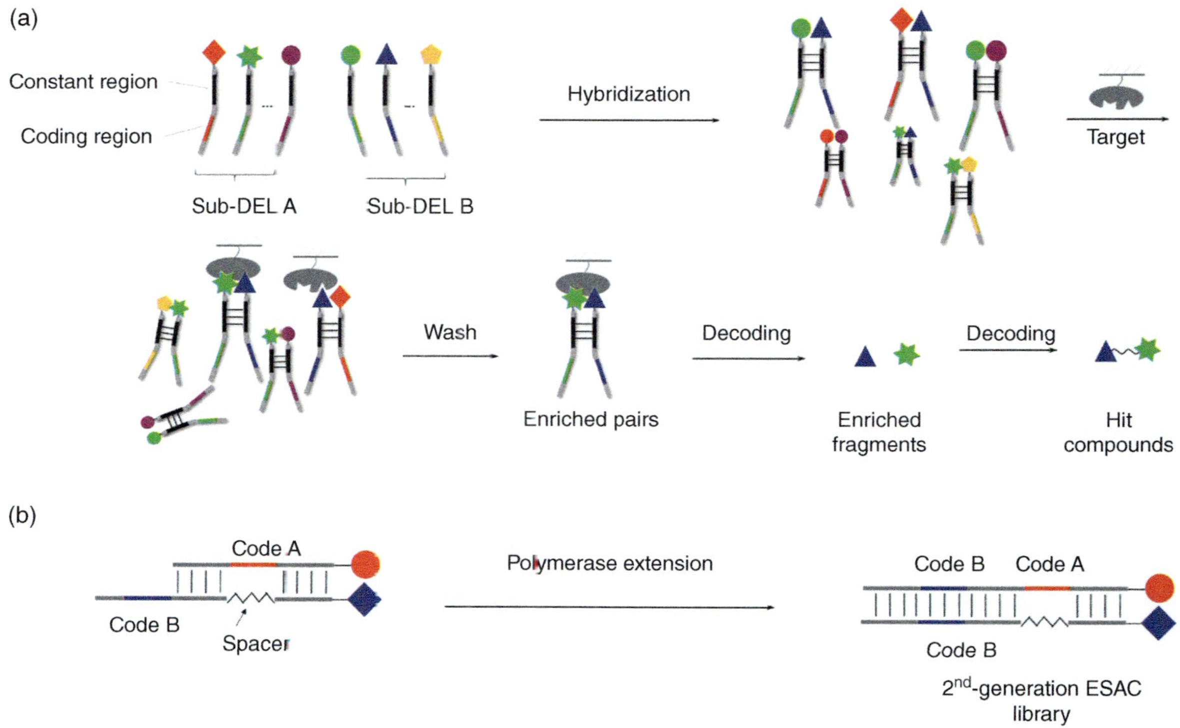

Figure 8.3 (a) The ESAC approach for encoding DEL. (b) The second-generation ESAC library. (c) The "Yocto-reactor"approach for encoding. (d) The DNA-BB conjugate assembly approach without DNA template. (e) PNA-encoded library can be displayed with DNA templates.

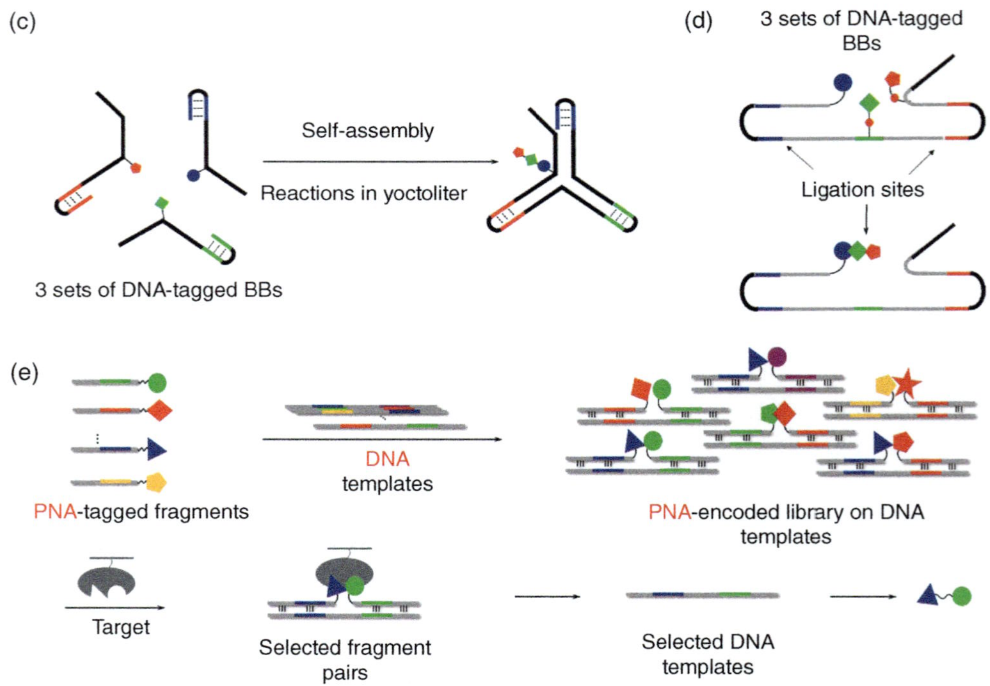

Figure 8.3 (*Continued*)

has realized the true dual-encoding and decoding by HTDS, and many ESAC libraries have been successfully created and selected against an array of biological targets [56].

The dual-pharmacophore ESAC libraries feature with two major advantages: (1) high library purity, since all DNA conjugates in sub-libraries can be individually HPLC purified and no further chemical reaction is involved in library assembly. Indeed, low library purity has been found to be harmful to the selection accuracy [57]; and (2) high degree of flexibility, compared to single-pharmacophore DEL, the arrangement of the BBs may more easily reach neighboring binding site on the protein surface. However, with dual-pharmacophore DELs, one needs to use an optimal linker (chemistry, length, rigidity, geometry need to be considered) to conjugate the selected binding pairs into a "full size binder" without DNA for further validation and hit optimization. This process is highly an empirical and largely laborious. In practice, Neri and co-workers have demonstrated that the selected binding pairs from ESAC libraries could be integrated into high-affinity binders by testing various linkers [43, 58, 59]. Finally, as an important future direction, it would be highly worthwhile to perform the simultaneous selection of not only the binding pair structures but also the linker structures directly from the library. In principle, this could be conceivable to DEL for selection and optimization of linker structures. For that purpose, application of affinity tests on locked nucleic acids (LNA) may facilitate the optimization of linkers [34].

Beyond the ESAC DEL, Zhang, Li, and their co-workers introduced the concept of dynamic combinatorial library (DCL) into dual-pharmacophore DELs and developed two kinds DEL named as DNA encoded dynamic libraries (hi-EDCCL) [44, 45] and DEDL [46, 47], respectively. Hi-EDCCL uses heating to reshuffle the DNA duplexes enriched from an ESAC library in the first round selection to generate a focused library, which can be further applied for the second round selection. Based on the very short hybridization region, the DEDL forms an unstable dynamic equilibrium dual-pharmacophore system; therefore, target addition moves the thermodynamic equilibrium and drives the formation of high-affinity duplexes, which were locked by the irreversibly photo-crosslinking for hit validation [46, 47].

These studies may integrate the advantages of classic DCL into DEL. In DCL selection, the target drives the thermodynamic equilibrium and improves the *in situ* synthesis of high-affinity binders at the expenses of non-binders; therefore, it may enjoy the numerical benefit and result in greater enrichment fold for more potency binders. These advantages have preliminarily been demonstrated in the hi-EDCCL technology with large size libraries [45]. Nevertheless, incorporating DNA encoding has solved a long-standing problem of small library size for classic DCL; therefore, it would be valuable to re-interrogate the protein targets previously tested with DCL but with much larger chemical diversities. Similar to ESAC, current dual-pharmacophore DEDL libraries are also limited to fragment discovery and a fragment linking/merging process is still necessary after selection campaign.

8.2.4. Direct DNA-BB Conjuate Assemly

In principle, combinatorially assembling several sets of DNA-BB conjugates should be a straightforward way to build and encode the library. This concept has been illustrated in the ESAC library with two sets of DNA-BB conjugates, which can be mixed and hybridized to form s table duplexes to present pairs of BBs with chemical diversity.

Scientists at Vipergen AS, located in Denmark, and the Gothelf group developed an approach of constructing DEL by DNA-BB conjugate assembly. This system has been named as "Yoctoliter Reactor™," since the DNA-directed reactions are restrained in a space at yoctoliter level (10^{-24} L) with multi-way DNA junctions formed by the DNA-BB conjugates (Figure 8.3c) [48]; and library encoding is achieved by enzymatic ligation among DNA-BB conjugates. In the self-assembly approaches developed by Li and co-workers, no DNA templates need to be prepared; without mismatch issue, the DNA sequence design is trivial (Figure 8.3d) [49].

8.2.5. PNA-Encoed Library

In principle, encoded metameric self-assembling libraries can also be created by using possible DNA surrogates, such as PNA and other analogy as tagging barcode. PNA is an artificially synthesized nucleic acid that can hybridize with DNA, RNA, and PNA itself, but with a synthetic neural and achiral peptidic backbone. The chemical structure of PNA allows the execution of reaction conditions in organic solvents, which may not always be possible with DNA; therefore, more BBs and reactions may be integrated in the PNA-encoded libraries. Winssinger and co-workers have extensively developed this technology; various libraries have been created and selected against different target proteins [26, 49, 50]. However, although more chemically stable, PNA is not adaptable with PCR amplification and DNA sequencing techniques. After the selection, the isolated PNA codes need to be deconvoluted with DNA microarray technology, resulted in significant limitation of the library size [26]. An alternative way is to transfer the PNA codes into DNA codes prior to PCR amplification and sequencing. For example, Bradley and co-workers have employed a selective nuclease digestion strategy to translate post-selection PNA sequences to DNA sequences [60]. Moreover, Winssinger and co-workers addressed this limitation by using a DNA/PNA hybridization strategy (Figure 8.3e). First, a pool of DNA templates was used to direct the combinatorial assembly of PNA-encoded BBs to form a dual-pharmacophore library; after selection, the enriched DNA templates were PCR-amplified and PCR products were subjected to assemble a focused library for next round of selection. The final pool of the enriched DNA templates was then applied to the standard deconvolution process.

As a useful alternative to standard DEL, PNA-encoded library is considered to access more types of chemistries to provide larger chemical diversities;

however, the hit deconvolution process must be systematically designed, since it will have significant impacts on the architectural design and the encoding sequences of the library.

8.3. CHEMICAL DIVERISTY OF DELs

After choosing an encoding method, one must systematically consider the library design, which directly influences the chemical diversity and properties of encoded compounds. The library design is essential for the success in isolation of the high-quality binders for further drug development.

8.3.1. Library Size

One of the most outstanding features of DEL is the size, ranging from hundreds of millions to billions to even trillions of compounds in pool collections, which is indeed remarkably huge compared with other chemical libraries. However, larger size is only related to but does not absolutely equal to higher chemical diversity and many other parameters should be considered. In addition, increasing the library size may raise an issue with poor library purity, which could result low selection accuracy [61], high false-negative rates [62], difficulties in decoding the sequencing data, etc. [63]. Furthermore, large libraries do not always associate with favorable discovery of novel binders [64]. Progressive evidences indicate that more focused and high purity libraries with compact scaffolds are more valuable [61, 64–67].

In practice, library size related with many other parameters including library architecture, synthetic chemistry for library, availability of BBs and/or scaffolds, and whether the library is an unbiased "general library" or a focused library biased toward one target or a specific type of target. For example, a general peptidic library with three to four sets of BBs can easily reach hundreds of million to billion scale due to the broad availability of the amino acids and the robustness of the amide formation reaction. In contrast, it is recently reported macrocycle DEL is designed to carry more backbone diversity, the laborious process to collect or produce the bi-functional backbone BBs became a limiting factor, and the library only has ~1/4-million compounds [68]. In addition, dual-pharmacophore libraries normally contain two sets of BBs; therefore, the library size (million compounds or less) is usually much smaller compared to single-pharmacophore libraries created with three or four sets of BBs. Furthermore, for the purpose of drug discovery, one fundamental criterion in DEL size is the number of BBs sets. Adding sets of BBs necessarily increases the compound's molecular weight; therefore, it is difficult to ensure the isolated binders fall into the chemical space of Lipinski's rule of 5, Veber descriptors, and other favorable physicochemical properties of drug molecules [3].

8.3.2. Building Blocks

From the logistical aspect, BB collections are one of the most important prerequisites for DEL synthesis, which cannot be undervalue. BBs dominate the chemical properties of library diversity. For unbiased "general libraries" intended to be selected against various types of targets, the choice of BBs is more flexible and usually depends on the library encoding method and the chemistries applied in library synthesis. For focused libraries, one may choose a set of BBs including not only the privileged structures for the specific type of targets as well as a scope of derivatives for diversification. In general, monofunctional BBs are mostly used as appendages on a central scaffold or as the end-capping BB in a linear library; bifunctional BBs may be used as linkers or at bifurcation point; and trifunctional BBs are highly useful scaffolds in branched library. However, BB availability decreases dramatically from mono- to bi- and trifunctional BBs. A report has shown that commonly available monofunctional BBs compatible with major types of DEL chemistries (see Section 3.3) at reasonable costs, including primary and secondary amines, carboxylic acids, aldehydes, aryl halides, boronates, activated acyl halides, alkynes, and sulfonyl halides, were estimated to be around ~13,000, while the number of bifunctional BBs decreased to ~3500 and the tri-functional BBs/scaffolds were only available a few [25]. In addition, physicochemical properties should also be considered in selecting library BBs with structural exclusion principles. For example, the incorporation of overly hydrophobic building blocks may end library hits with deficient for further medicinal chemistry optimization because they are far out the chemical space of Lipinski's rule of 5.

8.3.3. DEL-Compatible Reaction

In ESAC, DEDL, and hi-EDCCL, BBs are simply displayed on DNA and no chemistry is needed to connect the BBs. Otherwise, BBs may be conjugated or appended on a scaffold using chemical reactions, which inevitably influence the chemical diversity of the library. DEL construction requires compatible reactions with high conversion for most of the accessible BBs; moreover, such reactions should also be robust and easily applicable to an automated workflow for combinatorial synthesis process (Table 8.2) [90].

Historically, solid-phase peptide synthesis is the initial DEL chemistry [18–20]. The groups of Paegel and Kodadek have expanded the solid-phase-based DEL chemistry to include other reactions such as Suzuki-Miyaura reaction [91], Knoevenagel condensation [92], asymmetric Mannich reaction [93], and asymmetric Aldol and Mitsunobu reactions [94]. Recently, Brunschweiger and co-worker have expanded multicomponent reaction to solid-phase-coupled DNA such as Ugi and Ugi-azide [95]. The same group also reported a Yb(III)-mediated Castagnoli-Cushman reaction for the synthesis of isoquinolones and an Ag(I)-mediated cycloaddition for the synthesis highly substituted pyrrolidines on

TABLE 8.2 Summary and comparison of DEL selection methods

Selection methods		Features and advantages	Limitations	Refs.
Immobilized library, soluble targets		Compatible with non-immobilized and unmodified targets	Requires separation of target-bound beads from unbound ones.	[18–20, 52, 69, 70]
Solution library, immobilized targets		The most widely used selection method, well-validated in practice	Requires target modification and immobilization	[71]
Soluble library with soluble targets	Interaction dependent using unpurified proteins (IDUP)	Compatible with non-immobilized and endogenous targets in cell lysates	Targets need to be conjugated with a DNA tag; requires special library sequence design	[72–74]
	Ligand-directed crosslinking	Compatible with non-immobilized targets in cell lysates	Require additional steps in selection; not compatible with libraries encoded with dsDNA	[75–79]
	Binder trap enrichment (BTE)	Spatial separation of individual target/compound binding complexes	Requires target modification with a DNA tag; requires special library architecture	[80]
	Kinetic capillary electrophoresis	Target/compound binding complexes are separated from unbound targets with electrophoresis	Still in the early development stage	[81, 82]
Cell-based selections		Membrane proteins can be selected on live cells	Requires overexpressed target with few exceptions reported; future directions	[83–89]

Controlled Pore Glass (CPG) [96]. Pioneered by the groups of Baran and Dawson, Reversible Adsorption to Solid Support (RASS) technology has been examined as a promising method to increase DEL diversity by using standard organic synthesis procedures [97, 98]. Organic solvents and more choices of reagents could be applied to the solid-phase synthesis; in general, these reactions showed high conversion (85% to quantitative) in most cases.

In-solution DEL-compatible reactions are usually mild in an aqueous condition for solubility reason with neutral or slightly basic condition to avoid depurination. The in-solution DEL chemistry also started from peptide synthesis but has been expanded to become a diverse reaction toolbox, which has already been comprehensively reviewed [30, 53, 99–101]. Recently, many new reactions have been added to the toolbox. Fan *et al.* disclosed a ring opening reaction of on-DNA epoxides with amines assisted by zirconium tetrakis, thereby providing a new chemistry for using the abundantly available amines in DEL synthesis [102]. Lu and co-workers reported a ruthenium-promoted on-DNA ring-closing metathesis; by incorporating two olefins at different positions in a linear structure, this method provides a feasible way to access constrained macrocycles [103]. Brønsted acid catalysts, coinage transition metals, and oxidants furnish diverse drug-like structures from easily accessible BBs; however, these conditions were rarely considered as DEL-compatible because they may cause unacceptable DNA damage. To overcome this limitation, by using hexathymidine oligonucleotide adapter, Brunschweiger and co-workers developed acid- and gold-catalyzed reactions on DNA, which has been exploited to build the scaffold diversity at the initial step of library synthesis through cycloaddition [104] and three-component reactions [105]. In addition, Schreiber and co-workers have developed an elegant method to build DNA-conjugated polycyclic isoxazolidines via [3+2] nitrone–olefin cycloaddition [106]. To synthesize C3-alkylated indole DEL, Hitgen developed a two-step process via indolyl alcohol formation followed by metal-free transfer hydrogenation [107]. These reactions, although still limited by the BB availability, have reflected a trend in DEL-compatible chemistry research: development of reactions that can build drug-like core structures *in situ* during the library synthesis, rather than preparing them "off-line" and then conjugating to the DNA.

Traditionally, DELs tend to be deficient in C(sp^3) carbon centers, which is an important component of the pharmacologically relevant chemical spaces [108]. Recently, significant strides have been made on this aspect. Scientists from Pfizer reported a C(sp^3)-C(sp^3) coupling of DNA-tagged Michael acceptors and styrenes with α-amino acids by photoredox catalysis, this reaction tolerates a broad scope of radical precursors and the abundance of amino acids has made this method quite attractive in DEL synthesis [109]; optimized photoredox catalysis procedures have further been developed for on-DNA C(sp^2)-C(sp^3) coupling with various types of substrate types [110, 111]. Similarly, a radical-based Giese reaction has also been developed [112].

Additionally, Molander and co-workers have merged Ni/photoredox dual catalytic C(sp^2)-C(sp^3) cross-coupling, as well as photoredox-catalyzed alkylation reactions. Notably, these reactions install sp^3 carbon directly on the widely available alkyl halide BBs, which makes this reaction highly promising in building DELs with drug-like structures. In this work, more than 140 examples have been provided to demonstrate the broad applicability of this chemistry [113]. Later, the Molander group further explored this C(sp^2)-C(sp^3) cross-coupling with broader substrate scope including more available alkyl bromides and structurally diverse α-silylamines for creating versatile chemical foundation of DELs [114]. The groups of Zhong and Lu reported the first example of DEL-compatible C-H activation [115], a type of reactions that has been considered as a powerful strategy for constructing complex molecules from simple starting materials. In this work, simple and widely available aromatic carboxylic acids could be conjugated with Michael receptors catalyzed by Ruthenium. More recently, the Yu group showed the C(sp^3)–H arylation of aliphatic carboxylic acids, amides, and ketones with DNA-conjugated aryl iodides [116]. In addition, on-DNA reactions including photocatalytic [2+2] cycloaddition [117], cross-dehydrogenarative coupling (CDC) [118], and photoredox decarboxylation of amino acid coupling to DNA-linked carbonyl reaction [119] have been reported to increase C(sp^3) carbon diversity for DEL synthesis. Although the synthesis of large-sized DELs based on these reactions and the corresponding selections and hit discovery have yet to be reported, we expect to see wide utilities of these reactions in building DELs with more sp^3 contents, especially the ones with more drug-like features (e.g., more compact structure, less peptidic linkage). It may be useful to use these reactions to assemble a variety of central cores and then use the more "conventional DEL chemistry" to install appendages for further diversification. Moreover, the recently reported DEL-compatible reactions also include Suzuki/Suzuki-Miyaura reactions [120 124], Heck reactions [125], Sonogashira reactions [99, 126], nitro reduction [127, 128], in-solution Aldol reactions [129, 130], Michael reactions [129, 130], aromatic substitution [131, 132], C-N cross-coupling [133–135], C-S/Se/P cross-coupling [97, 136], electron-inversed Diels–Alder reactions [137, 138], and also biocatalytic reactions [139]. These reactions require specific types of BBs and may be more suitable for building more focused libraries. Nevertheless, collectively, they have further enriched the toolbox of DEL chemistry and provided researchers with more choices in accessing more chemical spaces in DELs.

The new DEL-compatible chemistry developments have the potential to revolutionize the DEL technology as they enable access to unprecedented molecular diversity. In the last few years, there have a significant progress in DEL-compatible chemistry; and in a sense, DEL chemistry is not that limited anymore. The actual limiting factor is the availability of BBs suitable for these reactions. Quite often, a reaction has been adapted to be compatible with DEL synthesis, but there are limited BBs available to build a sizable library.

8.3.4. BB Assembly Geometry

Similar to antibody-phage display, the assembly geometry of BBs has significant impact on the chemical diversity. In single-pharmacophore DELs, BBs could be assembled linearly, cyclically, or on a constant scaffold, while dual-pharmacophore DELs present the BBs on the DNA duplex with a flexible linker. Neri and co-workers developed a DEL with three sets of BBs displayed on a macrocyclic scaffold featured with β-sheets peptide, which has facilitated the generation of binders targeting poorly defined large surface of protein–protein interaction [35]; later, they reported a DEL design with two sets of BBs displayed on a smaller and more rigid stereoisomeric scaffold, which has also allowed de novo identification of various binders against different targets [140]. In an interesting study, the groups of Zhang and Keller used DNA origami structures to present pharmacophores, aiming to investigate protein–ligand interactions at a single-molecule level [141].

8.4. SELECTION METHODS OF DELs

DEL selection is mostly based on a binding assay (Table 8.3). In principle, the selection only delivers binders, disregarding where they bind and whether the binding is functionally relevant. Therefore, after the binders are isolated, they will be resynthesized "off-DNA" and further examined for their binding properties and biological activities. However, with a well-designed selection experiment itself, one could also guide the library selection to be biased for desired binding properties or biological functions.

8.4.1. Selection with Immobilized Library and Soluble Targets

In the early works, since DELs were prepared on solid supports [18–20], selections were performed by incubating the soluble target with the immobilized library. Later on, the majority of DELs are synthesized in solution and selected against immobilized targets. However, recently, the Paegel and Kodadek groups have reported a new generation of matrix-supported DELs that have been selected against soluble targets [52, 69, 70].

8.4.2. Selection In-Solution Library with Immobilized Targets

With in-solution DELs, selections were mostly performed with a purified protein immobilized on a matrix. The selection protocol is very simple: after library incubation with the target, non-binders are eliminated with repeated washing steps and the binders are then eluted under denaturing conditions [71]. Eluted compounds are collected and then subjected to the deciphering process. Due to the experimental nature of the washing procedures, it is suggested to use

TABLE 8.3 DEL-compatible reactions

Reaction	Scheme
(a) Suzuki [91, 120–124]	
(b) Knoevenagel [92]	
(c) Mannich [93]	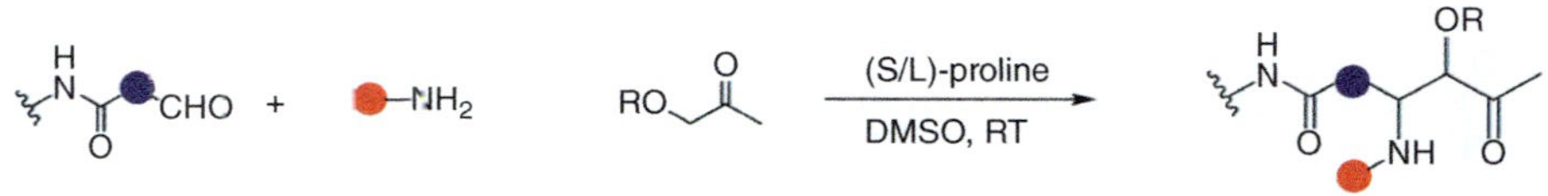
(d) Aldol [94, 129, 130]	
(e) Ugi [95]	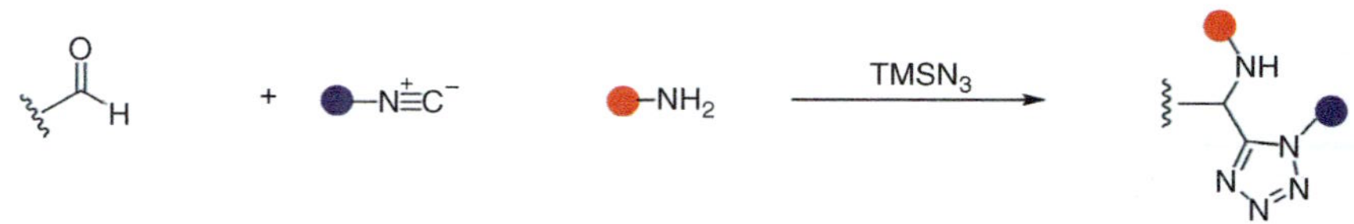

(*Continued*)

TABLE 8.3 (Continued)

	Reaction	Scheme
(f)	Castagnoli-Cushman [96]	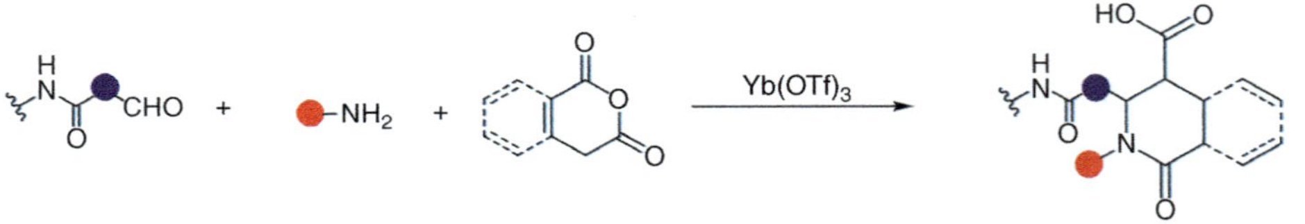 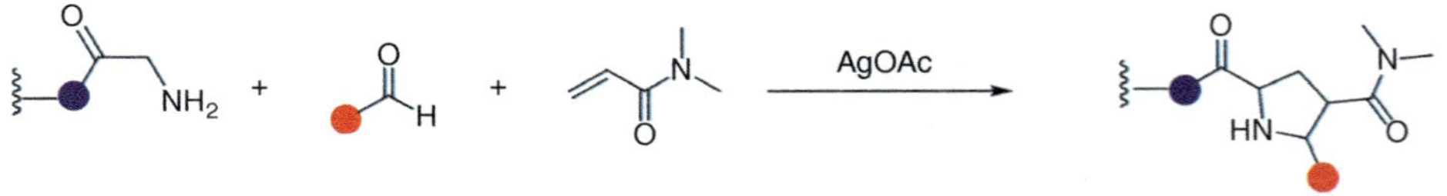
(g)	Ring-opening [102]	
(h)	Ring-closing metathesis [103]	
(i)	Three component cycloaddition [105]	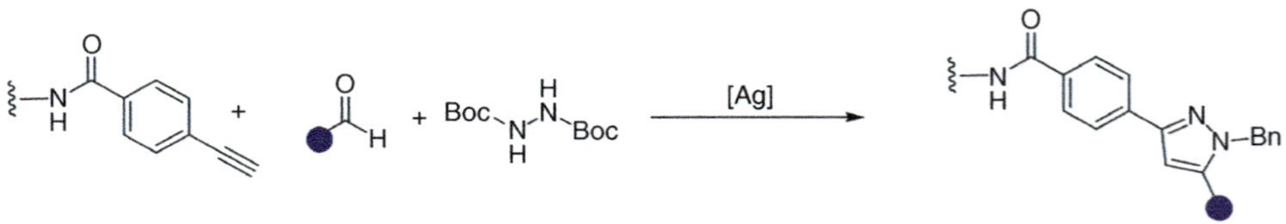

(j)	**[3+2] nitrone-olefin cycloadditions [106]**	
(k)	**C3 alkylation of indoles [107]**	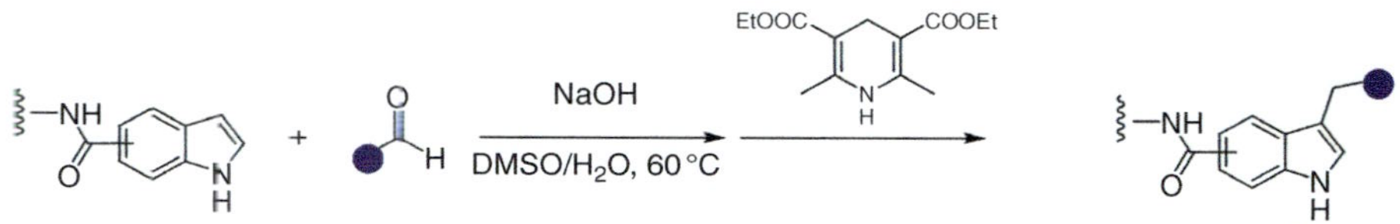
(l)	**Photoredox-mediated coupling [109–111, 113, 114]**	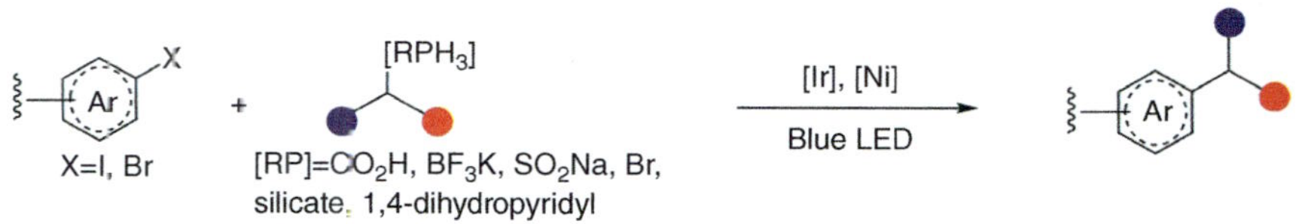
(m)	**Giese [112]**	
(n)	**Photocatalytic [2 + 2] cycloaddition [117]**	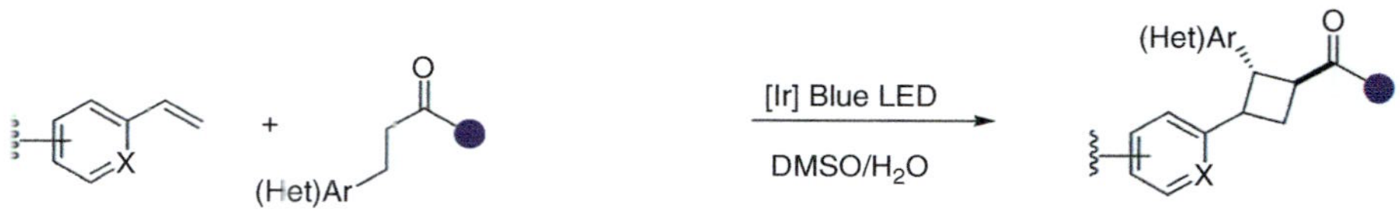

(Continued)

TABLE 8.3 (Continued)

	Reaction	Scheme
(o)	**C-H activation [115, 116]**	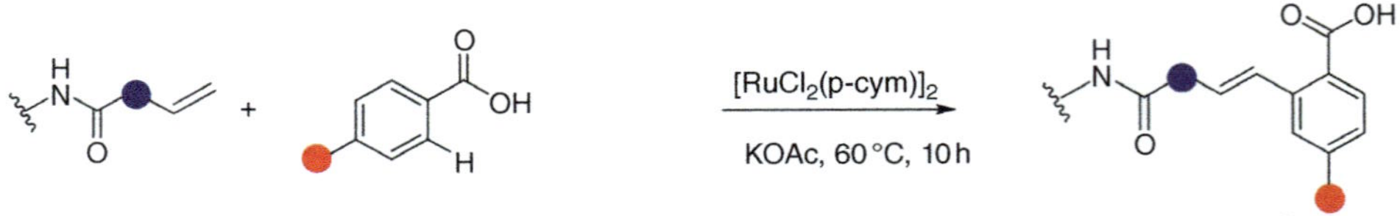

(p)	**Cross-dehydrogenative coupling [118]**	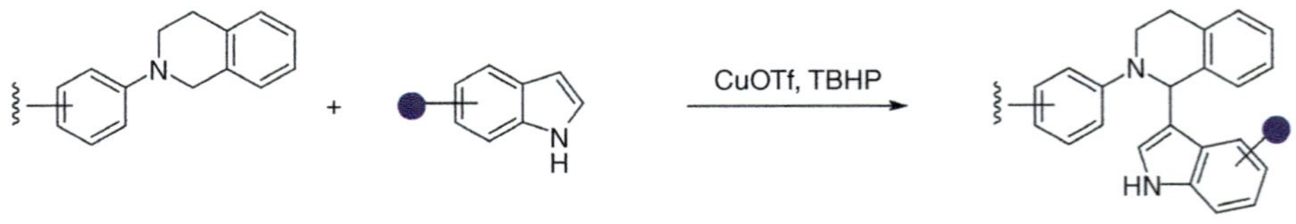
(q)	**Photoredox-mediated decarboxylative coupling [119]**	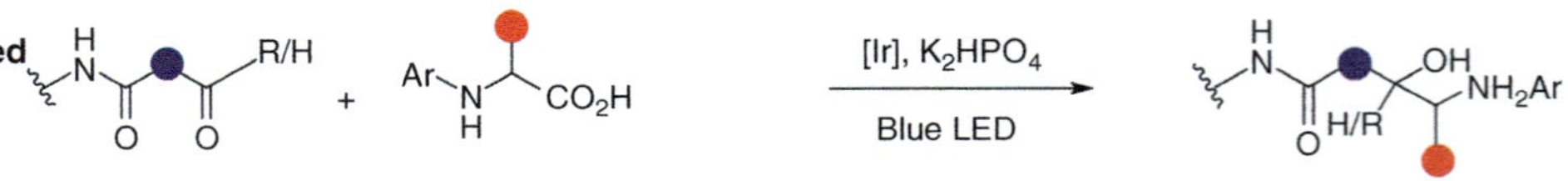
(r)	**Heck [125]**	

(s)	**Sonogashira** [99, 126]	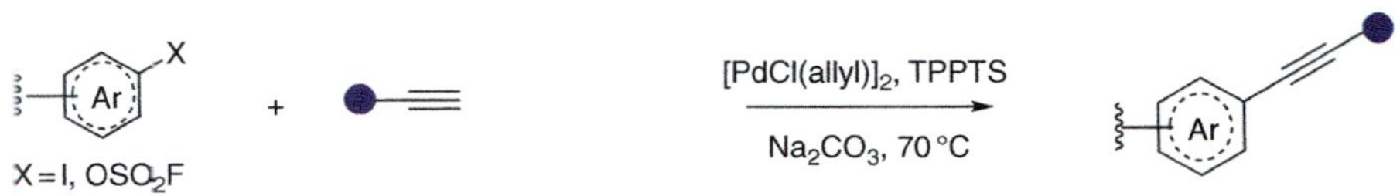
(t)	**Nitro reduction** [127, 128]	
(u)	**Michael** [129, 130]	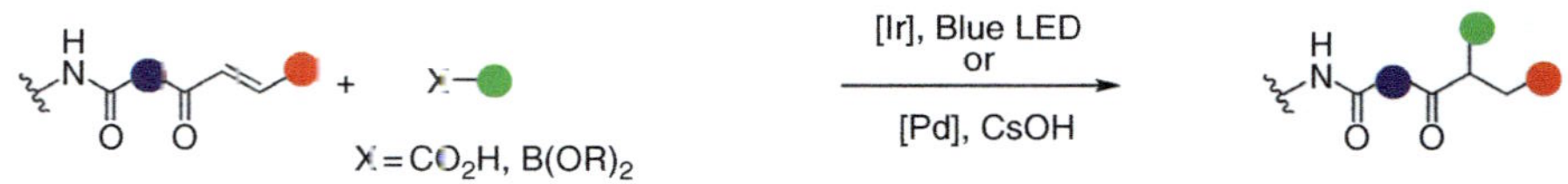
(v)	**Aromatic substitution** [131, 132]	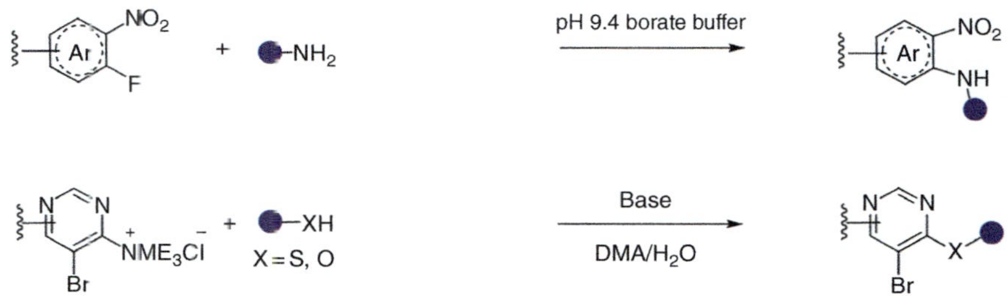

(Continued)

TABLE 8.3 (Continued)

	Reaction	Scheme
(w)	C-N cross-coupling [126, 133–135]	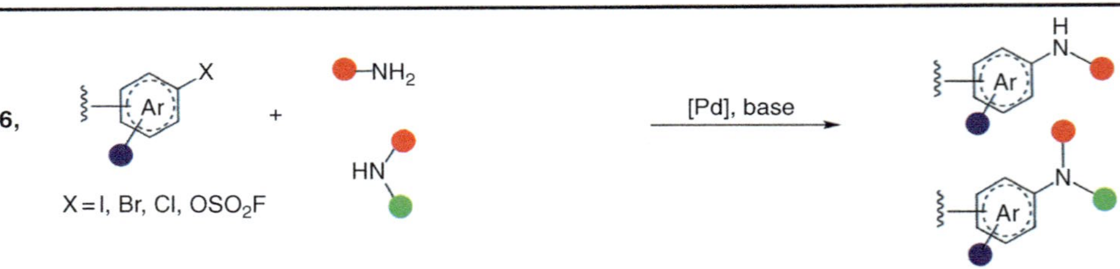
(x)	C-S/Se/P cross coupling [97, 136]	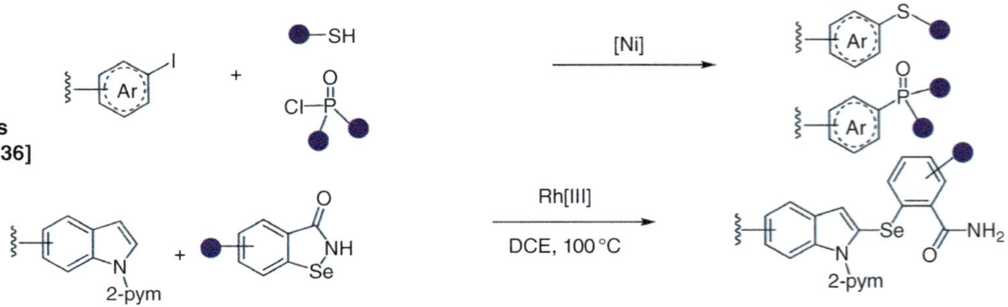
(y)	Electron-inversed D-A reaction [137, 138]	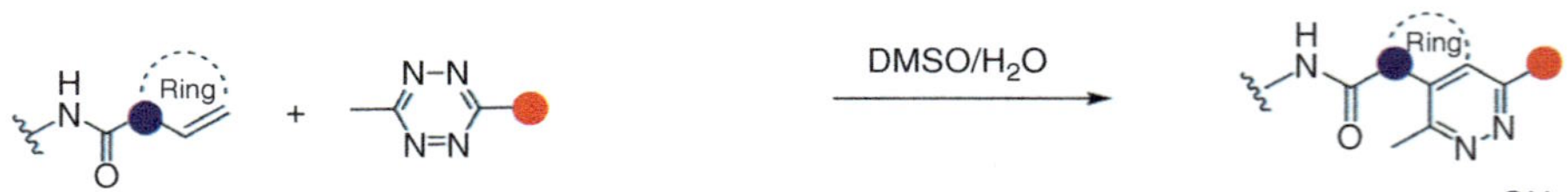
(z)	Biocatalytic reactions [139]	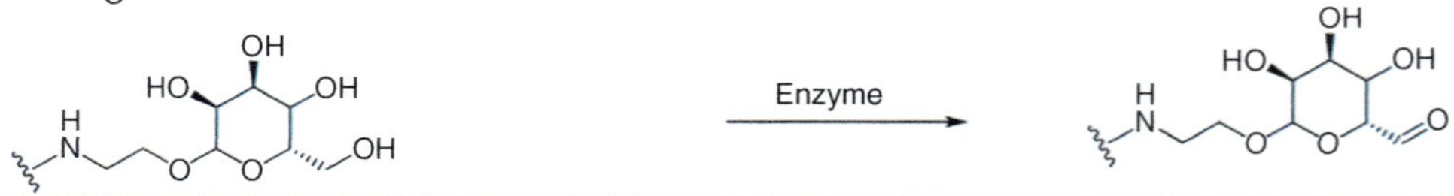

quantitatively PCR (qPCR) as a tool for selection quality control and include an internal standard extent possible [75, 142–145]. In addition, since DELs are selected at tiny amount (atmol~pmol), it is now already a routine process to apply multiple protein concentrations in DEL selections, since high-target concentration may identify both nM and μM binders, while low-target concentration preferably may identify high-affinity nM binders. In addition, running the selection over a range of concentrations may also help reduce false positives since the binders identified consistently in multiple selections are more likely to be the true hits. Furthermore, to control DNA-target interactions and other non-specific interactions, control selections with a "blank library" (no small molecule, only DNA tags) and an unrelated target are also necessary. Performing parallel selections with multiple target isoforms is beneficial in isolating binders with high selectivity. It is now quite common to include a control selection with bead only, a denatured target, or an unrelated protein, to compare with the target selection and identify specific binders.

8.4.3. Selection with Non-Immobilized Targets

Apparently, not all proteins can be easily purified and immobilized, and many proteins lose their structures and functions during immobilization. Complex targets such as protein complexes, membrane proteins, and live cells are more biologically relevant and hit compounds generated with these targets are more biologically relevant; however, they are not tractable to purification and immobilization. In addition, with purified proteins, certain desirable biological features are missing, such as post-translation modifications and endogenous binding partners [72]. Several approaches have been developed to enable selections with soluble proteins. The Interaction Determination using Unpurified Proteins (IDUP) method developed by Liu and co-workers modifies the target with a DNA tag prior to selection to enable selective PCR amplification of the binders in solution (Figure 8.4a) [72–74]. Vipergen developed a so-called Binder Trap Enrichment (BTE) method where selective PCR amplification is realized with a water–oil emulsion system to spatially separate binders from non-binders (Figure 8.4b) [80]. Alternatively, covalent crosslinking has been shown to be an effective approach to establish the link between the binders and the target, so that the binders could be isolated in the form of DNA–protein conjugates (Figure 8.4c). Li and co-workers used nuclease digestions to remove the non-binders [76]; in a later report, they gel-purified the DNA–protein conjugates for hit deconvolution [77, 78]. Also, Krusemark and co-workers employed affinity-pulldown to isolate the binders from the library after crosslinking (Figure 8.4d) [75]. Moreover, Krusemark and co-workers recently developed a substrate-mediated selection method to disclose compounds functions other than binding assay. This method includes two steps: (1) specific DNA-linked substrate selectively phosphorylated by a soluble protein tyrosine kinase

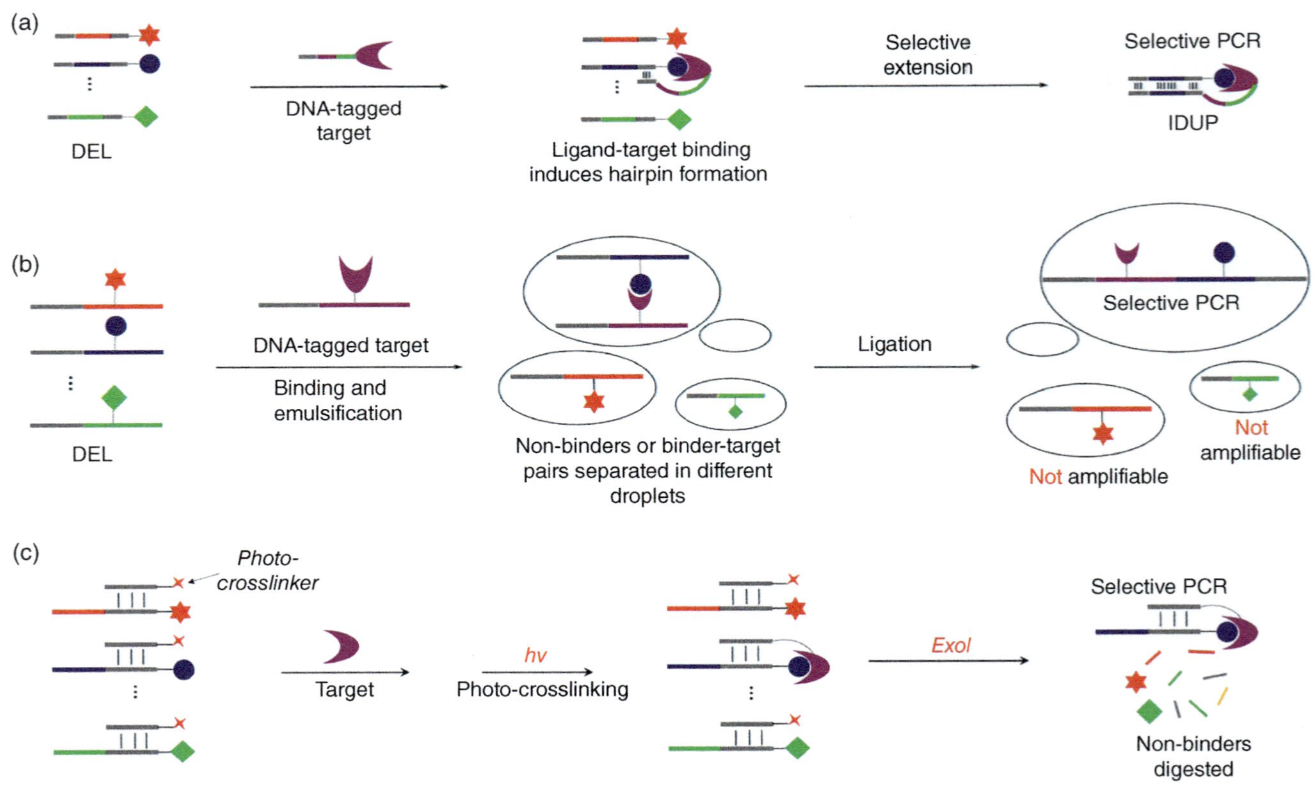

Figure 8.4 (a) IDUP. (b) Binder Trap Enrichment. (c) and (d) Two approaches based on ligand-directed photo-crosslinking strategy. (e) Substrate-mediated selection. (f) Selection method for targets within living cells. (g) Selection method for targets on surface of living cells.

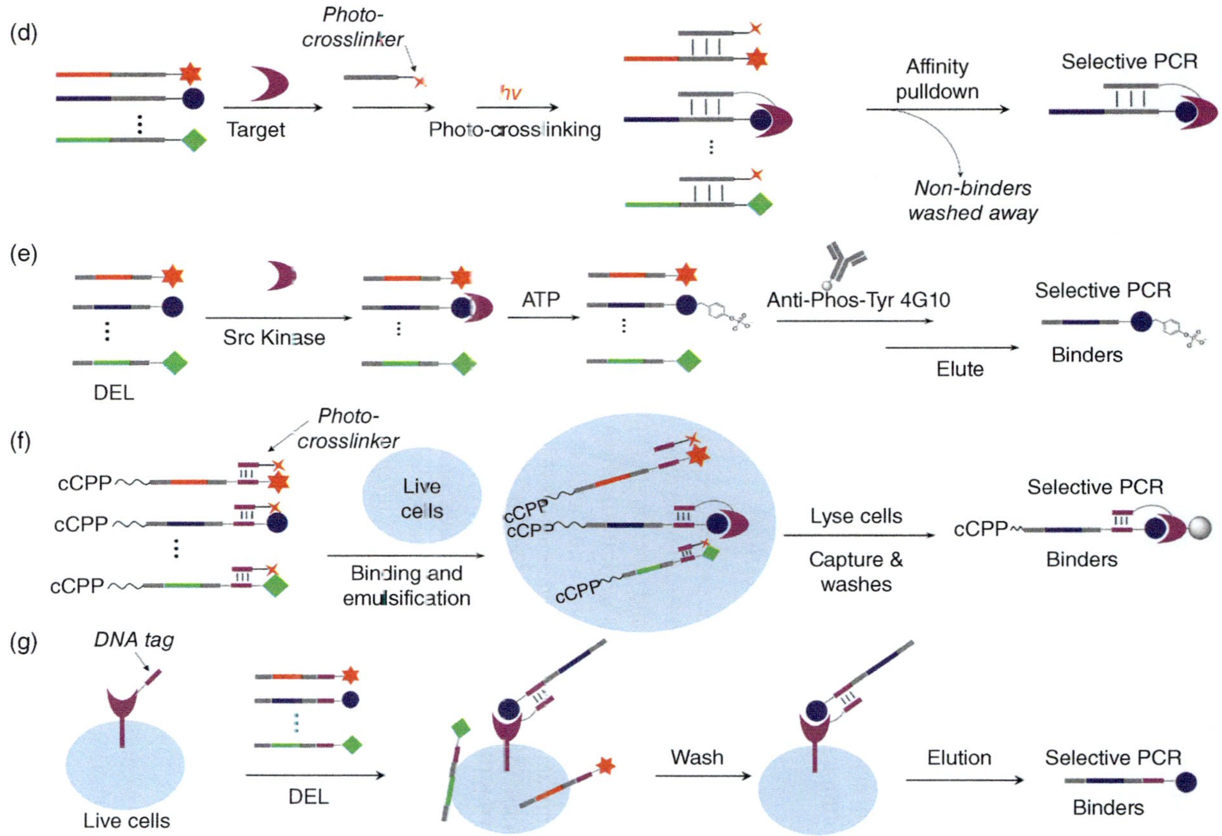

Figure 8.4 (*Continued*)

(c-Src), and (2) affinity-pulldown using specific antibody to identify DNA-linked phosphothiolates (Figure 8.4e) [79]. In addition, kinetic capillary electrophoresis has also been explored to partition the non-covalent ligand–protein complexes from non-binders [81, 82]. Finally, the dynamic DELs could be directly selected with soluble targets [46, 47].

Generally, selections with the immobilized target would be the method of choice for most researchers, provided the protein could be purified and modified without compromising its biological properties. For in-solution selections, one should consider whether the library in hand is architecturally compatible with the selection method. For example, the pre-tagging methods require special sequence design for the selective PCR amplification, while the crosslinking methods are less limited, since they utilize the common primer-binding site in the library. In addition, DELs encoded with dsDNAs cannot be directly used; conversion of the dsDNA tags to ssDNA is required prior to the selection in solution phase.

8.4.4. Cell-Based Selection

Cell surface membrane proteins are major classes of drug targets; however, they are also notoriously difficult to be expressed and purified while still maintaining the native structures and functions [146]. The ability to perform high-throughput screening with DELs against membrane protein targets on live cells is highly desired. DEL selections with the soluble domains [33, 83] and membrane proteins stabilized with detergent [83], nanodisc [84], or through genetic mutations [85] have been reported. In addition, GlaxoSmithKline selected a ~15-billion DEL against NK3 receptor [86] and Bradley and co-workers selected PNA-encoded libraries against chemokine receptor [87] and integrins [60] directly on live cells. Based on their previously reported two-step process including covalent crosslinking and affinity pulldown, Krusemark and co-workers reported a selection method of DEL against proteins within the cytosol and on the surface of live cells (Figure 8.4f) [88].

DEL selections on live cells face two issues: target specificity and target concentration. First, library interactions with other proteins or biomolecules on cell surface are inevitable; second, typically the effective concentration of membrane proteins on cell surface is in the low-nM range, significantly lower than the high-μM concentration usually required for DEL selection. Indeed, in the few examples of live-cell-based DEL selections, very high level of target expression is required, which may have limited the type of targets that can be selected. To address these issues, very recently, Li and co-workers developed a novel method to perform target-specific DEL selections against endogenous membrane proteins on live cells without overexpression or any other genetic manipulation, by specifically labeling protein of interest with a unique DNA anchor (Figure 8.4d) [89].

8.4.5. Selection for Covalent Inhibitors

Although a majority of selections aimed for non-covalent ligands, DELs could also be used to identify covalent inhibitors. In the early works by Winssinger, Schultz, and co-workers, PNA-encoded electrophilic probes were used to profile proteases in crude cell lysates [147–149]. Later on, the Winssinger group have employed PNA-encoded library to selected for covalent binders for kinases [150] and bromodomains [151]. The Neri group used the second-generation ESAC libraries to discover covalent inhibitors for JNK1 (c-Jun N-terminal kinase) [58]. With the in-solution selection method IDUP, Liu and co-workers discovered that ethacrynic acid, a known ligand for glutathione S-transferase, is also a covalent inhibitor for MAP2K6 kinase [74]. To discover covalent inhibitors, electrophilic "warheads," such as the α,β-unsaturated carbonyl motif, are often incorporated in the library to react with the nucleophilic side chains of lysines and cysteines, and selections are usually performed under stringent conditions to remove non-covalent binders, such as SDS washes [150] or heating [152].

8.5. APPLICATIONS OF DELS FOR HIT DISCOVERY

After selection, the DNA tags of the binders are PCR-amplified and surveyed by HTDS to determine fold of enrichment for each compound. With the ultra-large size of DELs, the automated processing of large sequencing data sets and data analysis tools become essential. In practice, each research group and organization seem to have their own methods in house, and there is no consensus on the general protocol. Recently, the Neri group has reported a valuable guide on data processing with great details [71]. The relative frequency of DNA tags is presumed to be corresponding to the relative abundance of the compounds; a comparison of the post-selection tag frequency with pre-selection (or control selection) tag frequency for all library compounds is statistically analyzed and plotted in a two- or three-dimension coordinate system (called "fingerprints") for hit picking. However, there is no generally accepted standards on how enrichment fold should be calculated and reported, criteria for hit picking and choosing the enrichment fold cut-off have been arbitrary, albeit the profound significance in finding high-affinity hits and avoiding artifacts. Very recently, Faver and co-workers reported a novel *z*-score enrichment metric, instead of simple enrichment fold, to set a more quantitative measure for DEL selection data analysis [153]. This method may solve a long-lasting issue and greatly improve the data analysis quality and hit deconvolution reliability in the field of DEL.

Figure 8.5 shows an example of the processed selection data for three sets of BBs in single-pharmacophore format [35]. The pre-selection HTDS data identify most of the library compounds with a uniform distribution of sequence counts, providing support for the homogeneity of the library. After the selection

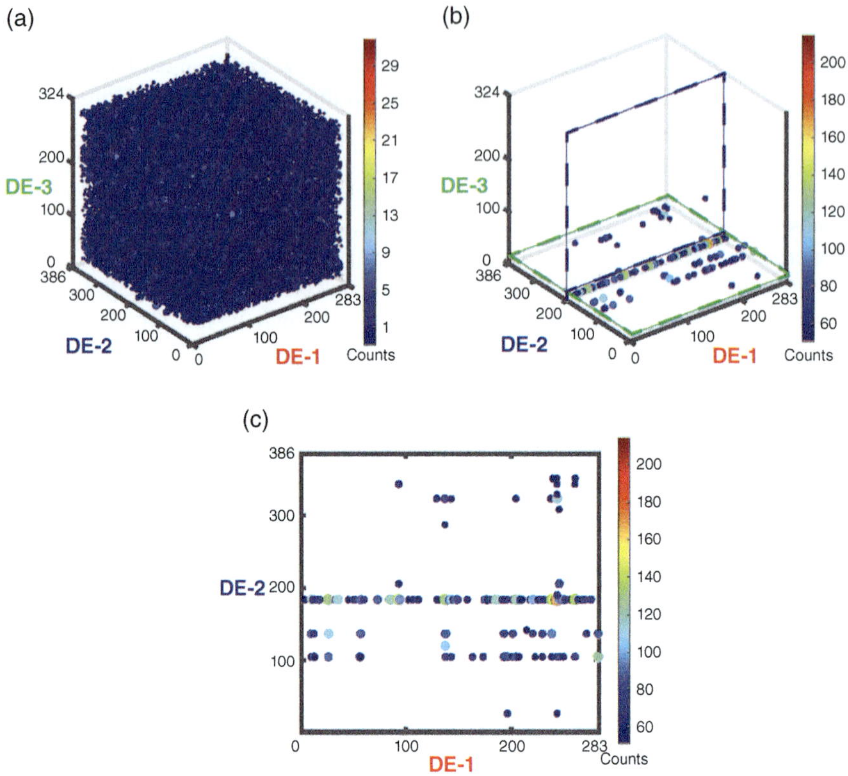

Figure 8.5 An example of DEL selection result analysis and plotted in "fingerprints." (a) Sequencing results of the pre-selection library. (b) and (c) are the same sequencing results of the post-selection library against CA II with 3-D or 2-D coordinate system. The axes represent the three BBs in the library (DE-1, DE-2, and DE-3); each dot represents a library member. Jet color scale represents the sequence counts.

against carbonic anhydrase 2 (CA II), the data plot clearly showed the binders and the BB distribution pattern. The relative enrichment fold is one of most important parameters for hits ranking; however, when picking the hits, one must also take consideration of other parameters, including SAR [65], target specificity [147], and purity of the library [57].

A wide variety of biological targets have been surveyed with DELs and numerous hits were isolated. Many of them are being further developed as drug candidates. Several reviews have comprehensively summarized the hits isolated from DELs in recent years [3, 27, 28, 154, 155]. Therefore, we update the hits isolated during the last two years from different designs of DELs or in different strategies of selections (Table 8.4).

As illustrated in initial proof-of-principle studies, the DEL seemed to be straightforward for building peptidic/peptidomimetic libraries and selecting against different protein surfaces and interfaces. The Krusemark group designed and screened various peptidic library DELs, and isolated peptide-like substrates/inhibitors to c-Src protein tyrosine kinase, CBX8 Chromodomain, and δ opioid receptor [79, 88, 156] (Table 8.4a–c). Recently, Brunschweiger and co-workers reported a DNA-encoded combinatorial peptoid library based on the Ugi four-component reaction by using indole side chains to probe the p53-binding domain of MDM2 and YAP-interacting domain of human TEAD4 (hTEAD4) (Table 8.4d) [157]. Using low-molecular-weight drug-like indole-based library, GSK developed a indoleamine 2,3-dioxygenase-1 (IDO1) inhibitor with moderate potency but high *in vivo* clearance (Table 8.4e) [158]. Recently, the Neri, Scheuermann, and co-workers reported a focused library based on a stereoisomeric scaffold with only 366,600 compounds and identified inhibitors against several pharmaceutically relevant proteins, including FANCD2-associated nuclease 1 (FAN1), human alpha-aminoadipic semialdehyde synthase (AASS), PI3K (phosphatidylinositol 3-kinase) (Table 8.4f) [140]. Researchers from Baylor college of medicine recently reported a 1-nM inhibitor of thrombin, which was identified from a protease-focused library featured with functional moieties such as guanidine, sulfonamide, urea, and carbamate (Table 8.4g) [159]. Based on three sets of BBs library design, Pfizer reported a potent and selective inhibitor to *N*-α -terminal acetyltransferase (Naa50) (an epigenetic regulator) (Table 8.4h) [160]. The Palzkill group developed a hit compound with submicromolar potency against oxacillinase-48 (OXA-48), which is one of problematic carbapenemases, from combinatorial triazine-based DEL within 162-million members (Table 8.4i) [161]. With follow-up structure optimization, Bassler and co-workers developed a catalytic inhibitor to PqsE, a thioesterase enzyme, which is vital for virulence of *Pseudomonas aeruginosa* that cause hospital-acquired infections, where the initial hits were isolated from a DEL of 550 million drug-like compounds (Table 8.4j) [162]. By combining fragment screening with DEL, researchers from GKS developed a bromodomain and extraterminal (BET) inhibitor with a low-predicted human dose through (Table 8.4k) [163]. Novel fragments with selectivity could be identified with dual-pharmacophore DELs for the affinity maturation purpose. The Neri group used the ESAC library and discovered a several new fragments for CREBBP bromodomain with selectivity against BRD4 (Table 8.4l) [164]. DEL appeared to be particularly suitable for building macrocycle libraries and selecting against large protein interfaces. Lim and co-workers isolated peptidic macrocycle binders against respiratory Skp2–Skp1 complex from a DEL with a DNA-encoded one-bead-one-compound library (Table 8.4m) [165]. A DEL resembling natural macrocycles and the retrosynthetic analysis strategy have also been described recently by group Gillingham (Table 8.4n) [68]. With their biological evolution, natural products have evolved with incredible structural diversities providing a fertile source for drug discovery. Yang, Lerner, and co-workers isolated inhibitors (luteoline and its analogue) for Poly(ADP–ribose) polymerase1 (PARP1), which binds to DNA in response to transient and plays

TABLE 8.4 A summary of identified hit compounds from recent DEL

Entry	institute(s)	Hit compound	Target	Affinity/Activity	Number of libraries	Total size	Structure	Cycles	Ref
(a)	Purdue University	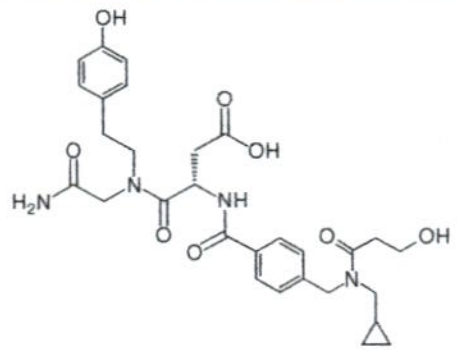	c-Src	IC_{50} = 48 μM by the inhibition of c-Src activity	1	5.5×10^5	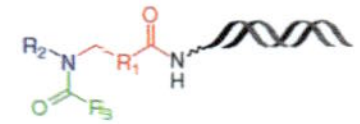	3	[79]
(b)	Purdue University	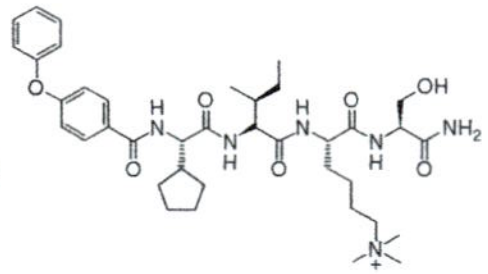	CBX8	K_d = 800 nM by FP	1	192	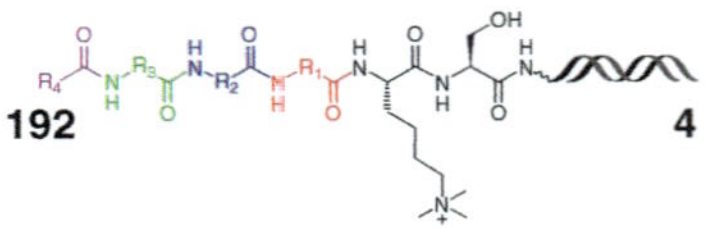	4	[156]
(c)	Purdue University	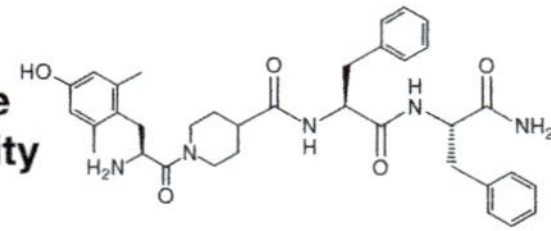	DOR	K_i = 2.7 nM by radioligand binding assay to DOR	1	96	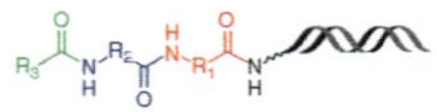	3	[88]
(d)	TU Dortmund University	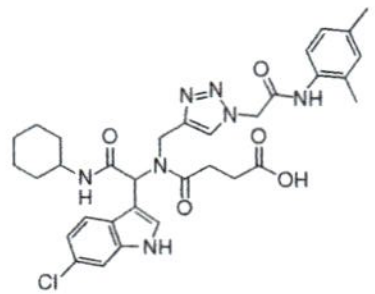	MDM2	K_d = 1.05 μM by SPR	1	8112	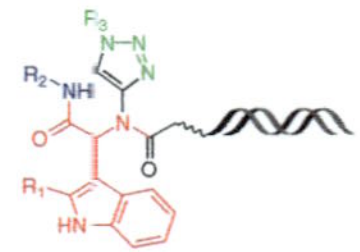	3	[157]

(e)	GlaxoSmith Kline	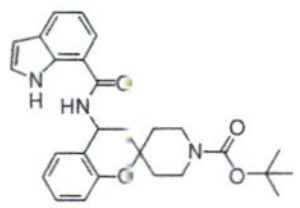	IDO1	IC_{50} = 251 nM by enzyme inhibition assays	1	3.6 million	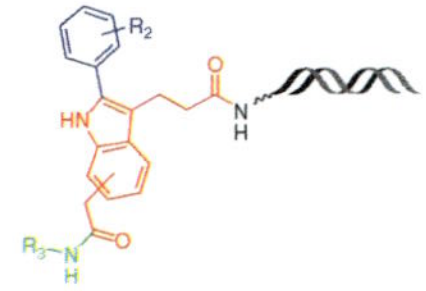	3	[158]
(f)	ETH Zürich	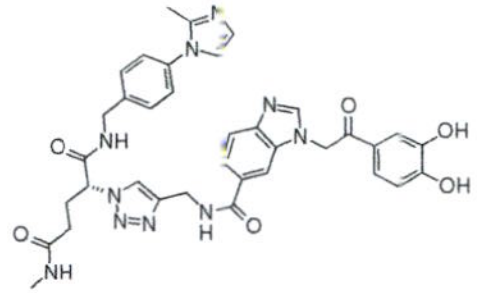	FAN1	K_d = 0.27 µM by ELISA binding assays					
		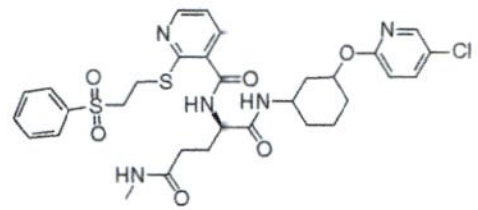	AASS	K_d = 0.7 µM by ELISA binding assays	1	366,600	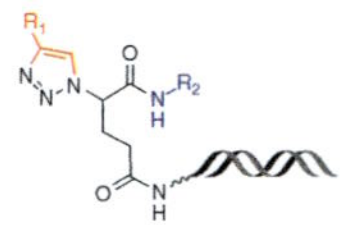	2	[140]
		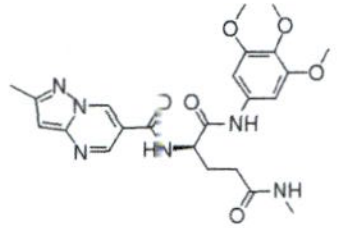	PI3K	K_d = 3.6 µM by ELISA binding assays					

(Continued)

TABLE 8.4 (Continued)

Entry	Institute(s)	Hit compound	Target	Affinity/Activity	Number of libraries	Total size	Structure	Cycles	Ref
(g)	Baylor College of Medicine	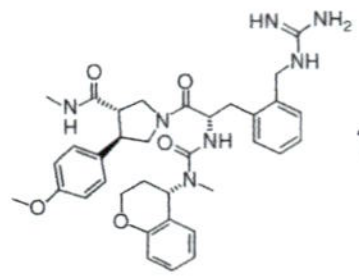	Thrombin	K_i= 1 nM in vitro inhibition assays	1	9.8 million	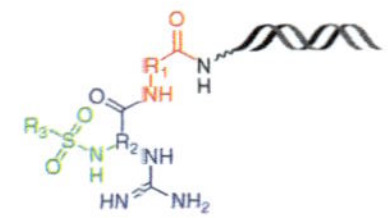	3	[159]
(h)	Pfizer and Hitgen	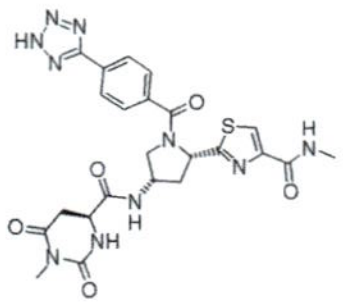	Naa50	IC_{50} = 7 nM by biochemical inhibitory activity	1	2.2 billion	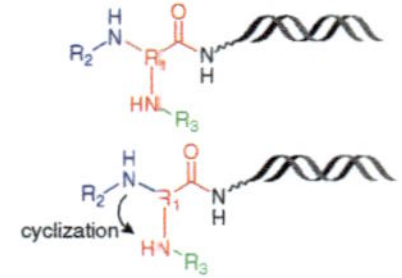	3	[160]
(i)	Baylor College of Medicine	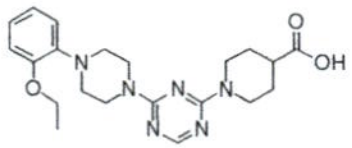	OXA-48	K_i= 053 µM in vitro inhibition assays	1	162 million	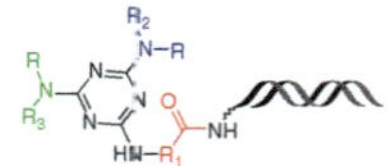	3	[161]
(j)	Princeton University	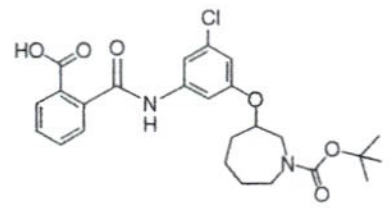	PqsE	IC_{50} = 5.6 nM by the ability to inhibit PqsE thioesterase activity	1	550 million	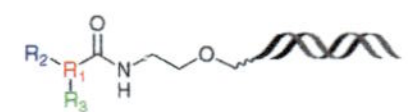	3	[162]
(k)	GlaxoSmith Kline	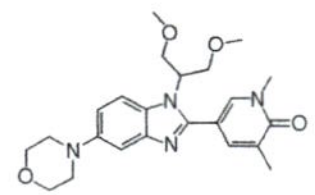	BRD4	pIC_{50} = 7.9 by FRET	1	117 million	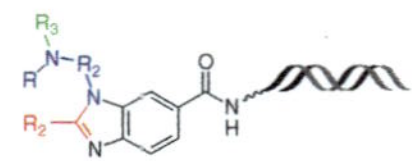	3	[163]

(l)	ETH Zürich	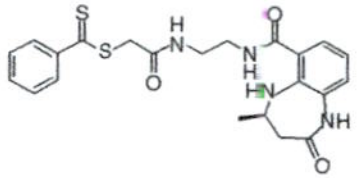	CREBBP Bromodomain	$K_d = 17\ \mu M$ by FP	1	788×425	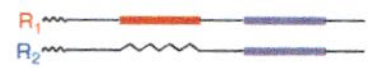	2	[164]
(m)	POSTECH	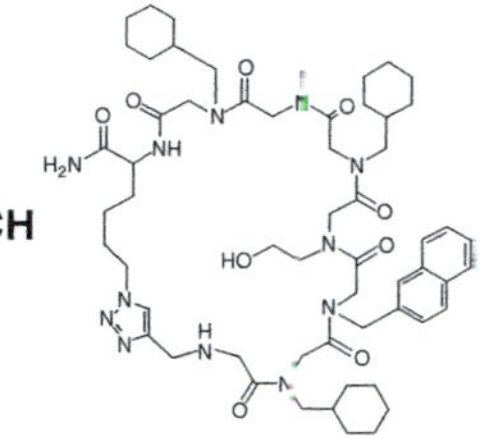	Skp2-Skp1	$K_d = 7.51\ \mu M$ by FA	1	11 million	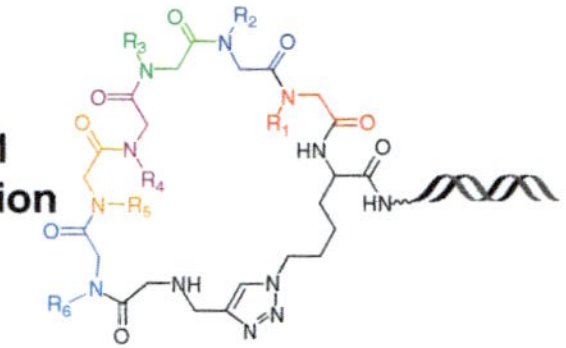	6	[165]
(n)	University of Basel	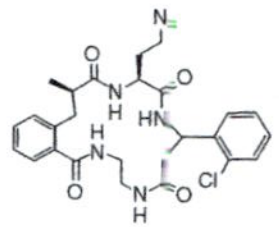	AGP	$K_d = 4.1\ \mu M$ by ITC	1	1.4×10^6	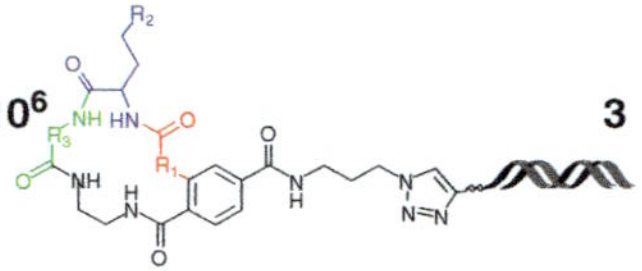	3	[68]
(o)	ShanghaiTech University	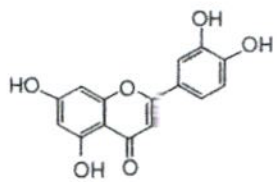	PARP 1	$IC_{50} = 7.5\ \mu M$ enzyme activity of PARP 1	1	10^4	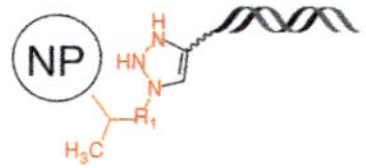	1	[51]
		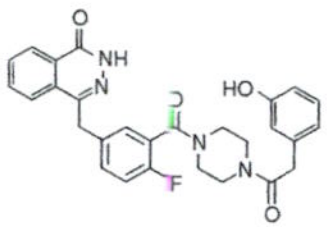	PARP 1	$IC_{50} = 2.5$ nM by enzyme activity of PARP 1					

(*Continued*)

TABLE 8.4 (Continued)

Entry institute(s)	Hit compound	Target	Affinity/Activity	Number of libraries	Total size	Structure	Cycles	Ref
(p) ShanghaiTech University	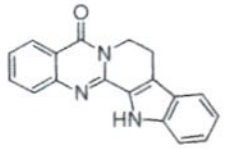	Insulin Receptor	$K_d = 14\,\mu M$ by SPR	1	10^4	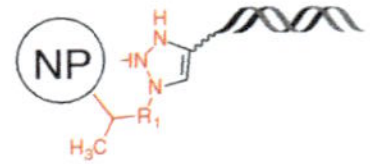	1	[166]
(q) GlaxoSmith Kline	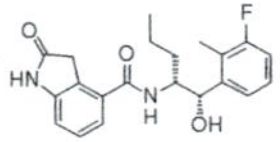	Z a1-antitrypsin	$pIC_{50} = 6.5$ by TR-FRET immunoassay	1	2×10^{12}	Not disclosed	1	[167]
	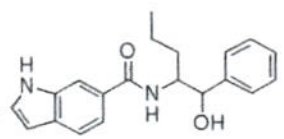	Z a1-antitrypsin	$pIC_{50} = 8.3$ by TR-FRET immunoassay					
(r) X-Chem	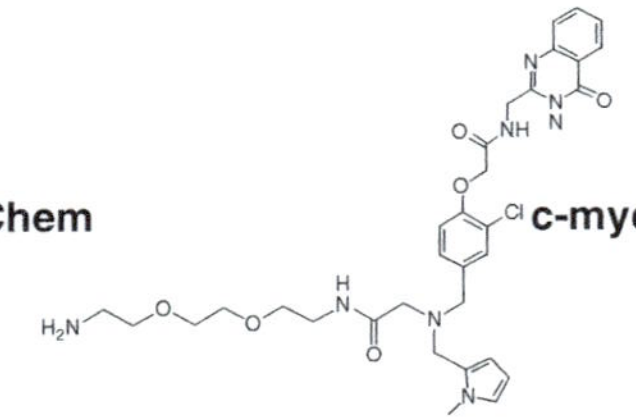	c-myc G4	$K_d = 1.05\,\mu M$ by SPR	33	120-billion	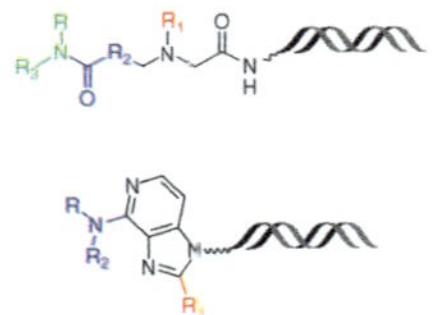	213	[168]

a significant role in a variety of biological processes from natural products enriched with DEL (Table 8.4o) [51]. With further selection, they reported a polycyclic nature product named Rutaecarpine, which is capable of both binding to and activating the insulin receptor (Table 8.4p) [166]. These studies have highlighted that DEL selection by using a relatively small but highly diverse library will also yield successful discovery of novel binders. In addition, in several studies where DEL structures were not disclosed, GSK reported the development of small molecules that correct misfolding and increases secretion of Z α_1-antitrypsin, based on the initial hit compound discovered form DEL (Table 8.4q) [167]. A large variety of protein targets have been interrogated with DELs; however, in principle other biomacromolecule of therapeutic interest could also be surveyed with DEL to isolate synthetic binders. X-Chem reported several novel nucleic-acid-binding small molecules, which identified from DEL selection against DNA targets (Table 8.4r) [168].

8.6. SUMMARY AND OUTLOOK

With around three decades' of development, DEL technology has reached a good level of maturity. It is especially satisfying to see the wide adoption of DEL by many major pharmaceutical companies. Experimental methods have been extensively developed and are well reported in the last few years. Nonetheless, we believe that progress in library design, DEL-compatible reactions, selection methods, and DNA decoding and hit-validation methodologies will surely open up new opportunities for DEL to exert its power and continue to make DEL more broadly accessible to the researchers in drug discovery. However, there still exist many important challenges. The proposal by Lerner and Brenner to make DEL an open source would certainly be a great way to explore this [17]; they propose a system where a pharmaceutical or a biotech company could provide DELs to an academic researcher through an intermediary who can facilitate the interactions between the library provider and the academic user. Companies usually have the resources to build large-scale DELs, whereas the academic users may have more novel targets or novel ways to use the libraries. To implementing this idea to practice, certain issues such as intellectual property arrangement, licensing relationship, material ownership and transfer, and profit sharing, etc. need to be addressed; however, putting them aside, an open-source DEL platform is expected to increase its efficiency in drug discovery and also facilitate the exploitation of its potential in academic research [169].

ACKNOWLEDGMENT

We thank all the research groups and companies mentioned for their work make the contents of this chapter possible.

REFERENCES

1. Shoichet, B. K., Walters, W. P., Jiang, H. et al. (2016). Advances in ComputationalMedicinal Chemistry: A Reflection on the Evolution of the Field and Perspective Going Forward. *Journal of Medicinal Chemistry* 59, 4033–4034.
2. Erlanson, D. A., Fesik, S. W., Hubbard, R. E. et al. (2016). Twenty Years on: The Impact of Fragments on Drug Discovery. *Nature Reviews Drug Discovery* 15, 605–619.
3. Goodnow, R. A., Jr., Dumelin, C. E. & Keefe, A. D. (2017). DNA-Encoded Chemistry: Enabling the Deeper Sampling of Chemical Space. *Nature Reviews Drug Discovery* 16, 131–147.
4. McCafferty, J., Griffiths, A. D., Winter, G. et al. (1990). Phage Antibodies: Filamentous Phage Displaying Antibody Variable Domains. *Nature* 348, 552–554.
5. Kang, A. S., Barbas, C. F., Janda, K. D. et al. (1991). Linkage of Recognition and Replication Functions by Assembling Combinatorial Antibody Fab Libraries Along Phage Surfaces. *Proceedings of the National Academy of Sciences of the United States of America* 88, 4363–4366.
6. Clackson, T., Hoogenboom, H. R., Griffiths, A. D. et al. (1991). Making Antibody Fragments Using Phage Display Libraries. *Nature* 352, 624–628.
7. Wilson, D. S., Keefe, A. D. & Szostak, J. W. (2001). The Use of mRNA Display to Select High-Affinity Protein-Binding Peptides. *Proceedings of the National Academy of Sciences of the United States of America* 98, 3750–3755.
8. Boder, E. T. & Wittrup, K. D. (1997). Yeast Surface Display for Screening Combinatorial Polypeptide Libraries. *Nature Biotechnology* 15, 553–557.
9. Hanes, J. & Pluckthun, A. (1997). in vitro Selection and Evolution of Functional Proteins by Using Ribosome Display. *Proceedings of the National Academy of Sciences of the United States of America* 94, 4937–4942.
10. Morimoto, J., Hayashi, Y., Iwasaki, K. et al. (2011). Flexizymes: Their Evolutionary History and the Origin of Catalytic Function. *Accounts of Chemical Research* 44, 1359–1368.
11. Deyle, K., Kong, X.-D. & Heinis, C. (2017). Phage Selection of Cyclic Peptides for Application in Research and Drug Development. *Accounts of Chemical Research* 50, 1866–1874.
12. Ng, S., Jafari, M. R. & Derda, R. (2012). Bacteriophages and Viruses as a Support for Organic Synthesis and Combinatorial Chemistry. *ACS Chemical Biology* 7, 123–138.
13. NCT02903966 GSK2982772 Study in Subjects with Ulcerative Colitis. https://clinicaltrials.gov/ct2/show/NCT02903966?term=GSK2982772&draw=2&rank=4 (accessed 20 August 2022).
14. Harris, P. A., Berger, S. B., Jeong, J. U. et al. (2017). Discovery of a First-in-Class Receptor Interacting Protein 1 (RIP1) Kinase Specific Clinical Candidate (GSK2982772) for the Treatment of Inflammatory Diseases. *Journal of Medicinal Chemistry* 60, 1247–1261.
15. NCT03486223 Soluble Epoxide Hydrolase Inhibition and Insulin Resistance. https://clinicaltrials.gov/ct2/show/NCT03486223?term=GSK2256294 (accessed 20 August 2022).

16. Cuozzo, J. W., Clark, M. A., Keefe, A. D. et al. (2020). Novel Autotaxin Inhibitor for the Treatment of Idiopathic Pulmonary Fibrosis: A Clinical Candidate Discovered Using DNA-Encoded Chemistry. *Journal of Medicinal Chemistry* 63, 7840–7856.
17. Lerner, R. A. & Brenner, S. (2017). DNA-Encoded Compound Libraries as Open Source: A Powerful Pathway to NewDrugs. *Angewandte Chemie-International Edition* 56, 1164–1165.
18. Brenner, S. & Lerner, R. A. (1992). Encoded Combinatorial Chemistry. *Proceedings of the National Academy of Sciences of the United States of America* 89, 5381–5383.
19. Needels, M. C., Jones, D. G., Tate, E. H. et al. (1993). Generation and Screening of an Oligonucleotide-Encoded Synthetic Peptide Library. *Proceedings of the National Academy of Sciences of the United States of America* 90, 10700–10704.
20. Nielsen, J., Brenner, S. & Janda, K. D. (1993). Synthetic Methods for the Implementation of Encoded Combinatorial Chemistry. *Journal of the American Chemical Society* 115, 9812–9813.
21. Kinoshita, Y. & Nishigaki, K. (1995). Enzymatic Synthesis of Code Regions for Encoded Combinatorial Chemistry (ECC). *Nucleic Acids Symposium Series*, 344, 201–202.
22. Melkko, S., Scheuermann, J., Dumelin, C. E. et al. (2004). Encoded Self-Assembling Chemical Libraries. *Nature Biotechnology* 22, 568–574.
23. Gartner, Z. J., Tse, B. N., Grubina, R. et al. (2004). DNA-Templated Organic Synthesis and Selection of a Library of Macrocycles. *Science* 305, 1601–1605.
24. Halpin, D. R. & Harbury, P. B. (2004). DNA Display I. Sequence-Encoded Routing of DNA Populations. *Plos Biology* 2, 1015–1021.
25. Goodnow, R. A., Jr. (2014). A Handbook for DNA-Encoded Chemistry: Theory and Applications for Exploring Chemical Space and Drug Discovery. Hoboken (NJ): John Wiley & Sons.
26. Zambaldo, C., Barluenga, S. & Winssinger, N. (2015). PNA-Encoded Chemical Libraries. *Current Opinion in Chemical Biology* 26, 8–15.
27. Franzini, R. M. & Randolph, C. (2016). Chemical Space of DNA-Encoded Libraries. *Journal of Medicinal Chemistry* 59, 6629–6644.
28. Neri, D. & Lerner, R. A. (2018). DNA-Encoded Chemical Libraries: A Selection System Based on Endowing Organic Compounds with Amplifiable Information. *Annual Review of Biochemistry* 87, 479–502.
29. Song, M. & Hwang, G. T. (2020). DNA-Encoded Library Screening as Core Platform Technology in Drug Discovery: Its Synthetic Method Development and Applications in DEL Synthesis. *Journal of Medicinal Chemistry* 63, 6578–6599.
30. Kunig, V., Potowski, M., Gohla, A. et al. (2018). DNA-Encoded Libraries – An Efficient Small Molecule Discovery Technology for the Biomedical Sciences. *Biological Chemistry* 399, 691–710.
31. Clark, M. A., Acharya, R. A., Arico-Muendel, C. C. et al. (2009). Design, Synthesis and Selection of DNA-Encoded Small-Molecule Libraries. *Nature Chemical Biology* 5, 647–654.
32. Buller, F., Zhang, Y., Scheuermann, J. et al. (2009). Discovery of TNF Inhibitors from a DNA-Encoded Chemical Library based on Diels-Alder Cycloaddition. *Chemistry & Biology* 16, 1075–1086.
33. Leimbacher, M., Zhang, Y., Mannocci, L. et al. (2012). Discovery of Small-Molecule Interleukin-2 Inhibitors from a DNA-Encoded Chemical Library. *Chemistry-a European Journal* 18, 7729–7737.

34. Wichert, M., Krall, N., Decurtins, W. et al. (2015). Dual-Display of Small Molecules Enables the Discovery of Ligand Pairs and Facilitates Affinity Maturation. *Nature Chemistry* 7, 241–249.
35. Li, Y., De Luca, R., Cazzamalli, S. et al. (2018). Versatile Protein Recognition by the Encoded Display of Multiple Chemical Elements on a Constant Macrocyclic Scaffold. *Nature Chemistry* 10, 441–448.
36. Litovchick, A., Clark, M. A. & Keefe, A. D. (2014). Universal Strategies for the DNA-Encoding of Libraries of Small Molecules Using the Chemical Ligation of Oligonucleotide Tags. *Artificial DNA, PNA & XNA* 5, e27896.
37. Tse, B. N., Snyder, T. M., Shen, Y. et al. (2008). Translation of DNA into a Library of 13,000 Synthetic Small-Molecule Macrocycles Suitable for I*n Vitro* Selection. *Journal of the American Chemical Society* 130, 15611–15626.
38. Usanov, D. L., Chan, A. I., Maianti, J. P. et al. (2018). Second-Generation DNA-Templated Macrocycle Libraries for the Discovery of Bioactive Small Molecules. *Nature Chemistry* 10, 704–714.
39. Li, Y., Zhao, P., Zhang, M. et al. (2013). Multistep DNA-Templated Synthesis Using a Universal Template. *Journal of the American Chemical Society* 135, 17727–17730.
40. Krusemark, C. J., Tilmans, N. P., Brown, P. O. et al. (2016). Directed Chemical Evolution with an Outsized Genetic Code. *Plos One* 11, e0154764.
41. Wrenn, S. J., Weisinger, R. M., Halpin, D. R. et al. (2007). Synthetic Ligands Discovered by *In vitro* Selection. *Journal of the American Chemical Society* 129, 13137–13143.
42. Melkko, S., Zhang, Y., Dumelin, C. E. et al. (2007). Isolation of High-Affinity Trypsin Inhibitors from a DNA-Encoded Chemical Library. *Angewandte Chemie-International Edition* 46, 4671–4674.
43. Scheuermann, J., Dumelin, C. E., Melkko, S. et al. (2008). DNA-Encoded Chemical Libraries for the Discovery of MMP-3 Inhibitors. *Bioconjugate Chemistry* 19, 778–785.
44. Reddavide, F. V., Lin, W., Lehnert, S. et al. (2015). DNA-Encoded Dynamic Combinatorial Chemical Libraries. *Angewandte Chemie-International Edition* 54, 7924–7928.
45. Reddavide, F. V., Cui, M., Lin, W. et al. (2019). Second Generation DNA-Encoded Dynamic Combinatorial Chemical Libraries. *Chemical Communications* 55, 3753–3756.
46. Li, G., Zheng, W., Chen, Z. et al. (2015). Design, Preparation, and Selection of DNA-Encoded Dynamic Libraries. *Chemical Science* 6, 7097–7104.
47. Zhou, Y., Li, C., Peng, J. et al. (2018). DNA-Encoded Dynamic Chemical Library and Its Applications in Ligand Discovery. *Journal of the American Chemical Society* 140, 15859–15867.
48. Hansen, M. H., Blakskjaer, P., Petersen, L. K. et al. (2009). A Yoctoliter-Scale DNA Reactor for Small-Molecule Evolution. *Journal of the American Chemical Society* 131, 1322–1327.
49. Daguer, J. P., Ciobanu, M., Alvarez, S. et al. (2011). DNA-Templated Combinatorial Assembly of Small Molecule Fragments Amenable to Selection/Amplification Cycles. *Chemical Science* 2, 625–632.
50. Daguer, J. P., Zambaldo, C., Ciobanu, M. et al. (2015). DNA Display of Fragment Pairs as a Tool for the Discovery of Novel Biologically Active Small Molecules. *Chemical Science* 6, 739–744.

51. Ma, P., Xu, H., Li, J. et al. (2019). Functionality-Independent DNA Encoding of Complex Natural Products. *Angewandte Chemie-International Edition* 58, 9254–9261.
52. MacConnell, A. B., McEnaney, P. J., Cavett, V. J. et al. (2015). DNA-Encoded Solid-Phase Synthesis: Encoding Language Design and Complex Oligomer Library Synthesis. *ACS Combinatorial Science* 17, 518–534.
53. Li, X. Y. & Liu, D. R. (2004). DNA-Templated Organic Synthesis: nature's Strategy for Controlling Chemical Reactivity Applied to Synthetic Molecules. *Angewandte Chemie-International Edition* 43, 4848–4870.
54. Weisinger, R. M., Wrenn, S. J. & Harbury, P. B. (2012). Highly Parallel Translation of DNA Sequences into Small Molecules. *Plos One* 7, e28056.
55. Weisinger, R. M., Marinelli, R. J., Wrenn, S. J. et al. (2012). Mesofluidic Devices for DNA-Programmed Combinatorial Chemistry. *Plos One* 6, e32299.
56. Scheuermann, J. & Neri, D. (2015). Dual-Pharmacophore DNA-Encoded Chemical Libraries. *Current Opinion in Chemical Biology* 26, 99–103.
57. Satz, A. L. (2016). Simulated Screens of DNA Encoded Libraries: The Potential Influence of Chemical Synthesis Fidelity on Interpretation of Structure Activity Relationships. *ACS Combinatorial Science* 18, 415–424.
58. Zimmermann, G., Rieder, U., Bajic, D. et al. (2017). A Specific and Covalent JNK-1 Ligand Selected from an Encoded Self-Assembling Chemical Library. *Chemistry-a European Journal* 23, 8152–8155.
59. Bigatti, M., Dal Corso, A., Vanetti, S. et al. (2017). Impact of a Central Scaffold on the Binding Affinity of Fragment Pairs Isolated from DNA-Encoded Self-Assembling Chemical Libraries. *Chemmedchem* 12, 1748–1752.
60. Svensen, N., Diaz-Mochon, J. J. & Bradley, M. (2011). Decoding a PNA Encoded Peptide Library by PCR: The Discovery of New Cell Surface Receptor Ligands. *Chemistry & Biology* 18, 1284–1289.
61. Satz, A. L. (2015). DNA Encoded Library Selections and Insights Provided by Computational Simulations. *ACS Chemical Biology* 10, 2237–2245.
62. Satz, A. L., Hochstrasser, R. & Petersen, A. C. (2017). Analysis of Current DNA Encoded Library Screening Data Indicates Higher False Negative Rates for Numerically Larger Libraries. *ACS Combinatorial Science* 19, 234–238.
63. Kuai, L., O'Keeffe, T. & Arico-Muendel, C. (2018). Randomness in DNA Encoded Library Selection Data Can Be Modeled for More Reliable Enrichment Calculation. *Slas Discovery* 23, 405–416.
64. Eidam, O. & Satz, A. L. (2016). Analysis of the Productivity of DNA Encoded Libraries. *Medchemcomm* 7, 1323–1331.
65. Franzini, R. M., Ekblad, T., Zhong, N. et al. (2015). Identification of Structure-Activity Relationships from Screening a Structurally Compact DNA-Encoded Chemical Library. *Angewandte Chemie-International Edition* 54, 3927–3931.
66. Franzini, R. M., Biendl, S., Mikutis, G. et al. (2015). "Cap-and-Catch" Purification for Enhancing the Quality of Libraries of DNA Conjugates. *ACS Combinatorial Science* 17, 393–398.
67. Yuen, L. H., Dana, S., Liu, Y. et al. (2019). A Focused DNA-Encoded Chemical Library for the Discovery of Inhibitors of NAD(+)-Dependent Enzymes. *Journal of the American Chemical Society* 141, 5169–5181.
68. Stress, C. J., Sauter, B., Schneider, L. A. et al. (2019). A DNA-Encoded Chemical Library Incorporating Elements of Natural Macrocycles. *Angewandte Chemie-International Edition* 58, 9570–9574.

69. Mendes, K. R., Malone, M. L., Ndungu, J. M. et al. (2017). High-Throughput Identification of DNA-Encoded IgG Ligands that Distinguish Active and Latent Mycobacterium Tuberculosis Infections. *ACS Chemical Biology* 12, 234–243.
70. Erharuyi, O., Simanski, S., McEnaney, P. J. et al. (2018). Screening One Bead One Compound Libraries Against Serum Using a Flow Cytometer: Determination of the Minimum Antibody Concentration Required for Ligand Discovery. *Bioorganic & Medicinal Chemistry Letters* 28, 2773–2778.
71. Decurtins, W., Wichert, M., Franzini, R. M. et al. (2016). Automated Screening for Small Organic Ligands Using DNA-Encoded Chemical Libraries. *Nature Protocols* 11, 764–780.
72. McGregor, L. M., Jain, T. & Liu, D. R. (2014). Identification of Ligand-Target Pairs from Combined Libraries of Small Molecules and Unpurified Protein Targets in Cell Lysates. *Journal of the American Chemical Society* 136, 3264–3270.
73. McGregor, L. M., Gorin, D. J., Dumelin, C. E. et al. (2010). Interaction-Dependent PCR: Identification of Ligand-Target Pairs from Libraries of Ligands and Libraries of Targets in a Single Solution-Phase Experiment. *Journal of the American Chemical Society* 132, 15522–15524.
74. Chang, A. I., McGregor, L. M., Jain, T. et al. (2017). Discovery of a Covalent Kinase Inhibitor from a DNA-Encoded Small Molecule Library x Protein Library Selection. *Journal of the American Chemical Society* 139, 10192–10195.
75. Denton, K. E. & Krusemark, C. J. (2016). Crosslinking of DNA-Linked Ligands to Target Proteins for Enrichment from DNA-Encoded Libraries. *Medchemcomm* 7, 2020–2027.
76. Zhao, P., Chen, Z., Li, Y. et al. (2014). Selection of DNA-Encoded Small Molecule Libraries Against Unmodified and Non-Immobilized Protein Targets. *Angewandte Chemie-International Edition* 53, 10056–10059.
77. Shi, B., Deng, Y. & Li, X. (2019). Polymerase-Extension-Based Selection Method for DNA-Encoded Chemical Libraries against Nonimmobilized Protein Targets. *ACS Combinatorial Science* 21, 345–349.
78. Shi, B., Deng, Y., Zhao, P. et al. (2017). Selecting a DNA-Encoded Chemical Library against Non-immobilized Proteins Using a "Ligate-Cross-Link-Purify" Strategy. *Bioconjugate Chemistry* 28, 2293–2301.
79. Kim, D., Sun, Y., Xie, D. et al. (2019). Application of a Substrate-Mediated Selection with c-Src Tyrosine Kinase to a DNA-Encoded Chemical Library. *Molecules* 24, 2764.
80. Blakskjaer, P., Christensen, A. B., Hansen, N. J., et al., (2012). A Method for Making an Enriched Library Patent. WO2012041633.
81. Bao, J., Krylova, S. M., Cherney, L. T. et al. (2016). Predicting Electrophoretic Mobility of Protein-Ligand Complexes for Ligands from DNA-Encoded Libraries of Small Molecules. *Analytical Chemistry* 88, 5498–5506.
82. Kochmann, S., Le, A. T. H., Hili, R. et al. (2018). Predicting Efficiency of NECEEM-Based Partitioning of Protein Binders from Nonbinders in DNA-Encoded Libraries. *Electrophoresis* 39, 2991–2996.
83. Kollmann, C. S., Bai, X., Tsai, C.-H. et al. (2014). Application of Encoded Library Technology (ELT) to a Protein-Protein Interaction Target: Discovery of a Potent Class of Integrin Lymphocyte Function-Associated Antigen 1 (LFA-1) Antagonists. *Bioorganic & Medicinal Chemistry* 22, 2353–2365.

84. Ahn, S., Pani, B., Kahsai, A. W. et al. (2018). Small-Molecule Positive Allosteric Modulators of the beta(2)-Adrenoceptor Isolated from DNA-Encoded Libraries. *Molecular Pharmacology* 94, 850–861.
85. Brown, D. G., Brown, G. A., Centrella, P. et al. (2018). Agonists and Antagonists of Protease-Activated Receptor 2 Discovered within a DNA-Encoded Chemical Library Using Mutational Stabilization of the Target. *Slas Discovery* 23, 429–436.
86. Wu, Z., Graybill, T. L., Zeng, X. et al. (2015). Cell-Based Selection Expands the Utility of DNA-Encoded Small-Molecule Library Technology to Cell Surface Drug Targets: Identification of Novel Antagonists of the NK3 Tachykinin Receptor. *ACS Combinatorial Science* 17, 722–731.
87. Svensen, N., Diaz-Mochon, J. J. & Bradley, M. (2011). Encoded Peptide Libraries and the Discovery of New Cell Binding Ligands. *Chemical Communications* 47, 7638–7640.
88. Cai, B., Kim, D., Akhand, S. et al. (2019). Selection of DNA-Encoded Libraries to Protein Targets within and on Living Cells. *Journal of the American Chemical Society* 141, 17057–17061.
89. Huang, Y., Meng, L., Nie, Q. et al. (2020). Selection of DNA-Encoded Chemical Libraries Against Endogenous Membrane Proteins on Live Cells. *Nature Chemistry* 13, 77–88.
90. Bobers, J., Skopic, M. K., Dinter, R. et al. (2020). Design of an Automated Reagent-Dispensing System for Reaction Screening and Validation with DNA-Tagged Substrates. *ACS Combinatorial Science* 22, 101–108.
91. Malone, M. L. & Paegel, B. M. (2016). What is a "DNA-Compatible" Reaction? *ACS Combinatorial Science* 18, 182–187.
92. Pels, K., Dickson, P., An, H. et al. (2018). DNA-Compatible Solid-Phase Combinatorial Synthesis of beta-Cyanoacrylamides and Related Electrophiles. *ACS Combinatorial Science* 20, 61–69.
93. Nam, T.-H. & Kodadek, T. (2018). Solid-Phase Synthesis of beta-Amino Ketones Via DNA-Compatible Organocatalytic Mannich Reactions. *ACS Combinatorial Science* 20, 55–60.
94. Shu, K. & Kodadek, T. (2018). Solid-Phase Synthesis of beta-Hydroxy Ketones Via DNA Compatible Organocatalytic Aldol Reactions. *ACS Combinatorial Science* 20, 277–281.
95. Kunig, V. B. K., Ehrt, C., Domling, A. et al. (2019). Isocyanide Multicomponent Reactions on Solid Phase Coupled DNA Oligonucleotides for Encoded Library Synthesis. *Organic Letters* 21, 7238–7243.
96. Potowski, M., Kunig, V. B. K., Losch, F. et al. (2019). Synthesis of DNA-Coupled Isoquinolones and Pyrrolidines by Solid Phase Ytterbium- and Silver-Mediated Imine Chemistry. *Medchemcomm* 10, 1082–1093.
97. Flood, D. T., Zhang, X., Fu, X. et al. (2020). RASS-Enabled S/P-C and S-N Bond Formation for DEL Synthesis. *Angewandte Chemie-International Edition* 59, 7377–7383.
98. Ruff, Y., Martinez, R., Pelle, X. et al. (2020). An Amphiphilic Polymer-Supported Strategy Enables Chemical Transformations under Anhydrous Conditions for DNA-Encoded Library Synthesis. *ACS Combinatorial Science* 22, 120–128.
99. Satz, A. L., Cai, J., Chen, Y. et al. (2015). DNA Compatible Multistep Synthesis and Applications to DNA Encoded Libraries. *Bioconjugate Chemistry* 26, 1623–1632.

100. Goette, K., Chines, S. & Brunschweiger, A. (2020). Reaction Development for DNA-Encoded Library Technology: From Evolution to Revolution? *Tetrahedron Letters* 61, 151889.
101. Shi, Y., Wu, Y.-r., Yu, J.-q. et al. (2021). DNA-Encoded Libraries (DELs): A Review of on-DNA Chemistries and Their Output. *RSC Advances* 11, 2359–2376.
102. Fan, L. & Davie, C. P. (2017). Zirconium(IV)-Catalyzed Ring Opening of on-DNA Epoxides in Water. *Chembiochem* 18, 843–847.
103. Lu, X., Fan, L., Phelps, C. B. et al. (2017). Ruthenium Promoted On-DNA Ring-Closing Metathesis and Cross-Metathesis. *Bioconjugate Chemistry* 28, 1625–1629.
104. Skopic, M. K., Salamon, H., Bugain, O. et al. (2017). Acid- and Au(I)-Mediated Synthesis of Hexathymidine-DNA-Heterocycle Chimeras, an Efficient Entry to DNA-Encoded Libraries Inspired by Drug Structures. *Chemical Science* 8, 3356–3361.
105. Skopic, M. K., Willems, S., Wagner, B. et al. (2017). Exploration of a Au(I)-Mediated Three-Component Reaction for the Synthesis of DNA-Tagged Highly Substituted Spiroheterocycles. *Organic & Biomolecular Chemistry* 15, 8648–8654.
106. Gerry, C. J., Yang, Z., Stasi, M. et al. (2019). DNA-Compatible 3+2 Nitrone-Olefin Cycloaddition Suitable for DEL Syntheses. *Organic Letters* 21, 1325–1330.
107. Cai, P., Yang, G., Zhao, L. et al. (2019). Synthesis of C3-Alkylated Indoles on DNA via Indolyl Alcohol Formation Followed by Metal-Free Transfer Hydrogenation. *Organic Letters* 21, 6633–6637.
108. Brow, D. G. & Bostrom, J. (2016). Analysis of Past and Present Synthetic Methodologies on Medicinal Chemistry: Where Have All the New Reactions Gone? *Journal of Medicinal Chemistry* 59, 4443–4458.
109. Kolmel, D. K., Loach, R. P., Knauber, T. et al. (2018). Employing Photoredox Catalysis for DNA-Encoded Chemistry: Decarboxylative Alkylation of alpha-Amino Acids. *Chemmedchem* 13, 2159–2165.
110. Kolmel, D. K., Meng, J., Tsai, M.-H. et al. (2019). On-DNA Decarboxylative Arylation: Merging Photoredox with Nickel Catalysis in Water. *ACS Combinatorial Science* 21, 588–597.
111. Kolmel, D. K., Ratnayake, A. S. & Flanagan, M. E. (2020). Photoredox Cross-Electrophile Coupling in DNA-Encoded Chemistry. *Biochemical and Biophysical Research Communications* 533, 201–208.
112. Wang, J., Lundberg, H., Asai, S. et al. (2018). Kinetically Guided Radical-Based Synthesis of C(sp(3))-C(sp(3)) Linkages on DNA. *Proceedings of the National Academy of Sciences of the United States of America* 115, E6404–E6410.
113. Phelan, J. P., Lang, S. B., Sim, J. et al. (2019). Open-Air Alkylation Reactions in Photoredox-Catalyzed DNA-Encoded Library Synthesis. *Journal of the American Chemical Society* 141, 3723–3732.
114. Badir, S. O., Sim, J., Billings, K. et al. (2020). Multifunctional Building Blocks Compatible with Photoredox-Mediated Alkylation for DNA-Encoded Library Synthesis. *Organic Letters* 22, 1046–1051.
115. Wang, X., Sun, H., Liu, J. et al. (2018). Ruthenium-Promoted C-H Activation Reactions between DNA-Conjugated Acrylamide and Aromatic Acids. *Organic Letters* 20, 4764–4768.
116. Fan, Z., Zhao, S., Liu, T. et al. (2020). Merging C(sp(3))-H activation with DNA-encoding. *Chemical Science* 11, 12282–12288.

117. Kolmel, D. K., Ratnayake, A. S., Flanagan, M. E. et al. (2020). Photocatalytic 2+2 Cycloaddition in DNA-Encoded Chemistry. *Organic Letters* 22, 2908–2913.
118. Wen, X., Duan, Z., Liu, J. et al. (2020). On-DNA Cross-Dehydrogenative Coupling Reaction toward the Synthesis of Focused DNA-Encoded Tetrahydroisoquinoline Libraries. *Organic Letters* 22, 5721–5725.
119. Wen, H., Ge, R., Qu, Y. et al. (2020). Synthesis of 1,2-Amino Alcohols by Photoredox-Mediated Decarboxylative Coupling of alpha-Amino Acids and DNA-Conjugated Carbonyls. *Organic Letters* 22, 9484–9489.
120. Li, J.-Y. & Huang, H. (2018). Development of DNA-Compatible Suzuki-Miyaura Reaction in Aqueous Media. *Bioconjugate Chemistry* 29, 3841–3846.
121. Gerry, C. J., Wawer, M. J., Clemons, P. A. et al. (2019). DNA Barcoding a Complete Matrix of Stereoisomeric Small Molecules. *Journal of the American Chemical Society* 141, 10225–10235.
122. Ding, Y., DeLorey, J. L. & Clark, M. A. (2016). Novel Catalyst System for Suzuki-Miyaura Coupling of Challenging DNA-Linked Aryl Chlorides. *Bioconjugate Chemistry* 27, 2597–2600.
123. Ding, Y., Franklin, G. J., DeLorey, J. L. et al. (2016). Design and Synthesis of Biaryl DNA-Encoded Libraries. *ACS Combinatorial Science* 18, 625–629.
124. Hunter, J. H., Prendergast, L., Valente, L. F. et al. (2020). High Fidelity Suzuki-Miyaura Coupling for the Synthesis of DNA Encoded Libraries Enabled by Micelle Forming Surfactants. *Bioconjugate Chemistry* 31, 149–155.
125. Wang, X., Sun, H., Liu, J. et al. (2019). Palladium-Promoted DNA-Compatible Heck Reaction. *Organic Letters* 21, 719–723.
126. Xu, H., Ma, F., Wang, N. et al. (2019). DNA-Encoded Libraries: Aryl Fluorosulfonates as Versatile Electrophiles Enabling Facile On-DNA Suzuki, Sonogashira, and Buchwald Reactions. *Advanced Science* 2019, 6, 1901551.
127. Du, H.-C. & Huang, H. (2017). DNA-Compatible Nitro Reduction and Synthesis of Benzimidazoles. *Bioconjugate Chemistry* 28, 2575–2580.
128. Du, H.-C., Simmons, N., Faver, J. C. et al. (2019). A Mild, DNA-Compatible Nitro Reduction Using B-2(OH)(4). *Organic Letters* 21, 2194–2199.
129. Liu, S., Qi, J., Lu, W. et al. (2021). Synthetic Studies toward DNA-Encoded Heterocycles Based on the On-DNA Formation of alpha,beta-Unsaturated Ketones. *Organic Letters* 23, 908–913.
130. Wu, R., Gao, S., Du, T. et al. (2020). Exploring Aldol Reactions on DNA and Applications to Produce Diverse Structures: An Example of Expanding Chemical Space of DNA-Encoded Compounds by Diversity-Oriented Synthesis. *Chemistry-an Asian Journal* 15, 4033–4037.
131. Ding, Y., Chai, J., Centrella, P. A. et al. (2018). Development and Synthesis of DNA-Encoded Benzimidazole Library. *ACS Combinatorial Science* 20, 251–255.
132. Wang, D.-Y., Wen, X., Xiong, C.-D. et al. (2019). Non-transition Metal-Mediated Diverse Aryl-Heteroatom Bond Formation of Arylammonium Salts. *Iscience* 15, 307–315.
133. de Pedro Beato, E., Priego, J., Gironda-Martinez, A. et al. (2019). Mild and Efficient Palladium-Mediated C-N Cross-Coupling Reaction between DNA-Conjugated Aryl Bromides and Aromatic Amines. *ACS Combinatorial Science* 21, 69–74.
134. Lu, X., Roberts, S. E., Franklin, G. J. et al. (2017). On-DNA Pd and Cu promoted C-N cross-coupling reactions. *Medchemcomm* 8, 1614–1617.

135. Chen, Y.-C., Faver, J. C., Ku, A. F. et al. (2020). C-N Coupling of DNA-Conjugated (Hetero)aryl Bromides and Chlorides for DNA-Encoded Chemical Library Synthesis. *Bioconjugate Chemistry* 31, 770–780.
136. Xu, H., Gu, Y., Zhang, S. et al. (2020). A Chemistry for Incorporation of Selenium into DNA-Encoded Libraries. *Angewandte Chemie-International Edition* 59, 13273–13280.
137. Li, H., Sun, Z., Wu, W. et al. (2018). Inverse-Electron-Demand Diels-Alder Reactions for the Synthesis of Pyridazines on DNA. *Organic Letters* 20, 7186–7191.
138. Xiong, H., Gu, Y., Zhang, S. N. et al. (2020). Iridium-Catalyzed C-H Amidation of s-Tetrazines. *Chemical Communications* 56, 4692–4695.
139. Thomas, B., Lu, X., Birmingham, W. R. et al. (2017). Application of Biocatalysis to on-DNA Carbohydrate Library Synthesis. *Chembiochem* 18, 858–863.
140. Bassi, G., Favalli, N., Vuk, M. et al. (2020). A Single-Stranded DNA-Encoded Chemical Library Based on a Stereoisomeric Scaffold Enables Ligand Discovery by Modular Assembly of Building Blocks. *Advanced Science* 7, 2001970.
141. Kielar, C., Reddavide, F. V., Tubbenhauer, S. et al. (2018). Pharmacophore Nanoarrays on DNA Origami Substrates as a Single-Molecule Assay for Fragment-Based Drug Discovery. *Angewandte Chemie-International Edition* 57, 14873–14877.
142. Belyanskaya, S. L., Ding, Y., Callahan, J. F. et al. (2017). Discovering Drugs with DNA-Encoded Library Technology: From Concept to Clinic with an Inhibitor of Soluble Epoxide Hydrolase. *Chembiochem* 18, 837–842.
143. Li, Y., Zimmermann, G., Scheuermann, J. et al. (2017). Quantitative PCR is a Valuable Tool to Monitor the Performance of DNA-Encoded Chemical Library Selections. *Chembiochem* 18, 848–852.
144. Denton, K. E., Wang, S., Gignac, M. C. et al. (2018). Robustness of in vitro Selection Assays of DNA-Encoded Peptidomimetic Ligands to CBX7 and CBX8. *Slas Discovery* 23, 417–428.
145. Sannino, A., Gabriele, E., Bigatti, M. et al. (2019). Quantitative Assessment of Affinity Selection Performance by Using DNA-Encoded Chemical Libraries. *Chembiochem* 20, 955–962.
146. Anson, L. (2009). Membrane Protein Biophysics. *Nature* 459, 343–343.
147. Franzini, R. M., Nauer, A., Scheuermann, J. et al. (2015). Interrogating Target-Specificity by Parallel Screening of a DNA-Encoded Chemical Library Against Closely Related Proteins. *Chemical Communications* 51, 8014–8016.
148. Machutta, C. A., Kollmann, C. S., Lind, K. E. et al. (2017). Prioritizing Multiple Therapeutic Targets in Parallel Using Automated DNA-Encoded Library Screening. *Nature Communications* 8, 10.
149. Winssinger, N., Ficarro, S., Schultz, P. G. et al. (2002). Profiling Protein Function with Small Molecule Microarrays. *Proceedings of the National Academy of Sciences of the United States of America* 99, 11139–11144.
150. Harris, J., Mason, D. E., Li, J. et al. (2004). Activity Profile of Dust Mite Allergen Extract Using Substrate Libraries and Functional Proteomic Microarrays. *Chemistry & Biology* 11, 1361–1372.
151. Urbina, H. D., Debaene, F., Jost, B. et al. (2006). Self-Assembled Small-Molecule Microarrays for Protease Screening and Profiling. *Chembiochem* 7, 1790–1797.
152. Zambaldo, C., Daguer, J. P., Saarbach, J. et al. (2016). Screening for Covalent Inhibitors Using DNA-Display of Small Molecule Libraries Functionalized with Cysteine Reactive Moieties. *Medchemcomm* 7, 1340–1351.

153. Faver, J. C., Riehle, K., Lancia, D. R., Jr. et al. (2019). Quantitative Comparison of Enrichment from DNA-Encoded Chemical Library Selections. *ACS Combinatorial Science* 21, 75–82.
154. Salamon, H., Skopic, M. K., Jung, K. et al. (2016). Chemical Biology Probes from Advanced DNA-Encoded Libraries. *ACS Chemical Biology* 11, 296–307.
155. Zhao, G., Huang, Y., Zhou, Y. et al. (2019). Future Challenges with DNA-Encoded Chemical Libraries in the Drug Discovery Domain. *Expert Opinion on Drug Discovery* 14, 735–753.
156. Wang, S. J., Denton, K. E., Hobbs, K. F. et al. (2020). Optimization of Ligands Using Focused DNA-Encoded Libraries To Develop a Selective, Cell-Permeable CBX8 Chromodomain Inhibitor. *ACS Chemical Biology* 15, 112–131.
157. Kunig, V. B. K., Potowski, M., Akbarzadeh, M. et al. (2020). TEAD-YAP Interaction Inhibitors and MDM2 Binders from DNA-Encoded Indole-Focused Ugi Peptidomimetics. *Angewandte Chemie-International Edition* 59, 20338–20342.
158. Kazmierski, W. M., Xia, B., Miller, J. et al. (2020). DNA-Encoded Library Technology-Based Discovery, Lead Optimization, and Prodrug Strategy toward Structurally Unique Indoleamine 2,3-Dioxygenase-1 (IDO1) Inhibitors. *Journal of Medicinal Chemistry* 63, 3552–3562.
159. Dawadi, S., Simmons, N., Miklossy, G. et al. (2020). Discovery of Potent Thrombin Inhibitors from a Protease-Focused DNA-Encoded Chemical Library. *Proceedings of the National Academy of Sciences of the United States of America* 117, 16782–16789.
160. Kung, P.-P., Bingham, P., Burke, B. J. et al. (2020). Characterization of Specific N-alpha-Acetyltransferase 50 (Naa50) Inhibitors Identified Using a DNA Encoded Library. *ACS Medicinal Chemistry Letters* 11, 1175–1184.
161. Taylor, D. M., Anglin, J., Park, S. et al. (2020). Identifying Oxacillinase-48 Carbapenemase Inhibitors Using DNA-Encoded Chemical Libraries. *ACS Infectious Diseases* 6, 1214–1227.
162. Valastyan, J. S., Tota, M. R., Taylor, I. R. et al. (2020). Discovery of PqsE Thioesterase Inhibitors for *Pseudomonas aeruginosa* Using DNA-Encoded Small Molecule Library Screening. *ACS Chemical Biology* 15, 446–456
163. Wellaway, C. R., Amans, D., Bamborough, P. et al. (2020). Discovery of a Bromodomain and Extraterminal Inhibitor with a Low Predicted Human Dose through Synergistic Use of Encoded Library Technology and Fragment Screening. *Journal of Medicinal Chemistry* 63, 714–746.
164. Catalano, M., Moroglu, M., Balbi, P. et al. (2020). Selective Fragments for the CREBBP Bromodomain Identified from an Encoded Self-assembly Chemical Library. *Chemmedchem* 15, 1752–1756.
165. Shin, M. H., Lee, K. J. & Lim, H.-S. (2019). DNA-Encoded Combinatorial Library of Macrocyclic Peptoids. *Bioconjugate Chemistry* 30, 2931–2938.
166. Xie, J., Wang, S., Ma, P. et al. (2020). Selection of Small Molecules that Bind to and Activate the Insulin Receptor from a DNA-Encoded Library of Natural Products. *Iscience* 23, 101197.
167. Lomas, D. A., Irving, J. A., Arico-Muendel, C. et al. (2020). Development of a small molecule that corrects misfolding and increases secretion of Z α_1-antitrypsin. *EMBO Molecular Medicine* 13, e13167.
168. Litovchick, A., Tian, X., Monteiro, M. I. et al. (2019). Novel Nucleic Acid Binding Small Molecules Discovered Using DNA-Encoded Chemistry. *Molecules* 24, 2026.
169. www.delopen.org

CHAPTER 9

NUCLEIC ACID–BASED NANOCARRIERS AS DRUG DELIVERY SYSTEM

SANGWOO HAN, DAJEONG KIM, YOONBIN JI, SUNGHYUN MOON, AND JONG BUM LEE
Department of Chemical Engineering, University of Seoul, Seoul, Republic of Korea

9.1. INTRODUCTION

9.1.1. Development of Nucleic Acid Nanotechnologies in Drug Delivery System

The intrinsic Watson–Crick base pairing property of complementary deoxyribonucleic acid (DNA) and ribonucleic acid (RNA) makes nucleic acids highly predictable and programmable polymers, which have attracted considerable attention in material science [1, 2]. These properties allow the materials to be precisely controlled, to achieve a rationally designed morphology, to fine-tune the structure, and to generate a systemic response to external stimuli [3]. Experts in the field of nanomaterial sciences exploit these advantages by using these biopolymers to build a variety of nanoscale structures in a bottom-up approach, based on their biological properties [4, 5]. The field of DNA nanotechnology has emerged and grown exponentially over the past three decades, since DNA branched junctions and lattices were reported in the 1980s by Nadrian Seeman [6]. Seeman's report was based on the concept of sequence-specific sticky ends, and their ability to produce self-assembled arrays. The cohesion of

Nucleic Acids in Medicinal Chemistry and Chemical Biology: Drug Development and Clinical Applications, First Edition. Edited by Lihe Zhang, Xinjing Tang, Zhen Xi, and Jyoti Chattopadhyaya.

sticky ends, wherein two DNA motifs containing complementary single-stranded extensions hybridize to form larger arrays, has allowed scientists to create a variety of DNA lattice structures [7]. To take a step further, increasing the number of branched sticky ends in a junction to over the four can allow multi-armed junctions to be made into tile-based 3D structures, such as DNA polyhedrons, including cubes, tetrahedrons, and octahedrons [8]. Beyond the tile-based DNA architecture, the report on DNA origami by Paul Rothemund in 2006 described an invention and lead to the rapid development of DNA nanotechnology [9]. DNA origami is a strategy in which a long single-stranded scaffold DNA can be folded into customized shapes, using hundreds of short complementary synthetic strands acting as DNA staples [3]. As a result, 2D DNA origami, honeycomb lattice structures, DNA origami with curvatures, and other complex shapes can successfully be synthesized. Although RNA and DNA share very similar properties, the development of RNA nanotechnologies has been hampered by its structural instability, caused by the extra hydroxyl group at the 2′ position of the sugar ring and by the complexity of its intra-strand double helices [10]. Despite the challenges in manipulating RNA, its complex biological functions have drawn attention toward the clinical applications of RNA nanotechnology, such as in the treatment of diverse genetic diseases and cancer [11].

Research on drug delivery systems has contributed to significant advances in the treatment of various diseases by allowing targeted drug delivery to the site of the disease [12, 13]. Novel drug delivery systems have aimed to overcome the issues associated with non-specific delivery, poor drug solubility, chemical or enzymatic degradation, drug toxicity and drug-resistance [14]. To achieve these, a drug delivery carrier should have the following properties: biocompatibility, efficient cargo loading, biodegradability, trigger-induced cargo release, high cellular uptake, and target specificity. Therefore, development of drug efficacy is necessary to improve therapeutic efficiency. Numerous drug delivery platforms have been developed to improve drug encapsulation and water solubility; to reduce nonspecific delivery; and to introduce synergistic effects with diagnostic imaging, or photothermal or photodynamic therapy. These include liposomes, viruses, polymer-based dendrimers, polymeric micelles, layer-by-layer-based nanoparticles, inorganic nanocarriers, and nanosponges, among others [12, 15–18]. Although these nanomaterials have considerably improved drug delivery technology, they do have some limitations [8]. In many cases, materials that enhance target efficiency or cellular uptake are inherently toxic [19]. In addition, it is difficult to accurately control the size, loading efficiency, or functionalization on a nanoscale. In this regard, nucleic acids have a huge advantage, since they are biocompatible, and their function can be finely controlled through programmable interactions of DNA and RNA [20]. This predictable and well-defined bottom-up approach is possible because the DNA and RNA strands can be chemically synthesized into customized sequences, facilitating the construction of various nano- or micro-sized geometries, suitable for stable drug delivery. In addition to the shape of the structures, the size of these nanoparticles can also be readily adjusted for efficient targeted delivery to most organs, including the brain, liver, tumors and lymph nodes [21].

9.1.2. Advantages of Nucleic Acid–Based Nanocarrier for Drug Delivery

Since chemical synthesis of arbitrary oligonucleotides sequences has become possible, the construction of nucleic acid–based nanomaterials has been widely explored [22]. The hybridization interaction of nucleic acids improves conformational rigidity with longer persistent length nucleic acids and allows the formation of excellent programmable materials. The programmability of nucleic acids has allowed many kinds of DNA- or RNA-based nanocarriers to be developed. There are five major advantages to their use as drug carriers (Figure 9.1). Firstly, any shape or size of nanocarriers can be precisely achieved [20], from

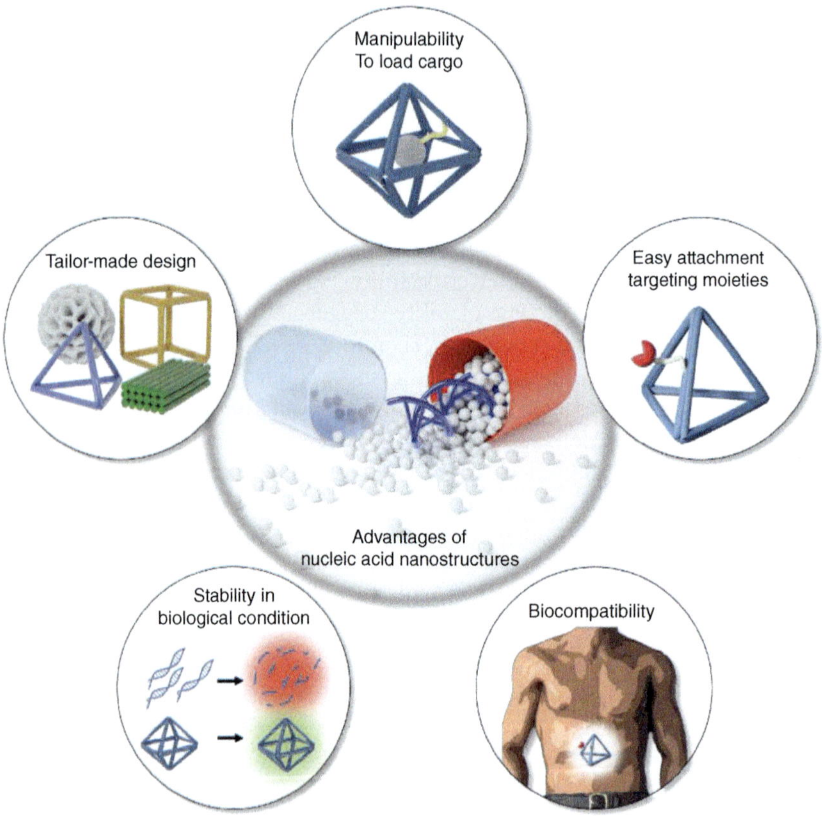

Figure 9.1 Advantages of nucleic acid–based carriers in drug delivery system.

simple two-dimensional (2D) structures, from periodic lattices, to complicated three-dimensional (3D) structures such as polyhedrons or DNA origami; numerous morphologies have been assembled to allow reliable drug delivery. Secondly, the cargo attachment can be readily tailored by sequence-specific hybridization, intrinsic intra-strand assembly, and chemical modification of the nucleotides [23–27]. Small molecules such as doxorubicin (DOX) and metal complexes can be intercalated within the DNA and RNA duplex, or encapsulated in DNA polyhedra. Some larger chemical drugs can be conjugated to nucleic acid nanostructures via chemical bonding with modified nucleotides. Thirdly, targeting moieties such as antibodies, peptides, and aptamers can easily be attached to the surface of nucleic acid–based structures. This can also be achieved by the hybridization of DNA and RNA through modified nucleotide-induced covalent or non-covalent bonds. Another advantage is that DNA and RNA nanostructures can protect embedded nucleic acid drugs, such as small-interfering RNA (siRNA), micro RNA (miRNA), messenger RNA (mRNA), or DNA CpG. Such nucleic acid drugs are vulnerable to the biological environment, due to their low structural resistance to various nucleases; encapsulation of these drugs in nucleic acid–based nanocarriers significantly improves the efficiency of their delivery to target cells or organs. Lastly, nucleic acids are highly biocompatible materials. This property is important for their application in drug delivery systems. The low cytotoxicity and universal functionalization properties of nucleic acid–based carriers have allowed them to be developed for drug delivery into cells and have introduced new approaches for clinical therapeutics.

9.2. DESIGN OF NUCLEIC ACID–BASED NANOSTRUCTURES FOR DRUG DELIVERY SYSTEM

9.2.1. Principle of Structural Nucleic Acid as a Building Block

Since the development of DNA tiles for nanoscale manipulability in the 1980s [6, 28], DNA nanotechnology, using DNA as a building block, has progressed research on DNA-based material engineering beyond its inherent genetic functions [29]. Nucleic acids are the key biopolymers for genetic function and are composed of nucleotides containing four distinct nitrogenous bases, represented by letters (Figure 9.2a) [30]. Watson–Crick base-pairing interactions, occurring via hydrogen bonding, offer nanoscale programmability with high predictability (Figure 9.2b) [22, 31]. In this section, we will discuss various DNA and RNA nanotechnologies for their bottom-up and top-down assembly, as well as the features of each structure for its potential as a drug nanocarrier.

9.2.2. Tile-Based Nanostructures

The most fundamental building motif is a tile-based structure consisting of tens of hybridized base pairs connecting each of the strands [29]. The longer

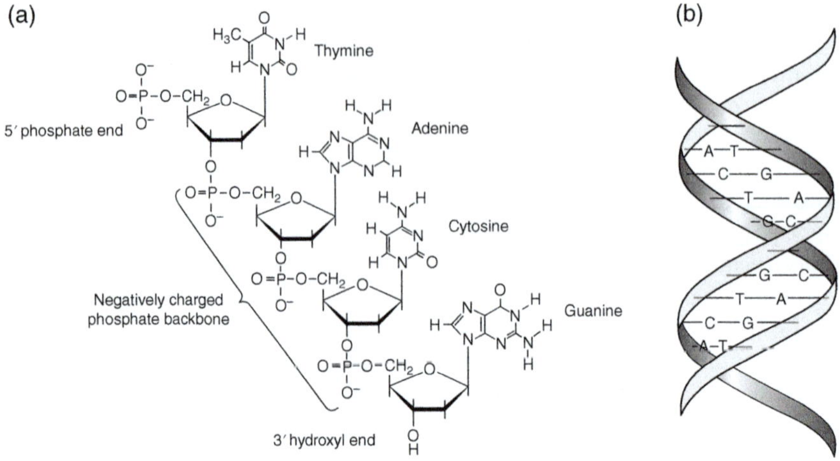

Figure 9.2 (a) Chemical structures of each nucleotide. (b) Watson–Crick base pairing of DNA. Source: Sinden [30] / with permission of Elsevier.

persistence length of double-stranded DNA (dsDNA) (~150 bp), compared to flexible single-stranded DNA (ssDNA) [32], makes it possible to assemble a multistranded junction (MJ), composed of the desired length and number of arms. In 1982, Seeman reported building the immobile DNA Holliday junction structure using asymmetric sequences, which are more stable than naturally occurring symmetric junctions (Figure 9.3a) [6]. These junctions could be synthesized by rationally designing sequences for desired base pairing and introducing covalent linkages using the sticky-ended ligation technique. More than five branched DNA MJs were constructed, which could then be assembled into 3D networks using the sticky ends (Figure 9.3b) [33, 34]. This would enable the direct deposition of biomolecules on 2D or 3D template structures with precise dimensions.

While MJs have been developed into higher-dimensional structures, their flexibility needs to be improved, in order to build stabilized and controlled nanostructures. A double-crossover (DX) DNA structure was developed using two Holliday junctions, connected by two double-helical domains (Figure 9.4a) [35, 36]. Using the 4-arm tile, in which each arm is composed of a DX, a two-dimensional lattice could be finely and stably prepared on a sub-nanoscale (Figure 9.4b) [7, 37]. To take a step further, these crossover-based multi-arm tiles could be developed into highly interconnected nanotubes (Figure 9.4c) [37]. Assisted by a T-shaped linkage at the end of the tiles, DNA polyhedrons such as tetrahedrons, octahedrons, and icosahedrons could successfully be self-assembled from these multi-arm tiles (Figure 9.4d) [38]. Moreover, triple-crossover (TX) bricks have been artificially designed using three dsDNA strands along with four individual strands (Figure 9.4e) [39]. This TX structure can be achieved by laying the three dsDNA

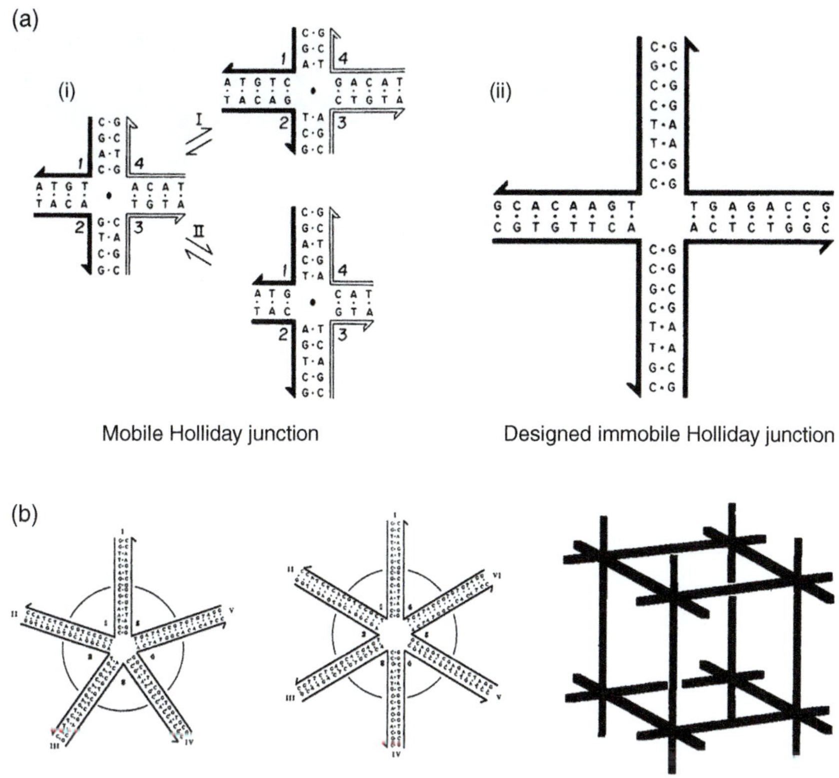

Figure 9.3 (a) DNA Holliday junction structure. (i) Mobile Holliday junction with symmetric sequences and (ii) designed immobile Holliday junction. Source: Seeman [6] / with permission of Elsevier. (b) Five- and six-arm junctions and conceptual cubic networks. Source: Wang *et al.* [33] / With permission of American Chemical Society.

on a plane and periodically extending them into a linear tube shape through aptamer–protein or protein–protein interactions (Figure 9.4f) [40, 41].

Precise control and design of RNA-based nanomaterials is challenging because RNA naturally forms complicated secondary or tertiary structures via inter-or intra-molecular interactions. To overcome this limitation, DNA and proteins have been used to fabricate well-defined RNA nanostructures. For example, Mao and co-workers reported a highly precise self-assembled RNA structure, built with the help of DNA strands, to form a well-defined nanostructure (Figure 9.5a) [42]. They designed several RNA–DNA hybrid branched nanomotifs and expanded them into two- or three-dimensional nanoarchitectures. Additionally, an RNA-based three-way junction (3WJ-RNA) was discovered through the addition of phi29 DNA packaging motors [43]. 3WJ-RNA exhibited high serum stability and was resistant to degradation. This improved stability is very beneficial in drug carriers. Moreover, six 3WJ-RNA building blocks,

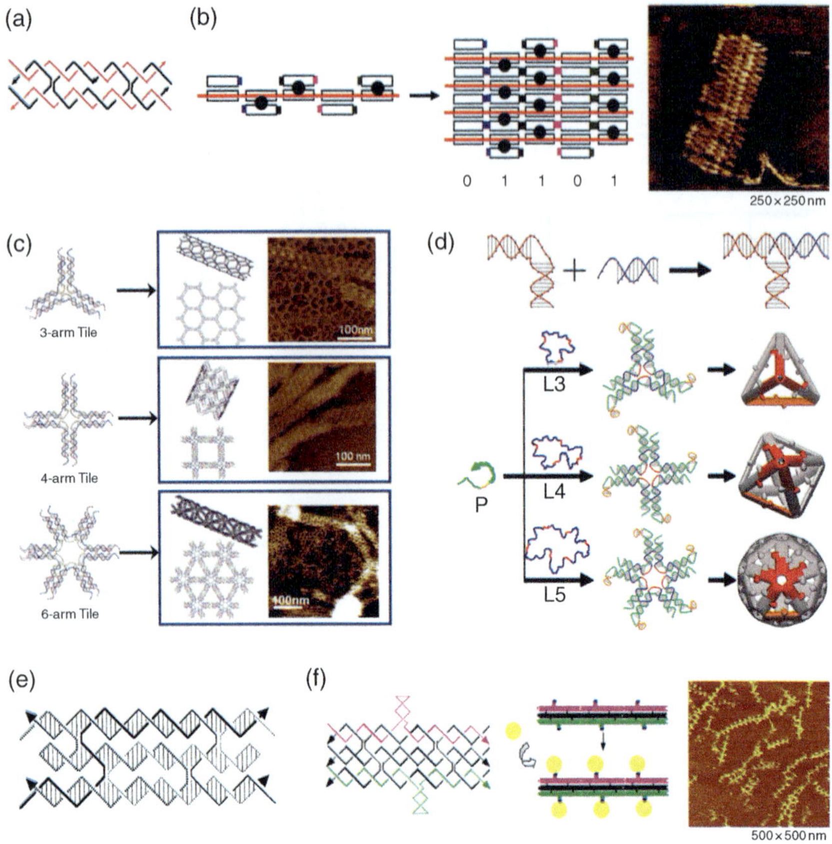

Figure 9.4 (a) Double-crossover (DX) DNA structure. (b) Two-dimensional lattice structure with DX tiles programmed through sticky ends. Black dots indicate the hairpin loop. Source: Yan *et al.* [36] / With permission of National Academy of Sciences. (c) Connected nanotube from multi-arm tile structure. Source: Shi *et al.* [37] / With permission of Royal Society of Chemistry. (d) DNA polyhedron with T-shape linkage. Source: Li *et al.* 38 / With permission of American Chemical Society. (e) Design of triple-crossover (TX). Source: LaBean *et al.* [39] / With permission of American Chemical Society. (f) Periodic linear nanotube by TX structure. Source: Li *et al.* [41] / With permission of American Chemical Society.

including siRNAs, could form kissing loops, permitting cross-linking between adjacent blocks, to form a closed RNA nanoring and enabling selective drug release (Figure 9.5b) [44].

The straightforward functionalization of the internal sequence and ends of the arms, as well as the ease of synthesis, make multistranded junctions a good choice for use in enhanced drug delivery systems [27, 45, 46]. Y-shaped DNA blocks with stretchable sticky ends could be elongated to a three-dimensionally branched

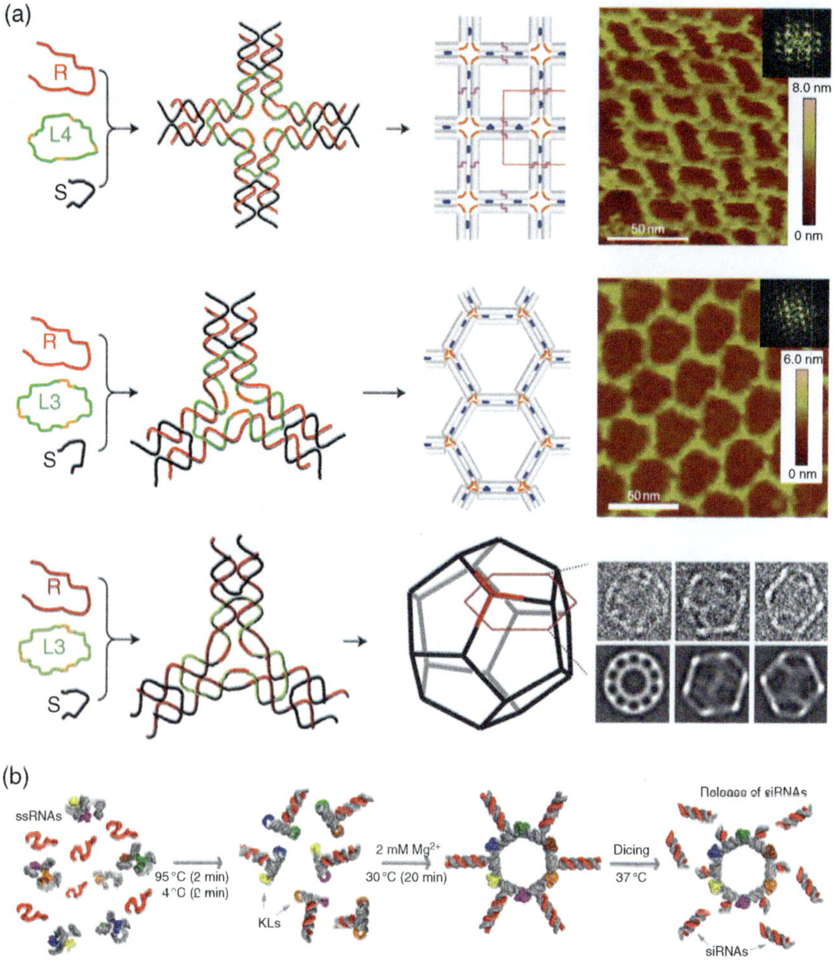

Figure 9.5 (a) Various RNA-based nanostructures by RNA–DNA hybrids. Source: Ko *et al.* [42] / with permission of Springer Nature. (b) RNA nanoring for the siRNA delivery system. Source: Afonin *et al.* [44] / with permission of American Chemical Society.

dendrimer, generating a number of end sites for further functionalization [45–48]. Further expansion could generate a macroscopic material, composed of interconnected sub-nanoscale bricks, which could be used for stimuli-responsive controlled drug release (Figure 9.6) [49, 50]. Altogether, the tile-based structures could form a wide range of architectures, from nanoscale to macroscopic, using nano-bricks or bottom-up assembly. This can allow the development of customized delivery system with controlled drug deposition properties at the molecular level.

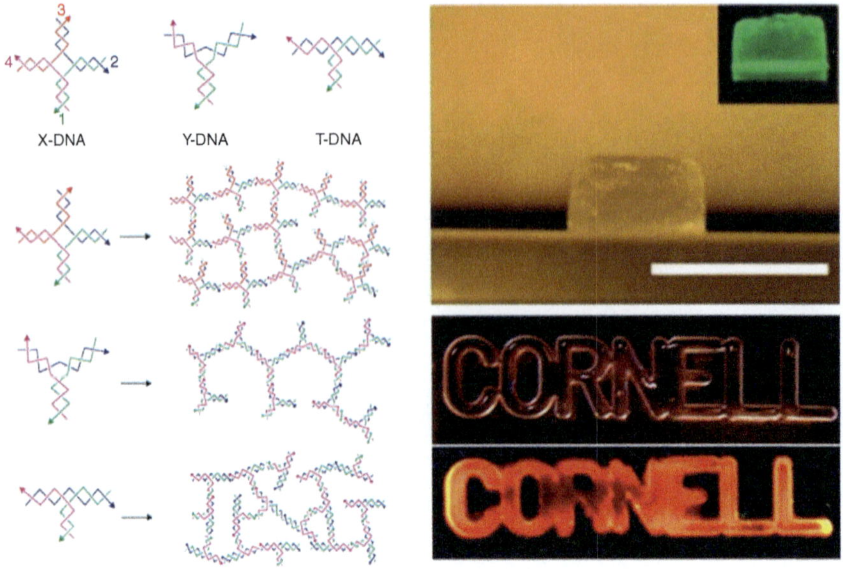

Figure 9.6 X-, Y-, and T-DNA blocks can be extended into macroscopic materials through the connection of sticky ends. Source: Um *et al.* [50] / with permission of Springer Nature.

9.2.3. Three-Dimensional Cages

The development of nucleic acid nanotechnology has enabled the assembly of complex polyhedrons or cages in desired shapes, sizes, and lengths. Unlike the tile-based three-dimensional nanostructures produced by the incorporation of repeated tile units, a more complex design of nucleic acid–based cage structures will be discussed in this section. In addition to the simple conjugation of tile-based units, the length, complementarity, and flexibility of the strands can also be applied to produce new types of structures. For example, eight individual strands, hybridized to neighboring strands, have been reported to build an octahedral structure (Figure 9.7a) [51]. This indicates that precise 3D nanostructures can be fabricated without the use of repeated tile-based units. Moreover, a complex design for cage structures has been developed by combining crossover units and the design of secondary structures of ssDNA. The ssDNA (1.7 kb) can be folded into a DNA octahedron via five DX and seven paranemic-crossover (PX) intermolecular interactions (Figure 9.7b) [52]. These studies imply that sequence variations, making variable loops on DNA scaffolds, could initiate the directed evolution of nucleic acid nanostructures. With the development of nucleic acid nanotechnology, partial stiffness and flexibility can be effectively controlled, resulting in structures that are more complex with high stability, for use as drug carriers.

(a)

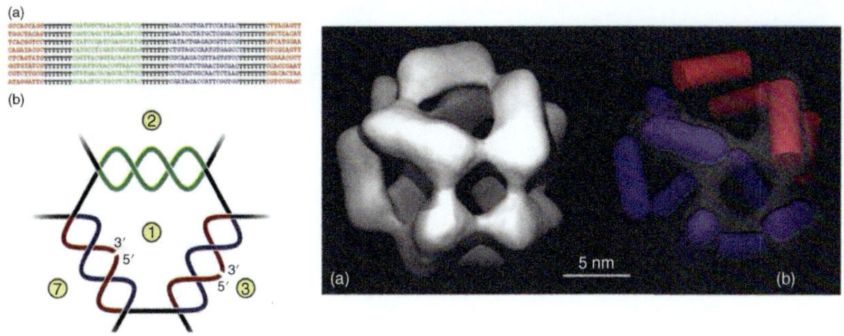

Assembly and structural analysis of a covalently closed nano-scale DNA cage

(b)

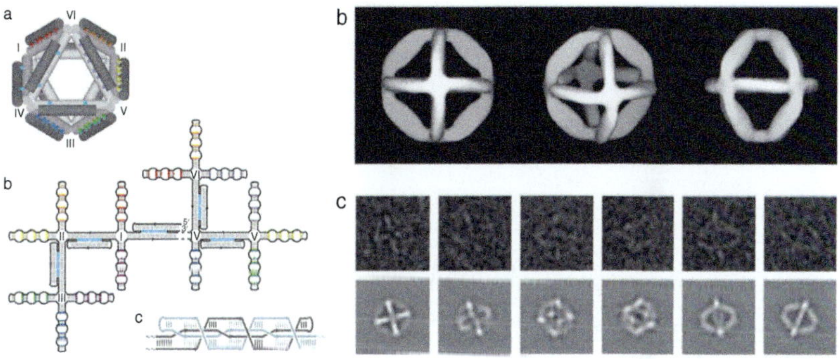

Figure 9.7 (a) Design of eight-stranded DNA octahedron cage (left) and 3D isodensity map of the cage based on cyro-TEM image. Source: Andersen *et al.* [51] / with permission of Oxford University Press. (b) Design of the DNA octahedron with DX and PX motifs and secondary branched folding structure of the intermediate (left). Three-dimensional map of the structure of DNA octahedron and its raw images of the structure by corresponding projection of computed map. Source: Shih *et al.* [52] / with permission of Springer Nature.

9.2.4. Nucleic Acid Origami

Tailorable hybridization has been extended to finely controlled structures. Staple strands can be used to help fold long DNA template strands and scaffolds, via guide-assisted assembly, into predicted shapes for use in enhanced drug delivery systems (Figure 9.8) [53]. In 2006, Rothemund developed a straightforward top-down approach to fabricate complex 2D domains, composed of dsDNA, by folding a single DNA strand (M13mp18) (Figure 9.9a) [9]. The piece of origami with the hairpin marker can mimic a jigsaw puzzle, assuming the shape of nanosized alphabets through interactions between the

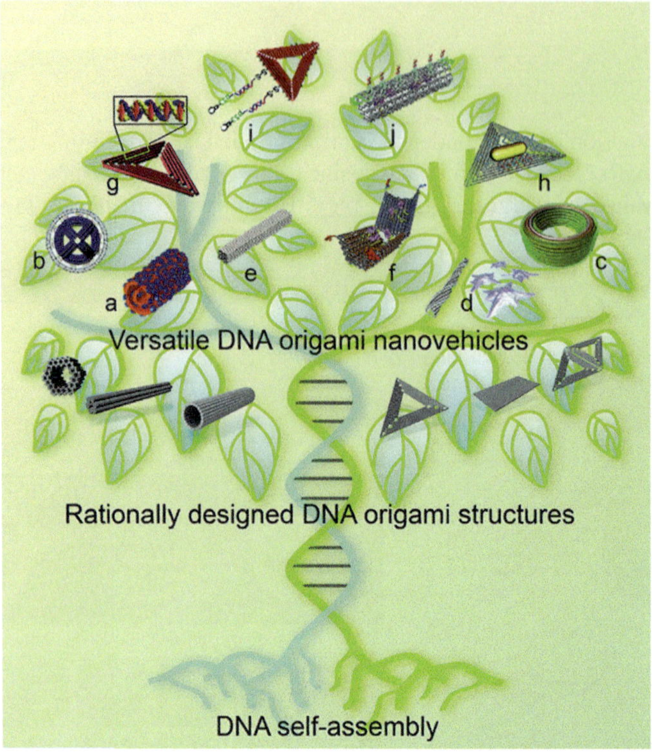

Figure 9.8 Rationally designed various DNA origami. Source: Jiang *et al.* [70] / with permission of John Wiley & Sons.

overhangs on each piece (Figure 9.9b) [54]. This could be further extended to create a layout and a pack of origami for fractal assembly, to produce large paintings such as the Mona Lisa (Figure 9.9c) [55].

Subtle manipulations of the nanoscale base unit, such as the addition and/or deletion of base pairs, could lead to compensatory deformation of the helices, resulting in the formation of curved structures (Figure 9.10a) [56–58]. The high controllability of the x and y dimensions and the high predictability have enabled the construction of 3D DNA origami structures. For example, a DNA strand was densely packed into a square lattice of definite length, as multiple layers of double helices [59]. Further regulation on the staple strand breakpoints facilitated the highly curved structures such as twisted or donut-shaped in three-dimensional scale (Figure 9.10b) [60, 61]. Moreover, the origami structure could be used as Lego bricks to build the desired shape by directional growth on a $10 \times 10 \times 10$-voxel 3D canvas [62].

Expanding beyond the static architectures, inherent characteristics of nucleic acids can be combined to allow origami to be programmed to produce dynamic

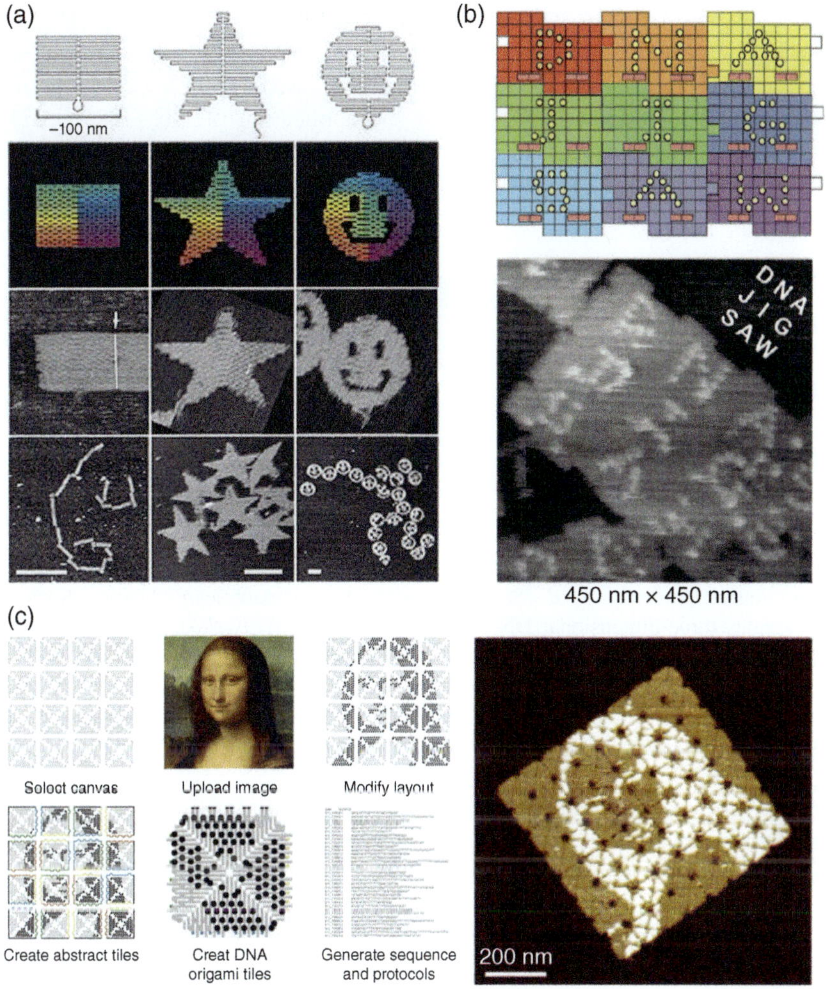

Figure 9.9 (a) Micrometer-scale DNA origami arrays with arbitrary shapes. Source: Rothemund *et al.* [9] / with permission of Springer Nature. (b) Jigsaw puzzles by DNA origami tiles with hairpin markers. Source: Rajendran *et al.* [54] / with permission American Chemical Society. (c) DNA painting assembled from a pack of origami tiles. Source: Tikhomirov *et al.* [55] / with permission of Springer Nature.

motion [63, 64]. In particular, reconfigurable 3D DNA origami tripod was devised of which interarms could be controlled by adjusting the lengths of connecting struts (Figure 9.11a) [64]. This reversible conformation was well controlled by adding or removing the locking strands, confirmed by gel electrophoresis and TEM images. These stimuli-responsive shape-changeable

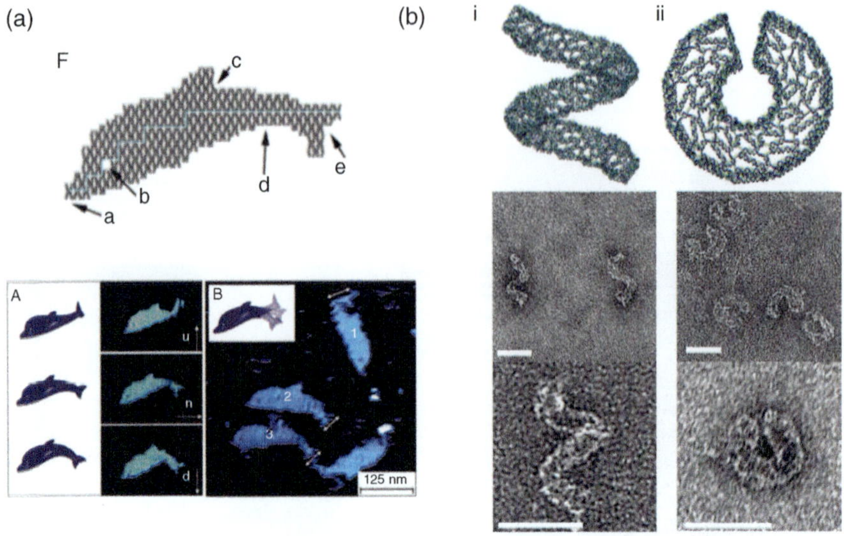

Figure 9.10 DNA origami with curvature. (a) Dolphin-shaped DNA origami structures. Source: Andersen *et al.* [58] / with permission of American Chemical Society. (b) Various three-dimensional DNA origami structures: (i) twisted, (ii) donut-shaped DNA origami. Source: Benson *et al.* [61] / with permission of Springer Nature.

origami structures could be beneficial in targeted drug delivery systems. Similarly, a locked DNA box was fabricated, where the "key" strands could open up the lid, resulting in the environmentally controlled release of internal components (Figure 9.11b) [65].

Since nucleic acids are highly programmable, single-stranded template DNA or RNA were rationally designed to self-fold into a single architecture without staple strands [66]. Using single-stranded nucleic acids, the production processes can feasibly be scaled-up, beyond the one-pot assembly of origami on a testing tube scale. Employing the DNA designed to encode the origami sequence, RNA strands, transcribed *in vitro*, could be self-folded into the intended shapes [67, 68]. The introduction of this method allows the process of preparation to be conserved, enabling large-scale synthesis, which is beneficial for drug delivery systems. More recently, an automatic designer program to synthesize tailor-made origami structures was developed to provide a general route for the fabrication of arbitrary shapes. This could accelerate the development of DNA origami for various applications [69]. Overall, the predictability and fine manipulability at the nano- to micro-scale, regardless of the size and shape, means that nucleic-based carriers can be used for targeted drug delivery and release [70].

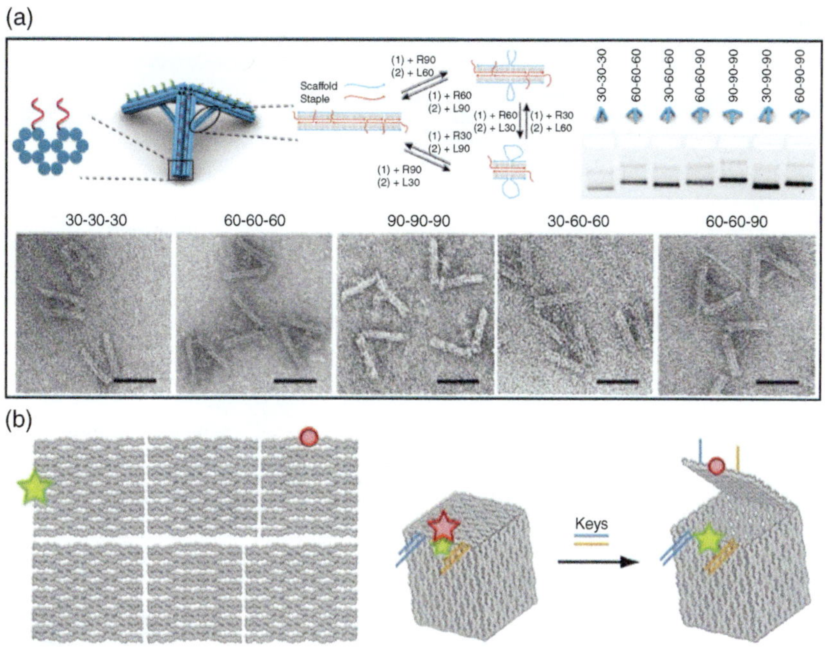

Figure 9.11 (a) Reconfiguration of DNA origami tripods and its confirmation with gel electrophoresis and TEM images. Source: Zhan *et al.* [64] / with permission of American Chemical Society. (b) DNA nanobox with controllable lid. Source: Andersen *et al.* [65] / with permission of Springer Nature.

9.2.5. Enzymatic Replication-Based Sponges

The rolling circle replication method is a powerful bottom-up technique for constructing nucleic acid–based nanostructures. Multimeric DNA is elongated by phi29 DNA polymerase, based on a circular DNA template, in a process called rolling circle amplification (RCA), and simultaneously self-assembled into sponge-like structures (Figure 9.12a) [71]. Densely packed RCA-based structures exhibit remarkable resistance to degradation by nucleases, which is a huge advantage for use as drug carriers. RNA can also be elongated from circular DNA using T7 RNA polymerase, in a process called rolling circle transcription (RCT). This results in a highly porous and compact spherical structure (Figure 9.12b) [18, 72]. RCT-mediated self-assembly can allow high cargo loading and enhance the physiological stability, further expanding its practical applications in medicine and pharmaceutics. Therefore, various functional RNAs, including siRNA, miRNA, mRNA, and ribozymes, have been embedded in the RCT-based RNA structures, and efforts have been made to increase drug efficacy, including tuning the size or sequences for effective drug release [73–75]. These sponge structures have several sites where loading

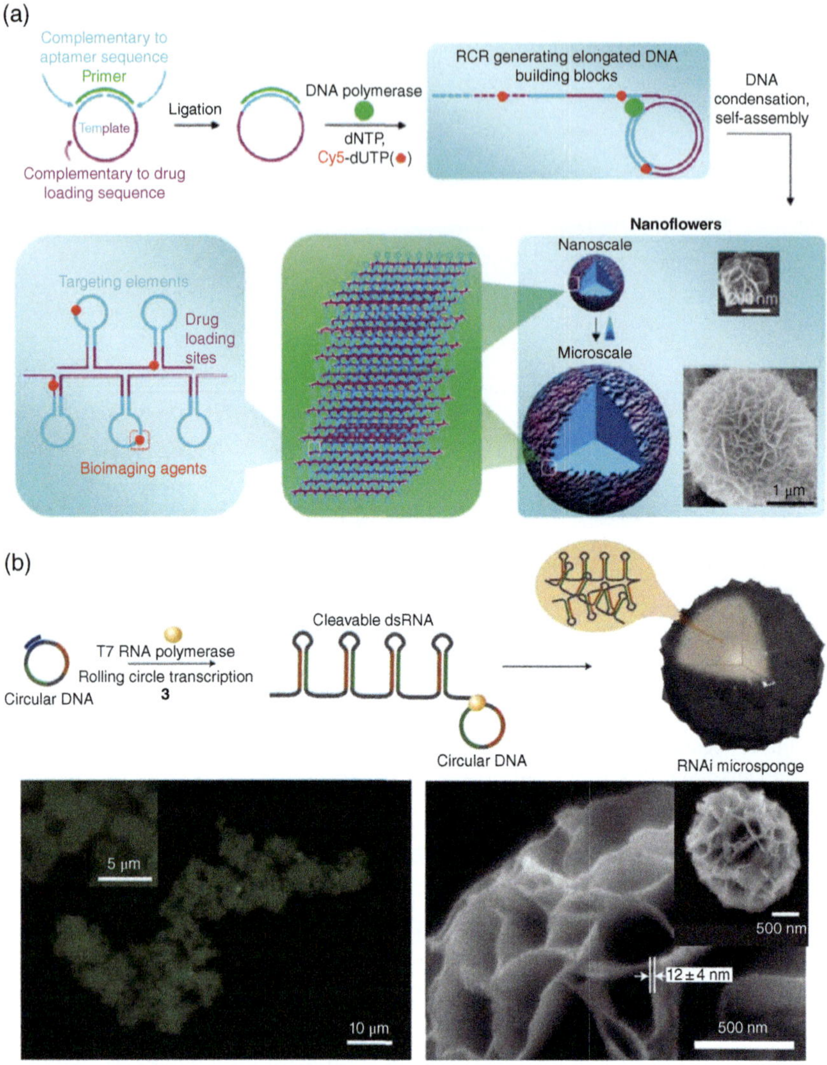

Figure 9.12 (a) Schematic illustration of RCT process and its microscopic structure. Source: Zhu *et al.* [71] / with permission of American Chemical Society. (b) RCT-based RNA microsponge. Source: Lee *et al.* [18] / with permission of Springer Nature.

of large quantities of drug can be achieved during the synthetic process. Also functional nucleic acid sequences, such as immunostimulants and anti-sense oligonucleotides, can be readily and efficiently incorporated into sponge structures, with rational sequence design [37, 76–78]. Secondly, the inorganic backbone provides a fascinating site for cation delivery (Figure 9.13a) [79, 80]. The individual nucleotides could also be modified with biomolecular moieties to

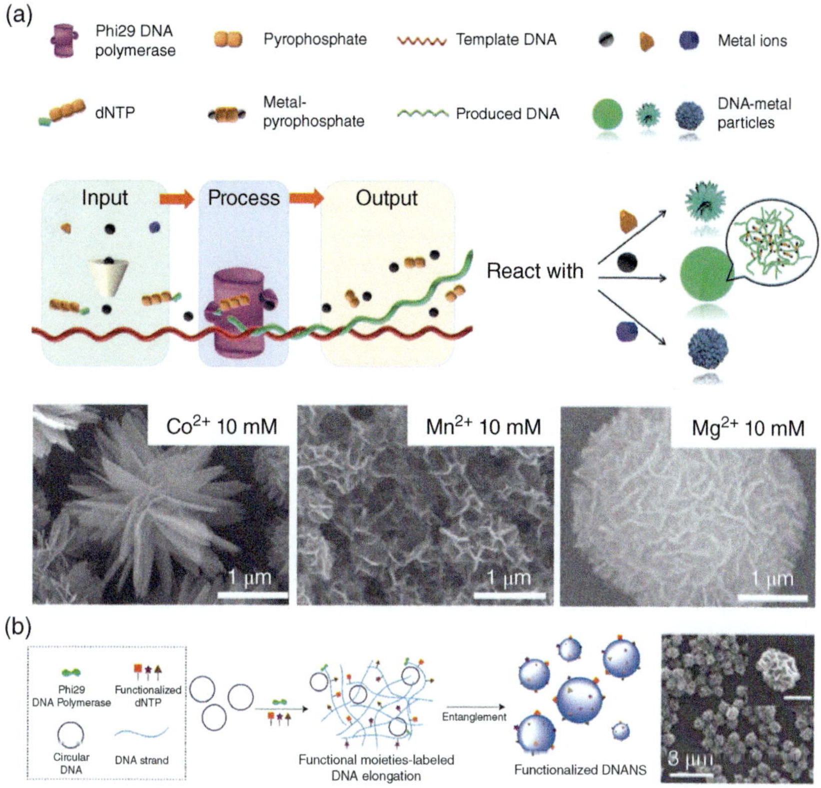

Figure 9.13 (a) Process of complexation between DNA and inorganic materials during RCA. Source: Lee *et al.* [80] / with permission of John Wiley & Sons. (b) Biomolecular functionalization on the elongated DNA. Source: Han *et al.* [82] / with permission of Royal Society of Chemistry.

achieve signal amplification and targeted drug delivery (Figure 9.13b) [81, 82]. Moreover, the biomolecules could be loaded in the sponges in a one-pot synthetic process, owing to their porous nature and large surface area [83]. Further protein–nucleic acid interactions can enable the active loading of proteins such as the Cas9 protein and thrombin [84, 85].

Based on these enzymes replication-based nucleic acid structures, multimeric DNA or RNA nanostructures can be incorporated into the whole structures (Figure 9.14a) [4, 86, 87]. Therefore, RCR could be used to achieve automated mass production of desirable structures, which is necessary for the clinical translation of nucleic acid–based drugs (Figure 9.14b). Furthermore, highly concentrated DNA/RNA hybrid nanostructures have also been fabricated via dual enzymatic polymerization, suggesting that multifunctional drug delivery may be possible (Figure 9.14c) [88, 89]. Overall, the architectures produced by enzymatic synthesis seem highly suitable for the delivery of biotherapeutic agents.

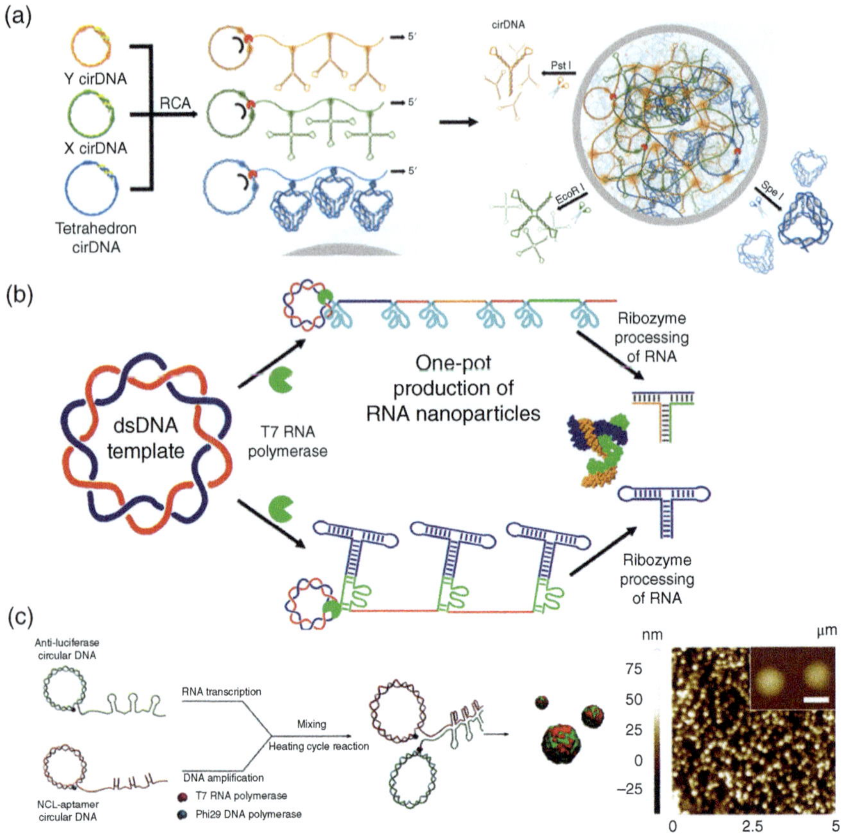

Figure 9.14 (a) Multimeric production of DNA nanostructures by RCA. Source: Ko *et al.* [4] / with permission of Elsevier. (b) One-pot synthesis of multimeric functional RNA by RCT. Source: Jasinski *et al.* [86] / With permission of American Chemical Society. (c) DNA–RNA hybrid nanostructures via dual enzymatic polymerization. Source: Park *et al.* [89] / with permission of American Chemical Society.

9.3. STRATEGIES FOR EFFICIENT DRUG DELIVERY WITH NUCLEIC ACID NANOSTRUCTURES

9.3.1. Advantages of Nucleic Acid for Functionalization

One of the main goals of drug delivery research is to develop technologies to achieve targeted drug delivery, in order to reduce the toxicity associated with nonspecific drug distribution [90]. Several active targeting strategies have been developed, using various moieties, including antibodies, peptides, proteins, aptamers, and biomolecules, for targeting specific receptors on the cells [3].

Bio-mimicking nanocarriers, such as virus capsid coatings, are also attractive strategies for enhancing the efficiency of cellular delivery [91]. Nucleic acids are good candidates to apply these strategies, since DNA and RNA are easy to functionalize, because of their inherent base-pairing properties and the synthetic accessibility of numerous functional group-modified nucleotides. In this chapter, we review diverse strategies for active molecular targeting and trigger-induced drug release, to achieve improved accumulation and localized delivery of drugs to specific cells or organs.

9.3.2. Molecular Targeting Strategies

The most common strategy for the recognition of a specific target is the use of an aptamer: a functional nucleic acid sequence that forms unique 3D structures and binds to specific molecules [92–95]. Their high specificity, strong binding affinity, and low immunogenicity make aptamers excellent targeting ligands. In addition, aptamers have the advantage of a large range of potential targets, including cells, proteins, and small molecules. Furthermore, aptamers can be easily integrated into 3D nucleic acid–based carriers without the need for chemical conjugation. Therefore, aptamers have been widely used in targeted drug delivery systems. Figure 9.15a shows an aptamer-based DNA dendrimer, as a clinically functional nanostructure. To achieve specific cancer cell recognition, sgc8 aptamer, which selectively targets the protein PTK7 expressed on the cell membranes, was assembled onto the DNA dendrimer by spontaneous hybridization. In this study, the target was a human T-cell acute lymphoblastic leukemia cell line (CCRF-CEM), overexpressing PTK7, and a human Burkitt's lymphoma cell line (Ramos), exhibiting low PTK7 expression, was used as a control. Successful specific cellular delivery was demonstrated by microscopic analysis (Figure 9.15a). While high fluorescence intensity was observed in the CCRF-CEM cell line, a weak fluorescent signal was observed with the Ramos cell line on treatment with the sgc8-conjugated DNA dendrimer (Figure 9.15b). In the absence of sgc8, no fluorescence signal was observed in either cell line, indicating that the DNA dendrimer was not taken up by the cells.

In addition to the integration of aptamers by the post-addition of target-specific oligonucleotides to nucleic acid–based nanostructures, there have been several studies on the one-step synthesis of aptamer-siRNA complex-embedded self-assembled structures. The Lee's group reported an enzymatic replication-based RNA nanovector, consisting of a DNA aptamer–siRNA complex [89]. This RNA nanovector was synthesized by stepwise dual enzyme polymerization (SDEP) employing RCA and RCT (Figure 9.16a). The RNA nanovector was assembled by hybridization of a polymerized DNA aptamer with siRNA. To target the overexpressed-nucleolin on the cancer cell membrane, the AS 1411 DNA aptamer was encoded. The drug delivery system exhibited specific cellular uptake, dependent on the presence of the aptamer (Figure 9.16b). This suggests that SDEP may serve as a platform for the development of carrier-free

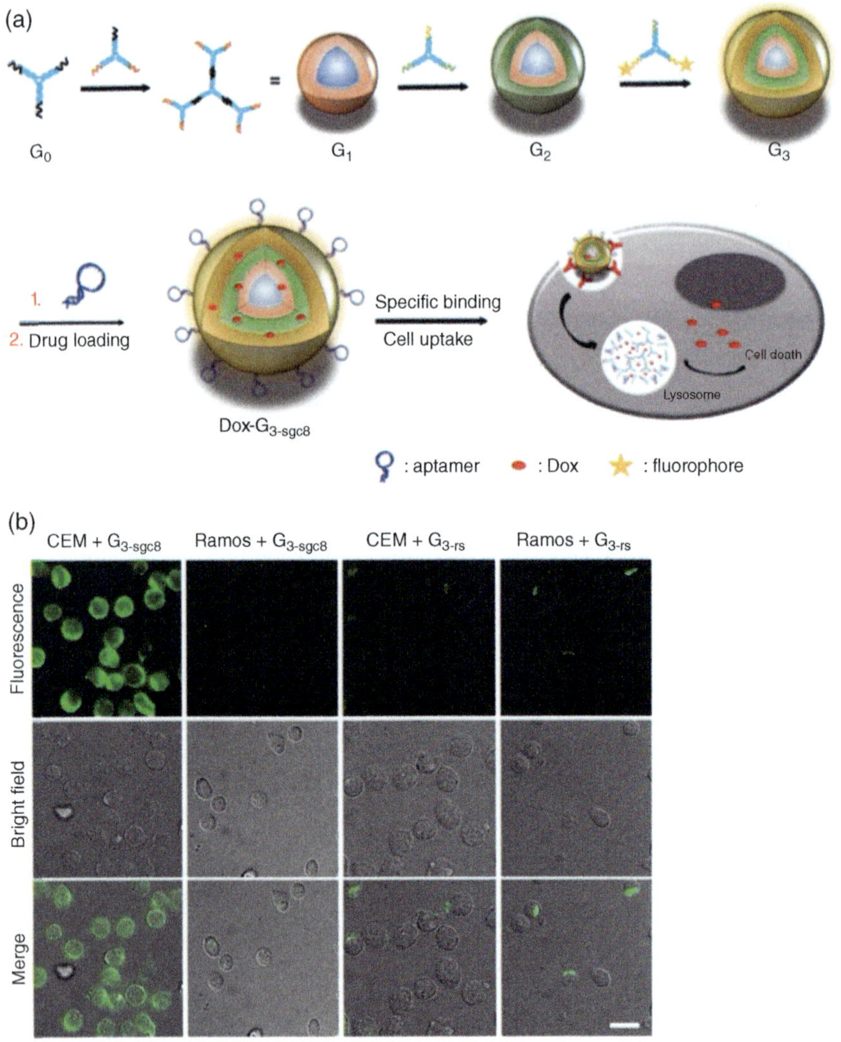

Figure 9.15 (a) Aptamer-conjugated DNA dendrimer and DOX delivery into target cells. (b) Target-cell-specific cellular delivery was confirmed by microscope images. Source: Zhang *et al.* [90] / Springer Nature / CC BY-4.0.

nucleic acid–based therapeutics. By designing various functional DNA and RNA sequences, this system has great potential for application in target-specific gene delivery.

Folic acid is also a well-known target molecule for folate receptor-dependent cellular uptake [96]. The folate receptor, a glycosylphosphatidylinositol membrane protein, is overexpressed in several malignant tumors and is absent in

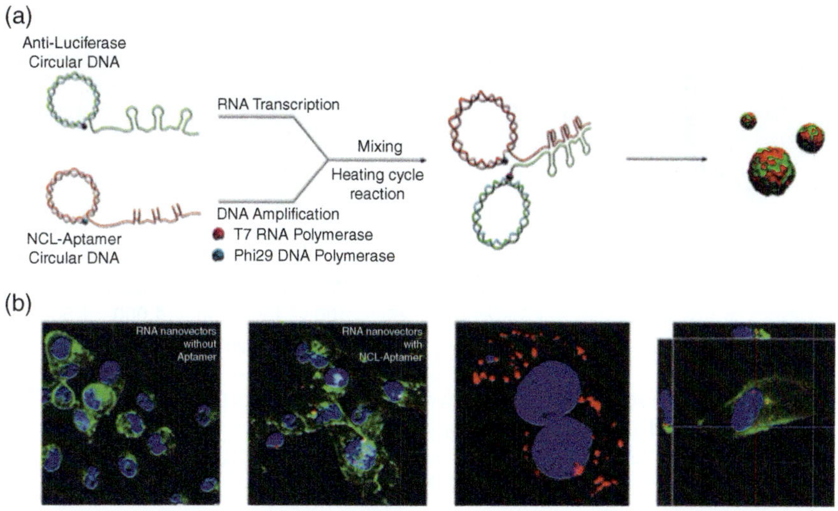

Figure 9.16 Enzymatic replication-based AS1411 apatamer-expressed RNA nanovector. (a) The RNA nanovector synthesis process. (b) AS1411 aptamer-dependent cellular uptake. Source: Park *et al.* [89] / With permission of American Chemical Society.

normal tissues [97]. In order to improve the effectiveness of targeted delivery of siRNA to human cancer cells, folic-acid-conjugated oligonucleotide nanoparticles (ONPs) were synthesized through base pairing (Figure 9.17). A tetrahedron formation was fabricated, using six single-stranded oligonucleotides, and folic acid was conjugated to the structures by hybridization. The pharmacokinetic profile and biodistribution of ONPs were investigated in nude-mice-bearing KB xenograft tumors. The folic-acid-conjugated ONPs were found to primarily accumulate in the tumors.

In addition to DNA or RNA hybridization, bio-orthogonal conjugation chemistry has also been applied to functionalize DNA nanocages with folic acid (Figure 9.18a) [97]. Biotinylated octahedral DNA nanocages were functionalized with folic acid, and DOX was successfully loaded into these DNA nanocages, for the treatment of cancer. Selective delivery, uptake, and internalization were achieved (Figure 9.18b).

RGD peptide (Arg-Gly-Asp), a tripeptide motif found in extracellular matrix proteins, is a unique biomaterial involved for cell adhesion and intracellular signal transduction through interactions with integrins expressed on the cell membrane [98, 99]. RGD may be used in drug delivery for cancer therapy, owing to its high efficacy for intracellular delivery of entrapped drugs to cancer cells. There have been many studies on the use of RGD-modified DNA nanostructures for efficient drug delivery into target cells. One of these studies used an RCA-induced DNA scaffold containing hundreds of repetitive sequence units, labeled with the RGD peptide by complementary DNA strands

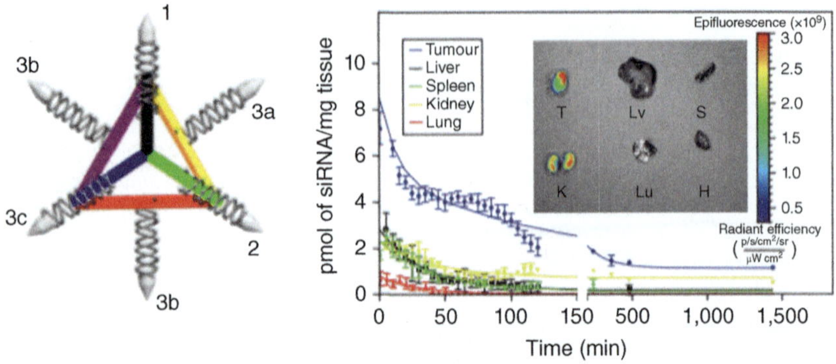

Figure 9.17 Folic acid expressed on DNA tetrahedron and its tumor-specific delivery. Source: Lee *et al.* [167] / With permission of Springer Nature.

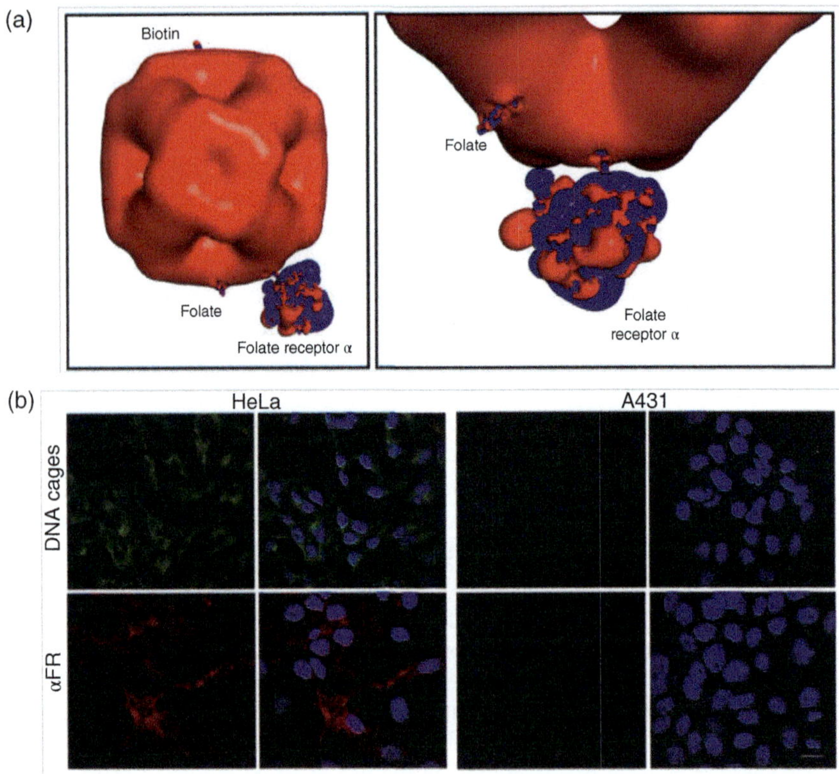

Figure 9.18 (a) Bio-orthogonal conjugation-induced functionalization of DNA nanocages. (b) Selective delivery of **αFR** into HeLa cells. Source: Raniolo *et al.* [97] / With permission of Elsevier.

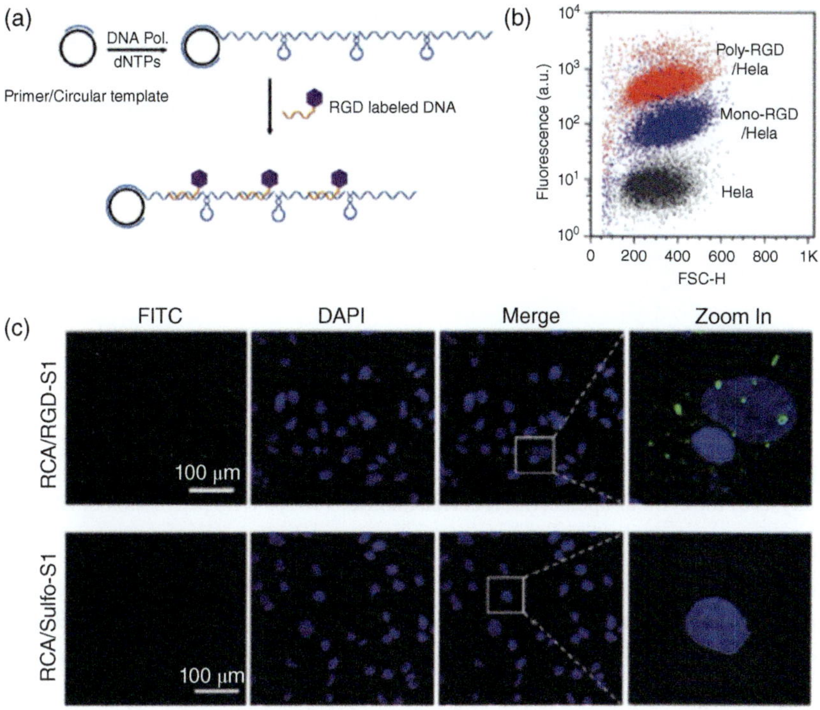

Figure 9.19 RGD-conjugated DNA nanostructure for the cellular uptake. (b) Quantification of the fluorescence intensity depending on the level of RGD modification. Source: Zhang *et al.* [99] / Royal Society of Chemistry / CC BY-3.0.

(Figure 9.19a). This RGD-conjugated DNA structure demonstrated high binding affinity to HeLa cells, as compared to the control group, showing an intense green fluorescence signal in the targeted cells (Figure 9.19b, c).

Antibody-drug carrier conjugates are promising systems for target-specific drug delivery [100]. Antibody-drug conjugation can improve the efficiency of drug delivery to target cells, overexpressing specific antigens on their cell membranes. This approach has attracted significant interest, owing to its successful tumor-selective delivery and low cytotoxicity. Classical systems for the delivery of therapeutic RNA, such as siRNA, have used cytotoxic positively charged polymers or liposome-based carriers [101]. These systems trigger an immune response and do not allow target specific delivery, resulting in low delivery efficiency [102–104]. Lee and co-workers suggested the use of an anti-epidermal growth factor receptor antibody (anti-EGFR Ab)-conjugated RNA nanoball to target diverse tumor cells [101]. The conjugation between the anti-EGFR Ab and the RNA nanoball was achieved through biotin-streptavidin (STA) interactions (Figure 9.20a). The RNA nanoball was labeled with biotin, using biotin-labeled UTP in the RCT process, followed by mixing with STA-anti-EGFR Ab.

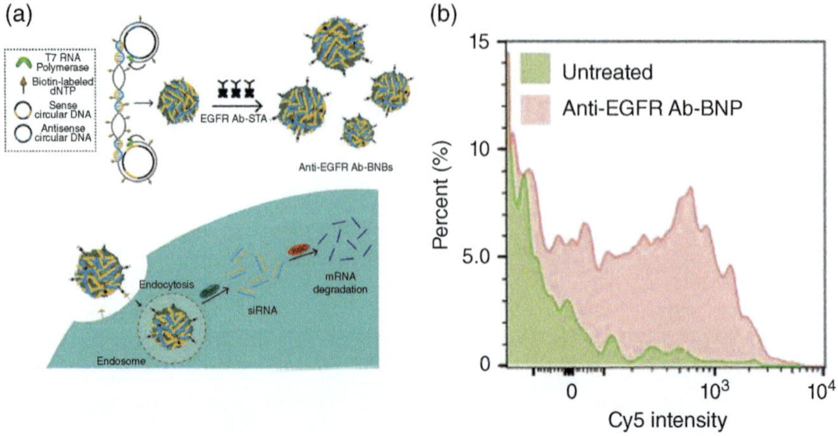

Figure 9.20 (a) Synthesis of anti-EGFR Ab-conjugated RNA nanoball. (b) Cytometric analysis of the cellular uptake. Source: Han and Bee [101] / With permission of Elsevier.

The uptake efficiency was confirmed by cytometry, which showed higher cyanine 5 intensity in the treated cells compared to untreated cells (Figure 9.20b). As a result, the GFP knockdown was successfully achieved, demonstrating this antibody conjugation strategy was efficient.

The covalent conjugation of antibodies to DNA nanostructures has also been reported to improve specific cellular drug delivery [105]. For the therapeutic delivery to EGFR-overexpressing breast cancer cells, a cetuximab (anti-EGFR antibody)-linked DNA tetrahedron was synthesized using a sulfo-SMCC crosslinker (Figure 9.21a). The DNA tetrahedron was formed by self-assembly using four complementary strands, and cetuximab was conjugated to the thiol groups on the tetrahedron. We confirmed that the enhanced uptake efficiency was dependent on the presence of cetuximab on the DNA nanostructure (Figure 9.21b). The studies suggest that target-specific antibody-linked nucleic acid–based nanostructures have huge potential as active targeting systems for efficient drug delivery.

A cell-penetrating peptide (CPP)-modified DNA nanostructure was also developed to improve endocytosis-mediated uptake into target cells [106]. CPP is a remarkable material, composed of short amino acid sequences, which are responsible for its cell-penetrating ability [107]. CPP may be covalently or non-covalently conjugated with nucleic acids; however, non-covalent assembly can cause the formation of aggregates with highly complex secondary and tertiary structures, resulting in low uptake efficiency [108, 109]. One study reported a CPP-labeled DNA tetrahedron, formed by a click reaction (Figure 9.22a) [106]. This nanocarrier demonstrated improved cellular uptake, and efficient delivery of DOX, by using a specific ligand, neuropilin-1, to target glioblastoma and endothelial cells. As shown in Figure 9.22b, the microscope images demonstrated

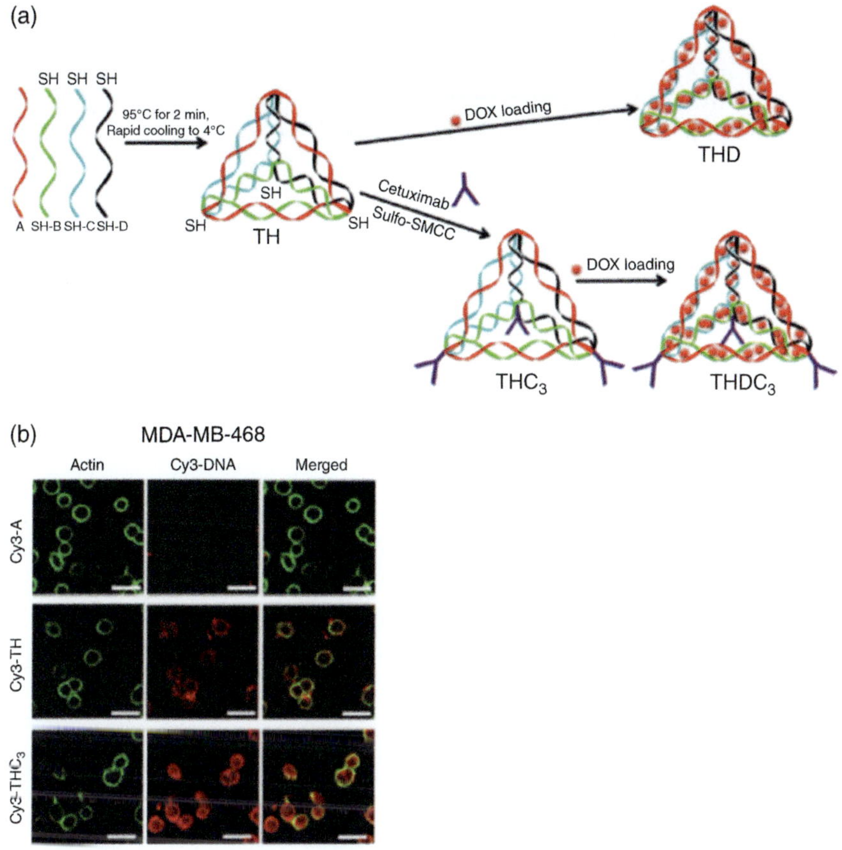

Figure 9.21 (a) Synthesis of cetuximab-expressed DNA tetrahedron nanostructure. (b) Microscope images for the confirmation of cetuximab-dependent enhanced uptake efficiency. Source: Setyawati *et al.* [105] / With permission of John Wiley & Sons.

that the CPP-modified DNA tetrahedron is the most effective nanostructure for targeting human primary glioblastoma cells (U87MG). This was further confirmed by flow cytometry profiles.

Biomimicking has also been reported as a strategy for the cellular delivery [91]. Virus capsid protein (CP) can bind to the surface of DNA origami structures through electrostatic interactions, and this complex shows improved transfection efficiency (Figure 9.23a). As shown in Figure 9.23b, internalization of the DNA origami–CP complex into kidney cells (HEK293) can be confirmed by the confocal microscope image. The transfection efficiency was enhanced by more than 13-fold compared to the non-complexed DNA origami. The study also reported that the transfection efficiency of the DNA–CP complex is much higher than that of the commercial transfection agent, lipofectamine 2000,

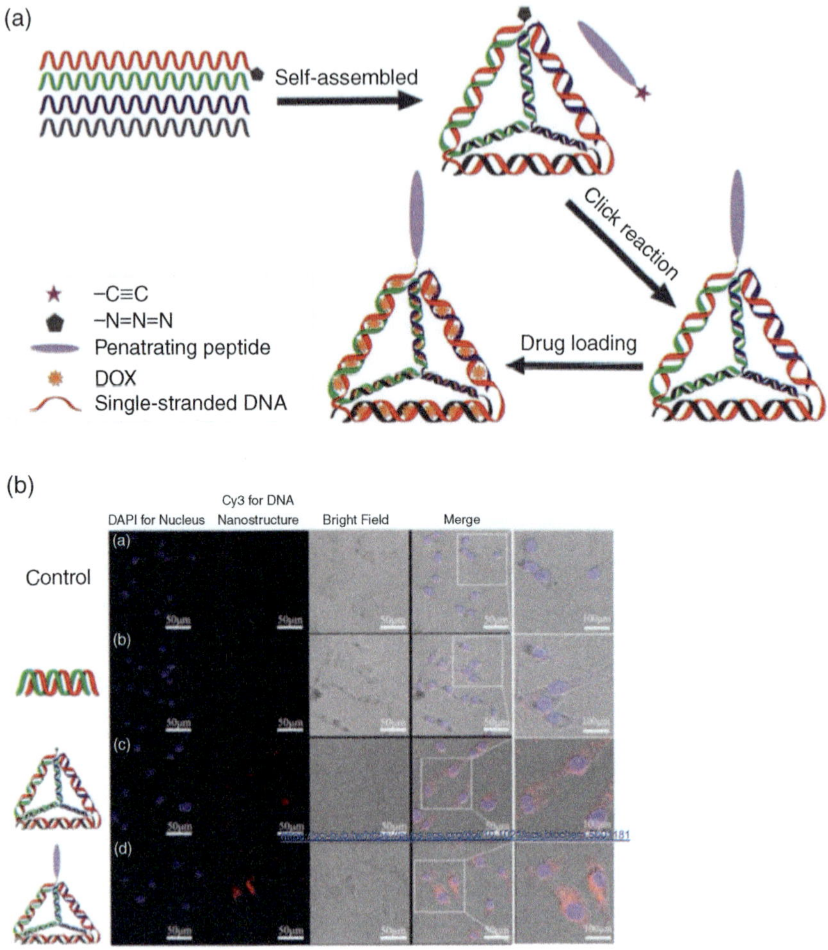

Figure 9.22 (a) Synthesis of CPP-labeled DNA tetrahedron through click chemistry. (b) NRP-specific delivery efficiency into U87MG cells. Source: Xia *et al.* [106] / With permission of American Chemical Society.

which was used as a control. Therefore, biomimicking-based transfection methods are expected to be improved by the addition of targeting moieties and ligands, or by the use of various viruses, to achieve low toxicity and enhanced cellular uptake.

9.3.3. Trigger-Induced Drug Delivery System

Beyond the traditional active targeting strategies, physiological environment–responsive DNA and RNA transformation-induced drug release has been widely

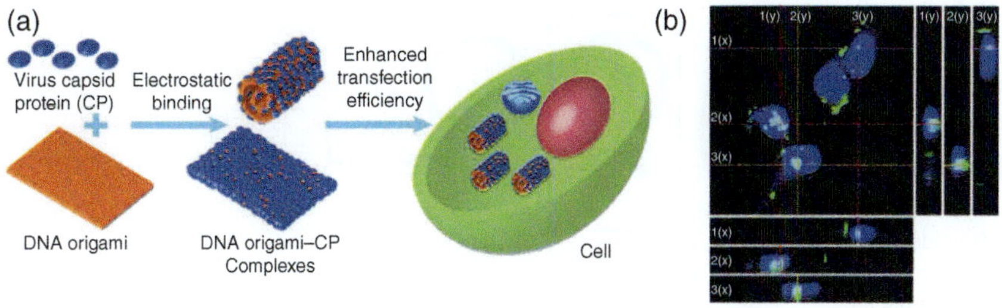

Figure 9.23 Virus capsid-encapsulated DNA origami-based delivery system. (a) Conceptual illustration of the synthesis of DNA origami-CP complex. (b) Confocal microscopy image for confirmation of cellular uptake. Source: Mikkila *et al.* [91] / With permission of American Chemical Society.

studied to help achieve biological specificity and therapeutic efficacy. These stimuli-responsive systems, also called "smart" or "intelligent" systems, are used in controlled and localized drug delivery and are able to minimize side effects. The drug is released from the carrier in response to triggers such as pH [110], temperature [111], light [112–115], target protein [116], ATP [117, 118], ultrasound [119], enzymatic activity [120], or pathogen-relevant miRNA molecules [121–123]. In this section, some trigger-induced drug delivery strategies are introduced.

9.3.3.1. pH-Responsive The pH-responsive transformation of DOX-loaded DNA nanoparticles between the i-motif quadruplex and DNA hairpin structure was reported (Figure 9.24a) [124]. In a normal physiological environment at pH 7.4, the internal structure maintained a hairpin shape, and DOX was intercalated in the duplex stem region. When the pH fell to 5.0, the hairpin transformed into an i-motif structure, releasing DOX from the DNA nanoparticle. This was confirmed by measuring the fluorescence intensity of DOX. When DOX is intercalated within the DNA duplex, the fluorescence intensity is very weak; when it is released from the duplex, the intensity increases. This system was tested on a CCRF-CEM cell line, human T lymphocytic leukemia cells, and Ramos cells, and on human B immunoblastic lymphoma cells as a control. DOX uptake efficiency was confirmed in each cell line. The CCRF-CEM cell line is known to have a low-pH in the cell organelles [125]. As shown in Figure 9.24b, high fluorescence intensity was recorded in in the CCRF-CEM cells, while a weaker intensity was recorded in the Ramos cells. Based on the uptake efficiency, the cell viability also differed between the CCRF-CEM and Ramos cell lines. Cell viability in the CCRF-CEM cells was lower than in the Ramos cells. This result indicates that the DNA nanostructures embedded in the i-motif quadruplex structure can be used as a means of pH-responsive targeted drug delivery.

In addition to the i-motif structure-based switching system, Hoogsteen base interactions also allow pH-responsive reversible changes in the conformation of

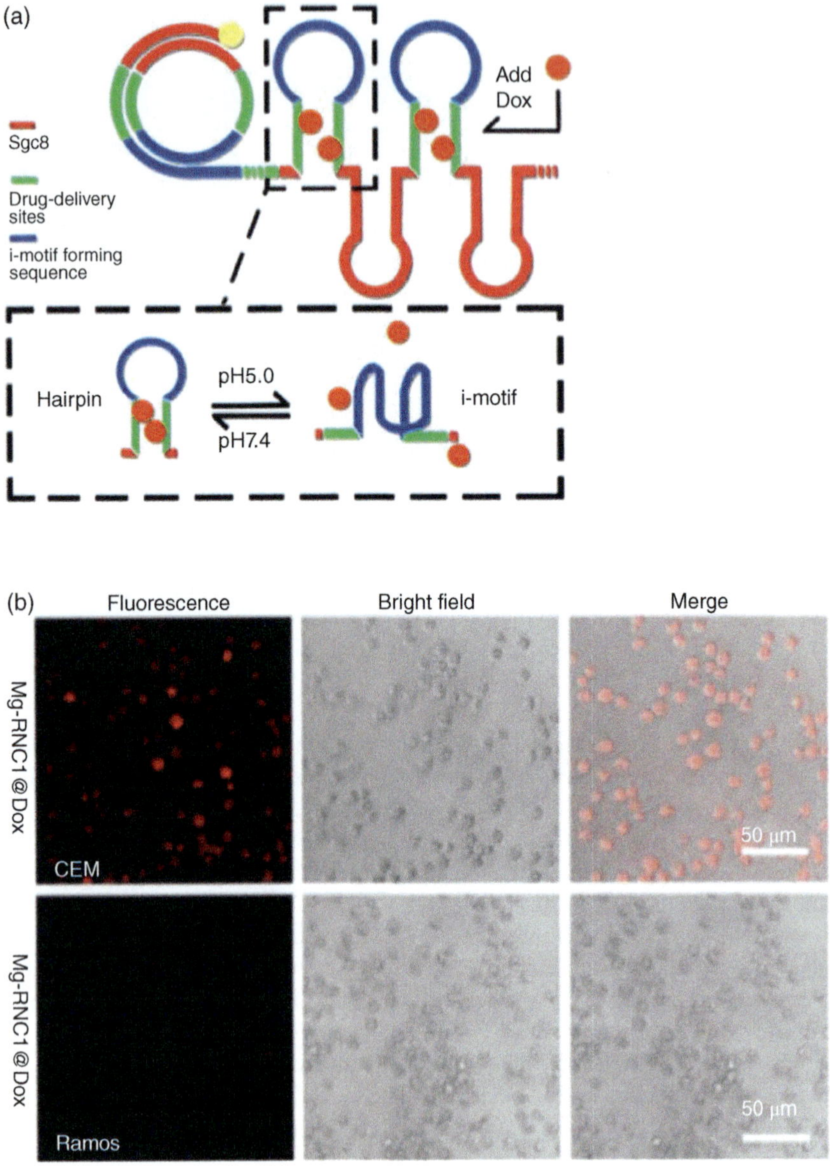

Figure 9.24 (a) pH-responsive transformation system between i-motif quadruplex and hairpin structures. (b) Cellular uptake efficiency depending on the target cells. Source: Zhao *et al.* [124] / With permission of American Chemical Society.

triplex DNA structures. Protonated CG·C$^+$ parallel domains favor triplex formation in acidic conditions, while TA·T parallel domains at neutral pH [110]. Hoogsteen base pairing can allow the rational design of pH-induced reversible DNA structures for the drug delivery.

9.3.3.2. Light-Responsive The use of light as a remote trigger is an attractive strategy for drug delivery [126]. Unlike pH-responsive systems, external irradiation with light serves as a remote trigger for light-sensitive molecules, causing the release of encapsulated drugs into target cells and organs. Since this strategy is dependent on external light, drug release can be rationally controlled by regulating the power or light sources [127] or by using reversible or irreversible transformable materials [113, 128]. In 2012, Tan and co-workers introduced photo-responsive DNA cross-linked hydrogels, containing the azobenzene moiety, which can transform between the *cis* and *trans* conformations [128]. Azobenzene-based derivatives have attracted considerable attention as photoswitchable materials, assuming the *cis* conformation upon UV irradiation at 350 nm, and the *trans* conformation upon irradiation with visible light [114, 129]. Azobenzene isomerization can reversibly denature and anneal DNA duplexes, depending on the light source. A photo-controllable DNA cross-linked hydrogel was formulated with an azobenezene-modified DNA linker (Azo-DNA linker) (Figure 9.25a). The Azo-DNA linker successfully connected DNA-polymer conjugates to form a drug-loaded network structure under visible light. UV irradiation caused this azobenzene moiety to switch to the *cis* state, leading to the denaturation of the hybridized DNA duplex, and subsequent drug release. It can be seen in Figure 9.25b that DOX was well encapsulated under visible light at 450 nm, while dramatic release was observed upon UV irradiation. This release could be observed with the naked eye, as shown in Figure 9.25c. The DOX-loaded DNA cross-linked hydrogel gradually degraded during the period of UV irradiation, releasing DOX from the matrix.

Sugiyama and co-workers also applied this concept to DNA nanocapsules, using photoirradiation to switch between an open and closed system for the smart delivery of the encapsulated drug [113]. The DNA-attached cargo was loaded into the DNA nanocapsules via the complementary strands, and it showed successful release by strand displacement, through a toehold system under UV irradiation.

Near-infrared (NIR)-sensitive DNA-based drug delivery systems have also been developed [112]. NIR is a highly attractive stimulus owing to its non-invasive nature and deep tissue penetrability. Farokhzad and co-workers introduced NIR-induced thermal chemotherapy, using gold nanoparticles coated with self-assembled DNA. NIR irradiation increases the temperature, resulting in DNA denaturation and the release of the intercalated DOX.

A wide range of approaches to photo-responsive DNA-based drug delivery have introduced a new class of smart materials, which are responsive to external stimuli. However, most light-responsive systems have been analyzed *in vitro* in

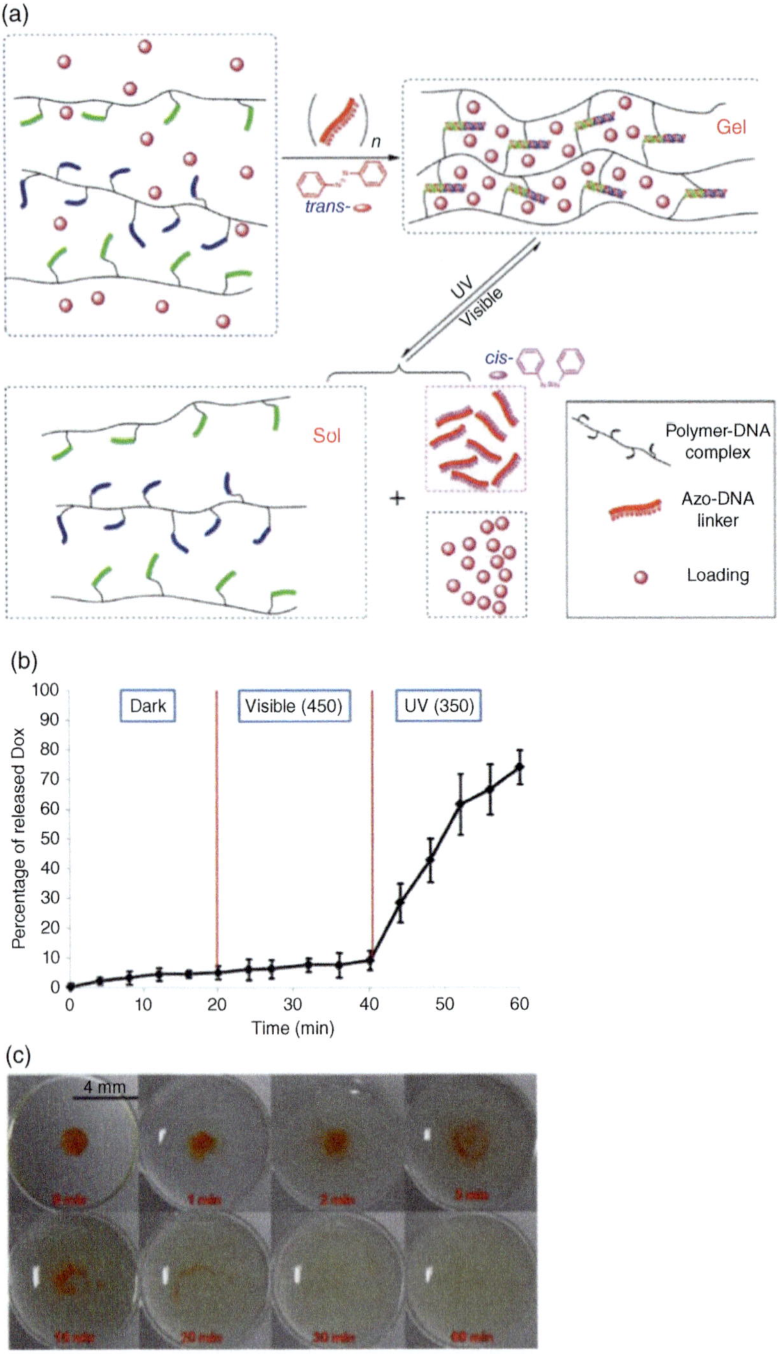

Figure 9.25 (a) UV and visible light-responsive cis-trans conformation-based drug delivery system. (b) Light-source-dependent DOX release profile. (c) Irradiation time-dependent DOX release. Source: Kang *et al.* [128] / With permission of American Chemical Society.

a controlled environment. Most studies have simply served as a proof-of-concept, and *in vivo* evaluation of photo-responsive nucleic acid–based materials remains a challenge [115]. *in vivo* evaluation of light responsiveness will make this an attractive advanced therapeutic strategy for precise and specific drug delivery [126].

9.3.3.3. Small Molecules-Responsive Biological triggers from target cells, such as metal ions, antigenic proteins, adenosine triphosphate (ATP), or miRNA can operate the smart lock system of nanocarriers used in targeted drug delivery [130]. DNA and RNA are powerful materials, due to their sequence specific binding with target molecules. Aptamers have been widely used in systemic drug delivery, since they are able to switch between an aptamer-complement duplex and an aptamer-target complex. In 2018, Li *et al.* reported DNA nanorobot for a cancer therapeutic in response to a molecular trigger [116]. Thrombin could be anchored in this DNA nanorobot by hybridization of poly-A and poly-T strands and closed by predesigned fastener strands consisting of specific antigen binding aptamer (AS1411) and complementary strand, which can be opened depending on the recognition of a target antigen (Figure 9.26a).

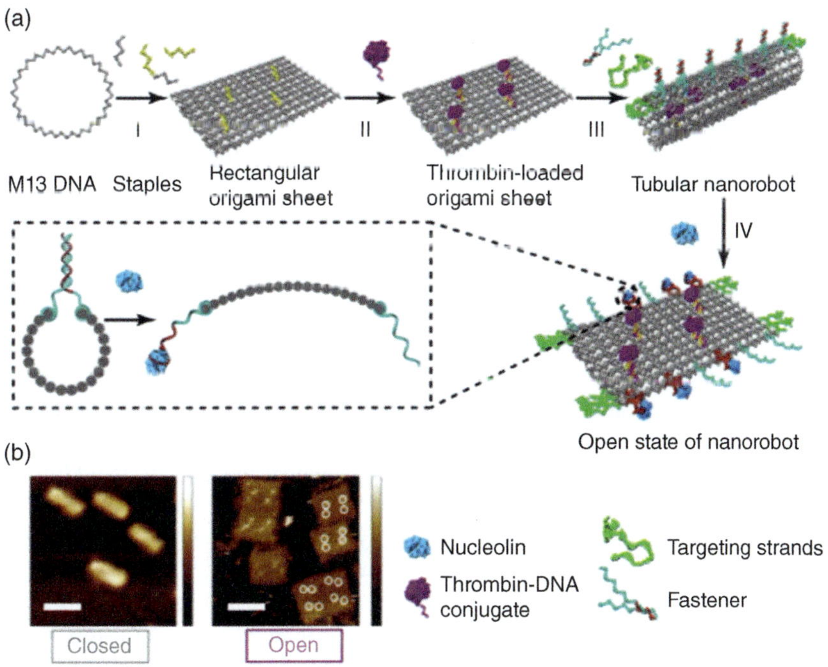

Figure 9.26 (a) Conceptual illustration of DNA nanorobot which is controllable by the presence of specific antigen. (b) AFM images of closed/open states of the DNA nanorobot. Source: Li *et al.* [116] / With permission of Springer Nature.

When the aptamer recognizes their targets, reconfiguration of the aptamer leads the dissociation of DNA duplex resulting in open state of the DNA origami cage. The payloads that are pre-conjugated inside the nanorobot can be successfully released by the antigen key. AFM images clearly shows the open and closed conformations of the DNA nanorobot (Figure 9.26b). This suggests that antigenic trigger-induced DNA origami-based systems may be used for specific cell-targeted delivery.

ATP is also widely used as a trigger for intracellular delivery of anticancer drugs. ATP, an energy transporter for cell metabolism, is present at low concentrations in the extracellular environment, and in relatively high concentrations within the cytosol [117, 131, 132]. This distinction makes it possible to enhance the intracellular drug release while keeping the drug encapsulated within the carrier in the extracellular environment. To achieve this, numerous studies have reported the incorporation of ATP-binding DNA aptamers into drug nanocarriers [117]. One example is a DOX-loaded ATP-responsive DNA scaffold, designed to improve the delivery of the chemotherapeutic agent to the site of tumor growth (Figure 9.27a). This nanocarrier design involved the transformation of an ATP aptamer into a tertiary structure in the presence of ATP, allowing it to dissociate from the DNA duplex. As a result, DOX, which was intercalated between the DNA duplexes, could be efficiently released. This drug delivery system enhanced the cytotoxicity and improved apoptosis of the target cells. As shown in Figure 9.27b, time-dependent release of DOX from the ATP-sensitive aptamer-encoded nanogel (DOX/NG) was confirmed in MDA-MB-231 cells, whereas the control nanogel (DOX/cNG), without ATP response, was unable to successfully release DOX. Moreover, the gradual accumulation of DOX in the nuclei can be seen in Figure 9.27c. Two hours after treatment with DOX/NG, DOX co-localized with late endosomes and lysosomes (left) within the cells. However, when the cells were incubated for 6 h, DOX was observed in the nuclei (right), indicating that the ATP-rich environment induced efficient DOX release from the DNA duplex. Therapeutic RNA and peptides could also be successfully co-delivered using the DNA nanostructure [118]. ATP triggered the disassembly of the DNA nanostructure in cancer cells and allowed the function of antimiR-21 to be restored, leading to successful inhibition of tumor growth. However, it is difficult to use the ATP-responsive system along, since ATP-responsiveness is generated after endosomal escape of the drugs, which means that an additional active targeting strategy is required for specific drug delivery.

MiRNAs are a class of small non-coding single-stranded RNA molecules, associated with the processing of cancer-relevant drug targets [121]. This endogenous biomolecule-regulated drug release system presents the advantage that the dose of the drug can be controlled by the intracellular expression of miRNA. A feedback system of the miRNA has a catalytic effect and boosts drug release [133, 134]. As shown in Figure 9.28a, miRNA-21 can catalyze DOX release, followed by disassembly of the DNA duplex. The catalytic reaction is initiated by the overexpression of miRNA-21 in the cytoplasm, and DNA fuel

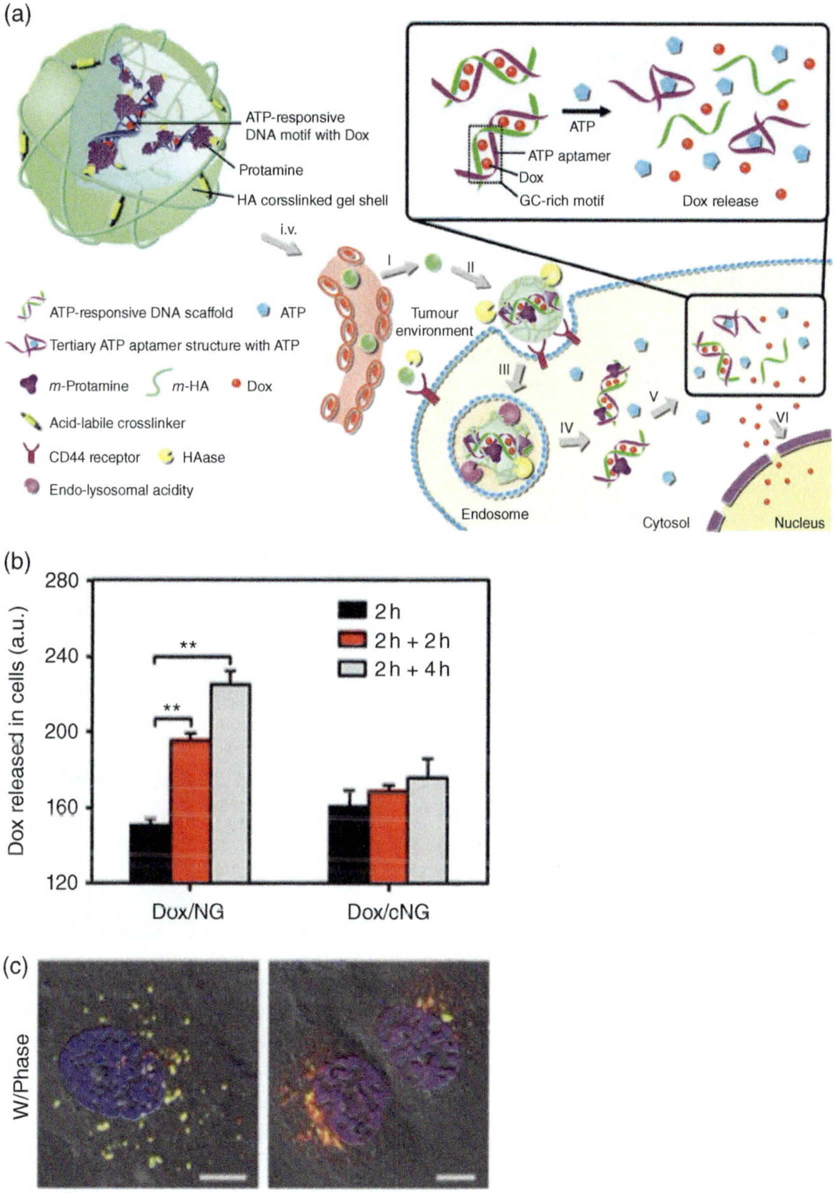

Figure 9.27 (a) ATP-responsive DOX release system. (b) Time-dependent release of DOX in presence of ATP and (c) accumulation of DOX in nuclei. Source: Mo *et al.* [117]/ With permission of Springer Nature.

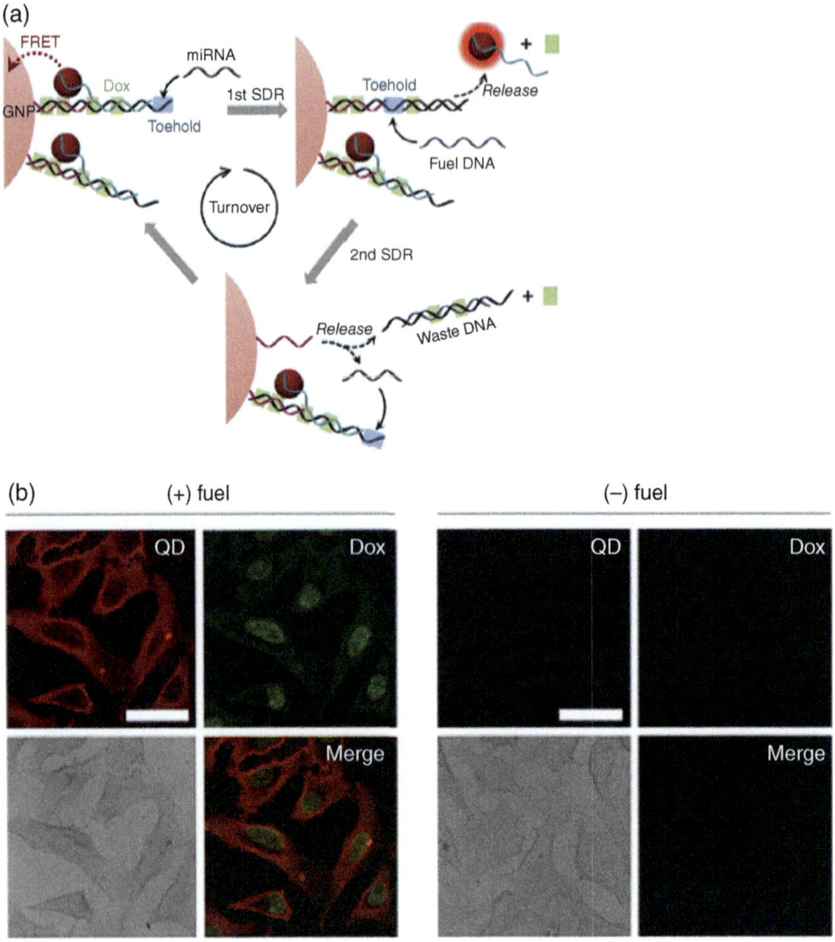

Figure 9.28 (a) Catalytic reaction and DOX release system, triggered by miRNA-21. (b) DOX release boosting in cytosol depending on the additional DNA fuel at low miRNA-21 concentration. Source: Luo *et al.* [121] / With permission of American Chemical Society.

maintains the turnover of catalytic activity, releasing the miRNA so that it can be reused to dissociate the neighboring DNA duplex for DOX release. This system can regulate the dose of DOX released, depending on the concentration of miRNA-21 in the cytoplasm. DNA fuel is required for boosting DOX release, even at low concentrations of miRNA-21, as demonstrated by the results shown in Figure 9.28b. The miRNA-responsive strategies have also been applied to RNA-based delivery approaches for gene editing systems [123]. Saito *et al.* developed a synthetic RNA-based, specific miRNA-responsive CRISPR-Cas9 (miR-Cas9) switch system for selective genome editing in mammalian cells. In

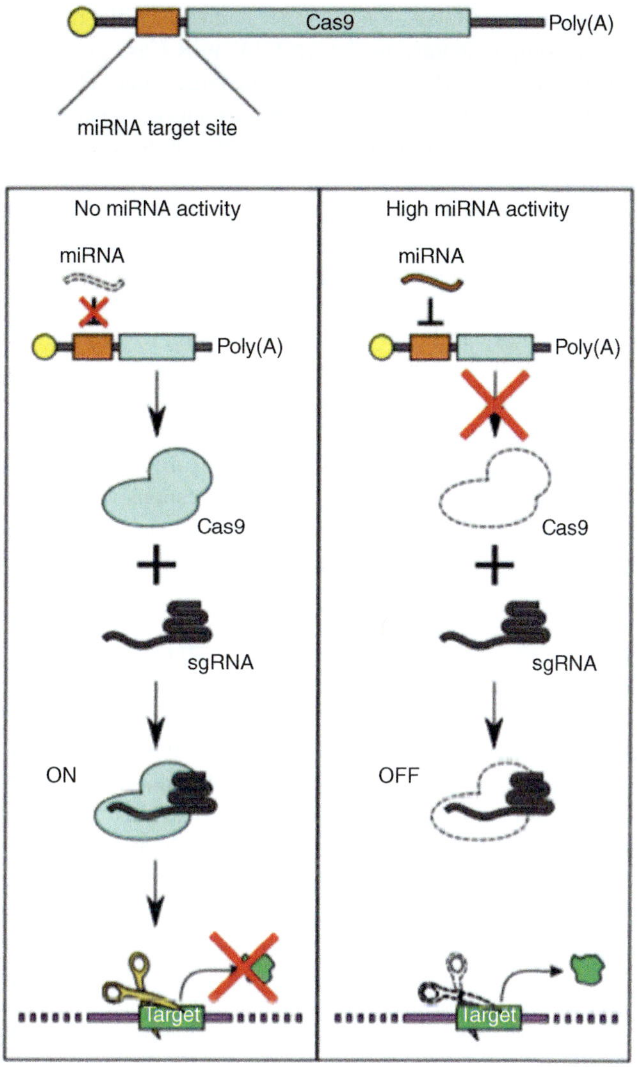

Figure 9.29 miRNA-responsive CRISPR-Cas9 switch system for selective genome editing. Source: Hirosawa *et al.* [123] / With permission of Oxford University Press.

this study, a novel miR-Cas9 switch system was designed to control Cas9 activity, depending on endogenous miRNA expression in the target human cells (Figure 9.29). The miRNA activity with the miR-Cas9 switch regulated the translation of the Cas9 protein and its complexation with sgRNA. When the miR-Cas9 switch did not interact with miRNA, Cas9 protein was translated from the miR-Cas9 switch and could form a Cas9-sgRNA complex, leading to the digestion of the target DNA. On the other hand, when the miR-Cas9 switch was completely complementary to the miRNA, miR-Cas9 could not be translated into Cas9, resulting in the expression of the target gene. This switch system successfully demonstrated the miRNA can be used as a trigger for smart therapeutic RNA delivery by selective engineering of genomes of target cells with endogenous miRNA.

Despite the significant developments in smart drug delivery systems, their clinical application has been limited by multidrug resistance [135, 136]. To overcome multidrug resistance, multi-stimuli-responsive delivery systems have been studied over the years. Guo *et al.* reported a nanocomplex for tri-stimuli-triggered DOX and pyronaridine (PND) delivery, using DNA and gold nanorods. DOX and PND-encapsulated dsDNA were conjugated on the gold nanorod, and the chemotherapeutic drugs were released in sequential steps: an NIR irradiation-induced photothermal effect, triggering endosomal escape; pH-responsive changes in DNA conformation; and cleavage of disulfide bonds through redox reactions (Figure 9.30) [137]. This multi-stimuli-responsive system has introduced a new route for sufficient and controllable drug payload release, allowing us to achieve efficient disease treatment.

9.4. APPLICATIONS OF DNA/RNA NANOSTRUCTURES FOR DRUG DELIVERY

We have discussed various kinds of the DNA- and RNA-based structures and targeting strategies for the efficient drug delivery. In this chapter, we will review the practical and in-depth application of the drug delivery with nucleic acid–based drug carrier. The application is detailed in three categories: chemical drug, protein drug, and nucleic acid drug delivery.

9.4.1. Small-Molecule Drug Delivery

One of the most commonly used drugs in the treatment of cancers is DOX. However, the properties of free DOX limits its clinical applications. Firstly, free DOX lacks tumor specificity and therefore is also cytotoxic toward healthy organs or cells. Secondly, being an anthracycline drug, DOX is composed of multiple rings, which makes it insoluble in water [138]. These factors significantly limit clinical translation and long-term drug storage. Therefore, it is important to develop an efficient delivery system to overcome these limitations. In this regard, DNA and RNA are highly versatile materials that can be used to overcome these limitations. Since DOX can be intercalated into the nucleic acid double-helix structure without the need for chemical modification or reaction, DNA and RNA are very good carriers for its delivery [112, 139]. Several nucleic acid–based nanostructures and delivery strategies have been discussed in previous chapters, and there have been numerous studies on targeted DOX delivery.

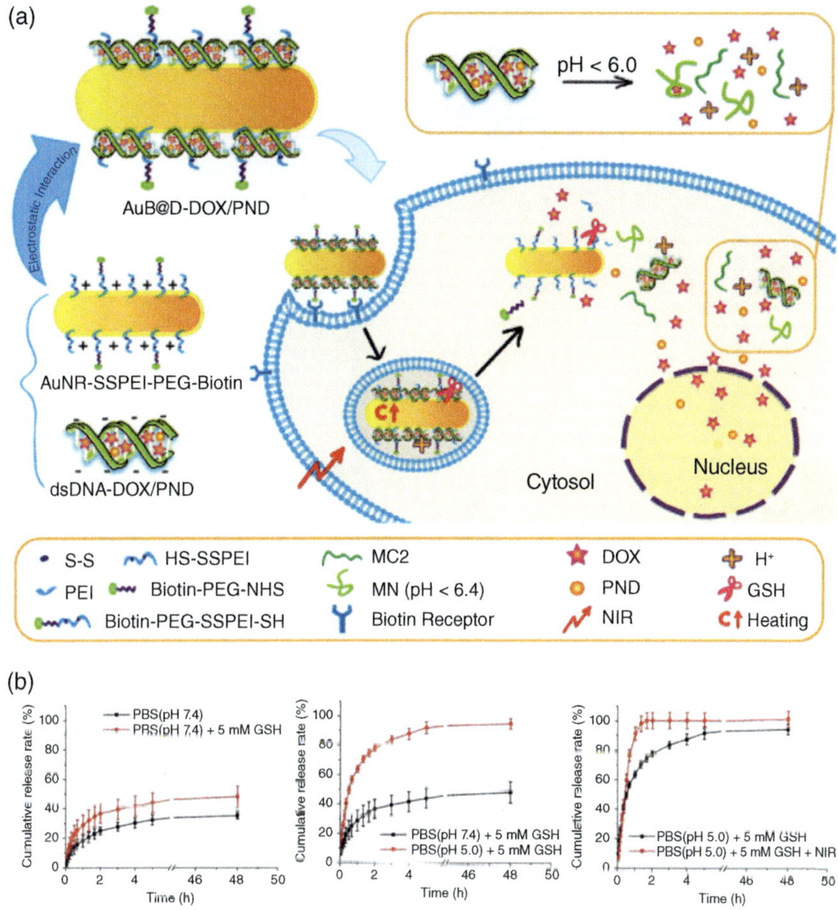

Figure 9.30 (a) Tri-stimuli-responsive DOX and PND delivery system with DNA and gold nanorod. This system is operated by NIR irradiation-induced photo thermal effect and pH-responsive conformation and cleavage. (b) GSH, pH, and NIR responsive release rate. Source: Wang *et al.* [137] / With permission of Elsevier.

As shown in Figure 9.31a, DOX-intercalated aptamer-tethered DNA nanotrains have been reported for tumor-targeted delivery. These self-assembled nanotrains could be synthesized from aptamers and short DNA building blocks, forming a chain of repeating "boxcars," where DOX could be intercalated. This model has a high DOX payload capacity and showed successful binding and internalization in the target cells (Figure 9.31b). Application of the aptamer sequence in this system was able to reduce off-target effects and achieve good antitumor efficacy. Moreover, DOX-encapsulated DNA nanocarriers have also been reported to help overcome drug resistance (Figure 9.31c) [25]. As shown in Figure 9.31d, a DNA tetrahedron was successfully internalized into both the

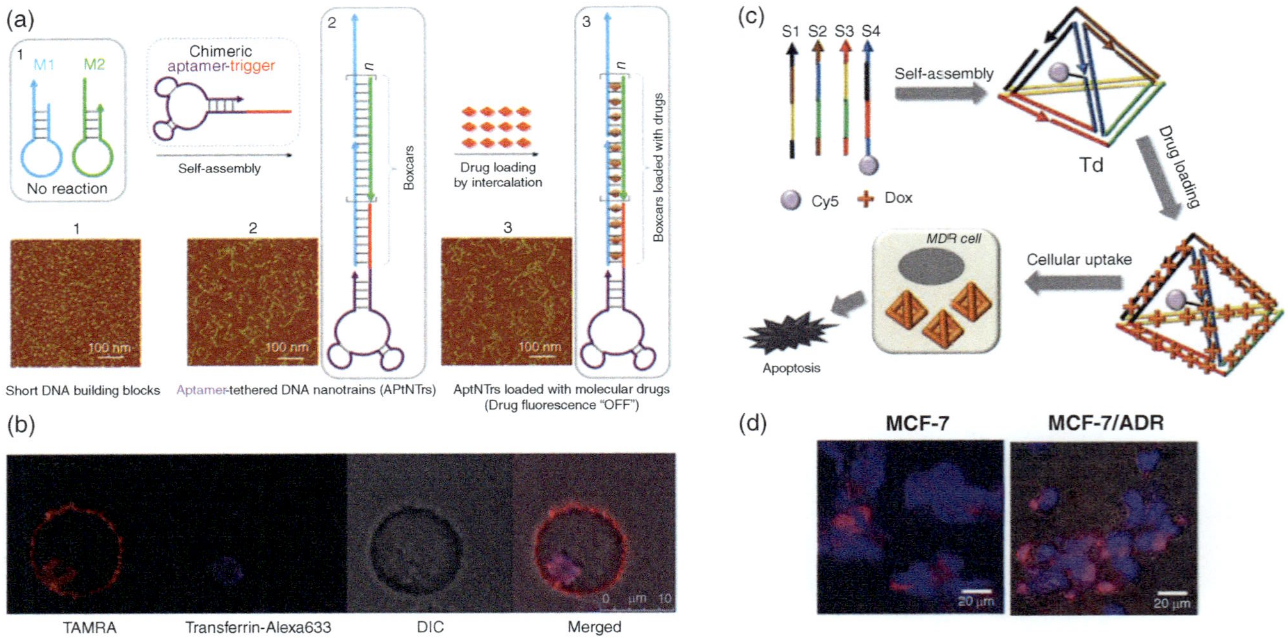

Figure 9.31 (a) Self-assembly of aptamer-tethered DNA nanotrain with DOX loading. (b) Confocal laser-scanning microscopy images for confirmation of cellular internalization. Source: Zhu *et al*. [181] / With permission of National Academy of Science. (c) Drug-loaded DNA tetrahedron and its apoptotic effect. (d) Fluorescence microscopic images of DOX delivery to drug-sensitive and resistant cells. Source: Kim *et al*. [25] / With permission of Royal Society of Chemistry.

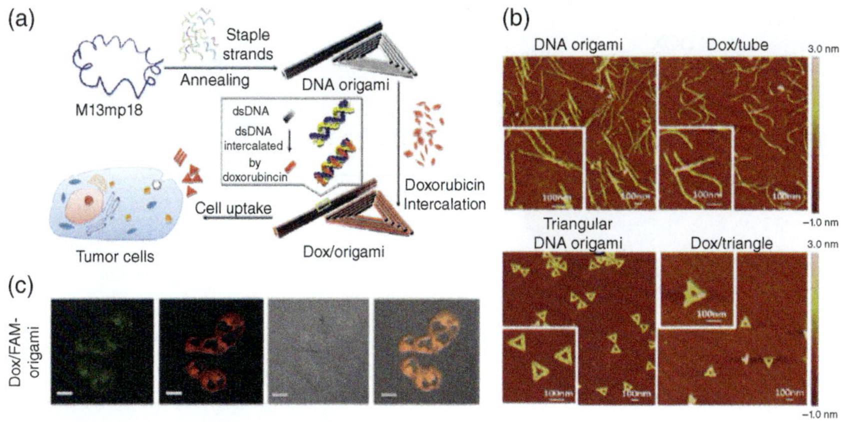

Figure 9.32 (a) DOX encapsulation in DNA origami and their cellular uptake. (b) AFM images of various DOX/DNA origami complex. (c) Confocal microscope images for confirmation of co-localization of DOX and DNA origami. Source: Jiang *et al.* [24] / With permission of American Chemical Society.

drug-sensitive (MCF-7) and drug-resistant breast cancer cells (MCF-7/ADR), demonstrating that DOX-DNA nanostructures can effectively inhibit the growth of the cancer cells. This concept was also applied to more elaborate DNA nanostructures, and DNA origami, to deliver DOX to drug-resistant cancer cells (Figure 9.32a, b) [24]. Yan, Ding, and co-workers evaluated the effects of DNA origami shapes on drug delivery efficiency. Tube and triangle-shaped DNA origami structures exhibited good cellular internalization (Figure 9.32c) and high biocompatibility, suggesting that DNA origami has huge potential as a drug carrier for anticancer drugs.

Enzymatically amplified DNA nanoclews were developed in 2014 and could be used along with a targeting moiety, as DOX carriers [26]. This self-assembled cocoon-like DNA nanocomposite could be fabricated by RCA, using folic acid for tumor targeting. An acid-degradable polymeric shell was expressed on the DNA nanoclew through DNA complementarity and electrostatic interactions (Figure 9.33a). Deoxyribonuclease (DNase) was encapsulated in the acid-degradable polymeric shell; therefore, the acidic environment in the tumors could induce its release, resulting in the degradation of the DNA nanoclews and release of DOX. As shown in Figure 9.33b and c, DOX was successfully released in at pH 5.4, and time-dependent DOX migration from the endosome to the nucleus was confirmed by confocal microscopy. These results show that RCA-based products could also be useful for DOX encapsulation and can be easily functionalized by other molecules for efficient target-specific delivery.

DOX can also be intercalated into RNA–DNA hybrid nanostructures [140]. The system was synthesized with a three-way junction (3WJ) motif from phi29 packaging RNA and annexin A2-targeting DNA aptamer (Endo28)

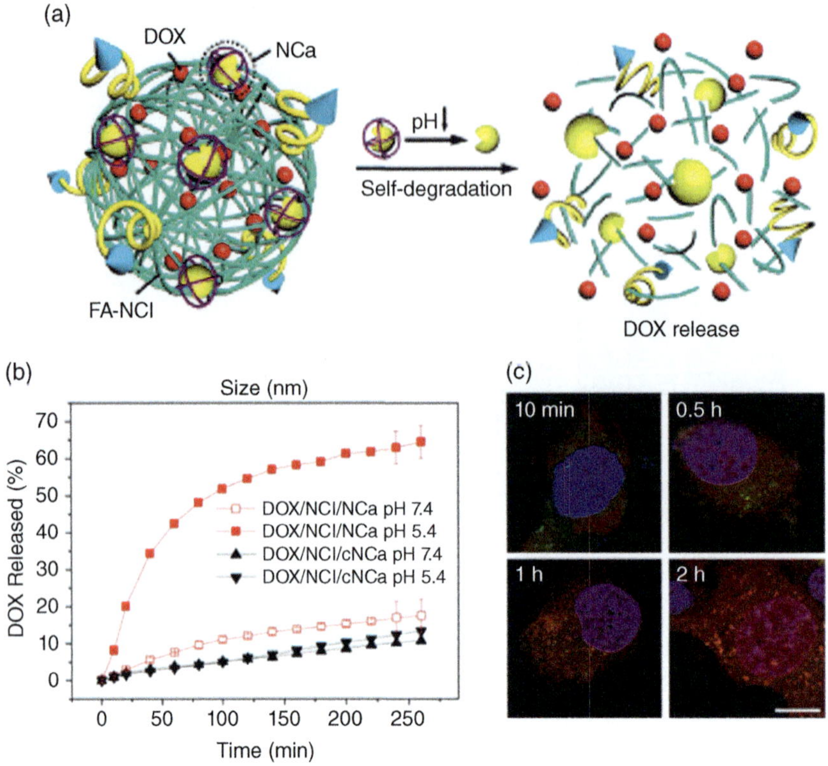

Figure 9.33 (a) Main components of the DNA nanoclew and DOX release with its degradable property. (b) DOX release profile depending on the system and pH. (c) Confocal laser scanning microscopy image of MCF-7 cells after the incubation with DNA nanoclew at different time points. Source: Sun *et al.* [26] / With permission of American Chemical Society.

(Figure 9.34a). The fluorescence spectra of the DOX-intercalated DNA–RNA hetero-duplex demonstrated good DOX loading efficiency, dependent on DOX concentration (Figure 9.34b). This 3WJ-based RNA–DNA hybrid nanoparticle allowed enhanced accumulation of DOX in ovarian cancer cells, with reduced distribution to healthy cells and organs (Figure 9.34c).

In addition to intercalation-based drug loading methods, modified nucleotides have also been widely used for drug loading by chemical or hydrophobic conjugation [27, 82, 141]. Chemical conjugation via click chemistry [142], carbonate bonds [143], disulfide bonds [144], and EDC/NHS chemistry [145], as well as non-covalent conjugation [146], such as hydrophobic interactions [154] with nucleic acids, have been developed through research on nucleotide modification [147, 148]. Guo and co-workers reported that paclitaxel (PTX)-labeled four-way junction (4WJ) RNA nanoparticles were able to improve the solubility of

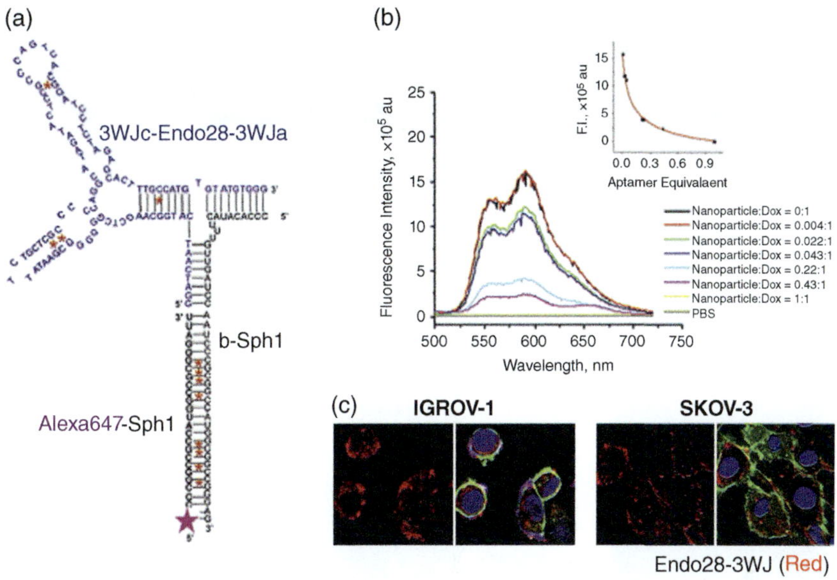

Figure 9.34 (a) Design of Endo28-3WJ. (b) Fluorescence spectra depending on the DOX concentration. (c) Selective cellular internalization of Endo28-3WJ. Source: Pi *et al.* [140] / With permission of Elsevier.

paclitaxel and could be used for targeted cancer therapy [27]. An alkyne handle was attached to the nucleotides in the RNA oligomers of the 4WJ RNA nanoparticles, and a PTX-azide (N_3) prodrug was prepared so that PTX-labeling could be done. The 4WJ-PTX complex was successfully formed by alkyne-azide click chemistry (Figure 9.35a). For the specific delivery to EGFR-overexpressing tumor cells, an anti-EGFR aptamer was incorporated into the 4WJ RNA nanoparticles and showed effective binding and tumor growth inhibition (Figure 9.35b, c).

With the designs of DNA and RNA nanoarchitectures, many research groups have found that nucleic acids are highly effective materials for regulating the loading efficiency and release kinetics of antitumor drugs. These studies have introduced new strategies for effective cancer treatment with reduced side effects, occurring because of nonspecific drug delivery.

9.4.2. Protein Drug Delivery

Proteins are among the most important biomolecules involved in biological activities. Their cellular uptake is limited by their inherent instability and charged surfaces, so functional protein delivery systems have been widely studied to overcome these challenges. To improve the efficiency of protein delivery, diverse approaches, including liposomes [149], polymers [150, 151], and other nanoparticle-based systems [152–154], have been developed to achieve high

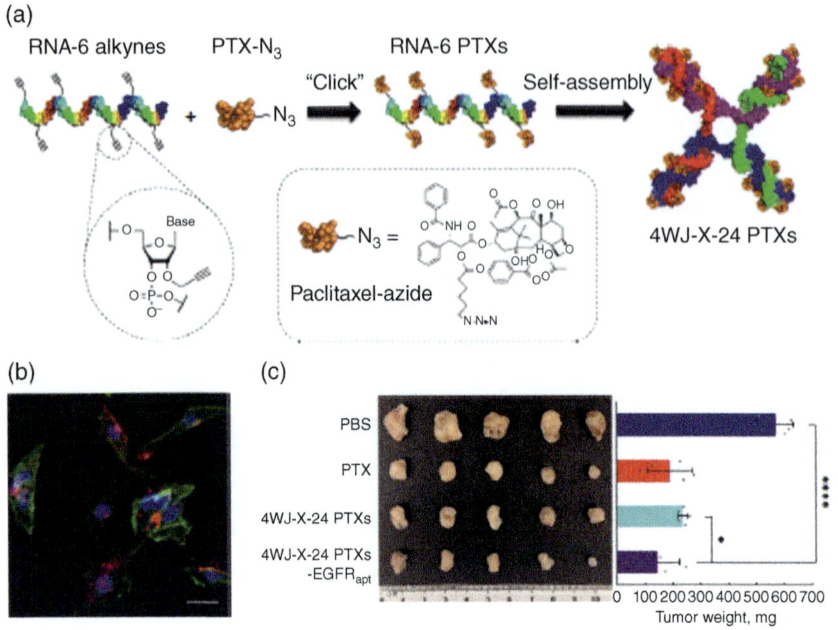

Figure 9.35 (a) Synthesis of PTX-4WJ RNA nanoparticle through click chemistry. (b) *in vitro* cellular binding of PTX-4WJ-EGFR apt nanoparticle. (c) *ex vivo* breast cancer tumors harvested from mice after the treatments. Source: Guo *et al.* [27] / Springer Nature / CC BY-3.0.

loading efficiency, to stabilize proteins, and to enhance their cellular internalization. Nucleic acids are particularly useful for protein delivery, since they can be readily coupled with peptides or proteins through various strategies such as EDC/NHS reaction [155], click chemistry [156], aptamer–protein interactions [40], RNA–protein interactions [85, 157], biotin–avidin linkage [158], or physical encapsulation [159].

In 2012, Chang, Yan and co-workers reported a vaccine delivery system involving programmable DNA nanostructures [159]. The DNA nanostructure assembled a model antigen, streptavidin (STV), and CpG adjuvants into a nanoscale complex and produced strong and long-lasting antibody responses *in vivo*. In this study, a tetrahedral DNA nanostructure was used as a scaffold to encapsulate the STV and CpG oligo-deoxynucleotides. This antigen–adjuvant–DNA complex was delivered to macrophages (Figure 9.36b), dendritic cells (DCs), and B cells, boosting the effective immune response compared to the free CpG and STV. Anti-STV IgGs were maintained at significantly higher levels (Figure 9.36c), and elevated levels of memory B cells were observed in the mice treated with the antigen–adjuvant–DNA complex (Figure 9.36d). These results demonstrate that the DNA nanostructure may be a good platform for vaccine delivery systems.

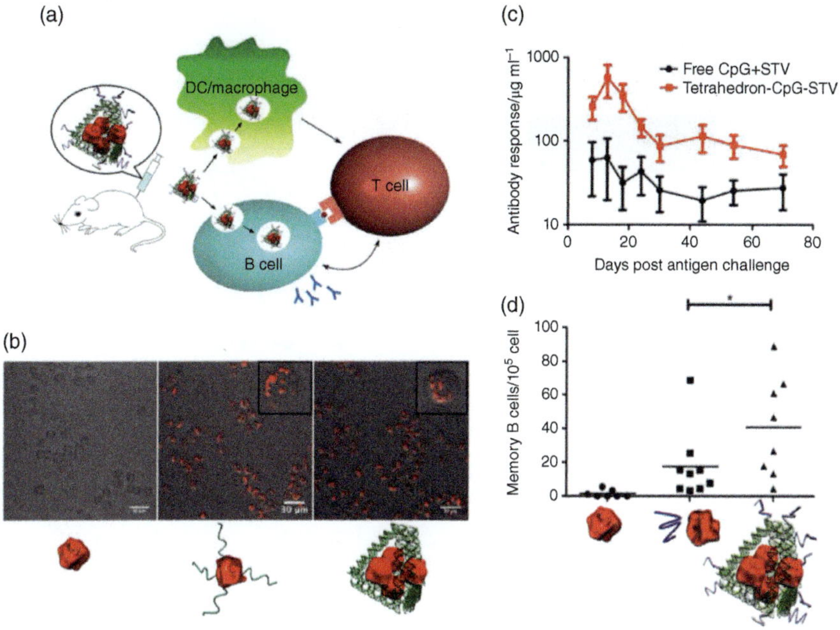

Figure 9.36 (a) Schematic illustration of antigen–adjuvant–DNA complex and its delivery mechanism to B cells, DCs, and macrophages. (b) Confocal microscopy images showing the internalized antigen–adjuvant–DNA complex. (c) Anti-STV IgG level and (d) memory B-cell response after the treatment of antigen–adjuvant DNA complex. Source: Liu *et al.* [159] / With permission of American Chemical Society.

Metal-histidine (His) tag affinity could also be used to aid the conjugation of nucleic acid–based structures with proteins. Gu *et al.* reported the development of cytokine-loaded DNA nanoclews, using Ni^{2+}-His tag affinity, to act as phospholipase A2 (PLA2)-responsive, transformable nanocarriers for cancer therapy [160]. Tumor necrosis factor-related apoptosis-inducing ligand (TRAIL) was used as a model cytokine, and the DNA nanoclews were made by RCA (Figure 9.37a). These TRAIL-loaded DNA nanoclews were designed to transform into nanofibers in a PLA2-rich microenvironment, facilitating the interaction of TRAIL and death receptors on the cell membrane, thereby increasing apoptotic signaling and reducing TRAIL internalization (Figure 9.37b). Figure 9.37c and d show the high binding capability of the fiber structure, transformed from a spherical DNA structure, and the increased apoptotic effect it produces. This study showed that cytokines can successfully be conjugated to DNA by metal–His tag affinity and demonstrated their therapeutic applications.

Mirkin and co-workers used spherical nucleic acid (SNA) nanoparticles, which consist of a nanoparticle core surrounded by oligonucleotides, for cellular delivery of functional enzymes [161]. In this study, β-galactosidase was used as a model protein and was modified with a polyethylene glycol-azide tag. DNA strands of

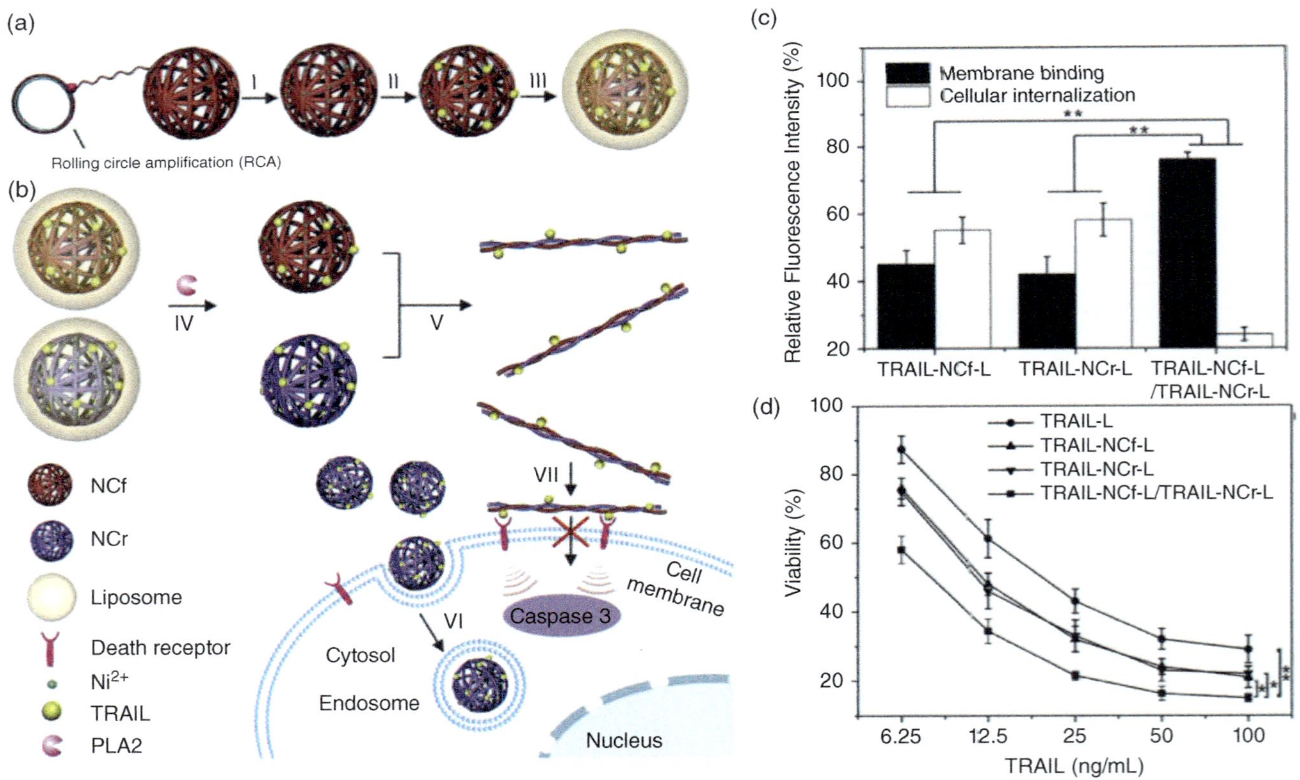

Figure 9.37 (a) Fabrication of TRAIL-DNA nanoclews complex by Ni^{2+}-His tag affinity. (b) Mechanism of PLA2-induced structural transformation and its binding and signaling process. (c) Increased binding affinity and the level of cellular internalization of each structures. (d) Apoptotic effect of morphological transformation of TRAIL-DNA nanoclews. Source: Sun *et al.* [160] / With permission of Elsevier.

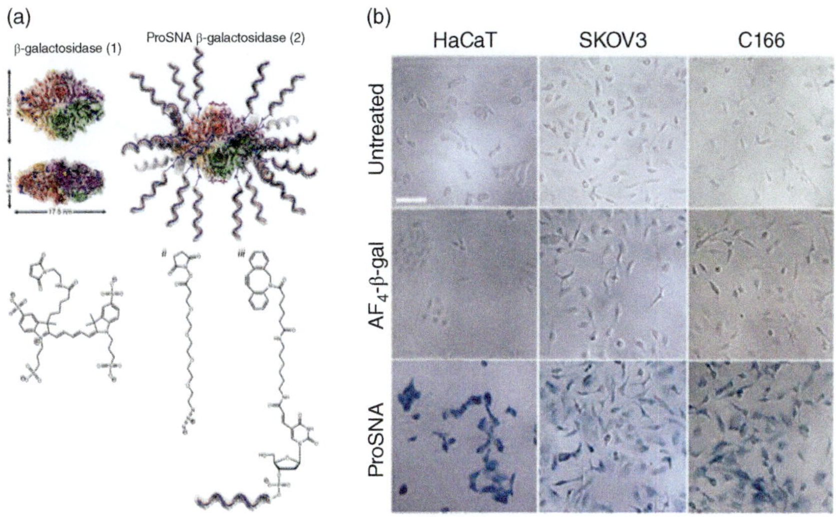

Figure 9.38 (a) Synthesis of **β**-gal functionalized SNA by click chemistry. (b) Confirmation of intracellular catalytic activity of **β**-gal SNA complex. Source: Brodin *et al.* [161] / With permission of American Chemical Society.

SNA were modified to incorporate dibenzocyclooctyne groups at the 5′-terminus, for conjugation with the azide groups through click chemistry. As a result, a chemically conjugated SNA–protein complex, exhibiting high intracellular catalytic activity, was synthesized (Figure 9.38a). Hydrolysis of the β-glycosidic linkage of 5-bromo-4-chloro-3-indolyl-β-D-galactopyranoside resulted in the release of an indole derivative, which could be visualized in blue (Figure 9.38b). This reports the development of a new strategy for immunomodulation, through the engineering of functional proteins into cell membrane-permeable systems.

Ahn and co-workers developed a biotin-streptavidin (STV)-induced enzyme delivery system, based on a DNA tetrahedron delivery platform [158]. STV is a biomolecule that can provide stoichiometric cargo loading sites, allowing a diverse range of functional enzymes to be loaded into the nanocarrier through STV–biotin interactions. In this study, biotinylated enantiomeric mirror DNA (L-DNA) was assembled into a DNA tetrahedron, and various biotinylated enzymes could be loaded onto this structure (Figure 9.39a, b). This hybrid platform demonstrated successful intracellular delivery of enzymes, including apoptotic enzymes, which can be used in cancer therapeutics (Figure 9.39c).

9.4.3. Nucleic Acid Drug Delivery

There are a range of functional nucleic acids, such as CpG, used for immunomodulation, and antisense oligonucleotides, siRNA, miRNA, and mRNA, that can be used for gene therapy [162]. In the case of functional nucleic acid

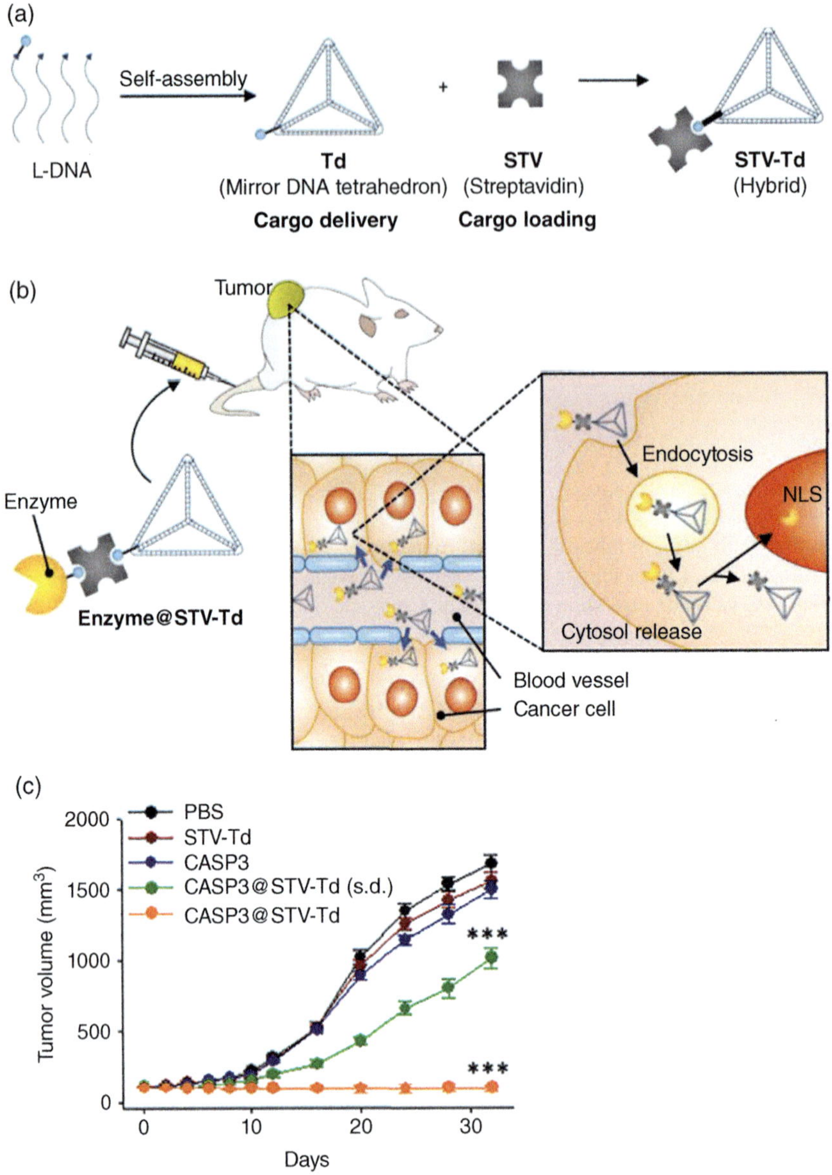

Figure 9.39 (a) Synthesis of enzyme-loaded DNA tetrahedron with L-DNA strands. (b) Universal enzyme-DNA tetrahedron hybrid system through biotin–STV interaction and its systemic delivery into tumor cells. (c) Growth inhibition of HeLa after the treatment of CASP3-DNA tetrahedron. Source: Kim *et al.* [158] / With permission of Elsevier.

delivery systems, the cargo and carrier are composed of the same material, making drug loading by sequence-specific hybridization very convenient [23]. Since functional nucleic acid loading in DNA- or RNA-based nanostructures does not require any chemical modification, organic solvents, or complicated synthetic processes, numerous studies on the use of nucleic acid carriers for the delivery of functional DNA and RNA have been conducted. In this section, we review various designs and strategies for the delivery of functional nucleic acids.

The simplest strategy for attaching nucleic acid cargoes to the carrier is through toeholds or sticky ends, where single-stranded DNA or RNA can be hybridized. With this sequence design, dendrimers, polyhedrons, or DNA origami-based nanocarriers have been used to deliver functional nucleic acids. For example, self-assembled DNA dendrimer nanoparticles have been used for the delivery of the CpG motif to enhance cytokine secretion. DNA dendrimers have been studied in the field of drug delivery because of their high stability, resulting from their branched structure [163]. In 2017, Ding and co-workers reported the use of DNA dendrimeric nanoparticles as a vehicle for the delivery of the immunostimulant, CpG. By designing X- and Y-shaped DNA structures with sticky ends, highly branched DNA dendrimers were successfully assembled, and CpG-encoded hairpin loops were expressed to trigger a strong immune response (Figure 9.40a). In this study, TAT, a typical cell-penetrating peptide, was also attached to the DNA dendrimer to enhance cellular internalization and increase cytokine production.

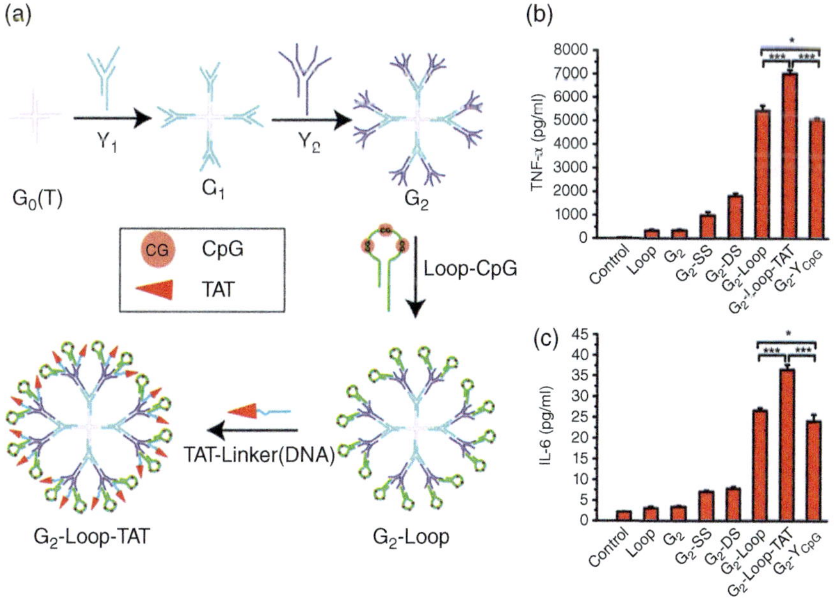

Figure 9.40 (a) Design of the CpG-expressed DNA dendrimer. (b, c) The level of TNF-**α** (b) and IL-6 (c) production after the treatment of CpG-expressed DNA dendrimer. Source: Qu *et al.* [163] / With permission of American Chemical Society.

Increased release of pro-inflammatory cytokines, including tumor necrosis factor α (TNF-α) and interleukin (IL) 6 from macrophages (RAW 264.7 cells) was detected on treatment with CpG-conjugated DNA dendrimers (Figure 9.40b, c). In contrast to loop-CpG and bare dendrimers, CpG-conjugated DNA dendrimers showed dramatically increased cytokine secretion, indicating that DNA dendrimers are highly effective in delivering CpG into macrophages.

Three-dimensional structures have also been widely used for CpG delivery. For instance, a DNA tetrahedron was used to deliver CpG oligonucleotides, noninvasively, to RAW 264.7 cells [164]. The CpG motifs were recognized by Toll-like receptor 9, which activates RAW264.7 cells, resulting in increased secretion of pro-inflammatory cytokines such as TNF-α, IL-6, and IL-12. As

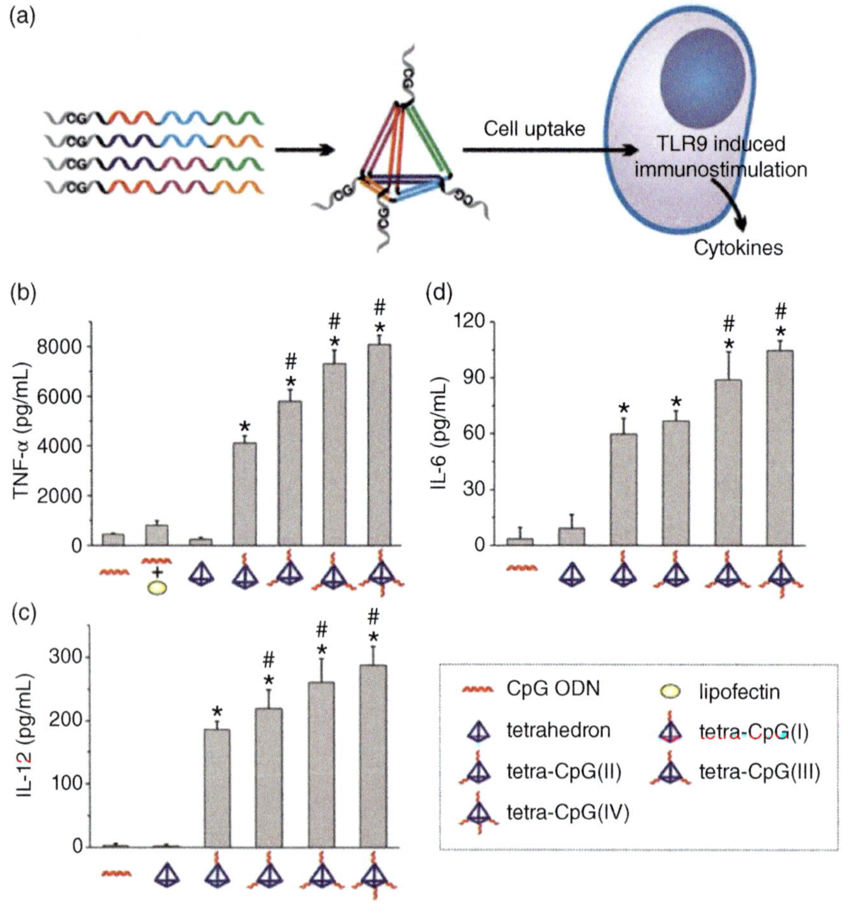

Figure 9.41 (a) Synthesis of CpG-expressed DNA tetrahedron and its cellular uptake through the TLR 9 pathway. (b–d) The level of pro-inflammatory cytokines release depending on the DNA nanostructures. Source: Li *et al.* [164] / With permission of American Chemical Society.

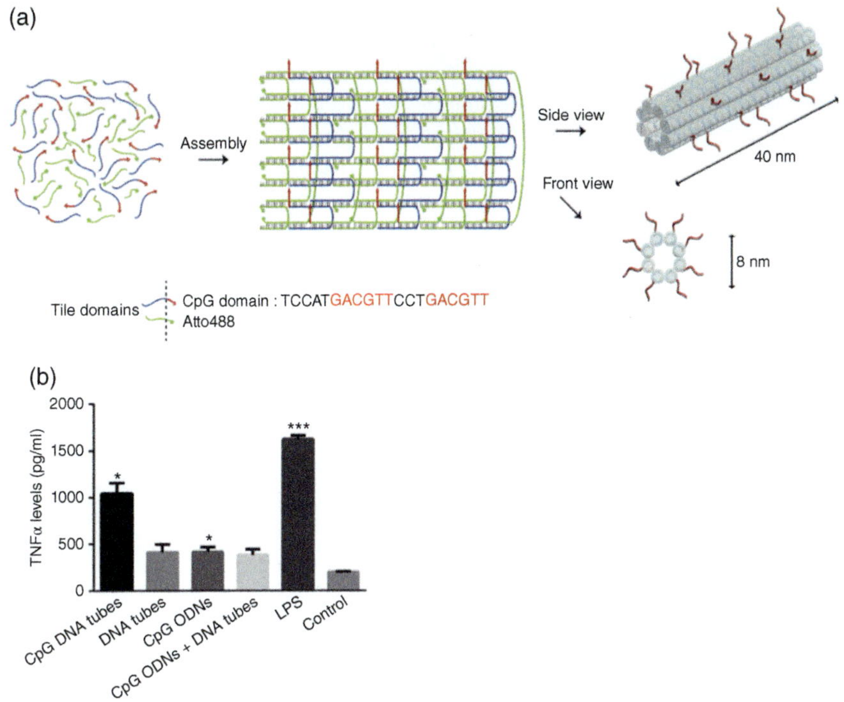

Figure 9.42 (a) Assembly of CpG-decorated DNA origami. (b) The level of TNF-α release. Source: Sellner *et al.* [165] / With permission of Elsevier.

shown in Figure 9.41a, the DNA tetrahedron was assembled from four designed DNA strands, expressing CpG oligonucleotides on each corner of the structure. The amount of CpG oligodeoxynucleotide (ODN) expressed was controlled by sequence design. Treatment with each structure produced distinct levels of cytokine secretion, indicating that CpG motifs were well delivered and the dose of CpG ODN affected immunostimulatory activity. The real-time internalization of CpG-labeled DNA origami nanocarriers and the immune response elicited are shown in Figure 9.42a [165]. Since DNA origami structures exhibit good nuclease resistance, they are thought to be a promising carrier system for functional nucleic acid delivery. As shown in Figure 9.42b, CpG sequence-decorated DNA nanotubes had an immunostimulatory effect on immune cells *in vivo*, increasing the release of TNF-α when injected into skeletal muscles of anesthetized mice.

Not only DNA nanostructures but also RNA-based structures have been utilized for the immune stimulation by hanging CpG ODN at the corners [166]. In 2014, Guo's group reported that the immune response could be controlled by the shape transition of the RNA polygons. The degree of immunostimulation is highly depended on the size, shape, and the number of CpG payload on each

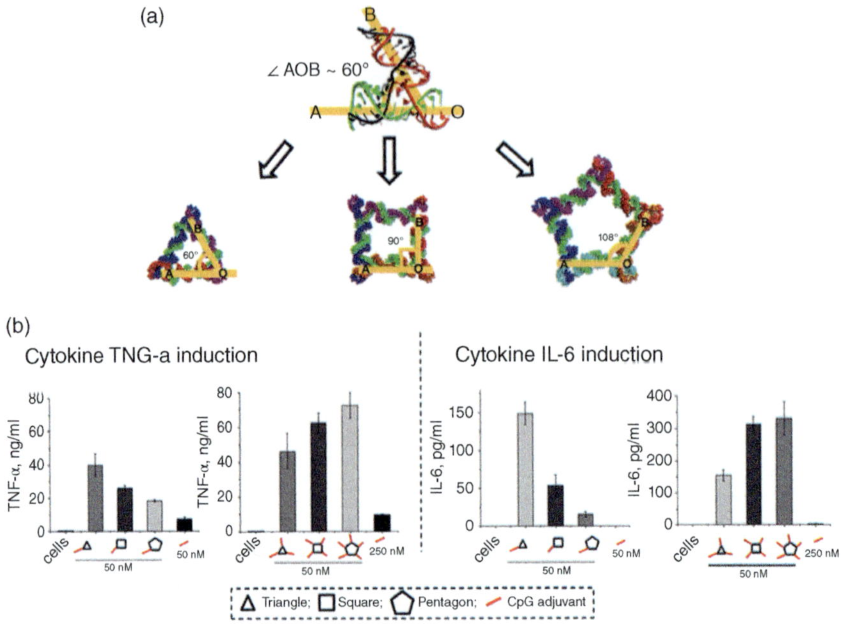

Figure 9.43 (a) Synthesis of triangle, square, and pentagon-shaped RNA structures. (b) The level of pro-inflammatory cytokines, TNF-**α**, and IL-6. Source: Khisamutdinov *et al.* [166] / Oxford University Press / CC BY-3.0.

nanostructure. The shapes of RNA polygons were controlled by modulating five external short strands, which can lead direct folding of the internal strand and the length of internal strands. Pro-inflammatory cytokines (TNF-α, IL-6) induction from RAW 264.7 cells showed the dependence of the shapes and the number of CpG per RNA polygons compared to the naked CpG ODN (Figure 9.43b). This research suggested that RNA polygon has the potential to serve as a various drug carrier not only CpG ODN, also therapeutic molecules such as functional RNA like miRNA or siRNA motifs and chemical drugs. Also the amount of payloads could be easily adjusted by tuning the shapes.

CpG-embedded DNA microparticles, produced by complementary RCA (cRCA), also brought about cytokine secretion [78]. This particle was designed to contain a double-stranded region that could be specifically cleaved by DNase II, and a mismatched region, called the "bubble," in which CpG could be embedded (Figure 9.44a). Repeated CpG motifs and cleavable regions could be produced by cRCA, allowing numerous CpG ODNs to be delivered simultaneously. This particle could be internalized into macrophages through phagocytosis and degraded in phagolysosomes by DNase II, leading to the release of CpG ODN (Figure 9.44b). As a result, pro-inflammatory cytokine release could be

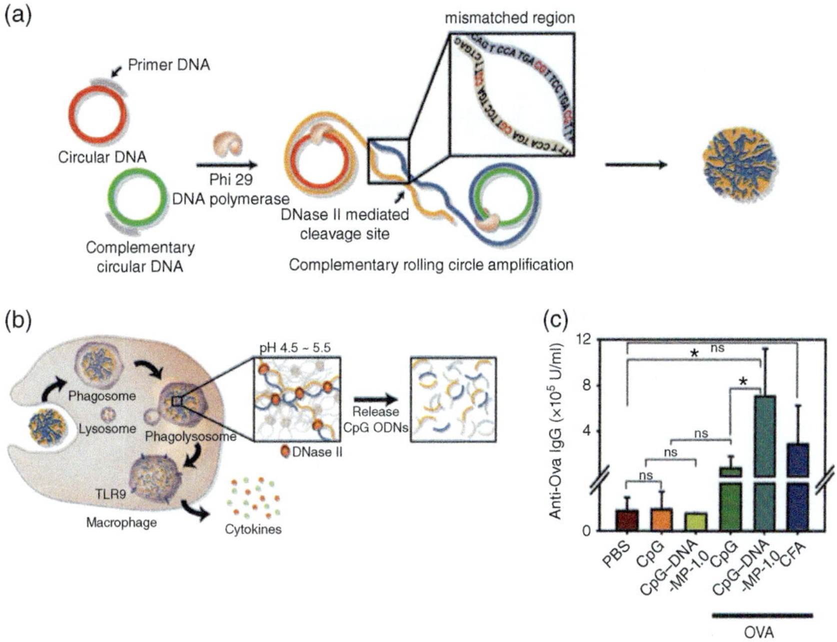

Figure 9.44 (a) Design and synthesis of the CpG ODN-embedded DNA microparticle by cRCA. (b) Phagocytosis of CpG-DNA microparticle and subsequential release of CpG ODN mediated by DNase II. (c) Quantification of the anti-OVA IgG in mouse serum after the treatment of CpG–DNA microparticle. Source: Jung *et al.* [78] / Royal Society of Chemistry / CC BY-3.0.

confirmed and elevated target antibody production could be achieved *in vivo*, with the model antigen ovalbumin (Figure 9.44c).

Variety of nucleic acid–based nanomaterials also have been widely explored for delivering siRNA – therapeutic RNA with 21–23 bases in length, which suppresses the expression of target genes by complex with RNA-induced silencing complex (RISC) resulting in the cleavage of a certain complementary mRNA [18, 167]. Conventional strategies such as liposomes or polymeric systems have caused several problems including the cytotoxicity, heterogeneity, immune response, or lack of encapsulation efficiency [19, 168, 169]. To overcome these limitations, finely controllable DNA and RNA have been attracted as a siRNA carrier with its hybridization or enzymatic replication methods.

In 2019, Zhang's group developed DNA tetrahedron-crosslinked nanogel, which can effectively embed the siRNA inside and protect the payloads [170]. Therapeutic siRNA with DNA tails could be assembled in nanogel as a linker between the DNA tetrahedrons by specific sticky end recognition (Figure 9.45a). This DNA nanogel showed good stability in 10% FBS-contained cell-culture medium for eight-hour incubation compared with the control group. This system

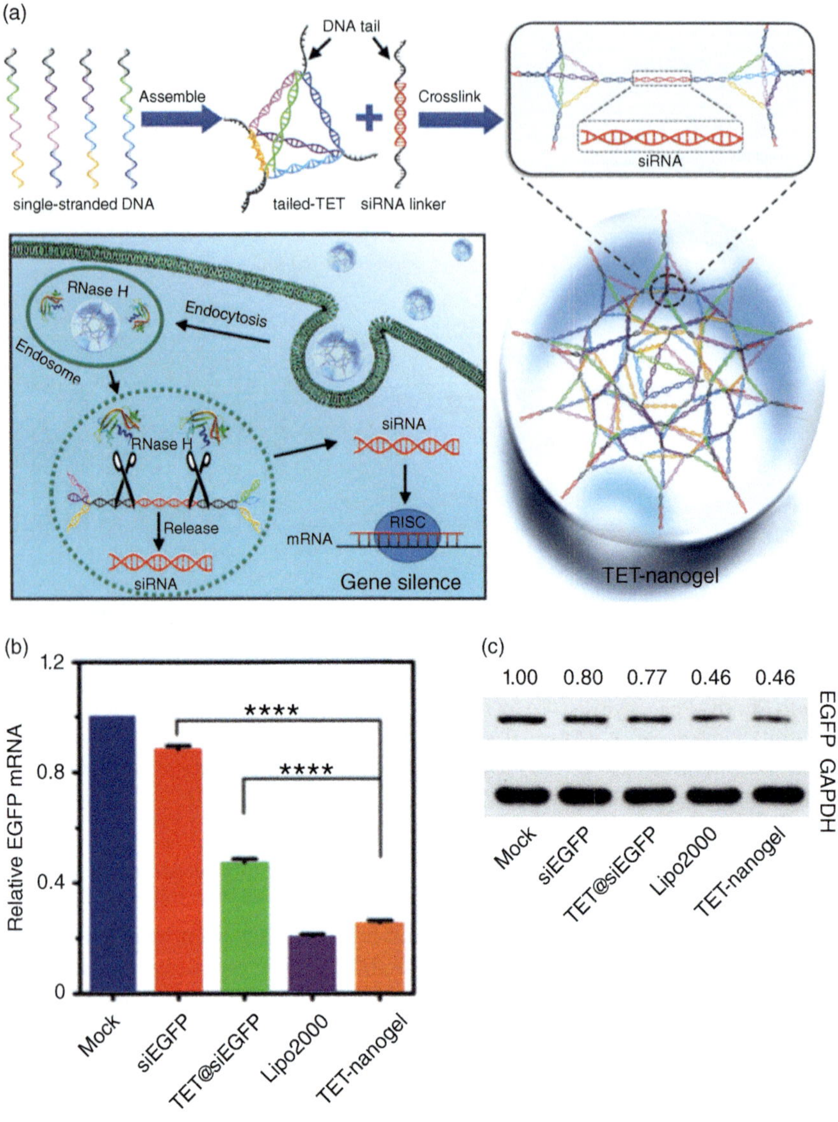

Figure 9.45 (a) Synthesis of DNA nanogel and the mechanism of siRNA release in cytoplasm. (b) The level of EGFP intensity after the treatment of DNA nanogel. (c) Western blot analysis of EGFRP expression. Source: Xue *et al.* [170] / With permission of Royal Society of Chemistry.

was designed to be efficiently cleaved in endosome by RNase H that exists in the cytoplasm, which can only cleave the RNA at DNA–RNA heteroduplex region. The sticky ends attached to siRNA linker could be cleaved, releasing functional siRNA by the treatment of RNase H. As a result, this nanogel dramatically decreased the mRNA expression of EGFP when it was transfected using lipofectamine 2000 (Figure 9.45b, c).

In 2012, Lee *et al.* reported self-assembled RNA microsponge for the stable siRNA delivery [18]. Instability of RNA and ineffectual loading efficiency of siRNA have remained critical challenges. To overcome both of limitations, rolling circle transcription (RCT) was applied to produce numerous multimeric siRNA units with one-step isothermal reaction. The RNA microsponge was composed of densely packed RNA strands embedding a huge amount of potential siRNA. To prepare this RNA microsponge, template DNA was designed to contain complementary sequences of the antisense and sense sequences. The transcripts can form the hairpin RNA structure that is cleavable by dicer to convert the hairpin RNA to siRNA after the cellular uptake (Figure 9.46a). The product of *in vitro* RCT has porous and wrinkled structures when it was observed with scanning electron microscopy (SEM) (Figure 9.46a). To enhance the cellular uptake of RNA microsponge, the synthetic polycation, polyethylenimine (PEI) was coated on the microsponge to express the positive charge on the surface. PEI-coated RNA

microsponge was well delivered into cancer cells, representing the effective suppression of luciferase expression *in vitro* (Figure 9.46c) and *in vivo* knockdown (Figure 9.45d). After the intratumoral injection of PEI–RNA microsponge into tumor in left thigh, decrease of luciferase expression was confirmed in contrast to the right thigh. This self-assembled RNA microsponge provided a promising strategy in siRNA delivery researches and it has been widely developed to improve the effectiveness of gene regulation. As one of them, in 2017, Lee's group suggested new design of the siRNA-embedded nanoparticles, bubble RNA-based cargo (BRC) [75]. BRC was synthesized by complementary rolling circle transcription (cRCT), which requires two partial complementary circular DNA to generate the sense and antisense strands for siRNA and nonhybridizing (bubble) region (Figure 9.47a). The approach was designed to improve the efficiency of Dicer cleavage which plays a critical role in RNA interference. As shown in Figure 9.47b, siRNA was efficiently generated from the BRC by the treatment of Dicer, followed by highly effective GFP gene knockdown confirmed by *ex vivo* image (Figure 9.47c). Overall, this developed design of RCT-based nanoparticle showed successful target-specific RNAi therapeutics with high efficiency and it was further applied to USE1 knockdown, which plays crucial roles in the progression of human lung cancer [74].

MicroRNA (miRNA) is a kind of short RNA that is related to the multiple intracellular and extracellulr biofunctions [171]. Many advanced strategies for tumor-suprressive microRNA (miRNA) delivery system have been developed. MicroRNA delivery into target cells was also hindered due to the structural vulnerability from the RNase degradation in physiological condition [172].

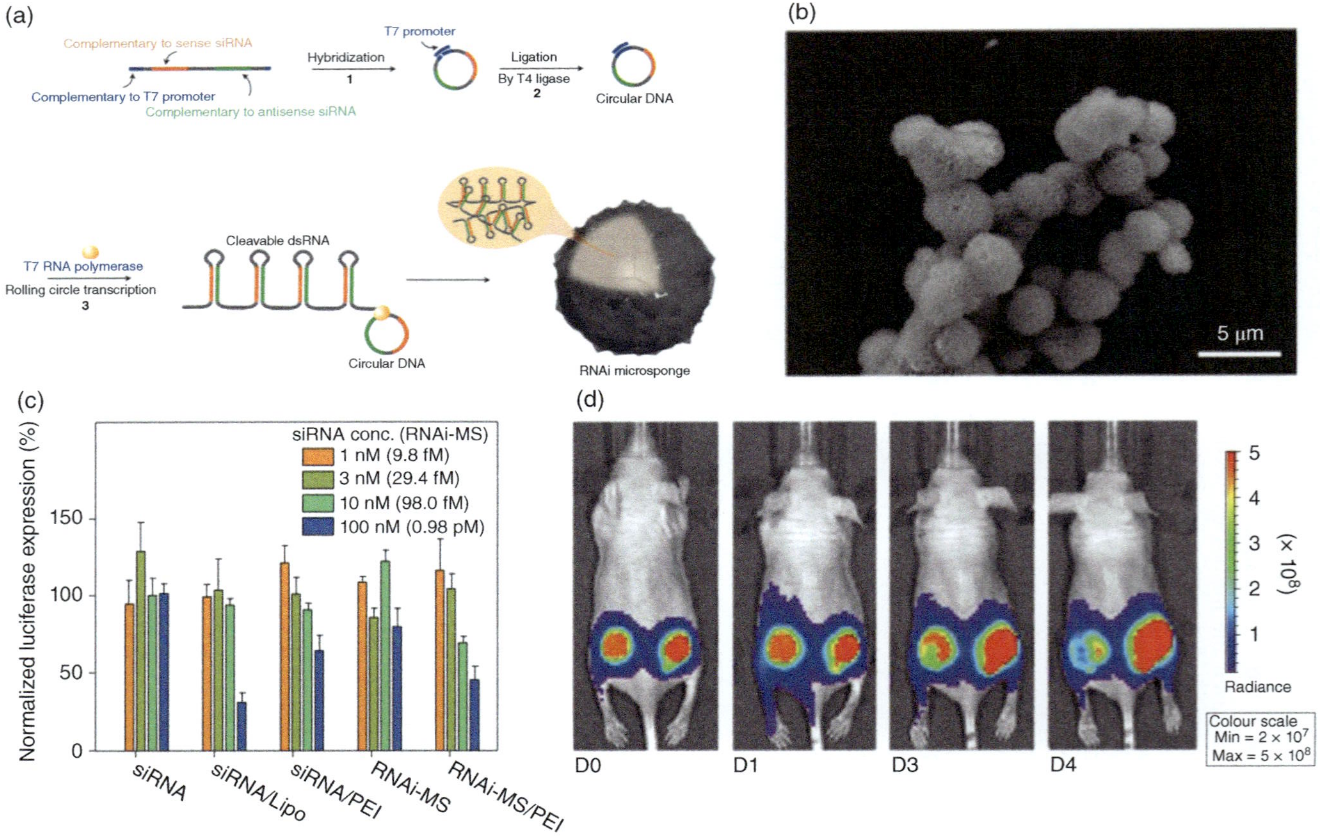

Figure 9.46 (a) Fabrication of siRNA-embedded RNA microsponge by RCT. (b) SEM image of RNA microsponge. (c, d) *in vitro* (c) and *in vivo* (d) gene knockdown efficacy of RNA microsponge. Source: Lee *et al.* [18]/ With permission of Sprigner Nature.

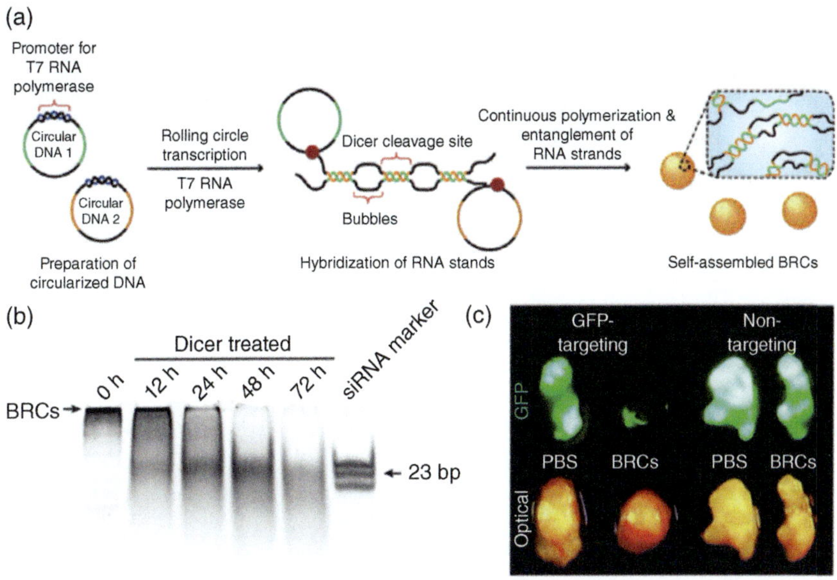

Figure 9.47 (a) Synthesis of BRC by cRCT. (b) Time-dependent siRNA generation after the dicer treatment. (c) Confirmation of GFP gene knockdown *ex vivo*. Source: Kim *et al.* [75] / John Wiley & Sons / CC BY-4.0.

To decreae the miRNA degradation caused by RNase, DNA nanostructures were introduced as a protector in harsh conditions, leading miRNA to endure for a long time, enhancing the tumor suppressive effect [173]. Qian *et al.* reported self-assembled DNA shuriken based on DNA star shape consisting of central strand (L), three of edge strands (M), and three of peripheral strands (S) (Figure 9.48a). Peripheral strands have the sites that can hybridize to miRNA and this hybridization lead to form a DNA shuriken. The effect of DNA shuriken exerted anticancer bioactivity. For untreated group, the spheroid have grown very fast and shown robust and uniform growth, while the DNA shuriken treated group has shown smaller size of spheroid after five days with spreaded out from the outer surface (Figure 9.48b). The level of cell proliferation was decreased when the DNA shuriken was treated to colorectal cancner cell line DLD-1 (Figure 9.48c).

Besides, variety of RNA-based structures such as RNA micelles by the hydrophobic assembly of the cholesterol-conjugated phi29 packaging RNA three-way junction (pRNA-3WJ) [174] or miRNA-embedded nanoparticle by RCT was utilzed for antiproliferation activity [175]. These developments of the miRNA delivery system have contributed to significantly improve the antitumoral thrapeutic efficacy and will provide more availability for delivery of various drugs such as immunemodulating adjuvants or antisense oligonucleotide.

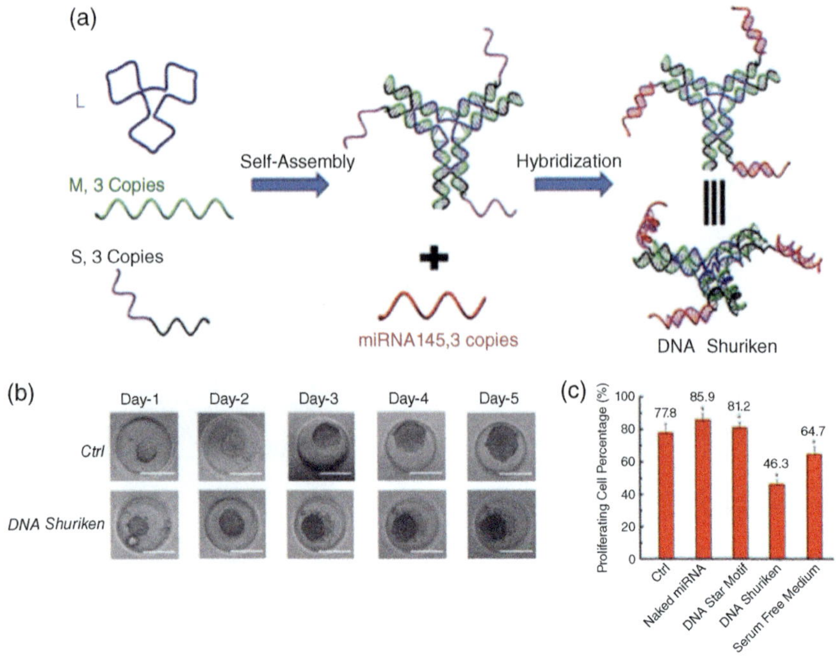

Figure 9.48 (a) Synthesis of DNA shuriken. (b) The effect of DNA shuriken on the tumor spheroid morphology. (c) Proliferation analysis of the DNA shuriken. Source: Qian *et al.* [173] / With permission of Royal Society of Chemistry.

9.4.4. Co-Delivery of Multiple Functional Drug Molecules

In this chapter, we will discuss co-delivery system for the synergistic effect of each therapeutic molecules. Co-delivery of multiple functional molecules enhance the therapeutic effect with improved delivery efficiency. In many cases, incorporation of therapeutic nucleic acid and proteins usaully have been reported as co-delivery system. For example, CpG motif and DOX with DNA hydrogel [176], CpG, and antigen with DNA polygon [159], two kinds of proteins with DNA origami [177] have been reported. Recently, the CRISPR-Cas9-mediated specific gene-editing system to permanently suppress the expression of the unwanted protein has been tremendously attracted and developed [178, 179]. Cas9 protein can recognize the target double-stranded DNA via the protospacer adjacent motif (PAM) sequence in the target gene and it can be complexed with single-stranded guide RNA (sgRNA). In 2017, Ahn and Lee's group considered a dual-function gene regulating system to improve the gene suppression efficiency by incorporating the Cas9 protein and siRNA [85]. This Cas9 and siRNA co-delivery system enabled to simultaneously cleave both target DNA and mRNA. Cas9-conjugated multimeric sgRNA and hairpin siRNA-embedded nanoparticles (poly-sgRNA/siRNA) were prepared by RCT process

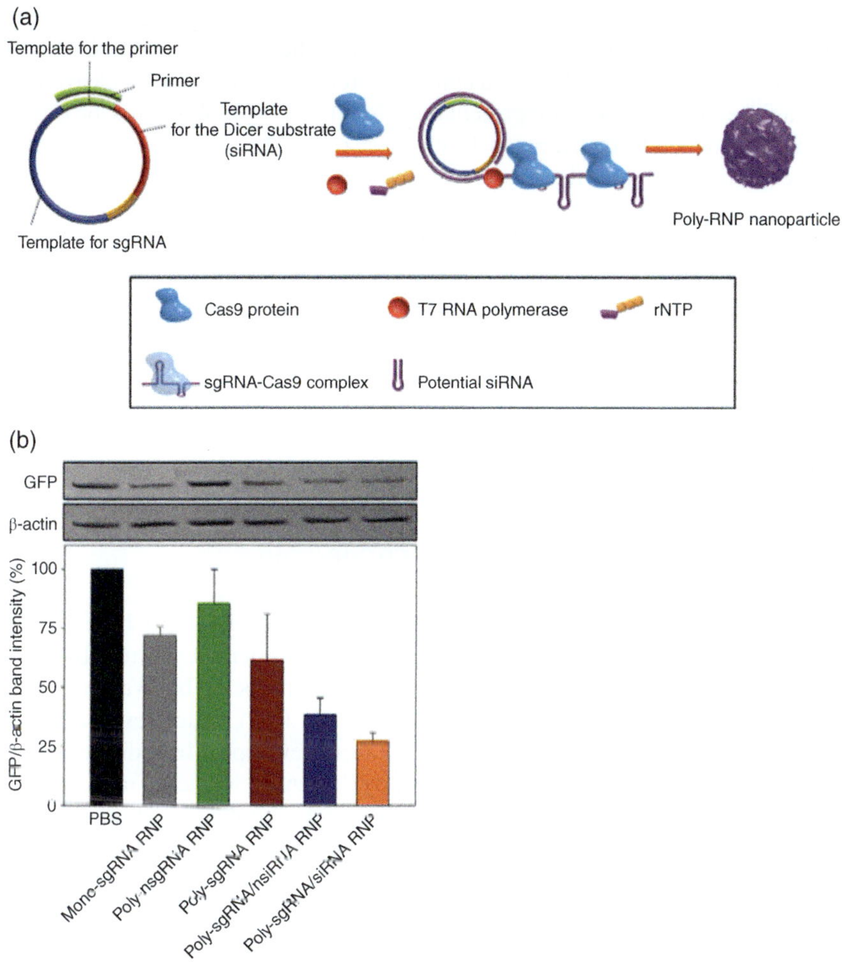

Figure 9.49 (a) Fabrication of Poly-sgRNA/siRNA nanoparticle. (b) GFP gene knockdown efficacy. Source: Ha *et al.* [85] / With permission of Elsevier.

(Figure 9.49a) and this material showed highly improved serum stability and cellular uptake compared to the monomeric RNA product. As a result, the poly-sgRNA/siRNA system demonstrated synergistic effect, showing the most effective knockout rates of the target gene both *in vitro* and *in vivo* (Figure 9.49b). Therefore, this concept of co-delivery system is expected to be emplyed to a number of target gene disruption in biomedical field of study.

One another example is the co-delivery of immune-boosting CpG ODN and anti-PD1 (aPD1) system [77]. Enzymatic replication, RCA, produced a spherical asssembled DNA structure, called DNA nano-cocoon, containing repeated CpG sequences and restriction enzyme-specific sites, which are capable to be digested (Figure 9.50a). Restriction enzyme-encapsulated triglycerol monostearate (TGMS) nanoparticles were attached on the aPD1-loaded DNA nanococoon. TGMS is amphiphilic and contains ester linkage that can be cleaved by esterases and matrix metalloproteinases (MMPs) [180]. With this property, inflammatory condition triggered the cleavage of TGMS, subsequetially releasing restriction enzyme, thereby converting the DNA nano-coccons into CpG ODNs, releasing aPD1 as well. Released CpG ODN activated denderitic cells to induce T-cell response with aPD1 to achieve synergistic effect for PD 1 blockade. As shown in Figure 9.50b and c, CpG ODN and aPD1 were specifically release in inflammatory condition with activated macrophages. The most effective antitumoral efficacy was confirmed when the TGMS-coated DNA nano-cocoons (G5) compared to other control groups (G1-G4) (Figure 9.49d). With this data, co-delivery of various functional molecules can be considered as a powerful strategy that can lead the synergistic potency in clinical application.

9.5. CONCLUSION AND FUTURE PERSPECTIVES

In Chapter 6, we briefly reviewed recent advances of nucleic acid–based structures for efficient drug delivery system. With unprecedented controllability in nanoscale of scope with excellent programmable property by Watson–Crick base pairing, a number of well-defined DNA and RNA structures from two-dimensioanl lattices to macrosized hydrogel were developed. In addition, inherent biocompatibility and biodegradability of nucleic acid make these developments of DNA and RNA structures to be applied to drug carriers. Nucleic acid engineering faciliatates the loading of the drugs and attaching the target-specific moieties with various strategies such as intercalation, physical entrapment, hydrogen bonding, or chemical modifications on DNA. Also, systemic cargoes delivery and release were also achieved as a smart drug delivery system with the unique modularity and flexibility of nucleic acid, which contributed the sequence replacement and change of shapes. With these advantages, DNA- and RNA-based structures were widely utilized for the delivery of all kinds of drugs including chemial drugs, protein, and functional nucleic acid.

Even though nucleic acid–based drug carriers have significant advantages, there are still certain hurdles that need to be addressed and improved to be more widely commercialized. First thing is the cost concern. Compared to the synthetic polymers, the production of purified functional DNA and RNA cost much higher. For the cost-effective large production that is necessary for practical applications, enzymatic amplification such as PCR, RCA, or RCT has been explored. These techniques can synthesize a number of functioanl DNA and RNA from the small amount of template DNA. Also effieicient purification technologies have been widely developed for the high yields. The second obstacles to be overcome is biosafety problem. Though DNA and RNA can be essential

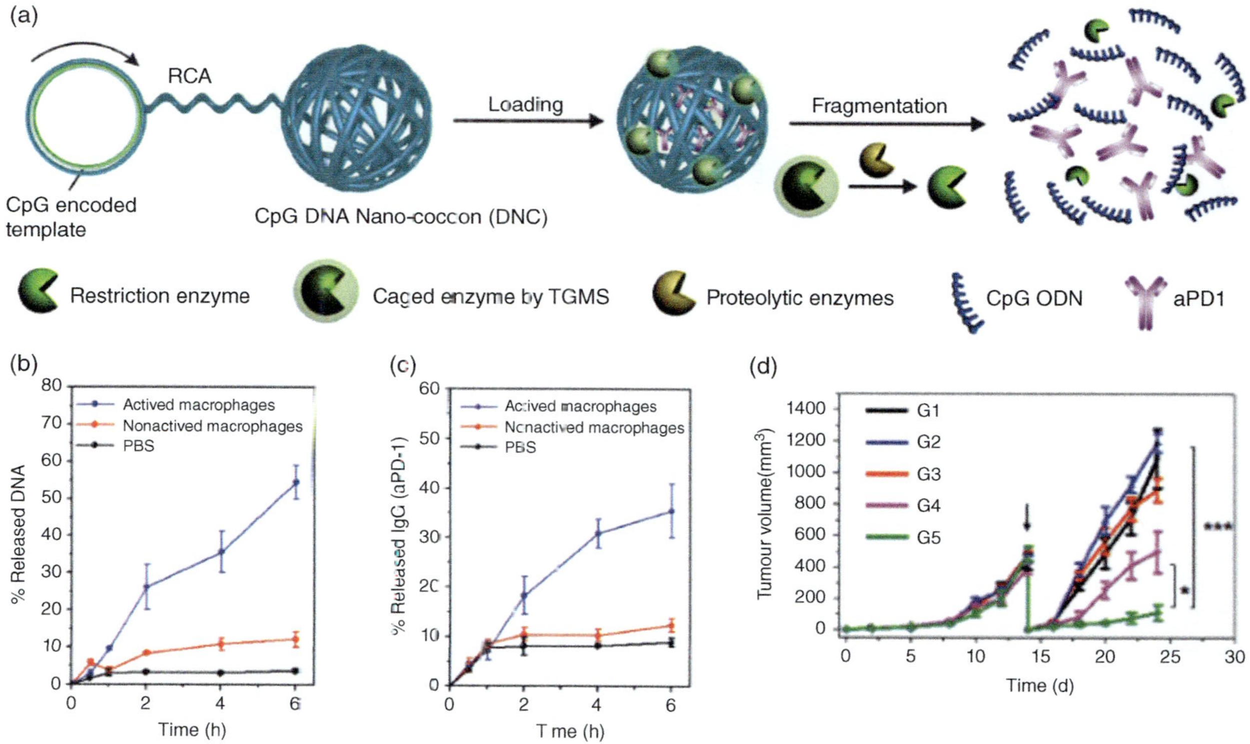

Figure 9.50 (a) aPD1-encapsulated and restriction enzyme-induced cleavable CpG DNA nano-cocoon. (b, c) The level of released DNA (b) and aPD1 (c) around activated macrophages. (d) Antitumoral efficacy of aPD1-encapsulated DNA nano-cocoon. Source: Wang *et al.* [77] / With permission of John Wiley & Sons.

treatment of the diseases, foreign DNA and RNA can cause the unexpected harmful genetic recombinationand or immunostimulation. In the future, these concerns should be clarified for nucleic acid nanotechnology to be more attractive in practical applications.

REFERENCES

1. N.C. Seeman, H.F. Sleiman, DNA nanotechnology, Nature Reviews Materials, 3 (2017) 1–23.
2. J.D. Watson, F.H. Crick, Molecular structure of nucleic acids, Nature, 171 (1953) 737–738.
3. B.R. Madhanagopal, S. Zhang, E. Demirel, H. Wady, A.R. Chandrasekaran, DNA nanocarriers: programmed to deliver, Trends in Biochemical Sciences, 43 (2018) 997–1013.
4. O. Ko, S. Han, J.B. Lee, Selective release of DNA nanostructures from DNA hydrogel, Journal of Industrial and Engineering Chemistry, 84 (2020) 46–51.
5. A.V. Pinheiro, D. Han, W.M. Shih, H. Yan, Challenges and opportunities for structural DNA nanotechnology, Nature Nanotechnology, 6 (2011) 763.
6. N.C. Seeman, Nucleic acid junctions and lattices, Journal of Theoretical Biology, 99 (1982) 237–247.
7. E. Winfree, F. Liu, L.A. Wenzler, N.C. Seeman, Design and self-assembly of two-dimensional DNA crystals, Nature, 394 (1998) 539–544.
8. Q. Hu, H. Li, L. Wang, H. Gu, C. Fan, DNA nanotechnology-enabled drug delivery systems, Chemical Reviews, 119 (2018) 6459–6506.
9. P.W. Rothemund, Folding DNA to create nanoscale shapes and patterns, Nature, 440 (2006) 297–302.
10. H. Kim, Y. Park, J. Kim, J. Jeong, S. Han, J.S. Lee, J.B. Lee, Nucleic acid engineering: RNA following the trail of DNA, ACS Combinatorial Science, 18 (2016) 87–99.
11. P. Guo, The emerging field of RNA nanotechnology, Nature Nanotechnology, 5 (2010) 833.
12. R. Langer, New methods of drug delivery, Science, 249 (1990) 1527–1533.
13. R. Langer, Drug delivery and targeting, Nature-London, 392 (1998) 5–10.
14. C. Alvarez-Lorenzo, A. Concheiro, Smart Materials for Drug Delivery, Royal Society of Chemistry 2013.
15. L. Steidler, W. Hans, L. Schotte, S. Neirynck, F. Obermeier, W. Falk, W. Fiers, E. Remaut, Treatment of murine colitis by Lactococcus lactis secreting interleukin-10, Science, 289 (2000) 1352–1355.
16. D.W. Pack, A.S. Hoffman, S. Pun, P.S. Stayton, Design and development of polymers for gene delivery, Nature Reviews Drug Discovery, 4 (2005) 581–593.
17. B. Duncan, C. Kim, V.M. Rotello, Gold nanoparticle platforms as drug and biomacromolecule delivery systems, Journal of Controlled Release, 148 (2010) 122–127.

18. J.B. Lee, J. Hong, D.K. Bonner, Z. Poon, P.T. Hammond, Self-assembled RNA interference microsponges for efficient siRNA delivery, Nature Materials, 11 (2012) 316–322.
19. H. Lv, S. Zhang, B. Wang, S. Cui, J. Yan, Toxicity of cationic lipids and cationic polymers in gene delivery, Journal of Controlled Release, 114 (2006) 100–109.
20. F. Zhang, J. Nangreave, Y. Liu, H. Yan, Structural DNA nanotechnology: state of the art and future perspective, Journal of the American Chemical Society, 136 (2014) 11198–11211.
21. A. Schroeder, D.A. Heller, M.M. Winslow, J.E. Dahlman, G.W. Pratt, R. Langer, T. Jacks, D.G. Anderson, Treating metastatic cancer with nanotechnology, Nature Reviews Cancer, 12 (2012) 39–50.
22. M.R. Jones, N.C. Seeman, C.A. Mirkin, Programmable materials and the nature of the DNA bond, Science, 347 (2015) 31.
23. J. Li, C. Fan, H. Pei, J. Shi, Q. Huang, Smart drug delivery nanocarriers with self-assembled DNA nanostructures, Advanced Materials, 25 (2013) 4386–4396.
24. Q. Jiang, C. Song, J. Nangreave, X. Liu, L. Lin, D. Qiu, Z.-G. Wang, G. Zou, X. Liang, H. Yan, DNA origami as a carrier for circumvention of drug resistance, Journal of the American Chemical Society, 134 (2012) 13396–13403.
25. K.-R. Kim, D.-R. Kim, T. Lee, J.Y. Yhee, B.-S. Kim, I.C. Kwon, D.-R. Ahn, Drug delivery by a self-assembled DNA tetrahedron for overcoming drug resistance in breast cancer cells, Chemical Communications, 49 (2013) 2010–2012.
26. W. Sun, T. Jiang, Y. Lu, M. Reiff, R. Mo, Z. Gu, Cocoon-like self-degradable DNA nanoclew for anticancer drug delivery, Journal of the American Chemical Society, 136 (2014) 14722–14725.
27. S. Guo, M. Vieweger, K. Zhang, H. Yin, H. Wang, X. Li, S. Li, S. Hu, A. Sparreboom, B.M. Evers, Ultra-thermostable RNA nanoparticles for solubilizing and high-yield loading of paclitaxel for breast cancer therapy, Nature Communications, 11 (2020) 1–11.
28. N.C. Seeman, Nucleic acid nanostructures and topology, Angewandte Chemie International Edition, 37 (1998) 3220–3238.
29. N.C. Seeman, DNA in a material world, Nature, 421 (2003) 427–431.
30. R. Sinden, Introduction to the structure, properties, and reactions of DNA, In Richard R. Sinden, Ed. DNA Structure and Function, Elsevier. (1994) 1–57.
31. J.D. Watson, F.H. Crick, Molecular structure of nucleic acids: a structure for deoxyribose nucleic acid, Nature, 171 (1953) 737–738.
32. H. Ramezani, H. Dietz, Building machines with DNA molecules, Nature Reviews Genetics, 21 (2020) 5–26.
33. Y. Wang, J.E. Mueller, B. Kemper, N.C. Seeman, Assembly and characterization of five-arm and six-arm DNA branched junctions, Biochemistry, 30 (1991) 5667–5674.
34. X. Wang, N.C. Seeman, Assembly and characterization of 8-arm and 12-arm DNA branched junctions, Journal of the American Chemical Society, 129 (2007) 8169–8176.
35. T.J. Fu, N.C. Seeman, DNA double-crossover molecules, Biochemistry, 32 (1993) 3211–3220.
36. H. Yan, T. H. LaBean, L. Feng, J. H. Reif, Directed nucleation assembly of DNA tile complexes for barcode-patterned lattices, PNAS, 100 (2003) 8103–8108.

37. X. Shi, X. Wu, T. Song, X. Li, Construction of DNA nanotubes with controllable diameters and patterns using hierarchical DNA sub-tiles, Nanoscale, 8 (2016) 14785–14792.
38. X. Li, C. Zhang, C. Hao, C. Tian, G. Wang, C. Mao, DNA polyhedra with T-linkage, ACS nano, 6 (2012) 5138–5142.
39. T.H. LaBean, H. Yan, J. Kopatsch, F. Liu, E. Winfree, J.H. Reif, N.C. Seeman, Construction, analysis, ligation, and self-assembly of DNA triple crossover complexes, Journal of the American Chemical Society, 122 (2000) 1848–1860.
40. Y. Liu, C. Lin, H. Li, H. Yan, Aptamer-directed self-assembly of protein arrays on a DNA nanostructure, Angewandte Chemie International Edition, 44 (2005) 4333–4338.
41. H. Li, S.H. Park, J.H. Reif, T.H. LaBean, H. Yan, DNA-templated self-assembly of protein and nanoparticle linear arrays, Journal of the American Chemical Society, 126 (2004) 418–419.
42. S.H. Ko, M. Su, C. Zhang, A.E. Ribbe, W. Jiang, C. Mao, Synergistic self-assembly of RNA and DNA molecules, Nature Chemistry, 2 (2010) 1050–1055.
43. D. Shu, Y. Shu, F. Haque, S. Abdelmawla, P. Guo, Thermodynamically stable RNA three-way junction for constructing multifunctional nanoparticles for delivery of therapeutics, Nature Nanotechnology, 6 (2011) 658–667.
44. K.A. Afonin, M. Viard, A.Y. Koyfman, A.N. Martins, W.K. Kasprzak, M. Panigaj, R. Desai, A. Santhanam, W.W. Grabow, L. Jaeger, Multifunctional RNA nanoparticles, Nano Letters, 14 (2014) 5662–5671.
45. C. Wu, D. Han, T. Chen, L. Peng, G. Zhu, M. You, L. Qiu, K. Sefah, X. Zhang, W. Tan, Building a multifunctional aptamer-based DNA nanoassembly for targeted cancer therapy, Journal of the American Chemical Society, 135 (2013) 18644–18650.
46. J.B. Lee, Y.H. Roh, S.H. Um, H. Funabashi, W. Cheng, J.J. Cha, P. Kiatwuthinon, D.A. Muller, D. Luo, Multifunctional nanoarchitectures from DNA-based ABC monomers, Nature Nanotechnology, 4 (2009) 430–436.
47. H.-M. Meng, X. Zhang, Y. Lv, Z. Zhao, N.-N. Wang, T. Fu, H. Fan, H. Liang, L. Qiu, G. Zhu, DNA dendrimer: an efficient nanocarrier of functional nucleic acids for intracellular molecular sensing, ACS Nano, 8 (2014) 6171–6181.
48. T. Zhou, P. Chen, L. Niu, J. Jin, D. Liang, Z. Li, Z. Yang, D. Liu, pH-Responsive size-tunable self-assembled DNA dendrimers, Angewandte Chemie International Edition, 51 (2012) 11271–11274.
49. Y. Xing, E. Cheng, Y. Yang, P. Chen, T. Zhang, Y. Sun, Z. Yang, D. Liu, Self-assembled DNA hydrogels with designable thermal and enzymatic responsiveness, Advanced Materials, 23 (2011) 1117–1121.
50. S.H. Um, J.B. Lee, N. Park, S.Y. Kwon, C.C. Umbach, D. Luo, Enzyme-catalysed assembly of DNA hydrogel, Nature Materials, 5 (2006) 797–801.
51. F.F. Andersen, B. Knudsen, C.L.P. Oliveira, R.F. Frøhlich, D. Krüger, J. Bungert, M. Agbandje-McKenna, R. McKenna, S. Juul, C. Veigaard, Assembly and structural analysis of a covalently closed nano-scale DNA cage, Nucleic Acids Research, 36 (2008) 1113–1119.
52. W.M. Shih, J.D. Quispe, G.F. Joyce, A 1.7-kilobase single-stranded DNA that folds into a nanoscale octahedron, Nature, 427 (2004) 618–621.
53. B. Saccà, C.M. Niemeyer, DNA origami: the art of folding DNA, Angewandte Chemie International Edition, 51 (2012) 58–66.

54. A. Rajendran, M. Endo, Y. Katsuda, K. Hidaka, H. Sugiyama, Programmed two-dimensional self-assembly of multiple DNA origami jigsaw pieces, ACS nano, 5 (2011) 665–671.
55. G. Tikhomirov, P. Petersen, L. Qian, Fractal assembly of micrometre-scale DNA origami arrays with arbitrary patterns, Nature, 552 (2017) 67–71.
56. H. Dietz, S.M. Douglas, W.M. Shih, Folding DNA into twisted and curved nanoscale shapes, Science, 325 (2009) 725–730.
57. A.M. Maier, W. Bae, D. Schiffels, J.F. Emmerig, M. Schiff, T. Liedl, Self-assembled DNA tubes forming helices of controlled diameter and chirality, ACS Nano, 11 (2017) 1301–1306.
58. E.S. Andersen, M. Dong, M.M. Nielsen, K. Jahn, A. Lind-Thomsen, W. Mamdouh, K.V. Gothelf, F. Besenbacher, J. Kjems, DNA origami design of dolphin-shaped structures with flexible tails, ACS Nano, 2 (2008) 1213–1218.
59. Y. Ke, S.M. Douglas, M. Liu, J. Sharma, A. Cheng, A. Leung, Y. Liu, W.M. Shih, H. Yan, Multilayer DNA origami packed on a square lattice, Journal of the American Chemical Society, 131 (2009) 15903–15908.
60. D. Han, S. Pal, J. Nangreave, Z. Deng, Y. Liu, H. Yan, DNA origami with complex curvatures in three-dimensional space, Science, 332 (2011) 342–346.
61. E. Benson, A. Mohammed, J. Gardell, S. Masich, E. Czeizler, P. Orponen, B. Högberg, DNA rendering of polyhedral meshes at the nanoscale, Nature, 523 (2015) 441–444.
62. Y. Ke, L.L. Ong, W.M. Shih, P. Yin, Three-dimensional structures self-assembled from DNA bricks, science, 338 (2012) 1177–1183.
63. T. Gerling, K.F. Wagenbauer, A.M. Neuner, H. Dietz, Dynamic DNA devices and assemblies formed by shape-complementary, non–base pairing 3D components, Science, 347 (2015) 1446–1452.
64. P. Zhan, P.K. Dutta, P. Wang, G. Song, M. Dai, S.-X. Zhao, Z.-G. Wang, P. Yin, W. Zhang, B. Ding, Reconfigurable three-dimensional gold nanorod plasmonic nanostructures organized on DNA origami tripod, ACS Nano, 11 (2017) 1172–1179.
65. E.S. Andersen, M. Dong, M.M. Nielsen, K. Jahn, R. Subramani, W. Mamdouh, M.M. Golas, B. Sander, H. Stark, C.L. Oliveira, Self-assembly of a nanoscale DNA box with a controllable lid, Nature, 459 (2009) 73–76.
66. D. Han, X. Qi, C. Myhrvold, B. Wang, M. Dai, S. Jiang, M. Bates, Y. Liu, B. An, F. Zhang, Single-stranded DNA and RNA origami, Science, 358 (2017) eaao2648.
67. M. Li, M. Zheng, S. Wu, C. Tian, D. Liu, Y. Weizmann, W. Jiang, G. Wang, C. Mao, in vivo production of RNA nanostructures via programmed folding of single-stranded RNAs, Nature Communications, 9 (2018) 1–9.
68. C. Geary, P.W. Rothemund, E.S. Andersen, A single-stranded architecture for cotranscriptional folding of RNA nanostructures, Science, 345 (2014) 799–804.
69. R. Veneziano, S.Ratanalert, K. Zhang, F. Zhang, H. Yan, W. Chiu, M. Bathe, Designer nanoscale DNA assemblies programmed from the top down, Science, 352 (2016) aaf4388.
70. Q. Jiang, S. Liu, J. Liu, Z.G. Wang, B. Ding, Rationally designed DNA-origami nanomaterials for drug delivery in vivo, Advanced Materials, 31 (2019) 1804785.
71. G. Zhu, R. Hu, Z. Zhao, Z. Chen, X. Zhang, W. Tan, Noncanonical self-assembly of multifunctional DNA nanoflowers for biomedical applications, Journal of the American Chemical Society, 135 (2013) 16438–16445.

72. H. Kim, Y. Park, J.B. Lee, Self-assembled messenger RNA nanoparticles (mRNA-NPs) for efficient gene expression, Scientific Reports, 5 (2015) 1–9.
73. D. Han, Y. Park, H. Nam, J.B. Lee, Enzymatic size control of RNA particles using complementary rolling circle transcription (cRCT) method for efficient siRNA production, Chemical Communications, 50 (2014) 11665–11667.
74. H. Kim, Y.K. Lee, K.H. Han, H. Jeon, I.-h. Jeong, S.-Y. Kim, J.B. Lee, P.C. Lee, BRC-mediated RNAi targeting of USE1 inhibits tumor growth in vitro and in vivo, Biomaterials, 230 (2020) 119630.
75. H. Kim, J. Jeong, D. Kim, G. Kwak, S.H. Kim, J.B. Lee, Bubbled RNA-based cargo for boosting RNA interference, Advanced Science, 4 (2017) 1600523.
76. Y.H. Roh, J.B. Lee, K.E. Shopsowitz, E.C. Dreaden, S.W. Morton, Z. Poon, J. Hong, I. Yamin, D.K. Bonner, P.T. Hammond, Layer-by-layer assembled antisense DNA microsponge particles for efficient delivery of cancer therapeutics, ACS Nano, 8 (2014) 9767–9780.
77. C. Wang, W. Sun, G. Wright, A.Z. Wang, Z. Gu, Inflammation-triggered cancer immunotherapy by programmed delivery of CpG and anti-PD1 antibody, Advanced Materials, 28 (2016) 8912–8920.
78. H. Jung, D. Kim, Y.Y. Kang, H. Kim, J.B. Lee, H. Mok, CpG incorporated DNA microparticles for elevated immune stimulation for antigen presenting cells, RSC Advances, 8 (2018) 6608–6615.
79. Y.R. Baker, J. Chen, J. Brown, A.H. El-Sagheer, P. Wiseman, E. Johnson, P. Goddard, T. Brown, Preparation and characterization of manganese, cobalt and zinc DNA nanoflowers with tuneable morphology, DNA content and size, Nucleic Acids Research, 46 (2018) 7495–7505.
80. J.S. Lee, H. Kim, C. Jo, J. Jeong, J. Ko, S. Han, M.S. Lee, H.Y. Lee, J.W. Han, J. Lee, Enzyme-driven hasselback-like DNA-based inorganic superstructures, Advanced Functional Materials, 27 (2017) 1704213.
81. R. Hu, X. Zhang, Z. Zhao, G. Zhu, T. Chen, T. Fu, W. Tan, DNA nanoflowers for multiplexed cellular imaging and traceable targeted drug delivery, Angewandte Chemie, 126 (2014) 5931–5936.
82. S. Han, J.S. Lee, J.B. Lee, Synthesis of a multi-functional DNA nanosphere barcode system for direct cell detection, Nanoscale, 9 (2017) 14094–14102.
83. E. Kim, L. Zwi-Dantsis, N. Reznikov, C.S. Hansel, S. Agarwal, M.M. Stevens, One-pot synthesis of multiple protein-encapsulated DNA flowers and their application in intracellular protein delivery, Advanced Materials, 29 (2017) 1701086.
84. J. Kim, D. Kim, J.B. Lee, DNA aptamer-based carrier for loading proteins and enhancing the enzymatic activity, RSC Advances, 7 (2017) 1643–1645.
85. J.S. Ha, J.S. Lee, J. Jeong, H. Kim, J. Byun, S.A. Kim, H.J. Lee, H.S. Chung, J.B. Lee, D.-R. Ahn, Poly-sgRNA/siRNA ribonucleoprotein nanoparticles for targeted gene disruption, Journal of Controlled Release, 250 (2017) 27–35.
86. D.L. Jasinski, D.W. Binzel, P. Guo, One-pot production of RNA nanoparticles via automated processing and self-assembly, ACS Nano, 13 (2019) 4603–4612.
87. T. Yata, Y. Takahashi, M. Tan, K. Hidaka, H. Sugiyama, M. Endo, Y. Takakura, M. Nishikawa, Efficient amplification of self-gelling polypod-like structured DNA by rolling circle amplification and enzymatic digestion, Scientific Reports, 5 (2015) 1–9.

88. G. Zhu, L. Mei, H.D. Vishwasrao, O. Jacobson, Z. Wang, Y. Liu, B.C. Yung, X. Fu, A. Jin, G. Niu, Intertwining DNA-RNA nanocapsules loaded with tumor neoantigens as synergistic nanovaccines for cancer immunotherapy, Nature Communications, 8 (2017) 1–13.
89. Y. Park, H. Kim, J.B. Lee, Self-assembled DNA-guided RNA nanovector via step-wise dual enzyme polymerization (SDEP) for carrier-free siRNA delivery, ACS Biomaterials Science & Engineering, 2 (2016) 616–624.
90. H. Zhang, Y. Ma, Y. Xie, Y. An, Y. Huang, Z. Zhu, C.J. Yang, A controllable aptamer-based self-assembled DNA dendrimer for high affinity targeting, bioimaging and drug delivery, Scientific Reports, 5 (2015) 10099.
91. J. Mikkila, A.-P. Eskelinen, E.H. Niemela, V. Linko, M.J. Frilander, P. Torma, M.A. Kostiainen, Virus-encapsulated DNA origami nanostructures for cellular delivery, Nano Letters, 14 (2014) 2196–2200.
92. M. Chang, C.-S. Yang, D.-M. Huang, Aptamer-conjugated DNA icosahedral nanoparticles as a carrier of doxorubicin for cancer therapy, ACS Nano, 5 (2011) 6156–6163.
93. S. Han, Y. Park, H. Kim, H. Nam, O. Ko, J.B. Lee, Double Controlled Release of Therapeutic RNA Modules through Injectable DNA–RNA Hybrid Hydrogel, ACS Applied Materials & Interfaces, 12 (2020) 55554–55563.
94. X. Li, Q. Zhao, L. Qiu, Smart ligand: aptamer-mediated targeted delivery of chemotherapeutic drugs and siRNA for cancer therapy, Journal of Controlled Release, 171 (2013) 152–162.
95. J.L. Litke, S.R. Jaffrey, Highly efficient expression of circular RNA aptamers in cells using autocatalytic transcripts, Nature Biotechnology, 37 (2019) 667–675.
96. M. Jang, J.H. Kim, H.Y. Nam, I.C. Kwon, H.J. Ahn, Design of a platform technology for systemic delivery of siRNA to tumours using rolling circle transcription, Nature Communications, 6 (2015) 1–12.
97. S. Raniolo, G. Vindigni, A. Ottaviani, V. Unida, F. Iacovelli, A. Manetto, M. Figini, L. Stella, A. Desideri, S. Biocca, Selective targeting and degradation of doxorubicin-loaded folate-functionalized DNA nanocages, Nanomedicine: Nanotechnology, Biology and Medicine, 14 (2018) 1181–1190.
98. Y. Sun, C. Kang, F. Liu, Y. Zhou, L. Luo, H. Qiao, RGD peptide-based target drug delivery of doxorubicin nanomedicine, Drug Development Research, 78 (2017) 283–291.
99. K. Zhang, R. Deng, Y. Sun, L. Zhang, J. Li, Reversible control of cell membrane receptor function using DNA nano-spring multivalent ligands, Chemical Science, 8 (2017) 7098–7105.
100. I. Dovgan, A. Ehkirch, V. Lehot, I. Kuhn, O. Koniev, S. Kolodych, A. Hentz, M. Ripoll, S. Ursuegui, M. Nothisen, On the use of DNA as a linker in antibody-drug conjugates: synthesis, stability and in vitro potency, Scientific Reports, 10 (2020) 1–9.
101. S. Han, J.B. Lee, Self-assembly of tumor-targeting RNA nanoball for carrier-free delivery of therapeutic RNA, Journal of Industrial and Engineering Chemistry, 64 (2018) 90–96.
102. K.E. Shopsowitz, Y.H. Roh, Z.J. Deng, S.W. Morton, P.T. Hammond, RNAi-microsponges form through self-assembly of the organic and inorganic products of transcription, Small, 10 (2014) 1623–1633.

103. M. Keller, Lipidic carriers of RNA/DNA oligonucleotides and polynucleotides: what a difference a formulation makes!, Journal of Controlled Release, 103 (2005) 537–540.
104. S. Werth, B. Urban-Klein, L. Dai, S. Höbel, M. Grzelinski, U. Bakowsky, F. Czubayko, A. Aigner, A low molecular weight fraction of polyethylenimine (PEI) displays increased transfection efficiency of DNA and siRNA in fresh or lyophilized complexes, Journal of Controlled Release, 112 (2006) 257–270.
105. M.I. Setyawati, R.V. Kutty, D.T. Leong, DNA nanostructures carrying stoichiometrically definable antibodies, Small, 12 (2016) 5601–5611.
106. Z. Xia, P. Wang, X. Liu, T. Liu, Y. Yan, J. Yan, J. Zhong, G. Sun, D. He, Tumor-penetrating peptide-modified DNA tetrahedron for targeting drug delivery, Biochemistry, 55 (2016) 1326–1331.
107. L.R. de Mello, I.W. Hamley, V. Castelletto, B.B.M. Garcia, S.W. Han, C.L.P. de Oliveira, E.R. da Silva, Nanoscopic structure of complexes formed between DNA and the cell-penetrating peptide penetratin, The Journal of Physical Chemistry B, 123 (2019) 8861–8871.
108. D.M. Copolovici, K. Langel, E. Eriste, U. Langel, Cell-penetrating peptides: design, synthesis, and applications, ACS Nano, 8 (2014) 1972–1994.
109. C. Bechara, S. Sagan, Cell-penetrating peptides: 20 years later, where do we stand?, FEBS Letters, 587 (2013) 1693–1702.
110. M. AngeláAleman-Garcia, pH-responsive and switchable triplex-based DNA hydrogels, Chemical Science, 6 (2015) 4190–4195.
111. K. Zhang, X. Zhu, F. Jia, E. Auyeung, C.A. Mirkin, Temperature-activated nucleic acid nanostructures, Journal of the American Chemical Society, 135 (2013) 14102–14105.
112. Z. Xiao, C. Ji, J. Shi, E.M. Pridgen, J. Frieder, J. Wu, O.C. Farokhzad, DNA self-assembly of targeted near-infrared-responsive gold nanoparticles for cancer thermo-chemotherapy, Angewandte Chemie, 124 (2012) 12023–12027.
113. T. Takenaka, M. Endo, Y. Suzuki, Y. Yang, T. Emura, K. Hidaka, T. Kato, T. Miyata, K. Namba, H. Sugiyama, Photoresponsive DNA nanocapsule having an open/close system for capture and release of nanomaterials, Chemistry–A European Journal, 20 (2014) 14951–14954.
114. A. Muzdalo, P. Saalfrank, J. Vreede, M. Santer, Cis-to-trans isomerization of azobenzene derivatives studied with transition path sampling and quantum mechanical/molecular mechanical molecular dynamics, Journal of Chemical Theory and Computation, 14 (2018) 2042–2051.
115. D.Y. Tam, X. Zhuang, S.W. Wong, P.K. Lo, Photoresponsive self-assembled DNA nanomaterials: design, working principles, and applications, Small, 15 (2019) 1805481.
116. S. Li, Q. Jiang, S. Liu, Y. Zhang, Y. Tian, C. Song, J. Wang, Y. Zou, G. J. Anderson, J.-Y. Han, Y. Chang, Y. Liu, C. Zhang, L. Chen, G. Zhou, G. Nie, H. Yan, B. Ding, Y. Zhao, A DNA nanorobot functions as a cancer therapeutic in response to a molecular trigger *in vivo*, Nature Biotechnology, 36 (2018) 258–264.
117. R. Mo, T. Jiang, R. DiSanto, W. Tai, Z. Gu, ATP-triggered anticancer drug delivery, Nature Communications, 5 (2014) 1–10.
118. K. Abnous, N.M. Danesh, M. Ramezani, M. Alibolandi, A. Bahreyni, P. Lavaee, S.A. Moosavian, S.M. Taghdisi, A smart ATP-responsive chemotherapy drug-free delivery system using a DNA nanostructure for synergistic treatment of breast cancer *in vitro* and *in vivo*, Journal of Drug Targeting, 28 (2020) 1–8.

119. S. Hernot, A.L. Klibanov, Microbubbles in ultrasound-triggered drug and gene delivery, Advanced Drug Delivery Reviews, 60 (2008) 1153–1166.
120. A. Biswas, K.-I. Joo, J. Liu, M. Zhao, G. Fan, P. Wang, Z. Gu, Y. Tang, Endoprotease-mediated intracellular protein delivery using nanocapsules, ACS Nano, 5 (2011) 1385–1394.
121. X. Luo, Z. Li, G. Wang, X. He, X. Shen, Q. Sun, L. Wang, R. Yue, N. Ma, MicroRNA-catalyzed cancer therapeutics based on DNA-programmed nanoparticle complex, ACS Applied Materials & Interfaces, 9 (2017) 33624–33631.
122. J. Shi, X. Yang, Y. Li, D. Wang, W. Liu, Z. Zhang, J. Liu, K. Zhang, MicroRNA-responsive release of Cas9/sgRNA from DNA nanoflower for cytosolic protein delivery and enhanced genome editing, Biomaterials, 256 (2020) 120221.
123. M. Hirosawa, Y. Fujita, C.J. Parr, K. Hayashi, S. Kashida, A. Hotta, K. Woltjen, H. Saito, Cell-type-specific genome editing with a microRNA-responsive CRISPR–Cas9 switch, Nucleic Acids Research, 45 (2017) e118.
124. H. Zhao, X. Yuan, J. Yu, Y. Huang, C. Shao, F. Xiao, L. Lin, Y. Li, L. Tian, Magnesium-stabilized multifunctional DNA nanoparticles for tumor-targeted and pH-responsive drug delivery, ACS Applied Materials & Interfaces, 10 (2018) 15418–15427.
125. Y. Chen, E.A. Arriaga, Individual acidic organelle pH measurements by capillary electrophoresis, Analytical Chemistry, 78 (2006) 820–826.
126. P. Rai, S. Mallidi, X. Zheng, R. Rahmanzadeh, Y. Mir, S. Elrington, A. Khurshid, T. Hasan, Development and applications of photo-triggered theranostic agents, Advanced Drug Delivery Reviews, 62 (2010) 1094–1124.
127. A.G. Skirtach, A.A. Antipov, D.G. Shchukin, G.B. Sukhorukov, Remote activation of capsules containing Ag nanoparticles and IR dye by laser light, Langmuir, 20 (2004) 6988–6992.
128. H. Kang, H. Liu, X. Zhang, J. Yan, Z. Zhu, L. Peng, H. Yang, Y. Kim, W. Tan, Photoresponsive DNA-cross-linked hydrogels for controllable release and cancer therapy, Langmuir, 27 (2011) 399–408.
129. M. Endo, Y. Yang, Y. Suzuki, K. Hidaka, H. Sugiyama, Single-molecule visualization of the hybridization and dissociation of photoresponsive oligonucleotides and their reversible switching behavior in a DNA nanostructure, Angewandte Chemie International Edition, 51 (2012) 10518–10522.
130. D. Wu, L. Wang, W. Li, X. Xu, W. Jiang, DNA nanostructure-based drug delivery nanosystems in cancer therapy, International Journal of Pharmaceutics, 533 (2017) 169–178.
131. M. Leist, B. Single, A.F. Castoldi, S. Kühnle, P. Nicotera, Intracellular adenosine triphosphate (ATP) concentration: a switch in the decision between apoptosis and necrosis, The Journal of Experimental Medicine, 185 (1997) 1481–1486.
132. M.W. Gorman, E.O. Feigl, C.W. Buffington, Human plasma ATP concentration, Clinical Chemistry, 53 (2007) 318–325.
133. L. Zhu, P. Kate, V.P. Torchilin, Matrix metalloprotease 2-responsive multifunctional liposomal nanocarrier for enhanced tumor targeting, ACS Nano, 6 (2012) 3491–3498.
134. N. Singh, A. Karambelkar, L. Gu, K. Lin, J.S. Miller, C.S. Chen, M.J. Sailor, S.N. Bhatia, Bioresponsive mesoporous silica nanoparticles for triggered drug release, Journal of the American Chemical Society, 133 (2011) 19582–19585.
135. T. Liu, L. Zeng, W. Jiang, Y. Fu, W. Zheng, T. Chen, Rational design of cancer-targeted selenium nanoparticles to antagonize multidrug resistance in cancer

cells, Nanomedicine: Nanotechnology, Biology and Medicine, 11 (2015) 947–958.

136. R.J. Kathawala, P. Gupta, C.R. Ashby Jr, Z.-S. Chen, The modulation of ABC transporter-mediated multidrug resistance in cancer: a review of the past decade, Drug Resistance Updates, 18 (2015) 1–17.
137. Y. Wang, Z. Zhang, S. Xu, F. Wang, Y. Shen, S. Huang, S. Guo, pH, redox and photothermal tri-responsive DNA/polyethylenimine conjugated gold nanorods as nanocarriers for specific intracellular co-release of doxorubicin and chemosensitizer pyronaridine to combat multidrug resistant cancer, Nanomedicine: Nanotechnology, Biology and Medicine, 13 (2017) 1785–1795.
138. M. Volkova, R. Russell, Anthracycline cardiotoxicity: prevalence, pathogenesis and treatment, Current Cardiology Reviews, 7 (2011) 214–220.
139. V. Bagalkot, O.C. Farokhzad, R. Langer, S. Jon, An aptamer–doxorubicin physical conjugate as a novel targeted drug-delivery platform, Angewandte Chemie International Edition, 45 (2006) 8149–8152.
140. F. Pi, H. Zhang, H. Li, V. Thiviyanathan, D.G. Gorenstein, A.K. Sood, P. Guo, RNA nanoparticles harboring annexin A2 aptamer can target ovarian cancer for tumor-specific doxorubicin delivery, Nanomedicine: Nanotechnology, Biology and Medicine, 13 (2017) 1183–1193.
141. J. Liu, L. Song, S. Liu, S. Zhao, Q. Jiang, B. Ding, A tailored DNA nanoplatform for synergistic RNAi-/chemotherapy of multidrug-resistant tumors, Angewandte Chemie International Edition, 57 (2018) 15486–15490.
142. Q. Mou, Y. Ma, F. Ding, X. Gao, D. Yan, X. Zhu, C. Zhang, Two-in-one chemogene assembled from drug-integrated antisense oligonucleotides to reverse chemoresistance, Journal of the American Chemical Society, 141 (2019) 6955–6966.
143. J. Zhang, Y. Guo, F. Ding, G. Pan, X. Zhu, C. Zhang, A camptothecin-grafted DNA tetrahedron as a precise nanomedicine to inhibit tumor growth, Angewandte Chemie, 131 (2019) 13932–13936.
144. X. Tan, X. Lu, F. Jia, X. Liu, Y. Sun, J.K. Logan, K. Zhang, Blurring the role of oligonucleotides: spherical nucleic acids as a drug delivery vehicle, Journal of the American Chemical Society, 138 (2016) 10834–10837.
145. M.P. Wickramathilaka, B.Y. Tao, Characterization of covalent crosslinking strategies for synthesizing DNA-based bioconjugates, Journal of Biological Engineering, 13 (2019) 1–10.
146. X. Zhou, X. Su, P. Pathak, R. Vik, B. Vinciguerra, L. Isaacs, J. Jayawickramarajah, Host–guest tethered DNA transducer: ATP fueled release of a protein inhibitor from Cucurbit [7] uril, Journal of the American Chemical Society, 139 (2017) 13916–13921.
147. M. Helm, Post-transcriptional nucleotide modification and alternative folding of RNA, Nucleic Acids Research, 34 (2006) 721–733.
148. A.H. El-Sagheer, T. Brown, Click chemistry with DNA, Chemical Society Reviews, 39 (2010) 1388–1405.
149. M.L. Tan, P.F. Choong, C.R. Dass, Recent developments in liposomes, microparticles and nanoparticles for protein and peptide drug delivery, Peptides, 31 (2010) 184–193.
150. X. Liu, W. Gao, In situ growth of self-assembled protein–polymer nanovesicles for enhanced intracellular protein delivery, ACS Applied Materials & Interfaces, 9 (2017) 2023–2028.

151. H. Zhao, Z.Y. Lin, L. Yildirimer, A. Dhinakar, X. Zhao, J. Wu, Polymer-based nanoparticles for protein delivery: design, strategies and applications, Journal of Materials Chemistry B, 4 (2016) 4060–4071.
152. Q. Gan, T. Wang, Chitosan nanoparticle as protein delivery carrier—systematic examination of fabrication conditions for efficient loading and release, Colloids and Surfaces B: Biointerfaces, 59 (2007) 24–34.
153. H.S. Park, C.W. Kim, H.J. Lee, J.H. Choi, S.G. Lee, Y.-P. Yun, I.C. Kwon, S.J. Lee, S.Y. Jeong, S.C. Lee, A mesoporous silica nanoparticle with charge-convertible pore walls for efficient intracellular protein delivery, Nanotechnology, 21 (2010) 225101.
154. F. Scaletti, J. Hardie, Y.-W. Lee, D.C. Luther, M. Ray, V.M. Rotello, Protein delivery into cells using inorganic nanoparticle–protein supramolecular assemblies, Chemical Society Reviews, 47 (2018) 3421–3432.
155. H. Pei, Y. Wan, J. Li, H. Hu, Y. Su, Q. Huang, C. Fan, Regenerable electrochemical immunological sensing at DNA nanostructure-decorated gold surfaces, Chemical Communications, 47 (2011) 6254–6256.
156. K. Zhang, L. Hao, S.J. Hurst, C.A. Mirkin, Antibody-linked spherical nucleic acids for cellular targeting, Journal of the American Chemical Society, 134 (2012) 16488–16491.
157. H. Ohno, T. Kobayashi, R. Kabata, K. Endo, T. Iwasa, S.H. Yoshimura, K. Takeyasu, T. Inoue, H. Saito, Synthetic RNA–protein complex shaped like an equilateral triangle, Nature Nanotechnology, 6 (2011) 116–120.
158. K.-R. Kim, D. Hwang, J. Kim, C.-Y. Lee, W. Lee, D.S. Yoon, D. Shin, S.-J. Min, I.C. Kwon, H.S. Chung, Streptavidin-mirror DNA tetrahedron hybrid as a platform for intracellular and tumor delivery of enzymes, Journal of Controlled Release, 280 (2018) 1–10.
159. X. Liu, Y. Xu, T. Yu, C. Clifford, Y. Liu, H. Yan, Y. Chang, A DNA nanostructure platform for directed assembly of synthetic vaccines, Nano Letters, 12 (2012) 4254–4259.
160. W. Sun, W. Ji, Q. Hu, J. Yu, C. Wang, C. Qian, G. Hochu, Z. Gu, Transformable DNA nanocarriers for plasma membrane targeted delivery of cytokine, Biomaterials, 96 (2016) 1–10.
161. J.D. Brodin, A.J. Sprangers, J.R. McMillan, C.A. Mirkin, DNA-mediated cellular delivery of functional enzymes, Journal of the American Chemical Society, 137 (2015) 14838–14841.
162. B. Draghici, M.A. Ilies, Synthetic nucleic acid delivery systems: present and perspectives, Journal of Medicinal Chemistry, 58 (2015) 4091–4130.
163. Y. Qu, J. Yang, P. Zhan, S. Liu, K. Zhang, Q. Jiang, C. Li, B. Ding, Self-assembled DNA dendrimer nanoparticle for efficient delivery of immunostimulatory CpG motifs, ACS Applied Materials & Interfaces, 9 (2017) 20324–20329.
164. J. Li, H. Pei, B. Zhu, L. Liang, M. Wei, Y. He, N. Chen, D. Li, Q. Huang, C. Fan, Self-assembled multivalent DNA nanostructures for noninvasive intracellular delivery of immunostimulatory CpG oligonucleotides, ACS Nano, 5 (2011) 8783–8789.
165. S. Sellner, S. Kocabey, K. Nekolla, F. Krombach, T. Liedl, M. Rehberg, DNA nanotubes as intracellular delivery vehicles *in vivo*, Biomaterials, 53 (2015) 453–463.
166. E.F. Khisamutdinov, H. Li, D.L. Jasinski, J. Chen, J. Fu, P. Guo, Enhancing immunomodulation on innate immunity by shape transition among RNA triangle, square and pentagon nanovehicles, Nucleic Acids Research, 42 (2014) 9996–10004.

167. H. Lee, A.K. Lytton-Jean, Y. Chen, K.T. Love, A.I. Park, E.D. Karagiannis, A. Sehgal, W. Querbes, C.S. Zurenko, M. Jayaraman, Molecularly self-assembled nucleic acid nanoparticles for targeted *in vivo* siRNA delivery, Nature Nanotechnology, 7 (2012) 389–393.
168. S.C. Semple, A. Akinc, J. Chen, A.P. Sandhu, B.L. Mui, C.K. Cho, D.W. Sah, D. Stebbing, E.J. Crosley, E. Yaworski, Rational design of cationic lipids for siRNA delivery, Nature Biotechnology, 28 (2010) 172–176.
169. K.A. Whitehead, R. Langer, D.G. Anderson, Knocking down barriers: advances in siRNA delivery, Nature Reviews Drug Discovery, 8 (2009) 129–138.
170. H. Xue, F. Ding, J. Zhang, Y. Guo, X. Gao, J. Feng, X. Zhu, C. Zhang, DNA tetrahedron-based nanogels for siRNA delivery and gene silencing, Chemical Communications, 55 (2019) 4222–4225.
171. D.P. Bartel, MicroRNAs: target recognition and regulatory functions, Cell, 136 (2009) 215–233.
172. L. Liang, J. Li, Q. Li, Q. Huang, J. Shi, H. Yan, C. Fan, Single-particle tracking and modulation of cell entry pathways of a tetrahedral DNA nanostructure in live cells, Angewandte Chemie International Edition, 53 (2014) 7745–7750.
173. H. Qian, C.Y. Tay, M. Setyawati, S. Chia, D. Lee, D. Leong, Protecting microRNAs from RNase degradation with steric DNA nanostructures, Chemical science, 8 (2017) 1062–1067.
174. H. Yin, H. Wang, Z. Li, D. Shu, P. Guo, RNA micelles for the systemic delivery of anti-miRNA for cancer targeting and inhibition without ligand, ACS Nano, 13 (2018) 706–717.
175. H. Kim, E. Lee, Y.Y. Kang, J. Song, H. Mok, J.B. Lee, Enzymatically produced miR34a nanoparticles for enhanced antiproliferation activity, Advanced Biosystems, 2 (2018) 1700158.
176. M. Nishikawa, Y. Mizuno, K. Mohri, N. Matsuoka, S. Rattanakiat, Y. Takahashi, H. Funabashi, D. Luo, Y. Takakura, Biodegradable CpG DNA hydrogels for sustained delivery of doxorubicin and immunostimulatory signals in tumor-bearing mice, Biomaterials, 32 (2011) 488–494.
177. Y. Fu, D. Zeng, J. Chao, Y. Jin, Z. Zhang, H. Liu, D. Li, H. Ma, Q. Huang, K.V. Gothelf, Single-step rapid assembly of DNA origami nanostructures for addressable nanoscale bioreactors, Journal of the American Chemical Society, 135 (2013) 696–702.
178. J.A. Doudna, E. Charpentier, The new frontier of genome engineering with CRISPR-Cas9, Science, 346 (2014) 1258096.
179. P.D. Hsu, E.S. Lander, F. Zhang, Development and applications of CRISPR-Cas9 for genome engineering, Cell, 157 (2014) 1262–1278.
180. T. Gajanayake, R. Olariu, F.M. Leclère, A. Dhayani, Z. Yang, A.K. Bongoni, Y. Banz, M.A. Constantinescu, J.M. Karp, P.K. Vemula, A single localized dose of enzyme-responsive hydrogel improves long-term survival of a vascularized composite allograft, Science Translational Medicine, 6 (2014) 249ra110.
181. G. Zhu, J. Zheng, E. Song, M. Donovan, K. Zhang, C. Liu, W. Tan, Self-assembled, aptamer-tethered DNA nanotrains for targeted transport of molecular drugs in cancer theranostics, PNAS, 110 (2013) 7998–8003.

CHAPTER 10

FLUORESCENCE HYBRIDIZATION IMAGING OF NEUROLOGICAL DISEASE MESSENGER RNAs

ERIC WICKSTROM
Laboratory of Molecular Design, Thomas Jefferson University, Philadelphia, PA, USA

ABBREVIATIONS

AEEA aminoethoxyethoxyacetyl, DAPI, 4′,6-diamidino-2-phenylindole
Htt huntingtin protein
MAO monoamine oxidase
MTT 3-(4,5-dimethylthiazol-2-yl) 2,5-diphenyltetrazolium bromide
PBS phosphate buffered saline
PNA peptide nucleic acid

10.1. INTRODUCTION

Neurological diseases become more prominent as the world population ages. Neurons malfunction due to mutations in neuronal genes that result in dysfunctional messenger RNAs (mRNAs) that encode malfunctioning or dysfunctioning neuronal proteins. Fluorescently labeled hybridization agents provide one diagnostic tool for assessing mRNA problems in neurologic diseases.

Nucleic Acids in Medicinal Chemistry and Chemical Biology: Drug Development and Clinical Applications, First Edition. Edited by Lihe Zhang, Xinjing Tang, Zhen Xi, and Jyoti Chattopadhyaya.

Here we consider two examples, imaging the mRNAs for monoamine oxidase A (MAO-A) [1] and huntingtin [2].

1. Monoamine oxidases (MAO) catalyze the oxidative deamination of many biogenic amines. MAO are expressed in various neuronal and nonneuronal cells in the central nervous system (CNS) and peripheral organs [3]. In the CNS, MAO carries out metabolic inactivation of released monoamine transmitters but also creates cytotoxic free radicals throughout aging and neurodegenerative diseases. MAO exists in two isoforms, MAO-A and MAO-B, which differ by their substrate specificity and more for inhibitor selectivity. MAO-A preferentially deaminates serotonin, dopamine, and norepinephrine in catecholaminergic neurons in the human brain, in particular adrenergic and noradrenergic neurons. On the other hand, MAO-B preferentially degrades exogenous bioamines ingested in the diet such as phenylethylamine and benzylamine. The catalytic activity of MAO-A, located in the mitochondria of presynaptic terminals in monoamine-releasing neurons, have a critical role in regulating the release and degradation of monoamines throughout the brain [4, 5].

Differential expression of the *MAOA* gene, called by some the "warrior gene," has been linked to numerous CNS disorders, particularly aggression [6–8], antisocial behavior, attention deficit hyperactivity disorder [9, 10], anxiety [11], anorexia nervosa [12, 13], bipolar disorder [14–16], drug abuse [17, 18], and Alzheimer's disease [19]. Elevated MAO-A levels in affect modulating regions are believed to exert an important monoamine-lowering process during key-depressive episodes of major depressive disorder. In postpartum depression, MAO-A levels and affinity increase sharply due to a decline (100- to 1000-fold) in estrogen levels, which creates a high risk for depressive episodes [20].

On the other hand, genetic deletion of *MAOA* produces aggressive phenotypes across species. Studies have found links between *MAOA* gene products and violent behavior. Lower MAO-A activity in cortical and sub-cortical brain regions results in higher self-reported trait aggression [21]. MAO-A also plays a critical role in alcohol [22, 23] and cocaine addiction [24]. A recent study showed upregulated MAO-A expression in alcoholic individuals, using whole genome sequencing of mRNA transcripts (RNA-Seq) [23]. In another study, *MAOA*-L (low repeat allele) carrier males with cocaine use disorders (CUD) revealed greater gray matter volume (GMV) reduction in the brain as compared to their healthy counterparts [24]. Over the last two decades, positron emission tomography (PET) has been used extensively for measuring MAO-A enzyme activity in different tissues [5]. ^{11}C-clorgyline is an example of compounds used for such studies [25]. However, no such method is available for noninvasive measurement of *MAOA* mRNA expression.

2. Huntington's disease (HD) is an autosomal-dominant neurodegenerative genetic disorder that is caused by the expansion of a polyglutamine-encoding CAG repeat within exon 1 of the huntingtin *HTT* gene [26]. HD is a fatal disorder that gradually damages brain cells and causes muscle uncoordination, cognitive decline, and psychiatric problems [27]. Striatal medium spiny neurons and cortical neurons are the primary neuronal substrates for disease symptoms, although other cell types, both neuronal and nonneuronal, contribute to the disease phenotype [28, 29]. Htt protein is expressed in all human and mammalian cells, including cells of the central nervous system and peripheral tissues, and occurs at particularly high levels in the brain [30–32].

The *HTT* CAG repeat is polymorphic in normal individuals, ranging in number from 6 to 35, while in affected individuals, the number of CAG repeats exceeds 37, with expansions as long as 200 observed in some early-onset cases [27]. The CAG expansion leads to production of a mutant, polyglutamine-expanded Htt protein, which is characterized by misfolding and aggregation [33]. Currently, therapies for HD include antidepressants and antipsychotics for psychiatric symptoms, and tetrabenazine or its deuterated form to treat choreic movements [34, 35], but there is no effective therapy to prevent or slow the progression of HD. Finding therapeutic agents to slow down or stabilize HD is thus a critical unmet need.

In recent years, investigators have applied various strategies to reduce *HTT* mRNA. Striatal injections of adeno-associated virus (AAV) expressing either a short hairpin RNA (shRNA) [36] or an artificial microRNA (miRNA) expression scaffold, mi2.4 [37], as well as intraventricular delivery of antisense oligonucleotides (ASOs) that target cellular mRNA transcripts via complementary base pairing [38, 39] have resulted in Htt protein knockdown efficacy in animal models of HD. Furthermore, the striatal infusion of synthetic small interfering RNA (siRNA) duplexes with cholesterol-conjugates-attenuated neuronal pathology and delayed the abnormal behavioral phenotypes in transgenic mouse models [40].

The development of therapeutic strategies that utilize *HTT* silencing requires appropriate, proximate biomarkers to ensure target engagement. The levels of mutant Htt protein or mutant *HTT* mRNA represent important proximate biomarkers to determine the efficacy of experimental knockdown therapies. Encouraging results from *HTT* knockdown studies in mice [38, 39] established a basis for a human trial with the same ASO [41]. Thus, the current human trial carries a requirement for a surrogate marker to determine efficacy of the therapeutic *HTT* knockdown agent in live patients in real time during therapy. An immunoassay of mutant Htt protein in cerebrospinal fluid withdrawn by spinal tap [42] has been utilized in the current human trial.

10.2. STRATEGY

10.2.1. *MAOA* mRNA

We conceived and pioneered PET imaging of mRNA levels in tumors [43] and fluorescence imaging of mRNA levels in cells, which showed equivalent results in both live and fixed cells [44]. To accomplish fluorescent imaging of *MAOA* mRNA, we designed an imaging agent composed of a complementary peptide

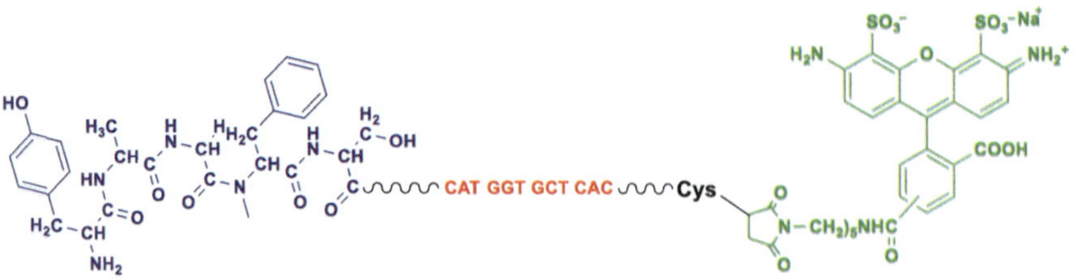

Figure 10.1 ChemDraw representation of human DAMGO-*MAOA* PNA-AF488, WT4879. Targeting peptide (blue) is linked to PNA (red) at N-terminus and to AlexaFluor488 (green) at C-terminus via thiol-maleimide linkage.

nucleic acid (PNA) labeled with a fluorescent dye for imaging and attached to a peptide moiety for receptor-mediated intracellular delivery (Figure 10.1).

PNA, an oligonucleotide analog where the sugar-phosphodiester backbone is replaced with a peptide-like aminoethylglycine backbone, has great potential for biomedical applications [45]. Owing to their achiral, uncharged, flexible backbone, PNAs hybridize with mRNA more strongly than normal RNA or DNA with mRNA [46]. They are resistant to enzymatic degradation and are stable over a wide range of pH. However, due to its uncharged nature, naked PNA is poorly taken up by mammalian cells [47]. As a result, the challenge of delivering PNA imaging agents into targeted cells must be overcome.

Fortunately, incorporation of a receptor-targeting peptide is an efficient way to deliver PNA imaging agents into targeted cells [43]. To specifically target our probes to neuronal cells, we extended a modified DAMGO μ-opioid targeting peptide, Tyr-D-Ala-Gly-NMePhe-Ser, from the N-terminus of PNA. DAMGO is an enkephalin analog that binds tightly to μ-opioid receptors on neuronal cells and induces internalization [48]. The phenolic hydroxyl (–OH) group of the *N*-terminal Tyr residue with a free cationic α-amino ($-NH_3^+$) group, similar to the tyramine moiety of morphine, and an aromatic amino acid separated by one or two residues are the core requirements for the binding of opioid peptides [49]. The replacement of the glycinol moiety with L-serine maintains receptor affinity (K_i value 0.56±0.006 nM for DAMGO with glycinol, 0.68±0.02 nM for DAMGO with L-serine) [50].

10.2.2. *HTT* mRNA

To accomplish fluorescent imaging of *HTT* mRNA as a real-time measurement of *HTT* mRNA knockdown, we designed a fluorescent *HTT* mRNA hybridization agent (Figure 10.2) to enable imaging and quantitation of *HTT* mRNA in Htt-expressing [51] HEK293T cells. The *HTT* mRNA imaging agent was composed of a complementary peptide nucleic acid (PNA) labeled on the *N*-terminus with a fluorescent dye for imaging and conjugated on the C-terminus to a peptide moiety for receptor-mediated intracellular delivery (Figure 10.2).

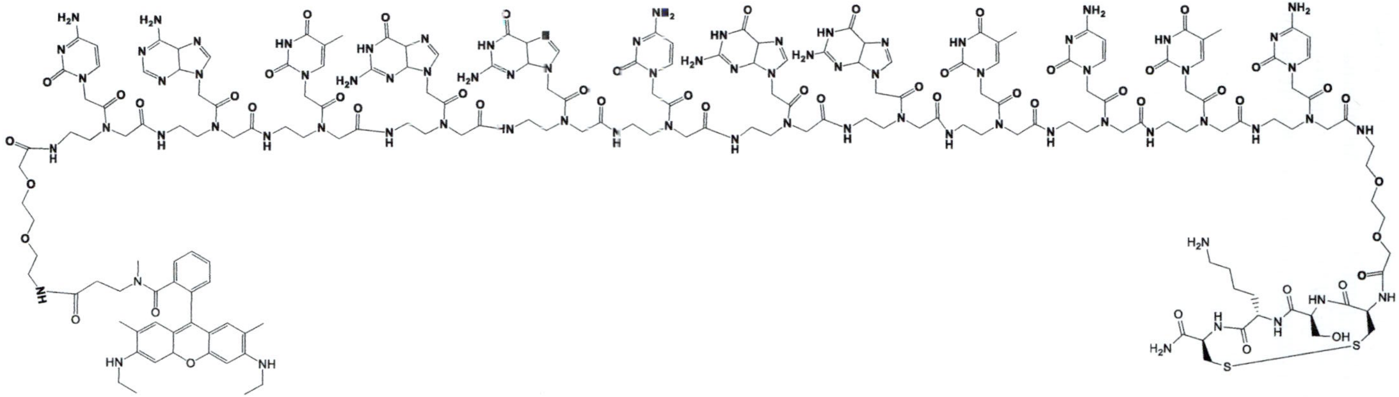

Figure 10.2 ChemDraw representation of human Cal560-AEEA-*HTT* PNA-AEEA-IGF1 tetrapeptide, Hs*HTT*.

One approach to the delivery of PNA imaging agents to cells utilizes the incorporation of a cell-penetrating peptide [52]. An alternative approach to target imaging agents specifically to *HTT*-expressing neuronal cells and to enable efflux of unbound agents is to incorporate a peptide ligand that enables receptor-mediated cellular uptake.

We previously showed that inclusion of a cyclized tetrapeptide fragment of insulin-like growth factor (IGF1) (Figure 10.2)-enabled IGF1 receptor-mediated cellular uptake [44], which was blocked in mice by excess IGF1 [53]. Radiolabeled or fluorescent agents are taken up by cells overexpressing high levels of the target receptor at a million or more copies per cell. In contrast, the mRNA targets are overexpressed at thousands of copies per cell. As a result, the target mRNAs are saturated with reporter agents. After excess reporter agents have effluxed, the remaining hybridized radiolabeled or fluorescent agents report the level of the target mRNAs [43]. Moreover, we previously found that reporter-PNA 12mers with a C-terminal IGF1 tetrapeptide stay in circulation by complexing with IGF1 binding proteins [54].

10.3. METHODS

10.3.1. *MAOA* mRNA

Here, we summarize the synthesis, characterization, and *in vitro* evaluation of peptide-PNA-fluorophore chimeras for *MAOA* mRNA imaging.

10.3.1.1. Fluorescent MAOA Hybridization Agent Synthesis A receptor-specific agent, Tyr-D-Ala-Gly-*N*-MePhe-Ser-AEEA-CAT GGT GCT CAC-AEEA-Cys-AlexaFluor488 (WT4879) (Figure 10.1), and a peptide mismatch control, Ala-D-Ala-Gly-*N*-MePhe-Ser-AEEA-CAT GGT GCT CAC-AEEA-Cys-AlexaFluor488 (WT4786), were synthesized by Fmoc-based solid-phase coupling (Scheme 10.1). The chimeras were purified using reverse-phase high-performance liquid chromatography (RP-HPLC) using C_{18} reversed-phase columns and characterized by mass spectrometry. AlexaFluor488-maleimide (Invitrogen, Carlsbad CA) was coupled to the C-terminal cysteine thiols in aqueous buffer under standard conditions.

10.3.2. Cell Culture

The human neuroblastoma cell-line SH-SY5Y (ATCC), which expresses both μ-opioid receptor and *MAOA* mRNA [55, 56], was used for *in vitro* studies of agent uptake. SH-SY5Y cells were maintained in a 1:1 (v/v) DMEM:F12 medium supplemented with 10% fetal bovine serum (FBS). Chinese hamster ovarian cells, CHO-K1 (ATCC), were used as a negative control. CHO-K1 cells were maintained in Ham's F12 medium containing 10% FBS. Cells were incubated at 37°C under 5% CO_2 in a humidified chamber.

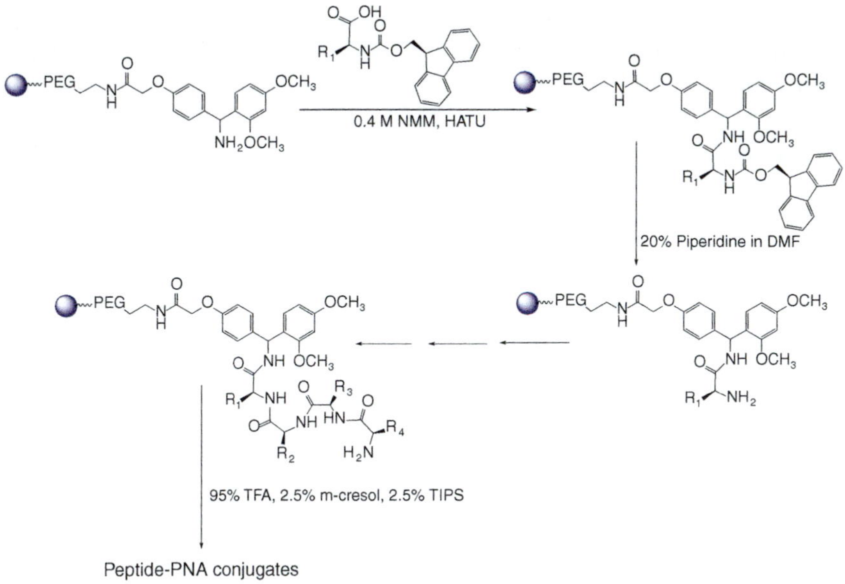

Scheme 10.1 Synthesis of peptide-PNA conjugates. Monomers were coupled in *N*-methylmorpholine with HATU as coupling reagent.

10.3.3. Cell Binding Assays

In brief, 4×10^4 SH-SY5Y cells/tube were incubated with a ramp of concentrations (0, 4, 8, 16, 32, 64, and 128 nM) of WT4879 (with and without 20× free DAMGO peptide, Set I and II) for 1 h at 4°C in the dark. After incubation, cells were washed twice with cold 1× calcium-magnesium free phosphate-buffered saline, CMF-PBS-5, containing 5% FBS, to remove the unbound probe. Subsequently, cells were fixed with CMF-PBS-5 + 1% paraformaldehyde for 10 min at 25°C in the dark, washed twice with CMF-PBS-5, resuspended in 300 μL of CMF-PBS-5, and analyzed on a FACS Calibur flow cytometer (BD Biosciences, San Jose, CA). In each case, the fluorescence intensities of 5000 vital cells were acquired. Vital cells were gated based on sideward and forward scatter. To calculate specific uptake, both non-specific uptake with 20× free DAMGO and background were subtracted from total uptake.

To evaluate the specificity of uptake, a ramp of concentrations (0.1, 0.2, 0.35, and 0.5 μM) of WT4879 (receptor-specific agent) and WT4786 (peptide mismatch agent) were prepared in 1:1 (v/v) DMEM:F12 medium. 2×10^5 SH-SY5Y cells/well were cultured in 6-well plates to ~70% confluence and the cells were incubated with imaging agent solutions for 6 h at 37°C under 5% CO_2 in a humidified incubator. After incubation, cells were washed with cold CMF-PBS-5 (3 mL) and detached using 0.05% trypsin. The cells were washed and fixed as described above and finally re-suspended in 300 μL of CMF-PBS-5 for flow cytometry; 25,000 vital cells were screened for each sample.

10.3.4. Cellular Uptake of Agents

Briefly, 6.5×10^4 SH-SY5Y cells were seeded in each well of a poly-D-lysine-coated 4-well micro-slide (BD Biosciences) one day before the experiment. Cells were incubated with 0.1 μM WT4879 for increasing times (0, 0.25, 0.5, 1, 2, 4, 8, and 12 h) at 37°C in a humidified chamber with 5% CO_2. After the incubation, cells were washed twice with CMF-PBS-5 and fixed using 4% paraformaldehyde in CMF-PBS-5 for 15 min. Prolong Gold Antifade (Invitrogen) with DAPI was used as mounting media for micro-slides. The slides were visualized using a LSM 510 Meta confocal microscope (Zeiss, New York NY) using the 488 nm line of an argon ion laser with a BP 505–550 nm band-pass filter. DAPI was visualized using a 405-nm diode laser. Following image acquisition, images were quantified with ImageJ software (http://rsbweb.nih.gov/ij).

10.3.5. Cellular Toxicity of Agents

The toxicity of WT4879 on SH-SY5Y cells was measured with a Vybrant MTT Cell Proliferation Assay Kit (Invitrogen). Briefly, cells were seeded in 96-well microtiter plates (2×10^4 cells/well) and cultivated overnight in phenol red-free DMEM:F12 medium (Invitrogen). Subsequently, SH-SY5Y cells were incubated with three concentrations (0.1, 0.5, and 1 μM) of WT4879 for 12 h at 37°C. Cells were then incubated with MTT dye at a concentration of 1 mg/mL for 4 h. The formazan product was solubilized with Me_2SO (50 μL). Cell viability was determined by measuring the absorbance of each sample at 540 nm using a microplate reader.

10.3.6. *HTT* mRNA

Here we summarize the synthesis, characterization, and cellular uptake of human Cal560-*HTT* PNA-IGF1 tetrapeptide in tumorigenic human embryonic kidney HEK293T cells.

10.3.6.1. Fluorescent HTT Hybridization Agent Synthesis Cal560-*HTT* PNA-peptides (Figure 10.2) (Table 10.1) were synthesized, purified, and analyzed as described [57]. Briefly, Fmoc-d-amino acids (Novabiochem, San Diego, CA) were composing the *N*-terminal peptide, Fmoc-AEEA linkers, and Fmoc-PNA monomers (Link Technologies, Bellshill, Scotland) and were first activated by *O*-(7-azabenzotriazole-1-yl)-1, 1, 3, 3-tetramethyluronium hexafluorophosphate (HATU) (Applied Biosystems, Foster City, CA), followed by coupling onto 0.2–0.3 mmol/g 9-Fmoc-amino-xanthen-3-yloxy NovaSyn TG

TABLE 10.1 Fluorophore-PNA-peptide sequences

Name	Sequence	Calc. mass	Exp. mass
Hs*HTT*	Cal560-AEEA-CATGGCGGTCTC-AEEA-D(CSKC)	4463.62 Da	4463.39 Da
Rr*htt*	Cal560-AEEA-CATG**a**CGG**c**CTC-AEEA-D(CSKC)	4432.61 Da	4432.38 Da

HsHTT complementary to initiation codon domain, nt 308–319 from NM_002111.7, 13,669 nt. Rat RrHTT differences from the human *HTT* sequence are shown in bold lowercase.

Sieber resin (Novabiochem) on a PS3 peptide synthesizer (Protein Technologies, Tucson, AZ). Then, the D-cysteine residues (where present) were deprotected, followed by cyclization with 0.1 M I_2 in $(CH_3)_2NCHO$ for 4 h on solid phase. After thorough rinsing of I_2 from the resin, Cal560 carboxylic acid (Biosearch Technologies, Petaluma, CA) was activated with HATU and coupled to the *N*-terminal AEEA amine in the synthesizer.

Following assembly, the Cal560-*HTT* PNA-peptides were deprotected and cleaved at room temperature with 85% CF_3CO_2H, 5% CH_2Cl_2, 9.5% m-cresol, and 0.5% Et_3SiH. Finally, the deprotected conjugates were purified by liquid chromatography on an Alltima reversed-phase C_{18} column (Alltech, Grace, Deerfield, IL) and analyzed on a 4800 matrix-assisted laser desorption/ionization time-of-flight mass spectrometer (MALDI-TOF MS) (Applied Biosystems, Foster City, CA) from an α-cyano-hydroxycinnamic acid matrix.

10.3.6.2. Cell Culture We used HEK293T cells (American Type Culture Collection, Manassas, VA), which express both *HTT* mRNA [51] and IGF1R [58], for *in vitro* studies of agent uptake. Thus, no further *HTT* DNA construction, transfection, or transformation was necessary for the cell culture model. HEK293T cells were maintained in Dulbecco's modified Eagle's medium (DMEM), 10% fetal bovine serum (FBS), 50 U/mL penicillin, 5 μg/mL streptomycin, and 2 mM glutamine, all from Thermo Fisher (Springfield Township, NJ) incubated at 37°C in humidified air containing 10% CO_2, according to the ATCC protocol.

10.3.6.3. siRNA Knockdown of HTT mRNA HEK293T cells grown in six-well plates (Thermo Fisher) were transfected with 2 μg/well of three unique 27mer siRNAs (#SR302082, OriGene Technologies, Rockville, MD) targeting human *HTT* mRNA, NM_002111.7 (Table 10.2), or non-targeting scrambled control siRNA (NTC, #SR30004) using Lipofectamine 2000 (Life Technologies, Carlsbad, CA); 2 μg/well of CMV-EGFP-C1 plasmid (Genentech, South San Francisco, CA) was included with the siRNAs to monitor the transfection rate. Cells were incubated at 37°C for 48 h after transfection, then trypsinized, pelleted, washed with PBS, resuspended, and titered; 5×10^4 cells were transferred

TABLE 10.2 Fluorophore-PNA-peptide sequences

Name	Sequence	Calc. mass	Exp. mass
Hs*HTT*	Cal560-AEEA-CATGGCGGTCTC-AEEA-D(CSKC)	4463.62 Da	4463.39 Da
Rr*htt*	Cal560-AEEA-CATG**a**CGG**c**CTC-AEEA-D(CSKC)	4432.61 Da	4432.38 Da

HsHTT complementary to initiation codon domain, nt 308-319 from NM_002111.7, 13,669 nt. Rat RrHTT differences from the human *HTT* sequence are shown in bold lowercase.

to 24-well plates with poly(D-lysine)-coated cover glasses (Thermo Fisher) for fluorescent hybridization uptake. The rest of the cells were left in the wells for RNA isolation.

10.3.6.4. HTT Fluorescent Hybridization Agent Uptake 48 h after transfection with siRNAs and indicator plasmids, the cells grown on coverslips were treated with 100 nM Cal560-PNA-peptides (Table 10.1) and incubated at 37°C for 4 h. Treated cells were washed twice for 15 min with culture medium at 37°C, then twice for 5 min with DPBS at room temperature. The cells were then fixed with 4% paraformaldehyde followed by three washes in DPBS. Nuclei were stained with Hoechst 22242 fluorescent stain (#28491-52-3, Thermo Fisher) for 10 min, followed by three 5-min washes with DPBS. Coverslips were mounted with Vectashield (Thermo Fisher) onto the glass slides. Three replicate treatments were carried out with both 100 and 200 nM Cal560-PNA-peptides.

10.3.6.5. Quantitative Polymerase Chain Reaction for HTT mRNA RNA was isolated from treated cells as described [43]. Briefly, 48 h post-transfection, cells from each well were homogenized with 1 mL Trizol reagent (#15596-026, Life Technologies) per well, then stabilized against RNase activity by addition of 0.2 mL $CHCl_3$, homogenized again, and sedimented. RNA in the aqueous phase was precipitated by addition of 0.5 mL of 100% isopropanol, then sedimented. The RNA pellet was washed with 1 mL of 75% EtOH, 25% water, then redissolved in 50 µL RNase-free water (Ambion, Austin, TX) at 55°C for 15 min, and quantitated by A_{260}.

Single-stranded cDNA was synthesized from 1 µg total RNA using a Taqman High Capacity cDNA Reverse Transcription Kit (#4368814, Applied Biosystems) by combining 2× reverse transcription (RT) master mix and total RNA with random primers in a volume of 20 µL reaction. Reverse transcription was performed for 2 h at 37°C in a thermocycler. To determine relative gene expression, 5′-FAM, 3′-TAMRA probe-based fluorescence qPCR was performed on the cDNA using Gene Expression Assays probes for human *HTT*

(#Hs00169273_m1, Applied Biosystems). Duplication cycle thresholds were normalized to a human GAPDH primer/probe set for determination of *HTT* mRNA knockdown vs. the experimental control probe.

10.3.6.6. Fluorescence Imaging of HTT mRNA Slides were mounted with Vectashield after PNA treatment and visualized by fluorescence microscopy (Leica DMRXA, #C8484-03). To define the nuclear or cytoplasmic localization of PNA:mRNA complexes, images were collected using an oil 60× objective lens. Images were captured with a Hamamatsu digital camera. The images that revealed an intact plasma membrane and cytosol free of bright puncta were used for quantification using Ivision software (Biovision Technologies, Exton, PA). Five images were acquired from each of three replicates for each treatment. The NTC sample was imaged first in each experiment in order to set the exposure times for rest of the siRNA-treated samples. All images within a treatment were acquired with identical exposure times.

Fluorescence intensities of imaged cells were measured and quantified using ImageJ (National Institutes of Health, Bethesda, MD) by the gray mean value (red images) obtained with the freehand selection tool. Ten background measurements were taken from each image to subtract an average background.

10.3.6.7. Statistical Analysis qPCR data were normalized to *HTT* mRNA levels in samples treated with the non-targeting control. Comparative levels were statistically evaluated using two-tailed unpaired Student's t-test. Analysis of fluorescent image intensity was performed by using one-way ANOVA. Post-hoc analyses were performed to assess for significant differences between individual groups. Pearson's correlation coefficient was determined from the qPCR and fluorescence quantification data obtained in three independent experiments, for all three siRNAs.

10.4. RESULTS

10.4.1. *MAOA* mRNA

10.4.1.1. Fluorescent MAOA Hybridization Agent Synthesis Fluorescent hybridization agents (Table 10.1) were synthesized at the 10 μmol scale by solid-phase synthesis. Representative reversed-phase HPLC (Figure 10.3) profiles are shown for the human *MAOA* mRNA agent, WT4879, and the peptide mismatch control agent, WT4786. Measured MALDI-TOF masses of each agent are shown in Table 10.1.

10.4.1.2. Cell Binding Assays In order to determine the K_d for WT4879 binding to its targeted μ-opioid receptor on SH-SY5Y cells, a competition binding assay was performed. The K_d for the entire agent found to be 11.6±0.8 nM, about 20-fold weaker than free DAMGO [50].

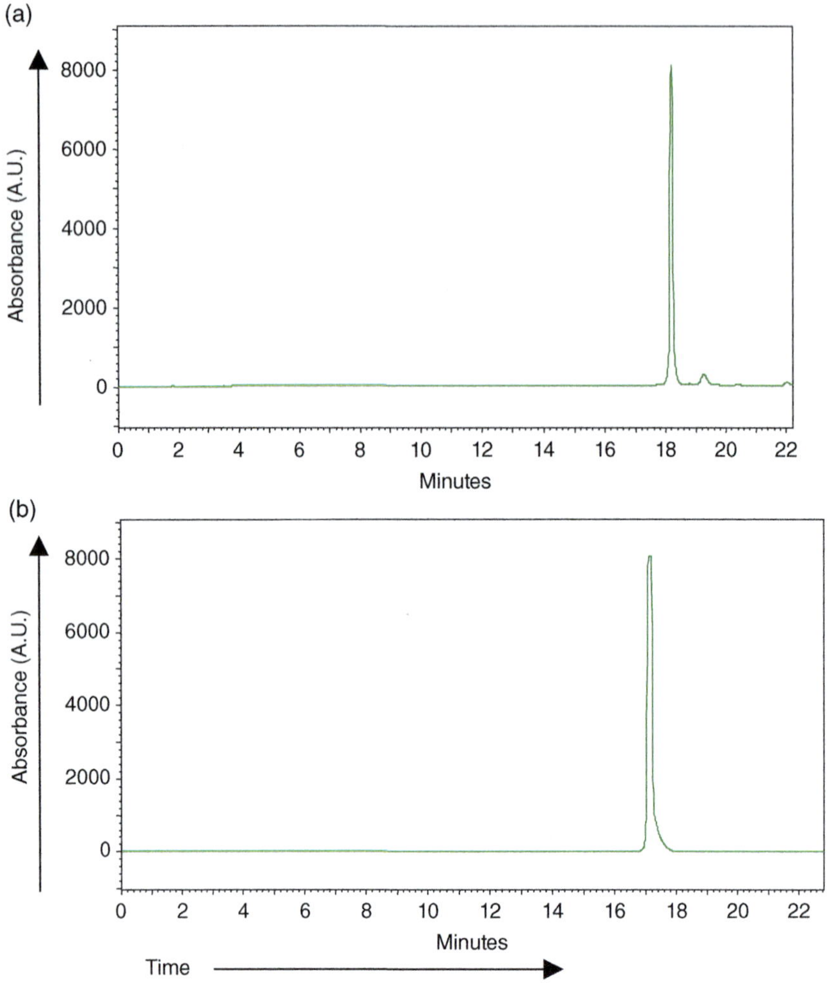

Figure 10.3 HPLC profiles of (a) Tyr-D-Ala-Gly-*N*-MePhe-Ser-AEEA-CAT GGT GCT CAC-AEEA-Cys-AlexaFluor488, WT4879 [DAMGO-MAOA-AlexaFluor488, m/z 4878.78 (calculated), 4879.20 (observed)]; and (b) Ala-D-Ala-Gly-*N*-MePhe-Ser-AEEA-CAT GGT GCT CAC-AEEA-Cys-AlexaFluor488, WT4786 [peptide-mismatch-MAOA-AlexaFluor488, m/z 4786.78 (calculated), 4786.08 (observed)] on a 10 × 250 mm Alltima C_{18} column, eluted at 2.5 mL/min with a 30-min gradient from 5 to 60% CH_3CN in aqueous 0.1% CF_3CO_2H, at 50°C, monitored at 488 nm.

10.4.1.3. Cellular Uptake of Agents A ratio of specific over nonspecific uptake was calculated and plotted against concentration (μM) (Figure 10.4).

The results revealed that specificity of uptake follows an inverse relationship to the concentration of the probe. At 0.1 μM, WT4879 showed ~5-fold higher uptake than WT4786, indicating high uptake specificity. A similar experiment

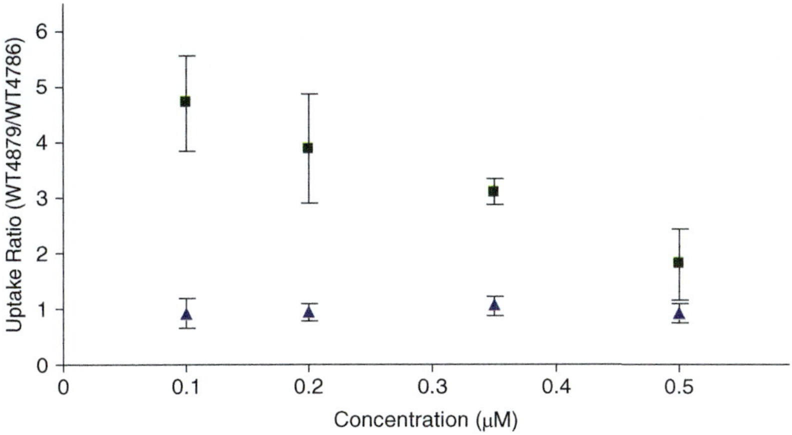

Figure 10.4 Specificity of uptake in SH-SY5Y (■) and CHO-K1 (▲) cells of WT4879 (μ-opioid receptor-specific agent) vs. WT4786 (peptide mismatch agent) over a ramp of concentrations measured by flow cytometry. Ratios of specific to nonspecific uptake, after background correction, were calculated and plotted against concentration (μM), averaged over three replicate experiments.

was performed using the CHO-K1 negative control cell line in Ham's F12 medium. The specificity ratio for WT4879 was found to be 1.0±0.3 at all the concentrations, implying no specific uptake in cells lacking the μ opioid receptor.

No significant uptake was observed at 0.25 and 0.5 h (data not shown). WT4879 was taken up strongly after 1 h and found to be uniformly distributed in the cytoplasm (Figure 10.5). The uptake intensity increased with time up to 4 h, leveling off thereafter.

In order to study the specificity for receptor-targeted delivery, a comparative uptake study was performed using CHO-K1 (1.5×10^4 cells/well) and SH-SY5Y (6.5×10^4 cells/well) by the same protocol, as described above. Cells were incubated for 4 h at 37°C with 0.1 μM WT4879 and WT4786 in respective growth media. Three fields were quantitated on each slide. The results revealed that negative control CHO-K1 cells took up ~21% as much receptor-specific WT4879 agent as did SH-SY5Y cells (Figure 10.6a,c). In SH-SY5Y cells, the peptide mismatch WT4786 agent showed ~20% uptake relative to the specific WT4879 (Figure 10.6b,c), consistent with the receptor-specific uptake observed by flow cytometry. In a structural control experiment, an agent with a fluorophore on the N-terminus, and DAMGO on the C-terminus, showed similar low uptake by SH-SY5Y cells (data not shown), consistent with the requirement for a free N-terminal cationic amine on DAMGO [49].

Following the specificity study, a live cell study was performed to evaluate the uptake mechanism of WT4879 in SH-SY5Y cells. Briefly, 2×10^5 cells/dish were seeded in 35 mm dishes (μ-dishes, ibidi GmbH, Germany) and transfected with CellLight BacMam reagent (Invitrogen), according to the manufacturer's protocol

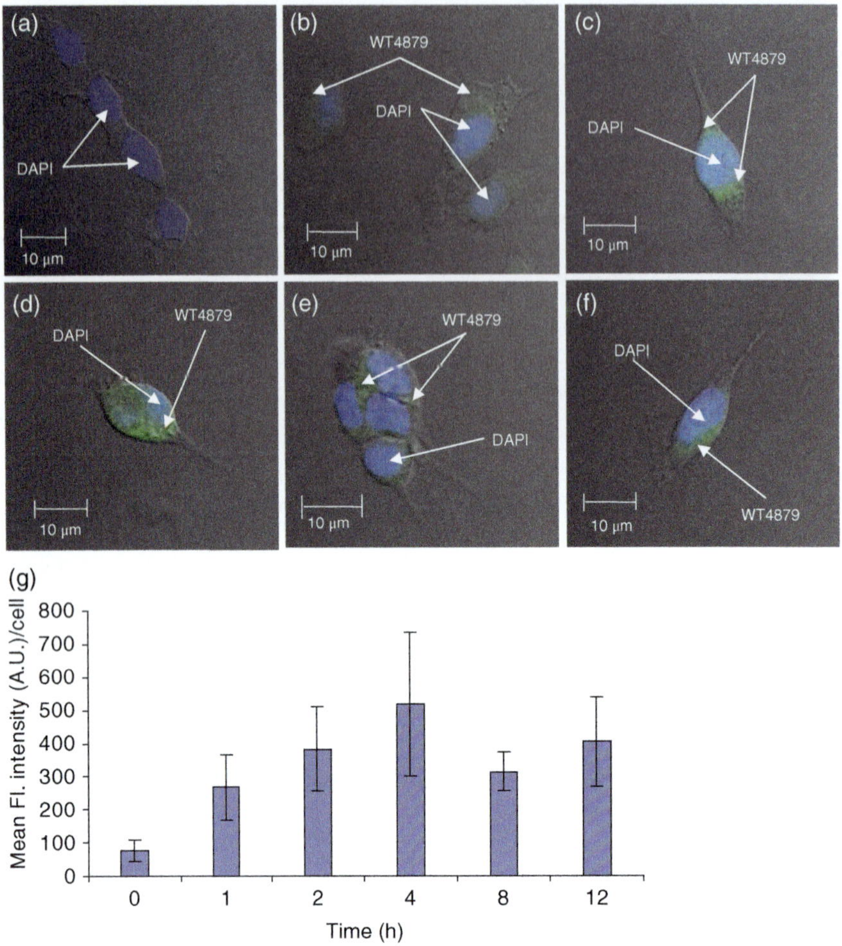

Figure 10.5 Confocal fluorescent images of WT4879 uptake in SH-SH5Y cells. (a) Control, (b) after 1 h, (c) 2 h, (d) 4 h, (e) 8 h, (f) 12 h, and (g) histogram showing mean fluorescence intensity (arbitrary units, A.U.) per cell measured at different times, averaged over three regions of interest.

(Invitrogen), to label early endosomes with RFP-labeled Rab5 protein. After the expression of RFP-labeled Rab5 protein, cells were incubated for 1 h at 37°C with 0.1 μM WT4879 in phenol red-free growth medium. The dishes were washed with phenol red-free medium (3 × 2 mL) and visualized using a heated stage on the LSM 510 Meta confocal microscope. The cells showed co-localization of WT4879 with Rab5 in early endosomes (Figure 10.7a–c). The same section of each dish was scanned after additional 1 h incubation at 37°C. The endosomes disappeared and WT4879 was found to be uniformly distributed in the cytoplasm (Figure 10.7d–f).

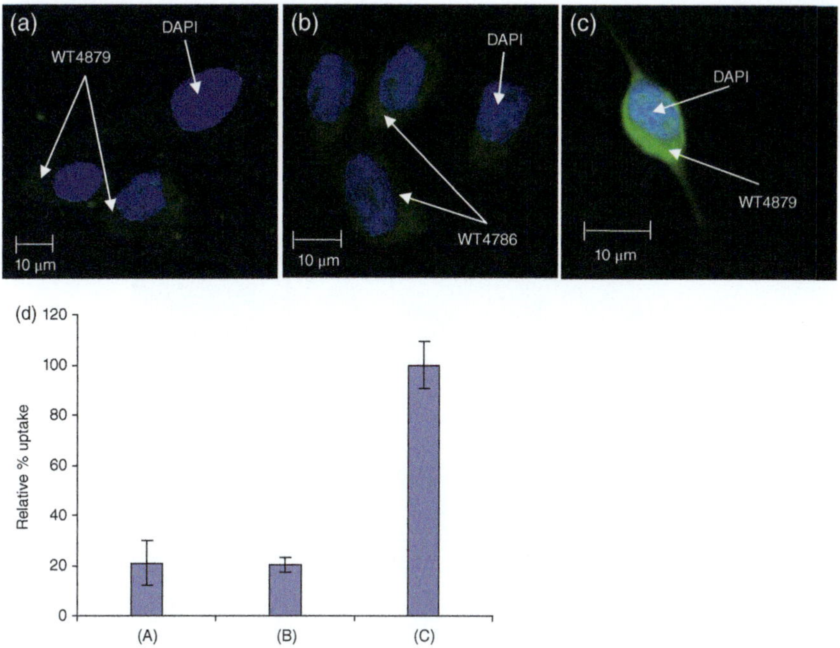

Figure 10.6 Confocal fluorescent images of (a) WT4879 uptake in CHO-K1 cells, (b) WT4786 uptake in SH-SH5Y cells, (c) WT4879 uptake in SH-SY5Y cells after 4 h of incubation at 37°C with 0.1 µM agent in respective growth media, and (d) histogram showing the relative percent of uptake for three fields on each slide.

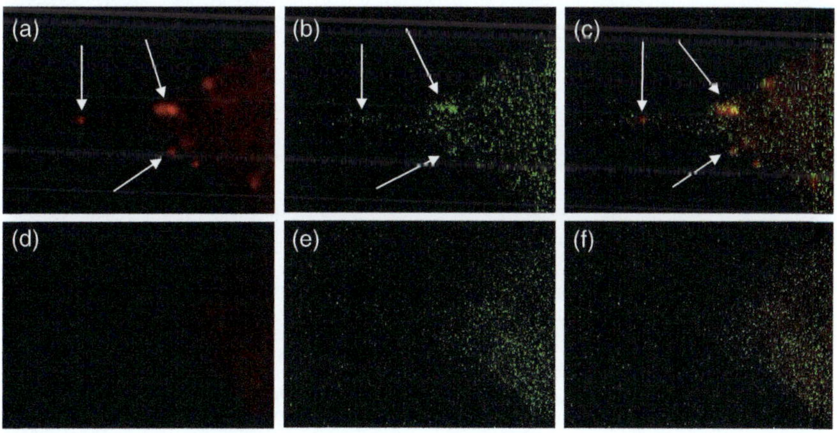

Figure 10.7 Confocal fluorescent images of live SH-SY5Y cells showing the (a) localization of RFP-labeled Rab5, (b) WT4879 in endosomes, (c) co-localization of RFP labeled Rab5 and WT4879, (d) RFP-labeled Rab5, (e) WT4879 distribution in cells after an additional 1 h incubation at 37°C, and (f) merged image of (d) & (e).

The control cells expressed RFP-labeled Rab5 but did not showed any green fluorescence. The intensities of RFP for labeled Rab5 and AlexaFluor 488 for WT4879 were measured along the x-axis using Amira 5 software, which showed >95% overlap (Figure 10.8). These results imply receptor-specific uptake of WT4879 into early endosomes, followed by cytosolic dispersal of cargo.

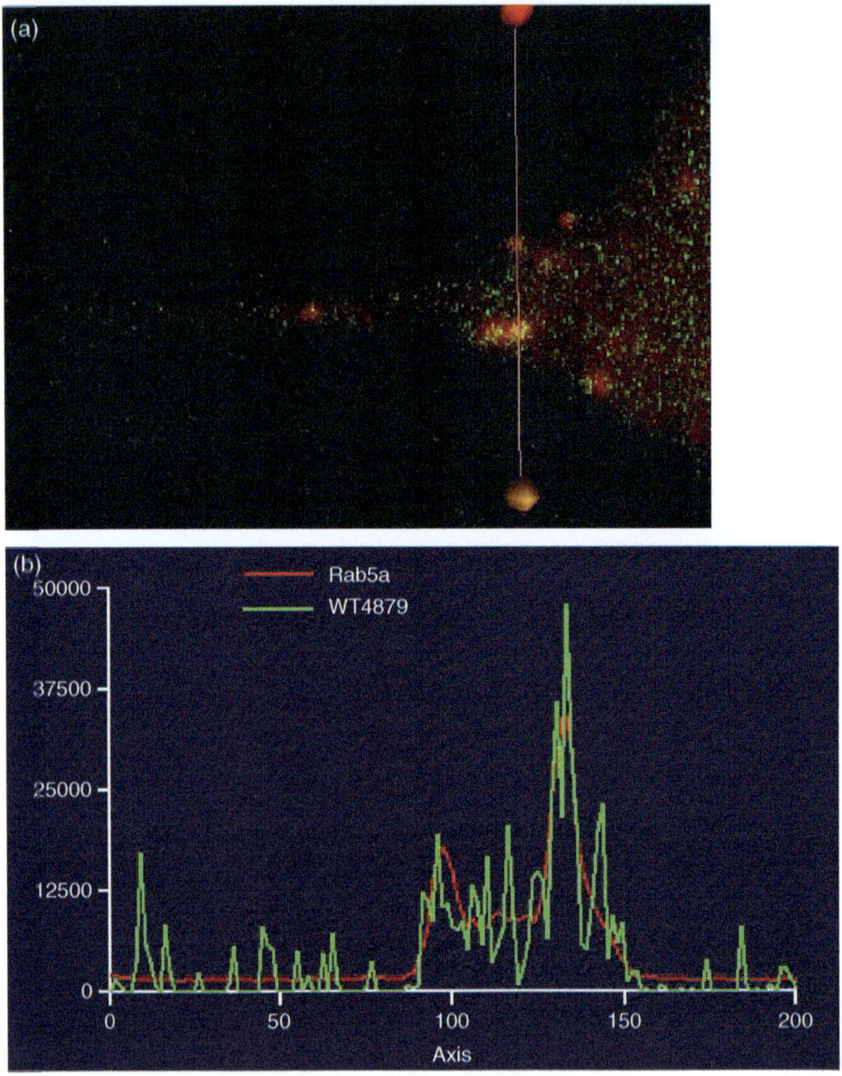

Figure 10.8 Quantization of RFP for labeled Rab5 and AlexaFluor 488 for WT4879 intensities along the x-axis in SH-SY5Y cells, using Amira 5 software. Pictures (a) showing line drawn along x-axis and (b) quantization of relative intensities showing co-localization in endosomes).

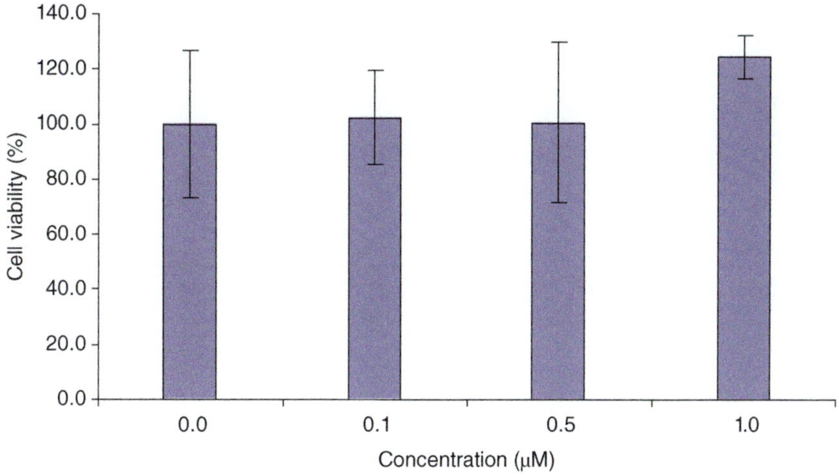

Figure 10.9 Cell viability analysis of SH-SY5Y cells using MTT assay for different concentrations of WT4879. Cells were incubated with probe solution for 12 h. No cellular toxicity was observed.

10.4.1.4. Cellular Toxicity by Agent The cells exposed to increasing concentrations of WT4879 survived just as well as control cells (Figure 10.9). No cellular toxicity was observed at any concentration tested, suggesting the biocompatibility of the synthesized probes.

10.4.2. *HTT* mRNA

10.4.2.1. Cal560-HTT PNA-Peptide Synthesis Fluorescent hybridization agents (Table 10.2) were synthesized at the 10 μmol scale by solid-phase synthesis. Representative reversed-phase HPLC (Figure 10.10) and MALDI-TOF MS (Figure 10.11) profiles are shown for the human *HTT* mRNA agent, HsHTT. Measured masses of each agent are shown in Table 10.2.

10.4.2.2. Validation of Cal560-HTT PNA-D(CSKC) as a Specific Imaging Agent for HTT mRNA Preliminary experiments with a fluorophore-PNA-HIV Tat cell-penetrating peptide revealed excellent cellular uptake, regardless of cell type or PNA sequence (not shown). However, although internalization via the Tat cell-penetrating peptide was efficient, little or no efflux of unbound agent occurred. Thus, the use of the Tat cell-penetrating peptide precluded mRNA-specific imaging. Therefore, in order to direct cytoplasmic uptake by human IGF1 receptor, as described [43], an IGF1 tetrapeptide was included at the C-terminus of the PNA sequence. IGF1R is expressed in the cortex and striatum [59–61], and in HEK293 cells [58]. For visualization, a Cal560 fluorophore was included at the N-terminus (Figure 10.2). HEK293T cells were incubated with 100 nM Cal560-Hs*HTT* PNA-IGF1 tetrapeptide in culture medium for 2 or 4 h at

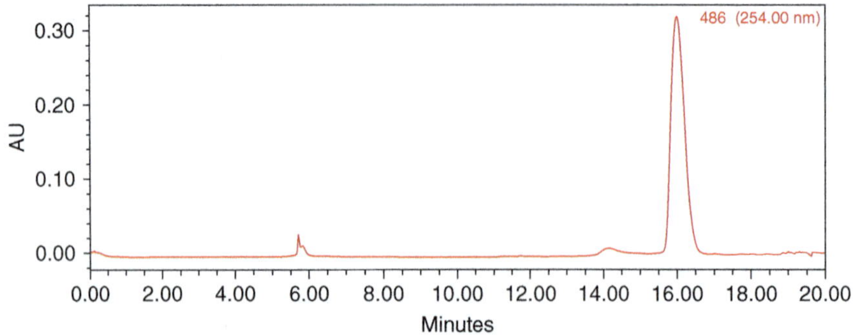

Figure 10.10 Representative analytical HPLC of *H. sapiens* Cal560-*HTT* PNA-D(CSKC) fluorescent imaging agent on a 10×250 mm Microsorb C_{18} column, eluted with a 35 min gradient from 0 to 60% CH_3CN in aqueous 0.1% CF_3CO_2H, at 1 mL/min, $\lambda = 260$ nm, at 50°C.

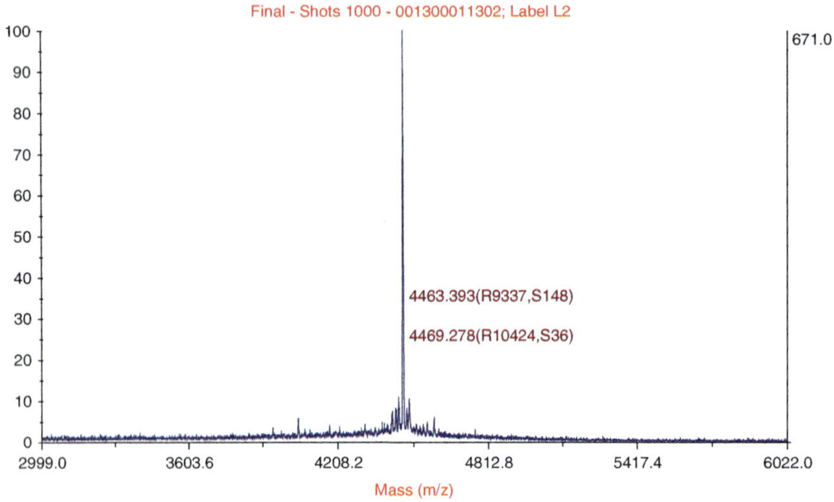

Figure 10.11 Representative MALDI-ToF mass spectrum of *H. sapiens* Cal560-*HTT* PNA-D(CSKC) fluorescent imaging agent. Calculated exact mass: 4463.62 Da; experimental mass: 4463.39 Da (M+H).

37°C to allow uptake while enabling efflux of unbound hybridization agent. Fluorescence microscopy revealed specific uptake of the human *HTT* agent into the cytoplasm, but not the nucleus (Figure 10.12). In contrast, incubation with 100 nM Cal560-Rr*htt* PNA-IGF1 tetrapeptide displayed a weak fluorescent signal (Figure 10.12). The weak fluorescence data imply that the two base difference between the Rr*htt* rat probe and the human *HTT* mRNA resulted in poor hybridization of the Rr*htt* rat probe to the human *HTT* mRNA.

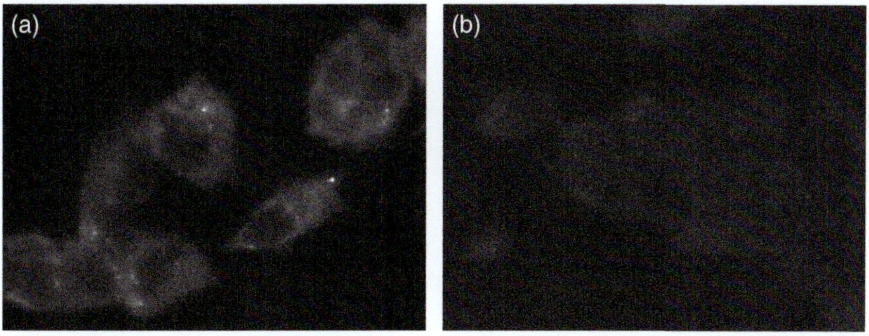

Figure 10.12 Fluorescent *HTT* imaging agent cellular uptake. HEK293T cells were incubated for 4 h at 37°C (a) with 100 nM HsHTT (*H. sapiens* Cal560-*HTT* PNA-D(CSKC) or (b) 100 nM RrHTT (*R. rattus* Cal560-*htt* PNA-D(CSKC). Cells were fixed and imaged by an individual blinded to the treatment condition using a Leica DMR fluorescence microscope. Fields used for imaging were selected based on viewing DAPI-stained nuclei with a blue filter and all images *within* an experiment were collected at the same exposure.

10.4.2.3. qPCR of HTT mRNA Following siRNA Knockdown The tri-silencer-27 siRNA kit containing three Dicer-Substrate duplexes (A, B, C) and a scrambled non-targeting control (NTC) (Table 10.3) were used to determine the most effective *HTT* mRNA knockdown reagent. No additional *HTT* gene expression was needed in the HEK293T cells because *HTT* is constitutively expressed in these cells.

siRNA transfection efficiency was determined using the co-transfected plasmid CMV-EGFP-C1. Following siRNA transfection, HEK293T cells were incubated for an additional 48 h, at which time, a portion of the cells were harvested to determine *HTT* mRNA levels.

The combined results of three independent siRNA knockdown experiments are shown in Figure 10.13. The percent knockdown of *HTT* mRNA for the type A siRNA was 85.7±4.5%, greater than type B at 72.5±9.5%, or type C at 57.6±3.9%. Three independent experiments confirmed that the type A siRNA

TABLE 10.3 Unique to human *HTT* siRNA 27mer duplexes

Type	27mer siRNA guide strands	Location, nt[a]
A	rGGAUAGUAGACAGCAAUAACUCGGT	3′UTR, 12836–12860
B	rGGGAUGUAGAGAGGCGUUAGUGGGC	3′UTR, 10534–10558
C	rCCUGUUACAACAAGUAAAUCCUCAT	Coding, exon 28, 3974–3998
NTC	rAUACGCGUAUUAUACGCGAUUAACGAC	Origene #SR302082

[a] from NM_002111.7, 13,669 nt; A =adenosine-3′-phosphate, 3′-adenylic acid; C =cytidine-3′-phosphate, 3′-cytidylic acid; G =guanosine-3′-phosphate, 3′-guanylic acid; T=ribothymidine-3′-phosphate, 3′-ribothymidylic acid; U =uridine-3′-phosphate, 3′-uridylic acid.

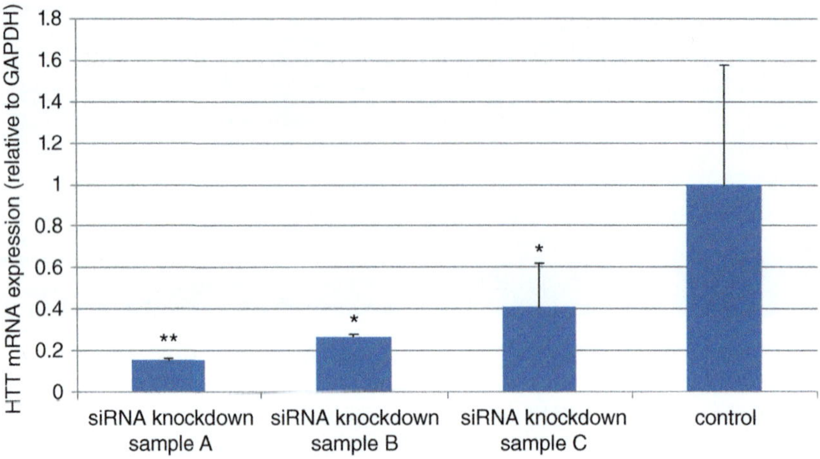

Figure 10.13 siRNA knockdown of *HTT* mRNA in HEK293T cells. Each bar represents the average of three replicates in each of three experiments. The two-tailed unpaired t test results are indicated by the asterisks (*for $p<0.05$; ** for $p<0.01$). Error bars represent SEM.

duplex most effectively reduced the level of *HTT* mRNA in HEK293T cells ($p<0.01$).

10.4.2.4. Fluorescence Imaging of HTT mRNA Following siRNA Knockdown After cells were transfected with siRNA knockdown sequences (Table 10.3) for 48 h, a portion of the cells was replated and incubated with match (Hs*HTT*) and mismatch (Rr*htt*) PNA agents (Table 10.2), in order to determine if changes in the PNA fluorescence intensity correlated with *HTT* RNA levels determined by qPCR.

Fluorescence images of HEK293T cells transfected with control (NTC) or specific (A, B, C) siRNAs (Table 10.3) were acquired after incubation with the human match hybridization agent, Hs*HTT*, or the rat mismatch agent, Rr*htt*. Five images were collected from each of three replicate slides in each of three experiments for image analysis and quantification. Six representative images of cells transfected with NTC or type A siRNA and incubated with Hs*HTT* or Rr*htt* are shown in Figure 10.14.

Quantification of these and additional images (Figure 10.15) was carried out using ImageJ. The percent reduction of cellular fluorescence intensity of *HTT* mRNA for the type A siRNA was 69±6%, greater than type B at 42±20%, or type C at 54±10%.

Analysis of the qPCR results (Figure 10.13) revealed that the type A siRNA knockdown duplex produced the lowest *HTT* mRNA level out of the three duplexes. Similarly, quantification (Figure 10.15) of fluorescence images, acquired following treatment with Hs*HTT* PNA imaging agents, revealed that,

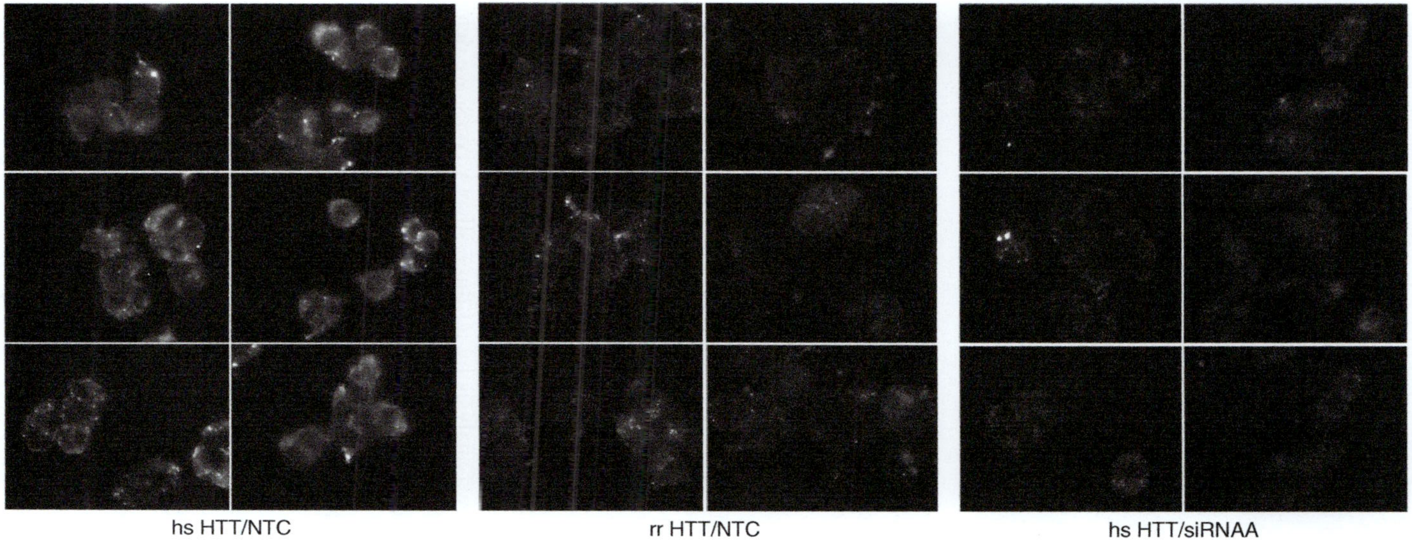

Figure 10.14 Fluorescence images of HEK293T cells transfected with siRNA, then incubated with Cal560-PNA-peptide. Left: NTC siRNA and human Hs*HTT*. Center: NTC siRNA and rat Rr*htt* differing from human by two bases. Right: *HTT* siRNA A and human Hs*HTT*.

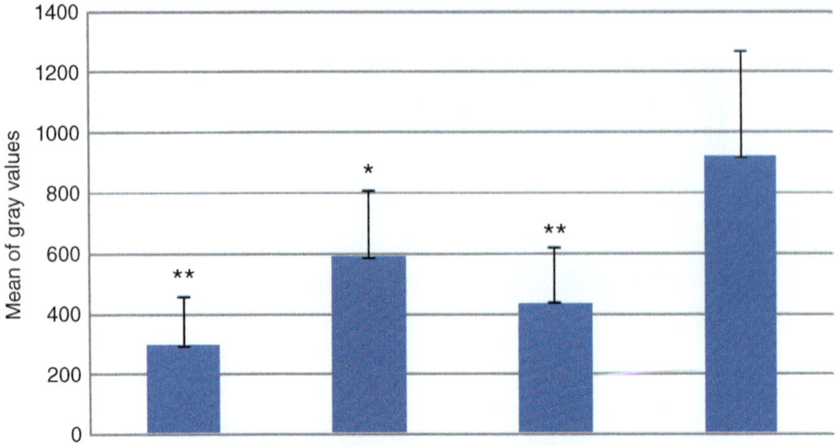

Figure 10.15 Average cell fluorescence intensities in HEK293T cells transfected with siRNAs, then incubated with human fluorescent agent Hs*HTT*. Image analysis of gray value data for fluorescence intensity was performed by using a one-way ANOVA, indicated by asterisks (* for $p<0.05$; **for $p<0.01$). Error bars represent SEM.

while all siRNA types produced a statistically significant reduction in PNA intensity, the type A siRNA resulted in the most substantial reduction, similar to what was observed by qPCR (Figures 10.14 and 10.15). These results illustrate both the efficacy of type A siRNA and the specificity of the Hs*HTT* imaging agent.

10.4.2.5. Statistical Comparison Finally, in order to determine if Cal560-PNA-IGF1 tetrapeptide fluorescence imaging could be used as a proxy for direct RNA knockdown analysis by all three siRNA types, we determined the Pearson correlation between the fluorescence and qPCR datasets. This analysis revealed a strong correlation between PNA fluorescence intensity and *HTT* mRNA levels for all three siRNAs (0.865), further supporting the development of this approach to quantify human *HTT* mRNA levels.

10.5. DISCUSSION

10.5.1. *MAOA* mRNA

A new *MAOA* mRNA imaging agent, coupled to a receptor-targeting peptide for specific intracellular delivery, was designed and synthesized. Initial *in vitro* studies demonstrated that WT4879 was specifically delivered to SH-SY5Y cells, which express μ-opioid receptors and *MAOA* mRNA, with high efficiency.

The receptor-specific probe (WT4879) showed ~five times higher uptake than the nonspecific probe (WT4786), in SH-SY5Y cells, using flow cytometry and confocal microscopy.

The control cell-line (CHO-K1) showed minimal nonspecific uptake of WT4879. Moreover, an MTT assay of WT4879 revealed high compatibility with SH-SY5Y cells. Such mRNA imaging agents, with near infrared fluorophores, might enable real-time imaging and quantitation of neuronal mRNAs in live animal models.

10.5.2. *HTT* mRNA

We quantitated *HTT* mRNA in HEK293T cells by qPCR with and without *HTT* siRNA knockdown, and also quantitated the accumulation of fluorescent human *HTT* hybridization agent in the same cells by fluorescence microscopic imaging. We asked whether the fluorescence intensity of treated cells was significantly different between the human *HTT* agent and the rat *htt* agent that differed by two bases, and whether the agents were stable in the cells.

First, the rat *htt* agent that differed by two bases should not hybridize stably enough to maintain its localization in the cytoplasm, even if the agent penetrates into the cells. Cells use passive and active transport mechanisms to import and export materials between cells and their environment. The unbound agents likely efflux by active transport activities [1]. The Hs*HTT* PNA agent that hybridizes with *HTT* mRNA should be stable and remain long enough to be detected by fluorescence microscopy. Elevated fluorescence intensity was expected to reflect elevated *HTT* mRNA copy number in the cytoplasm to bind with the introduced fluorescent agents, as we previously observed for oncogene mRNAs [43] and *MAOA* mRNA [1].

We found that the strongest siRNA sequence, type A, knocked down *HTT* mRNA in HEK293T cells by 86±5%. In parallel, we observed 69±6% reduction in cellular fluorescence intensity in type A siRNA-treated cells. These results suggest that quantitative fluorescence imaging of *HTT* mRNA in therapeutic animal models of HD before treatment and longitudinally during treatment can likely be achieved. Therapeutically, siRNA knockdown shows its limitations in these results, compared with the clinical success by single-stranded RNA analogs [41]. Ultimately, these fluorescence hybridization results suggest that PET imaging of *HTT* mRNA in HD patients undergoing *HTT* knockdown therapy could provide a noninvasive approach to quantitating this important proximate biomarker.

ACKNOWLEDGMENTS

This work was supported by NIH grant DA027746 to E.W. and M.L.T., as well as a pilot grant from the Hereditary Disease Foundation to E.W.

REFERENCES

1. Sethi, D., Chen, C.-P., Jing, R.-Y., Thakur, M.L., and Wickstrom, E. (2012) Fluorescent peptide-PNA chimeras for imaging monoamine oxidase A mRNA in neuronal cells. *Bioconjug Chem* **23**(2):158–163. PMID:22239616 3288683
2. Oh, E., Liu, Y., Sonar, M.V., Merry, D.E., and Wickstrom, E. (2018) Fluorescence imaging of huntingtin mRNA knockdown. *Bioconjug Chem* **29**(4):1276–82. PMID:29451775
3. Rendu, F., Peoc'h, K., Berlin, I., Thomas, D., and Launay, J.-M. (2011) Smoking related diseases: the central role of monoamine oxidase *Int J Environ Res Public Health* **8**:136–47.
4. Bolasco, A., Carradori, S., and Fioravanti, R. (2010) Focusing on new monoamine oxidase inhibitors. *Expert Opin Ther Pat* **20**(7):909–39. PMID:20553094
5. Fowler, J.S., Logan, J., Volkow, N.D., and Wang, G.J. (2005) Translational neuroimaging: positron emission tomography studies of monoamine oxidase. *Mol Imaging Biol* **7**(6):377–87. PMID:16265597
6. Brunner, H.G., Nelen, M., Breakefield, X.O., Ropers, H.H., and van Oost, B.A. (1993) Abnormal behavior associated with a point mutation in the structural gene for monoamine oxidase A. *Science* **262**(5133):578–80. PMID:8211186
7. Manuck, S.B., Flory, J.D., Ferrell, R.E., Mann, J.J., and Muldoon, M.F. (2000) A regulatory polymorphism of the monoamine oxidase-A gene may be associated with variability in aggression, impulsivity, and central nervous system serotonergic responsivity. *Psychiatry Res* **95**(1):9–23. PMID:10904119
8. Newman, T.K., Syagailo, Y.V., Barr, C.S., Wendland, J.R., Champoux, M., Graessle, M., Suomi, S.J., Higley, J.D., and Lesch, K.P. (2005) Monoamine oxidase A gene promoter variation and rearing experience influences aggressive behavior in rhesus monkeys. *Biol Psychiatry* **57**(2):167–72. PMID:15652876
9. Jiang, S., Xin, R., Lin, S., Qian, Y., Tang, G., Wang, D., and Wu, X. (2001) Linkage studies between attention-deficit hyperactivity disorder and the monoamine oxidase genes. *Am J Med Genet* **105**(8):783–8. PMID:11803531
10. Lawson, D.C., Turic, D., Langley, K., Pay, H.M., Govan, C.F., Norton, N., Hamshere, M.L., Owen, M.J., O'Donovan, M.C., and Thapar, A. (2003) Association analysis of monoamine oxidase A and attention deficit hyperactivity disorder. *Am J Med Genet B Neuropsychiatr Genet* **116B**(1):84–9. PMID:12497620
11. Tadic, A., Rujescu, D., Szegedi, A., Giegling, I., Singer, P., Moller, H.J., and Dahmen, N. (2003) Association of a MAOA gene variant with generalized anxiety disorder, but not with panic disorder or major depression. *Am J Med Genet B Neuropsychiatr Genet* **117B**(1):1–6. PMID:12555227
12. Urwin, R.E. and Nunn, K.P. (2005) Epistatic interaction between the monoamine oxidase A and serotonin transporter genes in anorexia nervosa. *Eur J Hum Genet* **13**(3):370–5. PMID:15523490
13. Urwin, R.E., Bennetts, B.H., Wilcken, B., Lampropoulos, B., Beumont, P.J., Russell, J.D., Tanner, S.L., and Nunn, K.P. (2003) Gene-gene interaction between the monoamine oxidase A gene and solute carrier family 6 (neurotransmitter transporter, noradrenalin) member 2 gene in anorexia nervosa (restrictive subtype). *Eur J Hum Genet* **11**(12):945–50. PMID:14508509
14. Gutierrez, M.J. and Scott, J. (2004) Psychological treatment for bipolar disorders—a review of randomised controlled trials. *Eur Arch Psychiatry Clin Neurosci* **254**(2):92–8. PMID:15146338

15. Preisig, M., Bellivier, F., Fenton, B.T., Baud, P., Berney, A., Courtet, P., Hardy, P., Golaz, J., Leboyer, M., Mallet, J., Matthey, M.L., Mouthon, D., Neidhart, E., Nosten-Bertrand, M., Stadelmann-Dubuis, E., Guimon, J., Ferrero, F., Buresi, C., and Malafosse, A. (2000) Association between bipolar disorder and monoamine oxidase A gene polymorphisms: results of a multicenter study. *Am J Psychiatry* **157**(6):948–55. PMID:10831475
16. Rubinsztein, D.C., Leggo, J., Goodburn, S., Walsh, C., Jain, S., and Paykel, E.S. (1996) Genetic association between monoamine oxidase A microsatellite and RFLP alleles and bipolar affective disorder: analysis and meta-analysis. *Hum Mol Genet* **5**(6):779–82. PMID:8776592
17. Huang, F., Khvorova, A., Marshall, W., and Sorkin, A. (2004) Analysis of clathrin-mediated endocytosis of epidermal growth factor receptor by RNA interference. *J Biol Chem* **279**(16):16657–61. PMID:14985334
18. Vanyukov, M.M., Maher, B.S., Devlin, B., Tarter, R.E., Kirillova, G.P., Yu, L.M., and Ferrell, R.E. (2004) Haplotypes of the monoamine oxidase genes and the risk for substance use disorders. *Am J Med Genet B Neuropsychiatr Genet* **125B**(1):120–5. PMID:14755456
19. Takehashi, M., Tanaka, S., Masliah, E., and Ueda, K. (2002) Association of monoamine oxidase A gene polymorphism with Alzheimer's disease and Lewy body variant. *Neurosci Lett* **327**(2):79–82. PMID:12098640
20. Sacher, J., Wilson, A.A., Houle, S., Rusjan, P., Hassan, S., Bloomfield, P.M., Stewart, D.E., and Meyer, J.H. (2010) Elevated brain monoamine oxidase A binding in the early postpartum period. *Arch Gen Psychiatry* **67**(5):468–74. PMID:20439828
21. Alia-Klein, N., Goldstein, R.Z., Kriplani, A., Logan, J., Tomasi, D., Williams, B., Telang, F., Shumay, E., Biegon, A., Craig, I.W., Henn, F., Wang, G.J., Volkow, N.D., and Fowler, J.S. (2008) Brain monoamine oxidase A activity predicts trait aggression. *J Neurosci* **28**(19):5099–104. PMID:18463263 PMC2430409
22. Mokrovic, G., Matosic, A., Hranilovic, D., Stefulj, J., Novokmet, M., Oreskovic, D., Balija, M., Marusic, S., and Cicin-Sain, L. (2008) Alcohol dependence and polymorphisms of serotonin-related genes: association studies. *Coll Antropol* **32** Suppl 1:127–31. PMID:18405071
23. Zhou, Z., Yuan, Q., Mash, D.C., and Goldman, D. (2011) Substance-specific and shared transcription and epigenetic changes in the human hippocampus chronically exposed to cocaine and alcohol. *Proc Natl Acad Sci U S A* **108**(16):6626–31. PMID:21464311 PMC3081016
24. Alia-Klein, N., Parvaz, M.A., Woicik, P.A., Konova, A.B., Maloney, T., Shumay, E., Wang, R., Telang, F., Biegon, A., Wang, G.J., Fowler, J.S., Tomasi, D., Volkow, N.D., and Goldstein, R.Z. (2011) Gene x disease interaction on orbitofrontal gray matter in cocaine addiction. *Arch Gen Psychiatry* **68**(3):283–94. PMID:21383264 PMC3127452
25. Fowler, J.S., MacGregor, R.R., Wolf, A.P., Arnett, C.D., Dewey, S.L., Schlyer, D., Christman, D., Logan, J., Smith, M., Sachs, H., Aquilonius, S. M., Bjurling, P., Halldin, C., Hartvig, P. K., Leenders, L., Lundqvist, H., Oreland, L., Stålnacke, C.-G., and Långström, B. (1987) Mapping human brain monoamine oxidase A and B with ^{11}C-labeled suicide inactivators and PET. *Science* **235**(4787):481–5. PMID:3099392
26. The Huntington's Disease Collaborative Research Group. (1993) A novel gene containing a trinucleotide repeat that is expanded and unstable on Huntington's disease chromosomes. *Cell* **72**:971–83.

27. Ross, C.A. and Tabrizi, S.J. (2011) Huntington's disease: from molecular pathogenesis to clinical treatment. *Lancet Neurol* **10**(1):83–98. PMID:21163446
28. Bradford, J., Shin, J.Y., Roberts, M., Wang, C.E., Li, X.J., and Li, S. (2009) Expression of mutant huntingtin in mouse brain astrocytes causes age-dependent neurological symptoms. *Proc Natl Acad Sci U S A* **106**:22480–5.
29. Benraiss, A., Wang, S., Herrlinger, S., Li, X., Chandler-Militello, D., Mauceri, J., Burm, H.B., Toner, M., Osipovitch, M., Jim, X.Q., Ding, F., Wang, F., Kang, N., Kang, J.S., Curtin, P.C., Brunner, D., Windrem, M.S., Munoz-Sanjuan, I., Nedergaard, M., and Goldman, S.A. (2016) Human glia can both induce and rescue aspects of disease phenotype in Huntington disease. *Nat Commun* **7**:11758.
30. Li, S.H., Schilling, G., Young III, W.S., Li, X.J., Margolis, R.L., Stine, O.C., Wagster, M.V., Abbott, M.H., Franz, M.L., Ranen, N.G., Folstein, S.E., Hedreen, J.C., and Ross, C.A. (1993) Huntington's disease gene (IT15) is widely expressed in human and rat tissues. *Neuron* **11**:985–93.
31. Strong, T.V., Tagle, D.A., Valdes, J.M., Elmer, L.W., Boehm, K., Swaroop, M., Kaatz, K.W., Collins, F.S., and Albin, R.L. (1993) Widespread expression of the human and rat Huntington's disease gene in brain and nonneural tissues. *Nat Genet* **5**:259–65.
32. Landwehrmeyer, G.B., McNeil, S.M., Dure IV, L.S., Ge, P., Aizawa, H., Huang, Q., Ambrose, C.M., Duyao, M.P., Bird, E.D., Bonilla, E., de Young, M., Avila-Gonzales, A.J., Wexler, N.S., DiFiglia, M., Gusella, J.F., MacDonald, M.E., Penney, J.B., Young, A.B., and Vonsattel, J.-P. (1995) Huntington's disease gene: regional and cellular expression in brain of normal and affected individuals. *Ann Neurol* **37**:218–30.
33. Scherzinger, E., Lurz, R., Turmaine, M., Mangiarini, L., Hollenbach, B., Hasenbank, R., Bates, G.P., Davies, S.W., and Wanker, E.E. (1997) Huntingtin-encoded polygolutamine expansions form amyloid-like protein aggregates in vitro and in vivo. *Cell* **90**:549–58.
34. Videnovic, A. (2013) Treatment of huntington disease. *Curr Treat Options Neurol* **15**(4):424–38. PMID:23417276 3677041
35. Huntington Study Group. (2016) Efffect of deutetrabenazine on chorea among patientis with Huntington disease: a randomized clinical trial. *JAMA* **316**:40–50.
36. Harper, S.Q., Staber, P.D., He, X., Eliason, S.L., Martins, I.H., Mao, Q., Yang, L., Kotin, R.M., Paulson, H.L., and Davidson, B.L. (2005) RNA interference improves motor and neuropathological abnormalities in a Huntington's disease mouse model. *Proc Natl Acad Sci U S A* **102**(16):5820–5. PMID:15811941 556303
37. McBride, J.L., Boudreau, R.L., Harper, S.Q., Staber, P.D., Monteys, A.M., Martins, I., Gilmore, B.L., Burstein, H., Peluso, R.W., Polisky, B., Carter, B.J., and Davidson, B.L. (2008) Artificial miRNAs mitigate shRNA-mediated toxicity in the brain: implications for the therapeutic development of RNAi. *Proc Natl Acad Sci U S A* **105**(15): 5868–73. PMID:18398004 2311380
38. Kordasiewicz, H.B., Stanek, L.M., Wancewicz, E.V., Mazur, C., McAlonis, M.M., Pytel, K.A., Artates, J.W., Weiss, A., Cheng, S.H., Shihabuddin, L.S., Hung, G., Bennett, C.F., and Cleveland, D.W. (2012) Sustained therapeutic reversal of Huntington's disease by transient repression of Huntingtin synthesis. *Neuron* **74**(6):1031–44.
39. Yu, D., Pendergraff, H., Liu, J., Kordasiewicz, H.B., Cleveland, D.W., Swayze, E.E., Lima, W.F., Crooke, S.T., Prakash, T.P., and Corey, D.R. (2012) Single-stranded RNAs use RNAi to potently and allele-selectively inhibit mutant Huntingtin expression. *Cell* **150**(5):895–08. PMID:22939619 3444165

40. DiFiglia, M., Sena-Esteves, M., Chase, K., Sapp, E., Pfister, E., Sass, M., Yoder, J., Reeves, P., Pandey, R.K., Rajeev, K.G., Manoharan, M., Sah, D.W., Zamore, P.D., and Aronin, N. (2007) Therapeutic silencing of mutant huntingtin with siRNA attenuates striatal and cortical neuropathology and behavioral deficits. *Proc Natl Acad Sci U S A* **104**(43):17204–9. PMID:17940007 2040405
41. Senthilingam, M. (2017) *Drug trial shows promising results to fight Huntington's disease*. CNN.
42. Wild, E.J., Boggio, R., Langbehn, D., Robertson, N., Haider, S., Miller, J.R., Zetterberg, H., Leavitt, B.R., Kuhn, R., Tabrizi, S.J., Macdonald, D., and Weiss, A. (2015) Quantification of mutant huntingtin protein in cerebrospinal fluid from Huntington's disease patients. *J Clin Invest* **125**(5):1979–86. PMID:25844897 4463213
43. Tian, X., Chakrabarti, A., Amirkhanov, N.V., Aruva, M.R., Zhang, K., Mathew, B., Cardi, C., Qin, W., Sauter, E.R., Thakur, M.L., and Wickstrom, E. (2005) External imaging of CCND1, MYC, and KRAS oncogene mRNAs with tumor-targeted radionuclide-PNA-peptide chimeras. *Ann NY Acad Sci* **1059**:106–44. PMID:16382049
44. Tian, X., Aruva, M.R., Qin, W., Zhu, W., Duffy, K.T., Sauter, E.R., Thakur, M.L., and Wickstrom, E. (2004) External imaging of *CCND1* cancer gene activity in experimental human breast cancer xenografts with Tc-99m-peptide-PNA-peptide chimeras. *Journal of Nuclear Medicine* **45**(12):2070–82. PMID:15585484
45. Nielsen, P.E. (2010) Peptide nucleic acids (PNA) in chemical biology and drug discovery. *Chem Biodivers* **7**(4):786–804. PMID:20397216
46. Egholm, M., Buchardt, O., Christensen, L., Behrens, C., Freier, S.M., Driver, D.A., Berg, R.H., Kim, S.K., Norden, B., and Nielsen, P.E. (1993) PNA hybridizes to complementary oligonucleotides obeying the Watson–Crick hydrogen-bonding rules. *Nature* **365**(6446):566–8. PMID:7692304
47. Gray, G.D., Basu, S., and Wickstrom, E. (1997) Transformed and immortalized cellular uptake of oligodeoxynucleoside phosphorothioates, 3′-alkylamino oligodeoxynucleotides, 2′-*O* methyl oligoribonucleotides, oligodeoxynucleoside methylphosphonates, and peptide nucleic acids. *Biochem Pharmacol* **53**(10):1465–76. PMID:9260874
48. Wang, Y., Van Bockstaele, E.J., and Liu-Chen, L.Y. (2008) in vivo trafficking of endogenous opioid receptors. *Life Sci* **83**(21–22):693–9. PMID:18930741 PMC2652254
49. Borics, A. and Toth, G. (2010) Structural comparison of mu-opioid receptor selective peptides confirmed four parameters of bioactivity. *J Mol Graph Model* **28**(6):495–505. PMID:20036591
50. Lowery, J.J., Yeomans, L., Keyari, C.M., Davis, P., Porreca, F., Knapp, B.I., Bidlack, J.M., Bilsky, E.J., and Polt, R. (2007) Glycosylation improves the central effects of DAMGO. *Chem Biol Drug Des* **69**(1):41–7. PMID:17313456
51. Shaw, G., Morse, S., Ararat, M., and Graham, F.L. (2002) Preferential transformation of human neuronal cells by human adenoviruses and the origin of HEK 293 cells. *FASEB J* **16**(8):869–71. PMID:11967234
52. Koppelhus, U., Awasthi, S.K., Zachar, V., Holst, H.U., Ebbesen, P., and Nielsen, P.E. (2002) Cell-dependent differential cellular uptake of PNA, peptides, and PNA-peptide conjugates. *Antisense Nucleic Acid Drug Dev* **12**(2):51–63. PMID:12074365
53. Tian, X., Aruva, M.R., Zhang, K., Cardi, C.A., Thakur, M.L., and Wickstrom, E. (2007) PET imaging of *CCND1* mRNA in human MCF7 estrogen receptor-positive

breast cancer xenografts with an oncogene-specific [^{64}Cu]DO3A-PNA-peptide radiohybridization probe. *J Nucl Med* **48**(10):1699–707. PMID:17909257

54. Opitz, A.W., Wickstrom, E., Thakur, M.L., and Wagner, N.J. (2010) Physiologically based pharmacokinetics of molecular imaging nanoparticles for mRNA detection determined in tumor-bearing mice. *Oligonucleotides* **20**(3):117–25. PMID:20406142 2966851
55. Toll, L. (1990) Mu-opioid receptor binding in intact SH-SY5Y neuroblastoma cells. *Eur J Pharmacol* **176**(2):213–7. PMID:1968847
56. Muller, I., Sarramegna, V., Renault, M., Lafaquiere, V., Sebai, S., Milon, A., and Talmont, F. (2008) The full-length mu-opioid receptor: a conformational study by circular dichroism in trifluoroethanol and membrane-mimetic environments. *J Membr Biol* **223**(1):49–57. PMID:18574543
57. Sonar, M.V., Wampole, M.E., Jin, Y.-Y., Chen, C.-P., Thakur, M.L., and Wickstrom, E. (2014) Fluorescence detection of *KRAS2* mRNA hybridization in lung cancer cells with PNA-peptides containing an internal thiazole orange. *Bioconjugate Chemistry* **25**(9):1697–708. PMID:25180641 PMC4166030
58. Kahlert, S., Nuedling, S., van Eickels, M., Vetter, H., Meyer, R., and Grohe, C. (2000) Estrogen receptor alpha rapidly activates the IGF-1 receptor pathway. *J Biol Chem* **275**:18447–53.
59. Fernandez A.M., and Torres-Alemán, I. (2012) The many faces of insulin-like peptide signalling in the brain. *Nat Rev Neurosci* **13**:225–39.
60. Bondy, C.A. and Cheng, C.M. (2004) Signaling by insulin-like growth factor 1 in brain. *Eur J Pharmacol* **490**(1–3):25–31. PMID:15094071
61. Mysoet, J., Canu, M.-H., Cieniewski-Bernard, C., Bastide, B., and Dupont, E. (2014) Hypoactivity affects IGF-1 level and PI3K/AKT signaling pathway in cerebral structures implied in motor control. *PLoS One* **9**:e107631.

CHAPTER 11

PROGRESS OF OLIGONUCLEOTIDE THERAPEUTICS TARGET TO RNA: REVIEW ON MODIFICATION, CONJUGATION, AND DELIVERY

YUFEI PAN[1], XINYANG ZHOU[1,2], AND ZHENJUN YANG[1]
[1] Key Laboratory of Natural and Biomimetic Drugs, School of Pharmaceutical Sciences, Peking University, Beijing, China
[2] School of Criminal Investigation & Forensic Science, People's Public Security University of China, Beijing, China

11.1. BACKGROUND

In the past three decades, especially after the discovery of the RNAi phenomenon, the field of RNA therapy for diseases has developed rapidly. Compared with traditional small-molecule drugs and protein drugs, oligonucleotide (ON) drugs have the advantages of defined structure, clear therapeutic target, simple preparation and synthesis process, etc., which make them a promising disease treatment option. The genetic roots of many common diseases have been identified owing to advances in gene sequencing technology [1]. Thus far, the sequence-based ON therapeutics have been applied to treatment of diseases such as cancer, neuropathy, immunopathy, and infectious diseases [2–5]. Historically, the FDA-approved drugs are two major categories of small molecule and protein drugs [6]. While the research and development of small-molecule drugs has reached a clear bottleneck, protein-based drugs (e.g., cytokines, antibodies, enzymes) also have their own problems, which strongly hinder their clinical promotion. There are

Nucleic Acids in Medicinal Chemistry and Chemical Biology: Drug Development and Clinical Applications, First Edition. Edited by Lihe Zhang, Xinjing Tang, Zhen Xi, and Jyoti Chattopadhyaya.

more than 10,000 proteins left in the human genome that cannot be targeted by small molecule and protein therapies up to now [7]. By contrast, ONs can target all protein-encoding genes theoretically and truly realize personalized gene therapies. Antisense oligonucleotides (ASOs) a type of ONs, by specifically bounding with target RNA via Watson–Crick base pairing, further represent an avenue to cure a genetic defect in addition to just treating it [8].

Over the past ten years, there are glimmers of light in ONs clinical application, serval ON-based therapies have been approved by FDA. However, there are still obstacles remaining to be solved for further improvement. Unlike antibodies, ONs have to pass through cell membranes to reach their RNA target. Naked ONs are susceptible to many hindering factors in the blood and inside cells. Among these barriers, translocation of ONs across the cell membrane and their escape from endocytic pathway are the two primary barriers to overcome [9]. As a result, how to deliver therapeutic ONs to desirable targets has always been the top priority of research in related field. Scientists have focused on the design of delivery methods and materials that will transport ON drugs to the site of interest. Judging from the current clinical progress, especially after the maturation of $GalNAc_3$ conjugation technology, it is relatively easy to realize liver targeting of nucleic acid drugs, but how to avoid liver accumulation and improve the targeting ability of other organs and cells has been a next field of competition in the development of RNA therapeutics. Conjugates have been widely explored in the study of nucleic acid drugs because of the diversity of conjugable molecules. On the basis of early chemical modification, functionalized molecular conjugate can continue to enhance *in vivo* biological properties of oligonucleotide drugs, such as optimizing transmembrane capabilities, improving cell and RNA targeting efficiencies, and improving tissue distributions. Carriers like liposomes are also common means to improve the efficiency of ON drug transfection. However, although a large number of literatures have reported various types of lipid structures so far, they are mainly cationic liposomes, and the toxicity problems seriously restrict their further development and application [10].

In this review, we will mainly put our emphasis on the delivery strategies used in ON research and development. The safety and specificity of ON drugs will also be discussed. Combined with the current advances in clinical development, we hope to provide some comprehensive picture of this topic that can serve as a useful reference for future development in ON therapy.

11.2. VARIETIES AND PROGRESS OF OLIGONUCLEOTIDE DRUGS IN CLINICAL TRIALS

Antisense technology has a history of 50 years. Since the first ASO drug was launched in 1998, a total of 8 ASO drugs and 4 siRNAs have been approved by FDA/EMA (Table 11.1), and a big number of them are in different stages of clinical trials, covering cardiovascular diseases, inflammation and autoimmune diseases, oncology, neurology, infectious diseases, etc. Especially after 2015, the

TABLE 11.1 Commercialized ONs approved by FDA/EMA (until January 2021)

Compd.	Name	Target	Modification	Vehicle	Marketing time	Delivery route
ASO	Fomivirsen (Vitravene)	CMV IE-2	PS	None	1998	Intraocular
	Mipomersen (Kynamro)	apo-B-100 mRNA	PS/2′-OMOE gapmer	None	2003	Subcutaneous
	Eteplirsen (Exondys 51)	exon 51	Morpholino	None	2016	Intravenous
	Nusinersen (Spinraza)	SMN2 pre-mRNA	PS/2′-OMOE gapmer	None	2016	Intrathecal
	Inotersen (Tegsedi)	TTR mRNA	PS/2′-OMOE gapmer	None	2018	Subcutaneous
	Volanesorsen (Waylivra)	apolipoprotein C-III mRNA	PS/2′-OMOE gapmer	None	2019	Subcutaneous
	Golodirsen (Vyondys 53)	exon 53	Morpholino	None	2019	Intravenous
	Viltolarsen (NS-065)	exon 53	Morpholino	None	2020	Intravenous
siRNA	Patisiran (Onpattro)	TTR mRNA	2′-OMe	LNP	2018	Intravenous
	Givosiran (Givlaari)	ALS1 mRNA	PS/2′-OMe/2′-F, GalNAc conjugate	None	2019	Subcutaneous
	Lumasiran (Oxlumo)	HAO1 mRNA	PS/2′-OMe/2′-F, GalNAc conjugate	None	2020	Subcutaneous
	Inclisiran (Leqvio)	PCSK9 mRNA	PS/2′-OMe/2′-F,GalNAc conjugate	None	2020	Subcutaneous

field of oligonucleotide medicine has ushered in a spurt of development, greatly arousing the enthusiasm of both scientific researchers and veterans of the big pharmaceutical companies.

11.2.1. ONs on the Market and in Clinical Studies

11.2.1.1. Commercialized ONs In 1998, FDA approved the first ASO drug Fomivirsen (ISIS2922) for the treatment of cytomegalovirus retinitis in AIDS patients. The sequence was totally modified with phosphorothioate (PS)-modification at a clinical dose of 165 μg/week for the first three weeks, followed by administration every two weeks, with an adverse effect of increased intraocular pressure [11].

In 2013, FDA approved a second ON drug, namely Mipomersen for the treatment of homozygous familial hypercholesterolemia (FoFH). Mipomersen is a phosphorothioated oligonucleotide. The clinical dosage is 200 mg subcutaneous injection once a week, and its hepatotoxicity should be monitored [12, 13].

In 2016, Exondys 51(Eteplirsen) was approved by the FDA for the treatment of Duchenne muscular dystrophy (DMD), which was the first FDA approval of a new drug for this indication. Exondys 51 is an antisense oligonucleotide modified by phosphoramidate morpholino oligonucleotide(PMO). It mediates an exon skipping event to bypass unwanted slicing and thus prevents the expression of disease-related mature mRNA sequences, thereby synthesizing the desired protein that can serve as the anti-muscular atrophy function. Its clinical dosage is 30 mg/kg in dwelling needle infusion, once a week [14].

In the same year, Spinraza (Nusinersen) was approved to treat spinal muscular atrophy (SMA) in children and adults, and it was also the first drug for this indication. The sequence was modified with 2′-*O*-methoxyethyl phosphorothioate (2′-OMOE, PS). The clinical administration concentration is 2.4 mg/mL, and the recommended dose is 12 mg per administration. The most common adverse reactions in treatment include upper respiratory tract infection, lower respiratory tract infection, and constipation [15].

Inotersen was developed by Ionis Pharmaceuticals for the treatment of hATTR (hereditary transthyretin-mediated amyloidosis) in 2018. Inotersen consists of gapmer design with five 2′-OMOE nucleotides present at each flank and PS modifications present throughout the sequence. hATTR is a rare genetic neurological disease. TTR amyloid is deformed and deposited in multiple organs and tissues throughout the body. The gradual accumulation of TTR amyloid deposits can lead to sensory, motor, and autonomic dysfunction. The principle of Inotersen is to inhibit the production of TTR protein and reduce the accumulation of TTR amyloidosis by binding to wild type and mutant TTR mRNA. Inotersen is also a naked oligonucleotide with a length of 20 nucleotides and is administered by subcutaneous injection. The phase III clinical results show that the patient's neurological function and quality of life can benefit from its treatment. But its side effects of thrombocytopenia and glomerulonephritis have been added with black-box warnings in the description [16].

Volanesorsen is used to treat FCS (familial chylomicronemia syndrome, familial hyperchylomicronemia). It was approved for marketing by EMA in 2019. It regulates triglyceride metabolism by inhibiting the production of apolipoprotein (ApoC-III). Phase III clinical results show that it can significantly reduce the fasting triglyceride level of patients. However, it will produce the side effect of lowering platelets, so it is necessary to monitor the patient's platelet level during treatment [17].

Golodirsen is also used to treat DMD, and Eteplirsen is the same product of Sarepta. Similar to Eteplirsen, Golodirsen also uses the chemical modification of PMO, which also produces a truncated but still functional anti-dystrophin by changing the splicing position of mRNA, except that the site of action is exon 53. Golodirsen also uses a naked oligonucleotide design, which is 25 nucleotides in length and is administered by intravenous infusion. The FDA also approved the listing of Golodirsen with accelerated review. Twenty-five patients were enrolled in this phase II clinical trial. The main clinical endpoint is the surrogate endpoint (anti-dystrophin level). The clinical results show that it can increase the expression of anti-dystrophin in patients. After 48 weeks of treatment, it is 0.10% from the baseline (relative to the level of healthy people) increase to 1.02% [18].

In 2018, the FDA approved the first siRNA drug ONPATTRO® (patisiran), which is a candidate therapy for hATTR (hereditary transthyretin amyloidosis) [19–21]. The safety of this siRNA drug is better than hATTR's ASO drug Inotesen (Tegsedi®) [22]. ONPATTRO is administered by intravenous injection at 0.3 mg/kg every three weeks.

In 2019, the FDA approved the second siRNA drug GIVLAARI™, which is the only effective treatment for acute hepatic porphyria other than intravenous heme [23–25]. In addition, heme is extremely difficult to obtain, and it is unstable in aqueous solution. Also, the long-term application is potentially toxic, and it is easy to cause various diseases including thrombophlebitis [26]. The dosage of Givosiran is 2.5 mg/kg subcutaneously injected monthly, and the peak plasma drug concentration is reached three hours after administration, the half-life is six hours, and the apparent clearance rate is 35 L/h [27]. In phase III clinical trials of the drug, the common side effects were nausea and injection site reactions.

Lumasiran was approved by the FDA in November 2020 for the treatment of primary type I hyperoxaluria in children and adults. The principle of Lumasiran is to induce gene silencing of mRNA encoding hydroxyacid oxidase 1 (HAO1) and reduce the production of glycolate oxidase (GO) enzymes in the body, thereby reducing the production of glycolate and oxalate. Lumasiran is 21/23 nucleotides in length. It uses chemical modification of PS, 2′-F, and 2′-OMe. It uses GalNAc coupling carrier to achieve targeted delivery to the liver. Its phase III clinical results show that it can continuously and significantly reduce the content of oxalic acid in the patient's urine [28].

Inclisiran was approved by the EMA in December 2020. Inclisiran can bind to PCSK9 mRNA and inhibit the synthesis of PCSK9. PCSK9 participates in the regulation of LDL-C levels by preventing the recycling of LDL receptors, so monoclonal antibodies targeting PCSK9 can play the role of lowering LDL-C. Inclisiran can also achieve the effect of reducing LDL-C by inhibiting the synthesis of PCSK9. PCSK9 is mainly synthesized in the liver, while Inclisiran uses GalNAc conjugate technology to achieve targeted delivery to the liver. With the chemical modification of PS, 2′-F, and 2′-OMe, unlike PCSK9 monoclonal antibody that needs to be injected every 2/4 weeks, Inclisiran only needs to be injected once every six months (subcutaneously). Its three phase III clinical trials show that it can significantly reduce the level of LDL-C in patients [29].

With the rapid development of genomics, the success of RNAi (RNA interference) drugs indicates the beginning of a new era of personalized therapy [3, 30, 31]. At present, all of siRNA drugs are developed by Alnylam Pharmaceuticals, which is a star company that promotes the listing of RNAi drugs. And multiple clinical trials in different countries/regions have shown that RNAi-based drugs have unique advantages over conventional therapies in terms of target specificity and safety. Many therapeutic siRNAs have entered clinical trials to treat multiple diseases. siRNA drugs currently in phase III clinical research are shown in Table 11.2 [42].

11.2.1.2. Case Studies of ASO vs siRNA The side-by-side approval of an ASO and an siRNA therapeutics in 2018 (inotersen and patisiran) to treat the same disease by inhibiting the same target makes it possible to directly compare the two major classes [9]. With the approval in July 2018 from the European Medicines Agency (EMA), inotersen became the world's first RNA-targeted therapeutic approved to treat hATTR. Robust efficacy was demonstrated in phase III study of inotersen. However, two significant adverse events (thrombocytopenia and renal dysfunction) that are related to this class of drugs were observed during the study and changes to the monitoring schedule were required. One patient with thrombocytopenia died because of intracranial hemorrhage (Ionis Pharmaceuticals Press Release in 2017). After changing to platelet monitoring, no additional serious sequelae secondary to thrombocytopenia have been observed. Inotersen was subsequently approved by the FDA and Health Canada in October 2018 (Akcea Therapeutics and Ionis Pharmaceuticals Press Releases). Only one month after the approval of inotersen, the FDA and the EMA granted the first-ever approval of an siRNA patisiran (ONPATTRO) to treat the same disease (Alnylam Pharmaceutical Press Releases). So far, there has been no significant evidence of safety concern related to thrombocytopenia, renal dysfunction, or liver enzyme elevations. This is a distinct advantage of patisiran over inotersen [20, 43]. However, pretreatment with antihistamines, nonsteroidal antihistamines, and a glucocorticoid are required to mitigate the pro-inflammatory effect of this formulation, because of patisiran in an LNP formulation. In addition, s.c. delivered inotersen provides convenience for patients compared with that administered through i.v. infusion [43, 44]. Nevertheless, the milestone of the treatment by patisiran marks the arrival of a new treatment approach for patients with rare and devastating diseases.

TABLE 11.2 The ON drugs in phase III clinical trial

Name	Target gene	Therapeutic indication	Sponsor	Modification	Formulation	Ref.
Tominersen	HTT	Huntington's disease	Ionis	PS, 2′-OMOE	Naked ASO	[32]
Tofersen	SOD1	Amyotrophic lateral sclerosis	Ionis	PS, 2′-OMOE	Naked ASO	[33]
AKCEA-TTR-LRx	TTR	ATTR cardio myopathy	Akcea/Ionis	PS, 2′-OMOE	GalNAc–ASO conjugate	[34]
AKCEA-APO(a)-LRx	apolipoprotein(a)	Cardiovascular disease	Akcea/Ionis/Novartis	PS, 2′-OMOE	GalNAc–ASO conjugate	[35]
Alicaforsen	ICAM-1	Pouchitis	Atlantic	PS	Naked	[36]
QPI-1002 (I5NP)	p53	Delayed graft function, other complication of kidney transplant	Quark Pharmaceuticals	2′-OMe	Naked	[37]
Vutrisiran (ALN-TTRSC02)	TTR	Amyloidosis	Alnylam Pharmaceuticals	PS, 2′-OMe, 2′-F	GalNAc–siRNA conjugate	[38]
Fitusiran (ALN-AT3SC)	AT	Hemophilia	Alnylam Pharmaceuticals	PS, 2′-OMe, 2′-F	GalNAc–siRNA conjugate	[39]
DCR-PHXC	LDHA	Primary hyperoxaluria	Dicerna Pharmaceuticals	Undisclosed	GalNAc–siRNA conjugate	[40]
Tivanisiran (SYL1001)	TRPV1	Dry eye disease, ocular pain	Sylentis, S.A	Undisclosed	Naked ASO	[41]

11.2.2. Other Promising Clinical Trials

11.2.2.1. GalNac Delivery Strategy There are at least 29 different GalNAc conjugates in clinical development (55% RNAi based and 45% ASO based), according to the information extracted from pipeline data on the websites of Alnylam, Ionis, Dicerna, Arrowhead Pharmaceuticals, Arbutus Biopharma, and Silence Therapeutics, etc. Arrowhead Pharmaceuticals have four additional siRNA conjugates using their targeted RNAi molecule (TRIM) platform in clinical trial focused on liver-related targets. Although it is very likely that they are targeting the ASGPR by GalNAc, all of these hepatocyte targeting ligand have not been publicly revealed. AMG-890 (ARO-LPA) is described as a GalNAc conjugate in clinical trial. Additionally, in a presentation on HBV drug using the platform TRIM in a phase II clinical trial of ARO-HBV/JNJ-3989, the drug was described as two siRNA triggered with GalNAc. Indeed, GalNAc conjugates targeting the mRNA of HBV proteins and thus HBV inactivation/depletion are being widely explored, with Ionis/GSK, Alnylam, and Dicerna/Roche also having candidates in phase I or II clinical development, with encouraging results being announced [45, 46].

In summary, after the first GalNAc-oligonucleotide conjugate is registered, there are five other GalNAc-oligonucleotide conjugates either submitted for approval or in phase III trials (Table 11.2), and there are at least a other 20 conjugates in earlier stage clinical trials. The tolerance of these conjugates so far appears to be good, with the most common adverse events in liver and injection site. It seems very likely that this will result in new robust of oligonucleotide therapeutics reaching the market in the next few years, with this number of agents in clinical development [34]. A future breakthrough for GalNAc oligonucleotides will be oral rather than subcutaneous administration, particularly for use for chronic disease and in broad patient populations. Progress on this new mode of delivery has recently been reported by Ionis and Alnylam, while most ligand-based approaches are utilizing GalNAc conjugation for delivery, other ligand-based delivery systems targeting oligonucleotides to extrahepatic tissues are emerging, such as a GLP1R agonist peptide can effectively target an ASO to the GLP1R on pancreatic insulin-secreting B cells [47]. Those have opened up the possibility to utilize ligand targeting of oligonucleotides to restore functional islet mass in type 2 diabetes. The transferrin receptor (TfR) is also gathering interest as a promising ligand target, and TfR-antibody-siRNA conjugates have been able to silence their target in skeletal and cardiac muscle after intravenous administration [48].

11.2.2.2. Lipid-Based Delivery Strategy Some siRNA drugs currently in clinical research also use lipid complexes for delivery (Table 11.3). At present, the lipid complex is still a mixed preparation method with cationic lipid as the main body. The already marketed ONPATTRO uses the SNALP delivery vehicle, and the siRNA drugs that are being developed by a number of companies to

TABLE 11.3 The siRNA drug using lipid-based delivery system in the clinical stage (until January 2021)

Drugs	Indications	Target	Target organ	lipids	N/P ratio	Sponsor	Status
ARB-001467	Chronic hepatitis B	HBsAg	Liver	SNALP	4/1	Arbutus Biopharma Corporation	Phase II completed
ALN-PCS02	Abnormal LDL cholesterol metabolism	PCSK9	Liver	SNALP	4/1	Alnylam Pharmaceuticals	Phase I completed
TKM-Ebola-Guinea	Ebola virus	polymerase L, VP24, VP35	Liver	SNALP	4/1	Arbutus Biopharma Corporation	Pan-African Clinical Trials Registry endorsement
ND-L02-s0201	Nonalcoholic steatohepatitis (NASH)/ liver cirrhosis	HSP47	Liver	Vitamin A/LNP	0.6/1	Bristol-Myers Squibb Pharmaceuticals	Phase II
siRNA-EphA2-DOPC	Malignant solid tumor	EphA2	Solid tumor	DOPC	4/1	M.D. Anderson Cancer Center	Phase I recruitment
Atu027	Metastatic pancreatic cancer	PKN3	Solid tumor	AtuPLEX®	12/1	Silence Therapeutics	Phase II completed

treat metabolic diseases and antiviral drugs also use this delivery strategy. The particle size of SNALP particles is generally 50–80 nm. The surface of the particles is uncharged or slightly positively charged. The ratio of nitrogen to phosphorus between lipids and nucleic acids is about 4/1. The liver is used as the target tissue for delivery, and a good gene silencing ability has been achieved by intravenous injection [49].

ARB-001467 is a siRNA drug developed by Arbutus Biopharma Corporation to treat chronic HBV (human hepatitis B virus) infection. The company, formerly known as Tekmira Pharmaceuticals, has a series of R&D product lines for anti-hepatitis B virus drugs and holds a patent for SNALP delivery vehicles. In a phase IIa clinical study, 0.4 mg/kg was administered every month, and a significant decrease in hepatitis B surface antigen (HBsAg) was observed after three consecutive administrations. However, the research and development of ARB-001467 has been terminated, and the company is developing the similar GalNAc-conjugated siRNA drug AB-729, which is still in preclinical research [50].

ALN-PCS02 is an siRNA drug developed by Alnylam Pharmaceuticals to treat excessive levels of low-density cholesterol (LDL) in the blood. Elevated levels of low-density cholesterol will directly increase the risk of cardiovascular and cerebrovascular diseases, and PCSK9 protein will bind to the LDL receptor (LDLR) on the surface of liver cells, causing it to be internalized and degraded in lysosomes [51, 52]. Therefore, free LDL cannot combine with LDLR, and its normal intracellular metabolic process cannot continue. Therefore, the level of LDL in the blood increases, which is likely to cause malignant cardiovascular and cerebrovascular diseases such as coronary heart disease and stroke. ALN-PCS02 in a rodent model, a single intravenous injection of 5 mg/kg lipid complex (the mass ratio of siRNA to lipid is 1/7.5), can reduce the expression of PCSK9 gene by 60–70%; this gene silencing effect can last for three weeks [53]. In the clinical phase I trial, the high-dose group (0.4 mg/kg) observed a significant reduction in plasma PCSK9 and LDL levels, but there were still multiple side effects, mainly skin rashes [54].

Another antiviral drug, TKM–Ebola–Guinea, developed by Arbutus Biopharma Corporation, can effectively inhibit virus replication in guinea pigs exposed to Ebola virus. The siRNA was partially 2′-OMe modified, and the rodent dose was 2 mg/kg [55]. After the Ebola virus attacks the human body, it first replicates in the mononuclear phagocyte system (MPS), and then the virus is secreted to various organs, of which the liver is the main site [56, 57]. Therefore, the target organs of TKM–Ebola–Guinea are MPS in the liver and blood. However, in clinical trials, the survival rate of patients treated with TKM–Ebola–Guinea did not improve [58, 59].

Another type of siRNA delivery carrier used in clinical practice is Vitamin A/LNP mixed carrier, which is used to deliver siRNA targeting HSP47 mRNA to the liver to treat steatohepatitis or cirrhosis (ND-L02-s0201). The formulation of Vitamin A/LNP is DC-6-14/cholesterol/dioleoylphosphatidylethanolamine/retinol (VA). The structure of the lipid is shown in Figure 11.1. The molar ratio of each lipid to siRNA is 11.5/8.625/8.626/23/1. The siRNA used in

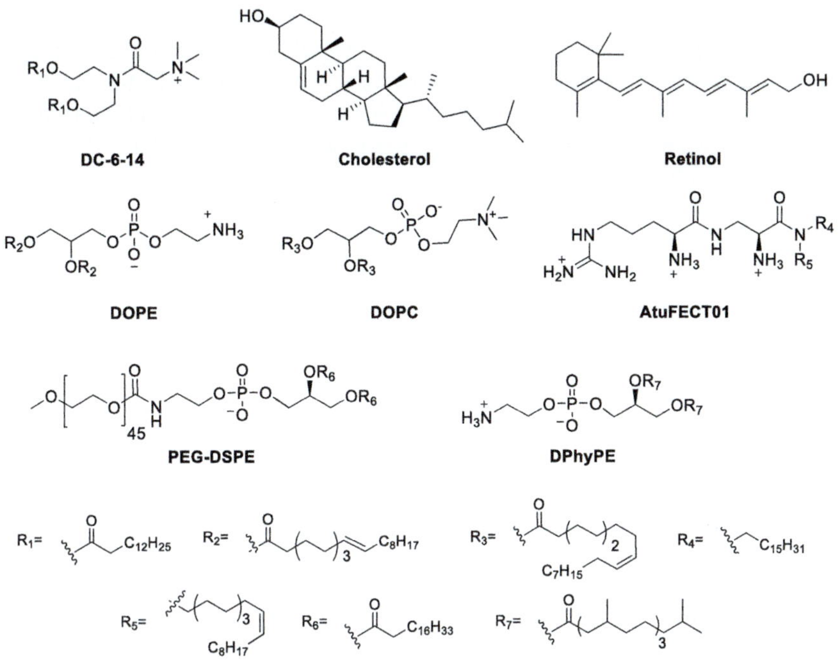

Figure 11.1 The lipid-based delivery system of siRNA in the clinical stage.

ND-L02-s0201 is 27 mer and has a 3′ end overhang, so the nitrogen to phosphorus ratio of lipid to siRNA is 0.6/1 [60]. The particle size of the nanoparticles formed by the mixed carrier and siRNA is about 150 nm, and the HSP47 protein can be almost completely knocked out at a concentration of 50 nM at the cellular level. In the rodent model, the dose is 0.75 mg/kg, which is administered by intravenous injection. The indication of ND-L02-s0201 is liver fibrosis, which is mainly related to hepatic stellate cells. Because vitamin A is incorporated in the preparation, nanoparticles can be specifically taken up by retinol binding protein on the surface of hepatic stellate cells.

siRNA-EphA2-DOPC was developed by MD Anderson Cancer Center in the United States. Its formulation is composed of siRNA targeting EphA2 and DOPC (Figure 11.1), a zwitterionic lipid material, with a nitrogen to phosphorus ratio of 4/1. It is used for the treatment of advanced solid tumors, which is administered by intravenous injection in rodent models at a dose of 0.25 mg/kg. The phase I clinical trial of siRNA-EphA2-DOPC started recruiting in 2012 and is expected to be completed in 2022.

Another siRNA-lipid complex used to treat tumors is Atu027, developed by Silence Therapeutics. Its target is PKN3, which is used to treat advanced or metastatic pancreatic cancer. Atu027 is delivered using the cationic lipid complex Atuplex. Its lipid composition is AtuFECT01 (cationic lipid)/DPhyPE/PEG-DSPE

(Figure 11.1), and its molar ratio to siRNA is 50/49/1/0.61. The particle size of Atu027 nanoparticles is about 118 nm, and the zeta potential is about 46 mV. In rodent models, 2.8 mg/kg is administered via tail vein injection to treat tumors. At present, the clinical phase I trial has been completed, and intravenous injection of 0.336 mg/kg siRNA preparation is safe enough. In phase II clinical trials, it is combined with gemcitabine, which may be due to the slightly lower efficacy of siRNA alone.

In general, compared with GalNAc conjugation, liposome delivery technique does not exhibit a strong development momentum. It is a great shame that none of the liposome-related delivery technologies have entered phase III trials since Patisiran's launch.

11.3. SAFETY AND SPECIFICITY OF OLIGONUCLEOTIDE DRUGS

11.3.1. Mechanisms

Since the concept of antisense was proposed in 1978, the action mechanism of RNA degradation has been extensively studied [7]. Unlike the traditional drug that acts on protein, oligonucleotide drugs can directly act on the target gene causing downregulation of oncogenes, interrupting the paracrine or autocrine signal loops [7]. The exact mechanism by which ASO/siRNA inhibit the target gene expression is not fully understood yet. According to the existing research, the mechanisms that have been elucidated include two aspects: active inhibition and physical blocking (Figure 11.2).

11.3.1.1. Active Inhibition Active inhibition refers to the process that RNA is cleaved by endogenous nuclease RNase H or argonaute-2 (Ago2) (Figure 11.2a) [61]. RNase H will digest the target mRNA, and the oligomer will be released after RNA cleavage at the site of antisense oligonucleotides. RNase H is an endonuclease ubiquitous in bacteria, animals, and humans. It can specifically degrade RNA chains in RNA–DNA double strands, and its efficiency is very high. In general, its inhibition efficiency is as high as 85–95%. The combination of ASOs and target mRNA activates RNase H and then degrade target mRNA. In addition, the abscisic ASOs can be combined with another target mRNA and again cause the degradation of this mRNA by RNase H [9]. One antisense oligonucleotide can cause the degradation of multiple mRNA. Therefore, the ASOs that activate RNase H can strongly inhibit its target. This is the most common means of regulating gene expression of ASOs. RNA interference (RNAi) was discovered 1998 by Fire *et al.*, which largely improved the development of siRNA. Ago2 enzyme mediates RNAi activity, promoting to form the catalytic core of the RNA-induced silencing complex. Different from ASO (usually single strand DNA), siRNA is commonly a double-stranded RNA molecule (a sense or passenger strand and an antisense or guide strand). Once

(a)

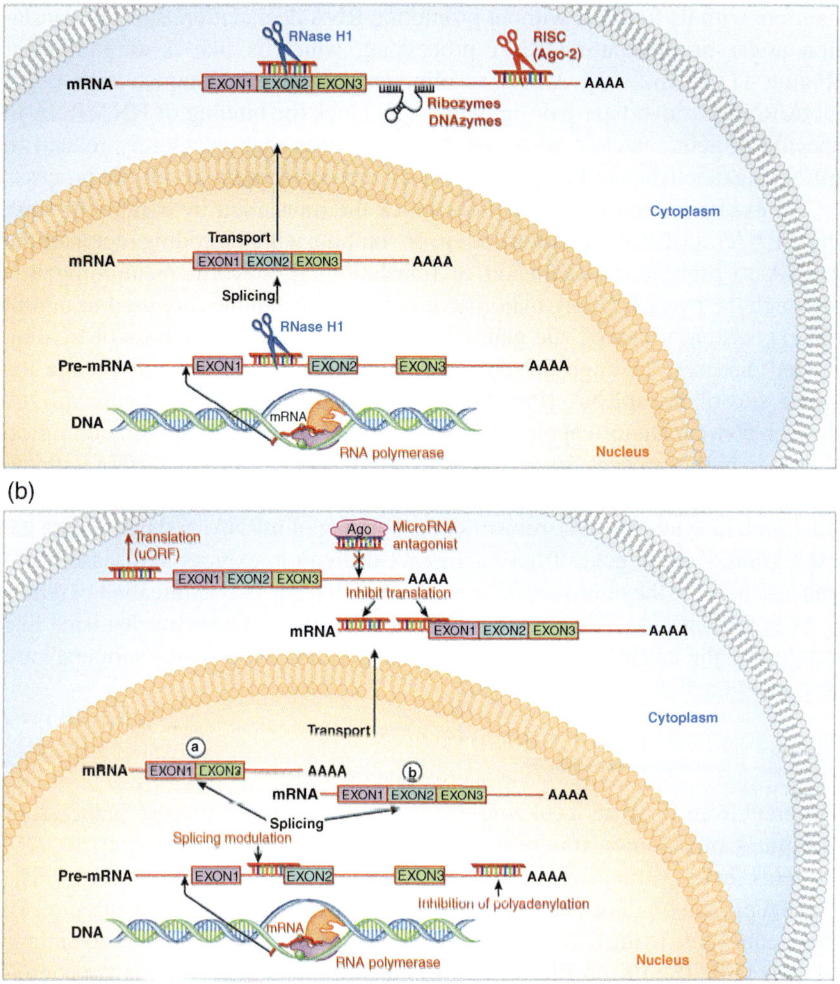

Figure 11.2 The mechanisms of ON *in vivo* (a) Active inhibition, (b) Physical blocking. Source: Bennett [61] / with permission of Annual Reviews.

inside the cell, the sense strand is eventually degraded, while the antisense strand recruits surrounding proteins including Dicer, Ago2, and TAR RNA-binding protein (TRBP) to form an active complex called "RISC." RISC then recognizes and binds to its target mRNA by sequence complementarity, resulting in Site-specific cleavage of the target RNA while retaining the antisense stand for a prolonged time period permitting it to bind additional target RNA. In some opinion, the main difference between the two mechanisms may lie in the oligonucleotides bind to mRNA and enzyme in different order [43, 61].

11.3.1.2. Physical Blocking In some circumstances, ONs bind to RNA and interfere with its function without promoting RNA degradation, such as translation arrest or modulating RNA processing, which is like a steric blocker (Figure 11.2b) [62, 63]. Antisense oligonucleotides can compete with target DNA/RNA through base pairing, which can block the binding of DNA/RNA to specific proteins, nucleic acids, or other cytokines that are closely related to mRNA metabolism and function, finally disrupting the normal cell life process.

For example, the oligonucleotides block the translation by combining with the mRNA translation initiation codon, or combine with the coding region of the mRNA to prevent the extension of translation. It is worth mentioning that although the overwhelming majority of antisense techniques are used to inhibit the expression of pathogenic genes in order to achieve the purpose of treating related diseases. Through the physical blocking of ASOs we can change the splice site of Pre-mRNA (Pre-mRNA can be processed into mature mRNA through 5′-wear hat (5′-capping), 3′-poly A tail (3′-polyadenylation), and intron shearing. In this process, some antisense oligonucleotides can interfere with the processing of Pre-mRNA, so that it cannot become a mature mRNA, thus inhibiting protein synthesis [64], reinstruct the splicing of mRNA, and upregulate the expression of some genes (these genes are difficult to express in disease state) and also achieve the purpose of disease treatment [65]. This application of ASOs is as important as the expression of suppressor genes. Other mechanisms like destroying the advanced structure of RNA [66] and serving as "antigene" are also mentioned in some articles [67].

11.3.2. Pharmacokinetics

Different from small molecules, ON therapeutics are not inhibitors, inducers, or substrates of common transporters and cytochrome P450 enzymes [68, 69]. Proteins serve as transporters for ONs rather than targets for small molecules. Plasma protein binding site won't be saturated because of the large amount. All these contribute to the low risk of drug interactions, which is quite an advantage of ONs [70]. The PK profiles of ON therapeutics largely depend on the chemical modification patterns, delivery materials, and conjugate strategies, and are mostly sequence independent [71–74]. Almost all oligonucleotide drugs are administered systemically or locally, only when they are administered together with permeation enhancers and only a modest efficacy can be observed [75]. In terms of absorption, ONs are readily absorbed after subcutaneous administration with Cmax achieved within serval hours and bioavailability ranging from 53 to nearly 100% [7, 76]. No matter through i.v. or s.c. administration, plasma concentrations decline rapidly in a multiphasic manner with a rapid distribution phase from plasma to tissues within a few minutes to hours. While the elimination phase may last long for weeks or months depending on the modification pattern of themselves [12, 71]. It has been discussed before that ONs are quite susceptible to ubiquitous nucleases and are easily cleared by liver and kidney.

PS-modified ONs are extensively bound to plasma proteins (85%) such as albumin, increasing the stability and bioavailability, and reducing urinary excretion and renal clearance [76]. As a comparison, ONs lack significant charges, such as peptide nucleic acids, morpholinos, and unmodified and unformulated siRNAs, and have low serum protein binding and are rapidly filtered by the kidney [77, 78]. So conjugates and carriers are being explored to enhance the distribution and cellular uptake on this basis. For example, patisiran, the siRNA already on the market. Naked patisiran barely bounds with plasma proteins and the terminal elimination half-life of patisiran is only three days, thus formulation must be conducted. Above all, the distribution of ONs is mainly concentrated in the liver and kidney, which account for over 80%. However, the means of targeting ONs to other tissues remain a spot of research, which are really meaningful. Nusinersen was approved to treat SMA by modifying the splicing of SMN2 pre-mRNA, an intended target in the CNS. In order to steer by the blood–brain barrier (BBB), it is directly administrated to the CSF by intrathecal injection. Following IT administration, they appear to distribute within the cerebrospinal fluid (CSF) and from CSF to spinal cord and brain tissues, with highest drug concentrations observed in spinal cord and cortical tissue and less drug found in deeper brain structures [76, 79]. Metabolism and extend elimination half-life also have a big relationship with chemical modification patterns. For example, eteplirsen (used to treat Duchenne muscular dystrophy) is an ASO fully modified with morpholino. Although it is metabolically stable and resistant to nuclease, it has a fast elimination due to poor protein-binding ability (60–70% of the dose cleared by renal within 24 hours through i.v. administration). In contrast, a combination of PS and 2′-OMOE modification has a better performance with a low renal clearance and much longer half-life [7]. In general, second-generation ONs with various sugar modifications are stabilized against endo- and exonuclease and are thus metabolized slowly over weeks (clearance half-lives around 2–4 weeks).

11.4. BARRIERS FOR ONs DELIVERY

Although ON drugs have many advantages, the inherent properties of oligonucleotides such as high molecular weight, high hydrophilicity, and high electronegativity lead to poor pharmacokinetic properties of these drugs *in vivo*. At present, nucleic acid drugs are mainly faced with the following problems in clinical application:

1. Poor serum stability. ONs (especially RNA) are easily recognized by ribozymes and nucleases, causing degradation and loss of efficacy [80].
2. Liver filtration. Macromolecule drugs are difficult to avoid opsonization mediated by reticuloendothelial system (RES), resulting in drug accumulation in liver, spleen, and other areas [81].

3. Immune response of the body. A variety of immune-related receptors can recognize specific oligonucleotide fragments (such as CpG) or special structures (such as dsRNA), thus showing certain immunogenicity [82, 83].
4. Renal clearance. Nano-drugs with small molecular size and naked can be easily excreted through glomerular filtration [84].
5. Intracellular metabolism. ONs can be hydrolyzed by intracellular nucleases.
6. Insufficient transmembrane capacity. Because ONs are electronegative macromolecules, they are easy to be rejected by the phospholipid membrane, which is also electronegative [85].
7. Poor endosomal release. Once after endocytosis, ONs can easily enter endosomes and be hydrolyzed by nucleases and acidic environment [86, 87].

To improve these properties, oligonucleotides must be chemically modified by changing the phosphate linkage, the ribose backbone, or the nucleobase. By manipulating their chemical structure, it is possible to enhance potency and limit toxicities [88, 89].

11.5. DELIVERY OF OLIGONUCLEOTIDE DRUGS

ON drugs face many obstacles in the treatment of diseases. Although chemical modification can make up for the deficiency of ON drugs to a certain extent, they are stagnant in clinical application due to high renal clearance rate, low target cell uptake, poor therapeutic effect, off-target effect, and other unintended side effects [90]. Therefore, the design of appropriate delivery strategies is particularly important in the diagnosis and treatment of diseases. The encapsulation strategy can protect ON drugs from the degradation of nuclease, avoid the adsorption of nonspecific proteins with serum, and improve the transmembrane efficiency and bioavailability through the design of the carrier system. ON delivery system can be divided into viral vectors and nonviral vectors. Viral vectors represent a big branch of research field for ON delivery. Nanocarriers headed by adeno-associated viruses are the most often used to carry nucleic acid cargo. Naturally, some viruses can release the virion contents into host cytoplasm directly through membrane fusion, which ensure 100% efficiency in delivery [91]. However, loading capacity and problem of immunogenicity severely limit their clinical translation. Membrane fusion of viruses is a complicated biological process, which so far remains impossible to imitate by artificial delivery systems. Nonviral vectors include protein and peptide nanomaterials, organic polymers, inorganic nanomaterials, lipid-based nanomaterials, etc. [92] Herein, the review will more focus on liposome and polymer-based nanovectors.

11.5.1. Liposomes

Liposomes have been widely studied for its good biological compatibility and easy modification. Liposome plays a very important role in the field of drug

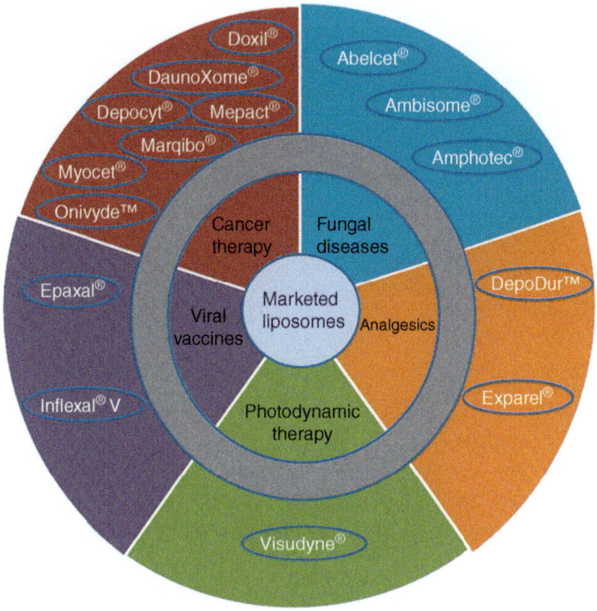

Figure 11.3 Therapeutic areas covered by liposome-based products. Source: Bulbake *et al.* [93] / MDPI / CC BY-4.0.

delivery and has been widely used in the transfection of oligonucleotide drugs (Figure 11.3). Despite many obstacles in the way of transfection, the pursuit of using liposome to deliver nucleic acid drugs has never stopped [94, 95].

11.5.1.1. Cationic Lipids Currently, cationic lipids are more often used to encapsulate nucleic acid drugs to improve the transfection efficiency in cells [96]. The so-called cationic lipids refer to the hydrophilic head containing cationic groups, which can adsorb negatively charged nucleic acids to the surface of the liposome or encapsulate them into the liposome through electrostatic interaction, thus improving the transfection efficiency of nucleic acid drugs and reducing the dose of drug administration (Figure 11.4) [97].

The surface of the complex has excess positive charge, which can adsorb on the surface of the negatively charged membrane, and endocytose into the cell to fuse with the lysosome [95]. Subsequently, the proton-sponge effect causes lysosomes to swell and release the complex into the cytoplasm [98]. However, it seems to be a common problem that ionizable lipid structure is the main factor affecting efficacy of the nanoparticle. All these cationic carriers exclusively take the route of endocytic pathway and less than 15% of the siRNAs internalized by cells can reach the cytosol. Although this obviously low transportation efficiency is acceptable to mediate pharmacological activity and knockdown of the disease-related gene at a least moderate level. It's specially worthy of pursuit to reduce

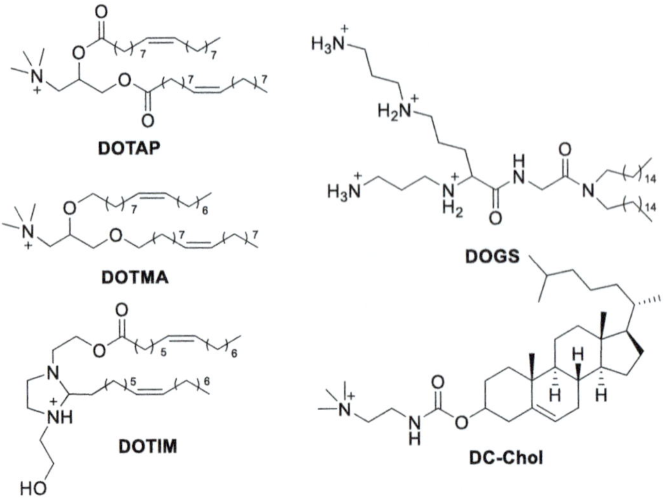

Figure 11.4 Common cationic lipid structure.

dosage with a better safety and lower cost [99]. The siRNA drug Patisiran launched in 2018 uses a liposome delivery strategy (Figure 11.5) [100]. The delivery system uses cationic liposome DLin-MC3-DMA as the core and adds zwitterionic liposome DSPC to enhance lysosomal escape ability; cholesterol improves membrane fluidity and reduces cationic toxicity. Adding lipidated PEG to increase the circulation time of nanoparticles in the body is an ideal nucleic acid transfection system [100]. Although cationic liposomes can efficiently deliver nucleic acid drugs into cells, the toxicity problem cannot be ignored: the excessive positive charge of the complex may also nonspecifically bind to the negative-point proteins and other substances in the plasma, thereby stimulating the body's immunity response effect and cause its own failure [101].

11.5.1.2. Nucleoside and Nucleotide Lipids To avoid toxic and side effects caused by excessive positive charges on the surface of the complex, the researchers focused on interactions like hydrogen bonding and π-π stacking by replacing electrostatic forces between positive and negative charges with these special forces between the bases. On this basis, researchers covalently connected hydrophobic tails on natural or modified nucleoside (acid) structures to form amphiphilic nucleoside (acid) lipids. In comparison with classical liposomes, nucleoside (acid) liposomes have a highly recognizable nucleoside (acid) head, long alkyl chain (such as oleic acid) tail, and ester or amide bonds as connection sites.

11.5.1.2.1. Cationic Nucleoside and Nucleotide Lipids In 2009, Barthélémy's research group reported a cationic nucleoside lipid with 3-nitropyrrole as the universal base, here referred to as TRN [102]. This lipid is the first reported nucleoside (acid) lipid material capable of transfecting siRNA.

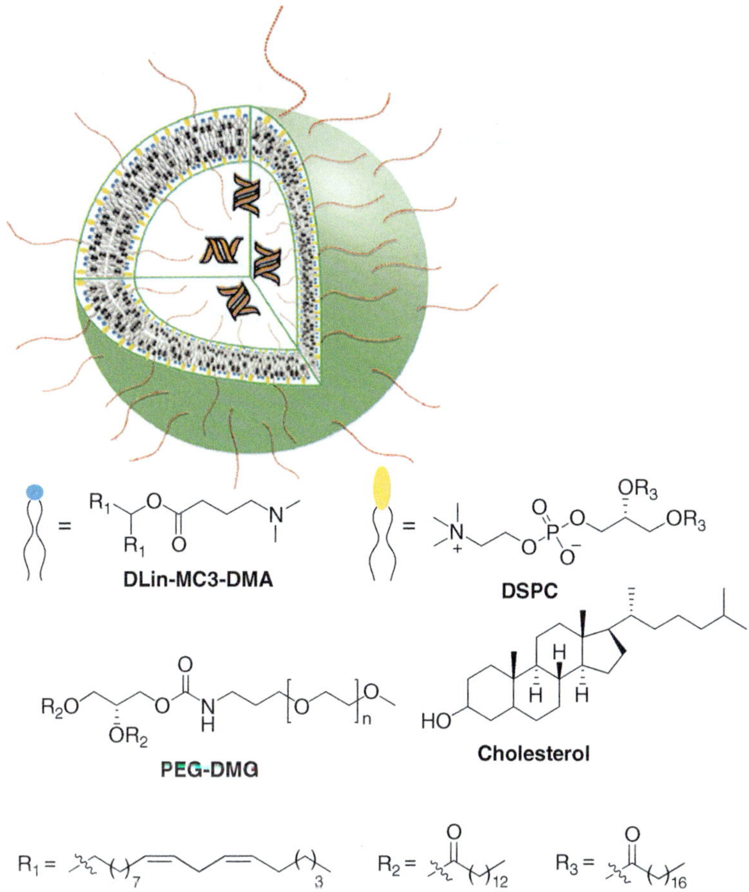

Figure 11.5 The formulation of Patisiran.

In 2013, the same group reported another nucleoside lipid material KNL with a ketal structure [103]. KNL can encapsulate siRNA to form a spherical complex with a particle size of about 100 nm when the N/P is 10 with a high delivery efficiency. The relatively stable conformation helps to form a stable complex, and the esterase is not sensitive to the ketal structure may be the reason for the improved delivery efficiency. KNL can effectively transfect siRNA and ASO and reduce the expression of Hsp27 (heat shock protein 27) in human prostate cancer cell PC-3 (Figure 11.6).

Anionic nucleotide lipids are negatively charged due to the presence of phosphate groups and will have a certain repulsive effect with the phosphate groups of the oligonucleotide skeleton. In theory, this repulsive force is unfavorable for the formation of complexes. Berti *et al.* [104, 105] designed and synthesized a nucleotide lipid molecule named POP-Ade (Figure 11.6). The lipid molecule is

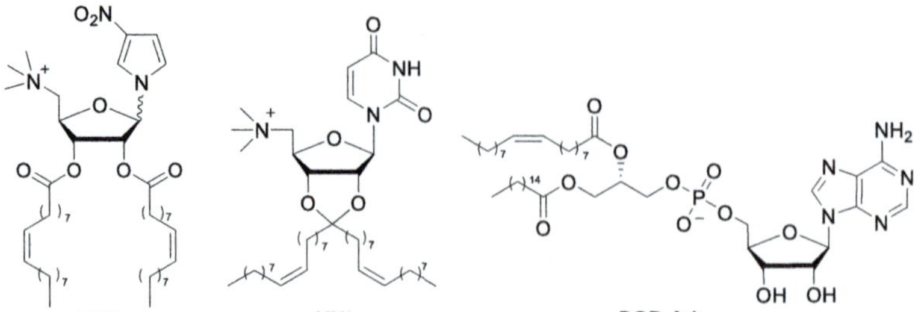

Figure 11.6 Structure of several ionic nucleoside and nucleotide lipids.

a typical anionic lipid. Ade is mixed with a certain amount of zwitterionic lipid POPE as an "auxiliary lipid" to improve the release ability of liposomes and reduce toxic and side effects. Studies have found that divalent cations (especially calcium ions) in POP-Ade/POPE play a key role in the formation of the single-stranded oligonucleotide complex. As the calcium ion concentration increases, the lysosomal escape of liposomes is significantly enhanced and the delivery efficiency is significantly improved. At the same time, calcium ion concentration is also one of the parameters that control the particle size of liposomes. A sharp drop in calcium ion concentration will induce the release of oligonucleotides, suggesting that calcium ion concentration can be used as one of the tools to control the release of liposomes. No interaction between the POPG/POPE mixed carrier and single-stranded oligonucleotides was observed at the same calcium ion concentration, which indicates that the bases in the POP-Ade structure also play an important role in the formation of the complex.

11.5.1.2.2. Neutral Nucleoside Lipids In order to reduce the dependence on a certain calcium ion concentration and simplify the structure of lipids, the researchers esterified the hydroxyl groups on the sugar ring of nucleosides with fatty acids and developed a class of neutral lipids based on nucleosides. Compared with anionic nucleotide lipids, this kind of neutral nucleoside lipids does not have phosphate groups. The emphasis is on the hydrogen bonding and π bonding between the bases, which is very distinctive. Lopez *et al.* designed and synthesized a simple neutral lipid material, named LU (Figure 11.7) [106]. When Poly A was encapsulated in the LU/DOTAP/DOPE mixed preparation, it was found that the neutral nucleoside lipid molecule can be strand DNA through the complementary pairing of bases and π-π stacking force [107]. The presence of LU will not affect the particle size and potential of the complex. ^{1}H NMR testing proves that the complex LU can be combined with Poly-A by hydrogen bonding.

Based on previous work, our group developed a series of neutral liposomes named DNCA, DNTA, DOCA, and DOTA, respectively, with base as the head, oleolenyl ether as the tail, ester bond or amide bond as the connecting arm

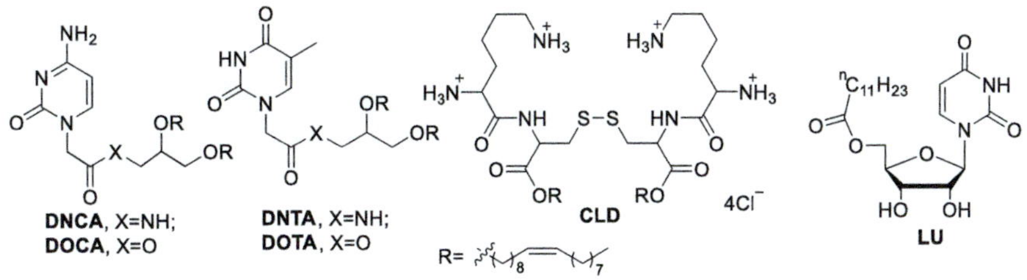

Figure 11.7 The structure of LU, DOTA, DOCA, DNTA, DNCA, and CLD.

(Figure 11.7) [107]. This kind of neutral base lipid can bind to oligonucleotide by hydrogen bonding and π-π stacking, pack and deliver various nucleic acid drugs such as aptamer, siRNA, antisense oligonucleotide, plasmid DNA, and mRNA into the cell. Among them, the cytosine derivative DNCA has great advantages in the delivery of aptamer (AS1411) and is an efficient neutral oligonucleotide delivery carrier with low toxicity *in vitro* and *in vivo*. CLD (Figure 11.7) is a cationic liposome with disulfide bond developed by our research group. With cysteine as the skeleton, two lysines as the head and two olefin alcohols as the tail, it has REDOX sensitivity [108]. CLD can efficiently deliver siRNA into the cell and break the disulfide bond under the action of intracellular glutathione, which improves the escape ability of lysosomes and accelerates the nucleic acid drug release process. CLD, as a classic cationic liposome, has the advantages of high encapsulation rate and high delivery efficiency, but its toxicity at high concentration is still an obstacle to its further application. In order to reduce the concentration of the cationic lipid and maintain a good delivery efficiency, our research group skillfully mixed the neutral lipid DNCA with the cationic lipid CLD in a certain proportion. The follow-up studies have proved the mixed lipids to be highly effective in encapsulating and delivering chemically modified ASO G3139 into cells with low toxicity, indicating that this strategy has great potential in advancing the further development of ON drugs [109].

11.5.2. Conjugates

As an alternative to nanoparticles, a more conceptually straightforward and chemically well-defined means of delivery is to directly conjugate a bioactive ligand to the RNA that will allow it to enter the cell of interest. Carbohydrate derivatives, lipid molecules, macromolecular polymers, peptides, antibodies, etc., have been widely reported (Figure 11.8) [110–112]. In clinical research, the molecular conjugation of nucleic acid drugs has begun to see light, showing a good development trend, especially the conjugation strategy led by GalNAc, which has been widely used in clinical research, provides a reference for nucleic acid drugs research [113, 114]. At present, compared to the current situation

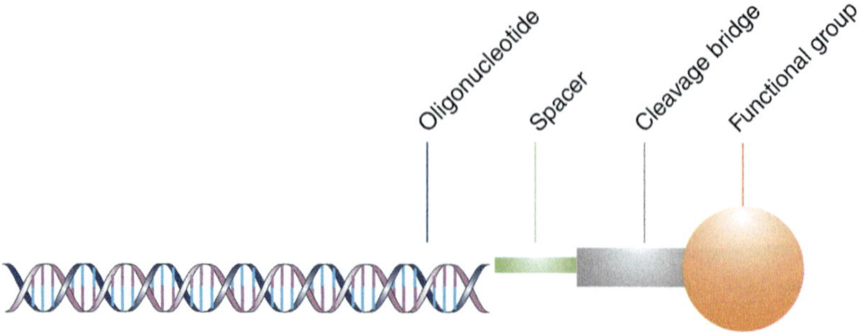

Figure 11.8 Oligonucleotide drug general conjugation strategy.

where the site-modified oligonucleotide drugs are mostly terminated in the first and second clinical phases, the GalNAc conjugation strategy surpasses the traditional site chemical modification strategy. More oligonucleotide drugs are pushed to the later stage of clinical research, increasing their application prospects [115].

11.5.2.1. Carbohydrates The covalent conjugation of carbohydrate derivatives and ONs has excellent conjugate performance, which can promote gene delivery and has broad application prospects. Carbohydrate derivatives are involved in a large number of biological processes, such as signal transmission through lectins, carbohydrate-binding proteins, and cell surface recognition [116, 117]. The synthesis of carbohydrate-ON conjugates has been intensively explored [118–120]. Galactose has become the most popular carbohydrate derivative used to increase the delivery of oligonucleotides, especially targeted delivery to hepatocytes. *N*-acetylgalactosamine is a high-affinity targeting group that can specifically bind to the asialoglycoprotein receptor in the liver (ASGPR) and is often used for liver targeting delivery of nucleic acid drugs [121–123], whose effect is remarkable [124]. Among them, the marketed siRNA drug Givosiran uses this conjugation strategy. Nair *et al.* prepared several CPG solid supports modified with protected divalent and trivalent GalNAc residues [123]. These CPG carriers are fully compatible with ON solid-phase synthesis and can synthesize a series of siRNA-(GalNAc) conjugates against apolipoprotein B100 (ApoB100) [125] and mouse transthyretin (mTTR). Interestingly, the authors observed that when siRNA is coupled with trivalent GalNAc, there will be a great impact on the biological activity, which is also confirmed by the high gene silencing level of the liver ApoB100 gene *in vitro* by cell uptake. In addition, the *in vivo* experiments targeted to ApoB100 gene confirmed that trivalent GalNAc conjugate can be effectively and specifically taken up by mouse liver in the absence of additional transfection reagents. The GalNAc conjugates used clinically are all trivalent, because if there is only a single GalNAc group, compared to the trivalent GalNAc conjugates used clinically, the performance is not as good as the trivalent GalNAc conjugates. It is reported in the literature that the activity will be reduced by about twice, and the target affinity will be reduced by about 10 times [126]. The pharmacokinetics of $GalNAc_3$-conjugated oligonucleotides in animals and humans have been well characterized. In general, $GalNAc_3$-oligonucleotides behave like prodrugs in that they preferentially deliver parental oligonucleotides to hepatocytes. GalNAc is stable in the blood, but after internalization in liver cells, the parent oligonucleotides are released by hydrolysis. The $GalNAc_3$ cluster itself is rapidly metabolized in liver cells and is rapidly excreted through the biliary and renal channels. This cluster or its related metabolites will not accumulate in the body, which is a characteristic of an ideal prodrug. At present, many pharmaceutical giants are scrambling to develop GalNAc-conjugated nucleic acid drugs for the treatment of liver diseases. Various forms of derivatives are also being developed (Figure 11.9) [114, 126–128].

11.5.2.2. Lipids Lipids are also one of the most commonly used modifications for the covalent introduction of nucleic acids. Such as bile and fatty acids, cholesterol and a large number of hydrophobic groups have been utilized to improve nucleic acid properties (such as cellular uptake, silencing activity, serum stability, and other pharmacokinetic characteristics), which is modified at the 3′- or 5′-end of siRNA to increase the transfection ability of siRNA [88, 129]. Khvorova *et al.* described the preparation of cholesterol-modified siRNA, which could be a library, targeting the genes of peptidylprolyl isomerase B (PPIB) or Huntingtin protein (HTT) [130, 131]. Therefore, this method can be used to screen out cholesterol-siRNA conjugate with high transfection efficiency, good stability and biodistribution, and low immune response *in vitro*. The authors observed that some cholesterol-modified siRNAs can promote the strong uptake of primitive neurons to cells and induce effective silencing of Huntingtin protein. Interestingly, the optimal Chol-siRNA candidates showed comparable inhibitory activity to unmodified siRNA (4 and 13 pmol/L, respectively). In most cases, the lipid tail chain can increase the binding capacity between the aptamer and serum albumin, thereby prolonging the half-life of the drug. This technology has been very mature in the development of hypoglycemic drugs [132–135]. Insulin detemir [136], insulin degludec [137], liraglutide [138], simaglutide [139], kumarutide [140], etc. all have fatty acid side chains in the drug structure (Figure 11.10a).

These lipid structures enable the drug to self-assemble and promote the reversible binding of the drug to albumin, thereby reducing the number of free drug molecules used for a long time. Lee *et al.* [141] synthesized a 16-mer oligonucleotide (ODN) with cholesterol modification at the 5′-end. Compared with unmodified ODN, the half-life of cholODN is greatly increased from less than 1 minute to about 10 minutes. The obtained cholODN can be further linked with low-density lipoprotein (LDL) to form a cholODN–LDL complex, which proves that the complex is stable against rat serum nuclease degradation.

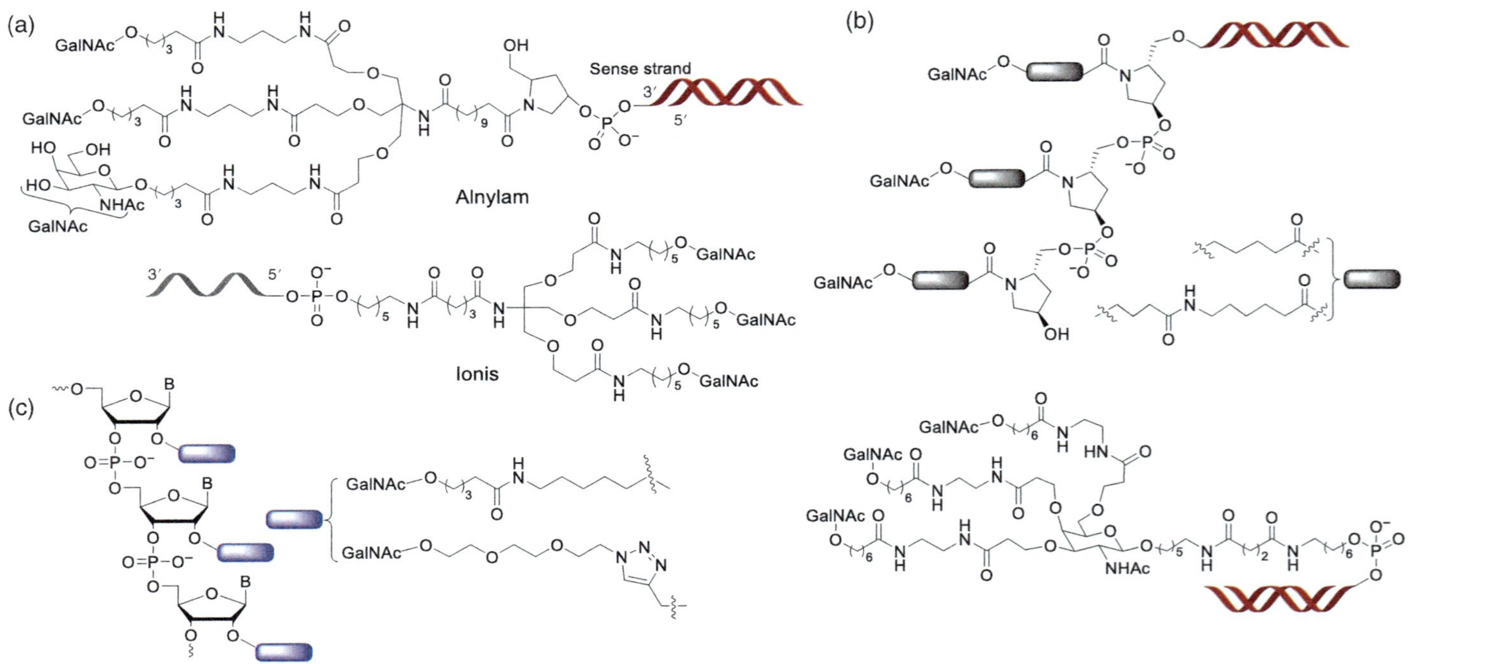

Figure 11.9 (a) The general structure of $GalNAc_3$ conjugate from Alnylam and Ionis company. (b) Trivalent GalNAc-siRNA conjugate. (c) Representative GalNAc-siRNA conjugates modifying at the 2′-position of nucleotide.

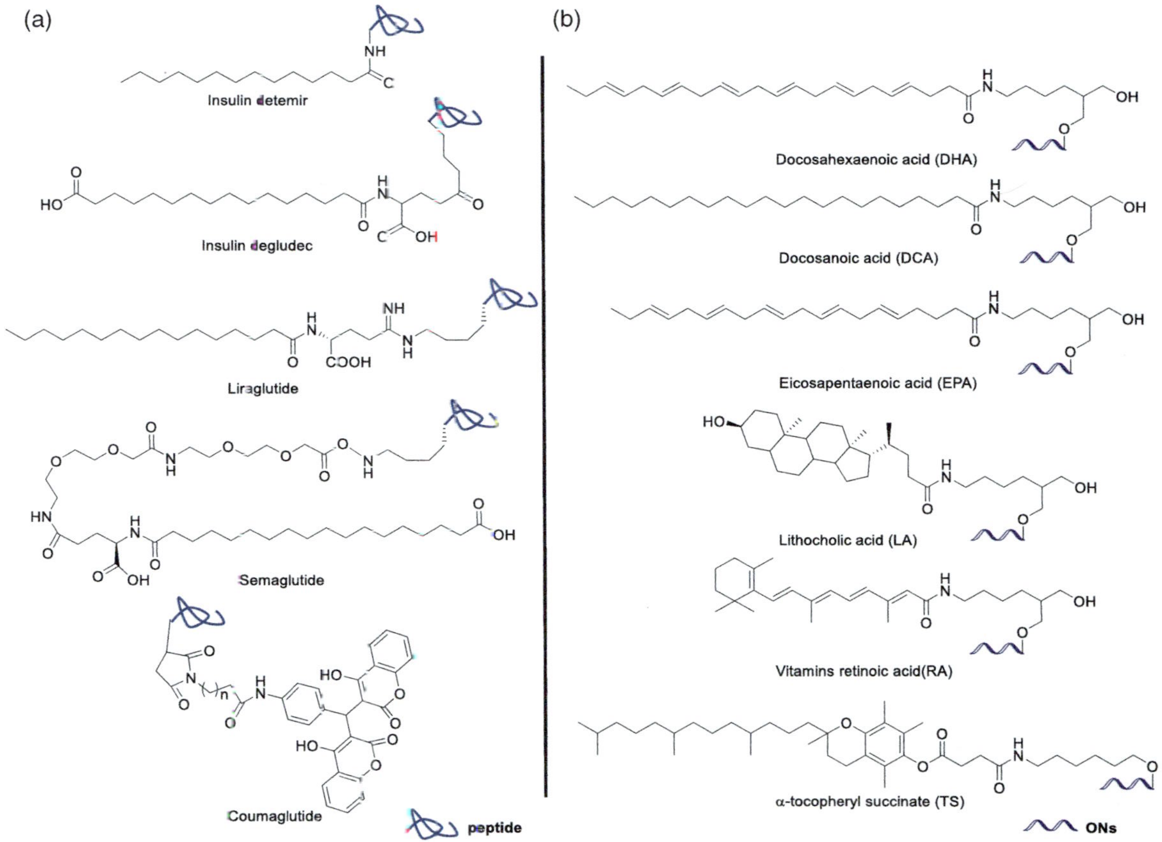

Figure 11.10 (a) Lipid molecules used for long-acting polypeptide modification; (b) Molecules used for nucleic acids conjugation.

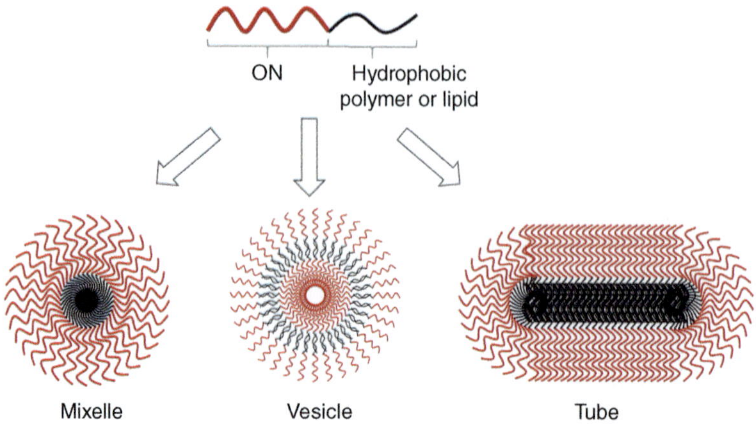

Figure 11.11 Self-assembly properties of lipid-conjugated nucleic acids.

In addition, many other lipid molecules were selected as oligonucleotide coupling candidates. Docosahexaenoic acid (DHA), docosahexaenoic acid (DCA), eicosapentaenoic acid (EPA), phenolic acid (LA), vitamin retinoic acid (RA), and alpha-tocopheryl succinate (TS) were used to construct ODN conjugates (Figure 11.10b). Experiments have shown that different lipid molecules can cause different tissue distribution in the body [142]. It is worth noting that compared with the above-mentioned types of conjugation strategies, lipid conjugation of oligonucleotides will give the conjugated molecules their own self-assembly properties (Figure 11.11). Taking advantage of this property, many basic researches have been carried out in the field of imaging and drug delivery [143–145]. Tan and his colleagues synthesized DPC with diacyl tails and found that by changing the length of DNA, micelles with a diameter of 8–36 nm can be formed [146]. After interacting with the cell membrane, the micelles decompose and undergo endocytosis mediated by the transferrin receptor into the cell. It was also found that the size of micelles affects cell uptake, and smaller micelles show faster endocytosis. These results, combined with studies of similar systems, have led to the idea that DNA lipid conjugates can be incorporated into cell membranes and endosomes [147, 148].

11.5.2.3. Polymer Polymers have become a common means to improve the uptake of ONs by cells and protect them from degradation by nucleases. Usually, the negatively charged oligonucleotide and the cationic polymer are electrostatically combined to form a complex, in many processes such as endosome escape, cytoplasmic transport, and nuclear uptake. Although other problems such as toxicity at this stage reduce the possibility of translating it to the clinic, these properties provide some ideas for promoting the internalization of ONs in cells [34]. The charge density and flexibility of polymers are important factors that determine the stability of the polymer. This improvement in stability is

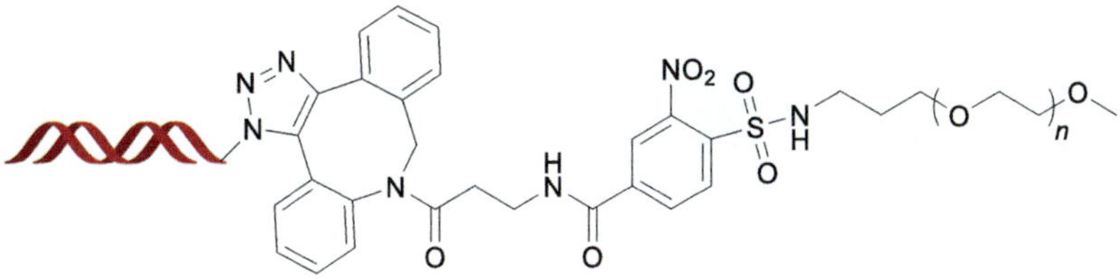

Figure 11.12 An environmentally sensitive and biodegradable PEG linker.

essential to improve the delivery of small drugs and nucleic acid therapies [149, 150]. In this regard, click chemistry and the use of disulfide bonds have become two well-known strategies [151]. Compared with polymers, the correct design of the degradable linking arm is very important, not only because it must maintain a certain degree of stability to resist the extracellular environment but also must be unstable to realize release activity after entering the cells. Huang *et al.* reportednonyl(2-nitrobenzenesulfonamidyl)-PEG-siRNAconjugates(Figure11.12). It showed resistance to acid and alkaline conditions and has the special feature of being partially cleaved by mercaptans. Further the above-mentioned cleavable nonyl group was functionalized with dibenzocyclooctyne-amine (DBCO), which has been widely used in biological coupling reactions between alkynes and modified oligonucleotides containing azide groups [152]. This covalent strategy provides a reference for the preparation of new polymer-siRNA constructs that have the potential environmental sensitivity under intracellular conditions.

11.5.2.4. Peptides Some studies have explored the potential of specific peptide sequences to promote tissue targeting in the context of oligonucleotide conjugates. However, it turns out that the limitation of this technology is that only a few targeting peptide–oligonucleotide conjugates have proven effective *in vivo* activity. Cell-penetrating peptides (CPPs) are a type of short cationic peptides, mainly used for the transportation of small and large molecules (such as antibodies, proteins, or nucleic acids) [153]. CPPs may come from natural or artificial synthesis and have the ability to cross biological barriers, such as cytoplasmic membrane and blood–brain barrier. This strategy was first used in the late 1990s [154], and it is still one of the best options for improving nucleic acid delivery [155]. The coupling of oligonucleotides and CPPs can overcome the risk of multiple dissociation in the bloodstream. In the absence of transfection reagents, the conjugate has good silencing activity *in vivo* and *in vitro*. In addition, Tat, atin, Transportan, and lytic melittin peptides have been selected as natural and artificial peptides for efficient delivery of siRNA [156–160]. Our group combined the peptide KALLAL with two siRNAs targeting MEK1 mRNA (siMek1) and mutant B-RAF mRNA (siMB3) [161]. According to the solid-phase method, the KALLAL peptide is introduced into the 3′-ends of the sense and antisense strands. KALLAL–siRNA conjugate can effectively silence the corresponding protein expression for a long time *in vitro* and shows good serum

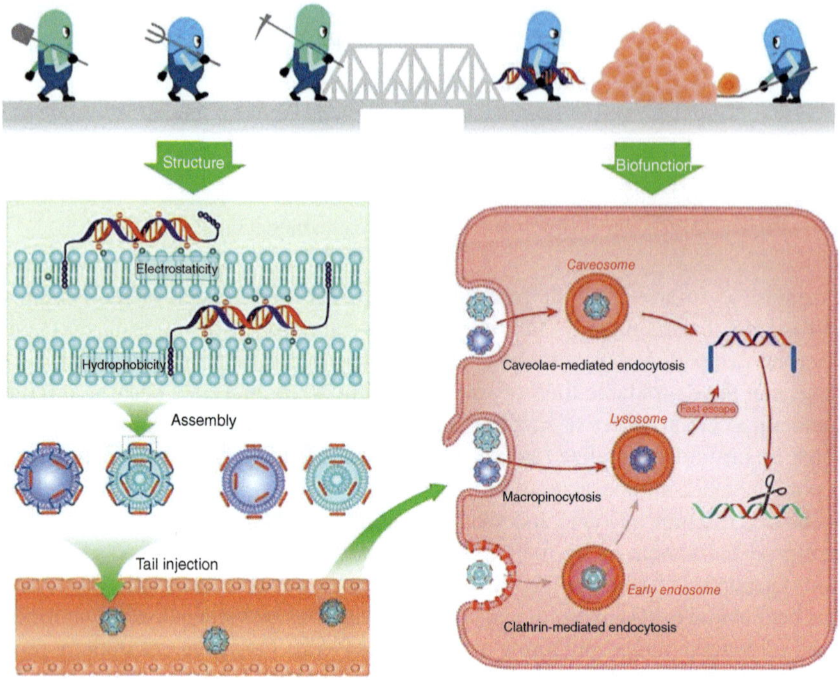

Figure 11.13 The uptake and endocytosis mechanism of 3′,3″-bis-peptide-siRNA conjugate.

stability (up to 80%). Compared with naked siRNA, the serum stability is increased by 21 times. Oligonucleotide–peptide conjugates can significantly inhibit the growth of transplanted tumors, prolong the circulation time, and reduce their accumulation in the kidney. Furthermore, we used a combination nanochemistry delivery strategy to deliver this 3′,3″-bis-peptide-siRNA conjugate using cationic lipids developed in the laboratory [162]. Significantly, compared to commercial liposomes, KALLAL-siRNA conjugates show better cell endocytosis performance (Figure 11.13). The proportion of caveolae-mediated endocytosis and macropinocytosis was much higher than that of clathrin-mediated endocytosis. These results show the potential of this modification in cancer treatment.

11.5.2.5. Antibodies As the antibody has the ability to specifically bind to extracellular proteins, using its target specificity, the antibody is used as an immunomodulator for direct pharmacological intervention, and antibody–drug conjugate (ADC) form is developed for drug delivery. The method also has more in-depth research in the field of nucleic acid drugs [163, 164]. Antibodies have a longer plasma half-life, which can improve the PK properties of the

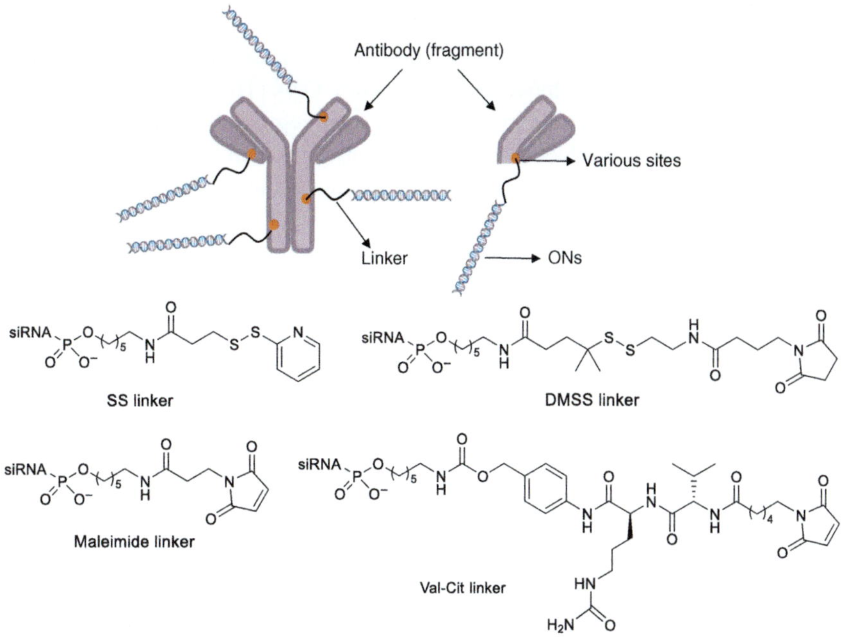

Figure 11.14 The strategy of antibody/ antibody fragment-ON conjugation.

therapeutic drugs carried [165]. Some early work in this field focused on attaching antibodies or antibody fragments to nanoparticles, and then developed to directly couple antibodies to nucleic acid drugs (Figure 11.14) [48, 149, 166]. Sugo *et al.* obtained CD71 (transferrin receptor, TfR) monomer antigen-binding fragment (Fab'), which was covalently linked to siRNA through different linking arms, and showed silencing activity even when the systemic dose was as low as 1 mg/kg. Compared with full-length antibodies, antibody fragments (including Fab' fragments) are smaller and provide greater dosage flexibility. In addition, Sugo confirmed that Fab' fragments have low binding affinity to TfR (compared to antibodies), which can make siRNA conjugates better release from their receptors, thereby achieving more efficient endosomal escape and subsequent efficacy [48]. However, other research groups have constructed coupling models of intact antibodies with siRNA and demonstrated *in vivo* efficacy in oncology models. Therefore, it is not clear whether Fab' fragments are truly superior to antibodies [149, 166]. In general, the field of antibody/antibody fragment coupling oligonucleotides is constantly developing. Combining the advantages of PK and cell specificity, the optimization of this targeting strategy has a good development prospect. However, the stability and scale preparation of antibody remains to be further improved (Figure 11.14).

11.5.3. Polymer-Based Delivery System

Polymers are a wide range of biocompatible macromolecular materials, which are broadly utilized in the delivery of therapeutic ONs [167, 168]. Under physiological environment, cationic polymers can form regular polyplexes with ONs spontaneously through electrostatic condensation, which shield from surface charges and enzyme attacks [169]. Amines are the main positively charged groups in polymers. Because of the abundant nitrogen atoms, they have a nice buffer capacity and are protonated at physiological pH (pK_a ~7.4). When it comes to interaction with cells, endocytosis or endocytosis-like mechanisms are mostly supported by scientists [170, 171]. Once inside the cells, polyplexes are trapped in endosomes, where they have a lower pH (~5.5). At the same time, some unprotonated amines groups accept protons that are pumped into endosomes, then followed by the passive influx of chloride ions and water. Osmotic pressure of endosomes increases, causing the swelling and rupture of endosome membrane and eventually the contents are released into cytosol. This process is the so-called proton sponge hypothesis [172, 173].

11.5.3.1. Polymers from Natural Sources Some naturally existing polymers such as cyclodextrin (CD), chitosan (CSN), etc. have been applied in ONs delivery for decades (Figure 11.15). Gonzalez *et al.* utilized CD to deliver

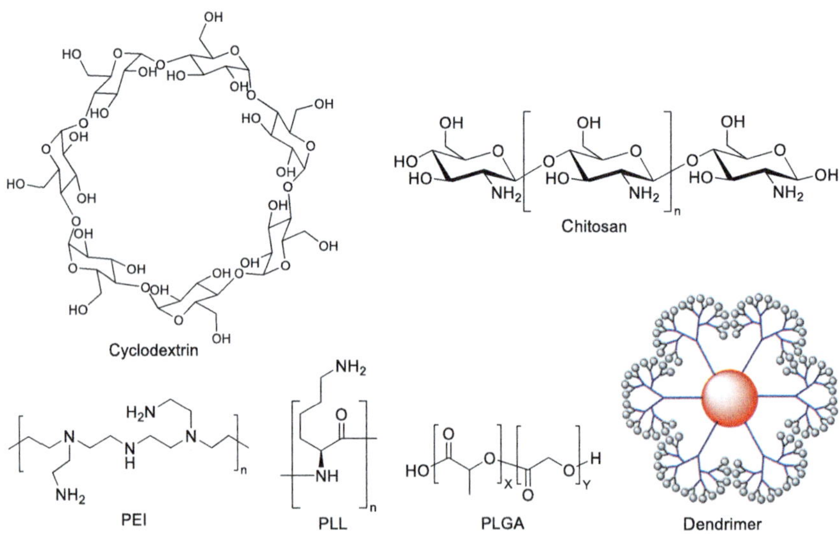

Figure 11.15 The chemical structure of CD, CSN, PEI, PLL, PLGA, and dendrimers (highly branched and precisely structured molecules synthesized from repeated growth reactions). Each cyclic reaction adds a branching layer, called a generation. It includes the main structure (kernel, branching unit, peripheral group) and the microenvironment (cavity).

plasmid DNA for the first time in 1999 [173, 174]. *In vivo* tests showed that unmodified CD polymers only were not enough to achieve an efficient delivery of ONs although the efficiency was high *in vitro*. Additional formulation elements were needed to form a stable and high- bioactivity complex [175–177]. Davis *et al.* constructed the first-targeted nanoparticle delivery platform with the help of CD polymer to transfect siRNA and pushed this technology into clinical trial (NCT00689065) (Figure 11.16). Apart from a linear CD polymer and therapeutic siRNA, the system also consisted of a human transferrin protein (TF) targeting ligand displayed on the exterior of the nanoparticle to engage TF receptors (TFR) on the surface of the cancer cells, a hydrophilic polymer (polyethylene glycol (PEG) used to promote nanoparticle stability in biological fluids. Robust and specific gene inhibition was observed in solid tumor in nonhuman primates treated with this kind of nanoparticle. However, it was discontinued in phase I clinical trial because of the obvious adverse effects [179, 180].

Chitosan (CSN) is the N-deacetyl product of chitin (Figure 11.15). Chitin and chitosan have many unique properties such as biodegradability, cell affinity, and biological effects, especially chitosan-containing free amino groups, which is the only basic polysaccharide among natural polysaccharides [181]. Raik *et al.* synthesized a chitosan derivative, *N*-[4-(*N*,*N*,*N*-trimethylammonium)benzyl]chitosan chloride (TMAB-CS). This CSN-based polyplexes mediated enhanced DNA delivery to target cells and protected it from nuclease degradation [182]. However, the high molecular weight of CSN (100–400 kDa) limits the dose *in vivo*. What's more, CSN can bound with DNA to form extremely stable polyplexes, causing a delayed release of therapeutic DNA [181].

11.5.3.2. Synthetic Polymers Synthetic polymers such as polyethyleneimine (PEI), poly-L-lysine (PLL), and poly (lactic-co-glycolic acid) (PLGA) have all been applied to ON delivery, with varying levels of success (Figure 11.15) [183–185]. PEI is one of the most commonly used nonviral polymers for ON transfection both *in vitro* and *in vivo* [173]. The amine groups of this nonviral vehicle can condense negatively charged DNA to form polyplexes with a desirable size and destabilize the endosome membrane, thus increasing the concentration of DNA in cytosol [186]. However, although the transfection efficiency increases with the degree of polymerization, the cytotoxicity also increases. This is a contradiction that needs to be resolved [187]. So, researchers modified PEI with different molecules to lower molecular weight of PEI. Certain ligands, like cyclodextrin (CD), chitosan (CSN), folic acid (FA), and 12-acryloyloxy dodecyl phosphorylcholine (ADPC), were all used to cross-link with PEI to decrease the surface charge of polyplexes. These reports demonstrated an improved gene transfection efficiency and low cytotoxicity [188–191]. In addition, there are some polymers specifically for ON delivery. Poly (β-amino esters) and its derivatives, in especial, have attracted widespread attention in ON delivery because of their easy synthesis process and biodegradability and have also presented an effective delivery of siRNA and mRNA [192, 193].

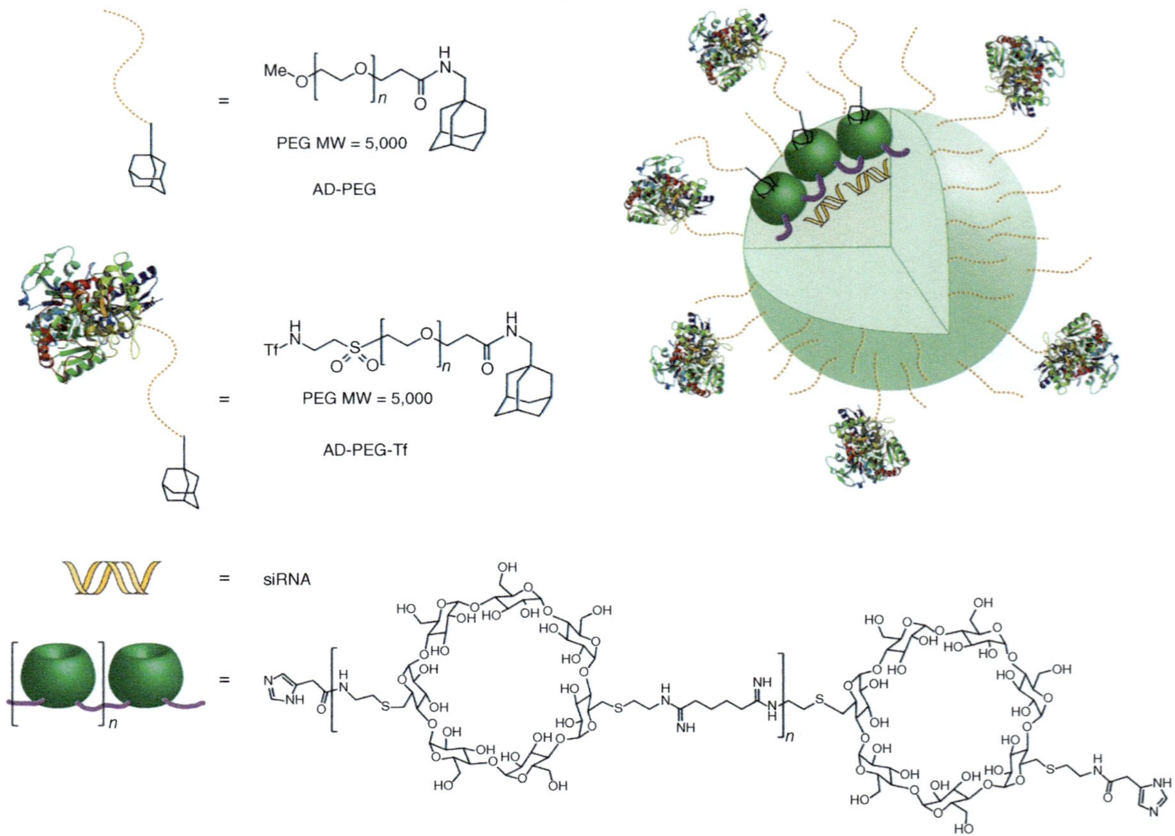

Figure 11.16 The composition of cyclodextrin polymer nanoparticles. Source: Kanasty *et al*. [178] / with permission of Springer Nature.

11.5.4. Dendrimers

Since the 1980s, a large variety of synthetic macromolecules with well-defined and highly branched peripheral chain ends have been developed (Figure 11.15). Dendrimers like poly (amidoamine) (PAMAM) dendrimers, poly (propyleneimine) (PPI) dendrimers, carbosilane dendrimers, and phosphorus dendrimers have been thoroughly studied. Dendrimers have a central core unit, while an exponential number of dendritic branches radiate from the core to form a dimensional core-shell architecture. Through modifying the repeating branching units, researchers can control the size of this nanostructure [173, 194]. Like many other polyplexes, rapid excretion from systemic circulation after administration and cytotoxicity are the common limitations of dendrimers. Varied chemical groups are applied to regulate their structure and improve biological properties [195, 196]. Take PAMAM for example, cholesterol, folic acid, and cell-penetrating peptide have been introduced to the surface of the structure to lower the toxicity [197–199]. Compared with unmodified PAMAM, hydrophobic moieties modified PAMAM had enhanced membrane interactions and cellular penetration, suggesting a higher delivery efficiency [194].

11.6. CHEMICAL MODIFICATION OF OLIGONUCLEOTIDE DRUGS

ON drugs in early clinical research mostly used chemical modification methods to increase the stability and nuclease resistance. Phosphate backbone, sugar ring, and base modification were the most common modification strategies. The strategies of using nucleic acid conjugation and carrier delivery to enhance the *in vivo* efficacy of ON drugs are all completed on the basis of chemical modification [200, 201]. Taking GalNAc-conjugated ON drugs as an example, in addition to the GalNAc conjugation at the 3′-end of sense strand, the marketed and clinically researched siRNA drugs also utilize phosphorothioate, 2′-F, 2′-OMe, and 2′-OMOE modification at multiple or all sites, so as to ensure the stability of oligonucleotide drugs *in vivo* [43, 114].

11.6.1. Phosphate Linkage Modifications

11.6.1.1. Phosphorothioate Replacement Phosphorothioate (PS) replacement (substitute a non-bridging phosphodiester oxygen with sulfur) is considered to be the most common way used in the nucleic acid modification, which is also well-known as the first generation of modification (Figure 11.17a). PS linkage directly increases stability in serum nucleases, contributing a lot to the half-life of DNA or RNA sequences. At the meantime, an increased binding affinity with proteins, especially serum proteins, enables ONs a longer time in circulation, slower removal rate by the kidney, and boosts the time available for uptake into target tissues [202]. Although originally incorporated to provide nuclease stability, the major impact of phosphorothioate modification has been on oligonucleotide trafficking and uptake. ASOs bearing phosphorothioate linkages are compatible with recruitment of RNase H, which cleaves the targets of ASOs [76, 203]. Moreover, phosphorothioates don't increase binding affinity with the RNA target or fully protect the ONs from degradation when used alone. Off-target effects and inflammatory responses are common at high doses.

11.6.1.2. Other Hydrophobic Modifications To avoid biological toxicity caused by phosphorothioates, Meade *et al.* proposed another way of masking negative charge by introducing lipophilic group onto the phosphate. This concept has previously proved feasible by several groups on single-stranded DNA. While inspired by a monophosphate nucleotide prodrug inhibitor of HIV and hepatitis B virus containing S-acyl-2-thioethyl (SATE) phosphotriester group, Meade *et al.* for the first time applied this kind of bioreversible phosphotriester linker into siRNA synthesis and obtained a good yield. The modification generates an siRNA prodrug that can be activated by cytoplasmic thioesterase and reverse into wild-type siRNA once entry into cytoplasm. However, despite that the modification on the phosphate groups perfectly shields negative charges of siRNA and prolongs the circulation time in blood, its poor membrane permeating capacity limits efficient delivery and a TAT peptide is necessary to be conjugated in the study. The reason may lie in weak hydrophobicity of tBu-SATE groups or due to the covalent linkage that prevent siRNA release from cell membrane once siRNA prodrug fuses into the membrane (Figure 11.17b) [204, 205].

11.6.1.3. The Chiral Problem As we all know, phosphodiester linkages are prochiral and each PS linkage has 2 stereoisomers (Figure 11.17c). Consequently, PS-modified oligonucleotides synthesized by standard methods are mixtures of 2^n diastereomers (n represents the number of PS linkages) [206]. In small molecule drugs, stereochemistry is strictly controlled to ensure potency and efficacy. While in ONs chemistry, it was considered nearly unfeasible to prepare stereopure ONs for clinical use. All oligonucleotide therapeutics approved to date are stereoisomeric mixtures [207]. It is recognized that the R_p and S_p stereoisomers differ in their binding affinity and susceptibility to degradation. However, there haven't a clear understanding or a thorough investigation on whether stereopure isomers can be more potent than stereorandom ONs or which stereoisomer can be better on cell level [207–209]. So it would be quite interesting to follow the development of stereopure ONs and publish data defining their potential as therapeutic agents [201]. Herein, Knouse *et al.* demonstrated a novel phosphorus(V)-based phosphorothioate reagent platform to overcome the feasibility hurdle of scalable synthetic process has been reported to yield stereopure oligonucleotides. The fundamentally different approach realizes diastereoselective phosphorus sulfur incorporation and can synthesize stereopure ONs with a high yield [210].

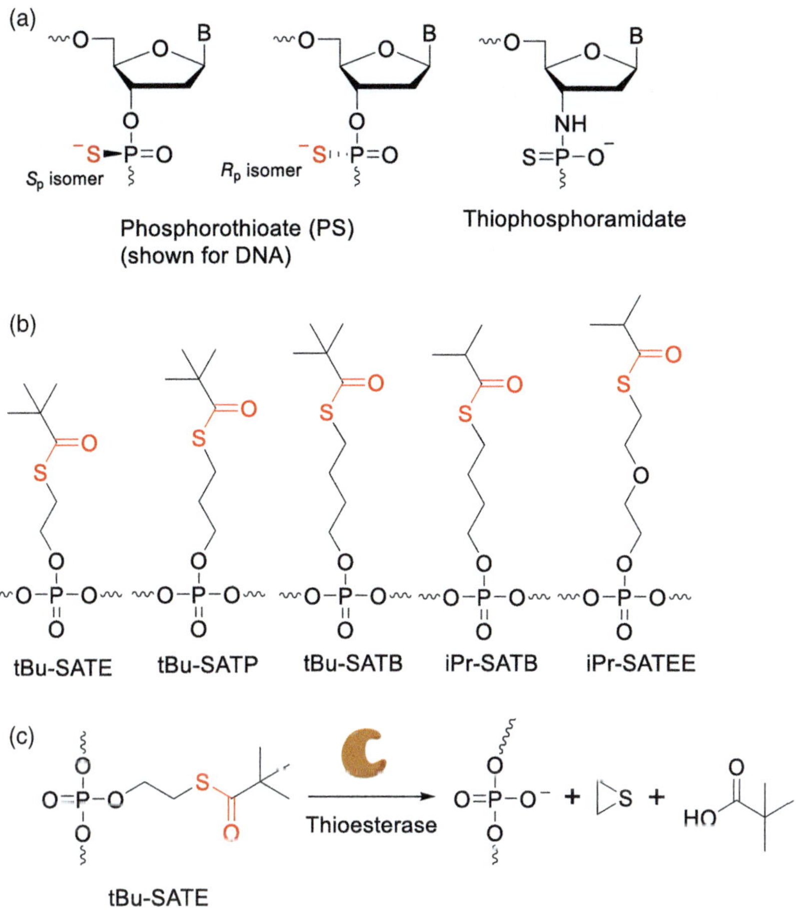

Figure 11.17 Examples of common chemical structure used in phosphate linkage modification. (a) The structure of phosphorothioates (S_p or R_p) and thiophosphoramidate. (b) The structure of tBu-SATE and its derivatives. (c) Degradation process of tBu-SATE *in vivo*.

11.6.2. Ribose Backbone Modification

In the past few decades, a number of chemical modifications to the ribose backbone have been made to improve antisense ON pharmacokinetic properties, tolerability profile, target binding affinity, and so on. For instance, phosphorodiamidate morpholino (PMO) and peptide nucleic acid (PNA) show better resistance to nucleases and enhanced uptake in cells. Tricyclo-DNAs (tcDNA) are conformationally constrained DNA analogues with increased potency and enhanced uptake in tissues after systemic administration. Ribose substitutions, including 2′-fluoro (2′-F), 2′-O-methyl (2′-OMe),

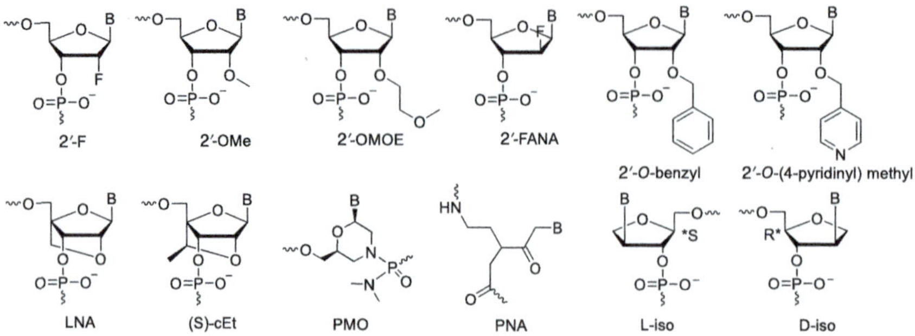

Figure 11.18 Nucleoside analogues could be applied on the modification of oligonucleotides.

2′-O-methoxyethyl (2′-OMOE), etc, are frequently used in combination to further increase stability, enhance target binding, and generally confer less toxicity than unmodified designs. Here are some detailed examples (Figure 11.18).

11.6.2.1. 2′-Ribose Modifications Among all ribose modifications, substitutions in the 2′-position most directly protect siRNAs against the action of serum nucleases, as the 2′-OH group is unstable itself under a nonneutral environment and participates in the cleavage of RNA by nucleases. This so-called second generation of modification is also widely used in siRNAs and ASOs chemistry. 2′-F, 2′-OMe, and 2′-OMOE represent the most commonly used replacement strategy in the clinic to date. The 2′-modifications increase stability toward digestion by nucleases by blocking the nucleophilic 2′-hydroxyl moieties. The modifications also increase the thermal stability of complementary hybridization, encouraging tighter binding and allowing the use of shorter ONs [44, 202]. 2′-Fluoro nucleosides improve binding affinity with target RNA to a larger degree than others do (ΔT_m 2.5°C per modified nucleotide), while >50% of the 2′-F in siRNA may cause toxicity [43]. The 2′-OMe modification (ΔT_m 0.5–1.5°C per modified nucleotide), which was discovered in nature, can also improve nuclease resistance, reduce immune stimulation, and stabilize 3′-endo ribose conformation [211]. Building on 2′-OMe chemistry, the binding affinity, and nuclease resistance were further improved by 2′-OMOE modification (ΔT_m 0.9–1.7°C per modified nucleotide).

Now, the combination of 2′-F, 2′-OMe, and phosphorothioate (PS) modifications has been applied for approved siRNA drugs and clinical studies [212]. With the introduction of GalNAc moiety, the ESC (enhanced stable chemistry) or ESC-plus platform has been established. The platform applied to the FDA-approved GIVLAARI®, OXLUMO®, and EMA-approved Leqvio® (Table 11.4). Thus, by avoiding the 2′-OH, the stability and nuclease resistance of siRNA improved, also the complexity of the synthesis process decreased. In addition, the fully modified single-stranded siRNA (ss-siRNA) is a newly developed strategy of siRNA drug, with enhanced stable 5′-(E)-vinyl

TABLE 11.4 The modification details and sequences of siRNA-GalNAc drugs

Name	Sequence (5′-3′)
GIVLAARI®	AS: UsAsA GAU GAG ACA CUC UUU CUGs GsU·22 Na+
(Givosiran)	S: CsAsG AAA GAG UGU CUC AUC UUA GalNAc·21 Na+
OXLUMO®	AS: UsAsU AUU UCC AGG AUG AAA GUCs CsA·22 Na+
(Lumasiran)	S: GsAsC UUU CAU CCU GGA AAU AUA GalNAc·21 Na+
Leqvio®	AS: AsCsA AAA GCA AAA CAG GUC UAG sAsA·22 Na+
(Inclisiran)	S: CsUsA GAC CUG UtU UGC UUU UGU GalNAc·21 Na+

Note: "B" represents 2′-*O*-methylation; "A, C, G, U" represents 2′-fluoronation; "s" represents phosphorothioation and "t" represents deoxythymidine.

phosphonate[5′-(E)-VP] at 5′-end, showing potent gene silencing activities *in vitro* and *in vivo* [213, 214]. The IC_{50} of this modified ss-siRNA in HeLa cells was similar to siRNA. The potency *in vivo* was potent with a whole 100 mg/kg dosage without a delivery system. Also, as contained no sense strand, the risk of undesirable off-target effects of ss-siRNA removed.

The approved antisense drug mipomersen, as well as numerous oligonucleotide drugs currently in clinical trials, carry the 2′-OMOE modification [215, 216]. Other modifications like 2′-*F*-arabino (2′-FANA), 2′-*O*-benzyl, and 2′-*O*-(4-pyridinyl) methyl [2′-*O*-CH_2Py(4)] oligonucleotide are also reported [217]. Unfortunately, 2′-modified nucleotides inhibit RNase H cleavage to some extent through "steric blocking" mechanisms (Figure 11.18) [218].

11.6.2.2. Constrained Nucleic Acids Nucleotides that are modified to limit conformation are referred to be locked nucleic acids (LNAs), which contain a methylene bridge between the 2′-oxygen and the 4′-carbon of the ribose. LNAs can increase target specificity by both DNAs and RNAs forming highly stable duplexes and reduced recognition by nucleases [219]. They can be incorporated into the synthesis of siRNAs [220] or gapmer ASOs supporting RNAse H activity, resulting in more potent gene silence *in vivo* than the 2′-OMOE analogs [221]. (S)-cET is also a 2′,4′-LNA, which is being tested in the clinic, and also can dramatically improve affinity and 5-fold to 10-fold in potency relative to a 2′-OMOE analog [222]. However, LNA and cEt modifications showed liver toxicity with sequence dependence and caused by off-target cleavage of mismatched transcripts [206, 223]. Tricyclo DNA (tcDNA) is another constrained nucleotide with binding affinity smaller than that of LNA but has shown much promise in splice-switching applications (Figure 11.18) [213, 224].

11.6.2.3. Morpholinos and Peptide Nucleic Acid Phosphoramidate morpholino oligomer (PMO) and peptide nucleic acid (PNA) are considered to be special modifications of oligonucleotides. PMO is a neutral ON, whose pentose is substituted by a morpholino ring, and the inter-nucleotide linkage is phosphoramidate instead of phosphodiester, with translational arrest mediated by steric interference of ribosomal assembly. PMO shows fewer nonspecific properties and lesser toxicity than PS but exhibits reduced cellular uptake. Conjugation with arginine-rich peptide (ARP) etc. can enhance its cellular uptake and antisense efficacy. Eteplirsen(30nt), for the treatment of Duchenne muscular dystrophy (DMD), represents the first FDA-approved morpholino-based ASO therapy for neurodegenerative disease [225–227]. PNA is a synthetic DNA, in which the deoxyribose phosphate backbone is replaced by polyamide linkages, has typically been used for modulating transcription and other applications not supporting RNase H activity. They are biologically stable and have good hybridization properties, but a PNA ASO targeting exon skipping in the DMD gene was poorly effective compared with an ASO with 2′-OMe modification (Figure 11.18) [228–230].

11.6.2.4. D-/L-Isonucleoside Isonucleosides are a type of nucleoside analogues. Generally, it can be considered that the nucleosides on the sugar ring have been transferred to the 2′ or 3′ position. This kind of modification has good chemical and enzymatic stability for there are no normal glycoside bonds in the structure [231]. Yang group applied single or multiple sites of D-/L-isonucleoside modification to three different aptamers, TBA, GBI-10, and AS1411 [232–234]. In all cases, some specific site (like loop areas) replaced with D-/L-isonucleoside showed a positive effect on stabilizing spatial conformation. The statistics of T_m value and circular dichroism (CD) spectra showed a certain correlation and proved the promotion in chemical stability. Serum stability test also demonstrated that the modified aptamers had a resistance against biodegradation. It's also worth noting that L-isonucleoside are more likely to enhance the biological activity of aptamers compared with D-isonucleoside (Figure 11.18).

11.6.3. Base Modifications

Modifications on the bases of nucleic acid are relatively rarely reported, but there are still some scientists who have made efforts to help us better understand the regulation pattern of gene silencing and contribute to alleviating the off-target effects caused by undesired binding with proteins and other mRNAs. Although new structures have been reported in succession (Figure 11.19), these modifications are at the

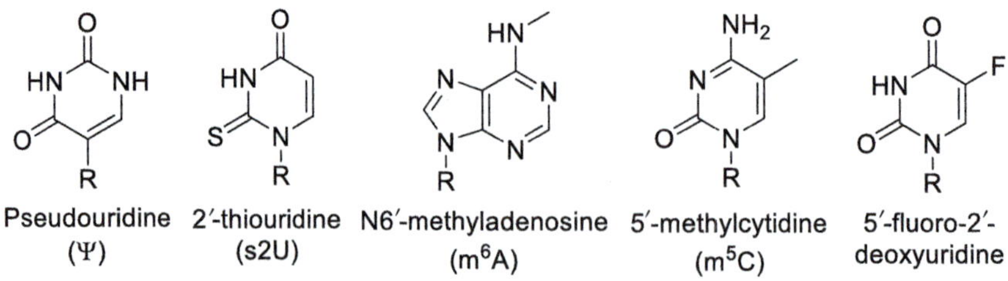

Figure 11.19 Modifications can be applied on nucleic acid base.

very early stage of research and development. While pharmaceutical corporations have reserved attitudes to these molecules because of the safety concerns that the metabolites of unnatural bases might be incorporated into human genome [212]. At the same time, base modifications may have a negative effect on the helix structure of siRNA, which will subsequently decrease the RNAi activity [235]. Hence, the presence of such modifications has not yet found wide application as similar effects can be obtained by using the two more common strategies mentioned above.

11.7. EPILOGUE

The current suite of chemical modifications provides good stability toward digestion by nucleases and achieves favorable binding to complementary sequences. Therefore, rather than achieving greater affinity or stability, the primary current needed for innovation is the development of improved methods for the delivery of oligonucleotides to disease-related tissues. In summary, the clinical translation of ON conjugates has been dominated by targeted GalNAc delivery. This revolutionary technology has promoted the growth and development of many companies and has also attracted the attention of the medical and investment communities. The next goal in this area should be to develop new ON conjugates that can treat diseases whose pathogenic sites are outside the liver. If an efficient and well-tolerated delivery system can be developed, it may lead to a change in the mode of disease intervention. This includes highly specific receptor-mediated uptake and endosomal escape, combined with improved protein binding to extend circulating half-life. Long-term plasma exposure is a key obstacle that needs to be resolved in the next generation of ON conjugates. The rapid clearance of these therapeutic agents reduces bioavailability and biodistribution, which at the same time reduces effective ON accumulation and distribution in extrahepatic target tissues. The use of different lipids or functional molecules as conjugation targeting moieties may be an effective way to confer the protein binding and PK properties required for extrahepatic delivery of ON. However, since the hepatotoxicity of heterologous lipophilic linkers may limit the dose, the tolerance of these lipid or functional molecule linkers needs to be considered. In this regard, the escape of the liver is also critical, and for this, innovative formulation strategies and shielding strategies may be required. All in all, a single chemical modification, nucleic acid conjugation, and vector delivery cannot perfectly solve all the problems of nucleic acid drug delivery *in vivo*. A combination of three strategies is needed to develop a new generation of efficient nucleic acid drug delivery system.

REFERENCES

1. Goodwin S, McPherson J D, McCombie W R. Coming of age: ten years of next-generation sequencing technologies. *Nat Rev Genet*, 2016, *17*(6): 333–351.

2. Stratton M R. Exploring the genomes of cancer cells: progress and promise. *Science*, 2011, *331*(6024): 1553–1558.
3. Kim J, Hu C, Moufawad El Achkar C, Black L E, Douville J, Larson A, Pendergast M K, Goldkind S F, Lee E A, Kuniholm A, Soucy A, Vaze J, Belur N R, Fredriksen K, Stojkovska I, Tsytsykova A, Armant M, DiDonato R L, Choi J, Cornelissen L, Pereira L M, Augustine E F, Genetti C A, Dies K, Barton B, Williams L, Goodlett B D, Riley B L, Pasternak A, Berry E R, Pflock K A, Chu S, Reed C, Tyndall K, Agrawal P B, Beggs A H, Grant P E, Urion D K, Snyder R O, Waisbren S E, Poduri A, Park P J, Patterson A, Biffi A, Mazzulli J R, Bodamer O, Berde C B, Yu T W. Patient-customized oligonucleotide therapy for a rare genetic disease. *New Engl J Med*, 2019, *381*(17): 1644–1652.
4. Jarvis J N, Frank M B. Functional genomics and rheumatoid arthritis: where have we been and where should we go? *Genome Med*, 2010, *2*(7): 44–48.
5. Deng Y B, Qin H J, Luo Y H, Liang Z R, Zou J J. Antiviral effect of hepatitis B virus S/C gene loci antisense locked nucleic acid on transgenic mice *in vivo*. *Genet Mol Res*, 2015, *14*(3): 10087–10095.
6. Kaczmarek J C, Kowalski P S, Anderson D G. Advances in the delivery of RNA therapeutics: from concept to clinical reality. *Genome Med*, 2017, *9*(1): 60–75.
7. Yin W, Rogge M. Targeting RNA: a transformative therapeutic strategy. *Clin Transl Sci*, 2019, *12*(2): 98–112.
8. Wilson R C, Doudna J A. Molecular mechanisms of RNA interference. *Annu Rev Biophys*, 2013, *42*(42): 217–239.
9. Dowdy S F. Overcoming cellular barriers for RNA therapeutics. *Nat Biotechnol*, 2017, *35*(3): 222–229.
10. Gujrati M, Vaidya A, Lu Z R. Multifunctional pH-sensitive amino lipids for siRNA delivery. *Bioconjug Chem*, 2016, *27*(1): 19–35.
11. Perry C M, Balfour J A. Fomivirsen. *Drugs*, 1999, *57*(3): 375–380.
12. Crooke S T, Geary R S. Clinical pharmacological properties of mipomersen (Kynamro), a second generation antisense inhibitor of apolipoprotein B. *Br J Clin Pharmacol*, 2013, *76*(2): 269–276.
13. McClorey G, Wood M J. An overview of the clinical application of antisense oligonucleotides for RNA-targeting therapies. *Curr Opin Pharmacol*, 2015, *24*: 52–58.
14. Young C S, Pyle A D. Exon skipping therapy. *Cell*, 2016, *167*(5): 1144.
15. Hoy S M. Nusinersen: first global approval. *Drugs*, 2017, *77*(4): 473–479.
16. Benson M D, Waddington-Cruz M, Berk J L, Polydefkis M, Dyck P J, Wang A K, Planté-Bordeneuve V, Barroso F A, Merlini G, Obici L, Scheinberg M, Brannagan T H, 3rd, Litchy W J, Whelan C, Drachman B M, Adams D, Heitner S B, Conceição I, Schmidt H H, Vita G, Campistol J M, Gamez J, Gorevic P D, Gane E, Shah A M, Solomon S D, Monia B P, Hughes S G, Kwoh T J, McEvoy B W, Jung S W, Baker B F, Ackermann E J, Gertz M A, Coelho T. Inotersen treatment for patients with hereditary transthyretin amyloidosis. *New Engl J Med*, 2018, *379*(1): 22–31.
17. Witztum J L, Gaudet D, Freedman S D, Alexander V J, Digenio A, Williams K R, Yang Q, Hughes S G, Geary R S, Arca M, Stroes E S G, Bergeron J, Soran H, Civeira F, Hemphill L, Tsimikas S, Blom D J, O'Dea L, Bruckert E. Volanesorsen and triglyceride levels in familial chylomicronemia syndrome. *New Engl J Med*, 2019, *381*(6): 531–542.

18. Frank D E, Schnell F J, Akana C, El-Husayni S H, Desjardins C A, Morgan J, Charleston J S, Sardone V, Domingos J, Dickson G, Straub V, Guglieri M, Mercuri E, Servais L, Muntoni F. Increased dystrophin production with golodirsen in patients with Duchenne muscular dystrophy. *Neurology*, 2020, *94*(21): e2270–e2282.
19. Ledford H. Gene-silencing technology gets first drug approval after 20-year wait. *Nature*, 2018, *560*(7718): 291–292.
20. Adams D, Gonzalez-Duarte A, O'Riordan W D, Yang C C, Ueda M, Kristen A V, Tournev I, Schmidt H H, Coelho T, Berk J L, Lin K P, Vita G, Attarian S, Plante-Bordeneuve V, Mezei M M, Campistol J M, Buades J, Brannagan T H, Kim B J, Oh J, Parman Y, Sekijima Y, Hawkins P N, Solomon S D, Polydefkis M, Dyck P J, Gandhi P J, Goyal S, Chen J, Strahs A L, Nochur S V, Sweetser M T, Garg P P, Vaishnaw A K, Gollob J A, Suhr O B. Patisiran, an RNAi therapeutic, for hereditary transthyretin amyloidosis. *New Engl J Med*, 2018, *379*(1): 11–21.
21. Kristen A V, Ajroud-Driss S, Conceicao I, Gorevic P, Kyriakides T, Obici L. Patisiran, an RNAi therapeutic for the treatment of hereditary transthyretin-mediated amyloidosis. *Neurodegener Dis Manag*, 2019, *9*(1): 5–23.
22. Titze-de-Almeida S S, Brandao P R P, Faber I, Titze-de-Almeida R. Leading RNA interference therapeutics part 1: silencing hereditary transthyretin amyloidosis, with a focus on Patisiran. *Mol Diagn Ther*, 2020, *24*(1): 49–59.
23. Bissell D M, Anderson K E, Bonkovsky H L. Porphyria. *New Engl J Med*, 2017, *377*(9): 862–872.
24. Sardh E, Harper P, Balwani M, Stein P, Rees D, Bissell D M, Desnick R, Parker C, Phillips J, Bonkovsky H L, Vassiliou D, Penz C, Chan-Daniels A, He Q L, Querbes W, Fitzgerald K, Kim J B, Garg P, Vaishnaw A, Simon A R, Anderson K E. Phase 1 trial of an RNA interference therapy for acute intermittent porphyria. *New Engl J Med*, 2019, *380*(6): 549–558.
25. de Paula Brandao P R, Titze-de-Almeida S S, Titze-de-Almeida R. Leading RNA interference therapeutics part 2: silencing delta-aminolevulinic acid synthase 1, with a focus on Givosiran. *Mol Diagn Ther*, 2020, *24*(1): 61–68.
26. Wang B, Rudnick S, Cengia B, Bonkovsky H L. Acute hepatic porphyrias: review and recent progress. *Hepatol Commun*, 2019, *3*(2): 193–206.
27. Scott L J. Givosiran: first approval. *Drugs*, 2020, *80*(3): 335–339.
28. Scott L J, Keam S J. Lumasiran: first approval. *Drugs*, 2021, https://doi.org/10.1007/s40265-020-01463-0.
29. Kosmas C E, Muñoz Estrella A, Skavdis A, Peña Genao E, Martinez I, Guzman E. Inclisiran for the treatment of cardiovascular disease: a short review on the emerging data and therapeutic potential. *Ther Clin Risk Manag*, 2020, *16*: 1031–1037.
30. Rancati G, Moffat J, Typas A, Pavelka N. Emerging and evolving concepts in gene essentiality. *Nat Rev Genet*, 2017, *19*: 34–49.
31. Woodcock J, Marks P. Drug regulation in the era of individualized therapies. *New Engl J Med*, 2019, *381*(17): 1678–1680.
32. Mühlbäck A, Lindenberg K S, Saft C, Priller J, Landwehrmeyer G B. [Gene-selective treatment approaches for Huntington's disease]. *Nervenarzt*, 2020, *91*(4): 303–311.
33. Chen J J. Overview of current and emerging therapies for amytrophic lateral sclerosis. *Am J Manag Care*, 2020, *26*(S9): S191–s197.

34. Debacker A J, Voutila J, Catley M, Blakey D, Habib N. Delivery of oligonucleotides to the liver with GalNAc: from research to registered therapeutic drug. *Mol Ther*, 2020, *28*(8): 1759–1771.
35. Alkhalil M. Lipoprotein(a) reduction in persons with cardiovascular disease. *New Engl J Med*, 2020, *382*(21): e65.
36. Greuter T, Rogler G. Alicaforsen in the treatment of pouchitis. *Immunotherapy*, 2017, *9*(14): 1143–1152.
37. Demirjian S, Ailawadi G, Polinsky M, Bitran D, Silberman S, Shernan S K, Burnier M, Hamilton M, Squiers E, Erlich S, Rothenstein D, Khan S, Chawla L S. Safety and tolerability study of an intravenously administered small interfering ribonucleic acid (siRNA) post on-pump cardiothoracic surgery in patients at risk of acute kidney injury. *Kidney Int Rep*, 2017, *2*(5): 836–843.
38. Janas M M, Zlatev I, Liu J, Jiang Y F, Barros S A, Sutherland J E, Davis W P, Liu J X, Brown C R, Liu X M, Schlegel M K, Blair L, Zhang X M, Das B, Tran C, Aluri K, Li J, Agarwal S, Indrakanti R, Charisse K, Nair J, Matsuda S, Rajeev K G, Zimmermann T, Sepp-Lorenzino L, Xu Y X, Akinc A, Fitzgerald K, Vaishnaw A K, Smith P F, Manoharan M, Jadhav V, Wu J T, Maier M A. Safety evaluation of 2′-deoxy-2′-fluoro nucleotides in GalNAc-siRNA conjugates. *Nucleic Acids Res*, 2019, *47*(7): 3306–3320.
39. Pasi K J, Rangarajan S, Georgiev P, Mant T, Creagh M D, Lissitchkov T, Bevan D, Austin S, Hay C R, Hegemann I, Kazmi R, Chowdary P, Gercheva-Kyuchukova L, Mamonov V, Timofeeva M, Soh C H, Garg P, Vaishnaw A, Akinc A, Sorensen B, Ragni M V. Targeting of antithrombin in hemophilia A or B with RNAi therapy. *New Engl J Med*, 2017, *377*(9): 819–828.
40. Dindo M, Conter C, Oppici E, Ceccarelli V, Marinucci L, Cellini B. Molecular basis of primary hyperoxaluria: clues to innovative treatments. *Urolithiasis*, 2019, *47*(1): 67–78.
41. Moreno-Montanes J, Bleau A M, Jimenez A I. Tivanisiran, a novel siRNA for the treatment of dry eye disease. *Expert Opin Investig Drugs*, 2018, *27*(4): 421–426.
42. Catuogno S, Esposito C L, Condorelli G, de Franciscis V. Nucleic acids delivering nucleic acids. *Adv Drug Deliv Rev*, 2018, *134*: 79–93.
43. Shen X, Corey D R. Chemistry, mechanism and clinical status of antisense oligonucleotides and duplex RNAs. *Nucleic Acids Res*, 2018, *46*(4): 1584–1600.
44. Crooke S T, Witztum J L, Bennett C F, Baker B F. RNA-targeted therapeutics. *Cell Metab*, 2018, *27*(4): 714–739.
45. Martinez M G, Villeret F, Testoni B, Zoulim F. Can we cure hepatitis B virus with novel direct-acting antivirals? *Liver Int*, 2020, *40* (S1): 27–34.
46. Javanbakht H, Mueller H, Walther J, Zhou X, Lopez A, Pattupara T, Blaising J, Pedersen L, Albæk N, Jackerott M, Shi T, Ploix C, Driessen W, Persson R, Ravn J, Young J A T, Ottosen S. Liver-targeted anti-HBV single-stranded oligonucleotides with locked nucleic acid potently reduce HBV gene expression *in vivo*. *Mol Ther Nucleic Acids*, 2018, *11*: 441–454.
47. Ämmälä C, Drury W J, 3rd, Knerr L, Ahlstedt I, Stillemark-Billton P, Wennberg-Huldt C, Andersson E M, Valeur E, Jansson-Löfmark R, Janzén D, Sundström L, Meuller J, Claesson J, Andersson P, Johansson C, Lee R G, Prakash T P, Seth P P, Monia B P, Andersson S. Targeted delivery of antisense oligonucleotides to pancreatic β-cells. *Sci Adv*, 2018, *4*(10): eaat3386.

48. Sugo T, Terada M, Oikawa T, Miyata K, Nishimura S, Kenjo E, Ogasawara-Shimizu M, Makita Y, Imaichi S, Murata S, Otake K, Kikuchi K, Teratani M, Masuda Y, Kamei T, Takagahara S, Ikeda S, Ohtaki T, Matsumoto H. Development of antibody-siRNA conjugate targeted to cardiac and skeletal muscles. *J Control Release*, 2016, *237*: 1–13.
49. Jayaraman M, Ansell S M, Mui B L, Tam Y K, Chen J, Du X, Butler D, Eltepu L, Matsuda S, Narayanannair J K, Rajeev K G, Hafez I M, Akinc A, Maier M A, Tracy M A, Cullis P R, Madden T D, Manoharan M, Hope M J. Maximizing the potency of siRNA lipid nanoparticles for hepatic gene silencing *in vivo*. *Angew Chem Int Ed Engl*, 2012, *51*(34): 8529–8533.
50. Lopatin U. Drugs in the pipeline for HBV. *Clin Liver Dis*, 2019, *23*(3): 535–555.
51. Grundy S M, Cleeman J I, Merz C N B, Brewer H B, Clark L T, Hunninghake D B, Pasternak R C, Smith S C, Stone N J, Cholesterol C C N. Implications of recent clinical trials for the national cholesterol education program adult treatment panel III guidelines. *Arterioscl Throm Vas*, 2004, *24*(8): E149–E161.
52. Horton J D, Cohen J C, Hobbs H H. Molecular biology of PCSK9: its role in LDL metabolism. *Trends Biochem Sci*, 2007, *32*(2): 71–77.
53. Frank-Kamenetsky M, Grefhorst A, Anderson N N, Racie T S, Bramlage B, Akinc A, Butler D, Charisse K, Dorkin R, Fan Y, Gamba-Vitalo C, Hadwiger P, Jayaraman M, John M, Jayaprakash K N, Maier M, Nechev L, Rajeev K G, Read T, Rohl I, Soutschek J, Tan P, Wong J, Wang G, Zimmermann T, de Fougerolles A, Vornlocher H P, Langer R, Anderson D G, Manoharan M, Koteliansky V, Horton J D, Fitzgerald K. Therapeutic RNAi targeting PCSK9 acutely lowers plasma cholesterol in rodents and LDL cholesterol in nonhuman primates. *P Natl Acad Sci USA*, 2008, *105*(33): 11915–11920.
54. Fitzgerald K, Frank-Kamonetsky M, Shulga-Morskaya S, Liebow A, Bettencourt B R, Sutherland J E, Hutabarat R M, Clausen V A, Karsten V, Cehelsky J, Nochur S V, Kotelianski V, Horton J, Mant T, Chiesa J, Ritter J, Munisamy M, Vaishnaw A K, Gollob J A, Simon A. Effect of an RNA interference drug on the synthesis of proprotein convertase subtilisin/kexin type 9 (PCSK9) and the concentration of serum LDL cholesterol in healthy volunteers: a randomised, single-blind, placebo-controlled, phase 1 trial. *Lancet*, 2014, *383*(9911): 60–68.
55. Geisbert T W, Lee A C H, Robbins M, Geisbert J B, Honko A N, Sood V, Johnson J C, de Jong S, Tavakoli I, Judge A, Hensley L E, MacLachlan I. Postexposure protection of non-human primates against a lethal Ebola virus challenge with RNA interference: a proof-of-concept study. *Lancet*, 2010, *375*(9729): 1896–1905.
56. Choi J H, Croyle M A. Emerging targets and novel approaches to Ebola virus prophylaxis and treatment. *Biodrugs*, 2013, *27*(6): 565–583.
57. Geisbert T W, Hensley L E, Kagan E, Yu E Y Z, Geisbert J B, Daddario-DiCaprio K, Fritz E A, Jahrling P B, McClintock K, Phelps J R, Lee A C H, Judge A, Jeffs L B, MacLachlan I. Postexposure protection of guinea pigs against a lethal Ebola virus challenge is conferred by RNA interference. *J Infect Dis*, 2006, *193*(12): 1650–1657.
58. Scott J T, Sharma R, Meredith L W, Dunning J, Moore C E, Sahr F, Ward S, Goodfellow I, Horby P, RAPIDE-TKM trial team. Pharmacokinetics of TKM-130803 in Sierra Leonean patients with Ebola virus disease: plasma concentrations exceed target levels, with drug accumulation in the most severe patients. *EBioMedicine*, 2020, *52*: 102601.

59. Dunning J, Sahr F, Rojek A, Gannon F, Carson G, Idriss B, Massaquoi T, Gandi R, Joseph S, Osman H K, Brooks T J, Simpson A J, Goodfellow I, Thorne L, Arias A, Merson L, Castle L, Howell-Jones R, Pardinaz-Solis R, Hope-Gill B, Ferri M, Grove J, Kowalski M, Stepniewska K, Lang T, Whitehead J, Olliaro P, Samai M, Horby P W, RAPIDE-TKM trial team. Experimental treatment of Ebola virus disease with TKM-130803: a single-arm phase 2 clinical trial. *PLoS Med*, 2016, *13*(4): e1001997.
60. Sato Y, Murase K, Kato J, Kobune M, Sato T, Kawano Y, Takimoto R, Takada K, Miyanishi K, Matsunaga T, Takayama T, Niitsu Y. Resolution of liver cirrhosis using vitamin A-coupled liposomes to deliver siRNA against a collagen-specific chaperone. *Nat Biotechnol*, 2008, *26*(4): 431–442.
61. Bennett C F. Therapeutic antisense oligonucleotides are coming of age. *Annu Rev Med*, 2019, *70*: 307–321.
62. Crooke S T. Molecular mechanisms of antisense oligonucleotides. *Nucleic Acid Ther*, 2017, *27*(2): 70–77.
63. Pirollo K F, Rait A, Sleer L S, Chang E H. Antisense therapeutics: from theory to clinical practice. *Pharmacol Ther*, 2003, *99*(1): 55–77.
64. Salomon W E, Jolly S M, Moore M J, Zamore P D, Serebrov V. Single-molecule imaging reveals that argonaute reshapes the binding properties of its nucleic acid guides. *Cell*, 2016, *166*(2): 517–520.
65. Lima W F, Murray H, Nichols J G, Wu H, Sun H, Prakash T P, Berdeja A R, Gaus H J, Crooke S T. Human Dicer binds short single-strand and double-strand RNA with high affinity and interacts with different regions of the nucleic acids. *J Biol Chem*, 2009, *284*(4): 2535–2548.
66. Shen W, Liang X H, Sun H, Crooke S T. 2′-Fluoro-modified phosphorothioate oligonucleotide can cause rapid degradation of P54nrb and PSF. *Nucleic Acids Res*, 2015, *43*(9): 4569–4578.
67. Zon G. Brief overview of control of genetic expression by antisense oligonucleotides and *in vivo* applications. Prospects for neurobiology. *Mol Neurobiol*, 1995, *10*(2–3): 219–229.
68. Geary R S. Antisense oligonucleotide pharmacokinetics and metabolism. *Expert Opin Drug Metab Toxicol*, 2009, *5*(4): 381–391.
69. Yu R Z, Warren M S, Watanabe T, Nichols B, Jahic M, Huang J, Burkey J, Geary R S, Henry S P, Wang Y. Lack of interactions between an antisense oligonucleotide with 2′-*O*-(2-methoxyethyl) modifications and major drug transporters. *Nucleic Acid Ther*, 2016, *26*(2): 111–117.
70. Watanabe T A, Geary R S, Levin A A. Plasma protein binding of an antisense oligonucleotide targeting human ICAM-1 (ISIS 2302). *Oligonucleotides*, 2006, *16*(2): 169–180.
71. Yu R Z, Kim T W, Hong A, Watanabe T A, Gaus H J, Geary R S. Cross-species pharmacokinetic comparison from mouse to man of a second-generation antisense oligonucleotide, ISIS 301012, targeting human apolipoprotein B-100. *Drug Metab Dispos*, 2007, *35*(3): 460–468.
72. Geary R S, Wancewicz E, Matson J, Pearce M, Siwkowski A, Swayze E, Bennett F. Effect of dose and plasma concentration on liver uptake and pharmacologic activity of a 2′-methoxyethyl modified chimeric antisense oligonucleotide targeting PTEN. *Biochem Pharmacol*, 2009, *78*(3): 284–291.
73. Yu R Z, Grundy J S, Geary R S. Clinical pharmacokinetics of second generation antisense oligonucleotides. *Expert Opin Drug Metab Toxicol*, 2013, *9*(2): 169–182.

74. Yu R Z, Grundy J S, Henry S P, Kim T W, Norris D A, Burkey J, Wang Y, Vick A, Geary R S. Predictive dose-based estimation of systemic exposure multiples in mouse and monkey relative to human for antisense oligonucleotides with 2′-*O*-(2-methoxyethyl) modifications. *Mol Ther Nucleic Acids*, 2015, *4*(1): e218.
75. Tillman L G, Geary R S, Hardee G E. Oral delivery of antisense oligonucleotides in man. *J Pharm Sci*, 2008, *97*(1): 225–236.
76. Geary R S, Norris D, Yu R, Bennett C F. Pharmacokinetics, biodistribution and cell uptake of antisense oligonucleotides. *Adv Drug Deliv Rev*, 2015, *87*: 46–51.
77. Cirak S, Arechavala-Gomeza V, Guglieri M, Feng L, Torelli S, Anthony K, Abbs S, Garralda M E, Bourke J, Wells D J, Dickson G, Wood M J, Wilton S D, Straub V, Kole R, Shrewsbury S B, Sewry C, Morgan J E, Bushby K, Muntoni F. Exon skipping and dystrophin restoration in patients with Duchenne muscular dystrophy after systemic phosphorodiamidate morpholino oligomer treatment: an open-label, phase 2, dose-escalation study. *Lancet*, 2011, *378*(9791): 595–605.
78. Solano E C, Kornbrust D J, Beaudry A, Foy J W, Schneider D J, Thompson J D. Toxicological and pharmacokinetic properties of QPI-1007, a chemically modified synthetic siRNA targeting caspase 2 mRNA, following intravitreal injection. *Nucleic Acid Ther*, 2014, *24*(4): 258–266.
79. Finkel R S, Chiriboga C A, Vajsar J, Day J W, Montes J, De Vivo D C, Yamashita M, Rigo F, Hung G, Schneider E, Norris D A, Xia S, Bennett C F, Bishop K M. Treatment of infantile-onset spinal muscular atrophy with nusinersen: a phase 2, open-label, dose-escalation study. *Lancet*, 2016, *388*(10063): 3017–3026.
80. Zhou Y, Zhang C, Liang W. Development of RNAi technology for targeted therapy – a track of siRNA based agents to RNAi therapeutics. *J Control Release*, 2014, *193*: 270–281.
81. Wang Y, Grainger D W. Barriers to advancing nanotechnology to better improve and translate nanomedicines. *Front Chem Sci Eng*, 2014, *8*(3): 265–275.
82. Krieg A M, Yi A K, Matson S, Waldschmidt T J, Bishop G A, Teasdale R, Koretzky G A, Klinman D M. CpG motifs in bacterial DNA trigger direct B-cell activation. *Nature*, 1995, *374*(6522): 546–549.
83. Alexopoulou L, Holt A C, Medzhitov R, Flavell R A. Recognition of double-stranded RNA and activation of NF kappaB by Toll-like receptor 3. *Nature*, 2001, *413*(6857): 732–738.
84. Whitehead K A, Langer R, Anderson D G. Knocking down barriers: advances in siRNA delivery. *Nat Rev Drug Discov*, 2009, *8*(2): 129–138.
85. Reischl D, Zimmer A. Drug delivery of siRNA therapeutics: potentials and limits of nanosystems. *Nanomedicine*, 2009, *5*(1): 8–20.
86. Jiang X, Wang G, Liu R, Wang Y, Wang Y, Qiu X, Gao X. RNase non-sensitive and endocytosis independent siRNA delivery system: delivery of siRNA into tumor cells and high efficiency induction of apoptosis. *Nanoscale*, 2013, *5*(16): 7256–7264.
87. Saw P E, Song E W. siRNA therapeutics: a clinical reality. *Sci China Life Sci*, 2020, *63*(4): 485–500.
88. Soutschek J, Akinc A, Bramlage B, Charisse K, Constien R, Donoghue M, Elbashir S, Geick A, Hadwiger P, Harborth J, John M, Kesavan V, Lavine G, Pandey R K, Racie T, Rajeev K G, Röhl I, Toudjarska I, Wang G, Wuschko S, Bumcrot D, Koteliansky V, Limmer S, Manoharan M, Vornlocher H P. Therapeutic silencing of an endogenous gene by systemic administration of modified siRNAs. *Nature*, 2004, *432*(7014): 173–178.

89. Morrissey D V, Lockridge J A, Shaw L, Blanchard K, Jensen K, Breen W, Hartsough K, Machemer L, Radka S, Jadhav V, Vaish N, Zinnen S, Vargeese C, Bowman K, Shaffer C S, Jeffs L B, Judge A, MacLachlan I, Polisky B. Potent and persistent *in vivo* anti-HBV activity of chemically modified siRNAs. *Nat Biotechnol*, 2005, *23*(8): 1002–1007.
90. Singh A, Trivedi P, Jain N K. Advances in siRNA delivery in cancer therapy. *Artif Cells Nanomed Biotechnol*, 2018, *46*(2): 274–283.
91. Kotterman M A, Schaffer D V. Engineering adeno-associated viruses for clinical gene therapy. *Nat Rev Genet*, 2014, *15*(7): 445–451.
92. Harrison S C. Viral membrane fusion. *Virology*, 2015, *479–480*: 498–507.
93. Bulbake U, Doppalapudi S, Kommineni N, Khan W. Liposomal formulations in clinical use: an updated review. *Pharmaceutics*, 2017, *9*(2): 12–44.
94. Alexander M Y, Akhurst R J. Liposome-medicated gene transfer and expression via the skin. *Hum Mol Genet*, 1995, *4*(12): 2279–2285.
95. Zhou J, Shum K T, Burnett J C, Rossi J J. Nanoparticle-based delivery of RNAi therapeutics: progress and challenges. *Pharmaceuticals (Basel)*, 2013, *6*(1): 85–107.
96. Agrawal M, Ajazuddin, Tripathi D K, Saraf S, Saraf S, Antimisiaris S G, Mourtas S, Hammarlund-Udenaes M, Alexander A. Recent advancements in liposomes targeting strategies to cross blood-brain barrier (BBB) for the treatment of Alzheimer's disease. *J Control Release*, 2017, *260*: 61–77.
97. Morille M, Passirani C, Vonarbourg A, Clavreul A, Benoit J P. Progress in developing cationic vectors for non-viral systemic gene therapy against cancer. *Biomaterials*, 2008, *29*(24-25): 3477–3496.
98. Creusat G, Rinaldi A S, Weiss E, Elbaghdadi R, Remy J S, Mulherkar R, Zuber G. Proton sponge trick for pH-sensitive disassembly of polyethylenimine-based siRNA delivery systems. *Bioconjug Chem*, 2010, *21*(5): 994–1002.
99. Choi C K K, Zhang L, Choi C H J. Efficient siRNA delivery with non-cationic carriers. *Nat Biomed Eng*, 2018, *2*(5): 275–276.
100. Titze-de-Almeida R, David C, Titze-de-Almeida S S. The race of 10 synthetic RNAi-based drugs to the pharmaceutical market. *Pharm Res*, 2017, *34*(7): 1339–1363.
101. Yang J P, Huang L. Overcoming the inhibitory effect of serum on lipofection by increasing the charge ratio of cationic liposome to DNA. *Gene Ther*, 1997, *4*(9): 950–960.
102. Ceballos C, Prata C A, Giorgio S, Garzino F, Payet D, Barthélémy P, Grinstaff M W, Camplo M. Cationic nucleoside lipids based on a 3-nitropyrrole universal base for siRNA delivery. *Bioconjug Chem*, 2009, *20*(2): 193–196.
103. Luvino D, Khiati S, Oumzil K, Rocchi P, Camplo M, Barthélémy P. Efficient delivery of therapeutic small nucleic acids to prostate cancer cells using ketal nucleoside lipid nanoparticles. *J Control Release*, 2013, *172*(3): 954–961.
104. Montis C, Milani S, Berti D, Baglioni P. Complexes of nucleolipid liposomes with single-stranded and double-stranded nucleic acids. *J Colloid Interface Sci*, 2012, *373*(1): 57–68.
105. Montis C, Gerelli Y, Fragneto G, Nylander T, Baglioni P, Berti D. Nucleolipid bilayers: a quartz crystal microbalance and neutron reflectometry study. *Colloids Surf B Biointerfaces*, 2016, *137*: 203–213.
106. Cuomo F, Ceglie A, Colafemmina G, Germani R, Savelli G, Lopez F. Polyadenylic acid binding on cationic liposomes doped with the non-ionic nucleolipid Lauroyl Uridine. *Colloids Surf B Biointerfaces*, 2011, *82*(2): 277–282.

107. Ma Y, Zhu Y J, Wang C, Pan D L, Liu S, Yang M Y, Xiao Z P, Yang X T, Zhao W T, Zhou X Y, Li Y D, Pan Y F, Sun J, Wang S H, Guan Z, Zhang L H, Yang Z J. Annealing novel nucleobase-lipids with oligonucleotides or plasmid DNA based on H-bonding or pi-pi interaction: assemblies and transfections. *Biomaterials*, 2018, *178*: 147–157.
108. Ma X F, Sun J, Qiu C, Wu Y F, Zheng Y, Yu M Z, Pei X W, Wei L, Niu Y J, Pang W H, Yang Z J, Wang J C, Zhang Q. The role of disulfide-bridge on the activities of H-shape gemini-like cationic lipid based siRNA delivery. *J Control Release*, 2016, *235*: 99–111.
109. Ma Y, Zhao W T, Li Y D, Pan Y F, Wang S H, Zhu Y J, Kong L X, Guan Z, Wang J C, Zhang L H, Yang Z J. Structural optimization and additional targets identification of antisense oligonucleotide G3139 encapsulated in a neutral cytidinyl-lipid combined with a cationic lipid *in vitro* and *in vivo*. *Biomaterials*, 2019, *197*: 182–193.
110. Deleavey G F, Damha M J. Designing chemically modified oligonucleotides for targeted gene silencing. *Chem Biol*, 2012, *19*(8): 937–954.
111. Zhang S, Zhao Y, Zhi D, Zhang S. Non-viral vectors for the mediation of RNAi. *Bioorg Chem*, 2012, *40*(1): 10–18.
112. Ku S H, Jo S D, Lee Y K, Kim K, Kim S H. Chemical and structural modifications of RNAi therapeutics. *Adv Drug Deliv Rev*, 2016, *104*: 16–28.
113. Benizri S, Gissot A, Martin A, Vialet B, Grinstaff M W, Barthélémy P. Bioconjugated oligonucleotides: recent developments and therapeutic applications. *Bioconjug Chem*, 2019, *30*(2): 366–383.
114. Weng Y, Xiao H, Zhang J, Liang X J, Huang Y. RNAi therapeutic and its innovative biotechnological evolution. *Biotechnol Adv*, 2019, *37*(5): 801–825.
115. Setten R L, Rossi J J, Han S P. The current state and future directions of RNAi-based therapeutics. *Nat Rev Drug Discov*, 2019, *18*(6): 421–446.
116. Varki A. Biological roles of oligosaccharides: all of the theories are correct. *Glycobiology*, 1993, *3*(2): 97–130.
117. Gabius H J, Siebert H C, André S, Jiménez-Barbero J, Rüdiger H. Chemical biology of the sugar code. *Chembiochem*, 2004, *5*(6): 740–764.
118. Mamidyala S K, Dutta S, Chrunyk B A, Préville C, Wang H, Withka J M, McColl A, Subashi T A, Hawrylik S J, Griffor M C, Kim S, Pfefferkorn J A, Price D A, Menhaji-Klotz E, Mascitti V, Finn M G. Glycomimetic ligands for the human asialoglycoprotein receptor. *J Am Chem Soc*, 2012, *134*(4): 1978–1981.
119. Zimmermann T S, Karsten V, Chan A, Chiesa J, Boyce M, Bettencourt B R, Hutabarat R, Nochur S, Vaishnaw A, Gollob J. Clinical proof of concept for a novel hepatocyte-targeting GalNAc-siRNA conjugate. *Mol Ther*, 2017, *25*(1): 71–78.
120. Huang Y. Preclinical and clinical advances of GalNAc-decorated nucleic acid therapeutics. *Mol Ther Nucleic Acids*, 2017, *6*: 116–132.
121. Østergaard M E, Yu J, Kinberger G A, Wan W B, Migawa M T, Vasquez G, Schmidt K, Gaus H J, Murray H M, Low A, Swayze E E, Prakash T P, Seth P P. Efficient synthesis and biological evaluation of 5′-GalNAc conjugated antisense oligonucleotides. *Bioconjug Chem*, 2015, *26*(8): 1451–1455.
122. Cedillo I, Chreng D, Engle E, Chen L, McPherson A K, Rodriguez A A. Synthesis of 5′-GalNAc-conjugated oligonucleotides: a comparison of solid and solution-phase conjugation strategies. *Molecules*, 2017, *22*(8): 1356–1367.

123. Kinberger G A, Prakash T P, Yu J, Vasquez G, Low A, Chappell A, Schmidt K, Murray H M, Gaus H, Swayze E E, Seth P P. Conjugation of mono and di-GalNAc sugars enhances the potency of antisense oligonucleotides via ASGR mediated delivery to hepatocytes. *Bioorg Med Chem Lett*, 2016, *26*(15): 3690–3693.
124. Shemesh C S, Yu R Z, Warren M S, Liu M, Jahic M, Nichols B, Post N, Lin S, Norris D A, Hurh E, Huang J, Watanabe T, Henry S P, Wang Y. Assessment of the drug interaction potential of unconjugated and GalNAc(3)-conjugated 2′-MOE-ASOs. *Mol Ther Nucleic Acids*, 2017, *9*: 34–47.
125. Higuchi K, Monge J C, Lee N, Law S W, Brewer H B, Jr., Sakaguchi A Y, Naylor S L. The human apoB-100 gene: apoB-100 is encoded by a single copy gene in the human genome. *Biochem Biophys Res Commun*, 1987, *144*(3): 1332–1339.
126. Prakash T P, Yu J, Migawa M T, Kinberger G A, Wan W B, Ostergaard M E, Carty R L, Vasquez G, Low A, Chappell A, Schmidt K, Aghajan M, Crosby J, Murray H M, Booten S L, Hsiao J, Soriano A, Machemer T, Cauntay P, Burel S A, Murray S F, Gaus H, Graham M J, Swayze E E, Seth P P. Comprehensive structure-activity relationship of triantennary *N*-acetylgalactosamine conjugated antisense oligonucleotides for targeted delivery to hepatocytes. *J Med Chem*, 2016, *59*(6): 2718–2733.
127. Roberts T C, Langer R, Wood M J A. Advances in oligonucleotide drug delivery. *Nat Rev Drug Discov*, 2020, *19*(10): 673–694.
128. Matsuda S, Keiser K, Nair J K, Charisse K, Manoharan R M, Kretschmer P, Peng C G, Kel'in A V, Kandasamy P, Willoughby J L, Liebow A, Querbes W, Yucius K, Nguyen T, Milstein S, Maier M A, Rajeev K G, Manoharan M. siRNA conjugates carrying sequentially assembled trivalent N-acetylgalactosamine linked through nucleosides elicit robust gene silencing *in vivo* in hepatocytes. *ACS Chem Biol*, 2015, *10*(5): 1181–1187.
129. Gallas A, Alexander C, Davies M C, Puri S, Allen S. Chemistry and formulations for siRNA therapeutics. *Chem Soc Rev*, 2013, *42*(20): 7983–7997.
130. Alterman J F, Hall L M, Coles A H, Hassler M R, Didiot M C, Chase K, Abraham J, Sottosanti E, Johnson E, Sapp E, Osborn M F, Difiglia M, Aronin N, Khvorova A. Hydrophobically modified siRNAs silence Huntingtin mRNA in primary neurons and mouse brain. *Mol Ther Nucleic Acids*, 2015, *4*(12): e266.
131. Haraszti R A, Roux L, Coles A H, Turanov A A, Alterman J F, Echeverria D, Godinho B, Aronin N, Khvorova A. 5′-Vinylphosphonate improves tissue accumulation and efficacy of conjugated siRNAs *in vivo*. *Nucleic Acids Res*, 2017, *45*(13): 7581–7592.
132. Havelund S, Plum A, Ribel U, Jonassen I, Vølund A, Markussen J, Kurtzhals P. The mechanism of protraction of insulin detemir, a long-acting, acylated analog of human insulin. *Pharm Res*, 2004, *21*(8): 1498–1504.
133. Ipp E, Genter P, Childress K. Semaglutide and cardiovascular outcomes in patients with type 2 diabetes. *New Engl J Med*, 2017, *376*(9): 890–891.
134. Lund A, Knop F K, Vilsbøll T. Glucagon-like peptide-1 receptor agonists for the treatment of type 2 diabetes: differences and similarities. *Eur J Intern Med*, 2014, *25*(5): 407–414.
135. Jonassen I, Havelund S, Hoeg-Jensen T, Steensgaard D B, Wahlund P O, Ribel U. Design of the novel protraction mechanism of insulin degludec, an ultra-long-acting basal insulin. *Pharm Res*, 2012, *29*(8): 2104–2114.

136. Plank J, Bodenlenz M, Sinner F, Magnes C, Görzer E, Regittnig W, Endahl L A, Draeger E, Zdravkovic M, Pieber T R. A double-blind, randomized, dose-response study investigating the pharmacodynamic and pharmacokinetic properties of the long-acting insulin analog detemir. *Diabetes Care*, 2005, *28*(5): 1107–1112.
137. Heise T, Hövelmann U, Nosek L, Hermanski L, Bøttcher S G, Haahr H. Comparison of the pharmacokinetic and pharmacodynamic profiles of insulin degludec and insulin glargine. *Expert Opin Drug Metab Toxicol*, 2015, *11*(8): 1193–1201.
138. Tamborlane W V, Barrientos-Pérez M, Fainberg U, Frimer-Larsen H, Hafez M, Hale P M, Jalaludin M Y, Kovarenko M, Libman I, Lynch J L, Rao P, Shehadeh N, Turan S, Weghuber D, Barrett T. Liraglutide in children and adolescents with type 2 diabetes. *New Engl J Med*, 2019, *381*(7): 637–646.
139. Gotfredsen C F, Mølck A M, Thorup I, Nyborg N C, Salanti Z, Knudsen L B, Larsen M O. The human GLP-1 analogs liraglutide and semaglutide: absence of histopathological effects on the pancreas in nonhuman primates. *Diabetes*, 2014, *63*(7): 2486–2497.
140. Han J, Sun L, Chu Y, Li Z, Huang D, Zhu X, Qian H, Huang W. Design, synthesis, and biological activity of novel dicoumarol glucagon-like peptide 1 conjugates. *J Med Chem*, 2013, *56*(24): 9955–9968.
141. Lee C H, Lee S H, Kim J H, Noh Y H, Noh G J, Lee S W. Pharmacokinetics of a cholesterol-conjugated aptamer against the Hepatitis C Virus (HCV) NS5B protein. *Mol Ther Nucleic Acids*, 2015, *4*(10): e254.
142. Osborn M F, Coles A H, Biscans A, Haraszti R A, Roux L, Davis S, Ly S, Echeverria D, Hassler M R, Godinho B, Nikan M, Khvorova A. Hydrophobicity drives the systemic distribution of lipid-conjugated siRNAs via lipid transport pathways. *Nucleic Acids Res*, 2019, *47*(3): 1070–1081.
143. Peterson A M, Heemstra J M. Controlling self-assembly of DNA-polymer conjugates for applications in imaging and drug delivery. *Wiley Interdiscip Rev Nanomed Nanobiotechnol*, 2015, *7*(3): 282–297.
144. Lee S, Yoon J H, Jo I S, Oh J S, Pine D J, Shim T S, Yi G R. DNA-functionalized 100 nm polymer nanoparticles from block copolymer micelles. *Langmuir*, 2018, *34*(37): 11042–11048.
145. Pokholenko O, Gissot A, Vialet B, Bathany K, Thiéry A, Barthélémy P. Lipid oligonucleotide conjugates as responsive nanomaterials for drug delivery. *J Mater Chem B*, 2013, *1*(39): 5329–5334.
146. Liu H, Zhu Z, Kang H, Wu Y, Sefan K, Tan W. DNA-based micelles: synthesis, micellar properties and size-dependent cell permeability. *Chemistry*, 2010, *16*(12): 3791–3797.
147. Alemdaroglu F E, Alemdaroglu N C, Langguth P, Herrmann A. DNA block copolymer micelles – a combinatorial tool for cancer nanotechnology. *Adv Mater*, 2008, *20*(5): 899–902.
148. Wu C, Chen T, Han D, You M, Peng L, Cansiz S, Zhu G, Li C, Xiong X, Jimenez E, Yang C J, Tan W. Engineering of switchable aptamer micelle flares for molecular imaging in living cells. *ACS Nano*, 2013, *7*(7): 5724–5731.
149. Bäumer S, Bäumer N, Appel N, Terheyden L, Fremerey J, Schelhaas S, Wardelmann E, Buchholz F, Berdel W E, Müller-Tidow C. Antibody-mediated delivery of anti-KRAS-siRNA *in vivo* overcomes therapy resistance in colon cancer. *Clin Cancer Res*, 2015, *21*(6): 1383–1394.

150. Luo C, Sun J, Sun B, He Z. Prodrug-based nanoparticulate drug delivery strategies for cancer therapy. *Trends Pharmacol Sci*, 2014, *35*(11): 556–566.
151. Fröhlich T, Edinger D, Russ V, Wagner E. Stabilization of polyplexes via polymer crosslinking for efficient siRNA delivery. *Eur J Pharm Sci*, 2012, *47*(5): 914–920.
152. Huang C H, Takemoto H, Nomoto T, Tomoda K, Matsui M, Nishiyama N. Utility of the 2-nitrobenzenesulfonamide group as a chemical linker for enhanced extracellular stability and cytosolic cleavage in siRNA-conjugated polymer systems. *ChemMedChem*, 2017, *12*(1): 19–22.
153. Fonseca S B, Pereira M P, Kelley S O. Recent advances in the use of cell-penetrating peptides for medical and biological applications. *Adv Drug Deliv Rev*, 2009, *61*(11): 953–964.
154. Wyman T B, Nicol F, Zelphati O, Scaria P V, Plank C, Szoka F C, Jr. Design, synthesis, and characterization of a cationic peptide that binds to nucleic acids and permeabilizes bilayers. *Biochemistry*, 1997, *36*(10): 3008–3017.
155. Lehto T, Ezzat K, Wood M J A, El Andaloussi S. Peptides for nucleic acid delivery. *Adv Drug Deliv Rev*, 2016, *106*(Pt A): 172–182.
156. Moschos S A, Jones S W, Perry M M, Williams A E, Erjefalt J S, Turner J J, Barnes P J, Sproat B S, Gait M J, Lindsay M A. Lung delivery studies using siRNA conjugated to TAT(48-60) and penetratin reveal peptide induced reduction in gene expression and induction of innate immunity. *Bioconjug Chem*, 2007, *18*(5): 1450–1459.
157. Ishihara T, Goto M, Kodera K, Kanazawa H, Murakami Y, Mizushima Y, Higaki M. Intracellular delivery of siRNA by cell-penetrating peptides modified with cationic oligopeptides. *Drug Deliv*, 2009, *16*(3): 153–159.
158. Muratovska A, Eccles M R. Conjugate for efficient delivery of short interfering RNA (siRNA) into mammalian cells. *FEBS Lett*, 2004, *558*(1-3): 63–68.
159. Meyer M, Dohmen C, Philipp A, Kiener D, Maiwald G, Scheu C, Ogris M, Wagner E. Synthesis and biological evaluation of a bioresponsive and endosomolytic siRNA-polymer conjugate. *Mol Pharm*, 2009, *6*(3): 752–762.
160. Lee S H, Castagner B, Leroux J C. Is there a future for cell-penetrating peptides in oligonucleotide delivery? *Eur J Pharm Biopharm*, 2013, *85*(1): 5–11.
161. Fan X M, Zhang Y F, Liu X J, He H Y, Ma Y, Sun J, Huang Y, Wang X F, Wu Y, Zhang L H, Yang Z J. Biological properties of a 3′,3″-bis-peptide-siRNA conjugate *in vitro* and *in vivo*. *Bioconjug Chem*, 2016, *27*(4): 1131–1142.
162. Sun J, Qiu C, Diao Y P, Wei W, Jin H W, Zheng Y, Wang J C, Zhang L H, Yang Z J. Delivery pathway regulation of 3′,3″-bis-peptide-siRNA conjugate via nanocarrier architecture engineering. *Mol Ther Nucleic Acids*, 2018, *10*: 75–90.
163. Han T H, Zhao B. Absorption, distribution, metabolism, and excretion considerations for the development of antibody-drug conjugates. *Drug Metab Dispos*, 2014, *42*(11): 1914–1920.
164. Hedrich W D, Fandy T E, Ashour H M, Wang H, Hassan H E. Antibody-drug conjugates: pharmacokinetic/pharmacodynamic modeling, preclinical characterization, clinical studies, and lessons learned. *Clin Pharmacokinet*, 2018, *57*(6): 687–703.
165. Ma Y, Kowolik C M, Swiderski P M, Kortylewski M, Yu H, Horne D A, Jove R, Caballero O L, Simpson A J, Lee F T, Pillay V, Scott A M. Humanized Lewis-Y specific antibody based delivery of STAT3 siRNA. *ACS Chem Biol*, 2011, *6*(9): 962–970.

166. Cuellar T L, Barnes D, Nelson C, Tanguay J, Yu S F, Wen X, Scales S J, Gesch J, Davis D, van Brabant Smith A, Leake D, Vandlen R, Siebel C W. Systematic evaluation of antibody-mediated siRNA delivery using an industrial platform of THIOMAB-siRNA conjugates. *Nucleic Acids Res*, 2015, *43*(2): 1189–1203.
167. Wilson D R, Sen R, Sunshine J C, Pardoll D M, Green J J, Kim Y J. Biodegradable STING agonist nanoparticles for enhanced cancer immunotherapy. *Nanomedicine*, 2018, *14*(2): 237–246.
168. Kaczmarek J C, Patel A K, Kauffman K J, Fenton O S, Webber M J, Heartlein M W, DeRosa F, Anderson D G. Polymer-lipid nanoparticles for systemic delivery of mRNA to the lungs. *Angew Chem Int Ed Engl*, 2016, *55*(44): 13808–13812.
169. Hess K L, Andorko J I, Tostanoski L H, Jewell C M. Polyplexes assembled from self-peptides and regulatory nucleic acids blunt toll-like receptor signaling to combat autoimmunity. *Biomaterials*, 2017, *118*: 51–62.
170. ur Rehman Z, Hoekstra D, Zuhorn I S. Mechanism of polyplex- and lipoplex-mediated delivery of nucleic acids: real-time visualization of transient membrane destabilization without endosomal lysis. *ACS Nano*, 2013, *7*(5): 3767–3777.
171. Tros de Ilarduya C, Sun Y, Düzgüneş N. Gene delivery by lipoplexes and polyplexes. *Eur J Pharm Sci*, 2010, *40*(3): 159–170.
172. Oupicky D. Design and development strategies of polymer materials for drug and gene delivery applications. *Adv Drug Deliv Rev*, 2008, *60*(9): 955–1094.
173. Li J, Xue S, Mao Z W. Nanoparticle delivery systems for siRNA-based therapeutics. *J Mater Chem B*, 2016, *4*(41): 6620–6639.
174. Gonzalez H, Hwang S J, Davis M E. New class of polymers for the delivery of macromolecular therapeutics. *Bioconjug Chem*, 1999, *10*(6): 1068–1074.
175. Bellocq N C, Pun S H, Jensen G S, Davis M E. Transferrin-containing, cyclodextrin polymer-based particles for tumor-targeted gene delivery. *Bioconjug Chem*, 2003, *14*(6): 1122–1132.
176. Mishra S, Webster P, Davis M E. PEGylation significantly affects cellular uptake and intracellular trafficking of non-viral gene delivery particles. *Eur J Cell Biol*, 2004, *83*(3): 97–111.
177. Bartlett D W, Davis M E. Physicochemical and biological characterization of targeted, nucleic acid-containing nanoparticles. *Bioconjug Chem*, 2007, *18*(2): 456–468.
178. Kanasty R, Dorkin J R, Vegas A, Anderson D. Delivery materials for siRNA therapeutics. *Nat Mater*, 2013, *12*(11): 967–977.
179. Davis M E. The first targeted delivery of siRNA in humans via a self-assembling, cyclodextrin polymer-based nanoparticle: from concept to clinic. *Mol Pharm*, 2009, *6*(3): 659–668.
180. Davis M E, Zuckerman J E, Choi C H, Seligson D, Tolcher A, Alabi C A, Yen Y, Heidel J D, Ribas A. Evidence of RNAi in humans from systemically administered siRNA via targeted nanoparticles. *Nature*, 2010, *464*(7291): 1067–1070.
181. Köping-Höggård M, Vårum K M, Issa M, Danielsen S, Christensen B E, Stokke B T, Artursson P. Improved chitosan-mediated gene delivery based on easily dissociated chitosan polyplexes of highly defined chitosan oligomers. *Gene Ther*, 2004, *11*(19): 1441–1452.
182. Raik S V, Poshina D N, Lyalina T A, Polyakov D S, Vasilyev V B, Kritchenkov A S, Skorik Y A. *N*-[4-(*N,N,N*-trimethylammonium)benzyl]chitosan chloride: synthesis, interaction with DNA and evaluation of transfection efficiency. *Carbohydr Polym*, 2018, *181*: 693–700.

183. Boussif O, Lezoualc'h F, Zanta M A, Mergny M D, Scherman D, Demeneix B, Behr J P. A versatile vector for gene and oligonucleotide transfer into cells in culture and *in vivo*: polyethylenimine. *Proc Natl Acad Sci U S A*, 1995, *92*(16): 7297–7301.
184. Guo J, Cheng W P, Gu J, Ding C, Qu X, Yang Z, O'Driscoll C. Systemic delivery of therapeutic small interfering RNA using a pH-triggered amphiphilic poly-l-lysine nanocarrier to suppress prostate cancer growth in mice. *Eur J Pharm Sci*, 2012, *45*(5): 521–532.
185. Khan A, Benboubetra M, Sayyed P Z, Ng K W, Fox S, Beck G, Benter I F, Akhtar S. Sustained polymeric delivery of gene silencing antisense ODNs, siRNA, DNAzymes and ribozymes: *in vitro* and *in vivo* studies. *J Drug Target*, 2004, *12*(6): 393–404.
186. Démoulins T, Milona P, Englezou P C, Ebensen T, Schulze K, Suter R, Pichon C, Midoux P, Guzmán C A, Ruggli N, McCullough K C. Polyethylenimine-based polyplex delivery of self-replicating RNA vaccines. *Nanomedicine*, 2016, *12*(3): 711–722.
187. Fischer D, Li Y, Ahlemeyer B, Krieglstein J, Kissel T. *in vitro* cytotoxicity testing of polycations: influence of polymer structure on cell viability and hemolysis. *Biomaterials*, 2003, *24*(7): 1121–1131.
188. Huang H, Yu H, Tang G, Wang Q, Li J. Low molecular weight polyethylenimine cross-linked by 2-hydroxypropyl-gamma-cyclodextrin coupled to peptide targeting HER2 as a gene delivery vector. *Biomaterials*, 2010, *31*(7): 1830–1838.
189. Ping Y, Liu C, Zhang Z, Liu K L, Chen J, Li J. Chitosan-graft-(PEI-β-cyclodextrin) copolymers and their supramolecular PEGylation for DNA and siRNA delivery. *Biomaterials*, 2011, *32*(32): 8328–8341.
190. Li J M, Wang Y Y, Zhang W, Su H, Ji L N, Mao Z W. Low-weight polyethylenimine cross-linked 2-hydroxypopyl-β-cyclodextrin and folic acid as an efficient and nontoxic siRNA carrier for gene silencing and tumor inhibition by VEGF siRNA. *Int J Nanomedicine*, 2013, *8*: 2101–2117.
191. Chen L, Wang H, Zhang Y, Wang Y, Hu Q, Ji J. Bioinspired phosphorylcholine-modified polyplexes as an effective strategy for selective uptake and transfection of cancer cells. *Colloid Surf B-Biointerfaces*, 2013, *111*: 297–305.
192. Dosta P, Segovia N, Cascante A, Ramos V, Borrós S. Surface charge tunability as a powerful strategy to control electrostatic interaction for high efficiency silencing, using tailored oligopeptide-modified poly(beta-amino ester)s (PBAEs). *Acta Biomater*, 2015, *20*: 82–93.
193. Tzeng S Y, Hung B P, Grayson W L, Green J J. Cystamine-terminated poly(beta-amino ester)s for siRNA delivery to human mesenchymal stem cells and enhancement of osteogenic differentiation. *Biomaterials*, 2012, *33*(32): 8142–8151.
194. Mignani S, Shi X, Zablocka M, Majoral J P. Dendrimer-enabled therapeutic antisense delivery systems as innovation in medicine. *Bioconjug Chem*, 2019, *30*(7): 1938–1950.
195. Luong D, Kesharwani P, Deshmukh R, Mohd Amin M C I, Gupta U, Greish K, Iyer A K. PEGylated PAMAM dendrimers: enhancing efficacy and mitigating toxicity for effective anticancer drug and gene delivery. *Acta Biomater*, 2016, *1*(43): 14–29.
196. Le T D, Nakagawa O, Fisher M, Juliano R L, Yoo H. RGD conjugated dendritic polylysine for cellular delivery of antisense oligonucleotide. *J Nanosci Nanotechnol*, 2017, *17*(4): 2353–2357.

197. Dung T H, Kim J S, Juliano R L, Yoo H. Preparation and evaluation of cholesteryl PAMAM dendrimers as nano delivery agents for antisense oligonucleotides. *Colloid Surf A-Physicochem Eng Asp*, 2008, *313–314*: 273–277.
198. Kang H, DeLong R, Fisher M H, Juliano R L. Tat-conjugated PAMAM dendrimers as delivery agents for antisense and siRNA oligonucleotides. *Pharm Res*, 2005, *22*(12): 2099–2106.
199. Kang C, Yuan X, Li F, Pu P, Yu S, Shen C, Zhang Z, Zhang Y. Evaluation of folate-PAMAM for the delivery of antisense oligonucleotides to rat C6 glioma cells *in vitro* and *in vivo*. *J Biomed Mater Res A*, 2010, *93*(2): 585–594.
200. Craig K, Abrams M, Amiji M. Recent preclinical and clinical advances in oligonucleotide conjugates. *Expert Opin Drug Deliv*, 2018, *15*(6): 629–640.
201. Khvorova A, Watts J K. The chemical evolution of oligonucleotide therapies of clinical utility. *Nat Biotechnol*, 2017, *35*(3): 238–248.
202. Schoch K M, Miller T M. Antisense oligonucleotides: translation from mouse models to human neurodegenerative diseases. *Neuron*, 2017, *94*(6): 1056–1070.
203. Eckstein F. Phosphorothioates, essential components of therapeutic oligonucleotides. *Nucleic Acid Ther*, 2014, *24*(6): 374–387.
204. Meade B R, Gogoi K, Hamil A S, Palm-Apergi C, van den Berg A, Hagopian J C, Springer A D, Eguchi A, Kacsinta A D, Dowdy C F, Presente A, Lönn P, Kaulich M, Yoshioka N, Gros E, Cui X S, Dowdy S F. Efficient delivery of RNAi prodrugs containing reversible charge-neutralizing phosphotriester backbone modifications. *Nat Biotechnol*, 2014, *32*(12): 1256–1261.
205. Zheng Y, Tai W. Insight into the siRNA transmembrane delivery – from cholesterol conjugating to tagging. *Wiley Interdiscip Rev Nanomed Nanobiotechnol*, 2020, *12*(3): e1606.
206. Kamola P J, Maratou K, Wilson P A, Rush K, Mullaney T, McKevitt T, Evans P, Ridings J, Chowdhury P, Roulois A, Fairchild A, McCawley S, Cartwright K, Gooderham N J, Gant T W, Moores K, Hughes S A, Edbrooke M R, Clark K, Parry J D. Strategies for *in vivo* screening and mitigation of hepatotoxicity associated with antisense drugs. *Mol Ther Nucleic Acids*, 2017, *8*: 383–394.
207. Iwamoto N, Butler D C D, Svrzikapa N, Mohapatra S, Zlatev I, Sah D W Y, Meena, Standley S M, Lu G, Apponi L H, Frank-Kamenetsky M, Zhang J J, Vargeese C, Verdine G L. Control of phosphorothioate stereochemistry substantially increases the efficacy of antisense oligonucleotides. *Nat Biotechnol*, 2017, *35*(9): 845–851.
208. Wan W B, Migawa M T, Vasquez G, Murray H M, Nichols J G, Gaus H, Berdeja A, Lee S, Hart C E, Lima W F, Swayze E E, Seth P P. Synthesis, biophysical properties and biological activity of second generation antisense oligonucleotides containing chiral phosphorothioate linkages. *Nucleic Acids Res*, 2014, *42*(22): 13456–13468.
209. Koziolkiewicz M, Krakowiak A, Kwinkowski M, Boczkowska M, Stec W J. Stereodifferentiation-the effect of P chirality of oligo(nucleoside phosphorothioates) on the activity of bacterial RNase H. *Nucleic Acids Res*, 1995, *23*(24): 5000–5005.
210. Knouse K W, de Gruyter J N, Schmidt M A, Zheng B, Vantourout J C, Kingston C, Mercer S E, McDonald I M, Olson R E, Zhu Y, Hang C, Zhu J, Yuan C, Wang Q, Park P, Eastgate M D, Baran P S. Unlocking P(V): reagents for chiral phosphorothioate synthesis. *Science*, 2018, *361*(6408): 1234–1238.

211. Monia B P, Johnston J F, Sasmor H, Cummins L L. Nuclease resistance and antisense activity of modified oligonucleotides targeted to Ha-ras. *J Biol Chem*, 1996, *271*(24): 14533–14540.
212. Hu B, Zhong L, Weng Y, Peng L, Huang Y, Zhao Y, Liang X J. Therapeutic siRNA: state of the art. *Signal Transduct Target Ther*, 2020, *5*(1): 101.
213. Lima W F, Prakash T P, Murray H M, Kinberger G A, Li W, Chappell A E, Li C S, Murray S F, Gaus H, Seth P P, Swayze E E, Crooke S T. Single-stranded siRNAs activate RNAi in animals. *Cell*, 2012, *150*(5): 883–894.
214. Prakash T P, Lima W F, Murray H M, Li W Y, Kinberger G A, Chappell A E, Gaus H, Seth P P, Bhat B, Crooke S T, Swayze E E. Identification of metabolically stable 5′- phosphate analogs that support single-stranded siRNA activity. *Nucleic Acids Res*, 2015, *43*(6): 2993–3011.
215. Prakash T P. An overview of sugar-modified oligonucleotides for antisense therapeutics. *Chem Biodivers*, 2011, *8*(9): 1616–1641.
216. Egli M, Minasov G, Tereshko V, Pallan P S, Teplova M, Inamati G B, Lesnik E A, Owens S R, Ross B S, Prakash T P, Manoharan M. Probing the influence of stereoelectronic effects on the biophysical properties of oligonucleotides: comprehensive analysis of the RNA affinity, nuclease resistance, and crystal structure of ten 2′-*O*-ribonucleic acid modifications. *Biochemistry*, 2005, *44*(25): 9045–9057.
217. Kenski D M, Butora G, Willingham A T, Cooper A J, Fu W, Qi N, Soriano F, Davies I W, Flanagan W M. siRNA-optimized modifications for enhanced *in vivo* activity. *Mol Ther Nucleic Acids*, 2012, *1*(1): e5.
218. Bennett C F, Baker B F, Pham N, Swayze E, Geary R S. Pharmacology of antisense drugs. *Annu Rev Pharmacol Toxicol*, 2017, *57*(1): 81–105.
219. Campbell M A, Wengel J. Locked vs. unlocked nucleic acids (LNA vs. UNA): contrasting structures work towards common therapeutic goals. *Chem Soc Rev*, 2011, *40*(12): 5680–5689.
220. Elmén J, Thonberg H, Ljungberg K, Frieden M, Westergaard M, Xu Y, Wahren B, Liang Z, Ørum H, Koch T, Wahlestedt C. Locked nucleic acid (LNA) mediated improvements in siRNA stability and functionality. *Nucleic Acids Res*, 2005, *33*(1): 439–447.
221. Swayze E E, Siwkowski A M, Wancewicz E V, Migawa M T, Wyrzykiewicz T K, Hung G, Monia B P, Bennett C F. Antisense oligonucleotides containing locked nucleic acid improve potency but cause significant hepatotoxicity in animals. *Nucleic Acids Res*, 2007, *35*(2): 687–700.
222. Hong D, Kurzrock R, Kim Y, Woessner R, Younes A, Nemunaitis J, Fowler N, Zhou T, Schmidt J, Jo M, Lee S J, Yamashita M, Hughes S G, Fayad L, Piha-Paul S, Nadella M V, Mohseni M, Lawson D, Reimer C, Blakey D C, Xiao X, Hsu J, Revenko A, Monia B P, MacLeod A R. AZD9150, a next-generation antisense oligonucleotide inhibitor of STAT3 with early evidence of clinical activity in lymphoma and lung cancer. *Sci Transl Med*, 2015, *7*(314): 314ra185.
223. Watts J K. Locked nucleic acid: tighter is different. *Chem Commun (Camb)*, 2013, *49*(50): 5618–5620.
224. Ittig D, Liu S, Renneberg D, Schümperli D, Leumann C J. Nuclear antisense effects in cyclophilin A pre-mRNA splicing by oligonucleotides: a comparison of tricyclo-DNA with LNA. *Nucleic Acids Res*, 2004, *32*(1): 346–353.

225. Summerton J. Morpholino antisense oligomers: the case for an RNase H-independent structural type. *Biochim Biophys Acta*, 1999, *1489*(1): 141–158.
226. Juliano R L, Ming X, Nakagawa O. Cellular uptake and intracellular trafficking of antisense and siRNA oligonucleotides. *Bioconjug Chem*, 2012, *23*(2): 147–157.
227. Crooke S T, Wang S, Vickers T A, Shen W, Liang X H. Cellular uptake and trafficking of antisense oligonucleotides. *Nat Biotechnol*, 2017, *35*(3): 230–237.
228. Kaihatsu K, Janowski B A, Corey D R. Recognition of chromosomal DNA by PNAs. *Chem Biol*, 2004, *11*(6): 749–758.
229. Malchère C, Verheijen J, van der Laan S, Bastide L, van Boom J, Lebleu B, Robbins I. A short phosphodiester window is sufficient to direct RNase H-dependent RNA cleavage by antisense peptide nucleic acid. *Antisense Nucleic Acid Drug Dev*, 2000, *10*(6): 463–468.
230. Aartsma-Rus A. FDA approval of nusinersen for spinal muscular atrophy makes 2016 the year of splice modulating oligonucleotides. *Nucleic Acid Ther*, 2017, *27*(2): 67–69.
231. Virgilio A, Varra M, Scuotto M, Capuozzo A, Irace C, Mayol L, Esposito V, Galeone A. Expanding the potential of G-quadruplex structures: formation of a heterochiral TBA analogue. *Chembiochem*, 2014, *15*(5): 652–655.
232. Li K F, Deng J L, Jin H W, Yang X T, Fan X M, Li L Y, Zhao Y, Guan Z, Wu Y, Zhang L H, Yang Z J. Chemical modification improves the stability of the DNA aptamer GBI-10 and its affinity towards tenascin-C. *Org Biomol Chem*, 2017, *15*(5): 1174–1182.
233. Li L Y, Hou J J, Liu X J, Guo Y J, Wu Y, Zhang L H, Yang Z J. Nucleolin-targeting liposomes guided by aptamer AS1411 for the delivery of siRNA for the treatment of malignant melanomas. *Biomaterials*, 2014, *35*(12): 3840–3850.
234. Cai B B, Yang X T, Sun L D, Fan X M, Li L Y, Jin H W, Wu Y, Guan Z, Zhang L R, Zhang L H, Yang Z J. Stability and bioactivity of thrombin binding aptamers modified with D-/L-isothymidine in the loop regions. *Org Biomol Chem*, 2014, *12*(44): 8866–8876.
235. Chiu Y L, Rana T M. siRNA function in RNAi: a chemical modification analysis. *RNA*, 2003, *9*(9): 1034–1048.

CHAPTER 12

CRISPR/CAS9: A THERAPEUTIC SOLUTION BY MANIPULATING NUCLEIC ACIDS

HONGKUI DENG[1], YULIN LIU[1], and SHICHENG SUN[2]
[1] School of Basic Medical Sciences, State Key Laboratory of Natural and Biomimetic Drugs, Peking University Health Science Center and the MOE Key Laboratory of Cell Proliferation and Differentiation, College of Life Sciences, Peking-Tsinghua Center for Life Sciences, Peking University, Beijing, China
[2] The Murdoch Children's Research Institute, Department of Paediatrics, University of Melbourne, Parkville, VIC, Australia

12.1. INTRODUCTION

Many human diseases are caused by gene mutations. In the 1970s, scientists realized that some human diseases could be cured by manipulating nucleic acids, the fundamental building blocks of genes [1]. In 1990, the FDA approved the first clinical trial of gene therapy to treat adenosine-deaminase-mediated severe combined immunodeficiency (ADA-SCID), a genetic disorder caused by a mutation of the gene encoding the ADA protein in T lymphocytes. In this trial, scientists isolated and genetically modified the T lymphocytes of patients, and then transferred the modified cells back into themselves. However, the expression level of ADA did not increase significantly [2]. Later in 1993, gene therapy was conducted for ADA-SCID patients using the hematopoietic stem cells (HSCs) [3]. Although no significant clinical benefits were observed, these early studies paved the way for the development of advanced gene-editing techniques with the aim of therapeutic applications.

Nucleic Acids in Medicinal Chemistry and Chemical Biology: Drug Development and Clinical Applications, First Edition. Edited by Lihe Zhang, Xinjing Tang, Zhen Xi, and Jyoti Chattopadhyaya.

Most clinical trials for gene therapy relied on the use of viral vectors to deliver an exogenous fragment of DNA into cells to replace defective DNA. The exogenous genes are either transiently expressed or integrated into the genome, depending on the types of viral vectors to be used. Concerns regarding the use of viral vectors have increased because of the risks of insertional mutagenesis associated with genome instability and toxicity. Cancer-like conditions or even death have been reported in several patients who received viral vector treatment [4, 5].

The clustered regularly interspaced short palindromic repeat (CRISPR)/CRISPR-associated protein 9 (Cas9) system has become an attractive technology for targeted gene therapy. Targeted gene editing is aimed at precise gene correction, deletion, or insertion at a specific locus in the human genome. In recent years, targeted gene editing has become quite attractive in clinical applications because it provides a tool for understanding gene functions by creating gene-modified animal models of human diseases as well as an opportunity for curing genetic disorders by correcting gene mutations. There are four types of genome-editing technologies available: meganucleases, zinc finger nucleases (ZFNs), transcription activator-like effector nucleases (TALENs), and CRISPR/Cas9. All of these gene-editing components include a guide element to locate the nucleus for a specific DNA sequence in the genome. Double-strand breaks (DSBs) are generated in specific location in the genome with a guide element for recognition and are then repaired by either homologous-direct recombination (HDR) or non-homologous end joining (NHEJ).

The CRISPR/Cas9 system has significant advantages of simplicity and technical feasibility for human cells. CRISPR sequences, which were first discovered in the immune system of prokaryotes, identify the nucleic acid information for an invasive pathogen and cleave reoccurring viral and plasmid DNAs [6]. Multiple proteins were discovered to be tightly associated with CRISPR sequences and were consequently named as CRISPR-associated (Cas) proteins [7]. Additional studies demonstrated the adaptive immune function required the hybridization of two small RNAs called CRISPR RNA (crRNA) and transactivating crRNA (tracrRNA) [8]. Unlike meganucleases, ZFNs and TALENs require to design specific nucleases that identify and bind to their target site through protein–DNA interactions, the CRISPR/Cas9 targets DNA via RNA–DNA interactions using a single-guide RNA (sgRNA) that can be easily made [9–12]. In addition, by introducing multiple sgRNAs with the same Cas nuclease, multiplexing gene editing can be achieved much easier with the CRISPR/Cas9 system [13]. These advantages suggest CRISPR/Cas9 is a simple and flexible tool with potential for clinical applications.

CRISPR/Cas9-based therapies have been tested in clinical trials. The first clinical trial examined whether CRISPR/Cas9-edited cells can be used for transplantation therapy. In this trial, CRISPR-Cas9 was used to modify donor HSPCs by deleting the HIV cell co-receptor gene CCR5, which recapitulated a natural mutant of CCR5 that is protective from HIV infection [14]. The transplantation of these CRISPR-edited stem cells was used to treat a leukemia patient with AIDS, demonstrating the feasibility and safety of CRISPR-Cas9 technology in stem cell therapy. Subsequent clinical trials have also demonstrated the notable therapeutic benefits and safety of CRISPR/Cas9 for treating human diseases, including SCD, β-thalassemia, and late-stage lung cancer [15, 16]. This chapter introduces the evolution of CRISPR-Cas9 from a gene-editing tool to a therapeutic approach and summarizes its recent clinical applications.

12.2. BASICS OF CRISPR/Cas9 FOR GENE THERAPY

CRISPR sequences were first discovered in the adaptive immune system of prokaryotes. These sequences can identify and cleave the nucleic acid of invasive pathogens [6]. In addition to Cas protein, the system requires the hybridization of crRNA and tracrRNA, which are used to guide the Cas nuclease to the location where DNA needs to be edited [8]. In 2012, the CRISPR/Cas9 can be programmed into a gene-editing tool to engineer bacterial genomes and the simplicity of this process has subsequently resulted in the extension of this tool to other applications, such as editing the genome of some polyploid plant cell types and mammalian cells [12, 17, 18].

12.2.1. Gene-Editing Components of CRISPR/Cas9 for Therapeutic Applications

To generate a double-stranded DNA break in human cells, CRISPR/Cas9 requires the expression of both Cas9 endonuclease and sgRNA that guides the Cas9 to the location where the DNA needs to be cut. Initially, a DNA fragment encoding the Cas9 nuclease and sgRNA were expressed by one plasmid or two separate plasmids [17, 19]. However, the use of exogenous DNA may trigger cellular immune responses and unwanted integration of Cas9-DNA segments into the genome [20, 21]. In addition, this method may induce the prolonged expression of Cas9 nuclease after the editing is completed, which raises the risk of preexisting adaptive immune response to Cas9 and off-target effects [22, 23]. To address this problem, an additional sgRNA targeting Cas9 was added to the construct to form a self-restricted CRISPR system, reducing the expression time for Cas9 and inducing the "suicide" of the Cas9 plasmid when the editing was completed ([24]).

Apart from using DNA to express Cas9, the endonuclease can be delivered in the form of mRNA into target cells [25]. In addition to the advantage of non-integration into the genome, this RNA strategy also hastens the onset of gene-editing efficiency because it does not require transcription from DNA. However, the clinical application of RNA transfection is limited by its instability and cellular immunogenicity. These limitations can be improved with the investigation of chemically modified nucleosides. RNAs contain more than 100 types of structurally distinguishable chemically altered nucleosides [26]. Some of the modifications were adopted *in vitro* transcribed (IVT) messenger RNA, including 5-methylcytidine (m5C), pseudouridine (Ψ), and methylpseudouridine (m1Ψ). *In vitro* transcribed chemically modified messenger RNAs have higher stability *in vivo* with increased resistance to RNases [27]. Certain combinations of nucleotide modifications reduce cellular immunogenicity by inhibiting interactions between the modified mRNA with Toll-like receptors and retinoid-inducible gene I [28]. Chemical modification of RNAs is a promising approach in transcript replacement therapy as the therapeutic efficacy was already confirmed in a mice model with fetal lung disease [29]. Recently, nucleotide modifications were also adopted in vaccine development for COVID [30]. In terms of CRISPR/Cas9 gene editing, modified nucleotides in Cas9 mRNA were screened to identify modifications that increase Cas9 activity. Some of the modifications also increased the CRISPR gene-editing efficiency [31].

The ribonucleoprotein (RNP) form of Cas9 combined with gRNA represents an attractive delivery approach, which has been used in published clinical trials [14–16]. Cas9 RNP cleaves the targeted DNA sequence immediately when RNP is in the cells, often taking only 3–24 hours to generate indels [32]. The entire editing process does not require the biosynthesis of proteins or RNA. In preclinical experiments, RNP has shown advantages in targeting primary human cells and human pluripotent stem cells, improved targeting efficiency, fewer off-target effects, and less stress for the cells [32, 33]. It was found that the sgRNA is very unstable in the absence of the Cas9 protein and therefore reduced the cleavage activity of CRISPR/Cas9 [34]. Chemical modification at three nucleotides on both terminals of the sgRNA can protect it from degradation and has improved the efficiency of genome editing in human primary cells and clinical trials [14, 35] (Figure 12.1). These important findings demonstrate the advantages of RNP as an optimal strategy in clinical applications.

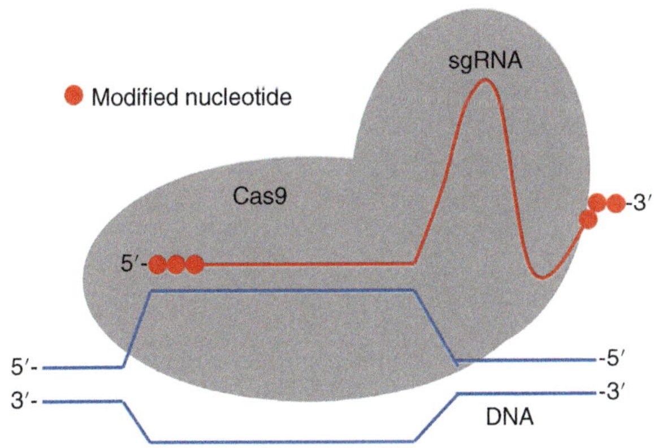

Figure 12.1 Schematic structure of Cas9 and sgRNA. Chemically modified nucleotides are shown in red dots.

12.2.2. Delivery of the CRISPR/Cas9 System *in vitro* and *in vivo*

The approach used to deliver CRISPR/Cas9 components into cells is an important factor involving gRNA and the Cas protein (Figure 12.2). The current strategy of CRISPR/Cas9-based therapy involves editing cells *in vitro*, and transfusing the edited cells into patients, which may circumvent the risks of *in vivo* gene editing. An often-used method for *in vitro* delivery is cell electroporation. The primary step is the isolation and culture of primary human cells *in vitro* for gene editing. Through electroporation, Cas9 RNP is directly delivered inside cells for gene editing and the gene-edited cells are transplanted into patients. *In vitro* gene editing also provides an opportunity to qualify the edited cells prior to transplantation, which can reduce the risks associated with gene editing.

The *in vivo* delivery of gene-editing tools is one of the biggest obstacles in gene editing therapy. Electroporation of plasmids, mRNA, or Cas9/sgRNA RNP is commonly used *in vitro* but is inefficient *in vivo*. Adeno-associated viruses (AAVs) have become important vectors for *in vivo* therapeutic gene delivery in recent years with a good safety profile [36]. DNA in an AAV vector cannot integrate into the host genome, which reduce the risks of DNA insertion into the host genome and mutagenesis. However, the low packaging capacity of AAVs (about 5 kb) limits the usage of AAVs to express Cas9. Lentivirus and

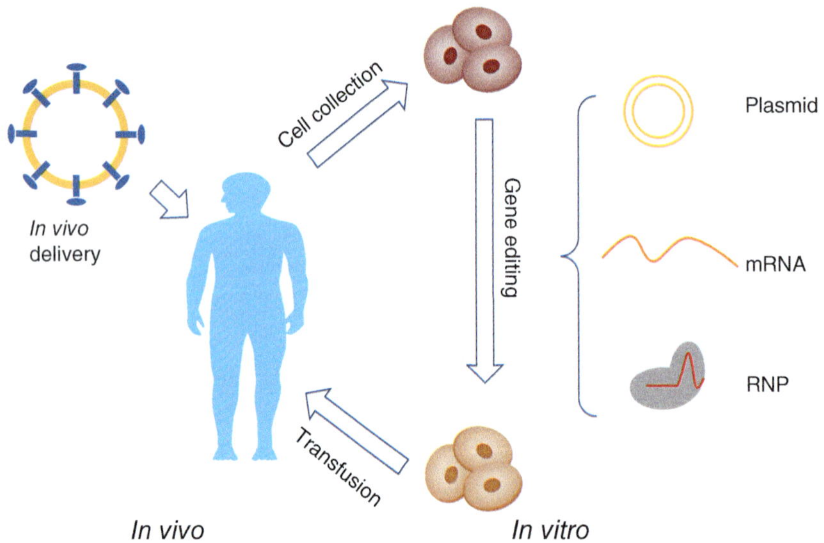

Figure 12.2 Delivery of gene-editing components. For *in vivo* gene editing, gene-editing components are delivered directly to the patients. For *in vitro* gene editing, cells were collected, modified with gene editing components including plasmids, mRNAs or RNPs, and transfused to the patients.

adenovirus vectors have larger packaging capacity, but also have their own limitations [37, 38]. Lentivirus vectors often insert DNA into the genome and the continuous expression of nucleases raises serious safety concerns about off-target effects. Adenovirus transiently expresses the nucleases but may cause severe immune responses in patients. Therefore, *in vivo* gene editing has been challenging and has only been performed in large animal models.

12.2.3. Targeted Cell and Disease Types by CRISPR/Cas Therapy

Current methods for gene therapy have been optimized for human cells *in vitro*, suggesting that CRISPR/Cas-based therapy is limited by cell types that can be collected, cultured, and engrafted *in vivo* after transplantation. The majority of cell types used in the currently documented clinical trials were hematopoietic cells, including T cells and HSCs, which can be cultured outside the body and transplanted into the patients (Table 12.1). Multiple genetic constructs, such as CCR5, TCR, and PD-1, have been targeted in hematopoietic cells to treat cancer, leukemia, AIDS, and other diseases.

12.3. CLINICAL TRIAL I, EDITING THE CCR5 GENE IN HUMAN STEM CELLS

12.3.1. Targeting CCR5 in HSPCs to cure HIV Infection

The first reported CRISPR/Cas9-based gene therapy valuated in a clinical trial was the transplantation of *CCR5*-edited HSPCs to treat a leukemia patient with HIV infection [14]. CCR5 is a chemokine receptor best known for its co-receptor function, which enables the HIV virus to enter human immune cells [39]. Some individuals carry a naturally occurring homozygous loss-of-function mutation of *CCR5* due to the deletion of 32 base pairs (Δ32 mutation), and these individuals show resistance to HIV-1 infection [40, 41]. In 2007, long-term eradication of the HIV-1 virus in an individual known as the "Berlin Patient" was achieved through an allogenic transplantation of HSPCs carrying a naturally occurring *CCR5* mutation and this was recognized as the first "functional cure" of HIV-1 infection [42]. Similarly, a recent report of the "London Patient" who exhibited long-term remission from HIV-1 after receiving a CCR5 Δ32 donor graft further demonstrated that a functional cure could be achieved through CCR5 Δ32 HSPC transplantation [43]. These findings suggest that deleting the *CCR5* gene from hematopoietic cells using gene-editing tools can replace HIV permissive cells in HIV patients and provide protection against HIV (Figure 12.3).

In 2014, a clinical trial using ZFNs to target the *CCR5* gene in human T cells was performed to treat AIDS [44]. Plasmids encoding a *CCR5* specific ZFN were electroporated to T cells collected from HIV patients and the T cells were transfused back into the patients. After transplantation, a decline in the frequency

TABLE 12.1 Ongoing clinical trials with *ex vivo* CRISPR gene editing

NCT	Sponsor	Disease	Target Cells	Target gene
NCT03164135	Affiliated Hospital to Academy of Military Medical Sciences	HIV-1-infection	Hematopoietic stem/ progenitor cell	CCR5
NCT03728322	Allife Medical Science and Technology Co., Ltd.	β-Thalassemia	iHSC	HBB correction
NCT04417764	Central South University	Advanced hepatocellular carcinoma	T cell	PD-1
NCT03747965	Chinese PLA General Hospital	Solid tumor	T cell	PD-1
NCT03398967	Chinese PLA General Hospital	Leukemia and lymphoma	T cell	CD19 and CD20/CD22
NCT03166878	Chinese PLA General Hospital	Leukemia and lymphoma	T cell	CD19
NCT03545815	Chinese PLA General Hospital	Mesothelin positive multiple solid tumors	T cell	TCR and PD-1
NCT04244656	CRISPR Therapeutics AG	Multiple myeloma	T cell	anti-B-cell maturation antigen (BCMA)
NCT04438083	CRISPR Therapeutics AG	Renal cell carcinoma	T cell	CD70-directed T-cell
NCT04035434	CRISPR Therapeutics AG	B-cell malignancy	T cell	CD19
NCT04502446	CRISPR Therapeutics AG	T-cell lymphoma	T cell	CD70
NCT04557436	Great Ormond Street Hospital for Children NHS Foundation Trust	B acute lymphoblastic leukemia	T cell	CD52 and TRAC
NCT04426669	Intima Bioscience, Inc.	Gastro-intestinal (GI) cancer	Tumor Infiltrating Lymphocytes (TIL)	CISH (Cytokine-induced SH2 protein)
NCT02793856	Sichuan University	Metastatic non-small cell lung cancer	T cell	PD-1
NCT03399448	University of Pennsylvania	Multiple myeloma	T cell	TCRα, TCRβ and PD-1
NCT03655678	Vertex Pharmaceuticals Incorporated	Transfusion-dependent β-Thalassemia	Hematopoietic stem/ progenitor cell	BCL11A
NCT03745287	Vertex Pharmaceuticals Incorporated	Sickle cell disease	Hematopoietic stem/ progenitor cell	BCL11A
NCT04037566	Xijing Hospital	$CD19^+$ leukemia or lymphoma	T cell	HPK1
NCT03044743	Yang Yang	Malignancies	T cell	PD-1

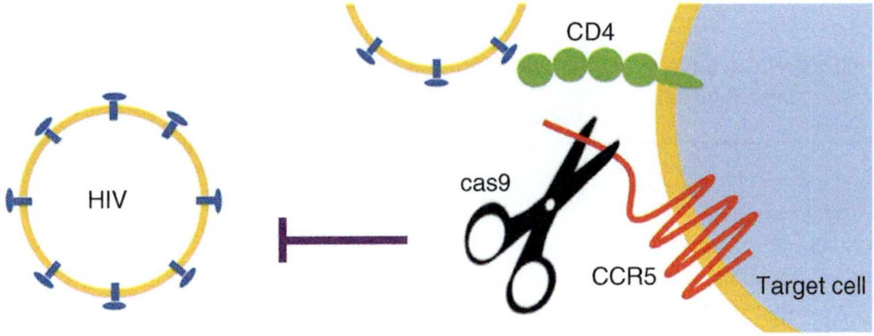

Figure 12.3 Targeting *CCR5* gene to cure HIV. The co-receptor help HIV-1 entering the cells can be deleted with CRISPR to provide protection against the virus.

of edited T cells was observed in the patients, possibly due to the low-level-differentiated T-cell proliferation. HSPCs are characterized by stable engraftment, self-renewal, and differentiation into all hematopoietic lineages after transplantation, suggesting they may serve as an optimal cell source for the production of CCR5-edited hematopoietic cells that can provide lifelong protection for patients. In 2019, Xu *et al.* used CRISPR/Cas9 to delete the *CCR5* gene from donor-derived CD34+ HSPCs. These cells showed durable engraftment and produced persistent CCR5-edited cells in the long term in a patient with both leukemia and HIV infection [14].

12.3.2. Pre-Clinical Investigations

An early attempt to investigate the therapeutic potential of *CCR5*-edited HSPCs were generated by ZFN, which involved the use of non-human primate HSPCs followed by transplantation. In the study, the frequency of *CCR5*-edited cells decreased dramatically from 64% *ex vivo* to 40% shortly after transplantation and 3–5% at six months post-transplantation [45]. A similar reduction was also reported in a study focusing on gene correction with CRISPR/Cas9, the targeting frequency decreased from 16% *ex vivo* to 3.5% after a 16-week engraftment [46]. This is possibly because that some HSPCs remained unmodified after the editing process, which can contribute to multilineage engraftment following transplantation in mice. This can also lead to a bias when analyzing the functional reconstitution by CCR5-edited HSPCs. The low frequency of modified cells in the whole engrafted human hematopoietic cells is a major challenge for therapy.

Prior to clinical trials, Xu *et al.* established a non-viral transfection CRISPR/Cas9 system to edit primary human HSPCs and reported high-editing efficiency and minimal off-target effects [47] (Figure 12.4). Gene-edited human CD34+ HSPCs retained their multilineage reconstitution, and the 31.2% CCR5 ablation rate was preserved in long-term reconstituted mice (ranging from 30 to 47 weeks post-transplantation). Most importantly, human blood cells from these primary

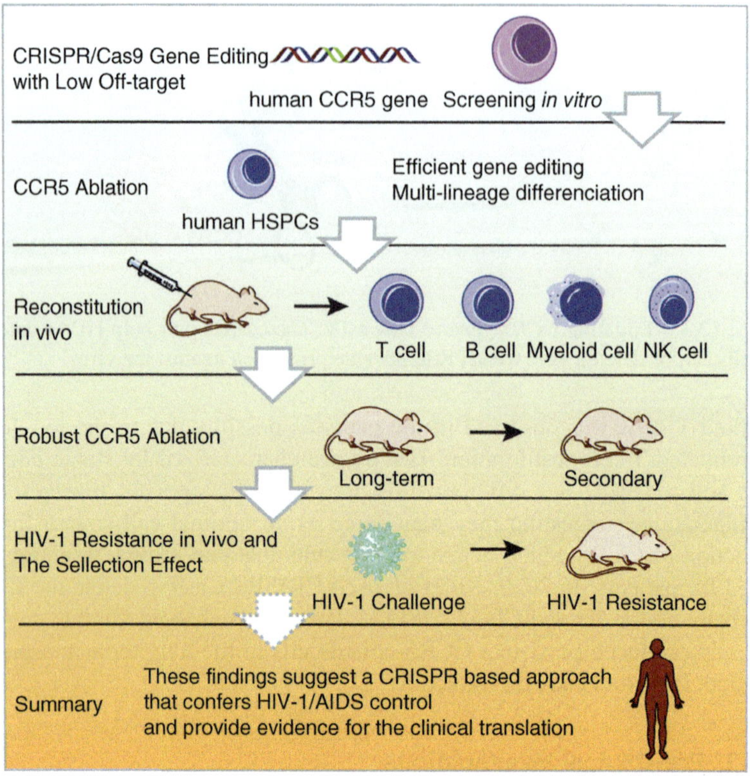

Figure 12.4 Pre-clinical CCR5 ablation with CRISPR in HSPCs (Source: Xu *et al.* [47]. Reproduced with permission of Deng). CCR5 was ablated with CRISPR in HSPCs. The ability of multilineage differentiation was tested *in vitro* and *in vivo*. The long-term persistence of CCR5 modified cells was confirmed in reconstituted mice model. HIV-1 resistance was observed in the HIV-1 challenge assay.

transplanted mice could be used for secondary transplantation, and CD34$^+$ HSPCs were present in the bone marrow of secondary transplanted mice, indicating that CCR5-modified human HSPCs retained their typical self-renew activity after transplantation.

Mice bearing CCR5-edited human HSPCs exhibited immune protection against HIV infection. After eight weeks of HIV-1 infection, a significant decrease of the viral load was observed in mice transplanted with CCR5-edited HSPCs, compared with mice transplanted with non-edited cells. In the study, the normal development of human CD4$^+$ T cells was observed unaffected by HIV infection, suggesting that CCR5-negative cells were enriched under the selection pressure of the HIV-1 virus. These findings demonstrate that CCR5 depletion in HSPCs by CRISPR/Cas9 provides HIV-1 resistance *in vivo*, setting the stage for clinical trials.

12.3.3. Clinical Trial: Patient Selection, Adverse Effects Monitoring, Gene Sequencing Analysis to Confirm Safety

The *CCR5* clinical trial was the first focusing on the therapeutic feasibility, clinical benefits and safety of CRISPR/Cas9 gene therapy [14]. The first recruited and tested patient was diagnosed with both HIV infection and acute leukemia and required immediate allogenic bone marrow transplantation. CD34$^+$ HSPCs were isolated from an unrelated human leukocyte antigen (HLA)-matched donor; one fraction of the cells was processed with gene editing, and the other fraction was reserved for co-transplantation with the edited cells in case the edited cells did not engraft in the patient. In this trial, isolated CD34$^+$ HSCs were treated with CRISPR gRNA and the Cas9 RNP to ablate the *CCR5* gene.

After transplantation, CCR5 ablation was confirmed in all major blood cell lineages of the treated patient for at least 19 months. To investigate the anti-HIV function of CCR5-deletion, related trial involving antiretroviral therapy (ART) interruption was performed for four weeks starting from 210 days post-transplantation. An increase of CCR5 ablation was found in T cells during ART interruption. During the trial, common adverse effects associated with HSPC transplantation were documented but were not considered to be related to the CRISPR/Cas9 gene editing. Whole-genome sequencing and deep sequencing were used to monitor off-targets of the CCR5 targeting and the results precluded all of the predicted hotspots for off-targets. Overall, this study provided the first line of evidence that the CRIPSR/Cas9 technology was safe and feasible for gene therapy and encouraged further and broader clinical trials for different of diseases.

12.4. CLINICAL TRIAL II FOR SCD, B-THALASSEMIA, AND CANCER

12.4.1. Targeting BCL11A in HSPCs to Cure SCD and b-Thalassemia

Another published clinical trial for CRISPR/Cas9-gene therapy involved the treatment of SCD and β-thalassemia. Both of these diseases are caused by the nonfunctional β-globin chain in hemoglobin [48, 49]. The nonfunctional β-globin chain can be replaced with high-level expression of fetal hemoglobin, which is downregulated after birth [50]. As the switch between fetal γ-globin and the adult β-globin after birth is mediated by the suppression of the γ-globin gene BCL11A, CRISPR/Cas9 was used to knockout the BCL11A gene in the HSCs of the patients. These results of the studies raised the possibility that a single-gene ablation of BCL11A can be a promising approach for the cure for SCD and β-thalassemia.

In 2020, Frangoul *et al.* used CRISPR/Cas9 to edit the erythroid-control region of BCL11A in HSCs and induced a high-level expression of fetal globin in red blood cells in patients with SCD and β-thalassemia [15].

12.4.2. CRISPR/Cas9 in CAR-T Cells to Treat Cancers

Genetic engineering of T cells is promising for cancer treatment. By editing T-cell receptors (TCRs) or a chimeric antigen receptor (CAR) T located on the surface, T cells can be reprogramed to specifically detect and kill tumor cells. T cells can be engineered to express a TCR that targets tumor cells [51]. This can be further improved by knocking down the immune checkpoint factor PD-1 and therapeutic effects have been shown in a mice model [52]. However, the competition for the expression of endogenous TCRs or mispairing may lead to low expression of transgenic TCR or lead to self-reactive TCRs, reducing the therapeutic effects. To address this limitation, TCRα and TCRβ can be deleted to reduce TCR mispairing. Using the CRISPR/Cas9 system to achieve multiplexing gene editing, TCRα, TCRβ, and PD-1 were targeted at the same time to enhance the expression of a synthetic, cancer-specific TCR transgene (NY-ESO-1). Compared to clinical trials where T cells were not edited for endogenous TCRs and endogenous PD-1, gene-edited T cells engrafted more durably in all patients for more than nine months [16].

12.5. CHALLENGES AND OPPORTUNITIES

Although the available evidence has clearly shown the clinical benefits and safety profile of CRISPR/Cas9 as a gene therapy tool, some challenges remain for the adoption of this technology as a routine therapeutic approach. A key safety concern is the off-target effect. Different assays have been employed to evaluate the off-target effects in clinical trials. In the clinical trial for CCR5-edited HSPCs, high-throughput whole-genome sequencing assay was used to evaluate the off-target effects using a pre-infusion cells and blood samples collected from the patient at three time points after transplantation. The hotspot candidates for the off-target loci were validated by deep sequencing. No off-target indels, chromosomal rearrangements, or long-range deletions were identified in any of the sequencing data sets [14].

In the clinical trial targeting *BCL11A*, off-target effects were evaluated based on the sequence similarity and ***g***enome-wide ***u***nbiased ***i***dentification of ***d***ouble-stranded breaks ***e***nabled by sequencing (GUIDE-seq), followed by hybrid capture [53]. No detectable off-target editing was observed in any of the four $CD34^+$ HSCs healthy donor samples. However, no off-target assay was performed on clinical samples to determine off-target editing appeared after engraftment [15]. In the multiplexing gene-editing clinical trial, the off-target effect was evaluated with a modification of the GUIDE-seq approach named iGUIDE [54]. Low-frequency off-target mutations were detected near the loci of the genes CLIC2, ZNF609, and LINC00377, but none of these off-targets showed cellular transformations during the pre-infusion quality examination. In addition, chromosomal translocations were detected in CRISPR/Cas9-edited T cells *in vitro*, but the frequency of these translocations declined after seven days during manufacturing and after transplantation, which may suggest that some off-targeted abnormal T cells will eventually die [16]. Off-targets of CRISPR/Cas9-edited HSPCs were reported at low frequency in the published results for all three clinical trials.

Given the flexibility of CRISPR technology for the manipulation of nucleic acids, the technology provides a broad therapeutic option in gene therapy. In addition to a knock-out strategy that has been tested clinically, this technology also permits gene correction by precisely changing nucleic acids without deactivating of the genes. For this purpose, emerging base-editing tools are more favorable for precision gene editing to change a single nucleic acid over subtle modification of the genome [55].

Apart from editing disease-causing genes, CRISPR/Cas9 has been used to modify HLA genes that govern the tissue compatibility of cells for transplantation. The likelihood of the rejection of such HLA-modified cells by the recipient's immune system is very low, and the cells are compatible for different recipients. As such, HLA-modified cells are expected to generalize cell therapies. The available clinical information for the CRISPR/Cas9 remains limited by the small number of patients tested and further research is required for understanding this technology in order to reduce off-target effects, improve gene editing efficiency, and minimize preexisting adaptive immune responses.

REFERENCES

1. Friedmann, T., Roblin, R. (1972). Gene therapy for human genetic disease? *Science* 175: 949–955.
2. Blaese, R.M., Culver, K.W., Miller, A.D. et al. (1995). T lymphocyte-directed gene therapy for ADA- SCID: initial trial results after 4 years. *Science* 270: 475–480.
3. Shpall, E.J., Jones, R.B., Bearman, S.I. et al. (1993). Positive selection of $CD34^+$ hematopoietic progenitor cells for transplantation. *Stem Cells* 11 (S3): 48–49.
4. Herzog, R.W. (2010). Gene therapy for SCID-X1: round 2. *Mol. Ther.* 18: 1891.
5. Sibbald, B. (2001). Death but one unintended consequence of gene-therapy trial. *Can. Med. Assoc. J.* 164: 1612.
6. Ishino, Y., Shinagawa, H., Makino, K. et al. (1987). Nucleotide sequence of the iap gene, responsible for alkaline phosphatase isozyme conversion in *Escherichia coli*, and identification of the gene product. *J. Bacteriol.* 69: 5429–5433.
7. Jansen, R., Embden, J.D., Gaastra, W. et al. (2002). Identification of genes that are associated with DNA repeats in prokaryotes. *Mol. Microbiol.* 43: 1565–1575.
8. Deltcheva, E., Chylinski, K., Sharma, C.M. et al. (2011). CRISPR RNA maturation by trans-encoded small RNA and host factor RNase III. *Nature* 471 (7340): 602–607.
9. Boch, J., Scholze, H., Schornack, S. et al. (2009). Breaking the code of DNA binding specificity of TAL-type III effectors. *Science* 326(5959): 1509–1512.
10. Deng, D., Yan, C., Pan, X. et al. (2012). Structural basis for sequence-specific recognition of DNA by TAL effectors. *Science* 335(6069): 6720–6723.

11. Durai, S., Mani, M., Kandavelou, K. et al. (2005). Zinc finger nucleases: custom-designed molecular scissors for genome engineering of plant and mammalian cells. *Nucleic Acids Res.* 33(18): 5978–5990.
12. Jinek, M., Chylinski, K., Fonfara, I. et al. (2012). A programmable dual-RNA-guided DNA endonuclease in adaptive bacterial immunity. *Science* 337(6096): 816–821.
13. Cong, L., Ran, F.A., Cox, D. et al. (2013). Multiplex genome engineering using CRISPR/Cas systems. *Science* 339(6121): 819–823.
14. Xu, L., Wang, J., Liu, Y. et al. (2019). CRISPR-edited stem cells in a patient with HIV and acute lymphocytic leukemia. *N. Engl. J. Med.* 381: 1240–1247.
15. Frangoul, H., Altshuler, D., Cappellini, M.D. et al. (2020). CRISPR-Cas9 gene editing for sickle cell disease and β-thalassemia. *N. Engl. J. Med.* 384: 252–260.
16. Stadtmauer, E.A., Fraietta, J.A., Davis, M.M. et al. (2020). CRISPR-engineered T cells in patients with refractory cancer. *Science* 367(6481): eaba7365.
17. Mali, P., Yang, L., Esvelt, K.M. et al. (2013). RNA-guided human genome engineering via Cas9. *Science* 339: 823–826.
18. Shan, Q., Wang, Y., Li, J. et al. (2013). Targeted genome modification of crop plants using a CRISPR-Cas system. *Nat. Biotechnol.* 31: 686–688.
19. Ran, F.A., Hsu, P.D., Wright, J. et al. (2013). Genome engineering using the CRISPR-Cas9 system. *Nat. Protoc.* 8: 2281–2308.
20. Ono, R., Yasuhiko, Y., Aisaki, K. et al. (2019). Exosome-mediated horizontal gene transfer occurs in double-strand break repair during genome editing. *Commun. Biol.* 2: 57. https://doi.org/10.1038/s42003-019-0300-2.
21. Sun, L., Wu, J., Du, F. et al. (2013). Cyclic GMP-AMP synthase is a cytosolic DNA sensor that activates the type I interferon pathway. *Science* 339: 786–791.
22. Charlesworth, C.T., Deshpande, P.S., Dever, D.P. et al. (2019). Identification of pre-existing adaptive immunity to cas9 proteins in humans. *Nat. Med.* 25: 249–254.
23. Merkle, F.T., Neuhausser, W.M., Santos, D. et al. (2015). Efficient CRISPR-Cas9-mediated generation of knockin human pluripotent stem cells lacking undesired mutations at the targeted locus. *Cell Rep.* 11: 875–883.
24. Chen, Y., Liu, X., Zhang, Y. et al. (2016). A self-restricted CRISPR system to reduce off-target effects. *Mol. Ther.* 24: 1508–1510.
25. Miller, J.B., Zhang, S., Kos, P. et al. (2017). Non-viral CRISPR/Cas gene editing *in vitro* and *in vivo* enabled by synthetic nanoparticle co-delivery of Cas9 mRNA and sgRNA. *Angew. Chem. Int. Ed.* 56(4): 1059–1063.
26. Machnicka, M.A., Kaja, M., Okan, O.O. et al. (2013). MODOMICS: a database of RNA modification pathways—2013 update. *Nucleic Acids Res.* 41 (D1): D262–D267.
27. Anderson, B.R., Muramatsu, H., Jha, B.K. et al. (2011). Nucleoside modifications in RNA limit activation of 2′-5′-oligoadenylate synthetase and increase resistance to cleavage by RNase L. *Nucleic Acids Res* 39(21): 9329–9338.
28. Katalin, K., Buckstein, M., Ni, H. et al. (2005). Suppression of RNA recognition by toll-like receptors: the impact of nucleoside modification and the evolutionary origin of RNA. *Immunity* 23(2): 165–175.
29. Kormann, M.S.D., Hasenpusch, G., Aneja, M.K. et al. (2011). Expression of therapeutic proteins after delivery of chemically modified mRNA in mice. *Nat. Biotechnol.*, 29(2): 154–157.
30. Walsh, E.E., Frenck, R.W., Falsey, A.R. et al. (2020). Safety and immunogenicity of two RNA-based Covid-19 vaccine candidates. *N. Engl. J. Med.* 383(25): 2439–3450.

31. Sriram, V., Azizian, K.T., Ashiqul, H.A.K.M. et al. (2018). Uridine depletion and chemical modification increase Cas9 mRNA activity and reduce immunogenicity without HPLC purification. *Mol. Ther. Nucleic Acids* 12: 530–542.
32. Kim, S., Kim, D., Cho, S.W. et al. (2014). Highly efficient RNA-guided genome editing in human cells via delivery of purified Cas9 ribonucleoproteins. *Genome Res.* 24: 1012–1019.
33. Schumann, K., Lin, S., Boyer, E. et al. (2015). Generation of knock-in primary human t cells using cas9 ribonucleoproteins. *Proc. Natl Acad. Sci.* 112(33): 10437–10442.
34. Ma, H., Tu, L.C., Naseri, A. et al. (2016). CRISPR-Cas9 nuclear dynamics and target recognition in living cells. *J. Cell Biol.* 29: 529–537.
35. Hendel, A., Bak, R.O., Clark, J.T. et al. (2015). Chemically modified guide RNAs enhance CRISPR-Cas genome editing in human primary cells. *Nat. Biotechnol.* 33: 985–989.
36. Drouin, L.M., Agbandje-Mckenna, M. (2013). Adeno-associated virus structural biology as a tool in vector development. *Future Virol.* 8(12): 1183–1199.
37. Li, C., Guan, X., Du, T. et al. (2015). Inhibition of hiv-1 infection of primary cd^{4+} t-cells by gene editing of ccr5 using adenovirus-delivered crispr/cas9. *J. Gen. Virol.* 96: 2381–2393.
38. Philippe, S., Sarkis, C., Barkats, M. et al. (2006). Lentiviral vectors with a defective integrase allow efficient and sustained transgene expression *in vitro* and *in vivo*. *Proc. Natl. Acad. Sci. U S A* 103(47):17684–17689.
39. Deng, H., Liu, R., Ellmeier, W. et al. (1996). Identification of a major co-receptor for primary isolates of HIV-1. *Nature* 381: 661–666.
40. Liu, R., Paxton, W.A., Choe, S. et al. (1996). Homozygous defect in HIV-1 coreceptor accounts for resistance of some multiply-exposed individuals to HIV-1 infection. *Cell* 86: 367–377.
41. Samson, M., Libert, F., Doranz, B.J. et al. (1996). Resistance to HIV-1 infection in Caucasian individuals bearing mutant alleles of the CCR-5 chemokine receptor gene. *Nature* 382: 722–725.
42. Hutter, G., Nowak, D., Mossner, M. et al. (2009). Long-term control of HIV by CCR5 Delta32/Delta32 stem-cell transplantation. *N. Engl. J. Med.* 360: 692–698.
43. Gupta, R.K., Abdul-Jawad, S., McCoy, L.E. et al. (2019). HIV-1 remission following CCR5Δ32/Δ32 haematopoietic stem-cell transplantation. *Nature* 568: 244–248.
44. Tebas, P., Stein, D., Tang, W.W. et al. (2014). Gene editing of CCR33 in autologous CD4 T cells of persons infected with HIV. *N. Engl. J. Med.* 370(10): 901–910.
45. Peterson, C.W., Wang, J., Norman, K.K. et al. (2016). Long-term multilineage engraftment of autologous genome-edited hematopoietic stem cells in nonhuman primates. *Blood* 127: 2416–2426.
46. Dever, D.P., Bak, R.O., Reinisch, A. et al. (2016). Crispr/cas9 β-globin gene targeting in human haematopoietic stem cells. *Nature* 539: 384–389.
47. Xu, L., Yang, H., Gao, Y. et al. (2017). Crispr/cas9-mediated ccr5 ablation in human hematopoietic stem/progenitor cells confers hiv-1 resistance *in vivo*. *Mol. Ther.* 25: 1782–1789.
48. Dean, J., Schechter, A.N. (1978). Sickle-cell anemia: molecular and cellular bases of therapeutic approaches (first of three parts). *N. Engl. J. Med.* 299(14): 752–763.
49. Olivieri, N.F. (1999). The beta-thalassemias. *N. Engl. J. Med.* 341: 99–109.

50. Pestina, T.I., Hargrove, P.W., Jay, D. et al. (2009). Correction of murine sickle cell disease using gamma-globin lentiviral vectors to mediate high-level expression of fetal hemoglobin. *Mol. Ther.* 17(2): 245–252.
51. Pulè, M.A., Straathof, K.C., Dotti, G. et al. (2005). A chimeric T cell antigen receptor that augments cytokine release and supports clonal expansion of primary human t cells. *Mol. Ther.* 12(5): 933–941.
52. Moon, E.K., Ranganathan, R., Eruslanov, E. et al. (2016). Blockade of programmed death 1 augments the ability of human T cells engineered to target NY-ESO-1 to control tumor growth after adoptive transfer. *Clin. Cancer Res.* 22: 436–447.
53. Tsai, S.Q., Zheng, Z., Nguyen, N.T. et al. (2015). GUIDE-seq enables genome-wide profiling of off-target cleavage by CRISPR-Cas nucleases. *Nat. Biotechnol.* 33: 187–197.
54. Nobles, C.L., Reddy, S., Salas-McKee, J. et al. (2019). iGUIDE: an improved pipeline for analyzing CRISPR cleavage specificity. *Genome Biol.* 20: 14. https://doi.org/10.1186/s13059-019-1625-3.
55. Anzalone, A.V., Randolph, P.B., Davis, J.R. et al. (2019). Search-and-replace genome editing without double-strand breaks or donor DNA. *Nature* 576: 149–157.

CHAPTER 13

APTAMER PROPERTIES, FUNCTIONS, AND APPLICATIONS

WEIHONG TAN[1,2], ZILONG ZHAO[2], and LIQIN ZHANG[3]
[1]The Cancer Hospital of the University of Chinese Academy of Sciences (Zhejiang Cancer Hospital), Institute of Basic Medicine and Cancer (IBMC), Chinese Academy of Sciences, Hangzhou, Zhejiang, China
[2]Molecular Science and Biomedicine Laboratory, State Key Laboratory of Chemo/Biosensing and Chemometrics, College of Chemistry and Chemical Engineering, College of Biology, Aptamer Engineering Center of Hunan Province, Hunan University, Changsha, Hunan, China
[3]State Key Laboratory of Natural and Biomimetic Drugs, School of Pharmaceutical Sciences, Peking University, Beijing, China

13.1. INTRODUCTION

Dating back to the discovery of target-specific therapy, the development of hybridoma technology leads to the rapid development of antibody drugs [1, 2]. However, antibody as a molecular probe is imperfect in performance, specifically in its chemical modification and stability, not to mention the complex selection and production process. Alternatively, aptamers, often termed chemical antibodies, have attracted extensive attention in bioanalysis and biomedicine since 1990 [3, 4]. In addition to their high binding affinity and specificity, aptamers present many advantages, including convenient synthesis, easy chemical modification, rapid tissue penetration, non-immunogenicity, long-term stability, and low variability between different synthetic batches. Comparison between aptamers and antibodies is shown in Table 13.1 to highlight their differences.

Aptamers with single-stranded oligonucleotide sequences typically contain defined structures, due to their propensity to form complementary base pairs. They can fold into various secondary structures (e.g. stem, loop, bugle,

Nucleic Acids in Medicinal Chemistry and Chemical Biology: Drug Development and Clinical Applications, First Edition. Edited by Lihe Zhang, Xinjing Tang, Zhen Xi, and Jyoti Chattopadhyaya.

TABLE 13.1 Comparison between aptamers and antibodies

	Aptamers	Antibodies
Stability	Withstand repeated rounds of denaturation/renaturation	Easy denatured
	Temperature resistant: stable at room temperature	Temperature sensitive and require refrigeration to avoid denaturation
	Long shelf life (several years)	Limited shelf life
	Can be lyophilized	Must be refrigerated for storage and transport
	Degradable by nucleases	Degradable by proteases
	Resistant to proteases	Resistant to nucleases
Production	*In vitro* SELEX takes only 2–8 weeks	Produced *in vivo*
	No batch-to-batch variation	More than 6 months
	Cheap to synthesis	Batch-to-batch variations
		Laborious and expensive
Target potential	From ions and small molecules to whole cells and live animals	Targets must cause a strong immune response for antibodies to be produced
Size	Small molecules	Relatively large by comparison
Modifiability	Aptamers can readily and easily be modified without affinity loss	Modifications often lead to reduced activity
Affinity	High and increased in multivalent aptamers	Dependent on the number of epitopes on the antigen
Specificity	Single point mutations identifiable	Different antibodies might bind the same antigen
Tissue uptake/ kidney filtration	Fast	Slow
Immunogenicity	None or low immunogenicity	High immunogenicity
		Increased immune reaction with repeated dosing
Patents/ distribution	Exclusive patents in SELEX technology	Expired protection or no early patents
	Limited initial distribution	More widespread distribution
Development/ market	The development pathway is less explored	Well-developed infrastructure
	Insufficient education and investment (R & D support)	Abundant supports from finance and education
	Commercialization has focused on diagnostic-based aptamer products	Rapid and sustained increase in medicine market share

pseudoknot, G-quadruplex, and kissing hairpin) [5]. These secondary structures further bend into unique three-dimensional (3D) structures for specifically recognizing their cognate targets through hydrophobic interaction, electrostatic interaction, hydrogen bond, van der Waals force, shape complementarity, and base stacking, or their combination [6]. Highly compared to the way antibodies bind to antigens, these 3D interactions drive the architecture of aptamers-target complexes [7]. Structure complexity and recognition mode diversity endow aptamers ability to distinguish closely related molecules, such as target molecules that there are different in functional groups [8, 9], conformational isomers [10], or even an amino acid mutation [11].

13.2. SELEX TECHNOLOGY

Aptamers are generated via an *in vitro* evolution process known as Systematic Evolution of Ligands by Exponential enrichment (SELEX), which was independently developed by Gold's group Szostak's group in 1990 [3, 4]. SELEX is a standard methodology of obtaining aptamers, where the targets of interest are first incubated with a 10^{14}–10^{16} single-stranded random oligonucleotides library with adequate structural diversity [12, 13]. Oligonucleotides in the SELEX pool typically consist of a random region with 20–60 oligonucleotides in the middle and fixed sequence regions with 18–24 oligonucleotides on both 5′ and 3′ ends for PCR primers binding. Subsequently, the oligonucleotide-target complexes are partitioned from the free oligonucleotides. The target-binding oligonucleotides are then recovered and amplified by PCR [14]. After several repeats of selection, the desired DNA sequences (aptamers) with high affinity are enriched in the library and sequenced (Figure 13.1, [15]). The selection procedure, of no matter DNA or RNA aptamers, for any targets of proteins, whole-cells, or living tissues, all require the above three pivotal steps: selection, partitioning, and amplification.

For RNA aptamers and DNA aptamers, there are some differences in the selection procedure. In terms of DNA SELEX, a library is usually obtained by solid-phase synthesis or by separating double-strandeded PCR products for each selection cycle. In the case of RNA aptamers, the random RNA library is transcribed from a DNA library or double-stranded DNA templates, which are reversely transcribed from evolved RNA sequences. Compared to RNA aptamers, DNA aptamers are inherently more stable and with lower manufacturing costs. However, based on diverse three-dimensional structures and strong intrastrand RNA–RNA interactions, RNA aptamers usually possess a better binding capacity and specificity than DNA aptamers [16].

According to the target kinds, SELEX can be classed into purified protein-based SELEX, whole cell-based SELEX, and live animal-based SELEX. Purified protein-based SELEX has produced many aptamers with high targeting ability toward purified recombinant proteins. However, these aptamers might be challenged by loss binding ability to recognize the native conformation of the target

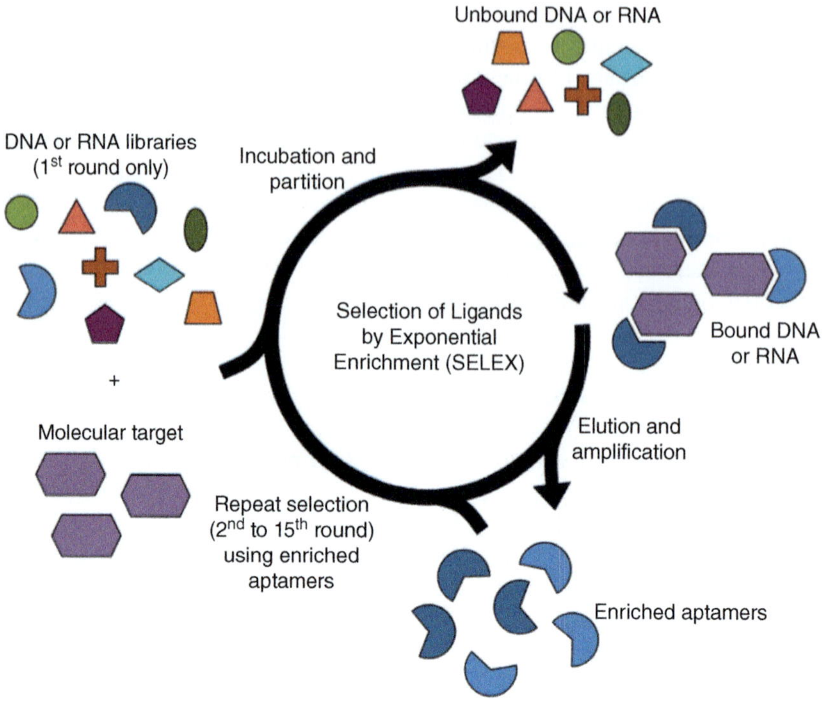

Figure 13.1 Schematic representation of SELEX technology. Source: Zhang *et al.* [15], p. 941/MDPI/CC BY-4.0.

proteins [17]. Cell-SELEX employs the intact cells as targets, which enable aptamers to selectively bind to the target protein in its native conformation, for example, the original glycosylation pattern of the extracellular domain. In addition, cell-SELEX can identify aptamers that internalized into the target cells. This increases the possibility of the selected aptamers being used directly for diagnostic and therapeutic applications [18–21]. However, the nonspecific binding of nucleic acids to dead cells impedes effective enrichment of target-specific nucleotide sequences or even result in SELEX failure [22]. To ensure the *in vivo* applications of aptamer, a live animal-based SELEX procedure has been developed to obtain tissue or organ-penetrating aptamers directly within a living animal model of the specific disease or condition. The live animal-based SELEX technology has proved its efficacy from aptamer selection to *in vivo* applications [23].

13.2.1. Recent Advances in SELEX Technology

Although conventional SELEX has demonstrated a promising and effective aptamer-selection procedure, time- and labor-consuming are significant challenges for the further development of aptamers [24, 25]. With advances in efficient separating and elution, accurate amplification and global data analysis, the overall performance of SELEX has been improved. For example, incorporation of a negative-SELEX or counter-SELEX can effectively eliminate oligonucleotides that bind to environmental components [26], such as immobilization matrix for partitioning support, or recognizing target analogs, improving selection efficacy and aptamer specificity. In addition, several partitioning technologies, e.g. capillary electrophoresis (CE) [27–29], atomic force microscopy (AFM) [30, 31], flow cytometry [5, 32], microfluidics [33], and surface plasmon resonance (SPR) [34], have been successfully adapted to the conventional SELEX procedure to improve selection efficacy of aptamers. Based on the advances in SELEX technology, aptamer selection can easily be completed in different laboratories with PCR machines.

13.3. PROPERTIES OF APTAMERS

In the last 30 years, there are plenty of researches on aptamers [35], which facilitate a comprehensive understanding of aptamers' properties and functions. Below we will introduce aptamers' general properties from structural characteristics, binding modes, target types, and chemical diversity.

13.3.1. Structural Characteristics

The distinctive structural features help understand the functionality of biomolecules. With the increase in the number of aptamers, their molecular structures' characteristics have gradually become apparent [36]. Among the aptamers reported in the early stage, the typical structure types include pseudoknots, hairpins, stem-loops/bulge, and G-quadruplexes (Figure 13.2a). For example, the aptamer against phage T4 DNA polymerase selected by Larry Gold's group in 1990 has a hairpin structure [3]. The aptamer against HIV-1 reverse transcriptase selected by the same group in 1992 has a pseudoknot structure [38]. In 1993, the ATP-binding aptamers selected by Mandana Sassanfar *et al.* showed a stem-loop structure [9]. The thrombin aptamer selected by Roman F. Macaya *et al.* in the same year presented a G-quadruplex structure [39]. Interestingly, these structures are also demonstrated by functional nucleic acid molecules found in nature, like human telomerase and riboswitches [40–42], and such similarity inspire researchers to explore aptamers' structures and properties. With the advancement in characterization methods of nucleic acid structure, the aptamers' structural details in the binding process are demonstrated. For example, Christoph Manz *et al.* characterized the SAM-1 riboswitch by single-molecule FRET microscopy and hidden Markov modeling analysis [37]. The result intuitively indicated that the SAM-1 riboswitch conformation kept shifting during several states when there exist or not the target molecule. Although aptamer structure can be simply predicted by some software. However, the accurate

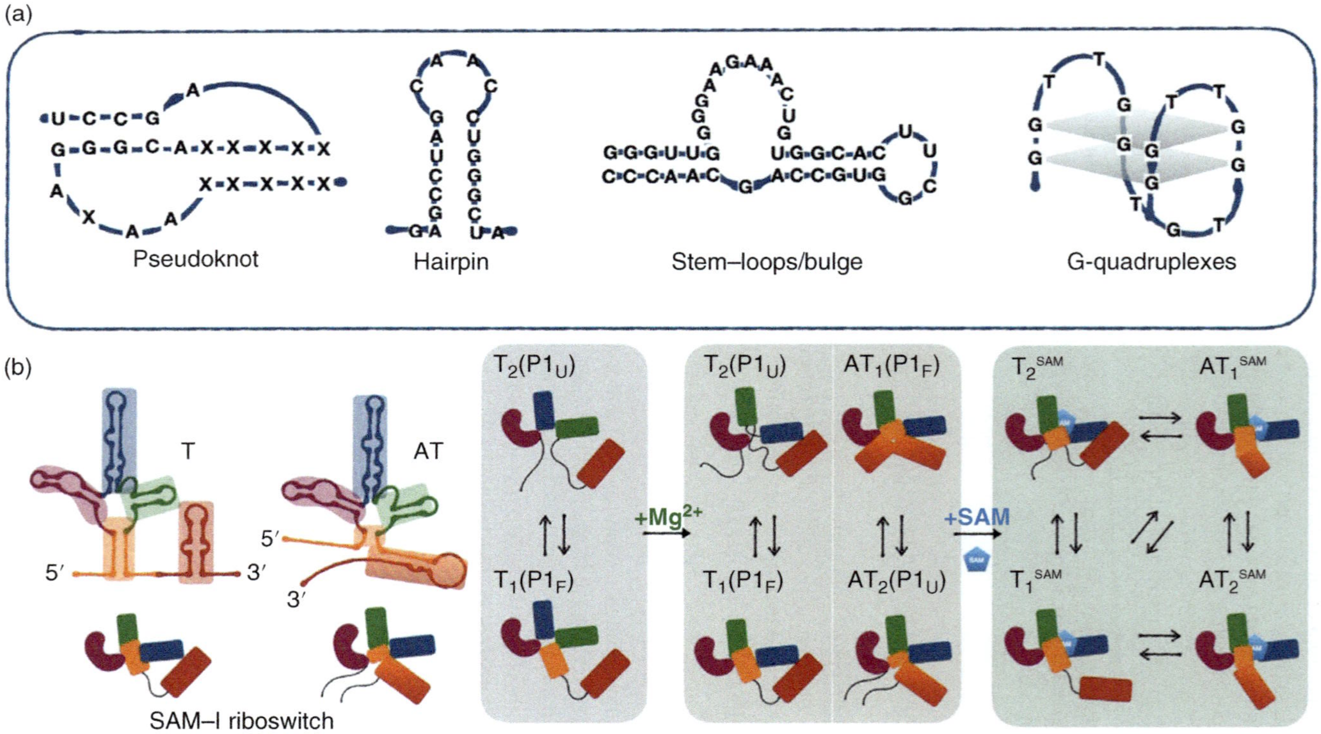

Figure 13.2 (a) The typical structure types of aptamers. It represents aptamer against HIV-1 reverse transcriptase, T4 DNA polymerase, ATP, and thrombin from left to right, respectively. (b) Different conformational states of SAM-1 riboswitch with or without the target. Source: Adapted from Manz *et al*. [37]. p. 1172.

structure information of aptamers is very high complicated and hard to explore by current techniques [43], which impedes the research and applications of aptamers.

13.3.2. Target Types

The very high diversity in sequences and structure in DNA/RNA library (10^{15}–10^{16}) ensure that aptamers can be selected against almost any targets theoretically by SELEX [35]. Since 1990, the target types of aptamers ranged from small molecules to macromolecules, viruses, virus, and live cells (Figure 13.3). Small molecules are an earlier target type involved in the field of aptamers. For example, in 1990, when Ellington *et al.* proposed aptamers' definition, the aptamer against small molecule, reactive blue 4 (B4, Figure 13.3a), were selected [4]. Subsequently, antibiotics (Figure 13.3b), duplex-stranded DNA (Figure 13.3c), and proteins (Figure 13.3d) are used as SELEX targets [44–46]. To obtain disease-specific aptamers and improve the *in vivo* application of aptamers, live cells (Figure 13.3e), virus (Figure 13.3f), and even live animals are used as SELEX targets [18, 47]. Therefore, the very wide target ranges of aptamer make aptamers hold a great perspective in analytical chemistry, biochemistry, biomedicine, and environmental monitoring.

13.3.3. Binding Modes

In general, aptamers interact with targets via hydrogen bonding, van der Waals force, base stacking interaction, electrostatic interaction, and molecular shape complementarity [48]. However, the binding modes between aptamers and targets are depended on the nature of targets. When aptamers interact with small molecules, such as aromatic compounds, hydrogen bonds are usually the main driven force [49, 50]. When aptamers interact with cationic ligands, the electrostatic interaction will be key factor [51, 52]. When aptamer interacts with the macromolecular target, like proteins, the molecular shape complementarity plays an important role in its interaction mode [6, 53]. It is worth noting that two or more binding modes are usually involved in the interactions between aptamers and targets. The interaction process between aptamers and targets is mediated by inducing fit and configuration selection (Figure 13.4a, [54]) and greatly affected by environmental factors, such as crowded environment, Mg^{2+}, and so on [55–57]. For example, Rode *et al.* found that crowded environments affect aptamers' recognition mechanism. The author took Flavin mononucleotide (FMN) aptamer for the research object and PEG200 to simulate crowded environments. In the analysis, when PEG200 was introduced, the binding process is changed from configuration selection mediated by Mg^{2+} to inducing fit (Figure 13.4b). Currently, although analytical techniques based on X-ray crystallography, nuclear magnetic resonance (NMR), surface plasmon resonance (SPR), quartz crystal microbalance (QCM), isothermal titration calorimetry (ITC), dynamic light scattering (DLS),

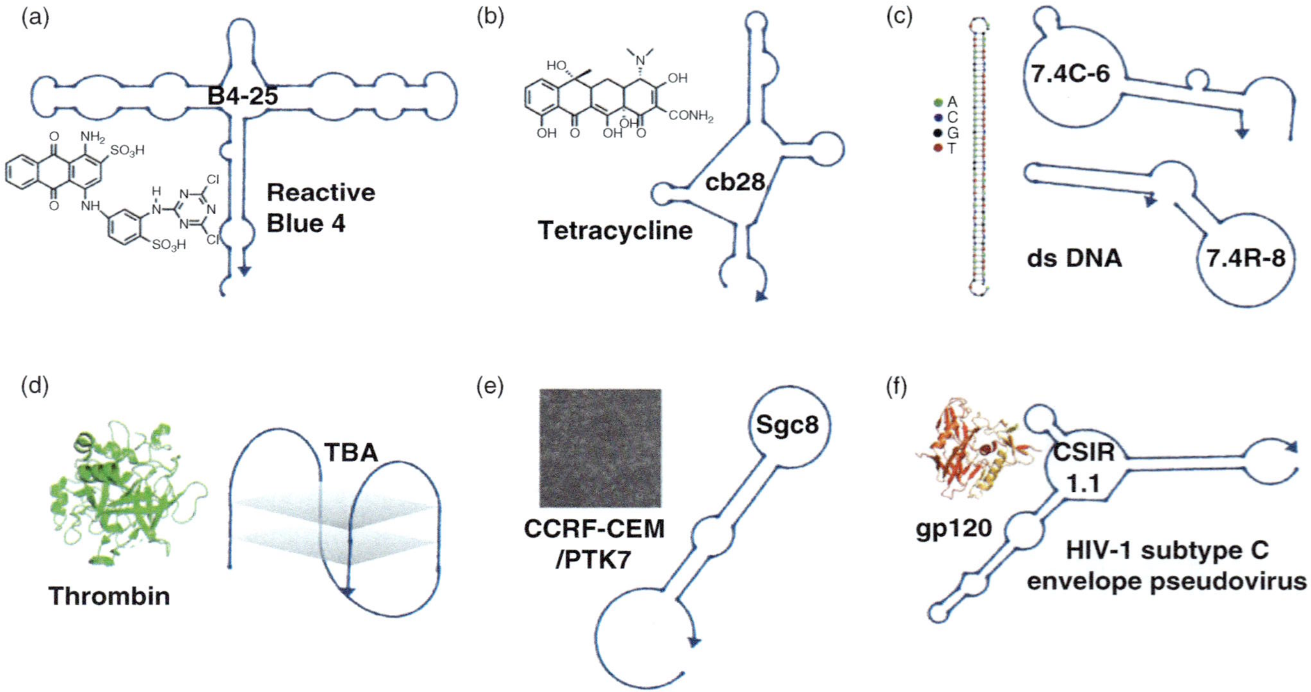

Figure 13.3 (a–f) Illustration of aptamers against different targets and their secondary structure. The types of targets are noted in the graph.

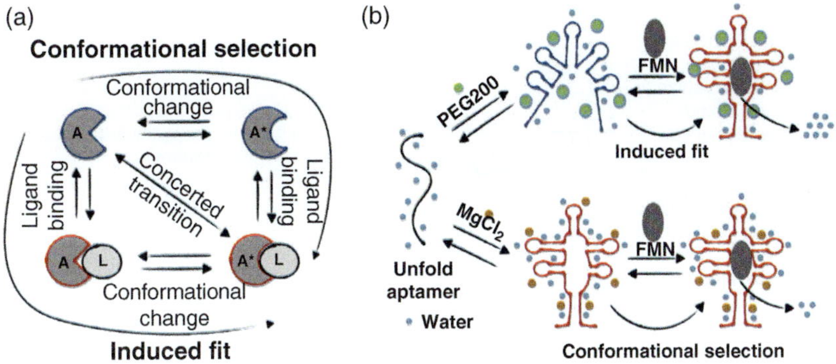

Figure 13.4 (a) Illustration of the molecular interaction mechanism. Source: Adapted from Galburt *et al.* [54], p. 11. (b) Recognition mechanism of FMN aptamer change by the environmental factors. Source: Adapted from Rode *et al.* [55], p. 6868.

circular dichroism (CD), and small-angle X-ray scattering (SAXS) have been used to study aptamer-target binding, however, the knowledge on the target-binding modes remain limited due to the lack of the detailed structural information of aptamers.

13.3.4. Chemical Diversity

Compared to the diversity carried by the side groups of amino acids, nucleotide bases in aptamers lack functional groups, which might affect the binding modes between aptamers and targets, and the binding affinity of aptamers [35]. However, aptamers are nucleic acid in nature. The robust molecular designability and accurate modification of nucleic acid can address the issue. In general, two strategies, post-SELEX modification and improved SELEX using a chemically modified nucleic acid library, are utilized to increase the chemical diversity of an aptamer. Post-SELEX modification strategy improve aptamer performance via introducing chemical groups in bases, sugar rings, phosphoric acid groups, or the entire nucleic acid backbone after aptamers are selected from DNA/RNA library (Figure 13.5a). Post-SELEX modification strategy is simple and play import roles in improving the anti-degradation of aptamers in biofluids [58]. However, post-SELEX modification probably causes the loss of binding affinity of aptamers. The improved SELEX using a chemically modified nucleic acid library are an alternative strategy to increase the chemical diversity of aptamers. For example, the artificial Ds:Px base-pair and Z:P base-pair have been utilized to prepare chemically modified nucleic acid library (Figure 13.5b, [59, 60]). Chemically modified nucleic acid library increases the sequence and structure diversity and also increases binding modes between aptamers and targets due to the introduction of functional groups. Therefore, such SELEX techniques more efficiently generate aptamers with high binding affinity than conventional SELEX. For example, Kwame Sefah

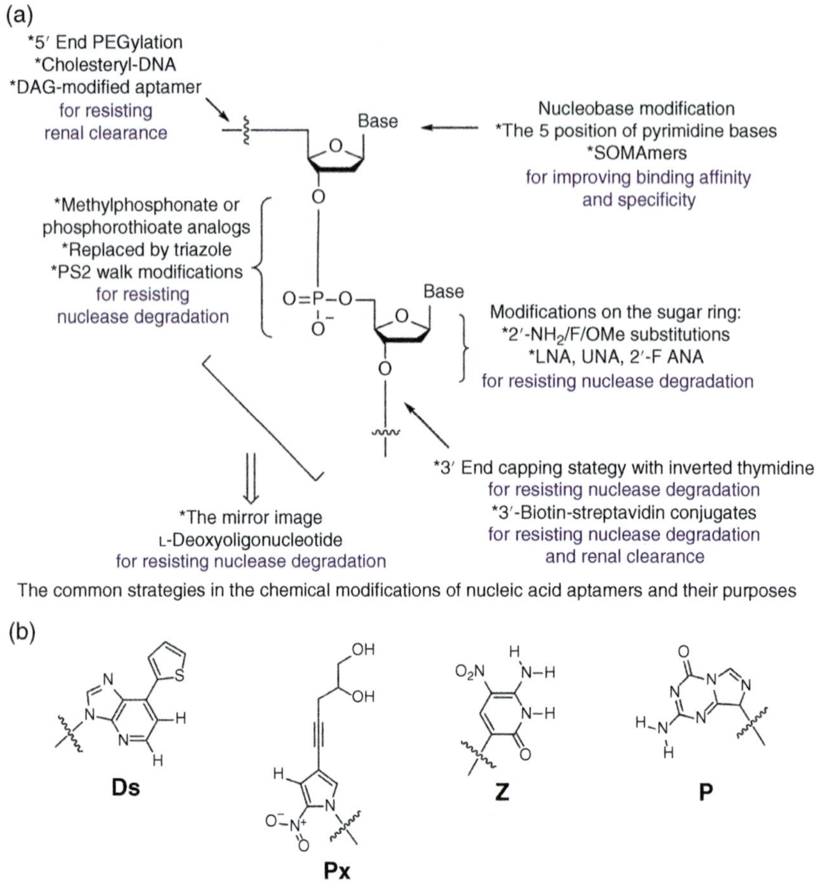

Figure 13.5 (a) Illustration of chemical modifications on aptamer. Source: Ni *et al.* [58], p. 1683/MDPI/CC BY-4.0. (b) Chemical structures of unnatural nucleic acid base-pair, including Z:P base-pair and Ds:Px base-pair.

successfully obtained high-affinity aptamers (Kd in the picomolar range) by using such library [60]. The main limitation of the strategy is that the chemically modified libraries need to be adapted to polymerases. Fortunately, advances in protein engineering and SELEX techniques make SELEX using chemically modified nucleic acid library popular, prompting the development and applications of aptamers.

13.4. APTAMER FUNCTIONS

Based on their high specificity and binding affinity, aptamers can directly bind to target proteins that were from different environments (*in vitro*/*in vivo* system, cell surface, intracellular environment, etc.) and thus interact with proteins to serve as agonist/antagonist, making them promising therapeutic agents for the application of different kinds of human disease. Usually, there are three interaction mechanisms for aptamers as agonist/antagonist: directly binding to the functional site of target proteins, preventing structural changes of target molecule, and regulating the distance or interaction between proteins. Herein, representative functions of aptamers are introduced.

13.4.1. Enzyme Inhibitor

So far, one of the most studied aptamer antagonists is thrombin-binding aptamers (TBAs) that were generated by directing a SELEX procedure against thrombin [12]. Thrombin is a multifunctional serine protease, which plays a key role in the coagulation process, thus serving as a vital target for the treatment of cardiovascular diseases [61]. Among TBAs, the 15-base TBA (TBA_{15}, 5′-GGTTGGTGTGGTTGG-3′) interacts with the exosite I of human alpha-thrombin [62], which is the binding site of fibrinogen, with a dissociation constant in the nano-molar range. Furthermore, TBA_{15} presents good specificity against thrombin and cannot recognize thrombin analogues. Therefore, the aptamer acts as an anti-coagulant agent inhibiting the activation of fibrinogen as well as platelet aggregation. Excellent performance enables the drug ARC183 developed based on TBA_{15} to be evaluated as an anticoagulant for potential use in acute cardiovascular disease in Phase I clinical trial conducted by the Archemix Corp. and Nuvelo Inc. [63]. Although the results of subsequent experiments are not ideal, it still provides valuable experience for the development of similar drugs.

13.4.2. Growth Factor Neutralization

In December 2004, the first aptamer drug, pegaptanib sodium (Macugen), was approved by the USA FDA for the treatment of all types of neovascular age-related macular degeneration (AMD) [64]. The pegaptanib sodium is a modified RNA aptamer that recognizes the 165 isoform of vascular endothelial growth factor (VEGF), which is responsible for vascular permeability and pathological ocular neovascularization. Pegaptanib works as an antagonist of VEGF. When it is injected into the eyes, Pegaptanib blocks the actions of VEGF and then reduces the growth of the blood vessels located within the eye. The clinical success of pegaptanib sodium is undoubtedly encouraging, which has laid a certain foundation for the development of nucleic acid drugs. For example, the *in vivo* half-lives of natural aptamers are generally less than half an hour because of rapid nuclease-mediated degradation [65]. In the design of pegaptanib sodium, 2′-O-methylated purines and 2′-fluorine-modified pyrimidines are introduced to improve aptamer ability to resist enzyme digestion, which provides molecular design solutions for many nucleic acid drugs used in clinical testing [66].

13.4.3. Immune Manipulation

Many important processes of the immune system require signal transmission mediated by target molecules on the cell surface. Therefore, as one of the most important molecular tools that can dynamically regulate the function of cell surface molecules (especially proteins), many aptamers have been screened and developed for immune system regulation and further used for immunotherapy of malignant tumors [67]. For example, cytotoxic T-cell antigen 4 (CTLA-4) acts as a "brake" switch for T-cell proliferation, the inhibition of CTLA-4 can therefore attenuate T-cell-mediated immune responses. In 2003, Santulli-Marotto *et al.* reported RNA aptamers that recognize and bind CTLA-4 with high affinity and specificity [68]. In addition to inhibit CTLA-4 function *in vitro*, these aptamers can be used in tumor research of mice models to enhance tumor immunity, thus manipulating the immune system for therapeutic application. Another strategy of immunotherapy is to stimulate T-cell activation. By manipulating costimulatory signaling molecules (such as CD28, 4-1BB, and OX40) on the surface of T cells, aptamers screened by researchers against target molecules can initiate costimulatory signaling to boost T cell survival [69–71]. These aptamers have to form dimeric and/or multimeric aptamer structures to induce a receptor oligomerization, efficiently stimulating T cells and inhibiting tumor growth as compared to agonistic antibodies. In addition to directly regulating the costimulatory signal pathways, aptamers that can block related receptors (such as IL10R, IL4Rα, and IL6) have also been developed for regulating the activity of corresponding immune cells [72–74]. As for other similar strategies to further regulate its function by manipulating the activity of immune cell surface receptors, please refer to other reviews [75].

13.4.4. Antivirus

Virus infection can induce serious diseases in humans and animals. Similar to antibodies, another potential application of aptamers is to resist/defense against viral infections. A large number of aptamers have been generated and utilized to antagonize various viruses, such as HIV-1, HCV, HBV, SCoV, HPV, HSV, influenza, and virus [76, 77]. In these cases, aptamers were mainly used to block the binding of the virus to host receptors or to inhibit the activity of related transcriptional replicase proteins. For example, surface glycoprotein (gp120) is an essential surface receptor for virus entry into cells and several aptamers selected against gp120 have been demonstrated to show antiviral activity [78]. In 2003, Khati *et al.* reported the isolation of 2-F-modified RNA aptamers that bind to the gp120 of HIV-1 that potently neutralized HIV-1 infectivity in human cells [79]. HIV-1 reverse transcriptase, a multifunctional enzyme during the retroviral life cycle, which can catalyze the conversion of single-stranded genomic viral RNA into double-stranded DNA, was also set as an ideal target since Tuerk and Gold identified RNA aptamers against HIV-1 reverse transcriptase from an

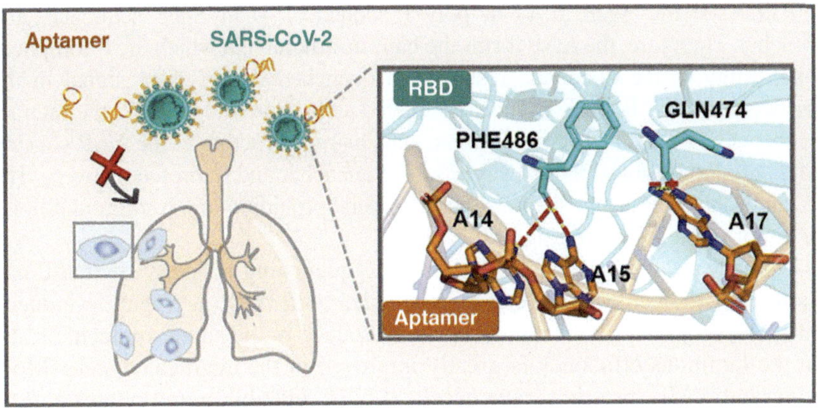

Figure 13.6 The mechanism of aptamer blocking SARS-CoV-2 infection.

RNA pool in 1992 [38]. Since then, various aptamers have been raised against HIV-1 reverse transcriptase and to be explored for the inhibition of virus replication. For example, a single-stranded DNA aptamer (ODN112) for the RNase H domain of HIV-1 RT reverse transcriptase was isolated to this regard [80]. More recently, SARS-CoV-2 has greatly affected the health of all human in the world. In terms of virus-targeted therapy, inhibiting the interaction between the S protein receptor-binding domain of SARS-CoV-2 and the ACE2 receptor of human epithelial cells is considered an efficient and specific treatment way. Sun *et al.* reported the development of an aptamer blocking method to inhibit the SARS CoV-2 virus infection [81], which provides a new direction for the development of aptamer-based drugs for the COVID-19 and other emerging infectious diseases (Figure 13.6).

13.5. APTAMER-BASED ANALYSIS METHOD

13.5.1. Aptamer-Based Targeted Imaging

Even without prior knowledge about diseased cells, aptamers can differentiate the differences between diseased cancer cells and normal cells. In addition, based on the programmable self-assembly and excellent molecular computing ability of DNA, aptamer are desired molecular recognition tools to explore the basic laws of life and study various diseases [35, 82]. Up to now, various aptamer-based biosensors with highly controllable structures and functions have been rationally constructed to detect a very wide range of targets, including metal ions, small organic molecules, proteins, cells, tissues, and so on [66, 83]. For example, Wang *et al.* designed an activatable aptamer probe for tumor imaging. In this study, the AAP (activatable aptamer probe) is composed of five parts from 5′ to 3′ terminal: fluorescent dyes (e.g. FAM, Cy5), an 8-nt sequence complementary to the 3′-terminal of PTK7-specific Sgc8, a 17-nt poly-T sequence, 41-nt Sgc8 sequence, and quencher. Therefore, the AAP forms the hairpin structure, in which the fluorophore is proximity to the quencher, resulting in quenched fluorescence signal in the absence of PTK7. In the presence of targeted cells expressing PTK7, the interaction between PTK7 and Sgc8 can change the hairpin structure of the AAP, causing the separation of fluorophore away from the quencher and fluoresce recovery. The design of the APP is general for other aptamers, providing low background noise for tumor fluorescent imaging [84].

Contrary to the aggregated fluorescence quenching property of traditional fluorescent molecules, fluorescent molecules with AIE (aggregation-induced luminescence) effect have almost no fluorescence in the single-molecule state, but the luminous efficiency is greatly improved in the aggregated state. More importantly, AIE materials have excellent biocompatibility and photostability, which make them have obvious advantages in biological detection and imaging applications [85]. Thus, if aptamer can be combined with AIE materials effectively and easily, the design of fluorescent sensor based on aptamer will be greatly expanded. Given this, Tang *et al.* developed a simple and universal method to modify AIE molecules with aptamers for tumor-targeted imaging. As shown in Figure 13.7, taking AS1411 aptamer as an example, hydrophilic aptamer molecule will transform into an amphiphilic molecule by modifying cholesterol molecule at the end of AS1411 aptamer. Thus, cholesterol-modified aptamer could be anchored on the surface of the liposome by the strong interaction between cholesterol and hydrophobic cavities formed by phospholipid self-assembly. Moreover, hydrophobic AIE molecules will also be spontaneously

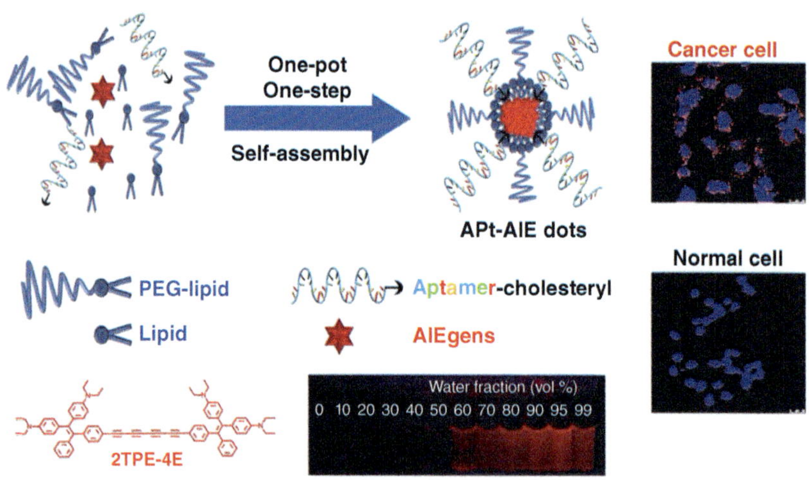

Figure 13.7 Schematic illustration of aptamer-anchored AIE-dots (Apt-AIE dots) preparation by one-pot self-assembly method. Source: Zhang *et al.* [86] / with permission of American Chemical Society.

encapsulated in the liposome in the self-assembly process, thus forming a state of aggregation, and the fluorescence is greatly enhanced. Cell targeting experiments showed that the AS1411 aptamer modified on AIE liposome still keeps its molecular recognition ability and can specifically recognize tumor cells with high nucleolin expression (such as breast cancer cell line MCF-7 and small-cell lung cancer cell line A549), but there is no obvious fluorescence signal on normal cells. Although only AS1411 aptamers were used in this paper, this method should also be easily extended to other aptamers. Therefore, it provides a shortcut for the construction of AIE smart probes with excellent molecular recognition ability [86].

With the development of precision medicine, people are increasingly aware that the occurrence of the disease is essentially derived from the changes at the molecular level. However, due to the complexity of the occurrence of diseases, the information differences of a single molecule are difficult to be completely and directly translated into the development of the disease. Therefore, molecular profiling of multiple molecular information related to diseases at the same time is expected to provide a more accurate way to achieve a precise diagnosis of diseases. Sun's group reported an aptamer-based thermophoretic technique to profile extracellular-vesicle (EV) surface proteins for the early detection and classification of cancers. In this study, fluorescent dye-labeled aptamers first incubated EVs for 2 h. The mixture was then applied on a microfluidic chip irradiated by a 1480-nm laser. After irradiation, a temperature gradient is generated in fluids. Based on thermophoresis, EVs of 30–1000 nm tend to quickly converge to the center of the chip, where EVs can be enriched 1400 times, while the free aptamer remains dispersed. After EVs enrichment, the fluorescence signal from aptamers is in direct proportion to protein expression on EVs surface. Thus, the EV membrane proteomics information could be easily obtained by a panel of aptamers. By using a set of seven aptamers, combined with a machine learning algorithm, EV membrane proteomics information was used for early automatic typing detection of six types of cancer, prostate tumor liquid biopsy, and prostate cancer recurrence monitoring. Overall, this work establishes a low-cost blood test to analyze EV surface proteins in serum, which may potentially translate into standard clinical practice for cancer diagnosis [87].

Besides, molecular imaging technology can provide real-time and noninvasive imaging information at the molecular level of the target site, which is the frontier field of imaging research, and the key lies in the selection of molecular probes. The nucleic acid aptamer is a kind of oligonucleotide with high affinity and specificity to corresponding target molecules, which is superior to traditional antibodies in targeting, stability, and safety, and has shown a good application prospect in the fields of targeted drug transport and disease diagnosis, so it is an ideal molecular probe. Nucleic acid aptamers can mediate optical quantum dots, magnetic nanoparticles, and ultrasound contrast agents to realize optical, magnetic resonance, ultrasound, and multimodal multifunctional imaging of target sites, which further promotes the application of molecular imaging in the field of disease diagnosis and treatment [88, 89].

13.5.2. Protein-Specific Modification Based on Aptamer

Protein-specific modification is the latest development trend of protein chemistry, and it is also a key step to manipulate protein functions. The genetic engineering protein developed in the past is time-consuming and laborious. To achieve precisely and efficiently chemical modification of the protein, Tan's research group reported a "recognition-then-reaction" strategy based on aptamer, which can realize the site-selective recognition and efficient conjugation of protein with DNA in a one-step reaction [90]. In this study, DNA aptamer extended with 20-nt DNA sequence, which hybridizes complementary DNA sequences with an amine-reactive group, was designed. When aptamer interacts with the target protein, resulting in the amine-reactive group close to the protein. Therefore, the amine-reactive group can be efficiently with amine groups in protein. After removing aptamer by using complementary, the protein–DNA conjugate was left for further utilization. The strategy can make the protein be selectively modified by DNA sites in a mild and biocompatible strategy for protein precise modification.

Furthermore, as shown in Figure 13.8, Tan's group designed a new chemical method, which used aptamer-oriented bioorthogonal reaction to couple different molecules on the target protein, to carry out multicolor imaging and cell surface nano-engineering, providing a comprehensive strategy for exploring cell activity [91]. In this process, the aptamers guide the functional molecules

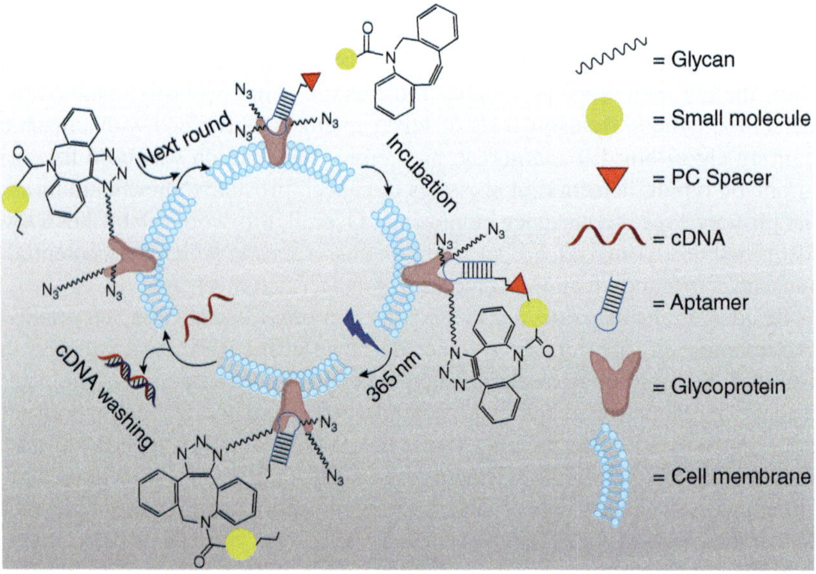

Figure 13.8 Aptamer-induced protein-specific bioorthogonal modification technology to deliver and tag functional molecules onto a specific protein. Source: Cheng *et al.* [91], p.37845 / with permission of American Chemical Society.

(e.g. fluorescent dye) and dibenzocyclooctyne to the specific glycoprotein labeled with azide. The interaction between aptamers and targets makes dibenzocyclooctynes proximate to azides and triggers click reaction, resulting in the covalent crosslinking of the functional molecules and targets. Because a photocleavable (PC) spacer is inserted between the aptamer and the functional molecule, aptamers can be removed from the cell surface by complementary DNA strands after UV irradiation, which cleaves the PC linker. Finally, only the functional molecules are successfully covalently crosslinked on the target glycoprotein. This method provides a new strategy to couple multiple functional molecules on specific proteins for multicolor protein imaging, cell membrane nano-engineering, and determining therapeutic targets.

13.5.3. Biological Separation Based on Aptamer

In recent years, the liquid biopsy of circulating targets including circulating tumor cells (CTCs) in blood and other body fluids has been considered as one of the frontiers in malignant tumor diagnosis [92, 93]. Because CTCs generally exist in billions of normal cells with several to dozens of cells, clinical analysis and identification of CTCs are like looking for a needle in a haystack, which is a great challenge. Given this, Yang's group proposed a new strategy of multiscale collaborative capture and enrichment of microfluidic chips based on hydrodynamic separation and bionic multivalent recognition [94]. The bionic octopus microfluidic chip designed in this study consists of tens of thousands of microcolumns modified with aptamer probes that can specifically capture CTCs. These microcolumn arrays can capture cancer cells, but not normal cells based on the differences in hydromechanics on the micron scale. CTCs with large sizes collide with microcolumns repeatedly, which increases the probability of being caught by aptamers. While normal cells rarely collide with microcolumns and move along the direction of liquid flow because of their small size, thus avoiding nonspecific capture. It is worth noting that aptamers are immobilized on the surface of each microcolumn via gold nanoparticles as bridge substrates. It is calculated that hundreds of aptamer probes are modified on each gold nanoparticle. Therefore, these aptamers can cooperate to capture CTCs like octopus tentacles. The capture efficiency is increased by three times compared with monovalent aptamers. After adding glutathione, aptamers fall off the gold nanoparticle interface, ensuring the lossless release of CTC for subsequent cell culture or gene mutation analysis.

13.6. APTAMER-BASED BIOMARKER DISCOVERY

The characteristic molecules, typically termed as biomarkers, on the cell membrane are actively involved in a myriad of biological events in cells, such as environmental condition sensing, signal transduction, as well as the downstream pathway activation [95–97]. In particular, serving as the targets of 66% of approved drugs in the DrugBank database, such biomarkers are of substantial biomedical significance [98]. The in-depth interpretation of the molecular codes on the cell surface will absolutely benefit the understanding of the onset and progression of diseases and offer new insights for the therapeutic responses, thereby favoring the clinical strategy of precision medicine and personalized therapy. However, currently suitable molecular tools for the discovery of the biomarkers are still lacking, despite the emerging assistance from deep learning and big data analysis [99, 100], yet some biased artifacts that originate from the mismatch between acquired peptide sequences and protein fragments cannot be clearly avoided.

Cell membrane proteins play key roles in various physiological functions. Dysregulation and abnormal expression of membrane proteins usually take place in cancer cells. Thus, membrane proteins are important kind of potential disease biomarkers. However, the identification of low-abundance proteins in complex sample with wide dynamic range of abundance of membrane proteins, such as cell lysate, remains a formidable challenge for proteome analysis. Aptamer, emerging as one of the artificial molecular tools, offers great advantages to decipher the codes on the cell surface without requirement for the prior knowledge on the biomarkers [18, 101]. The specific molecular recognition between aptamers and their targets makes it possible to achieve the identification of both quantitative and qualitative differences between normal and diseased proteomes. In addition, the negative selection with the normal proteome in cell-SELEX and aptamer-based affinity purification facilitates the identification of less abundant biomarkers that are usually missed by most mass-spectrometric techniques due to the presence of more abundant proteins. Therefore, aptamer-based biomarker discovery strategy has great promise in biomarker discovery.

To discovery unknown biomarker by aptamer, cell-SELEX, target identification and biological function analysis are usually involved (Figure 13.9). Cell-SELEX generates evolved nucleic acid library or aptamers differentiate the molecular difference between disease cells and normal cells (left panel in Figure 13.9). Then the evolved nucleic acid library or aptamers are incubated diseased cells or cell lysate, the bound molecules can be pulled down and purified by mass spectroscopy (Middle panel in Figure 13.9). In the following steps, the biological function and clinical significance of the characteristic target will be evaluated for further study (Right panel in Figure 13.9).

Pioneered by Tan's lab, the aptamer-based biomarker discovery strategy undergoes rapid progress in the recent decades [18, 101]. Up till now, a variety of excellent potential biomarkers on the cells have been discovered in his lab, for example PTK7 in leukemia cells [102], STIP1 in ovarian cancer [103], CD109 in nasopharyngeal carcinoma [104], and CD71 in pancreatic cancer [105]. Additionally, it has also been proved to be an excellent approach by more than 40 labs worldwide, such as the CD 63 in vemurafenib-resistant melanoma [106], PrPc in prostate cancer [107], integrinα4 in leukemia [108], and so

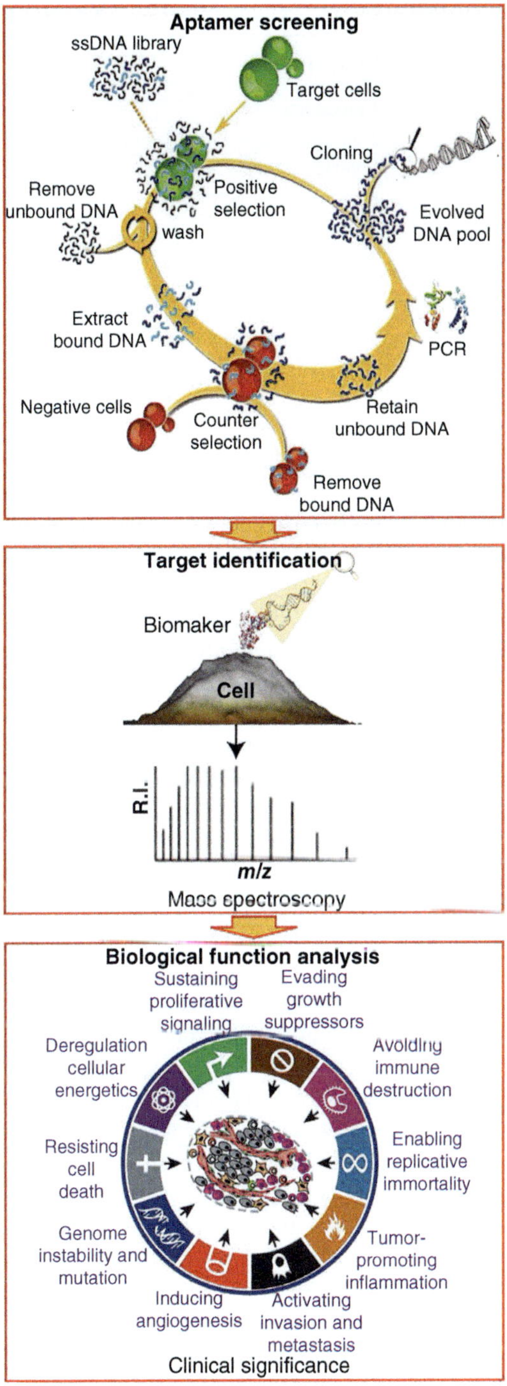

Figure 13.9 The strategy of aptamer-based biomarker discovery in cells.

TABLE 13.2 The biomarkers have been identified by DNA aptamers

Biomarker	Cell type	Aptamer
PTK7	Leukemia	Sgc8
CD109	Nasopharyngeal carcinoma	S3
STIP1	Ovarian cancer	TOV6
CD71	Pancreatic cancer	XQ-2d
CD63	Vemurafenib-resistant melanoma	LL-4A
L1CAM [109]	Neuroblastoma	YLY-12
PrPc	Prostate cancer	Wy-5a
AP dimers [110]	--	BG2
Integrin α4	Leukemia	Sgc4e

on, as summarized in Table 13.2. Notably, above mentioned biomarkers have been proved to be promising in diagnosis and therapy, therefore reinforced the significance of aptamers in biomarker discovery [87, 111, 112].

In the treatment of cancer, heterogeneity is a knotty problem lying ahead, because the molecular signatures of cancer cells vary significantly in different locations and in different patients. As a result, for cancer treatment, it calls urgently for precision medicine and personalized therapy [113, 114]. Based on the discussion above, the aptamer-based biomarker discovery strategy can provide a promising approach to overcome the problems ahead. Recently, tumor tissue-based and *in vivo*-based aptamer screening techniques have been developed [115, 116]. Consequently, the information of the specific biomarkers in heterogeneous tumor cells can be received. Thus, such an approach offers a new insight to combat cancer.

13.7. APTAMER–BASED TARGETED THERAPY

Cancer has been one of the severe lethal diseases all over the world. Among limited effective therapeutic methods, chemotherapy has been adopted generally for cancer therapy and presents modest survival benefits [117]. However, nonnegligible limitations associated with poor tissue selectivity have caused an unwanted effect, where patients succumb to the adverse effects of myelosuppression and cardiotoxicity far earlier than the elevation of the tumor burden [118]. Therefore, it's a critical step to improve the selectivity of therapeutic drugs for maximizing the advantages of chemotherapy.

Current strategies for targeted therapy include screening small-molecule drugs that specifically bind to biomarkers to inhibit their biofunctions and conjugating drug with targeted ligands, such as an antibody. Antibody-drug conjugates (ADCs) have existed for a long time and some have been approved for clinical treatment [119]. However, aptamers exhibit several excellent performances compared to antibodies, such as simple aptamer screening and synthesis, a rich variety of target molecules, faster tissue penetration, and so on [111]. Hence, aptamer-drug conjugates (ApDCs) are more attractive for targeted therapy. Some representative ApDCs in recent years will be described in the section below, revealing that aptamers can improve chemotherapeutic efficacy and promote the cancer targeted therapy progress.

13.7.1. Aptamer-Drug Noncovalent Conjugates

By advantage of the chemical properties of oligonucleotides and small-molecule drugs, ApDCs can be achieved through noncovalent conjugation. In 2006, the Farokhzad group reported the first ApDCs in the real sense, where DOX was intercalated into a PSMA-targeting RNA aptamer, A10 (Figure 13.10a, [120]). Dox and other closely associated drugs, such as doxorubicin, have been shown to be intercalated within the DNA GC sequence due to the presence of flat aromatic rings in these molecules. A10 was an RNA aptamer that binds with PSMA. Upon conjugated to A10, ApDCs targeted PSMA-expression LNCaP cells but not PSMA-non-expression PC3 cell, while free DOX readily diffused through the plasma membrane of LNCaP and PC3 cells indistinguishably. Cell cytotoxicity results showed selective toxicity for ApDCs between targeted and untargeted cells. Subsequent researches demonstrated that upon internalization into targeted cells, apatemr-DOX conjugates were trafficked into lysosomes, where the low pH microenvironment triggered a structural change of ApDCs and rapid release of DOX payload exclusively within cells, leading to cell death. Importantly, ApDCs rapidly released their DOX payload when the pH dropped below 6.0 and reached maximum release at pH 5.0 (Figure 13.10b, [121]) Beyond that, DOX unloading from internalized ApDC was also facilitated by other intracellular factors such as ionic environment and nuclease degradation [124]. All these studies suggested that the targeting ability of aptamer can improve therapeutic efficiency tremendously.

In the above-mentioned PSMA-DOX conjugate, ApDCs showed an approximately 1:1.2 molar equivalence of DOX to the aptamer. To obtain a higher drug-loading capacity and achieve more excellent therapy outcome, a series of researches based on aptamer-DOX conjugates were carried out. Zu group selected aptamer #1S which could specifically recognize cellular CD38 proteins and CD38-expressing cells [121]. For drug delivery, they added a CG-repeat structure at the end of an aptamer to intercalate more DOX. They measured the formed ApDCs was at a 1:5 molar ratio of an aptamer to DOX. Zhu *et al.* took advantage of DNA nanotechnology to exploit an aptamer-tethered DNA nanotrain (aptNTr) [122]. The nanotrain was self-assembled simply from two relatively short DNA building blocks upon initiation of aptamer-tethered trigger probes through hybridization chain reaction (HCR). The cavity structures could be loaded with DOX through intercalation. Aptamer served as a locomotive to guide a series of "boxcars" carrying drug toward target cells. The long linear "boxcars" provided enough spatially addressable sites, allowing high-capacity

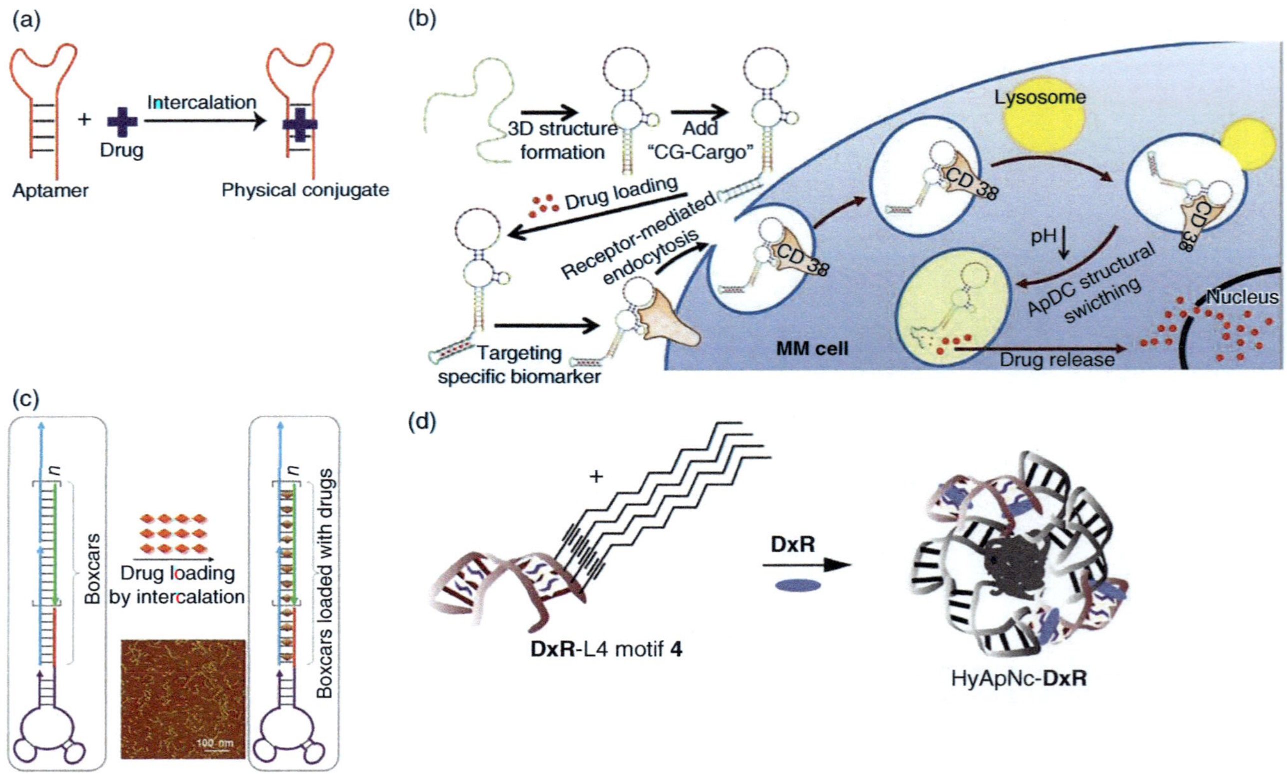

Figure 13.10 Noncovalent conjugation of aptamer-DOX. (a) Physical-conjugate formation between an aptamer and a model drug. Source: Bagalkot *et al*. [120], p.8149 / with permission of John Wiley & Sons. (b) A unique ApDC was formulated by synthesizing a CD38-specific aptamer with a CG-cargo structure. Source: Wen *et al*. [121], p.987 / with permission of Springer Nature. (c) The long nanotrains were tethered with aptamers working as locomotives on one end, with multiple repetitive "boxcars" to be loaded with molecular drugs. Source: Zhu *et al*. [122] / with permission of National Academy of Sciences. (d) Lipidated versions of a cMet-binding aptamer and a separate lipidated GC-rich DNA hairpin motif loaded with intercalated doxorubicin self-assemble into spherical hybrid-nanoconstructs that specifically bind cMet. Source: Prusty *et al*. [123], p.535 / with permission of Springer Nature/CC BY-4.0.

loading of DOX. In this design, the sgc8–NTr/drug molar ratio of 1/50 was achieved. Sgc8–NTrs were verified to have the selective recognition ability between target CEM cells and nontarget Ramos cells and potent cytotoxicity was confirmed in CEM cells, but much less cytotoxicity in Ramos cells. Based on the intercalation of DOX in GC base pair, Prusty *et al.* developed supramolecular aptamer nano-constructs for receptor-mediated cancer-cell-targeted therapy (Figure 13.10d, [123]). The researches demonstrate that aptamer-drug nonvalent conjugate provide a kind of simple and efficient targeted drugs for cancer therapy.

13.7.2. Aptamer-Drug Conjugates (ApDC)

As mentioned above, accurate chemical modification can be easily achieved in solid-phase synthesis, providing various functional groups, such as -SH, NH_2, -COOH, -DBCO, and -N_3, for conjugating drug molecules. Therefore, aptamer-drug conjugates can be easily prepared via covalent bonds. Compared to oligonucleotides could be modified with numerous active groups during solid-phase synthesis and these modified aptamers possess the chemical activity to react with various drugs to form stable chemical bonds. Huang *et al.* reported the first aptamer-drug conjugate through covalent conjugation [125]. They chose a hydrazone linker to conjugate a PTK7 targeting aptamer, sgc8c, with DOX (Figure 13.11a). When aptamer recognize targeting CEM cell with PTK7 high expression, ApDCs were internalized and located in the endosome. In an ultra-acidic environment of endosome (pH 4.5–5.5), the hydrazone linker was cleaved and DOX was released from ApDCs. The cell cytotoxicity results showed that ApDCs exhibited the same toxicity to CEM cells with free DOX while the control group, TDO5-DOX conjugate, where TDO5 had no targeting ability to CEM cells, was similar to free sgc8c. Meanwhile, ApDCs was much more toxic to CEM cells than nontargeted NB-4 cells and Ramos cells.

After this successful conjugation, a series of ApDCs with novel covalent conjugations and much more highly toxic drugs were well designed, and their physiological properties were systematically explored. Zhang's group utilized a cathepsin B-labile valine-citrulline dipeptide linker to conjugate nucleon binding aptamer, AS1411, and the first-line drug for many kinds of cancers, paclitaxel [124]. The authors selected SKOV3 cells as targeted cells that highly expressed nucleon on the cell membrane. ApDCs were postulated to be intact during blood circulation and PTX lost their antitumor activity as the activity site was occupied by the linker. Once they were delivered to tumor cells, AS1411 recognized nucleon and receptor-mediated endocytosis occurred. ApDCs trafficked into the endosome, where the high distribution of cathepsin B cleavage valine-citrulline dipeptide linker to be hydrolyzed and release active PTX. For the cell growth inhibition experiment, ApDCs exhibited higher cytotoxicity to SKOV3 cells than to normal cells. In the human ovarian cancer xenografted mouse model, ApDCs were prior to accumulating to tumor site than other normal

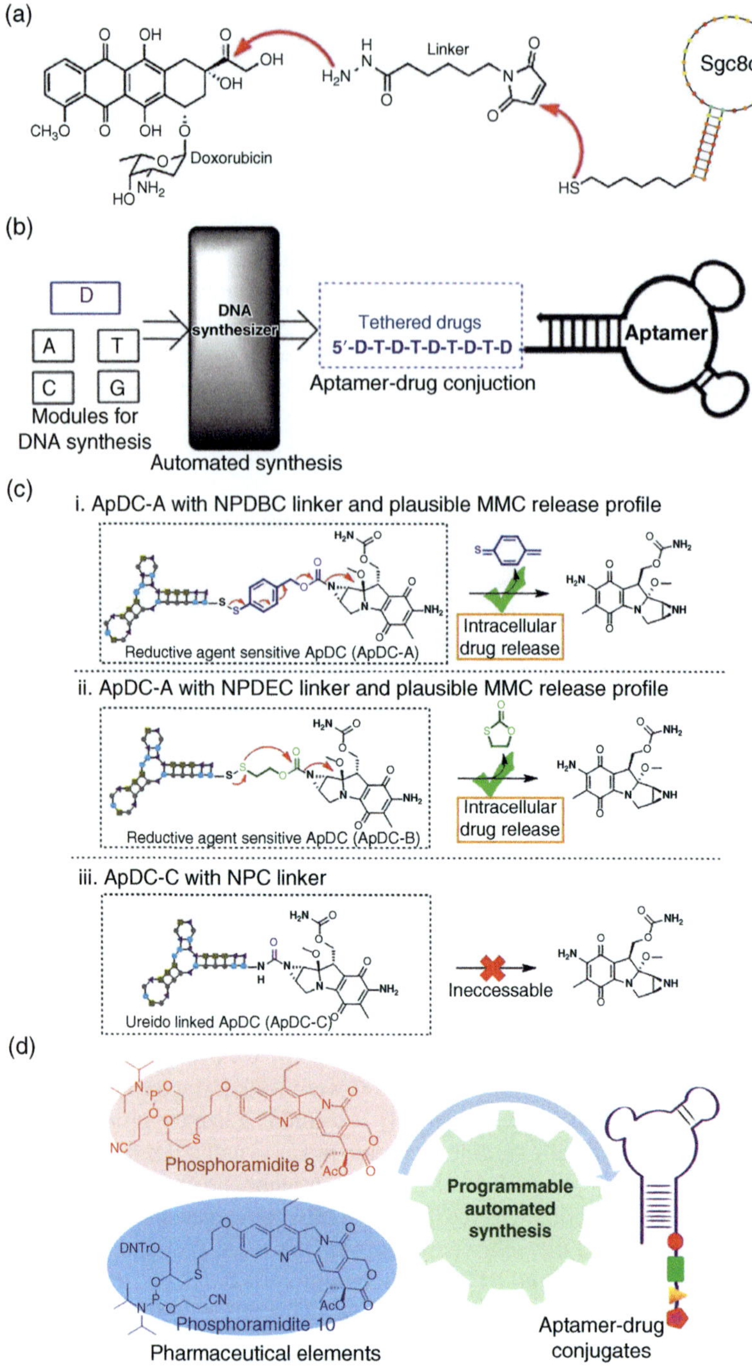

Figure 13.11 Covalent conjugation of aptamer-drug conjugation. (a) Sgc8c-DOX conjugates linked by a hydrazone linker. Source: Huang *et al.* [125], p. 862/John Wiley & Sons. (b) Synthesis of three aptamer–MMC conjugates with different linkers. Source: Yang *et al.* [126], p. 2532/American Chemical Society. (c) Automated modular synthesis of aptamer–drug conjugates. Source: Wang *et al.* [127] / with permission of American Chemical Society. (d) Repurposing CPT as drug phosphoramidites for automated modular synthesis of aptamer–drug conjugates. Source: Adapted from Wang *et al.* [128], p. 2000074.

organs and the antitumor ability of PTX was tremendously enhanced due to the conjugation of the aptamer. Beyond the excellent stability and targeting ability, ApDCs also performed the lack of immunogenicity and toxicity to other organs, making a solid foundation for the clinical translation. Yang *et al.* focused on the impact of linkers on the toxicity of aptamer-drug conjugates [126]. They synthesized three different XQ-2d-Mitomycin C (MMC) conjugates linked by nitrophenyl 4-(2-pyridyldithio) benzyl carbonate (NPDBC), (4-nitrophenyl 2-(2-pyridyldithio) ethyl carbonate (NPDEC), and 4-nitrophenyl chloroformate separately (Figure 13.11b). The active site of MMC was occupied when they were conjugated with linkers and the drug efficacy was counteracted. The former two conjugates were GSH responsive and when they were trafficked into targeted cancer cells, drug release occurred rapidly under intercellular reductant microenvironment. The last one only could become druggable when they were digested by endosome enzymes. The cell growth inhibition results demonstrated reductant-responsive ApDCs exhibited much higher toxicity to tumor cells than nonresponsive ApDC and a higher release rate resulted in a more toxic effect for the two responsive ApDCs. This study suggested that the cytotoxicity of newly developed ApDCs could be tuned using rational cross-linking chemistry.

13.7.3. Automated Modular Synthesis of ApDCs

Except for the traditional covalent conjugation methods, ApDCs can be automatically synthesized by a DNA synthesizer. Wang *et al.* first synthesized a drug phosphoramidite for ApDCs automated synthesis [127]. In this case, the antitumor drug fluorouracil (5-FU) was chemically modified into a modular phosphoramidite (Figure 13.11c). This drug phosphoramidite can be integrated automatedly and controllably into an aptamer at any predesigned position through a DNA solid-phase synthesizer. To better control with drug release, the photocleavable linker (PC-linker) was introduced in the backbone of oligonu cleotides. The 5-FU-phosphoramidite-based ApDCs exhibited similar toxicity as the naked drug, but better selectivity to targeted tumor cells. Based on the desired cellular cytotoxicity and synthesis convenience, other therapeutic drugs, such as camptothecin (CPT) and Combretastatin A4 (CA4), have been utilized to synthesize drug phosphoramidites as drug models for preparing ApDCs (Figure 13.11d, [128]) To further prompt the applications of ApDCs, Wang *et al.* summarized the cytotoxicity mechanism was a five-step interaction between ApDCs and target cells: receptor-mediated binding, macropinocytosis endocytosis, endo/lysosomal escape, destabilizing tubulin, and inducing apoptosis. One of the molecular events, drug release, was triggered by abundant endo/lysosomal and cytoplasmic phosphatases. Once the phosphatases activity was inhibited, a little CA4 signal in HPLC spectra was monitored. Besides, the *in vivo* therapeutic and safety evaluation indicated the specificity and reliability of solid-phase ApDCs, prompting the development of aptamer-mediated targeted therapy and precise medicine.

13.7.4. Aptamer–Synergistic Drug Conjugates for Improving Targeted Therapy Index

Combination therapy simultaneously inhibits multiple physiological activities, efficiently lowers the dosage of individual drugs at specific molar ratios and overcomes resistance mechanisms, thus providing a new strategy for efficient chemotherapy. Most ApDCs were in the form of one kind of drug per aptamer, which limited their therapeutic efficacy. Pusuluri *et al.* reported a novel design conjugating aptamer with peptides where drug combination at optimized molar ratio was preloaded (Figure 13.12a, [129]). They used DOX and CPT as drug models. The conjugate inhibited tumor growth *in vivo* at extremely low doses of DOX (500 μg/kg/dose) and CPT (350 μg/kg/dose). The dosage was approximately 8-fold and 21-fold lower than what was required for DOX and CPT individually to achieve similar effects, respectively, suggesting the successful construction of a synergistic drug carrier. Other than using peptides as the linkers for the precise loading of discrete drugs to the aptamer, Zhou *et al.* constructed a combination therapy system by directly chemically conjugating different drugs to aptamer bases [112]. They adopted a circle aptamer to load two different drugs. Two XQ-2d monomers were extended with 13-nucleotide complementary sequences at 5′-end. They were modified with CPT or PTX respectively and ligated by DNA ligase to obtain circular bivalent aptamer–drug conjugates (cb-ApDCs), which could be tunable on the drug ratios for optimal therapeutic efficacy (Figure 13.12b). The cell growth inhibition result indicated that when the molar ratio of CPT to PTX was 3:1, the combination index was the smallest and performed the best synergistic effect. Undoubtedly, molecular engineering of ApDCs with combined drug dosage could provide a novel strategy for forwarding the clinical research of ApDCs.

13.8. APTAMER/APTAZYME-BASED RIBOSWITCHES

Riboswitches are typically located in the 5′/3′-untranslated region (5′-UTR) and usually consist of an evolutionarily conserved ligand-binding domain and expression platform, which regulate the expression of the downstream coding sequences in a ligand-responsive manner. The first aptamer-based riboswitch was reported by Werstuck and Green in 1998 [130]. They inserted Hoechst 33258-targeting RNA aptamer in the 5′-UTR of a reporter gene. After adding the dye, the expression of the reporter gene was substantially repressed in CHO cells. The elegant work even predated the report of natural bacterial riboswitches. Because RNA aptamer can be generated against a variety of compounds by the SELEX technique. In addition, RNA aptamers have a small molecular weight and lack immunogenicity. Therefore, various aptamer/aptazyme-based artificial riboswitches have been engineered [131–133]. Based on adaptability to diverse ligand input and regulatory targets, tunable regulatory

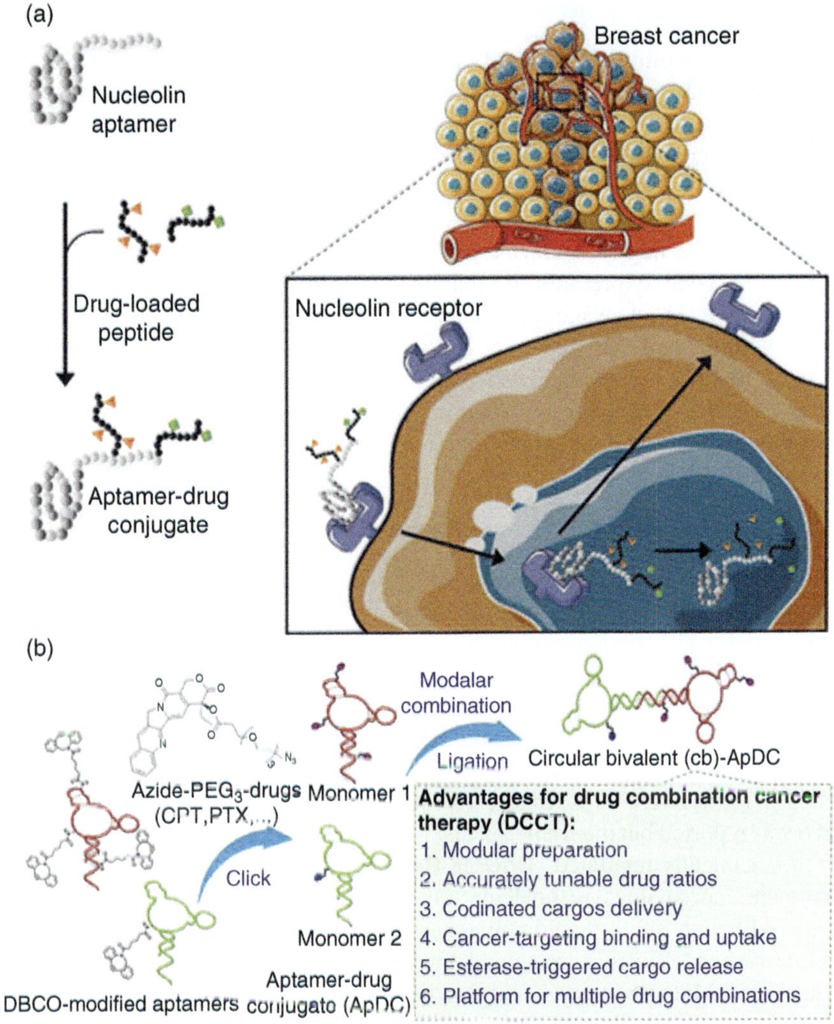

Figure 13.12 Aptamer–synergistic drug conjugates. (a) Conjugation of an aptamer with a carrier peptide pre-loaded with drug combinations at defined molar ratios. Source: Pusuluri *et al.* [129], p. 1437 / with permission of American Chemical Society. (b) Circular bivalent aptamer–drug conjugates with accurate tunability of drug ratios for drug combination cancer therapy. Source: Zhou *et al.* [112], p. 11787 / with permission of American Chemical Society John Wiley & Sons.

stringency, and design modularity, the synthetic riboswitches hold a great promise in the ligand-controlled gene-regulatory system. Herein, the applications and the underlying regulatory mechanisms of synthetic riboswitches are introduced.

13.8.1. Controlling mRNA Cleavage

Based on the combination of catalytic RNA (e.g. ribozyme) with ligand-specific aptamers, aptazymes are a kind of synthetic riboswitches that harness the self-cleavage of a ribozyme in a ligand-responsive manner. The first cell-free aptazyme was reported by Tang and Breaker *et al.* by integrating ATP-binding aptamer and hammerhead ribozyme [134]. In 2004, Yen demonstrated that a hammerhead ribozyme can regulate gene expression in the cell and *in vivo*, which pave the way to aptazyme-based riboswitches inserted in 5′/3′ UTR for regulating gene expression in mammalian cells [135]. Based on ligand-responsive RNA cleavage, aptazyme-based riboswitches have been used for constructing synthetic gene networks and regulating transgene expression (Figure 13.13a). For example, Chen *et al.* engineered a T-cell proliferation regulatory system by a synthetic aptazyme-based riboswitch, which can prompt cytokine expression in a theophylline-responsive manner [136].

13.8.2. Controlling Ribosomal Frameshifting

Hsu *et al.* first controlled programmed-1 ribosomal frameshifting (-1PRF) in mammalian cells by designing aptamer-based riboswitches with theophylline and S-adenosyl-L-homocysteine (SAH) aptamers [137]. -1PRF element is composed of a 7-nt slippery sequence X XXY YYZ followed by a stable secondary structure such as a pseudoknot or a stem-loop. By inserting RNA aptamer in the stable secondary structure domain, -1PRF can result in an alternative polypeptide translated from an mRNA by changing the translation reading frame from XXY YYZ to XXX YYY (Figure 13.13b). In the first system, an ON/OFF ratio of 6 was reported but the frameshifting efficiency was low. Recently, Matsumoto *et al.* chemically regulated -1PRF in HeLa cells by using naphthyridine carbamate tetramer analogs as molecular glue to mediate the formation of a pseudoknot. Although about 9-fold activation was observed, however, the maximum frameshifting efficiency remained low (3.2%) in the presence of molecular glue [138]. Matsumoto *et al.* suggest that the toxicity of the ligand might affect the frameshifting efficiency in HeLa cells.

13.8.3. Controlling MicroRNA Maturation Pathway

Besides controlling translation via mRNA cleavage or frameshifting, aptamer-based riboswitches can control the microRNA (miRNA) processing pathway to achieve the regulation of mRNA expression (Figure 13.13c). An *et al.* first demonstrated that an aptamer-based riboswitch chemically regulates RNA interference (RNAi) by modulating the Dicer mediated cleavage of a short-hairpin RNA (shRNA) [139]. Smolke group has demonstrated an alternative strategy by inserting RNA aptamers in proximity to the Drosha cleavage site to control Drosha cleavage ability and upregulate gene expression [140]. For example, the

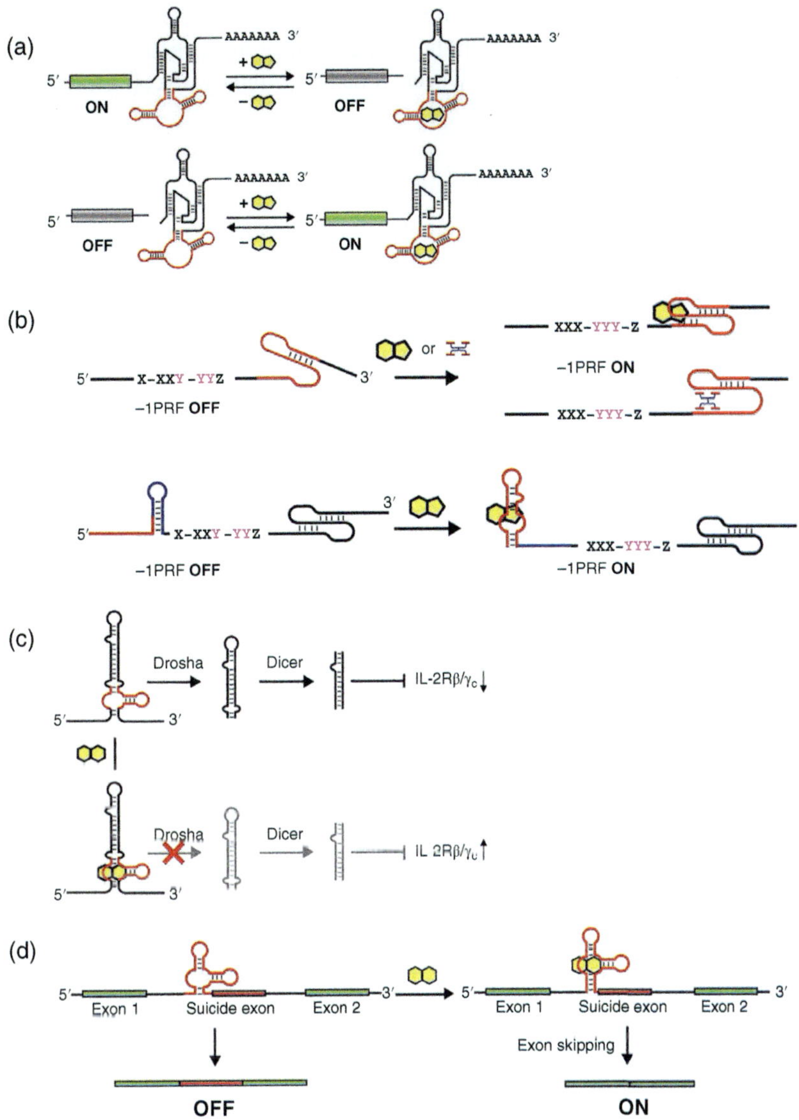

Figure 13.13 Regulatory mechanisms of synthetic aptamer/aptazyme riboswitches. (a) Aptazymes riboswitches regulate the integrity of mRNA via ligand-dependent cleavage. (b) Aptamer riboswitches control ribosomal frameshifting via ligands or molecular glues. (c) Interfering with the cleavage of pre-miRNA by aptamer riboswitches. (d) Aptamer riboswitches mediates exon skipping via a ligand. Source: Yokobayashi *et al.* [133], p. 73/With permission of Elsevier.

group used a (6R)-folinic acid (FA)-targeting RNA aptamer to control miRNA targeting IL-2 receptors. Although the modulation of the targeted gene expression was modest in this case, however, robust regulation of cell proliferation was observed.

13.8.4. Controlling Exon Skipping Processes

In addition, aptamer-based riboswitches can mediate gene expression by controlling exon skipping (Figure 13.13d). For example, an aptamer-based riboswitch engineered by inserting a tetracycline RNA aptamer near the 3′ splice site along with a suicide exon can control exon skipping and cause 5-fold activation of gene expression in HEK 293 cells in the presence of ligand [141]. By integrating with an aptazyme device, about 7-fold activation can be achieved in HEK 293 cells.

Although the promising perspective for cellular function regulation, the limited number of available aptamers and the challenges of integrating aptamers and expression platforms to work in concert caused by the unpredictable nature of the relationship between RNA sequence, structure, folding, and function remain to be addressed. In addition, the safety issue *in vivo* is also an important issue.

13.9. CHALLENGES AND SOLUTIONS

13.9.1. PCR Bias

Conventional SELEX obeys the law of "survival of the fittest" and is *in vitro* evolution procedure, by which very rare sequences displaying high-affinity to a target of interest are amplified and enriched from a highly diversified nucleic acid library via multiple rounds of PCR. To obtain the fittest target-binding nucleic acid sequences, equal amplification of each sequences in the enriched nucleic acid library during iterative selection is the precondition. However, unlike homogeneous templates, the heterogeneous templates in the highly diverse nucleic acid library are very varied in GC content, secondary structure, melting temperature, and their ability to adapt to polymerase, primer, and PCR conditions. In addition, compared to high-affinity sequences, sequences displaying low affinity to target are usually over-represented in the enriched nucleic acid library. Therefore, apart from the well-known PCR by-products from primer–product and product-product hybridization, the sequences in the enriched nucleic acid pool are never equally amplified, causing significant loss of sequence information during SELEX process and resulting in PCR bias. Usually, GC- and AT-rich sequences, as well as other sequences as well as highly structured oligonucleotides are amplified with low efficiency compared to other sequences.

Conventional SELEX usually take 8–15 rounds to enrich target-binding sequences. However, the diversity of a library can decrease by up to 50% after single PCR amplification, as demonstrated by Levay *et al.* [142]. Therefore, without proper control in the conventional SELEX, the sequences are likely to be selected to fit best to PCR systems and not to target of interest. Recently, emulsion PCR (ePCR), in which each oligonucleotide sequence is encapsulated into an individual PCR droplet surrounded by a hydrophobic organic phase, can substantially reduce the formation of by-products and avoid PCR bias. By using ePCR, highly structural sequences can be amplified and the library diversity can be maintained throughout the SELEX procedure, improving the chance to obtain high-affinity aptamer fitting best to targets.

To completely eliminate PCR bias and improve the selection efficacy of aptamer, Prof. Weihong Tan put forward a pioneering scientific concept of "single round selection of aptamer" based on the advances in sequencing technology. In this scientific concept, aptamer will be obtained by a single round selection. Therefore, aptamer – the best fittest to the target of interest – will be obtained from a highly diverse nucleic acid library.

13.9.2. Selection Procedure in "Black Box"

Conventional SELEX is essentially operated in many unknown conditions. By iterative selections, a final enriched nucleic acid pool will be obtained. The final pool is then subjected to cloned and sequenced to determine aptamer candidates for further detailed characterization. The procedure is similar to the operation in the "black box" until a few aptamers are identified. During the procedure, researchers experientially apply experimental conditions, resulting in low selection efficacy and loss of high-affinity aptamers.

Therefore, making the SELEX procedure operation in a "transparent box" is pursued by researchers. Fortunately, high-throughput sequencing (HTS) technology and the robust bioinformatic tool can make the SELEX procedure visible by revealing the global and real-time sequence information of each enriched nucleic acid pool, including nucleotide composition and frequency. By analyzing the sequence information of each round of selection, the dynamic changes of each enriched nucleic acid pool throughout the SELEX procedure can be quantitatively assessed. Therefore, HTS and the bioinformatic tool can present a dynamic global landscape of SELEX procedure, in which the evolution process of sequences can be clearly acknowledged, facilitating elucidating aptamer structure, properties, functions, and aptamer–target interactions. In addition, high-throughput sequencing (HTS) technology and the robust bioinformatic tool can avoid PCR bias and efficiently find the fittest aptamers, which are usually lost in conventional SELEX procedures.

13.9.3. Structure and Functions

As mentioned above, aptamers can fold into a wide variety of tertiary structure, such as hairpins, bulges, pseudoknots, or G-quadruplexes, to match the structure of targets, forming aptamer–target complexes by a combination of the hydrogen bond, hydrophobic interaction, van der Waals interaction, aromatic stacking. Crystal structures of aptamer-target complexes can shed light on the molecular basis of the interactions, providing a comprehensive knowledge of the aptamer-target binding mode and aptamer functions. However, the formation of the crystal structure of aptamer-target complexes is subject to DNA parameters (e.g. purity, length, and sticky end), target parameters (e.g. protein purity and polydispersity), and crystallization parameters (e.g. pH, ionic strength, temperature, protein concentration, the presence of various salts, additives, the type of precipitant, and the crystallization method). Therefore, only a few crystal structures of aptamer-target complexes are available, and the structures of most aptamer are unknown.

The scarcity of co-crystal structure of aptamer–target complexes hampers not only the exploring of interaction mode [143–146], but also the studies on aptamer functions. Currently, the *in vivo* applications of therapeutic aptamers are based on their occupancy-driven pharmacology. For example, Macugen block VEGF-A_{165} binding to VEGFR1. To achieve the therapeutic effect, occupancy-driven-based drugs need to be maintained at high concentration at the disease site, which usually easily cause adverse side effects on normal tissues. The limited knowledge of aptamer functions impedes the *in vivo* applications of aptamers. Therefore, revealing aptamer functions, for example triggering protein degradation, will facilitate the *in vivo* applications of aptamers.

Although the studies on the co-crystal structure of aptamer-target complex have proven difficult. Technology advances, the pursuit of scientists, and the demand of the market will drive us to obtain atomic resolution structures of aptamer–protein complexes to elucidate the aptamer structures and functions.

13.10. CONCLUSION

Based on the elegant properties and functions, aptamers provide a kind of powerful molecular tool for disease analysis and therapy biosensor. In the last 30 years, we are witnessing the substantial advances in the aptamer field. For example, the first aptamer-based drug Macugen was approved by the US FDA for the treatment of neovascular age-related macular degeneration in 2004. Another aptamer-based drug discovered by Professor Zhang Ge from Hong Kong Baptist University was approved as an orphan drug by the US FDA. The proteomics company SomaLogic based on aptamers has attracted substantial capital. Although there remain challenges in the aptamer field. However, it can be believed that we will continue to witness the further development of the field based on the improvement of selection methods, biostability of aptamers, knowledge about the structure, and function of aptamers.

REFERENCES

1. Strebhardt, K., Ullrich, A. (2008) Paul Ehrlich's magic bullet concept: 100 years of progress. *Nature Reviews Cancer* 8 (6): 473–480.
2. Kohler, G. Milstein, C. (1975). Continuous culture of fused cells secreting antibody of predefined specificity. *Nature* 256 (5517): 495–497.
3. Tuerk, C., Gold, L. (1990) Systematic evolution of ligands by exponential enrichment: RNA ligands to bacteriophage T4 DNA polymerase. *Science* 249 (4968): 505–510.
4. Ellington, A. D. Szostak, J. W. (1990). in vitro selection of RNA molecules that bind specific ligands. *Nature* 346 (6287): 818–822.
5. Mayer, G. (2010). The chemical biology of aptamers. *Angewandte Chemie International Edition* 48 (15): 2672–2689.
6. Gelinas, A. D., Davies, D. R. Janjic, N. (2016). Embracing proteins: structural themes in aptamer-protein complexes. *Current Opinion in Structural Biology* 36: 122–132.
7. Gold, L. (1995). Oligonucleotides as search, diagnostic, and therapeutic agents. *Journal of Biological Chemistry* 270 (23): 13581–13584.
8. Jenison, R. D., Gill, S. C., Pardi, A. et al. (1994). High-resolution molecular discrimination by RNA. *Science* 263 (5152): 1425–1429.
9. Sassanfar, M., Szostak, J. W. (1993) An RNA motif that binds ATP, *Nature* 364 (6437): 550–553.
10. Geiger, A., Burgstaller, P., von der Eltz, H. (1996). RNA aptamers that bind L-arginine with sub-micromolar dissociation constants and high enantioselectivity. *Nucleic Acids Research* 24 (6): 1029–1036.
11. Chen, L., Rashid, F., Shah, A. et al. (2015). The isolation of an RNA aptamer targeting to p53 protein with single amino acid mutation. *Proceedings of the National Academy of Sciences of the United States of America* 112 (32): 10002–10007.
12. Bock, L. C., Griffin, L. C., Latham, J. A. et al. (1992). Selection of single-stranded DNA molecules that bind and inhibit human thrombin. *Nature* 355 (6360): 564–566.
13. Robertson, D. L., Joyce, G. F. (1990) Selection in vitro of an RNA enzyme that specifically cleaves single-stranded DNA. *Nature* 344 (6265): 467–468.
14. O'Connell, D., Koenig, A., Jennings, S. et al. (1996). Calcium-dependent oligonucleotide antagonists specific for L-selectin. *Proceedings of the National Academy of Sciences of the United States of America* 93 (12): 5883–5887.
15. Zhang, Y., Lai, B., Juhas, M. (2019) Recent advances in aptamer discovery and applications. *Molecules* 24 (5): 941.
16. Shu, Y., Pi, F., Sharma, A. et al. (2014) Stable RNA nanoparticles as potential new generation drugs for cancer therapy. *Advanced Drug Delivery Reviews* 66: 74–89.
17. Pestourie, C., Cerchia, L., Gombert, K. et al. (2006). Comparison of different strategies to select aptamers against a transmembrane protein target. *Oligonucleotides* 16 (4): 323.
18. Shangguan, D., Li, Y., Tang, Z. et al. (2006) Aptamers evolved from live cells as effective molecular probes for cancer study. *Proceedings of the National Academy of Sciences of the United States of America* 103 (32): 11838–11843.
19. Guo, K. T., Ziemer, G., Paul, A. et al. (2008). CELL-SELEX: novel perspectives of aptamer-based therapeutics. *International Journal of Molecular Sciences* 9 (4): 668–678.

20. Ohuchi, S. (2012) Cell-SELEX technology. *BioResearch Open Access* 1 (6): 265–272.
21. Cerchia, L., Giangrande, P. H., McNamara, J. O. et al. (2009). Cell-specific aptamers for targeted therapies. *Methods in Molecular Biology* 535: 59–78.
22. Avci-Adali, M., Metzger, M., Perle, N. et al. (2010). Pitfalls of cell-systematic evolution of ligands by exponential enrichment (SELEX): existing dead cells during in vitro selection anticipate the enrichment of specific aptamers. *Oligonucleotides* 20 (6): 317–323.
23. Cheng, C., Chen, Y. H., Lennox, K. A. et al. (2013). in vivo SELEX for identification of brain-penetrating aptamers. *Molecular Therapy Nucleic Acids* 2 (1): e67.
24. Ozer, A., Pagano, J. M., Lis, J. T. (2014). New technologies provide quantum changes in the scale, speed, and success of SELEX methods and aptamer characterization. *Molecular Therapy Nucleic Acids* 3 (8): e183.
25. Darmostuk, M., Rimpelova, S., Gbelcova, H. et al. (2015). Current approaches in SELEX: an update to aptamer selection technology. *Biotechnology Advances* 33 (6): 1141–1161.
26. Ellington, A. D. Szostak, J. W. (1992). Selection in vitro of single-stranded DNA molecules that fold into specific ligand-binding structures. *Nature* 355 (6363): 850–852.
27. Drabovich, A. P., Berezovski, M., Okhonin, V. et al. (2006). Selection of smart aptamers by methods of kinetic capillary electrophoresis. *Analytical Chemistry* 78 (9): 3171–3178.
28. Roman, Y., Svetlana, M. K., Christine, B. et al. (2015) Emulsion PCR significantly improves nonequilibrium capillary electrophoresis of equilibrium mixtures-based aptamer selection: allowing for efficient and rapid selection of aptamer to unmodified ABH2 protein. *Analytical Chemistry* 87 (2): 1411–1419.
29. Mosing, R. K., Mendonsa, S. D. Bowser, M. T. (2005). Capillary electrophoresis-SELEX selection of aptamers with affinity for HIV-1 reverse transcriptase. *Analytical Chemistry* 77 (19): 6107–6112.
30. Yusuke, M., Nobuaki, S., Chiaki, O. et al. (2010) Selection of DNA aptamers using atomic force microscopy. *Nucleic Acids Research* 38 (4): e21.
31. Peng, L., Stephens, B. J., Bonin, K. et al. (2007). A combined atomic force/fluorescence microscopy technique to select aptamers in a single cycle from a small pool of random oligonucleotides. *Microscopy Research Technique* 70 (4): 372–381.
32. Raddatz, M. L., Dolf, A., Endl, E. et al. (2008) Enrichment of cell-targeting and population-specific aptamers by fluorescence-activated cell sorting. *Angewandte Chemie International Edition* 47 (28): 5190–5193.
33. Cho, M., Xiao, Y., Nie, J. et al. (2010). Quantitative selection of DNA aptamers through microfluidic selection and high-throughput sequencing. *Proceedings of the National Academy of Sciences of the United States of America* 107 (35): 15373–15378.
34. Sayer, N., Ibrahim, J., Turner, K. et al. (2002) Structural characterization of a 2′F-RNA aptamer that binds a HIV-1 SU glycoprotein, gp120. *Biochemical & Biophysical Research Communications* 29 (3): 924–931.
35. Dunn, M. R., Jimenez, R. M., Chaput, J. C. (2017). Analysis of aptamer discovery and technology. *Nature Reviews Chemistry* 1: 0076.
36. McGown, L. B., Joseph, M. J., Pitner, J. B. et al. (1995). The nucleic acid ligand. A new tool for molecular recognition. *Analytical Chemistry* 67(21): 663A–668A.
37. Manz, C., Kobitski, A. Y., Samanta, A. et al. (2017). Single-molecule FRET reveals the energy landscape of the full-length SAM-I riboswitch. *Nature Chemical Biology* 13 (11): 1172–1178.

38. Tuerk, C., MacDougal, S., and Gold, L. (1992) RNA pseudoknots that inhibit human immunodeficiency virus type 1 reverse transcriptase. *Proceedings of the National Academy of Sciences of the United States of America* 89 (15): 6988–6992.
39. Macaya, R. F., Schultze, P., Smith, F. W. et al. (1993). Thrombin-binding DNA aptamer forms a unimolecular quadruplex structure in solution. *Proceedings of the National Academy of Sciences of the United States of America* 90 (8): 3745–3749.
40. Staple, D. W., Butcher, S. E. (2005) Pseudoknots: RNA structures with diverse functions. *PLOS Biology* 3 (6): e213.
41. Breaker, R. R. (2011). Prospects for riboswitch discovery and analysis. *Molecular Cell* 43(6): 867–879.
42. Haider, S. M., Parkinson, G. N., Neidle, S. (2003). Structure of a G-quadruplex-ligand complex. *Journal of Molecular Biology* 326 (1): 117–125.
43. Carothers, J. M., Oestreich, S. C., Davis, J. H. et al. (2004) Informational complexity and functional activity of RNA structures. *Journal of the American Chemical Society* 126(16): 5130–5137.
44. Berens, C., Thain, A., Schroeder, R. (2001). A tetracycline-binding RNA aptamer. *Bioorganic & Medicinal Chemistry* 9 (10): 2549–2556.
45. Soukup, G. A., Ellington, A. D., Maher, J. L. (1996) Selection of RNAs that bind to duplex DNA at neutral pH. *Journal of Molecular Biology* 259 (2): 216–228.
46. London, G. M., Mayosi, B. M., Khati, M. (2015). Isolation and characterization of 2′-F-RNA aptamers against whole HIV-1 subtype C envelope pseudovirus. *Biochemical and Biophysical Research Communications* 456 (1): 428–433.
47. Mi, J., Liu, Y., Rabbani, Z. N. et al. (2010). in vivo selection of tumor-targeting RNA motifs. *Nature Chemical Biology* 6 (1): 22–24.
48. Hermann, T., Patel, D. J. (2000). Adaptive recognition by nucleic acid aptamers. *Science* 287 (5454): 820–825.
49. Lin, C. H., Patei, D. J. (1997). Structural basis of DNA folding and recognition in an AMP-DNA aptamer complex: distinct architectures but common recognition motifs for DNA and RNA aptamers complexed to AMP. *Chemistry & Biology* 4 (11): 817–832.
50. Nonin-Lecomte, S., Lin, C. H., Patel, D. J. (2001). Additional hydrogen bonds and base-pair kinetics in the symmetrical AMP-DNA aptamer complex. *Biophysical Journal* 81 (6): 3422–3431.
51. Hermann, T., Westhof, E. (1998). Aminoglycoside binding to the hammerhead ribozyme: a general model for the interaction of cationic antibiotics with RNA. *Journal of Molecular Biology* 276 (5): 903–912.
52. Ye, H., Zhou, Y., Ma, P., et al. (2020) Analysis of the anti-inflammatory effect of the aptamer LA27 and its binding mechanism. *International Journal of Biological Macromolecules* 165: 308–313.
53. Krauss, R. I., Merlino, A., Giancola, C., et al. (2011) Thrombin-aptamer recognition: a revealed ambiguity. *Nucleic Acids Research* 39 (17): 7858–7867.
54. Galburt, E. A., Tomko, E. J. (2017). Conformational selection and induced fit as a useful framework for molecular motor mechanisms. *Biophysical Chemistry* 223: 11–16.
55. Rode, A. B., Endoh, T., Sugimoto, N. (2018) Crowding shifts the FMN recognition mechanism of riboswitch aptamer from conformational selection to induced fit. *Angewandte Chemie International Edition* 57 (23): 6868–6872.
56. Hetzke, T., Vogel, M., Gophane, D. B., et al. (2019). Influence of Mg^{2+} on the conformational flexibility of a tetracycline aptamer. *RNA* 25 (1): 158–167.

57. Suddala, K. C., Wang, J., Hou, Q. et al. (2015) Mg^{2+} shifts ligand-mediated folding of a riboswitch from induced-fit to conformational selection. *Journal of the American Chemical Society* 137 (44): 14075–14083.
58. Ni, S., Yao, H., Wang, L., et al. (2017). Chemical modifications of nucleic acid aptamers for therapeutic purposes. *International Journal of Molecular Sciences* 18 (8): 1683
59. Kimoto, M., Yamashige, R., Matsunaga, K. I. et al. (2013). Generation of high-affinity DNA aptamers using an expanded genetic alphabet. *Nature Biotechnology* 31 (5): 453–457.
60. Sefah, K., Yang, Z., Bradley, K. M. et al. (2014) In vitro selection with artificial expanded genetic information systems. *Proceedings of the National Academy of Sciences of the United States of America* 111 (4): 1449–1454.
61. Huntington, J. A., Baglin, T. P. (2003). Targeting thrombin–rational drug design from natural mechanisms. *Trends in Pharmacological Sciences* 24 (11): 589–595.
62. Tsiang, M., Jain, A. K., Dunn, K. E. et al. (1995) Functional mapping of the surface residues of human thrombin. *Journal of Biological Chemistry* 270 (28): 16854–16863.
63. Musumeci, D., Montesarchio, D. (2012). Polyvalent nucleic acid aptamers and modulation of their activity: a focus on the thrombin binding aptamer. *Pharmacology & Therapeutics* 136 (2): 202–215.
64. Ng, E. W., Shima, D. T., Calias, P. et al. (2006). Pegaptanib, a targeted anti-VEGF aptamer for ocular vascular disease. *Nature Reviews Drug Discovery* 5 (2): 123–132.
65. Healy, J. M., Lewis, S. D., Kurz, M. et al. (2004). Pharmacokinetics and biodistribution of novel aptamer compositions. *Pharmaceutical research* 21 (12): 2234–2246.
66. Zhuo, Z., Yu, Y., Wang, M. et al. (2017) Recent advances in SELEX technology and aptamer applications in biomedicine. *International Journal of Molecular Sciences* 18 (10): 2142.
67. Nozari, A., Berezovski, M. V. (2017). Aptamers for CD antigens: from cell profiling to activity modulation. *Molecular Therapy Nucleic Acids* 6: 29–44.
68. Santulli-Marotto, S., Nair, S. K., Rusconi, C. et al. (2003) Multivalent RNA aptamers that inhibit CTLA-4 and enhance tumor immunity. *Cancer Research* 63 (21): 7483–7489.
69. Pastor, F., Soldevilla, M. M., Villanueva, H. et al. (2013). CD28 aptamers as powerful immune response modulators. *Molecular Therapy Nucleic Acids* 2 (6): e98.
70. McNamara, J. O., Kolonias, D., Pastor, F. et al. (2008). Multivalent 4-1BB binding aptamers costimulate $CD8^+$ T cells and inhibit tumor growth in mice. *The Journal of Clinical Investigation* 118 (1): 376–386.
71. Dollins, C. M., Nair, S., Boczkowski, D. et al. (2008). Assembling OX40 aptamers on a molecular scaffold to create a receptor-activating aptamer. *Chemistry & Biology* 15 (7): 675–682.
72. Berezhnoy, A., Stewart, C. A., Mcnamara II, J. O. et al. (2012). Isolation and optimization of murine IL-10 receptor blocking oligonucleotide aptamers using high-throughput sequencing. *Molecular Therapy* 20 (6): 1242–1250.
73. Roth, F., Adriana, C., Vella, J. L. et al. (2012) Aptamer-mediated blockade of IL4Rα triggers apoptosis of MDSCs and limits tumor progression. *Cancer Research* 72 (6): 1373–1383.
74. Gupta, S., Hirota, M., Waugh, S. M. et al. (2014). Chemically modified DNA aptamers bind interleukin-6 with high affinity and inhibit signaling by blocking its interaction with interleukin-6 receptor. *Journal of Biological Chemistry* 289 (12): 8706–8719.

75. Zhou, G., Wilson, G., Hebbard, L. et al. (2016) Aptamers: a promising chemical antibody for cancer therapy. *Oncotarget* 7 (12): 13446–13463.
76. González, V. M., Martín, M. E., Fernández, G. et al. (2016). Use of aptamers as diagnostics tools and antiviral agents for human viruses. *Pharmaceuticals* 9 (4): 78.
77. Zou, X., Wu, J., Gu, J. et al. (2019) Application of aptamers in virus detection and antiviral therapy. *Frontiers in Microbiology* 10: 1462.
78. Musumeci, D., Riccardi, C., Montesarchio, D. (2015). G-quadruplex forming oligonucleotides as anti-HIV agents. *Molecules* 20 (9): 17511–17532.
79. Khati, M., Schüman, M., Ibrahim, J. et al. (2003). Neutralization of infectivity of diverse R5 clinical isolates of human immunodeficiency virus type 1 by gp120-binding 2′ F-RNA aptamers. *Journal of Virology* 77 (23): 12692–12698.
80. Andreola, M. L., Pileur, F., Calmels, C. et al. (2001). DNA aptamers selected against the HIV-1 RNase H display in vitro antiviral activity. Biochemistry 40 (34): 10087–10094.
81. Sun, M., Liu, S., Wei, X. et al. (2021) Aptamer blocking strategy inhibits SARS-CoV-2 virus infection. *Angewandte Chemie*. https://doi.org/10.1002/anie.202100225.
82. Wang, T., Chen, C., Larcher, L. M. et al. (2019) Three decades of nucleic acid aptamer technologies: lessons learned, progress and opportunities on aptamer development. *Biotechnology Advances* 37 (1): 28–50.
83. Famulok, M., Mayer, G. (2014). Aptamers and SELEX in chemistry & biology. *Chemistry & Biology* 21(9): 1055–1058.
84. Shi, H., He, X., Wang, K. et al. (2011) Activatable aptamer probe for contrast-enhanced *in vivo* cancer imaging based on cell membrane protein-triggered conformation alteration. *Proceedings of the National Academy of Sciences of the United States of America* 108 (10): 3900–3905.
85. Shi, L., Liu, Y., Wang, Q. et al. (2018) A pH responsive AIE probe for enzyme assays. *Analyst* 143 (3): 741–746.
86. Zhang, P., Zhao, Z., Li, C. et al. (2017) Aptamer-decorated self-assembled aggregation-induced emission organic dots for cancer cell targeting and imaging. *Analytical Chemistry* 90 (2): 1063–1067.
87. Liu, C., Zhao, J., Tian, F. et al. (2019). Low-cost thermophoretic profiling of extracellular-vesicle surface proteins for the early detection and classification of cancers. *Nature Biomedical Engineering* 3 (3): 183–193.
88. Savla, R., Taratula, O., Garbuzenko, O. et al. (2011) Tumor targeted quantum dot-mucin1 aptamer-doxorubicin conjugate for imaging and treatment of cancer. *Journal of Controlled Release* 153 (1): 16–22.
89. Ko, H. Y., Lee, J. H., Kang, H. et al. (2010). A nucleolin-targeted multimodal nanoparticle imaging probe for tracking cancer cells using an aptamer. *Journal of Nuclear Medicine* 51 (1): 98–105.
90. Cui, C., Zhang, H., Wang, R. et al. (2017). Recognition-then-reaction enables site-selective bioconjugation to proteins on live-cell surfaces. *Angewandte Chemie* 129 (39): 12116–12119.
91. Chen, X., Qiu, L., Cai, R., et al. (2020). Aptamer-directed protein-specific multiple modifications of membrane glycoproteins on living cells. *ACS Applied Materials & Interfaces* 12 (34): 37845–37850.
92. Crowley, E., Di Nicolantonio, F., Loupakis, F. et al. (2013). Liquid biopsy: monitoring cancer-genetics in the blood. *Nature Reviews Clinical Oncology* 10 (8): 472–484.

93. Heitzer, E., Ulz, P., Geigl, J. B. (2015). Circulating tumor DNA as a liquid biopsy for cancer. *Clinical Chemistry* 61 (1): 112–123.
94. Song, Y., Shi, Y., Huang, M. et al. (2019) Bioinspired engineering of a multivalent aptamer-functionalized nanointerface to enhance the capture and release of circulating tumor cells. *Angewandte Chemie* 58 (8): 2236–2240.
95. Hedin, L. E., Illergård, K.Elofsson, A. (2011). An introduction to membrane proteins. *Journal of Proteome Research* 10 (8): 3324–3331.
96. Nwamba, O. C. (2020). Membranes as the third genetic code. *Molecular Biology Reports* 47 (5): 4093–4097.
97. Kang, M. K., Tullman-Ercek, D. (2018). Engineering expression and function of membrane proteins. *Methods* 147: 66–72.
98. Bausch-Fluck, D., Goldmann, U., Müller, S. et al. (2018). The in silico human surfaceome. *Proceedings of the National Academy of Sciences of the United States of America* 115 (46): 10988–10997.
99. Bailey, M. H., Tokheim, C., Porta-Pardo, E. et al. (2018). Comprehensive characterization of cancer driver genes and mutations. *Cell* 173 (2): 371–385.
100. Carnielli, C. M., Macedo, C. C. S., De Rossi, T. et al. (2018). Combining discovery and targeted proteomics reveals a prognostic signature in oral cancer. *Nature Communications* 9 (1): 3598.
101. Jin, C., Qiu, L., Li, J. et al. (2016). Cancer biomarker discovery using DNA aptamers. *Analyst* 141 (2): 461–466.
102. Shangguan, D., Cao, Z., Meng, L., et al. (2008) Cell-specific aptamer probes for membrane protein elucidation in cancer cells. *Journal of Proteome Research* 7 (5): 2133–2139.
103. Van Simaeys, D., Turek, D., Champanhac, C., et al. (2014) Identification of cell membrane protein stress-Induced phosphoprotein 1 as a potential ovarian cancer biomarker using aptamers selected by cell systematic evolution of ligands by exponential enrichment. *Analytical Chemistry* 86 (9): 4521–4527.
104. Jia, W., Ren, C., Wang, L., et al. (2016). CD109 is identified as a potential nasopharyngeal carcinoma biomarker using aptamer selected by cell-SELEX. *Oncotarget* 7 (34): 55328–55342.
105. Wu, X., Liu, H., Han, D., et al. (2019) Elucidation and structural modeling of CD71 as a molecular target for cell-specific aptamer binding. *Journal of the American Chemical Society* 141 (27): 10760–10769.
106. Li, H., Liu, J., Xiao, X. et al. (2019). A novel aptamer LL4A specifically targets vemurafenib-resistant melanoma through binding to the CD63 protein. *Molecular Therapy Nucleic Acids* 18: 727–738.
107. Bing, T., Wang, J., Shen, L., et al. (2020). Prion protein targeted by a prostate cancer cell binding aptamer, a potential tumor Mamrker? *ACS Applied Bio Materials* 3 (5): 2658–2665.
108. Bing, T., Shangguan, D., Wang, Y. (2015). Facile discovery of cell-surface protein targets of cancer cell aptamers. *Molecular & Cellular Proteomics* 14 (10): 2692–2700.
109. Wang, L., Bing, T., Liu, Y., et al. (2018) Imaging of neurite network with an anti-L1CAM aptamer generated by neurite-SELEX. *Journal of the American Chemical Society*, 140 (51): 18066–18073.
110. Bing, T., Shen, L., Wang, J. et al. (2019) Aptameric probe specifically binding protein heterodimer rather than monomers. *Advanced Science* 6 (11): 1900143.

111. Xuan, W., Peng, Y., Deng, Z., et al. (2018) A basic insight into aptamer-drug conjugates (ApDCs). *Biomaterials* 182: 216–226.
112. Zhou, F., Wang, P., Peng, Y., et al. (2019) Molecular engineering-based aptamer-drug conjugates with accurate tunability of drug ratios for drug combination targeted cancer therapy. *Angewandte Chemie* 131 (34), 11787–11791.
113. Tsimberidou, A. M., Fountzilas, E., Nikanjam, M. et al. (2020) Review of precision cancer medicine: evolution of the treatment paradigm. *Cancer Treatment Reviews* 86: 102019.
114. Jackson, S. E., Chester, J. D. (2015). Personalised cancer medicine. *International Journal of Cancer* 137 (2): 262–266.
115. Mi, J., Ray, P., Liu, J., et al. (2016). in vivo selection against human colorectal cancer xenografts identifies an aptamer that targets RNA helicase protein DHX9. *Molecular Therapy Nucleic Acids* 5 (4): e315.
116. Li, S., Xu, H., Ding, H., et al. (2009). Identification of an aptamer targeting hnRNP A1 by tissue slide-based SELEX. *The Journal of Pathology* 218 (3): 327–336.
117. Siegel, R. L., Miller, K. D., Jemal, A. (2020). Cancer statistics, 2020. *CA: A Cancer Journal of Clinicians* 70: 7–30.
118. Mahato, R., Tai, W., Cheng, K. (2011). Prodrugs for improving tumor targetability and efficiency. *Advanced Drug Delivery Reviews* 63 (8): 659–670.
119. Scott, A. M., Wolchok, J. D., Old, L. J. (2012) Antibody therapy of cancer. *Nature Reviews Cancer* 12 (4): 278–287.
120. Bagalkot, V., Farokhzad, O. C., Langer, R. et al. (2006). An aptamer-doxorubicin physical conjugate as novel targeted drug-delivery platform. *Angewandte Chemie International Edition* 45 (48): 8149–8152.
121. Wen, J., Tao, W., Hao, S. et al. (2016). A unique aptamer-drug conjugate for targeted therapy of multiple myeloma. *Leukemia* 30 (4): 987–991.
122. Zhu, G.; Zheng, J.; Song, E. et al. (2013) Self-assembled, aptamer-tethered DNA nanotrains for targeted transport of molecular drugs in cancer theranostics. *Proceedings of the National Academy of Sciences of the United States of America* 110 (20): 7998–8003.
123. Prusty, D. K., Adam, V., Zadegan, R. M. et al. (2018). Supramolecular aptamer nano-constructs for receptor-mediated targeting and light-triggered release of chemotherapeutics into cancer cells. *Nature Communications* 9 (1): 535.
124. Li, F.; Lu, J.; Liu, J. et al. (2017). A water-soluble nucleolin aptamer-paclitaxel conjugate for tumor specific targeting in ovarian cancer. *Nature Communications* 8: 1390.
125. Huang, Y., Shangguan, D., Liu, H. et al. (2009). Molecular assembly of an aptamer-drug conjugate for targeted drug delivery to tumor cells. *ChemBioChem* 10 (5): 862–868.
126. Yang, Q., Deng, Z., Wang D. et al. (2020). Conjugating aptamer and Mitomycin C with reductant-responsive linker leading to synergistically enhanced anticancer effect. *Journal of the American Chemical Society* 142 (5): 2532–2540.
127. Wang, R., Zhu, G., Mei, L. et al. (2014). Automated modular synthesis of aptamer-drug conjugates for targeted drug delivery. *Journal of the American Chemical Society* 136: 2731–2734.
128. Wang, D., Peng, Y., Deng, Z. et al. (2020) Modularly engineered solid-phase synthesis of aptamer-functionalized small molecule drugs for targeted cancer therapy. *Advanced Therapeutics* 3: 2000074.

129. Pusuluri, A., Krishnan, V., Lensch, V. et al. (2019). Treating tumors at low drug doses using an aptamer–peptide synergistic drug conjugate. *Angewandte Chemie International Edition* 58 (5): 1437–1441.
130. Werstuck, G., Green, M. R. (1998). Controlling gene expression in living cells through small molecule-RNA interactions. *Science* 282 (5387): 296–298.
131. Ausländer, S., Fussenegger, M. (2017). Synthetic RNA-based switches for mammalian gene expression control. *Current Opinion in Biotechnology* 48: 54–60.
132. Lee, C. H., Han, S. R., Lee, S. W. (2016). Therapeutic applications of aptamer-based riboswitches, *Nucleic Acid Therapy* 26 (1): 44–51.
133. Yokobayashi, Y. (2019) Aptamer-based and aptazyme-based riboswitches in mammalian cells. *Current Opinion in Chemical Biology* 52: 72–78.
134. Tang, J., Breaker, R. R. (1997) Rational design of allosteric ribozymes. *Chemistry & Biology* 4 (6): 453–459.
135. Yen, L., Svendsen, J., Lee, J. S. et al. (2004) Exogenous control of mammalian gene expression through modulation of RNA self-cleavage. *Nature* 431 (7007): 471–476.
136. Chen, Y., Jensen, M. C., Smolke, C. D. (2010). Genetic control of mammalian T-cell proliferation with synthetic RNA regulatory systems. *Proceedings of the National Academy of Sciences of the United States of America.* 107 (19): 8531–8536.
137. Hsu, H. T., Lin, Y. H., Chang, K. Y. (2014). Synergetic regulation of translational reading-frame switch by ligand-responsive RNAs in mammalian cells. *Nucleic Acids Research* 42 (22): 14070–14082.
138. Matsumoto, S., Caliskan, N., Rodnina, M. V. et al. (2018). Small synthetic molecule-stabilized RNA pseudoknot as an activator for −1 ribosomal frameshifting. *Nucleic Acids Research* 46 (16): 8079–8089.
139. An, C. I., Trinh, V. B., Yokobayashi, Y. (2006). Artificial control of gene expression in mammalian cells by modulating RNA interference through aptamer-small molecule interaction. *RNA* 12 (5):710–716.
140. Wong, R. S., Chen, Y. Y., Smolke, C. D. (2018) Regulation of T cell proliferation with drug-responsive microRNA switches. *Nucleic Acids Research* 46:1541–1552.
141. Vogel, M., Weigand, J. E., Kluge, B. et al. (2018) A small, portable RNA device for the control of exon skipping in mammalian cells. *Nucleic Acids Research* 46 (8): e48.
142. Levay, A., Brenneman, R., Hoinka, J. et al. (2015). Identifying high-affinity aptamer ligands with defined cross-reactivity using high-throughput guided systematic evolution of ligands by exponential enrichment, *Nucleic Acids Research* 43 (12): e82.
143. Padmanabhan, K., Padmanabhan, K. P., Ferrara, J. D. et al. (1993). The structure of alpha-thrombin inhibited by a 15-mer single-stranded DNA aptamer. *The Journal of Biological Chemistry* 268 (24): 17651–17654.
144. Huang, D. B., Vu, D., Cassiday, L. A. et al. (2003). Crystal structure of NF-κb (p50)2 complexed to a high-affinity RNA aptamer. *Proceedings of the National Academy of Sciences of the United States of America* 100 (16): 9268–9273.
145. Nomura, Y., Sugiyama, S., Sakamoto, T. et al. (2010). Conformational plasticity of RNA for target recognition as revealed by the 2.15 Å crystal structure of a human IgG-aptamer complex. *Nucleic Acids Research* 38 (21): 7822–7829.
146. Convery, M. A., Rowsell, S., Storehouse, N. J. et al. (1998) Crystal structure of an RNA aptamer-protein complex at 2.8 Å resolution. *Nature Structural Biology* 5 (2): 133–139.

CHAPTER 14

G-QUADRUPLEX AS A PLATFORM FOR NEW PERSPECTIVES IN NUCLEIC ACID TARGETING FOR THERAPEUTIC APPLICATIONS

MANLIO PALUMBO AND CLAUDIA SISSI
Department of Pharmaceutical and Pharmacological Sciences, University of Padova, Padova, Italy

14.1. INTRODUCTION

The peculiar properties exhibited by guanine-rich nucleic acid sequences were first described in the early 1960s, but it took over 40 years to make sure that the three-dimensional structures they originate, the so-called G-quadruplexes (G4), were not just a scientific curiosity to play with, but elements endowed with vital functions in living organisms. As soon as the genome sequences became available, the massive presence of nonrandomly repeated G-rich components pointed to a functional significance of these nucleic acid sequences and their related noncanonical structures. However, only later, proper bioanalytical protocols supported their formation in living cells. Interestingly, the largest occurrence of these species was located at the tips of chromosomes (telomeres) and at the promoter regions of oncogenes.

G4s consist of four guanine strands held together by Hoogsteen hydrogen bonding. Despite a generally conserved arrangement of the G-tetrads core, it soon became clear that G4s can adopt a variety of conformations, in which the base sequences can fold depending upon sequence context and environmental conditions [1]. The energy

Nucleic Acids in Medicinal Chemistry and Chemical Biology: Drug Development and Clinical Applications, First Edition. Edited by Lihe Zhang, Xinjing Tang, Zhen Xi, and Jyoti Chattopadhyaya.

barriers to switch from one conformation to another are relatively small (1–3 kcal/mol). Hence, the various forms can be fully exploited to trigger a biological event because of a given change. In cells, this folding process can occur starting from a single-stranded polynucleotide like at the 3′ protruding telomere ends. Conversely, at double-stranded DNA (dsDNA) tracks like at activating/deactivating replication or transcription sites, the assistance of enzymatic partners or of low molecular weight effectors can be required to favor G4 formation. Last but not least, these events can involve DNA- or RNA-based sequences or hybrid mixtures thereof [2]. As a result, a world of functions can be imagined under this huge variety of conditions [3].

G4s can be properly generated to perform/control physiological functions but they can also improperly operate in the case of pathological conditions with cancer representing a leading example. Hence, a pharmacological intervention might counteract the progress of disease. Consistently, many potential new drugs presently under investigation ideally target G4 structures to interfere with undesired G4-related events like, for example, uncontrolled cell growth. Drugs can affect the crucial equilibrium mixture of possible G4 forms by selectively binding to one of them in a mutual adaptation form (induced fit) or can disrupt a pre-existing interaction (i.e. with proteins) by displacing it from the complex or by forming a ternary system. Of course, DNA (or RNA) binding proteins that play a primary role in the folding–unfolding processes involving G4s surely represent important co-targets, yet hardly druggable, for successful treatment [4].

In principle, with such potent weapons in our hands, it should be possible to regulate G4 formation/disruption at any selected nucleic acid sequence for medicinal purposes. This enthusiastic conclusion unfortunately compares very poorly with everyday experience, as we impotently witness the death of thousands of patients, while none of the G4-binding drugs reached the clinic. This stimulates all of us to a deeper commitment in our scientific mission. In fact, things are not so simple as we depicted them. The main problem has to do with the exceedingly high complexity of a living organism. We need accurate guesses from simplified model systems to safely treat a human patient. Notwithstanding the amazing progresses of the current (bio)technologies, translating quantitative information from the bench to the bed side is nevertheless too big of a jump. We usually treat biological systems using the classical laws of thermodynamics as if they were compartments at equilibrium in a steady state [5]. On the contrary, a continuous flow of matter and energy occurs within nonhomogeneous cellular compartments, which clearly requires substantial kinetics corrections and drastic theoretical simplifications. In addition, it is a long journey for a drug molecule from the release site to the action site, during which the drug interacts (and reacts) with a large number of species according to the familiar ADMET (Absorption, Distribution, Metabolism, Elimination, and Toxicity) acronym. In addition, if the drug is targeted at the cell nucleus, severe crowding conditions occur, which may substantially change the dielectric properties and activity coefficients of the partner species with substantial changes in the thermodynamic parameters, including local pH. Other specific limitations to accurately simulate G4 conformational equilibria and their ligand targeting include possible overexpression of K^+-related channels with associated decrease of intranuclear metal ion concentration; variations in effective ionic strength with changes in G4 stability; complexity of physiological conditions; high sensitivity to small pH changes; off-target binding to highly charged species; interference by non-aqueous polar solvents (dimethylsulfoxide, DMSO); epigenetic events mediated by G4; processing of the nucleic acids by helicases and other nuclear proteins; formation of DNA higher order structures that formally reduce DNA accessibility. Finally, yet importantly, interactions with RNA G-quadruplexes must be considered carefully. Indeed, they are even in the cytosol where they affect distinct cellular functions with reference to those controlled by DNA G4s and, noteworthy, they can reduce the free drug concentration available for DNA recognition [2, 6]. Lastly, we cannot forget that the active agent is metabolized and the number of derived species potentially entering in the cell multiplied.

Hence, one must be extremely cautious when considering *in vitro* data transferable to *in vivo* studies. There is a wealth of work to be done to fill in the gap between theory and practice starting from *in silico* studies to wet chemistry experiences to investigations in living cells, to end up with clinical studies and hopefully new drugs' approval. These steps are strictly interdependent and require synergic intervention of scientists with different expertise. "Unity is strength" says an old Italian proverb. In fact, the more people collaborate to achieve a shared purpose, the greater the likelihood of successful outcome. Indeed, because of joint international concerted efforts, vaccines against SARS-CoV-2 became available a few months after the outbreak of pandemics.

Can we cope with the serious limitations mentioned above? Where are we now in the field of G4 binders as possible drugs? What new directions are being taken? Several excellent reviews published in the last year or so deal with these topics extensively, and we refer the reader to them [7–10]. Still, a lot of interesting new results became lastly available, which we would like to discuss in the present chapter, to focus on the main strategies that are currently under development starting from computer-aided studies, proceeding to *in vitro* and *in vivo* chemical and biological investigations, to end with pharmacological perspectives.

14.2. IS THE LIGAND DESIGN CHANGING?

The overall structural differences between dsDNA and G4 promoted the search for small molecules able to selectively bind the tetrahelical structures even before the final assessment of their occurrence *in cells* and of the disclosure of their regulatory roles along multiple functional pathways. Unveiling these concerns opens bright new therapeutic applications and pushed up the production of a vast number of potential hits and leads with the ultimate goal to identify efficient and selective binders. Still, this process is not complete since, currently, no compound is approved for clinical use yet. So, it is crucial to merge the information that the already available data provide us to properly build up new perspectives.

14.2.1. The Aromatic Core

So far, small ligands were considered to target different G4 domains, i.e. loops, grooves, or G4 core. The emerging picture indicates the G-tetrads and, in particular, the solvent accessible external ones as the most represented binding site used to drive a small molecule toward preferentially G4 vs dsDNA. Therefore, up to date most tested derivatives refer to an extended planar aromatic core designed to stack on a G-tetrad. In this connection, quantitative relationships between the ligands aromatic surface and their affinity (and selectivity) for G4 have been reported [11]. Based on this correlation, the planar aromatic domain of novel hits was frequently enlarged through the link or conjugation of multiple aromatic cycles. However, this strategy is not always suitable when looking for pharmacological applications. Indeed, it leads to the synthesis of derivatives frequently characterized by a poor ADMET profile. Thus, it is important to focus on the key requirements behind the G-tetrads recognition.

A crucial feature to be kept in mind while working on the ligand aromatic surface is the need to preserve a shape complementary to the interacting G-tetrad partner. Several examples can help in addressing this point. An interesting one is represented by indole derivatives (Figure 14.1a). This scaffold has been widely exploited as potential pharmacophore for G4 recognition and, in particular, bis-indole derivatives show very promising affinity [12]. Among them, a new series took into account for the first time the link of the two indole moieties at position 3. This connection allows the new compounds to fully overlap with the G-tetrads. To proficiently implement the interaction properties with the

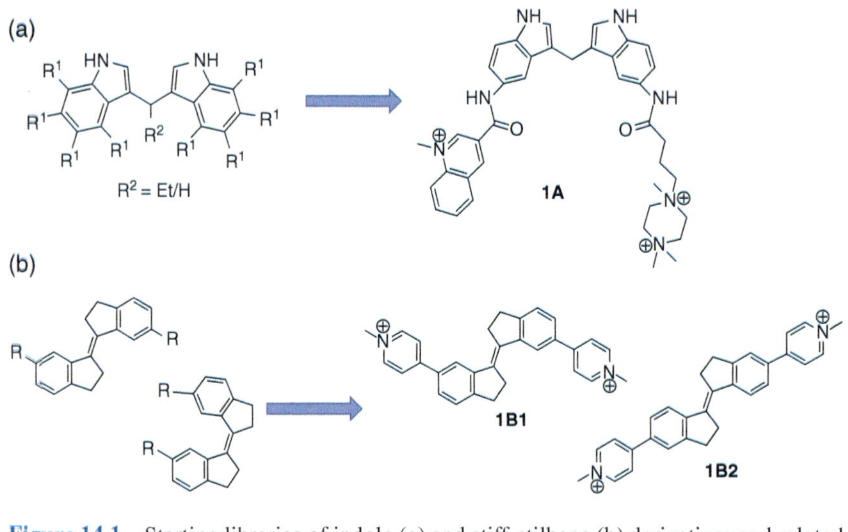

Figure 14.1 Starting libraries of indole (a) and stiff-stilbene (b) derivatives and related selective active compounds.

nucleic acid, the resulting core has been also functionalized with protonable side chains located according to both symmetrical and nonsymmetrical patterns.

The position of the substituents was crucial for both selectivity and affinity: their introduction at positions 5–5′ retains the ability to adopt a conformation that properly places them in the right position to bind the G-tetrad. Noteworthy, the insertion of a methylated quinoline in the two side chains at position 5 generated the most efficient bis-indoles in terms of G4 binding and stabilization. The methylation of the quinoline was mandatory for efficient recognition. Remarkably, molecular dynamics studies evidenced that the positive effect of this substitution is not simply due to charge–charge interaction. Indeed, one quinolone unit actually stacks on the G-tetrad system. Consistently, replacement of one methylated quinoline with a methylated piperazine provides an asymmetric derivative (**1A**) that fully retains a good G4 binding profile. Thus, the right balance of the shape of the core and relative orientation of side groups emerged.

It is to note that these compounds are quite flexible since the best binders have methylene groups to connect the two indoles, thus allowing mutual rotations of the two domains. This can explain why these compounds are good in discriminating between G4 and dsDNA, whereas they can bind different G4 topologies with comparable efficiency.

A strategy to overcome this drawback is to reduce the space potentially explored by the ligand by introducing proper structural constraints that retain the active distribution of functional groups. A preliminary study considered a series of stiff-stilbene ligands (Figure 14.1b) and allowed to dissect the structural features that govern their binding to nucleic acid targets in terms of optimal molecular configuration, electronics, and substitution pattern [13]. Again, the relative orientation of the aromatic domains with reference to the cationic terminals confirmed to be critical to grant an efficient G4 interaction (**1B1**). However, whereas the best binder can explore also the G4 grooves as binding sites, the constrained geometry allows the one compound of the series (**1B2**) to force G4 into a single topology.

Noteworthy, the new-stilbene derivatives demonstrate high toxicity both to parasites and cancerous cell lines. In the cellular models, *in vitro* cytotoxicity to cancer cells strongly correlates with G4-binding activity, and dose-response times indicate long-term mechanisms of action that may include G4-mediated pathways, such as telomerase inhibition. Meanwhile, they remain up to two orders of magnitude less toxic for a nontumoral model. This study represents an intriguing proof-of-concept for the geometry-activity connection and, along with the possible further improvement of this emergent highly G4 selective chemotype, indicates stiff-stilbene as an interesting lead scaffold for DNA-targeted drug development.

The advantages related to the preferential target of the external G-tetrads are not based on their solvent accessibility only. Indeed, if compared to the interaction occurring between neighboring base pairs in dsDNA, the stacking interactions between subsequent G-tetrads are much stronger. As a result, the energetic cost to disrupt them is expected to be too large to be efficiently promoted by a small intercalating ligand. This actually turned out not to be the case when DNA quadruplexes are intrinsically characterized by a sufficient conformational freedom. This was elucidated by NMR titration of the alternative tetrahelical structures assumed by a G-rich sequence corresponding to a domain of the regulatory region of the PLEKHG3 human gene (VK2: d(G_3AGCGAG_3AGCGAG_3AGCGAG_3AGCG)) with the G4 binder 360A [14]. The solved high-resolution structure of the **VK2–360A** complex represents the first report of intercalation of a heterocyclic ligand into a quadruplex structure (Figure 14.2). The detailed structural insights into interactions of 360A with AGCGA-quadruplexes uncover how the enthalpic–entropic balance is granted by a proper combination of flexibility and alternative structural elements of the quadruplex structure. Indeed, upon 360A complexation, the hydrophobic core is extended, thanks to the interaction of the ligands with GAGA and GCGC-quartet platforms.

14.2.2. The Charged Side Chains

Beside the hydrophobic interactions supported by the stacking of the ligand core on the G-tetrads, the reported examples highlight the advantages provided by the presence of cationic moieties in the side chains. As we discussed above, they actively complement the selective recognition of G4, increase the ligand affinity for the nucleic acid, and potentially implement the bioavailability of the final products. Metal complexes that have been extensively considered for G4 recognition can similarly provide these benefits. As additional features, they are

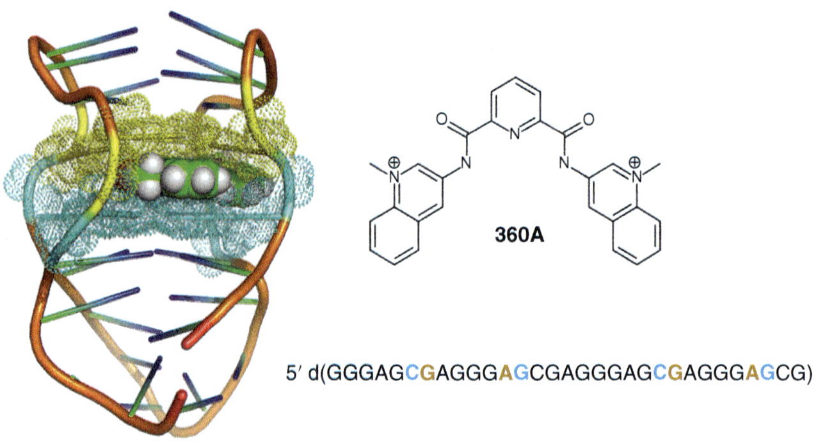

Figure 14.2 Model of the intercalation of **360A** into the tetrahelical folded **VK2** sequence (PDB ID:6SX3). The DNA platforms interacting with the ligand are highlighted in cyan (GCGC) and yellow (GAGA).

frequently derivatives with unique properties (i.e. spectroscopic or redox) and allow exploring different geometries including chirality. Thus, besides the therapeutic potential we can foresee their applications as sensors or probes for G4.

According to the metal complex design, two main divergent approaches are possible. The first one aims at localizing the metal ion in the center of an aromatic surface. In this way, the cation results locked within the central negatively charged channel of G4, thus causing an increment of the binding affinity and a concomitant reduction of the ligand binding modes. Porphyrin metal complexes are actually pioneering models for this approach but they generally retain ADMET issue [15].

A novel development considers the use of small self-assembly domains. In this regard, metallacrowns have an enormous potential (Figure 14.3a). Among them, the comparison of the Cu(II) vs Ln(II) metal complexes highlighted a better profile for the first one. This is largely based on the better interaction with the G4 core supported by the planar square-based geometry and the overall net positive charge of the metallacrown [16]. Worth of note, beside the presence of multiple cations, the 1:1 complex stoichiometry was confirmed.

We can consider this result in connection with those obtained along an alternative experimental design that considers metal complexes as substituents of a neutral aromatic scaffold. A wise selection of the two components is expected to finely tune both affinity and selectivity. As an example, recently trisubstituted fan-shaped trinuclear Pt(II) complexes were synthesized and characterized (Figure 14.3b) [17]. These compounds showed a remarkably high selectivity for G4 vs dsDNA with a better kinetic profile for the tetrahelix binding process. Interestingly, when tested on a telomeric sequence, they shifted the equilibria toward the antiparallel arrangement. This property was rationalized by assuming an interaction of these branched molecules within the large groove provided by this G4 topology and supported by the possible rotation of the Pt(II) substituents. Noteworthy, these fan-shaped trinuclear Pt(II) complexes inhibited telomerase activity. These preliminary data are encouraging the development of novel anticancer drugs firmly based on structure–function relationships.

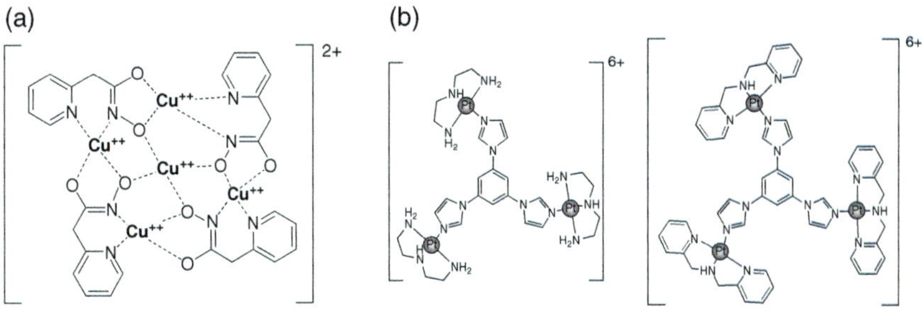

Figure 14.3 Copper metallacrown (a) and trisubstituted fan-shaped trinuclear Pt(II) complexes (b).

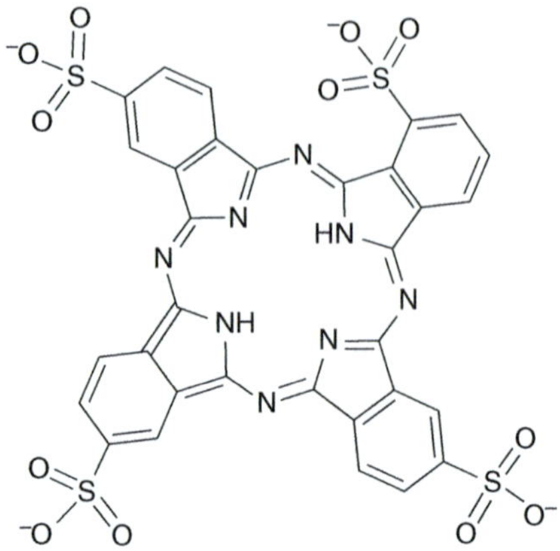

Figure 14.4 Molecular structure of **APC**.

Finally, it must be reminded that whenever we translate the *in-vitro*-derived G4-small-molecule interaction model within a complex physiological environment, we have to reconsider all restriction rules. A key example is represented by the effective telomerase inhibition exhibited by the negatively charged phthalocyanine 3,4′,4″,4‴-tetrasulfonic acid (**APC**) with the telomeric G4 (Figure 14.4) [18]. In this case, the reported experimental stabilization of the telomeric hybrid G4 and the consequent telomerase inhibition pairs with the concurrent recruitment of physiological metal ions like K^+ that is required to support G4 formation and the recruitment of the negatively charged ligand. Although this binding mechanism and the lack of positive charges should increment the selectivity toward the tetra-helix, actually this compound binds the double helix and the G4 as well. However, as suggested by a combination of Molecular Docking and Molecular Dynamics simulations, hybrid I telomeric G4 and dsDNA can accommodate **APC** within the grooves, whereas only the hybrid form II of the telomeric G4 provides a G-tetrad arrangement suitable to engage efficient end-stacking interactions with the ligand. In perspective, this rational for preferential binding of a single topology must be further developed for the design of novel selective G4 stabilizers.

14.2.3. Pharmacophores Conjugation with Advanced Properties

The above reported examples refer to a central core substituted with variegate functions to implement the affinity of the resulting molecules for the G4 target. By keeping in mind that the side chains actively participate in the proper recognition of the G4 structure as well as on its delivery at specific positions, a smart combination of these elements can largely amplify the potential applications of the final product.

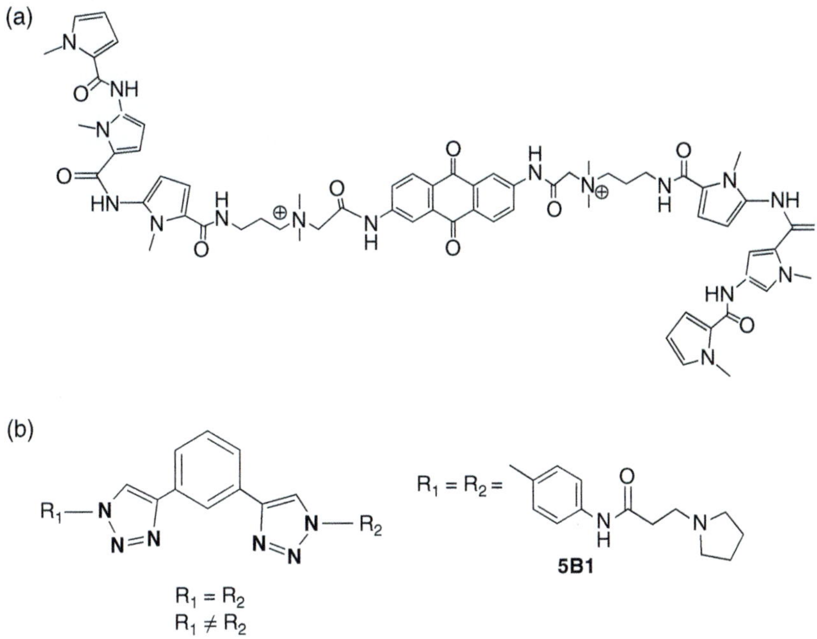

Figure 14.5 Chemical structures of the most efficient anthraquinone-(pyrrol)carboxamide conjugate (a) and symmetric/asymmetric phenyl ditriazoles (b).

Anthraquinone was one of the first scaffolds to be extensively explored for the design of binders able to stabilize the telomeric G4 causing telomerase inhibition [19]. As well, the grooves of G4 were identified as promising binding sites for minor groove binders like as distamycin [20]. Thus, the conjugation of the anthraquinone pharmacophore with two oligopyrrole carboxamide chains of growing length was exploited (Figure 14.5a) [21]. The increment of N-methyl pyrrole units from 1 to 3 was linearly related to the increment of the affinity for G4, with a strong preference for c-myc and kit G4 vs dsDNA. Remarkably, Molecular Dynamics analyses on its binding with c-myc G4 showed that the anthraquinone only partly stacks at the 3′ G-tetrads of the folded DNA. This localization is driven by the side chains that lie within the grooves, thus raising the stability of the complex up to $\Delta G = -34.6$ kcal/mol. Remarkably, these compounds showed an appealing cytotoxic activity for cancer cells (HeLa and HEK293T) that was associated to the induction of apoptotic pathways, thus supporting their potential interest for further development.

Other interesting entities used to functionalize G4 binders are carbohydrates. In addition to increase the affinity and/or selectivity for the target G4, they contribute to tune the solubility of the final product and, taking advantage of the glucose transporters (GLUT) that are overexpressed in cancer cells they can optimize the drug accumulation at the desired site, thus reducing side effects. The final achievement obtained with the introductions of these substituents largely depends upon the chemistry of the selected sugar moieties and their relative location on the planar scaffold. The drawback is represented by the difficulties related to the control of the reaction for conjugation. However, click chemistry was successfully applied to prepare small libraries of these conjugates to rationally describe their outcomes. In particular, starting from a phenyl ditriazole central core (Figure 14.5b), this protocol was applied to prepare a new family of symmetrical and dissymmetrical carbohydrate conjugates [22]. When compared to the control leading compound (**5B1**, Figure 14.5), the G4 affinity of all conjugates was reduced, likely due to a lowered net charge of the final products. Nevertheless, within the series, the asymmetric compounds were much stronger G4 binders with a preference for the human telomeric sequence that paired to an implemented G4 vs dsDNA selectivity. Unfortunately, only a low antiproliferative activity was observed *in cell*. However, the reason was related to a poor stability and/or low uptake, thus putting the basis for further developments.

Alternatively, the conjugation project can start from sugar and the planar ligands known to bind nucleic acid but with modest affinity with the aim to realize a synergic effect of two moieties [23]. This was the case of neomycin, a carbohydrate with several basic amino groups and known to bind ribosomal RNA. This antibiotic was reported to fit in the G4 groves with a very low affinity. As well, phenanthroline (Phen) is a poor end-stacking agent due to its relatively small aromatic surface. Their conjugation at position 5 of Phen (Figure 14.6, **6A**) produced a compound with much better recognition properties toward telomeric G4, thus opening the interest in testing more variegate combinations. At first, the work focused on the contribution of the planar system, so it analyzed the consequences of the progressive increment of this surface [24]. Indeed, using a different conjugation chemistry in a new series of compounds, the Phen unit was substituted with an anthraquinone moving forward to a pyrene, a naphthalene diimide, and, finally, BQQ (Figure 14.6, **6B**).

Consistently with the starting hypothesis, the BQQ-neomycin conjugate **6B** was the best binder with affinity constants for different G4s ranging between 10^7 and 10^8 M^{-1}. When tested on the Oxytricha Nova G4, a 1:1 stoichiometry was observed, indicative of a single site for BQQ–neomycin complexation. Noteworthy, the molecular details of the G4 complex formation by these conjugates evidence that all of them take advantage of the Neomycin structural flexibility. Indeed, docking analyses showed that in the G4 complex, upon changing of the planar ligand, the antibiotic rearranges its poses to maximize the bonding network with the oligonucleotide. This explains how the conjugation increases the targeting of G4 while, remarkably, preventing the recognition of the rRNA A-site by the antibiotic, thus suppressing its antibiotic activity.

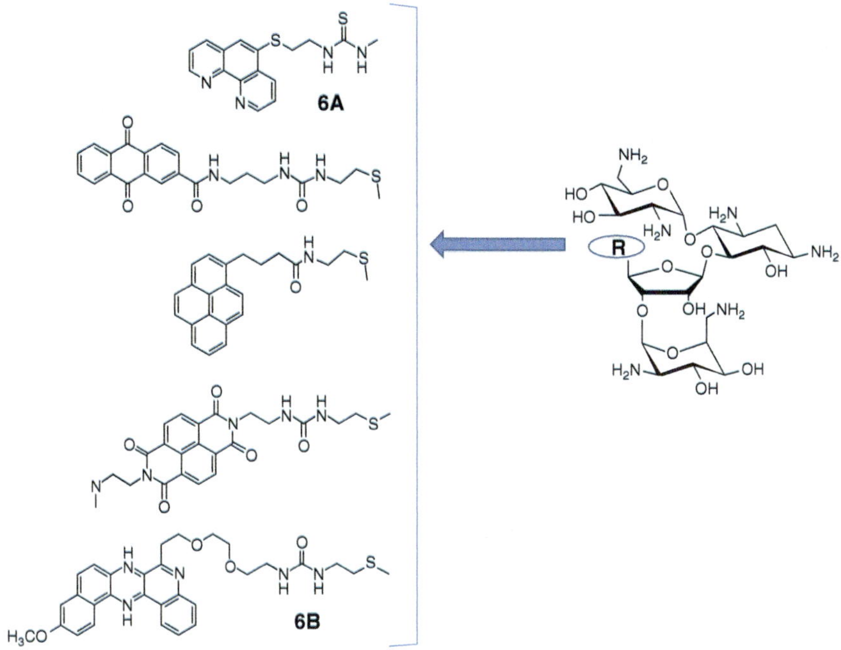

Figure 14.6 Examples of studied Neomycin conjugates.

14.3. THERAPEUTIC EFFICIENCY: SELECTIVE vs NOT SELECTIVE?

14.3.1. G4 vs dsDNA: The NDI Model

Clearly, the huge synthetic and analytical efforts devoted to produce good G4 binders is motivated by their potential use *in cells* and, hopefully, in humans. To reach this goal, it is important to assess the selectivity the new derivatives. In particular, the potential competition with the much more abundant dsDNA is still a matter of concern. Originally, an extended planar surface was considered sufficient to increase the binding to the G4 and, simultaneously, to reduce dsDNA recognition for steric reasons. However, this is not always the case. As an example, one of the most widely studied class of ligands, the NDI, efficiently intercalate between the base-pairs of the double helix. For this scaffold, the possibility to introduce substituents at positions 2, 3, and even 4 has been extensively exploited. Indeed, the use of multiple side chains further increases the stability of the final complex through interactions with the grooves. Moreover, since the double-stranded DNA contains only two grooves, it represents an efficient strategy to implement the selectivity for G4 vs dsDNA. Examples in this context are multiple [11, 25]. Among the most recent, we mention **CM03**, a tri-substituted analogue active in Gemcitabine-resistant cell lines (Figure 14.7a) [26]. As evidenced by RNA-seq data analyses, it simultaneously downregulates multiple

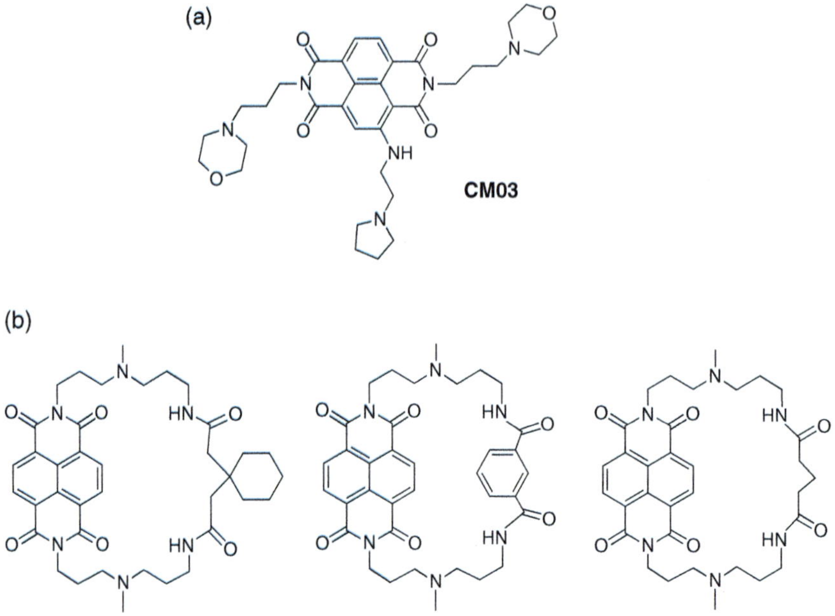

Figure 14.7 Examples of tri-substituted (**CM03**, (a) and cyclic (b) naphthalene diimide derivatives.

pathways related to cancer growth and sustenance. It follows that **CM03** activity is not related to the binding of one single G4 but of panel of them.

To further impair the recognition of the double helix, the cyclization of one side chain between two NDI positions was originally proposed (Figure 14.7b) [27, 28]. Remarkably, these derivatives bind G4 and stabilize them both in the presence of limited K^+ concentration and under crowding conditions. This evidence indicates that this class of NDI-derivatives can preserve their G4-binding properties also in the complex physiological environment.

14.3.2. Selectivity: A Matter of Shape?

The above commented studies nicely describe the general features of a small molecule that can be proficiently exploited to implement G4 recognition. They leave open the question of the selectivity among different G4s. Since G4 topologies are highly variegate, it was considered possible to realize ligands highly selective for each single G4 (or family of G4 structures). If we consider that a quite important number of G4 structures have been solved, a rational design supported by computational tools (*de novo* design, docking, Molecular Dynamics) should be expected to properly drive this process. However, the issue is more complex. Indeed, as widely reported for the "simple" monomeric telomeric sequence, each G-rich sequence often folds according to multiple intramolecular topologies (basket, chair, hybrid, propeller), thus making it difficult to choose the correct starting model. Moreover, the conformational distribution at the equilibrium conditions can significantly change according to minimal environmental variations (i.e. DNA concentrations, metal ions, pH), thus making it hard to identify the most favorable target to obtain a well-defined therapeutic response. Another poorly studied critical point concerns the kinetic evolution of the G4 folding and the characterization of the structural features of the involved intermediates. Indeed, for each G4 forming sequence the thermodynamically most stable G4 state can be reached through the formations of kinetic intermediates along pathways that are not conserved for all systems [29]. Noteworthy, in some instances their lifetime can be compatible with physiological processes. In this scenario, it is thus demanding to assess how the presence of a single ligand actually alters all these parameters. This behavior is not easy to monitor and, even less to predict since, intriguingly, structurally closely related compounds can often drive divergent effects. This was shown using the two perylene derivatives (Figure 14.8). **K20** and **PIPER** share a great structure similarity, common duplex-DNA recognition properties, but they diverge on the preference for telomeric G4 topology. This property was proficiently tested also on a sequence (kit2) present in the proximal promoter of *KIT*. Solution studies indicated it as a system with a complex folding pathway covering both kinetically and thermodynamically favored species [30, 31]. Both derivatives resulted good binders for the folded kit2, but it emerged that only K20 was able to recognize and stabilize a two-quartet anti-parallel G4 corresponding to a transient kinetic intermediate. As a result, this species, that is totally absent under equilibrium conditions, is conformationally selected by **K20**. This result opens the opportunity to consider these ligands as useful tools to assess if, under physiological conditions, the different topologies occurring at a single G4-forming site can actually exploit common or different functions [25].

14.3.3. Target One or Multiple G4

Despite these problems in rationally approaching the G4 vs G4 selectivity issue, several model studies deal with derivatives screened to target with high-efficiency

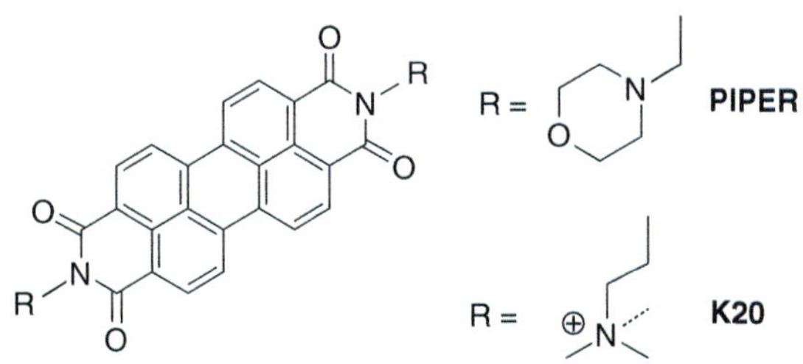

Figure 14.8 Chemical structures of perylene derivatives tested for their G4 conformational selection along kit2 folding pathway.

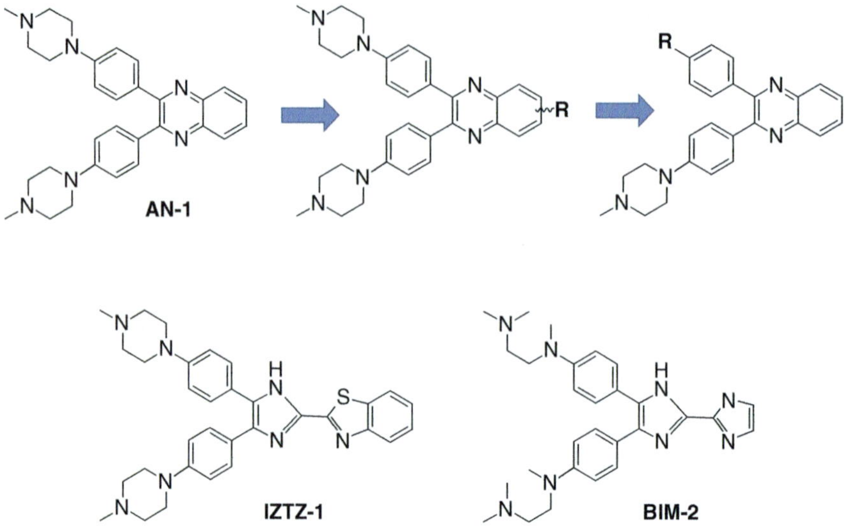

Figure 14.9 **AN-1** and structurally related derivatives.

one-specific sequence. In this context, the focus has been devoted to ligands that can act at oncogene promoters although the selectivity issue can be extended to all other genomic positions where G4 can form. A good model scaffold is quinoxaline, a system that, thanks to the wide possible derivatization, has been extensively used as a platform to design new G4 binders (Figure 14.9). Noteworthy, several members have been screened for their interaction and stabilization of G4 at the promoter of *c-MYC* [32, 33]. All these compounds showed interesting selectivity for G4 vs dsDNA. This property was supported by cellular studies that revealed that **AN-1** was able to reduce the oncogene transcriptions only of those mRNAs under the control of a G4 forming site. Ultimately, these compounds also suppressed the growth of c-myc-dependent cancer cell lines, i.e. melanoma and Triple Negative breast cancer. As the authors clearly evidenced, these data supported a promising application of the derivative to all cancers that require high c-myc level to support cell proliferation and migration. Although *in vivo* studies on these compounds are not available yet, they present an interesting drug-likeness profile, thus suggesting that their optimization might be easily addressed.

For **AN-1** and **IZTZ**, the selective targeting of one single site have been considered as the best strategy to target such aggressive cancers with minimal off-target effects. However, in a complex pathology like cancer, cell response to a single treatment can rapidly change. Thus, a smart strategy can actually be based on the simultaneous target of a subset of G4s with related functions within the cells. This would be expected to provide an amplified therapeutic response upon drug exposure. Moreover, it should retain the activity even in the case of mutation of a single target. This rational justifies why now-a-day several screening programs are run in parallel on multiple G4 forming sequences and the binders with the most transversal profile are selected for deeper characterization. This general workflow allowed to select for example a new drug-like diimidazole chemotype that simultaneously targets effectively two different G4s occurring at the promoter of *c-MYC* and of *BCL2*, respectively, with an efficiency much larger in comparison to the one observed on other G4s [33]. After the selection of the most promising candidate, (**BIM-2**) *in cell* data confirmed its ability to downregulate both *c-MYC* and *BCL-2* expression in acute myeloid leukemia, thus causing effective inhibition of cell growth.

The initial screening step for the identification of dual binders can be run also by virtual screening protocols eventually of wide or focused libraries (Figure 14.10a) [34]. For example, five indolo-naphthyridine derivatives were selected along a project starting from a previously identified promising HTS/c-MYC G4s dual binder (**10A1**). The members of the new series showed different preference for several G4 forming sites but one derivative (**10A2**) that, on the contrary, was quite selective for kit1 G4s. This preferential recognition was directly related to the formation of two additional hydrogen bonds between the morpholino-propyl side chain and the DNA backbone. This result further highlights how the energetic barriers that separate a selective vs. a dual binder can be extremely low and, thus, difficult to predict.

With the same aim, indole was also conjugated to other scaffolds and, in one example, simultaneously with multiple entities (Figure 14.10b) [35]. Although its evaluation is still ongoing, the resulting active compound is chemically interesting since it comprises in one single small molecule, an unusual spiro-indole with a coumarin, a thiazolidine, and a pyrazole moiety.

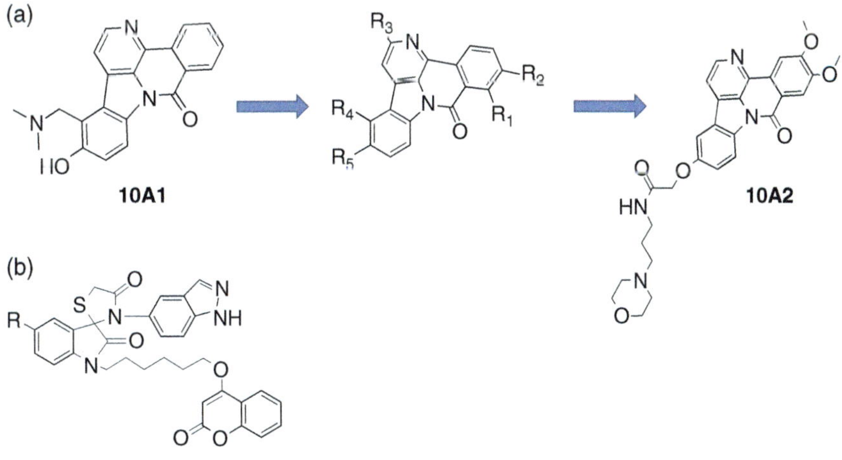

Figure 14.10 Scheme of the substitution pattern considered for indolo-naphthyridine derivatives (a) and of spiro-indole coumarin conjugates (b).

14.3.4. Recognition of G4 Repeats

Telomeres are characterized by a large number of conserved repeats, in humans corresponding to the (TTAGGG) sequence. This composition enables the formation of multiple G4s that can form close one to each other. Consequently, the conformational distribution of the telomeric G4 units can be affected.

In some instances, the ligand itself can "trap" preferentially one single conformation, thus leading to a conformational selection of the most suitable target G4. This was observed when the telomeric sequence was treated with **BMPQ** (Figure 14.11a), a compound that favors the (3+1) conformation rather than the monomeric G4 formed by four telomeric repeats and on longer sequences where, as it occurs in cells, multiple G4 units are simultaneously present [36]. This conformational selectivity results in a high affinity for these long chains in comparison to all tested monomeric G4s. Consistently, *in cell* BMPQ caused an increment in G4 that localized at telomere, thus leading to a disruption of their physiological architecture and to the activation of the DNA damage response. This occurred much more efficiently in comparison to other telomeric G4 binders not selective for multiple telomeric G4s.

We must keep in mind that the localization of multiple closely located G4 units can drive the preferential topology of each single unit and, in addition, generate G4–G4 interactions according to different overall architectures that, on a general basis, can be clustered into a beads-on-a-string or a cross-talking models. Noteworthy, besides the telomere-linked issues, several human pathologies are related to the presence of long stretches of G-repeats occurring in the coding regions [37]. Moreover, even a subset of oncogenes contains multiple vicinal G4 forming sites [38]. These domains actually represent additional levels of tridimensional organization of noncanonical structures, with more complex

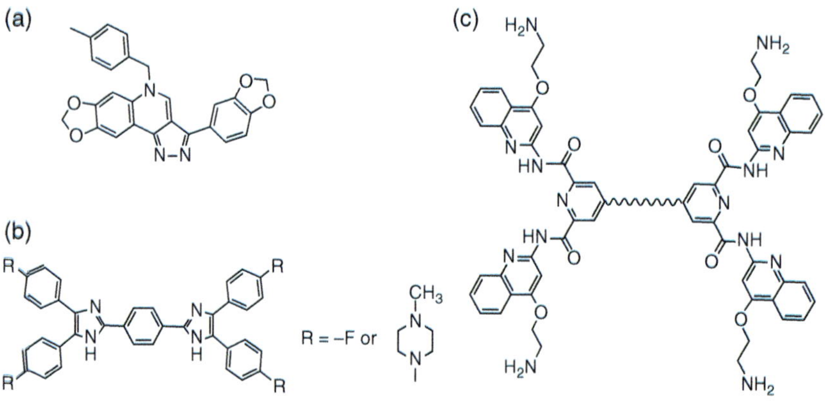

Figure 14.11 Different scaffolds and substitution patterns used for the selective recognition of multimeric G4. **BMPQ,** aryl-substituted imidazoles and pyridostatin derivatives are reported in Panel a, b and c, respectively.

architectures in comparison to the isolated G4. It derives that they are good candidates to finely tune the selective delivery of ligands at these sites.

A common strategy is to target the G4–G4 interaction surface. The energy cost is expected to be much lower in comparison to intercalation. In addition, an increment of the planar surface can grant a suitable stacking surface for the recruitment of one G4 at each side. A successful example of this strategy was obtained by dimerization of aryl-substituted imidazoles [39]. Although the aromatic surface corresponds to a homodimeric imidazole, the functionalization of the side chains allowed testing a certain degree of asymmetry in these series of derivatives (Figure 14.11b). This was important because, among the tested compounds only the one with the two fluorine and the two methyl-piperazine substituents *in trans* was selective for multimeric vs monomeric G4s. This ligand binds and stabilizes multimeric G4, possibly by intercalating at the G4–G4 interface. The resulting activity profile *in cell* was consistent with a good G4 selectivity. Indeed, it inhibited growth in an ALT cancer line, whereas it did not reduce the transcription of genes containing monomeric G4. Notably, its cytotoxicity was extremely reduced in normal cells. This points to the value of this targeting approach.

An alternative strategy for the recognition of G4–G4 platform refers to the connection of two independent G4 binding units through a properly designed linker. The advantage of this approach is that through appropriate linker design and functionalization, we can target dimeric but also multimeric G4s. Moreover, the inserted G4 binders can been selected to have different binding properties (i.e. binding mode, G4 topology preference, etc.,) thus making them tunable platforms that can be easily modified according to the nucleic acid sequence to be targeted. For these compounds, the length and the chemical composition of the linker must be carefully optimized since they must grant the best complementarity of the conjugate on the multiple G4. Interestingly, a study performed using pyridostatin (**PDS**) as ligand and polyether linker chain of variable length showed how this parameter was actually driving the preferential recognition of antiparallel (n =1) vs parallel (n=2) telomeric G4 (Figure 14.11c) [40]. By keeping in mind that these two conformations are quite different, it is clear that when the two G4 units are consecutively occurring on a single string, the relative distance between the two PDS binding sites will change according to the DNA folding topology. Thus, at least for the telomeric G4s that have very reduced energetic barriers to shuffle between these two folding topologies, the conformational selection can be simply controlled by the linker length that best matches the binding sites distance. Similar outcomes resulted using acridine or tri-functionalized naphthalene diimides as monomeric planar scaffold [41, 42].

14.3.5. Dual Binders: G4 and Other Targets

All above reported data actually balance between the better performance of ligands for one single G4 vs multiple G4s either if separated or in a run. If we consider the G4 formation from G-rich double-stranded sites, we know that we have to deal with a high energetic cost to melt the paired strand. To balance this cost, we can simply increase the affinity of the ligand for the G4. However, we can take advantage of the plastic behavior of the complementary strand that, being C-rich, can fold into an I-motif (iM). As a result, the two noncanonical motifs co-localize and thus can be simultaneously targeted. To obtain a synergic combination in terms of affinity and selectivity, the two of them must be simultaneously stabilized, thanks to the conjugation of the designed ligands. This model has been generally tested on the *c-MYC* oncogene promoter by conjugating two binding units of a single ligand (an acridine) previously selected for its ability to stabilize either G4 and iM [43]. An optimization work was exploited to identify the best linker to connect the two units to implement G4 and iM recognition but to prevent dsDNA binding (Figure 14.12a). The most promising compound (**12A1**) identified was thus tested in cell lines. Beside its remarkable antiproliferative effects, it was confirmed to cause an extensive inhibition of c-*MYC* transcription and expression. Thus, it can be considered a promising approach although not yet a proof-of concept of the advantage derived from simultaneous stabilization of the two structures *in cells*. Indeed, it was already reported that their occurrence at a single site should be mutually exclusive for steric reasons.

This was elegantly proved *in cells* thanks to a comparative analysis of two ellipticine derivatives, one selective for G4 and one for iM structures (Figure 14.12b) [44]. Remarkably, the two tested ligands were both ellipticine derivatives: the close structural relationships grant avoiding bias due to recognition of differential alternative binding sites. When used independently, both

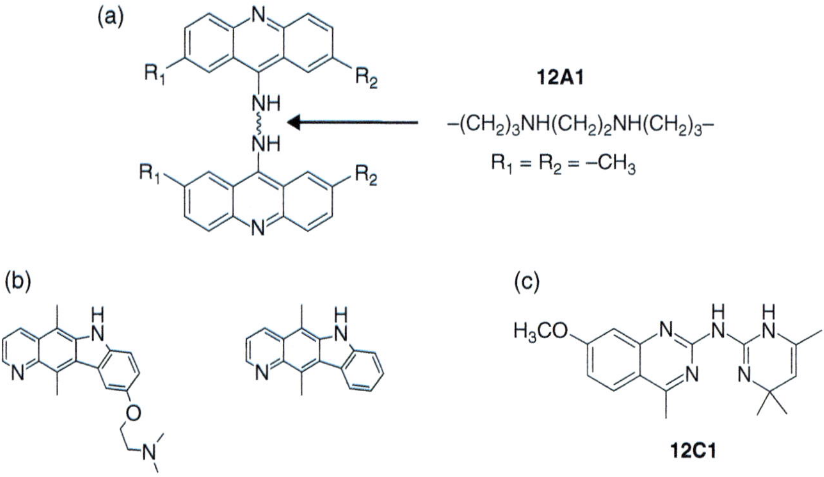

Figure 14.12 Examples of small molecules designed to simultaneously recognize G4 and different targets. Acridine, ellipticine and quinazoline derivatives are reported in Panel a, b and c, respectively.

compounds confirmed their ability to selectively induce G4 or iM *in cell* nuclei. However, on synchronized MCF-7, the simultaneous addition of the two ligands did not exert a synergic effect but, conversely, confirmed that they work competitively. This result appears as totally in contrast with the previous one. However, Waller and co-workers also evidenced how in untreated cells the number of G4 and iM structural foci distinctly change along the cell cycle with G4 being maximally present at early S phase, and iM at G1. Thus, it is conceivable that a diffuse stabilization of the two of them along the whole cell cycle, although temporally independent, can cause a more prominent damage to the cells.

Within the frame of selectivity issue, it is finally demanding to put the light that the issue of off-target effects for all these G4 binders is not limited to the interaction with DNA or, more precisely, with nucleic acids. Indeed, we must be aware that, since they preserve small-molecule features, even other not related biomolecules can act as binding sites and once bound they can eventually modulate different physiological/pathological functions. Off-target events are difficult to predict, and they are likely to account for poor performance of these drugs *in vivo*. However, it is worth to be mentioned that not always they represent a source of undesirable toxicity. This was recently exploited by studying a novel series of quinazoline derivatives [45]. The project was based on the G4 stabilizing properties of a leading compound (**12C1**, Figure 14.12c) and on its structural similarity with STAT3 inhibitors. The new library was preliminary screened for G4 affinity and selectivity vs dsDNA. Both the lead compounds and two related congeners were confirmed to be good G4 binders both *in vitro* and in human cell lines where they induced the hallmarks associated with DNA damage and, ultimately apoptosis. As well, the observed reduction of STAT3 in treated cell was confirmed to be associated with their direct interaction with the protein (?). As the G4 recognition, also this interaction was specific, since STAT18 was not affected by the ligands. Clearly, the rational optimization of these derivatives will be complex due to the need to balance several targeting processes. Nevertheless, this study opens a promising perspective: a single small molecule that works according to two synergic mechanisms of action can help in obtaining more efficient drugs with reduced side effects.

14.4. PROTEIN-BASED LIGANDS FOR G-QUADRUPLEXES

As above mentioned, an ideal situation occurs when a small ligand can establish an effective interaction with the target, eliminating or substantially reducing off-target effects. Establishing effective "bullets" can be achieved by imitation of mother nature, which developed numerous selective nucleic-acid binding proteins. Once defined the peptide sequence responsible for selectivity, this could be incorporated into the drug producing a highly effective tethered medicine. The approach is apparently simple. However, an isolated short peptide is in general much more flexible than in the protein context. The shorter the

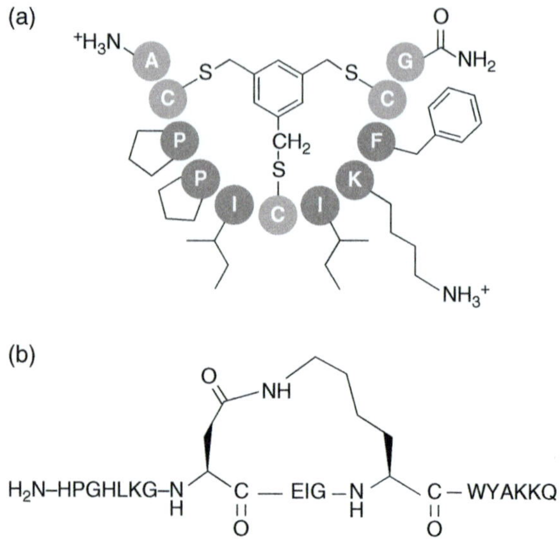

Figure 14.13 Example of cyclic (a) and staple (b) peptides.

peptide and the less structured it is. Losing conformational stability renders the binding poorly selective. One possibility is trying to "freeze" the short peptide in a favorable conformation (stapling), by producing cyclic derivatives with the aid of an external scaffold covalently linked to the peptide by bonding to reactive amino acid side chains. Another approach consists in producing a cyclization by intramolecular link, for example, through the reaction of the side chain of a lysine with a glutamic acid residue. Of course, these reactions should involve amino acids essential for nucleic acid recognition. Two recent examples deal with staple peptides as G4 recognition elements (Figure 14.13) [46, 47]. One describes the production of a library of bicyclic peptides supported on a three-valent scaffold, obtained by a phage display technique (Figure 14.13a), while the other generates a monocyclic stapled system by intramolecular reaction of RHAU peptidyl side chains (Figure 14.13b). In both cases, a substantial increase in G4 affinity is observed in comparison with the linear peptide. Furthermore, stapled systems are characterized by higher chemical stability.

14.5. METHODS FOR G4 IDENTIFICATION AND CHARACTERIZATION

14.5.1. Characterization of G4-Ligands in Solution

Detection of the conformation-dependent ligand-nucleic acid complexes under physiological condition is highly desirable both in terms of selectivity and for distinguishing between naturally occurring species as opposed to ligand-induced

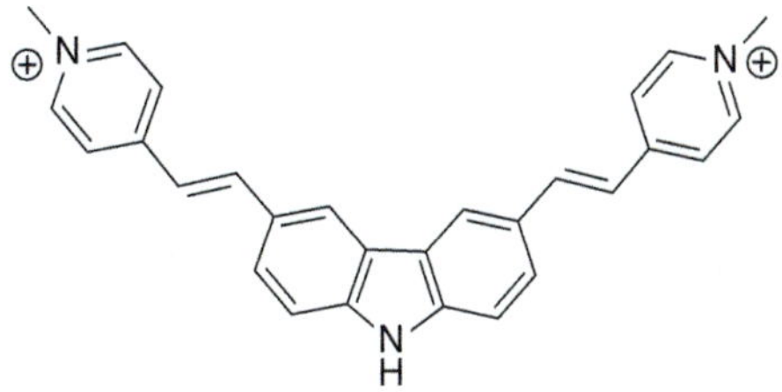

BMVC

Figure 14.14 **BMVC**, a G4 ligand studied by G4 microarray.

forms. This should apply at early screening stage up to the *in cell* models. Thus, new methodological tools are required.

The set-up of G-rich microarrays represents a successful high-throughput combinatorial method to assess the selectivity of G4 ligands [48]. The large number of sequences, and thus structures, that can be simultaneously tested provides a huge amount of information that can be usefully integrated to better define the sequence features that drive the G4 formation and the selective recognition by a ligand. Using **BMVC** (Figure 14.14) as a probe, the microarray system output well compared with NMR and spectroscopic results and allow addressing it as a selective binder for parallel G4, thanks to the recognition of thymine in the 3′-flanking region. The final emerging picture is a detailed recognition model that can be proficiently applied to derive differential binding profiles of multiple potential binders.

Clearly, the increasing number of high-resolution structures of G4 and G4-ligands complex is of great help for finely tuning the synthetic efforts related to the production of well-performing ligands. However, X-ray and NMR both suffer of several experimental limitations. Thus, it is worth mentioning that other spectroscopic methods are currently under development to support our comprehension of the molecular details behind the ligand-G4 complex formation. A mention is deserved by Ultraviolet Resonance Raman, UVRR, spectroscopy that finely addresses, which chemical groups of ligand and DNA are involved in the interaction and nicely identify the π interactions. As well, SAXS can provide valuable details on the overall conformation of the complex [49, 50]. Thus, a proper combination of all these techniques can provide a highly detailed description of the structural and molecular features of G4s and their complexes even under experimental conditions not suitable for high-resolution methods.

14.5.2. Detection of G4 in Cell

While studying G4 in solution, we try to adjust our experimental conditions to mimic as much as possible the cellular environment. Nevertheless, this strategy will always provide a biased simplified model with reference to the *in cell* system.

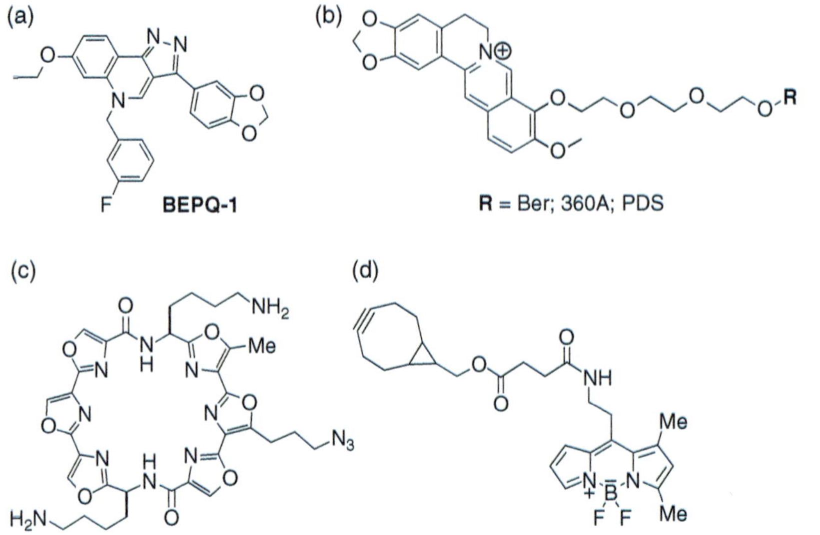

Figure 14.15 Example of fluorescent G4 ligands (a and b) and of ligand (c) and fluorophore (d) used for post-target click chemistry.

To properly follow a G4 ligand *in cell,* we can refer to molecules that light up a strong fluorescence upon G4 binding (Figure 14.15) [51–53]. Alternatively, we can label a G4 ligand with a fluorescent probe. Hopefully, the fluorescence properties of the fluorophore and the binding profile of the ligands will be preserved *in cell*. However, as above discussed, even a modest chemical modification can remarkably alter the G4 recognition pattern of a small molecule. A possible alternative is to proceed with a post-target-binding visualization strategy [54]. Click chemistry is of a great help in this kind of application since it can be run in "mild" conditions with fluorophores that are homogenously distributed inside the cell.

Another critical issue is related to the limited knowledge of the fate of G4 in a living system. So, we need ligands specifically designed not only for therapeutic applications but also as G4 molecular probes *in cells*.

To fulfil this need, multiple small molecules have been designed. In this case, they are expected to be not cytotoxic and to be highly environment-sensitive. In addition, we can design compounds that equally distribute within all the intracellular space or that preferentially accumulate into one specific compartment (i.e. cytosol and nuclei). This was the case of some alkynyl-gold (I) complexes (Figure 14.16) [55]. By analyzing several substituents at Au(I), it was possible to identify a fluorescent noncytotoxic G4 binder that efficiently localizes in the cytoplasm of intact melanoma cells but cannot enter the nuclei.

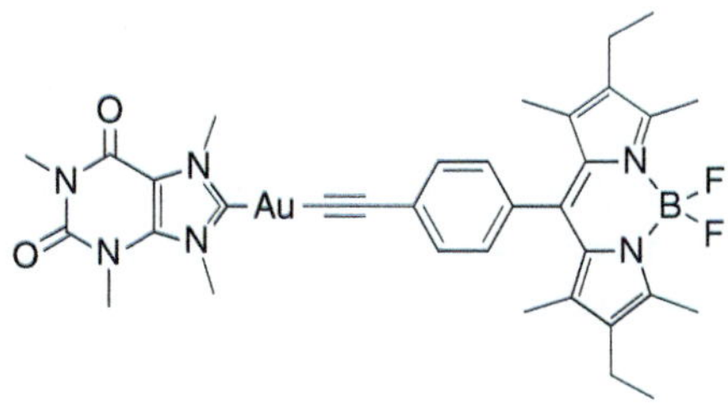

Figure 14.16 Noncytotoxic Au(I) complex.

14.6. CONCLUDING REMARKS

G4 structures represent possible significant targets for therapeutic intervention, given their multiple and essential roles in practically all cellular processes. This is corroborated by the impressive number of G-rich sequences able to fold into noncanonical structures found in the genome. They principally belong to regulatory genes like c-myc and to telomere-related structures located at the tips of chromosomes and processed by telomerase. At the beginning, G4s were considered easily druggable targets and a large number of synthetic and natural compounds have been shown to interact with them. However, it was soon realized that the intricate interconnection between thermodynamics, kinetics, and structure(s) of ligand-G4 adduct(s) renders rational design quite elusive. Only now, after 40 years of poorly successful efforts, we begin to understand the ABC of a complicated drug-target recognition process, as we discussed earlier.

Moreover, still we did not properly focus the different clinical strategies that should be applied when the target is an oncogenic sequence as opposed to a telomeric sequence [56]. In fact, in the latter case, the therapeutic goal is the silencing of telomerase activity, selectivity is especially important and off-target interactions should be avoided or drastically reduced. Conversely, in the case of oncogene-related target, the goal is to produce the maximum silencing of cell signaling, like in a classical chemotherapy approach: accordingly, multiple targeting could be accepted, if not preferred. Hence, aiming at improved selectivity, including simultaneous recognition of G4 dimers, might be expected to be preferentially restricted to telomeric targeting. Incidentally, structure–activity relationships should become more reliable getting closer to the physiological cellular conditions including crowding.

Meanwhile, novel investigation techniques are disclosed, which permit us more sensitive cellular studies to locate and quantitatively determine the true sites of G4 formation and their operation mechanisms. New active compounds are rendered available for biological testing, each showing an individual fingerprinting in terms of intensity and sites of recognition. Given the induced fit for mutual adaptation, this means that the final conformation of G4 in the complex will differ (by a low or large extent) from the original one. Of course, the less energy is spent in conformational changes, the more stable the complex will be.

Recent studies demonstrate that G4 forming sequences are found in many organisms including viruses (see for example the recent literature on the Nipah virus and Covid-2) and protozoa, which confirms the conservation of G4 arrangements during evolution [57–59]. Hence, the general importance of noncanonical structures stimulates further work in the field. In this connection, targeting RNA G4 arrangements opens a (almost) new chapter for medicinal chemistry to deal with. RNA G4s are structurally similar but distinguishable from DNA G4s in terms of stability, hydration sphere, and flanking structures, which gives unique opportunities for new nucleic-acid-directed drug design and development [60, 61].

ACKNOWLEDGMENTS

This work was supported by University of Padova, Italy/grant # SISS_SID19_01 and by AIRC grant # IG 2021—ID. 26474.

REFERENCES

1. Yuan, W. *et al.* (2020) The influencing factors and functions of DNA G-quadruplexes, *Cell Biochemistry and Function*. John Wiley and Sons Ltd, *38*(5), pp. 524–532. https://doi.org/10.1002/cbf.3505.
2. Kharel, P. *et al.* (2020) The role of RNA G-quadruplexes in human diseases and therapeutic strategies, *WIREs RNA*. Blackwell Publishing Ltd, *11*(1), p. e1568. https://doi.org/10.1002/wrna.1568.
3. Spiegel, J., Adhikari, S. and Balasubramanian, S. (2020) The structure and function of DNA G-quadruplexes, *Trends in Chemistry*. Cell Press, *2*, pp. 123–136. https://doi.org/10.1016/j.trechm.2019.07.002.
4. Lansdorp, P. and van Wietmarschen, N. (2019) Helicases FANCJ, RTEL1 and BLM act on guanine quadruplex DNA in vivo, *Genes*. MDPI AG. https://doi.org/10.3390/genes10110870.
5. Tan, Z., Hao, Y. and Zheng, K. (2020) Kinetics, conformation, stability, and targeting of G-quadruplexes from a physiological perspective, *Biochemical and Biophysical Research Communications*. Elsevier B.V., *531*(1), pp. 84–87. https://doi.org/10.1016/j.bbrc.2020.04.047.
6. Masai, H. and Tanaka, T. (2020) G-quadruplex DNA and RNA: Their roles in regulation of DNA replication and other biological functions, *Biochemical and Biophysical Research Communications*. Elsevier B.V., *531*(1), pp. 25–38. https://doi.org/10.1016/j.bbrc.2020.05.132.
7. Asamitsu, S., Bando, T. and Sugiyama, H. (2019) Ligand design to acquire specificity to intended G-quadruplex structures, *Chemistry - A European Journal*. Wiley-VCH Verlag, pp. 417–430. https://doi.org/10.1002/chem.201802691.
8. Carvalho, J. *et al.* (2020) G-quadruplex, friend or foe: The role of the G-quartet in anticancer strategies, *Trends in Molecular Medicine*. Elsevier Ltd, pp. 848–861. https://doi.org/10.1016/j.molmed.2020.05.002.

9. Ma, Y., Iida, K. and Nagasawa, K. (2020) Topologies of G-quadruplex: Biological functions and regulation by ligands, *Biochemical and Biophysical Research Communications*. Elsevier B.V., *531*(1), pp. 3–17. https://doi.org/10.1016/j.bbrc.2019.12.103.
10. Nakanishi, C. and Seimiya, H. (2020) G-quadruplex in cancer biology and drug discovery, *Biochemical and Biophysical Research Communications*. Elsevier B.V., *531*(1), pp. 45–50. https://doi.org/10.1016/j.bbrc.2020.03.178.
11. Zhirov, A. M. *et al.* (2020) Diazapyrenes: Interaction with nucleic acids and biological activity, *Chemistry of Heterocyclic Compounds*. Springer, *56*(6), pp. 674–693. https://doi.org/10.1007/s10593-020-02717-1.
12. Prasad, B. *et al.* (2020) The relation between position and chemical composition of bis-indole substituents determines their interactions with G-quadruplex DNA, *Chemistry - A European Journal*. Wiley-VCH Verlag, *26*(43), pp. 9561–9572. https://doi.org/10.1002/chem.202000579.
13. O'Hagan, M. P. *et al.* (2020) Stiff-stilbene ligands target G-quadruplex DNA and exhibit selective anticancer and antiparasitic activity, *Chemistry - A European Journal*. Wiley-VCH Verlag, *26*(28), pp. 6224–6233. https://doi.org/10.1002/chem.201905753.
14. Kotar, A., Kocman, V. and Plavec, J. (2020) Intercalation of a heterocyclic ligand between quartets in a G-rich tetrahelical structure, *Chemistry - A European Journal*. Wiley-VCH Verlag, *26*(4), pp. 814–817. https://doi.org/10.1002/chem.201904923.
15. Pipier, A. *et al.* (2019) G-Quadruplex binding optimization by gold(iii) insertion into the center of a porphyrin, *Dalton Transactions*. Royal Society of Chemistry, *48*(18), pp. 6091–6099. https://doi.org/10.1039/c8dt04703k.
16. Rajczak, E. and Juskowiak, B. (2020) Conformational rearrangements of G-quadruplex topology promoted by Cu(II) 12-MCCu(II)PyrAcHA-4 metallacrown, *International Journal of Biological Macromolecules*. Elsevier B.V., *156*, pp. 1258–1269. https://doi.org/10.1016/j.ijbiomac.2019.11.163.
17. Xu, C. X. *et al.* (2020) Two novel fan-shaped trinuclear Pt(ii) complexes act as G-quadruplex binders and telomerase inhibitors, *Dalton Transactions*. Royal Society of Chemistry, *49*(27), pp. 9322–9329. https://doi.org/10.1039/d0dt01767a.
18. Wang, Z. *et al.* (2020) Molecular insight into the selective binding between human telomere G-quadruplex and a negatively charged stabilizer, *Clinical and Experimental Pharmacology and Physiology*. Blackwell Publishing, *47*(5), pp. 892–902. https://doi.org/10.1111/1440-1681.13249.
19. Sun, D. *et al.* (1997) Inhibition of human telomerase by a G-Quadruplex-Interactive compound, *Journal of Medicinal Chemistry*. American Chemical Society, pp. 2113–2116. https://doi.org/10.1021/jm970199z.
20. Randazzo, A. *et al.* (2002) Interaction of distamycin a and netropsin with quadruplex and duplex structures: A comparative 1H-NMR study, *Nucleosides, Nucleotides and Nucleic Acids*, *21*(8–9), pp. 535–545. https://doi.org/10.1081/NCN-120015067.
21. Roy, S. *et al.* (2020) Specific stabilization of promoter G-Quadruplex DNA by 2,6-disubstituted amidoanthracene-9,10-dione based dimeric distamycin analogues and their selective cancer cell cytotoxicity, *European Journal of Medicinal Chemistry*. Elsevier Masson SAS, *195*. https://doi.org/10.1016/j.ejmech.2020.112202.

22. Arévalo-Ruiz, M. *et al.* (2020) Symmetric and dissymmetric carbohydrate-phenyl ditriazole derivatives as DNA G-quadruplex ligands: Synthesis, biophysical studies and antiproliferative activity, *Bioorganic Chemistry*. Academic Press Inc., *99*, p. 103786. https://doi.org/10.1016/j.bioorg.2020.103786.
23. Singh, M. *et al.* (2020) Use of neomycin as a structured amino-containing side chain motif for phenanthroline-based G-quadruplex ligands and telomerase inhibitors, *Chemical Biology & Drug Design*. Blackwell Publishing Ltd, *96*(5), pp. 1292–1304. https://doi.org/10.1111/cbdd.13741.
24. Ranjan, N. *et al.* (2020) Surface dependent dual recognition of a G-quadruplex DNA with neomycin-intercalator conjugates, *Frontiers in Chemistry*. Frontiers Media SA, *8*. https://doi.org/10.3389/fchem.2020.00060.
25. Amjadi Oskouie, A. and Abiri, A. (2021) Refining our methodologies for assessing quadruplex DNA ligands; selectivity or an illusion of selectivity?, *Analytical Biochemistry*. Academic Press Inc., *613*, p. 113744. https://doi.org/10.1016/j.ab.2020.113744.
26. Ahmed, A. A. *et al.* (2020) A G-quadruplex-binding compound shows potent activity in human gemcitabine-resistant pancreatic cancer cells, *Scientific Reports*. Nature Research, *10*(1), p. 12192. https://doi.org/10.1038/s41598-020-68944-w.
27. Esaki, Y. *et al.* (2014) Design of tetraplex specific ligands: Cyclic naphthalene diimide, *Chemical Communications*. The Royal Society of Chemistry, *50*(45), pp. 5967–5969. https://doi.org/10.1039/c4cc01005a.
28. Zou, T. *et al.* (2020) The interaction of cyclic naphthalene diimide with G-quadruplex under molecular crowding condition, *Molecules*. MDPI AG, *25*(3). https://doi.org/10.3390/molecules25030668.
29. Grün, J. T. *et al.* (2020) Conformational dynamics of strand register shifts in DNA G-quadruplexes, *Journal of the American Chemical Society*. American Chemical Society, *142*(1), pp. 264–273. https://doi.org/10.1021/jacs.9b10367.
30. Ceschi, S. *et al.* (2020) A two-quartet G-quadruplex topology of human KIT2 is conformationally selected by a perylene derivative, *Biochimie*. Elsevier B.V., *179*, pp. 77–84. https://doi.org/10.1016/j.biochi.2020.09.015.
31. Rigo, R. *et al.* (2017) Conformational profiling of a G-rich sequence within the c-KIT promoter, *Nucleic Acids Research*. Oxford University Press, *45*(22), pp. 13056–13067. https://doi.org/10.1093/nar/gkx983.
32. Hu, M. H. *et al.* (2019) New substituted quinoxalines inhibit triple-negative breast cancer by specifically downregulating the c-MYC transcription, *Nucleic Acids Research*. NLM (Medline), *47*(20), pp. 10529–10542. https://doi.org/10.1093/nar/gkz835.
33. Wu, T. Y. *et al.* (2020) A drug-like imidazole-benzothiazole conjugate inhibits malignant melanoma by stabilizing the c-MYC G-quadruplex, *Bioorganic Chemistry*. Academic Press Inc., *99*. https://doi.org/10.1016/j.bioorg.2020.103866.
34. Catalano, R. *et al.* (2019) Targeting multiple G-quadruplex–forming DNA sequences: Design, biophysical and biological evaluations of indolo-naphthyridine scaffold derivatives, *European Journal of Medicinal Chemistry*, *182*. https://doi.org/10.1016/j.ejmech.2019.111627.
35. Khanna, L. *et al.* (2020) Spiro-Indole-Coumarin hybrids: Synthesis, ADME, DFT, NBO studies and in silico screening through molecular docking on DNA G-quadruplex, *ChemistrySelect*. Wiley-Blackwell, *5*(11), pp. 3420–3433. https://doi.org/10.1002/slct.201904783.
36. Gao, C. *et al.* (2020) BMPQ-1 binds selectively to (3+1) hybrid topologies in human telomeric G-quadruplex multimers, *Nucleic Acids Research*. Oxford University Press, *48*(20), pp. 11259–11269. https://doi.org/10.1093/nar/gkaa870.

37. Khristich, A. N. and Mirkin, S. M. (2020) On the wrong DNA track: Molecular mechanisms of repeat-mediated genome instability, *Journal of Biological Chemistry*. American Society for Biochemistry and Molecular Biology Inc., pp. 4134–4170. https://doi.org/10.1074/jbc.REV119.007678.
38. Berselli, M., Lavezzo, E. and Toppo, S. (2020) QPARSE: Searching for long-looped or multimeric G-quadruplexes potentially distinctive and druggable, *Bioinformatics*. Oxford University Press, *36*(2), pp. 393–399. https://doi.org/10.1093/bioinformatics/btz569.
39. Hu, M. H. *et al.* (2020) Dimeric aryl-substituted imidazoles may inhibit ALT cancer by targeting the multimeric G-quadruplex in telomere, *European Journal of Medicinal Chemistry*. Elsevier Masson s.r.l., *186*, p. 111891. https://doi.org/10.1016/j.ejmech.2019.111891.
40. Ma, T. Z. *et al.* (2020) Dimers formed with the mixed-type G-quadruplex binder pyridostatin specifically recognize human telomere G-quadruplex dimers, *Organic and Biomolecular Chemistry*. Royal Society of Chemistry, *18*(5), pp. 920–930. https://doi.org/10.1039/c9ob02470k.
41. Paluszkiewicz, E. *et al.* (2020) Design, synthesis and high antitumor potential of new unsymmetrical bisacridine derivatives towards human solid tumors, specifically pancreatic cancers and their unique ability to stabilize DNA G-quadruplexes, *European Journal of Medicinal Chemistry*. Elsevier Masson s.r.l., *204*. https://doi.org/10.1016/j.ejmech.2020.112599.
42. Platella, C. *et al.* (2020) Trifunctionalized naphthalene diimides and dimeric analogues as g-quadruplex-targeting anticancer agents selected by affinity chromatography, *International Journal of Molecular Sciences*. MDPI AG, *21*(6). https://doi.org/10.3390/ijms21061964.
43. Kuang, G. *et al.* (2020) Syntheses and evaluation of new bisacridine derivatives for dual binding of G-quadruplex and i-Motif in regulating oncogene c-myc expression, *Journal of Medicinal Chemistry*. American Chemical Society, *63*(17), pp. 9136–9153. https://doi.org/10.1021/acs.jmedchem.9b01917.
44. King, J. J. *et al.* (2020) DNA G-quadruplex and i-Motif structure formation is interdependent in human cells, *Journal of the American Chemical Society*. American Chemical Society, *142*(49), pp. 20600–20604. https://doi.org/10.1021/jacs.0c11708.
45. Jamroskovic, J. *et al.* (2020) Quinazoline ligands induce cancer cell death through selective STAT3 inhibition and G-quadruplex stabilization, *Journal of the American Chemical Society*. American Chemical Society, *142*(6), pp. 2876–2888. https://doi.org/10.1021/jacs.9b11232.
46. Liu, K. C. *et al.* (2020) Affinity-selected bicyclic peptide G-quadruplex ligands mimic a protein-like binding mechanism, *Journal of the American Chemical Society*. American Chemical Society (ACS), *142*(18), pp. 8367–8373. https://doi.org/10.1021/jacs.0c01879.
47. Yaneva, M. Y. *et al.* (2020) Stapling a G-quadruplex specific peptide, *Biochemical and Biophysical Research Communications*. Elsevier B.V., *531*(1), pp. 62–66. https://doi.org/10.1016/j.bbrc.2020.02.144.
48. Wu, G. *et al.* (2020) Custom G4 microarrays reveal selective G-quadruplex recognition of small molecule BMVC: A large-scale assessment of ligand binding selectivity, *Molecules*. MDPI AG, *25*(15). https://doi.org/10.3390/molecules25153465.

49. Comez, L. *et al.* (2020) Polymorphism of human telomeric quadruplexes with drugs: A multi-technique biophysical study, *Physical Chemistry Chemical Physics*. Royal Society of Chemistry, *22*(20), pp. 11583–11592. https://doi.org/10.1039/d0cp01483d.
50. Di Fonzo, S. *et al.* (2020) Ligand binding to G-quadruplex DNA: New insights from ultraviolet resonance Raman spectroscopy, *Physical Chemistry Chemical Physics*. Royal Society of Chemistry, *22*(15), pp. 8128–8140. https://doi.org/10.1039/d0cp01022g.
51. Dhamodharan, V. and Pradeepkumar, P. I. (2019) Specific recognition of promoter G-quadruplex DNAs by small molecule ligands and light-up probes, *ACS Chemical Biology*. American Chemical Society, *14*(10), pp. 2102–2114. https://doi.org/10.1021/acschembio.9b00475.
52. Liao, T. C. *et al.* (2020) Human telomere double G-quadruplex recognition by berberine-bisquinolinium imaging conjugates in vitro and in cells, *International Journal of Biological Macromolecules*. Elsevier B.V., *158*, pp. 1299–1309. https://doi.org/10.1016/j.ijbiomac.2020.04.171.
53. Zhao, J. *et al.* (2020) Specific recognition of telomeric multimeric G-quadruplexes by a simple-structure quinoline derivative, *Analytica Chimica Acta*. Elsevier B.V., *1132*, pp. 93–100. https://doi.org/10.1016/j.aca.2020.07.017.
54. Yasuda, M. *et al.* (2020) Target identification of a macrocyclic hexaoxazole G-quadruplex ligand using post-target-binding visualization, *Chemical Communications*. Royal Society of Chemistry, *56*(85), pp. 12905–12908. https://doi.org/10.1039/d0cc04957c.
55. Meier-Menches, S. M. *et al.* (2020) Comparative biological evaluation and G-quadruplex interaction studies of two new families of organometallic gold(I) complexes featuring N-heterocyclic carbene and alkynyl ligands, *Journal of Inorganic Biochemistry*. Elsevier Inc., *202*. https://doi.org/10.1016/j.jinorgbio.2019.110844.
56. Bryan, T. M. (2020) G-quadruplexes at telomeres: Friend or foe?, *Molecules*. MDPI AG. https://doi.org/10.3390/molecules25163686.
57. Guillon, J. *et al.* (2020) Design, synthesis, and antiprotozoal evaluation of new 2,4-bis[(substituted-aminomethyl)phenyl]quinoline, 1,3-bis[(substituted-aminomethyl)phenyl]isoquinoline and 2,4-bis[(substituted-aminomethyl)phenyl]quinazoline derivatives, *Journal of Enzyme Inhibition and Medicinal Chemistry*. Taylor and Francis Ltd, *35*(1), pp. 432–459. https://doi.org/10.1080/14756366.2019.1706502.
58. Majee, P. *et al.* (2020) Identification and characterization of two conserved G-quadruplex forming motifs in the Nipah virus genome and their interaction with G-quadruplex specific ligands, *Scientific Reports*. Nature Research, *10*(1), p. 1477. https://doi.org/10.1038/s41598-020-58406-8.
59. Zhao, C. *et al.* (2021) Targeting RNA G-quadruplex in SARS-CoV-2: A promising therapeutic target for COVID-19?, *Angewandte Chemie International Edition*. Wiley-VCH Verlag, *60*(1), pp. 432–438. https://doi.org/10.1002/anie.202011419.
60. Groaz, E. and De Jonghe, S. (2021) Overview of Biologically Active Nucleoside Phosphonates, *Frontiers in Chemistry*. Frontiers, *8*, 616863. https://doi.org/10.3389/fchem.2020.616863.
61. Doboszewski, B. et al. (2013) Synthesis of Phosphonoglycine Backbone Units for the Development of Phosphono Peptide Nucleic Acids, *European Journal of Organic Chemistry*. Chemistry Europe, *2013*(22), pp. 4804–4815. https://doi.org/10.1002/ejoc.201300523.

INDEX

Note: Page numbers followed by *f* or *t* indicate material in figures or tables respectively

Nucleic Acids in Medicinal Chemistry and Chemical Biology: Drug Development and Clinical Applications, First Edition. Edited by Lihe Zhang, Xinjing Tang, Zhen Xi, and Jyoti Chattopadhyaya.

Wiley Series in Drug Discovery and Development

Binghe Wang, Series Editor

Drug Delivery: Principles and Applications
Edited by Binghe Wang, Teruna Siahaan, and Richard A. Soltero

Computer Applications in Pharmaceutical Research and Development
Edited by Sean Ekins

Glycogen Synthase Kinase-3 (GSK-3) and Its Inhibitors: Drug Discovery and Development
Edited by Ana Martinez, Ana Castro, and Miguel Medina

Aminoglycoside Antibiotics: From Chemical Biology to Drug Discovery
Edited by Dev P. Arya

Drug Transporters: Molecular Characterization and Role in Drug Disposition
Edited by Guofeng You and Marilyn E. Morris

Drug-Drug Interactions in Pharmaceutical Development
Edited by Albert P. Li

Dopamine Transporters: Chemistry, Biology, and Pharmacology
Edited by Mark L. Trudell and Sari Izenwasser

Carbohydrate-Based Vaccines and Immunotherapies
Edited by Zhongwu Guo and Geert-Jan Boons

ABC Transporters and Multidrug Resistance
Edited by Ahcène Boumendjel, Jean Boutonnat, and Jacques Robert

Drug Design of Zinc-Enzyme Inhibitors: Functional, Structural, and Disease Applications
Edited by Claudiu T. Supuran and Jean-Yves Winum

Kinase Inhibitor Drugs
Edited by Rongshi Li and Jeffrey A. Stafford

Evaluation of Drug Candidates for Preclinical Development: Pharmacokinetics, Metabolism, Pharmaceutics, and Toxicology
Edited by Chao Han, Charles B. Davis and Binghe Wang

HIV-1 Integrase: Mechanism and Inhibitor Design
Edited by Nouri Neamati

Carbohydrate Recognition: Biological Problems, Methods, and Applications
Edited by Binghe Wang and Geert-Jan Boons

Chemosensors: Principles, Strategies, and Applications
Edited by Binghe Wang and Eric V. Anslyn

Medicinal Chemistry of Nucleic Acids
Edited by Li He Zhang, Zhen Xi, and Jyoti Chattopadhyaya

Plant Bioactives and Drug Discovery: Principles, Practice, and Perspectives
Edited by Valdir Cechinel Filho

Dendrimer-Based Drug Delivery Systems: From Theory to Practice
Edited by Yiyun Cheng

Cyclic-Nucleotide Phosphodiesterases in the Central Nervous System: From Biology to Drug Discovery
Edited by Nicholas J. Brandon and Anthony R. West

Drug Transporters: Molecular Characterization and Role in Drug Disposition, 2nd Edition
Edited by Guofeng You and Marilyn E. Morris

Drug Delivery: Principles and Applications, 2nd Edition
Edited by Binghe Wang, Longqin Hu, and Teruna J. Siahaan

Carbon Monoxide in Drug Discovery: Basics, Pharmacology, and Therapeutic Potential
Edited by Binghe Wang and Leo E. Otterbein

Drug Transporters: Molecular Characterization and Role in Drug Disposition, Third Edition
Edited by Guofeng You and Marilyn E. Morris

Hydrogen Sulfide: Chemical Biology Basics, Detection Methods, Therapeutic Applications, and Case Studies, First Edition
Edited by Michael D. Pluth

Nucleic Acids in Medicinal Chemistry and Chemical Biology: Drug Development and Clinical Applications, First Edition
Edited by Lihe Zhang, Xinjing Tang, Zhen Xi, and Jyoti Chattopadhyaya

Printed and bound by CPI Group (UK) Ltd, Croydon, CR0 4YY
17/09/2026
14964320-0002